Enzyme Catalysis Today and the Chemistry of the 21st Century

Gertz I. Likhtenshtein

Enzyme Catalysis Today and the Chemistry of the 21st Century

Springer

Gertz I. Likhtenshtein
Department of Chemistry
Ben-Gurion University of the Negev
Beersheba, Israel

Federal Research Center of Problems
of Chemical Physics and Medicinal
Chemistry
Russian Academy of Sciences
Moscow, Russia

ISBN 978-3-031-51551-4 ISBN 978-3-031-51552-1 (eBook)
https://doi.org/10.1007/978-3-031-51552-1

This Springer imprint is published by the registered company Springer Nature Switzerland AG
The registered company address is: Gewerbestrasse 11, 6330 Cham, Switzerland

Preface

If we imagine that natural enzymes were to cease their functioning for one moment, this would mean an immediate collapse of all life on earth. Indeed, due to unique catalytic properties of enzymes such as precise specificity, selectivity, high rate of chemical reaction, and regulatory capacity, modern enzymology "stretches its arms wide" to all areas of wildlife and plays an important role in biotechnology, medicine, physiology, and the daily life of a human being. Classical and modern physical chemistry, chemical kinetics, organic, inorganic, and quantum chemistries provide an arsenal of physical methods and establish a basis for the investigation of structure and action mechanism of enzymes. The enzymes' outstanding properties are possible owning to the multifunctional active centers, by the unique structure of protein globules, possessing both rigidity and flexibility and by the formation of catalytic ensembles. The rapid progress in knowledge of the structure and mechanism of action of enzymes, on the one hand, makes it possible to control and practical use of enzymatic processes, and on the other hand opening the door to a new chemical revolution through appropriate mimicry (naturally inspired design). Enzyme mimics can have the important advantages including tunable structures and catalytic efficiencies, stability in experimental conditions, lower cost, and simple synthetic manufacturing routes to their preparation. Modern synthetic chemistry presents an almost unlimited possibility of preparing model catalysts of various structures.

This book is a viewpoint on a view of enzyme catalysis, and it mimics by a physical chemist with long-term experience in the investigation of structure and action mechanism of biological catalysts. This book is not intended to provide an exhaustive survey of each topic but rather a discussion of their theoretical and experimental background, and recent developments. The literature of enzyme catalysis is so vast, and many scientists have made important contributions in the area that it is impossible in the space allowed for this book to give a representative set of references. The author apologizes to those he has not been able to include.

Chapter 1 of the book is a brief overview of the general information about the main classes of enzymes and its applications. Methods of analysis in enzyme processes are described in Chap. 2. Methods of enzyme kinetics are the subject of Chap. 3. Chapters 4–6 form the theoretical and experimental background for elementary mechanisms of enzyme reactions including radical (Chap. 5) and multielectron redox (Chap. 6) processes. In Chap. 7, a general survey is made on molecular dynamics of proteins and their functional activity. Chapter 8 reviews theoretical and experimental data on enthalpy–entropy relationships in enzyme reactions. Chapters 9–13 gather and analyze information on recent progress made in the areas of artificial and nanoenzymes (Chap. 9), artificial photosynthesis (Chap. 10), modeling nonhemin iron proteins and enzymes (Chap. 11), artificial hydroxylases and hydrolases (Chap. 12), and miscellaneous biomimetic processes (Chap. 13).

The monograph is intended for scientists working on enzyme catalysis and adjacent areas such as chemical modeling of biological processes, homogeneous catalysis, biomedical research, and biotechnology. The book can be used as a subsidiary manual for instructors, and graduate and undergraduate students of university biochemistry and chemistry departments.

Beersheba, Israel Gertz I. Likhtenshtein

Acknowledgments The author is extremely grateful to the Department of Chemistry, Ben-Gurion University of the Negev and Federal Research Center of Problems of Chemical Physics and Medicinal Chemistry, Russian Academy of Sciences for providing ideal conditions for working on the book.

Contents

1 Enzymatic Processes as a Basis for the Creation of Advanced Chemical Catalysts 1
1.1 Introduction 1
1.2 Enzymes in the Processes of Solar Energy Conversion and Stored Energy Through the Formation of Energy-Rich Molecules 4
1.2.1 Solar Energy Conversion in Chemical Energy 4
1.2.2 Another Source of Synthesis of the Reach-Energy Molecules 6
1.3 Enzymes in Metabolism 8
1.3.1 Synthesis of Small Molecules 9
1.3.2 Synthesis of Biological Polymers 11
1.3.3 Degradation of Biological Macro Molecules 15
1.4 Miscellaneous Enzymes 16
1.5 Enzymes in Biotechnology and Industry 19
1.5.1 Introduction 19
1.5.2 Miscellaneous Applications of Enzymes 20
1.6 Enzymes in Medicine 25
1.6.1 General 25
1.6.2 Enzymes in Cancer 26
References 31

2 Methods of Analysis in Enzyme Processes 43
2.1 Optical Methods 43
2.1.1 Introduction 43
2.1.2 Absorption (UV/Vis) 45
2.1.3 Raman and Infrared Spectroscopy 47
2.1.4 Fluorescence Spectroscopy 48
2.1.5 Dual Fluorophore-Nitroxides (FNRO) as Redox Sensors and Spin Traps 51
2.1.6 Terahertz Spectroscopy 54

2.2 Electrochemical Methods 55
2.2.1 Introduction 55
2.2.2 Amperometry 56
2.2.3 The Oxygen Electrode 56
2.2.4 Voltammetry 56
2.2.5 Polarography 58
2.2.6 Cyclic Voltammetry 58
2.2.7 Potentiometry 59
2.2.8 Chronoamperometry 60
2.2.9 Coulometry 62
2.2.10 Electrochemical Impedance Spectroscopy 62
2.3 Enzyme Biosensors 63
2.4 Magnetic Resonance 67
2.4.1 Grounds 67
2.4.2 Nuclear Magnetic Resonance-Based Methods 68
2.4.3 Nitric Oxide Analysis by ESR 70
2.5 Microdialysis 72
2.6 Nitric Oxide Analysis 74
2.6.1 Introduction 74
2.6.2 NO Luminescence Analysis 75
2.6.3 Nitric Oxide Analysis by Electrochemistry 79
2.6.4 Nitric Oxide Analysis Based on Utilization of Dual Fluorescent-Nitroxide Probes 79
2.7 Calorimetry 83
2.8 Mass Spectrometry-Based Methods 85
References 88

3 Methods of Enzyme Kinetics 99
3.1 Introduction 99
3.2 Artificial Intermediates 100
3.3 Jump Methods [10–30] 101
3.4 Kinetic Isotope Effect (KIE) 105
3.5 Method of Transition State Analogs [59–89] 119
3.6 Theoretical Methods 125
3.6.1 Foundation of Theoretical Approaches 125
3.6.2 Combined Quantum Mechanics/Molecular Mechanics (QM/MM) Methods in Computational Enzymology 127
3.6.3 Examples Illustrating Theoretical Approaches 130
References 133

4 Elementary Mechanisms of Enzyme Reactions (Part 1) 143
4.1 Proximity and Oriental Effects 144
4.2 Concerted Chemical Processes 146
4.2.1 Theoretical Consideration 146
4.2.2 Simple Molecules Reactions 147
4.2.3 Enzymatical Reaction 148

4.3 Principle of "Optimum Motion" and Mechanism of Enzymatic Reactions 151
4.4 Electron Transfer 154
4.4.1 First Grounds 154
4.4.2 Long-Range Electron Transfer 155
4.4.3 Photosynthetic Light Energy Conversion 160
4.5 Magnetic Isotope Effects 169
References 171

5 Enzymes Mechanisms. (Part 2) Radical Processes 177
5.1 Cytochromes P450 177
5.1.1 General Properties and Action Mechanism 177
5.1.2 Crystal Structure 178
5.1.3 Mechanism and Active Cite Structure 181
5.2 Nitric Oxide Synthase 192
5.2.1 General 192
5.2.2 Structure of NOS and It Active Cite 194
5.3 Methane Mono Oxygenase 201
5.3.1 General 201
5.3.2 Soluble Methane Monooxigenase 201
5.3.3 Particulate MMO 205
References 210

6 Multielectron Redox Mechanisms 217
6.1 Two Electron Processes. Hydride Transfer in Hydrogenases 218
6.2 Multielectron Processes. Four-Electron Reduction and Oxidation 220
6.2.1 General 220
6.2.2 Nitrogen Fixation 222
6.3 Photosynthetic Water Oxidation Complex 232
References 243

7 Molecular Dynamics of Proteins and Their Functional Activity 247
7.1 Introduction 247
7.2 Early Studies 251
7.2.1 Hypothesizes and First Experiments 251
7.2.2 Early Mechanisms of Protein Dynamics 253
7.3 Further Developments 255
7.3.1 Proteins Dynamics Computations 255
7.3.2 Miscellaneous Dynamic Properties of Proteins Studied by Physical Methods 261
7.3.3 Protein Dynamics and Their Functional Activity and Evolution 265
7.3.4 Role Solvation Dynamics 270
References 273

8 Enthalpy–Entropy Relationships in Enzyme Reactions 279
8.1 Introduction 279
8.2 Early Theoretic Considerations 280
8.3 Statistical Artifacts at Apparent EEC 282
8.4 General Theoretical Considerations 286
8.5 Miscellaneous Examples 287
References 297

9 Artificial and Nanoenzymes 301
9.1 Introduction 302
9.2 Enzyme Mimetic Catalysts 303
9.2.1 Modeling Enzymes Active Centers Structure and Activity 303
9.2.2 Nanoenzymes 303
9.3 Molecular Imprinting 324
References 328

10 Artificial Photosynthesis 333
10.1 Introduction 334
10.2 Light Harvesting and Charge Separation in Artificial Photosynthetic Systems 335
10.2.1 Artificial Light Harvesting Antenna 335
10.2.2 Charge Separation in Artificial Donor–Acceptor Pairs 337
10.2.3 Photophysical and Photochemical Processes in Dual Flourophore-Nitroxide Molecules 350
10.2.4 Artificial Photoenzymes on the Base of Model Vesicles 354
10.3 Oxygen Evolution Complex (OEC) Mimics 359
10.4 Carbon Dioxide Utilization 368
References 372

11 Modeling Nonhemin Iron Proteins and Enzymes 379
11.1 Iron–Sulfur Clusters 379
11.2 Nitrogenase Chemical Models 381
11.3 Hydrogenases Chemical Models 394
11.3.1 General 394
11.3.2 [Fe–Fe] Hydrogenase Models 394
11.3.3 Nickel–Iron and Iron Hydrogenases 402
References 404

12 Artificial Hydroxylases and Hydrolases 409
12.1 Introduction 410
12.2 Cytochrome P450 412
12.3 Methane Monooxygenase Mimicking 421
12.4 Artificial Hydrolases 427
References 435

13 Miscellaneous Biomimetic Processes 439
13.1 Nitric Oxide Producing Activity 439
13.1.1 S-Nitrosoglutathione as NO Donor 440
13.1.2 NO Complexes Donors 441
13.2 Catalytic Antibodies 446
13.2.1 Fundamentals 446
13.2.2 Typical Examples 447
13.3 Miscellaneous Artificial Enzymes [70–88] 455
References 460

Curriculum Vitae

Name: Likhtenshtein, Gertz I.

Academic Background

1976	Institute of Chemical Physics, USSR Academy of Science, Moscow, Professor of Chemical Physics.
1972	Institute of Chemical Physics, USSR Academy of Science, Moscow, Physical Chemistry, Doctor of Chemical Science.
1959–1963	Institute of Chemical Physics, USSR Academy of Science, Moscow, Physical Chemistry, Ph.D.
1951–1956	Gorky State University, Gorky, USSR, Faculty of Chemistry. Physical Chemistry, M.Sc.

Present and Previous Employment

2006–2007	Kyoto University, Graduate School of Human and Environmental Studies (visiting professor)
2003–till now	Ben-Gurion University of the Negev, Beer-Sheva, Israel Department of Chemistry, Professor Emeritus
1992-2003	Ben-Gurion University of the Negev, Beer-Sheva, Israel Department of Chemistry, Full Professor
1966–1992	Institute of Chemical Physics, USSR Academy of Science, Moscow Chief of Department of Enzyme Catalysis,
1963–1966	Institute of Molecular Biology, USSR Academy of Science, Moscow Senior Researcher,
1959–1963	Institute of Chemical Physics, USSR Academy of Science, Moscow Post-graduate student,
1956–1959	Institute of Chemistry, Dushanbe, USSR Junior Researcher,

Awards

2018	Foreign Member of the Academy of Science, Republic of Tadjikistan
2015	Academician N. M. Emmanuel Golden Medal on Chemical Biophysics
2008	Medal from the President of Academy of Science, Republic of Tadjikistan
2007	Academician V. V. Voevodsky Price on Chemical Physics
2000	Diploma of Israel Chemical Society
1977	USSR State Price
1979	Diploma of Discovery, USSR
1968	Medal of the Exhibition of Economic Achievement, USSR

Membership in Scientific Societies

International ESR Society
American Biophysical Society
Israel Chemical Society
Israel Electron Spin Resonance Society

Research Interest

Long distance electron transfer in proteins and model systems: role of orbital overlap, local polarity and dynamics. Multielectron and synchronous processes.

Linear free-energy relationships in photochemistry and photophysics of stilbenes as models for intramolecular charge transfer. Distribution of electrostatic potential around molecules of biological importance. Biophysical labeling methods in molecular biology. Novel fluorescence biosensing: fluidity of biomembranes, express immunoassay, analysis of nitric oxide, superoxide in antioxidants in bioobjects.

Teaching Experience

Lectures for graduated students

Kinetcs and mechanism of biological processes

Electron Spin Resonance in chemistry and biology
Biophysical labeling methods in molecular biology
Metalloenzymes
Enzyme Catalysis Today and Chemistry of 21th Century

Laboratories for graduated students

1. General and Analytical Chemistry
2. General Chemistry
3. Physical Chemistry

List of Publications

Books

1. Likhtenshtein GI (1974) Spin Labeling Method in Molecular Biology. Nauka, Moscow
2. Likhtenshtein GI (1976) Spin labeling method in molecular biology. Wiley Interscience, New York, NY
3. Likhtenshtein GI (1979) Multinuclear redox metalloenzymes. Nauka, Moscow
4. Likhtenshtein GI (1988) Chemical physics of redox metalloenzymes. Springer-Verlag, Heidelberg
5. Likhtenshtein GI (1993) Biophysical labeling methods in molecular biology. Cambridge University Press, Cambridge N.Y.
6. Denisov ET, Sarkisov OM, Likhtenshtein GI (2000) Chemical Kinetics, Khymia, Moscow
7. Likhtenshtein GI (2003) New Trends in enzyme catalysis and mimicking chemical reactions. Kluwer Academic/ Plenum Publishers, New York

8. Denisov ET Sarkisov OM, Likhtenshtein GI (2003). Chemical kinetics: fundamentals and recent developments, Elsevier Science
9. Likhtenshtein GI, Yamauchi J, Nakatsuji S, Smirnov A., Tamura R (2008) Nitroxides: application in chemistry, biomedicine, and materials science. Wiley-VCH, Weinheim
10. Likhtenshtein GI (2009) Stilbenes: Application in Chemistry, Life Science and Material Science WILEY-VCH, Weinheim).
11. Likhtenshtein GI (2012) Solar energy conversion. Chemical aspects. Wiley-VCH, Weinheim
12. Likhtenshtein GI (2016) Electron spin in chemistry and biology: fundamentals, methods, reactions mechanisms, magnetic phenomena, structure investigation. Springer
13. Likhtenshtein GI (2018) Nitric oxide in chemistry and biology–physicochemical aspect: fundamental and recent developments. NOVA Science Publishers
14. Likhtenshtein GI (2020) Nitroxides: brief history and recent developments. Springer
15. Likhtenshtein GI (2021) Biological water: physicochemical aspect. Springer Nature
16. Likhtenshtein GI Enzyme Catalysis Today and Chemistry of 21th Century, Springer

International Recognition

1. **Who's Who in the World**, 14th Editions. A Who's Who in America Publication. (1997) p.853.
2. **The International Directory of Distinguished Leadership**. 7th Edition. American Bibliographic Institute Inc. (1998) p.119.
3. **International Who's Who of Intellectuals**. 13th Edition. International Biographical Centre. Cambridge, England (1999) p.433.
4. **Who's Who in the World**. 25th Editions. Silver Anniversary (2008).

Chapter 1
Enzymatic Processes as a Basis for the Creation of Advanced Chemical Catalysts

Abstract This chapter is a brief overview of the general information about the main classes of enzymes and its applications. The key enzymatic systems involved in the conversion of solar energy into chemical energy, the formation of energy-rich molecules, carbon dioxide, and atmospheric nitrogen fixation, the synthesis and degradation of amino acids, lipids, proteins, nucleotides, and nucleic acids. Enzymes provide vital chemical reactions like oxidative phosphorylation, Krebs cycle, Glycolysis Calvin-Benson cycle. Several paragraphs are devoted to miscellaneous enzymes including cytochromes P450, methane monooxygenases, hydrolases, kinases, and phosphatases. Enzymes play a vital role for all living things and control all normal and pathological processes. Enzymes in healthcare and medicine are used for treating disorders, assisting metabolism, medical device cleaning, the manufacture of medicines, regulation of digestion, detection of diseases, addressing food quality, treating heart, brain, lung, digestive tract, and mental diseases. Separate sections refer to the crucial role of enzymes in cancer and in COVID-19. Enzymes are widely used in such practically important areas as pharmaceutical industries, yogurt, beer and cheese production, apple wine fermentation, baking, brewing, production of cakes, preparation of digestive aids, fruit juices, and starch syrups and addition to the dough of bread, biodiesel production, enzymatic degumming, textile and paper industry, textile desizing, laundry detergents, the enzyme-mediated ceramic synthesis, skincare and cosmetics industry, the enzyme-mediated ceramic synthesis, the enzymatic decomposition of organic matter in soil. Thus, modern enzymology represents an inexhaustible source of knowledge and ideas for modern chemistry to initiate and bring about the real industrial revolution of the twenty-first century.

1.1 Introduction

Almost all key chemical and physical processes in wildlife (human and animal bodies, reptiles, plants, algae, and microorganisms) are controlled by enzymes [1–28]. The effective, close to ideal, course of these processes under ambient conditions

G. I. Likhtenshtein, *Enzyme Catalysis Today and the Chemistry of the 21st Century*,
https://doi.org/10.1007/978-3-031-51552-1_1

is provided by such unique properties of enzymes as high speed, selectivity, selectivity, and ability to be regulated. Thus, enzymatic catalysis is at the center of interest of biochemists, biophysicists, molecular biologists, and biomedical scientists. From the point of view of the strategy for the development of basic and applied chemistry, the creation of model catalysts based on knowledge of the structure and mechanism of action of enzymes can lead to a new chemical revolution in the industry and contribute positively to everyday human life.

This section briefly describes the key enzymatic systems involved in the conversion of solar energy into chemical energy, the formation of energy-rich molecules, carbon atmospheric nitrogen fixation, the synthesis of amino acids, lipids, proteins, nucleotides, and nucleic acids. Currently, more than 4000 enzymes have been isolated and characterized. According to Recommendations of the Nomenclature Committee of the International Union of Biochemistry and Molecular Biology on the Nomenclature and Classification of Enzymes by the Reactions they Catalyse (https://iubmb.qmul.ac.uk/enzyme/World Wide Web version prepared by G.P. Moss), a list of the enzymes is: (1) Oxidoreductases catalyze the transfer of electrons from electron donor to the electron acceptor; (2) transferases catalyze the transfer of specific functional groups from donor to another molecule; (3) hydrolases catalyze the cleavage of a covalent bond using water; (4) lyases are the enzymes responsible for catalyzing addition and elimination reactions, breaking the bond between a carbon atom and another atom such as oxygen, sulfur, or another carbon atom; (5) isomerases convert one isomer to another, meaning that the end product has the same molecular formula; (6) ligases catalyze the joining (ligation) of two large molecules by forming a new chemical bond; and (7) translocases are enzymes that assists in moving another molecule, catalyzing for example the movement of ions or molecules across membranes or their separation within membranes.

The molecules upon which enzymes may act are called substrates, and the enzymes convert the substrates into different molecules known as products. In enzymology, turnover number is defined as the number of substrate molecules transformed per minute by a single enzyme molecule when the enzyme is the rate-limiting factor [27].

A substrate binds to the active site of the enzyme to produce an enzyme-substrate complex ES and is transformed into an enzyme-product complex EP and from there to product P, via a transition state ES*:

$$E + S \rightleftarrows ES \rightleftarrows ES* \rightleftarrows EP \rightleftarrows E +$$

The Michaelis–Menten equation [27–30] describes how the (initial) reaction rate v_0 depends on the enzyme concentration [E], the position of the substrate-binding equilibrium, and the rate constant k_2 (Fig. 1.1). If the enzyme concentration is much less than the substrate concentration—the rate of product formation is given by

$$v_0 = \frac{V_{\max}[S]}{[S] + K_m} \tag{1.1}$$

where the Michaelis constant

$$K_m = \frac{k_{-1} + k_2}{k_1} \tag{1.2}$$

and the maximal velocity

$$V\max = k_2[E] \tag{1.3}$$

Enzymes can have ratios of k_2/K_m near 10^8 to 10^9 $M^{-1}s^{-1}$ that is close to the maximum allowed by the rate of diffusion. Most enzymes have specificity constants from 10^9 to 10^0s (see Fig. 1.2 and Table 1.1) [31].

Fig. 1.1 Michaelis–Menten graph. https://en.wiki

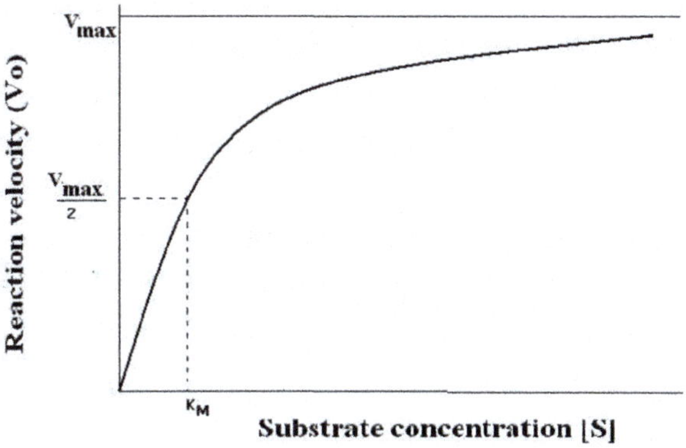

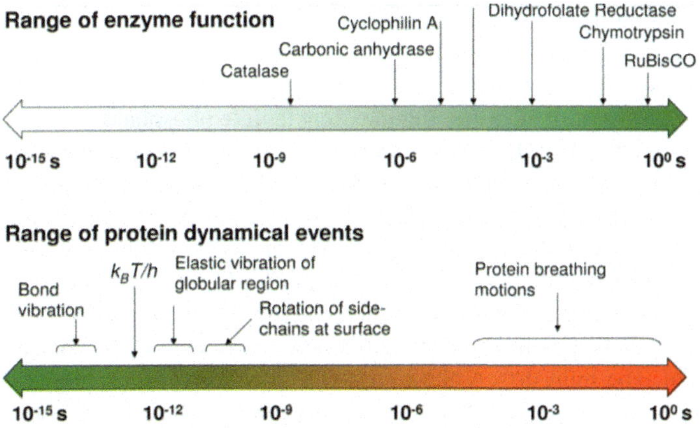

Fig. 1.2 Range of timescales involved in substrate turnover step of enzyme-catalyzed reactions and internal protein dynamics are similar. Note the universal frequency factor (k_BT/h), which is commonly used in transition state theory; k_B is Boltzmann's constant, T represents the ambient temperature, and h is Planck's constant [32]. Distributed under the terms of the Creative Commons CC BY license from Springer Nature

Table. 1.1 Parameter values vary widely between enzymes: https://en.wikipedia.org/wiki/Michaelis-Menten_kinetics

Enzyme	K_M (*M*)	\\k_2 (s^{-1})	K_2/K_M ($M^{-1}s^{-1}$)
Chymotrypsin	1.5×10^{-2}	0.14	9.3
Pepsin	3.0×10^{-4}	0.50	1.7×10^3
tRNA synthetase	9.0×10^{-4}	7.6	8.4×10^3
Ribonuclease	7.9×10^{-3}	7.9×10^2	1.0×10^5
Carbonic anhydrase	2.6×10^{-2}	4.0×10^5	1.5×10^7
Fumarase	5.0×10^{-6}	8.0×10^2	1.6×10^8

1.2 Enzymes in the Processes of Solar Energy Conversion and Stored Energy Through the Formation of Energy-Rich Molecules

1.2.1 Solar Energy Conversion in Chemical Energy

In plant photosynthesis, six carbon dioxide molecules and six water molecules are converted by light energy captured by chlorophyll into a sugar molecule and six oxygen molecules [33–38]:

$$6CO_2 + 6H_2O \rightarrow C_6H_{12}O_6 + 6O_2.$$

This process is carried out with the help of Photosystem I and Photosystem II and Calvin cycle. In the Photosystem I and Photosystem II systems, light energy is used to convert the initial stable molecules into energy-rich molecules, which are further used in a series of reactions leading to the conversion of CO_2 into glucose (Calvin cycle). The net effect of these steps is to convert light energy into chemical energy in the form of ATP and nicotine adenine dinucleotide phosphate (NADPH).

The Photosystem I (PSI, or plastocyanin–ferredoxin oxidoreductase) absorbs light energy creating photoseparated charges (strong electron donor D^- and strong electron acceptor (A^+)). The latter receives electrons from plastocyanin or cytochrome c_6, while D^- drives the synthesis of NADPH. Within Photosystem II (water-plastoquinone oxidoreductase), enzymes capture photons of light with formation photoseparated donor–acceptor pair. Then electron from the photoreduced donor is transferred through a variety of coenzymes and cofactors to reduce plastoquinone to plastoquinol. The light-driven photosynthetic electron transport from water in Mn cluster to the secondary quinone electron acceptor (Q) results in water oxidation most likely by the Mn cluster by one-step four-electron mechanism.

The Calvin-Benson cycle [39–41] (Fig. 1.3), in which carbon of CO_2 is reduced and involves the formation of intermediate sugar phosphates in a cyclic sequence. The overall reaction can be represented as follows:

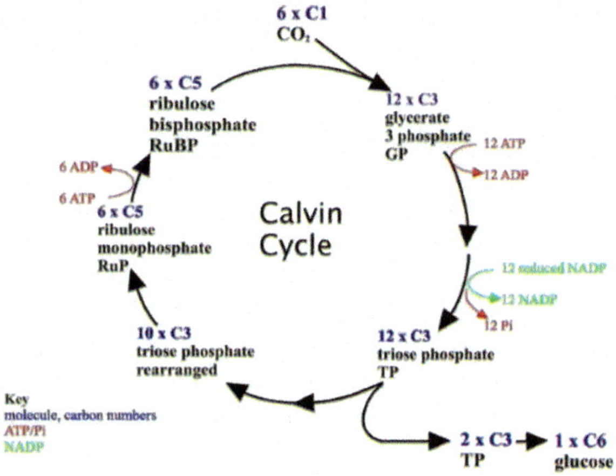

Fig. 1.3 Calvin cycle. Ribulose 1,5-bisphosphate (RuBP) I is the principal CO_2 acceptor in plant. https://www.quora.colvin-cycle-more-superior-than-glycolysis

$$3\ CO_2 + 6\,NADPH + 6\,H^+ + 9\,ATP + 5\,H_2O \rightarrow \text{glyceraldehyde} - 3 - \text{phosphate (G3P)} + 6\,NADP^+ + 9\,ADP + 8\,P_i \quad (P_i = \text{inorganic phosphate})$$

The cycle incorporates three molecules of carbon dioxide and produces one molecule of the three-carbon glyceraldehyde-3-phosphate (Gal3P), sugar phosphates, amino acids, sucrose, and carboxylic acids.

Calvin cycle is operated by 11 different enzymes that catalyze 13 reactions (Fig. 1.3). The regulatory enzymes are RuBisCO, FBPase, TKL, and PRK. The process runs in the following steps: (1) RuBisCO catalyzes the carboxylation of ribulose-1,5-bisphosphate (RuBP); (2) FBPase changes the fructose-1,6-bisphosphate into fructose 6-phosphate, the enzyme catalyzes the reverse reaction by phosphofructokinase; (3) PRK catalyzes the ATP dependent phosphorylation of ribulose-5-phosphate (RuP) into RuBP; (4) TKL catalyzes the reverse reaction, the conversion of sedoheptulose-7-P and glyceraldehyde-3-P to pentoses, d-ribose-5-phosphate (aldose) and d-xylulose-5-P (ketose); (5) in the key reactions of the Calvin cycle, carbon from CO_2 is added to a five-carbon molecule Ribulose 1,5-bisphosphate (RuBP) using the enzyme, RuBP oxygenase-carboxylase(rubisco). These reactions use chemical energy from NADPH and ATP that were produced in the light and other reactions. The final product of the Calvin cycle is glucose [29].

1.2.2 *Another Source of Synthesis of the Reach-Energy Molecules*

1.2.2.1 Oxidative Phosphorylation

Oxidative phosphorylation [42–48] refers to the metabolic pathway in which the energy released by nutrients during oxidation is utilized to generate ATP through electrical transport chain (Fig. 1.4). The ATP synthesis is coupled to the movement of electrons through the mitochondrial electron transport chain and the associated consumption of oxygen. The chain of redox reactions driving the flow of electrons through the electron transport chain, from electron donors such as NADH to electron acceptors such as oxygen and hydrogen (protons), is an exergonic process, whereas the synthesis of ATP is an endergonic process. Within proteins, electrons are transferred between flavin cofactors, iron–sulfur clusters, and cytochromes.

In Fig. 1.4, Complex I, NADH-coenzyme Q oxidoreductase, catalyzes the two-electron oxidation of NADH by coenzyme Q10 or ubiquinone:

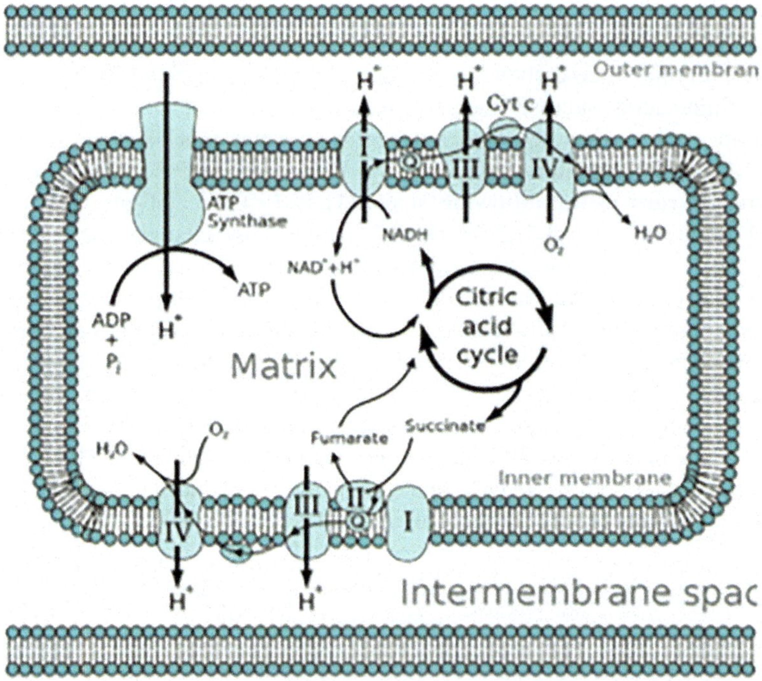

Fig. 1.4 Electron transport chain in the cell is the site of oxidative phosphorylation. The NADH and succinate generated in the citrm/How-is-the-Caic acid cycle are oxidized, releasing the energy of O_2 to power the ATP synthase. https://en.wikipedia.org/wiki/Oxidative_phosphorylation

$$\text{NADH (matrix)} + \text{Q10} + 5\,\text{H}^+ \rightarrow \text{NAD} + \text{QH}_2 + 4\text{H}^+\text{(intermembrane)}.$$

Complex II, Succinate-Q oxidoreductase, consists of four protein subunits and contains a bound flavin adenine dinucleotide (FAD) cofactor and iron–sulfur clusters and participates in electron transfer to coenzyme Q (Succinate + Q ➔ Fumarate + Q_2). Electron transfer flavoprotein-ubiquinone oxidoreductase, ETF-Q oxidoreductase (Complex III), containing a flavin and a [4Fe–4S] cluster accepts electrons from electron-transferring flavoprotein in the mitochondrial matrix, and uses these electrons to reduce ubiquinone:

$$4\,\text{Cytc(ox)} + \text{O}_2 + 8\text{H}^+\text{(matrix)} \rightarrow 4\,\text{Cytc(ox)} + 2\text{H}_2\text{O} + 4\,\text{H}^+\text{(intermediate)}$$

ATP synthase, complex V, is the final enzyme in the oxidative phosphorylation pathway. The enzyme uses the energy stored in a proton gradient across a membrane to drive the synthesis of ATP from ADP and phosphate (P_i). A number of protons required to synthesize one ATP have ranged from three to four:

$$\text{ADP} + \text{Pi} + 4\text{H}^+\text{(intermembrane)} \rightarrow \text{ATP} + \text{H}_2\text{O Pi} + 4\,\text{H}^+\text{(matrix)}$$

This phosphorylation reaction is an equilibrium, which can be shifted by altering the proton-motive force. ATP synthase consists of two regions F_O and F_1 which create a pathway for protons movement across the inner mitochondrial membrane. F_O causes rotation of F_1. F_1 has a water-soluble part that can hydrolyze ATP while F_O has mainly hydrophobic regions.

1.2.2.2 Krebs Cycle and Glycolysis

The citric acid cycle (CAC), TCA cycle (tricarboxylic acid cycle), or the Krebs cycle (Fig. 1.5) [49–55] are a series of chemical reactions to release stored energy through the oxidation of acetylCoA derived from carbohydrates, fats, and proteins in the help of corresponding enzymes. Sequence of these reactions provides energy for storage in NADH, Guanosine-5'-triphosphate (GTP), and ubiquinone Q2.

(https://en.wikipedia.org/wiki/Citric_acid_cycle#/media/File:Citric_acid_cycle_with_aconitate_2.svg)

Glycolysis is a cytoplasmic pathway which breaks down glucose into two three-carbon compounds and generates the high-energy molecules NADH and ATP. The final product of glycolysis is pyruvate in aerobic settings and lactate in anaerobic conditions. The overall process of glycolysis is:

$$\text{Glucose} + 2\,\text{NAD}^+ + 2\,\text{ADP} + 2\,\text{P}_1 \rightarrow 2\,\text{pyruvate} + 2\,\text{NADH} + 2\text{H}^+ + 2\,\text{ATP}$$

The three key enzymes of glycolysis are hexokinase, phosphofructokinase, and pyruvate kinase. The first step is phosphorylation of glucose by a family of enzymes

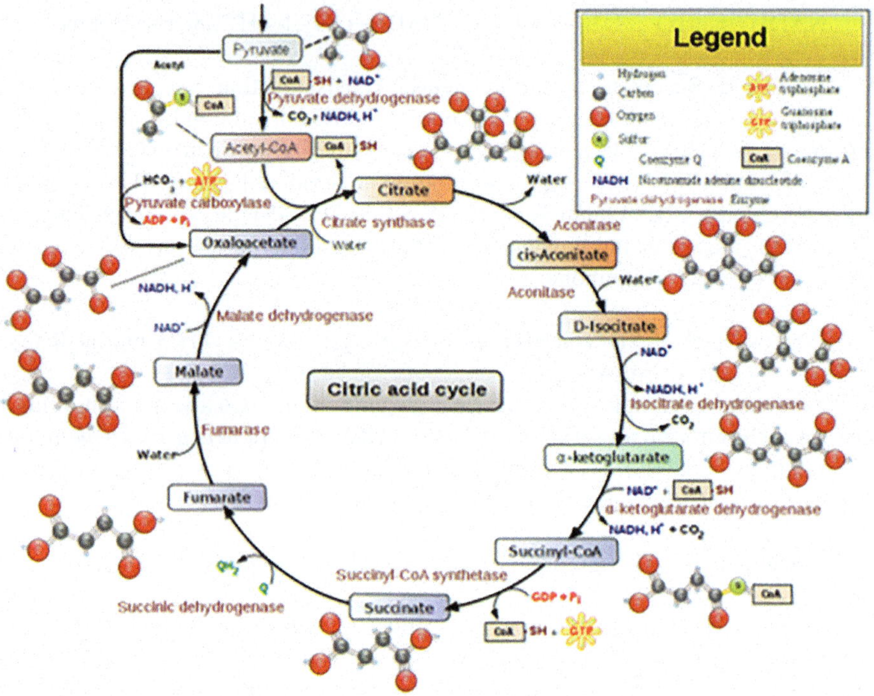

Fig. 1.5 Overview of the citric acid cycle. https://en.wikipedia.org/wiki/Citric_acid_cycle. Published under the terms of the GNU Free Documentation License

called hexokinases to form glucose 6-phosphate (G6P). This reaction consumes ATP. The second stage involves breaking down the glucose into two three-carbon molecules and the third stage involves harvesting the energy in the chemical bonds of glucose to form a few ATP molecules as well as pyruvate and NADH molecules.

1.3 Enzymes in Metabolism

Metabolism is the chemical reactions in the body's cells that change food into energy using a cascade of enzymes (Fig. 1.6). Metabolism includes catabolism (breakdown of molecules to obtain energy) and anabolism (synthesis of compounds needed by the cells). The three main purposes of metabolism are: the conversion of the energy in food to energy available to run cellular processes, the conversion of food to building blocks for proteins, lipids, nucleic acids, and carbohydrates; and the elimination of metabolic wastes.

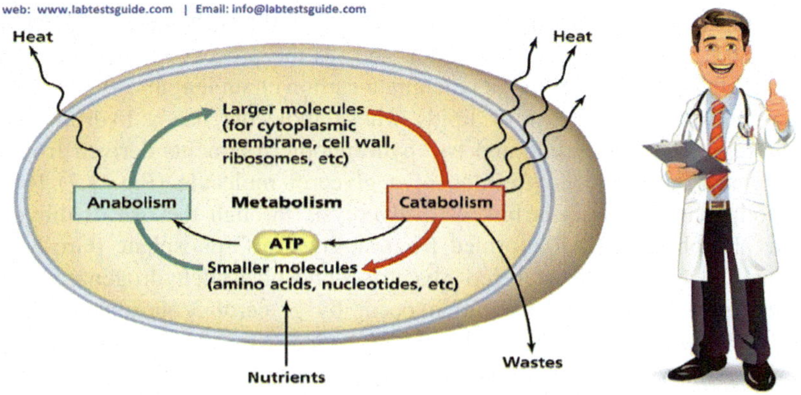

Fig. 1.6 Schematic presentation of role of anabolism and catabolism in cells. https://www.labtestsguide.com/metabolism

1.3.1 Synthesis of Small Molecules

In amino acids enzymatic synthesis, the enzymes glutamate dehydrogenase and glutamine synthetase convert ammonia to glutamate and glutamine, the universal donor of amine groups [56–67]. Amino group from glutamine then can be transferred to other carbon backbones by transamination reactions to make amino acids such as alanine and aspartate, asparagine. Similarly, asparagine and glutamine are synthesized from aspartate and glutamate, respectively, by an amidation reaction step. The process of enzymatic transamination was discovered in the mid-1930s by A.E. Braunstein and associates [58]. Over the next 50 years, they and other authors made important discoveries in the field of aminotransferases and other pyridoxal 5'-phosphate enzymes.

Fatty acid synthesis is the creation of fatty acids from acetylCoA and NADPH through the action of fatty acid synthases. The acetylCoA

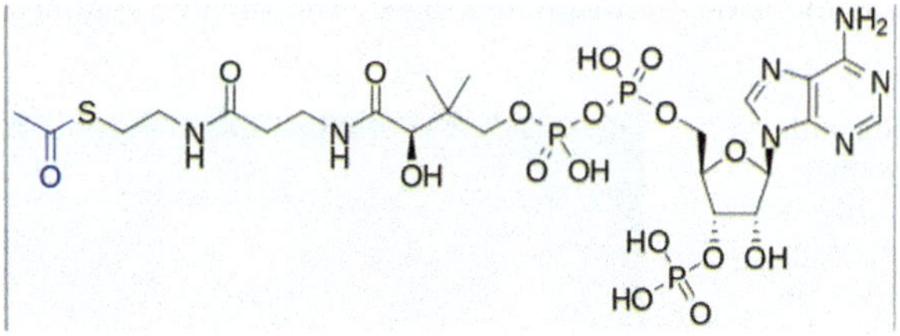

is derived from carbohydrates via the glycolytic pathway. For example, straight-chain fatty acid synthesis of the 16-carbon palmitic acid occurs via the six recurring reactions catalyzed by fatty acid synthase II (FASII), which contain multiple enzymes: (1) The conserved acyl carrier protein (ACP) acts as the mobile domain responsible for shuttling the intermediate fatty acid substrates to various catalytic sites; (2) following each round of elongation the beta keto group is reduced to the fully saturated carbon chain by the sequential action of a ketoreductase (KR), dehydratase (DH), and enoyl reductase (ER); and (3) the process is completed by the action of a thioesterase (TE) upon reaching a carbon chain length.

Phospholipids are a class of lipids whose molecule has a hydrophilic part containing a phosphate group and two hydrophobic fragments derived from fatty acids, joined by an alcohol residue of a glycerol molecule (Fig. 1.7) [68–72]. Phospholipid biosynthesis is linked to glycolysis through the use of dihydroxyacetone phosphate (DHAP), formed from sn-glycerol-3-phosphate (Gro-3-P) via its reduction by NADH, catalyzed glycerol-3-phosphate dehydrogenase .Various phospholipid synthesis and allocation occur by glycerol 3-phosphate acyltransferase (GPAT), acylglycerol-3-acyltransferase (AGPAT), phosphatidic acid phosphatase (PAP) cytidine diphosphate diacylglycerol synthase (CDS), lysophosphatidic acid acyltransferases (LPAAT), phosphatase and choline phosphotransferase, and floppase.

Nucleotide units are made up of a nitrogenous base (cytosine, guanine, adenine, or thymine), pentose sugar (deoxyribose), and phosphate group [73–80]. Nucleotides consist of three chemical components, one of the five bases, a ribose moiety in RNA and a deoxyribose in DNA, and one to three phosphate groups linked together and to the ribose. There are two routes for nucleotide biosynthesis in living cells, the *de novo* and salvage pathways. Carbamoyl phosphate synthetase catalyzes the first step in pyrimidine *de novo* synthesis. The second enzyme in pyrimidine *de novo* synthesis is aspartate carbamoyltransferase (ATCase) [74]. ATCase catalyzes the committed step in the *de novo* synthesis of uridine-5'-monophosphate (UMP), from which all other pyrimidine nucleotides are made. Nucleotide salvage pathways are used to recover bases and nucleosides that are formed during degradation of RNA and DNA. For example, uridine triphosphate can be synthesized in following reactions: (1) Uridine phosphorylase or pyrimidine-nucleoside phosphorylase substitutes the anomeric-carbon-bonded phosphate of ribose 1-phosphate for the free base uracil, forming the nucleoside uridine, (2) Uridine kinase phosphorylates the 5'-carbon of this nucleoside into uridine monophosphate (UMP), and (3) UMP/CMP kinase phosphorylates UMP into uridine diphosphate, which can be phosphorylated into uridine triphosphate by nucleoside diphosphate kinase. The salvage synthesis of nucleotides

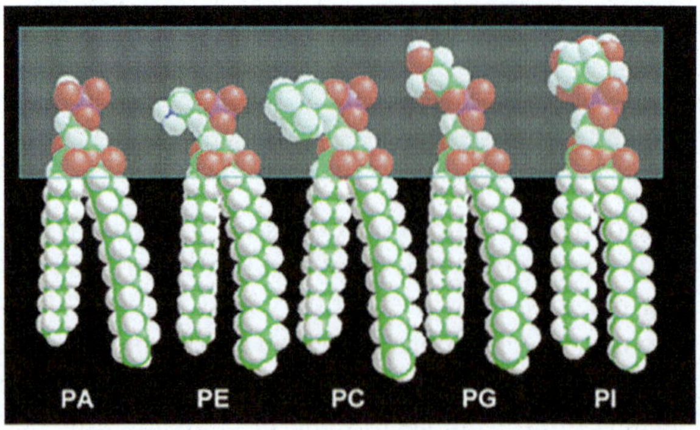

Fig. 1.7 Protrusion of the groups esterified to the phosphate in the surface of the membrane [68]. With permission from Elsevier

occurs from uracil using a reaction catalyzed by a phosphoribosyltransferase enzyme:

$$\text{Uracil} + \text{P}-\text{Rib - PP}$$
$$\rightarrow \text{uridine5'}-\text{monophosphate} + \text{PPi}$$

1.3.2 *Synthesis of Biological Polymers*

Cellulose is a polysaccharide composed of linear glucan chains that are linked together by β-1,4-glycosidic bonds with cellobiose residues. The repeating unit at degree of polymerization ranged from several hundreds to over ten thousands [82, 83]. The individual glucose units are connected via glycosidic bonds between their C1 and C4 positions, and the anomeric C1 carbon adopts the β-configuration. Cellulose is produced by membrane-integrated cellulose synthase (CeS) complexes, containing at least 18 catalytic CESA protein subunits, which synthesize a linear β-1,4-glucan from cytosolic UDP-activated glucose (UDP-Glc, the donor sugar). UDP-glucose pyrophosphorylase (UGP) is the enzyme capable of activating glucose-1-phosphate (Glc-1-P) to UDP-glucose (UDP-Glc).

In both plants and bacteria, UDP-glucose (UDP-Glc) is the key intermediate in cellulose synthesis. The cellulose synthase complex uses the glucose moiety from UDP-Glc transports the glucose across the cell membrane or cell wall and adds the monomer to the nascent extracellular cellulose chain (Fig. 1.8). UDP-glucose is used in nucleotide sugar metabolism as an activated form of glucose, a substrate for enzyme UDP-glucose pyrophosphorylase (UGP).

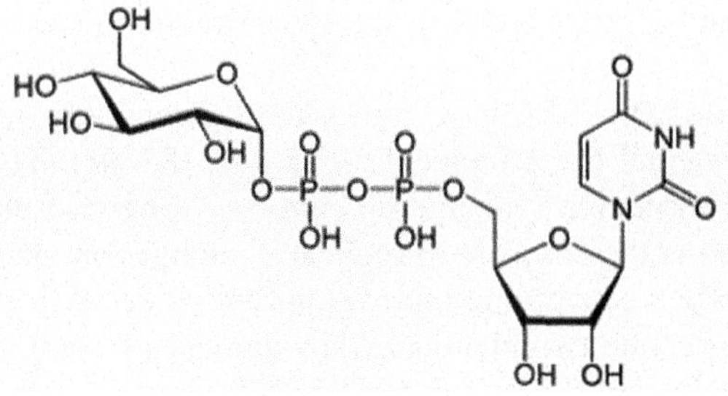

Starch or amylum is a polymeric carbohydrate consisting of amylose (a linear alpha(1–4)-glucan) and amylopectin (an alpha(1–4)-glucan with alpha(1–6)-branch points) in which numerous glucose units are joined by glycosidic bonds [84–88, 83–86]. Starch biosynthesis requires coordinated activities among multiple

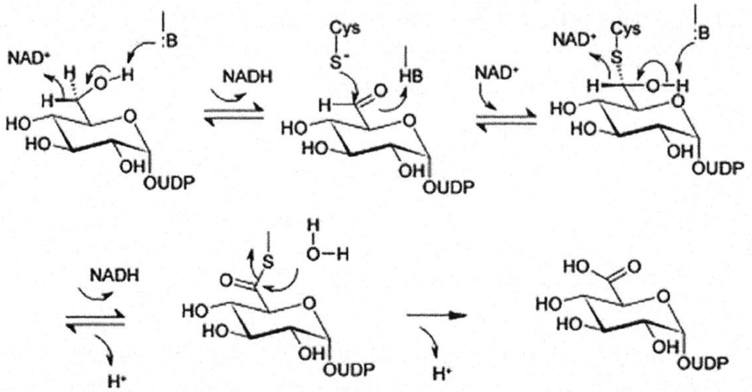

Fig. 1.8 Proposed three-step enzymatic transformation of UDP-Glc into UDPGlcUA catalyzed by [81]. Published under the terms of the Creative Commons CC-BY license from Elsevier

enzymes, including adenosine diphosphoglucose (ADP)-glucose pyrophosphorylase (AGPase), starch synthase (SS), starch branching enzyme (SBE) and starch debranching enzyme (DBE). Glucans and amylopectins) are synthesized via the ADP-glucose pathway by three key enzymes: ADP-glucose pyrophosphorylase, starch synthases, and starch branching enzymes.

DNA synthesis is the creation of deoxyribonucleic acid (DNA) macromolecules, made up of nucleotide units. Nucleotide units are in turn made up of a nitrogenous base (cytosine, guanine, adenine, or thymine), pentose sugar (deoxyribose) and phosphate group. Specific base pairing (adenine A and thymine T, guanine G and cytosine C) occurs when hydrogen bonds form between the nucleotide bases. Two sugar-phosphate chains are paired through hydrogen bonds between A and T and between G and C, thus forming the twin-stranded double helix of the DNA molecule.

DNA replication is the biological process of producing two identical replicas of DNA from one original DNA molecule (Fig. 1.9) [87–91]. The process occurs in four major steps: formation ("unzipping" into two single strands), primer binding (the leading strand is the simplest to replicate), elongation, and termination. DNA replication ends when converging replication forks meet. All in all, these enzymes working in groups create two identical DNA duplexes from a single original DNA duplex. During DNA replication, primer, which is short strands of RNA, is required for DNA synthesis to initiate polynucleotide synthesis [88–91]. Primases are special RNA polymerases that synthesize short-lived oligonucleotides. The following key enzymes are involved in the DNA replication [88]: (1) DNA polymerase, the enzyme responsible for catalyzing the addition of nucleotide substrates to DNA in the 5′ to 3′ direction during the reaction deoxynucleoside triphosphate ($DNA_n \rightleftharpoons$ pyrophosphate + DNA_{n+1}); (2) topoisomerase relaxes the DNA from its supercoiled nature; (3) helicase separates the two strands of DNA at the replication fork behind the topoisomerase; and (4) DNA ligase re-anneals the semiconservative strands and joins Okazaki fragments of the lagging strand. Okazaki fragments are the short lengths of DNA that are produced by the discontinuous replication of the lagging strand. Thus, at the replication fork, many replication enzymes assemble on the DNA into a complex molecular machine called the replisome (Fig. 1.9).

Protein synthesis proceeds into two phases—transcription and translation (Fig. 1.10). Synthesis of RNA (transcription) is catalyzed by an enzyme—RNA polymerase—using DNA as a template (Fig. 1.10). Steps of the protein synthesis are: (1) Initiation of transcription begins with the binding of the enzyme to a promoter (the promotor is a sequence of DNA to which proteins bind to initiate transcription of a single RNA transcript); (2) the DNA double helix is unwound by the helicase activity of the enzyme; and (3) the enzyme then synthesizes a complementary RNA molecule with elongation occurring in the 5' to 3' direction. The single-stranded ribonucleic acid (RNA) product synthesized by transcription of DNA, processes to yield messenger ribonucleic acids (mRNAs), transfer RNA (tRNAs), and ribosomal RNA (rRNAs) and can be modified by enzymes after transcription. There are also a number of RNA-dependent RNA polymerases that use RNA as their template for synthesis of a new strand of RNA.

Transfer RNA (tRNA), composed, typically from 76 to 90 nucleotides in length carries an amino acid to the protein synthesizing machinery of a cell called the ribosome (Fig. 1.11). [96, 97]. Ribosomes consist of two major components: the small and large ribosomal subunits which in turn consists of ribosomal RNA (rRNA) molecules and ribosomal proteins. Using the mRNA as a template, the ribosome traverses each codon (3 nucleotides) of the mRNA, pairing it with the appropriate amino acid provided by an aminoacyl-tRNA. Aminoacyl-tRNA contains a complementary anticodon on one end and the appropriate amino acid on the other. Complementation of a 3-nucleotide codon in a messenger RNA (mRNA) by a 3-nucleotide anticodon of the tRNA results in protein synthesis based on the mRNA code.

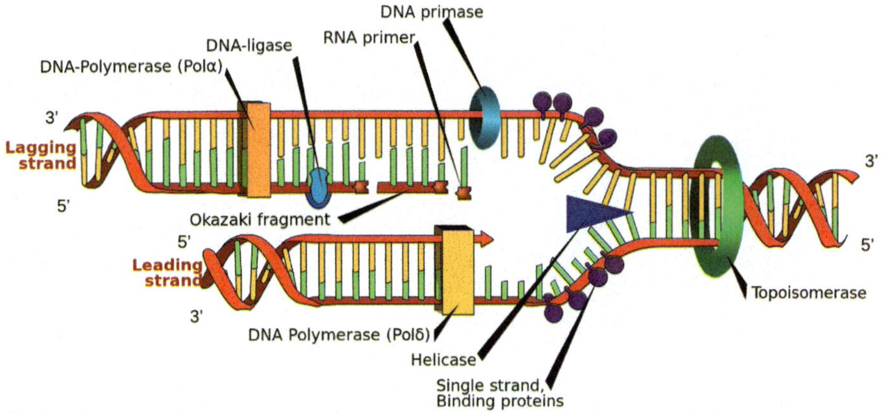

Fig. 1.9 Mechanisms of DNA replication termination. Reproduced under Lady of Hats grants

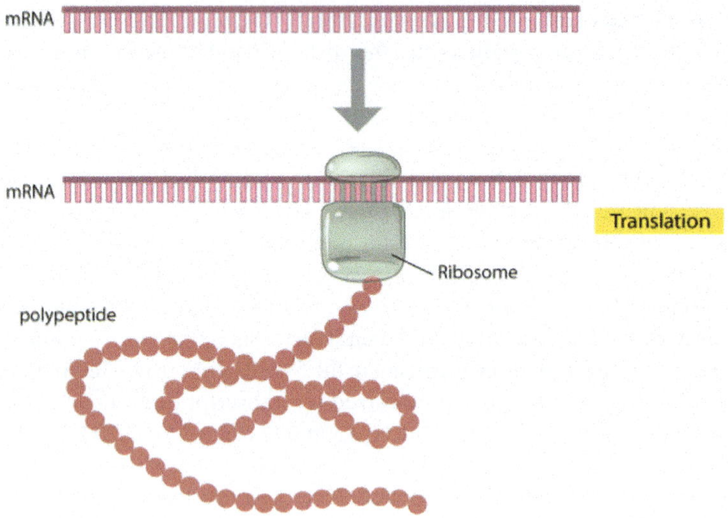

Fig. 1.10 A gene is expressed through the processes of transcription and translation. During transcription, the enzyme RNA polymerase (green) uses DNA as a template to produce a pre-mRNA transcript (pink). The pre-mRNA is processed to form a mature mRNA molecule that can be translated to build the protein molecule (polypeptide) encoded by the original gene [94] delete[93]. With permission from Springer Nature

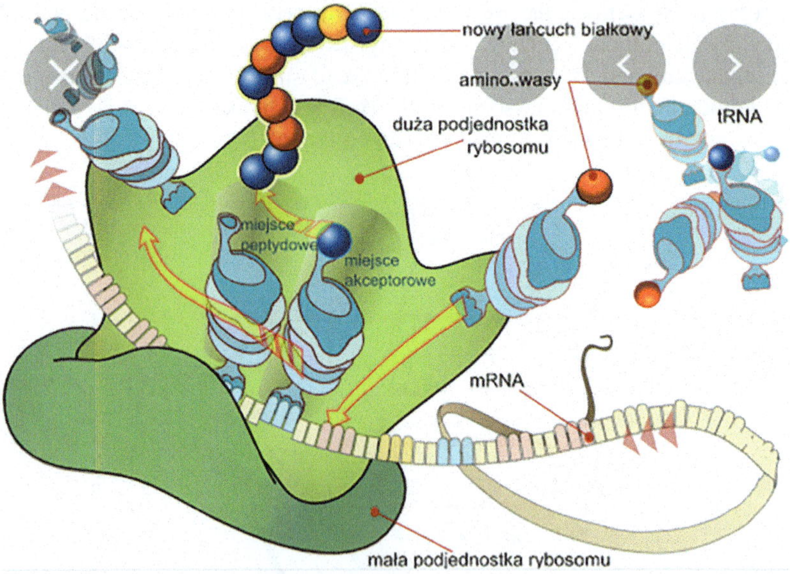

Fig. 1.11 Diagram showing the translation of mRNA and the synthesis of proteins by a ribosome. Published under the term of Lady of Hats grants

Translation proceeds in several steps: [96, 97]: (Fig. 1.10) (1) The ribosome assembles around the target mRNA; (2) the first tRNA is attached at the start codon; (3) the last tRNA is accommodated the small ribosomal subunit (accommodation) transfers the amino acid; (3) it carries to the large ribosomal subunit which binds it to the one of the preceding admitted tRNA (transpeptidation); (4) the ribosome creates an amino acid chain by moving to the next mRNA codon; (translocation); (5) when a stop codon is reached, the ribosome releases the polypeptide; and (6) the remained intact ribosomal complex moves on to the next mRNA to be translated. Crystal structures of the ribosome have shown that the peptidyl-transferase activity is catalyzed by the 28-S RNA in the major subunit of rRNA [98]. For recognition of the appropriate tRNA, the ribosome utilizes large conformational changes and the polypeptide later folds into an active protein

1.3.3 Degradation of Biological Macro Molecules

Degradation of biological molecules protein includes the following catabolic reactions: protein→ amino acid → AcetylCoA, polysaccharides → monosaccharides → AcetylCoA, fats → fatty acid → AcetylCoA. In turn, AcetylCoA [100].is involved in citric acid cycle and oxidative phosphorylation. Lipolysis is the metabolic pathway through which lipid triglycerides are hydrolyzed into a glycerol and three fatty acids by lipases [100, 102–[104]. The six separate steps of hydrolysis involved in the lipolysis process: (1) triacylglycerol is hydrolyzed to make diacylglycerol; (2) this is catalyzed by adipose triglyceride lipase (ATGL); (3) diacylglycerol is hydrolyzed to make monoacylglycerol; (4) this is catalyzed by hormone-sensitive lipase (HSL); (5) monoacylglycerol is hydrolyzed to make glycerol; and (6) this is catalyzed by monoacylglycerol lipase (MGL).

Mitochondrial beta-oxidation of fatty acids proceeds several steps, to produce three energy storage molecules per round of oxidation, including one NADH, one FAD(H2), and one acetylCoA molecule [105, 106]The steps are: (1) acyl CoA dehydrogenase involves in the handling of fatty acids; (2) enoyl CoA hydratase performs a hydration step of the double bond between the alpha and beta carbons; (3) beta-hydroxyl acyl CoA dehydrogenase enzyme catalyzed the chemical reaction

$$\begin{aligned}&(\mathrm{S}) - 3 - \mathrm{hydroxyacyl} - \mathrm{CoA}+ \rightarrow \mathrm{NAD}^{+}\\&= 3 - \mathrm{oxoacyl} - \mathrm{CoA} + \mathrm{NADH} + \mathrm{H}^{+};\end{aligned}$$

and (4) the process's final step involves cleavage of the bond between the alpha and beta carbon by CoASH (Коэнзим A)

Proteases or peptidase are enzymes that catalyze proteolysis, breaking down proteins into smaller polypeptides and eventually into amino acids by hydrolysis [108–110], 108–111]. Based on the chemical nature of group, attacking a substrate in the enzyme active site and mechanism of catalysis, proteases are classified into six distinct classes, namely, aspartic, glutamic, metalloproteases, cysteine, serine, and threonine proteases. The first three classes utilize an activated water molecule as a nucleophile to attack the peptide bond of the substrate, whereas in other enzymes, the nucleophile is an amino acid residue (Cys, Ser, or Thr, respectively). A metalloproteinase, or metalloprotease, is protease enzyme whose catalytic mechanism involves a metal (zinc, cobalt). The ligands coordinating the metal ion can vary with histidine, glutamate, aspartate, lysine, and arginine. The fourth coordination position is taken up by an activated labile water molecule.

Amylase enzymes (salivary amylase and pancreatic amylase) catalyze the breakdown of starch, which is a mixture of amylose and amylopectin polysaccharide polymers, into sugars. α-Amylase acts randomly along the starch chain hydrolyzing the α-(1→4) glycosidic bonds to produce a combination of maltotriose, maltose, and limit dextrins. β-Amylase breaks down starch into the disaccharide maltose. Oligosaccharidases catalyze the disaccharide hydrolyzing. Cellulases break down the cellulose molecule into monosaccharides such as beta-glucose, or shorter polysaccharides and oligosaccharides [111–114, 112–114]. Cellulases cleave the β-1,4-glucan bonds in cellulose, belonging to the glycoside hydrolases family of enzymes which catalyzes cleave glycosidic linkages in carbohydrates. The three types of reaction catalyzed by cellulases are: (1) Breakage of the noncovalent interactions present in the amorphous structure of cellulose (endocellulase); (2) Hydrolysis of chain ends to break the polymer into smaller sugars (exocellulase); and (3) Hydrolysis of disaccharides and tetrasaccharides into glucose (beta-glucosidase). Five general types of cellulases based on the type of reaction catalyzed are: (1) Endocellulases randomly cleave internal bonds at amorphous sites that create new chain ends; (2) exocellulases or cellobiohydrolases cleave two to four units from the ends of the exposed chains produced by endocellulase, resulting in tetrasaccharides [or disaccharides, such as cellobiose; (3) cellobiases or beta-glucosidases hydrolyze the exocellulase product into individual monosaccharides; (4) oxidative cellulases depolymerize cellulose by radical reactions; and (5) cellulose phosphorylases depolymerize cellulose using phosphates instead of water. Three types of cellulases are secreted by fungi that are endo-β-1, 4-glucanases, cellobiohydrolase, and cellobioses or β-glucosidases

1.4 Miscellaneous Enzymes

Hydroxylases are enzymes which hydroxylate organic compounds. In mammals, these enzymes oxidize steroids, fatty acids, and xenobiotics[115–117]. The action of these systems on compounds having unsaturated bonds may result in their epoxidation or demethylation with methylamines. These enzymes are involved in a number of vital processes including the biosynthesis of steroids, degradation of xenobiotics, drug metabolism, and carcinogenesis. For example, phenylalanine hydroxylase (PAH) is an enzyme that catalyzes the hydroxylation of the aromatic side chain of phenylalanine to generate tyrosine. PAH belongs to a class of monooxygenase that uses tetrahydrobiopterin (BH_4, a pteridine cofactor) and a nonheme iron for catalysis.

Cytochromes P450 containing protoheme IX, powerful enzyme hydroxylation systems of organic substrates, i.e., steroids, hydrocarbons, organic acids, alcohols, and amines, are operatives in animal and plant tissues and bacteria [117–126]. The most common reaction catalyzed by cytochromes P450, containing protocheme IX is a monooxygenase reaction, e.g., insertion of one atom of oxygen into the aliphatic position of an organic substrate (RH), while the other oxygen atom is reducedto water:

$$RH + O_2 + NADPH + H^{+\infty} \rightarrow ROH + H_2O + NADP^+$$

The reduction of the enzyme can involve flavin reductases and electron carriers, such as adrenodoxin, rubredoxin, and cytochrome b_5. Dioxygen, being a weak one-electron oxidant, is activated after the two-electron reduction in the enzyme heme coordination sphere. Ferryl rebound mechanisms suggest the formation of the ferryl-oxene(Por•FeIV-O) [126, 127]. This mechanism proposed by Groves and McClusky involves initial hydrogen abstraction from the alkane by (Por•FeIV-O) followed by the alkyl radical rebound on the ferryl-hydroxo intermediate with formation alcohol. Current evidences strongly support Fe(IV)=O as the hydroxylating intermediate.

Nitrogen fixation is a chemical process by which molecular nitrogen, with a strong triple covalent bond, is converted into ammonia. Biological nitrogen fixation (BNF) occurs when atmospheric nitrogen is reduced to ammonia by a nitrogenase enzyme [14, 128–135, 138, 139]. The overall reaction for BNF is:

$$N_2 + 8H^+ + 8e^- + 16\,MgATP \rightarrow 2NH_3 + H_2 + 16\,MgADP + 16P_i$$

Nitrogenase consists of two proteins, a catalytic MoFe protein having of two clusters: $[Fe_4S_3]$ and $[MoFe_3S_3]$, which are linked by three sulfide ions and a reducing iron-only protein carrying two $[Fe_4S_3]$ cubes (P-clusters) linked by a central carbon atom. The mechanism proceeds via a series of protonation and reduction steps wherein the FeMoco active site hydrogenates the N_2 substrate. The function of the Fe protein, the dinitrogenase reductase, is to transfer electrons from a reducing agent, such as ferredoxin or flavodoxin to the MoFe protein nitrogenase protein. The electron transfer is coupled with the reaction of ATP hydrolysis the chemical energy of which is used to increase the reduction potential of the P cluster. The Lowe–Thorneley kinetic model suggests the eight correlated proton and electron transfers required throughout the reaction [14] and references therein. At the key stage of the nitrogenase process, after the accumulation of four reducing equivalents, the reduction of the complex N_2 in cluster $[MoFe_3S_3]$, leads to the formation of an intermediate hydrazine derivative by one-step four-electronic mechanism (four-electron mechanism) [14, 133–135]. The transfer of electrons requires an input of chemical energy which comes from the binding and hydrolysis of ATP. The hydrolysis of ATP also causes a conformational change within the nitrogenase complex, bringing the Fe protein and MoFe protein closer together for easier electron transfer.

Methane monooxygenases (MMOs) are the enzymes that catalyze the oxidation of methane via incorporation of one hydroxyl group into substrate:

$$CH_4 + O_2 + NAD(P)H + H- > CH_3OH + NAD(P)^+ + H_2O$$

The enzyme can also cooxidize a wide range of hydrocarbons and chlorinated hydrocarbons. Methane monooxygenase belongs to the class of oxidoreductase enzymes, nonheme diiron enzymes [130–132]. The reaction is shown to occur via a bound-radical mechanism. There are two forms of MMO: the well-studied soluble form (sMMO) and the particulate form (pMMO). The active site in sMMO contains a diiron center bridged by an oxygen atom (Fe-O-Fe), whereas the active site in pMMO utilizes copper. Structures of both proteins have been determined by X-ray crystallography [132].

A kinase is an enzyme that attaches a phosphate group to a protein, while a phosphatase is an enzyme that removes a phosphate group from a protein [130, 132–146]. A protein kinase catalysis is the transfer of phosphate from ATP (or GTP) to its protein substrates. All kinases require a Mg^{2+}–nucleotide complex as one of the enzyme substrates. Specifically, the enzymes catalyze phosphoryl transfer from MgATP to Ser/Thr and Tyr side chains in proteins. Currently, the following types of kinases have been identified and are being intensively studied: cyclin-dependent kinases, mitogen-activated protein kinases phosphatidylinositol kinases, sphingosine kinases, riboflavin kinase and thymidine kinase. A coordination between highly conserved protein kinases and the family of phosphoprotein phosphatases (PPP) establishes the phosphorylation state of proteins, which controls their function. Recent studies have shown that protein phosphatases are specific and selective enzymes and that their activity can be tightly regulated. Protein kinases and protein phosphatases regulate each other during mitosis (Fig. 1.12) [141]. Inhibition of enzyme PP2A/B55 as well as other members of the PPP family is directly or indirectly controlled by Cdk1/cyclin B.

Kinetically, kinases can be classified into two general types, those that follow ternary complex mechanisms and those that follow ping-pong mechanisms [132, 143]. Ternary complex mechanisms typically involve direct reaction between the two substrates to afford the two products, whereas ping-pong mechanisms proceed through a covalent enzyme intermediate, which in the case of kinases would be a phosphoenzyme species. Mechanistic scheme of associative and dissociative transition states of phosphoryl transfer was described. Figure 1.13 illustrates the kinases role in signal transduction pathways.

Phosphatases have been classified into four major classes based on their substrate preference, inhibitor sensitivity, and catalytic mechanism [145–151]. These classes include phosphoprotein phosphatases (PPP), Mg^{2+}/Mn^{2+}-dependent protein phosphatases (PPM), phosphotyrosine phosphatases (PTP), and aspartate-based protein phosphatases. For example, cysteine-dependent phosphatases (CDPs) catalyze the hydrolysis of a phosphoester bond via a phospho-cysteine intermediate. Metallo-phosphatases (PP2C) coordinate catalytically essential metal ions within their active site. The free cysteine nucleophile forms a bond with the phosphorus atom of the phosphate moiety, and the P-O bond links the phosphate group to the tyrosine [150]. In addition to the major serine/threonine-specific phosphoprotein phosphatase, Mg^{2+}-dependent phosphoprotein phosphatase, and protein tyrosine phosphatase families,

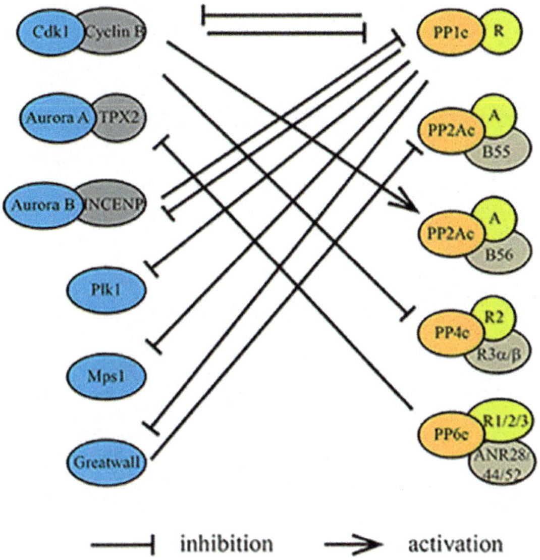

Fig. 1.12 Protein kinases and protein phosphatases regulate each other during mitosis. Protein kinases and protein phosphatases coordinate with each other through underlying dynamic phosphorylation changes within kinase/phosphatase catalytic or regulatory subunits. Cdk1/cyclin B, directly or indirectly, inhibits the phosphatase activity of PP1, PP2A-B55, and PP4. Conversely, PP1 suppresses the kinase activity of AURKB through its regulatory protein Sds22 (PPP1R7), Plk1 through Mypt1 (PPP1R12A) and Gwl through PPP1R3B. Phosphorylation by Cdk1/cyclin B within or near the PP2A-B56 binding LxxIxE motif in substrates increases the affinity of PP2A-B56 interactions. PP2A-B55 activity is inhibited during mitosis by Gwl phosphorylation of the inhibitory proteins ENSA and ARPP-19. PP1 inactivates Gwl at mitotic exit, thereby activating PP2A-B55. PP6 is the T-loop phosphatase for AURKA, thereby decreasing its activity directly [130, 141].

there are novel protein phosphatases, including enzymes with aspartic acid-based catalysis and subfamilies of protein tyrosine phosphatases [145].

1.5 Enzymes in Biotechnology and Industry

1.5.1 Introduction

Biotechnology is characterized as the exploitation of natural procedures for industrial and varied purposes such as pharmaceuticals, bio-fuels, paper and pulp, food and brewing, agricultural industries, and others [152–204]. Enzymes are of great importance in the development of industrial bioprocesses. Many industrial processes, such as cheese manufacturing, have traditionally used impure enzyme sources often

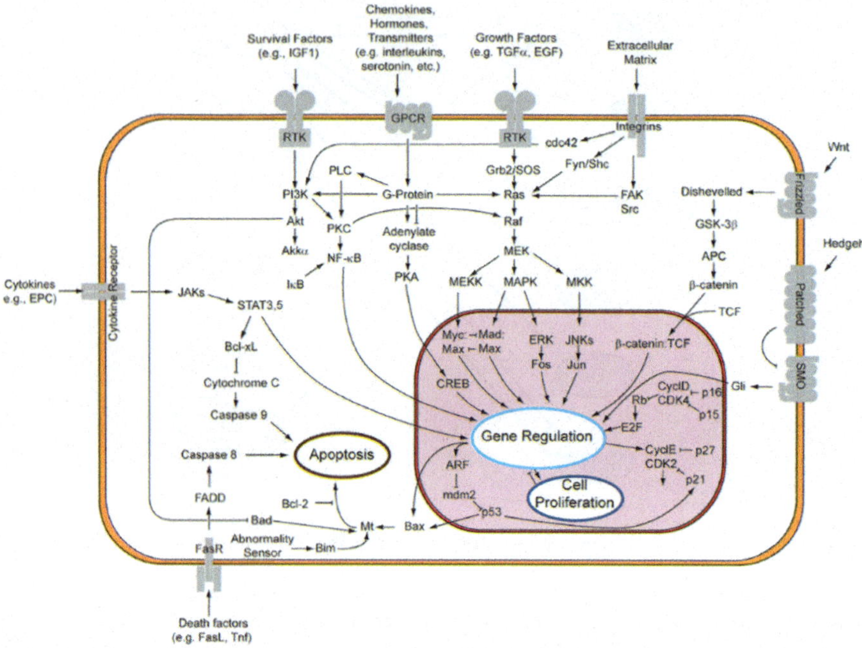

Fig. 1.13 Overview of signal transduction pathways. Many of the proteins involved are kinases, including protein kinases (such as MAPK and JAK) and lipid kinases (such as PI3K) https://en.wikipedia.org/wiki/Kinase

from animals or plants. Much of modern industrial enzymology with the commercial exploitation of microbial enzymes were developed in Table 1.2. Industrial enzymes have many applications such as detergent, food, animal feed, cosmetics, biofuel, medication, pharmaceuticals, and technical use. Thermophilic and hyperthermophilic enzymes represent a superior source for industrial applications owning their stability. The global market of enzymes is estimated in 2015 to be about 3.7 billion USD with a 10% expansion. Commercially available microbial enzymes are about 200 out of almost 4000 enzymes known.

1.5.2 *Miscellaneous Applications of Enzymes*

Microorganisms such as bacteria, yeast, and fungi have been used in food fermentation since ancient times and fermentation processes are still applied in the preparation of many of the food items [155–157]. Microbial enzymes are the preferred source to plants or animals due to several advantages such as easy, cost-effective, and consistent production. Microbial cellulases find applications in utilization of biomass for

Table 1.2 Enzymes as a unit operation https://wikious.com/en/Industrial_enzymes

Enzyme	Industry	Application
Palatase	Food	Enhance cheese flavor
Lipozyme TL IM	Food	Interesterification of vegetable oil
Lipase AK Amano	Pharmaceutical	Synthesis of chiral compounds
Lipopan F	Food	Emulsifier
Cellulase	Biofuel	Class of enzymes that degrade cellulose to glucose monomers
Amylase	Food/biofuel	Class of enzymes that degrade starch to glucose monomers
Xylose isomerase	Food	High-fructose corn syrup production
Rasinase	Paper	Pitch control in paper processing
Penicillin amidase	Pharmaceutical	Synthetic antibiotic production
Amidase	Chemical	Class of enzymes used for nonproteinogenic enantiomerically pure amino acid production

fuel production employing cellulases, because enzymatic hydrolysis of cellulose is an effective approach for obtaining fermentable sugars.

Lipases serve important roles in human practices as yogurt and cheese fermentation from ancient time. In more modern applications, lipases are being exploited as baking, laundry detergents, for the removal of size lubricants, manufacturing of sausage, modifying the food flavor by synthesis of esters of short-chain fatty acids and alcohols, and for enhancing the aroma and speeding up the fermentation of apple wine [158–163]. The α-amylases have been used in baking, brewing, production of cakes, preparation of digestive aids, fruit juices, and starch syrups and addition to the dough of bread in order to degrade the starch into smaller dextrins [154–157]. Enzymatic biodiesel is becoming an increasingly popular topic in bioenergy literature. An innovative system for enzymatic biodiesel production involving a combination of a lipase and two phospholipases was developed. This allowed performing the enzymatic degumming and transesterification in a single step, using crude soybean oil as feedstock, and converting part of the phospholipids into biodiesel. The application of alkaline pectinases in the textile industry for the retting and degumming of fiber crops, production of good quality paper, fermentation of coffee and tea, and in bringing down the cloudiness and bitterness of fruit juices has been reported. Employment of lipases as biocatalysts in the transesterification of triacylglycerides allows mild reaction conditions and easy recovery of glycerol [170].

Lignocellulose-degrading enzymes, belonging to glycoside hydrolases (GH), polysaccharide lyases (PL), carbohydrate esterases (CE), lignin oxidases (LO), and lignin-degrading auxiliary enzymes (LDA) families, are used as industrial biocatalysts for fuels and chemicals [165–166]. Pectinases break down complex polysaccharides of plant tissues into simpler molecules like galacturonic acids [160, 167].

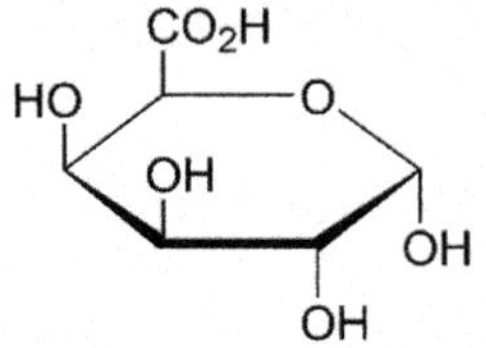

Peroxidases are oxidoreductases produced by a number of microorganisms and plants and catalyze a variety of practical reactions in the presence of peroxides such as hydrogen peroxide [168] including (1) oxidative polymerization of phenols and aromatic amines, conducted by horseradish peroxidase (HRP) in water and water-miscible organic solvents; (2) soil detoxification; while HRP, soybean, and turnip peroxidases have been applied; (3) bioremediation of waste waters by HRP, soybean and turnip peroxidases contaminated with phenols, cresols and chlorinated phenols; and (4) biopulping and biobleaching in the paper industry using lignin peroxidase (LiP) and manganese peroxidase (MnP).

Fungal naringinase is used industrially in large amounts due to its activities to hydrolyze Naringenine-7-rhamnosidoglucoside which include the production of rhamnose

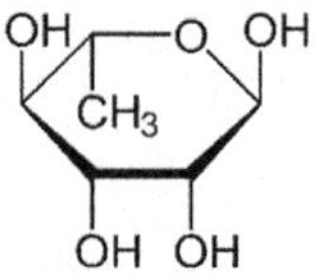

and prunin

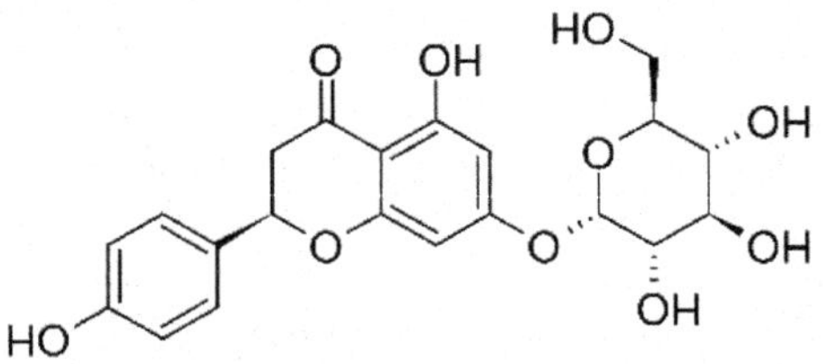

and debittering of citrus fruit juices.

Yeasts invertase converts sucrose to glucose and fructose, while zymase converts it finally to ethanol and CO_2. The enzyme zymase was used as biocatalyst to improve ethanol production. Various operational and nutritional parameters were investigated and analyzed to determine the optimum ethanol production [171]. Proteases, amylases, lipases, cellulases, and mannanases are enzymes that are used as laundry detergent being the largest application of industrial enzymes [174–185]. They catalyze the reaction between stains and the water solution, thus aiding stain removal and improving efficiency. Specifically [174] (1) Proteases degrade stains comprised mainly of protein, such as grass, blood, egg, and others, giving clothes a clean appearance; (2) Amylases break down starch-based stains commonly caused by pasta, potatoes and baby food, which are commonly found on dishes; (3) lipases target fat-based stains such as butter, oil, and human sebum; (4) for cotton fabrics, cellulases improve overall cleanness by reducing redeposition of particulate soils such as soot, clay, and rust during the wash; cellulases also provide fabric and color care; (6) mannanases degrade stains containing mannans; and (7) pectate lyases act on pectin-based stains from fruits and vegetables, jams and other food containing thickeners.

Untreated cotton fibers can break easily when being woven into textures. Amylase enzymes are generally utilized in desizing (the process of removing the size material from warp yarns after a textile fabric is woven). The studies [175, 176] explored the potential of amylase for textile desizing extracted from indigenous environmental *Aspergillus tubingensis* isolate. The sized cotton fabric was enzymatically desized, scoured (clean or brighten), and bleached with an enzyme mixture of starch-degrading enzymes, pectinases [176–178]. Glucose oxidases were added to generate hydrogen peroxide. In the one-step process for enzymatic desizing and bioscouring of cotton fabrics, a mixture of α-amylase (Am) and polygalacturonase (PG) enzymes from *Trichoderma harzianum* induced with orange peel was produced [179]. The enzymes used in the textile field are amylases, catalase, and laccase which are employed to removing the starch, degrading excess hydrogen peroxide, bleaching textiles, and degrading lignin [183]. The cellulases are applied for denim (a sturdy cotton warp-faced textile) finishing, and lactases are used for decolorization of textile effluents and textile bleaching. The use of highly specific and efficient enzymes, working under mild conditions, results in reduced process times, energy and water savings, improved product quality. Biopolishing is a finishing process that improves fabric quality by mainly reducing fuzziness and pilling property of cellulosic fiber, bringing the fabric a cleaner surface, a cooler feel, luster, and softer feel [184, 185]. The aims of the study [186] were to analyze the treatment conditions for the use of acid and neutral cellulase enzymes and evaluate the changes in the various properties of the treated fabrics. In this study, biopolishing was carried out on bleached fabric using acid cellulase (Mega PK) and neutral cellulase. On the basis of the observed results, it was suggested that Mega PK showed a better result when using 1% concentration at 55 °C for 40 min. On the other hand, a comparatively better result was found using 1% Mega L at 55 °C for 50 min.

Currently industries such as textile and pulp and paper shift away from toxic, carcinogenic chemical bleaching processes to hydrogen peroxide-based bleaching because of its nontoxic and eco-friendly characteristics [180] For example, in an effort to replace these compounds, hydrogen peroxide was generated as a bleaching compound by means of enzymatic reactions [181] Three different pathways were investigated: (1) the H_2O_2 production from glucose; (2) the production of H_2O_2 from carboxymethylcellulose (CMC) by the action of both cellulase, which promotes the hydrolysis of the polymeric chain, and glucose oxidase, which oxidizes the smaller fractions to produce H_2O_2; and (3) H_2O_2 was obtained from ethanol, which is present in liquid detergents, by the action of the enzyme alcohol oxidase. The carbohydrate oxidase from *Microdochium nivale* (CAOX), heterologously expressed in *Aspergillus oryzae*, and cellobiose dehydrogenase from *Myriococcum thermophilum* (MtCDH), were assessed for their ability to generate bleaching species at a pH suitable for liquid detergents. Both enzymes were able to produce an amount of H_2O_2 sufficient to decolorize the pure anthocyanins

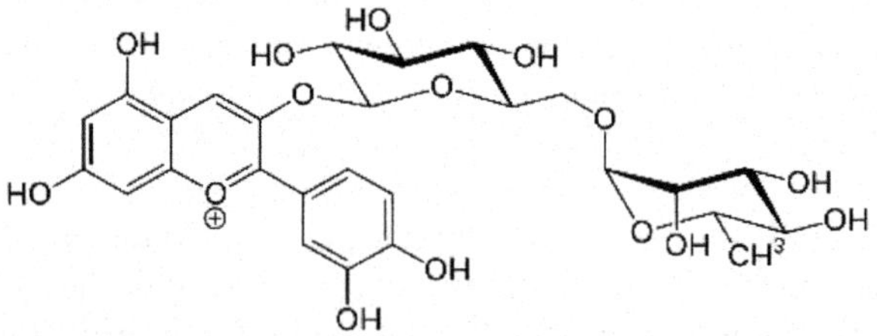

within 2 h and showed significant cleaning benefits on the stains [182].

Enzymes used in the skincare and cosmetics industry are: coenzyme Q10, superoxide dismutase (SOD), proteolytic enzymes, diacylglycerol acyltransferase, lysyl and prolyl hydroxylases, and peroxidase enzymes [190–192]. In research [189], SOD is combined with catalase in skin formulations to prevent wrinkles, liver spots, and fine lines and also helps in scar tissue and keloid reduction and quicker healing of wounds. Pepsin, bromelain, and papain are work by breaking the skin proteins into peptides and then amino acids. Lysyl and prolyl hydroxylases when combined with vitamin C, increase skin collagen which firms the skin and prevents wrinkles and fine lines. Peroxidase preserves the skin creams by preventing bacterial growth. Superoxide dismutase and peroxidase help to protect the skin by neutralizing dangerous substances that can be generated by sun exposure and pollution and reduce UV-induced erythema [187]

Sole enzymes play a role in the decomposition of organic matter and their activities are related to decomposition, microbial activity, soil properties, and biomass. The soil enzymes include amylase, phosphatase, urease, cellulase, protease, glycosidase, phenol oxidase, peroxidase, and arylsulfatase [194, 195]. These enzymes can act in skincare and cosmetics industry. The effects of manure and tillage (the preparation of land for growing crops) practices on activities of enzymes that are key to biogeochemical cycling such as β-glucosidase (C biogeochemical cycling), α-galactosidase (C cycling), β-glucosaminidase (C and N cycling) and phosphomonoesterases (P cycling) were evaluated [195] Results of the study indicated significant responses in C and P cycling enzyme activities to manure applications. This finding suggested potential benefits to soil biogeochemical cycling essential for the productivity of agroecosystems [195]

Polylactic acid (PLA), a thermoplastic polyester with backbone formula,

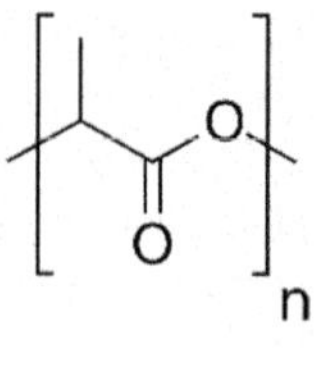

is a promising biomass-derived polymer [192–194, 197–200]. PLA-based materials might be mixed with fleece (the woolly covering of a sheep or goat) or cotton. Production of PLA homopolymer and its copolymer, poly(3-hydroxybutyrate-co-lactate), P(3HB-co-LA), by direct fermentation of metabolically engineered Escherichia coli was reported [188, 189, 197, 198] Synthesis of PLA and P(3HB-co-LA) in E. coli was also performed involving engineered propionate CoA-transferase and polyhydroxyalkanoate (PHA) synthase for the efficient generation of lactyl-CoA and incorporation of lactyl-CoA into the polymer, respectively [199]. The research focused on the study of enzymatic copolymerization of LA with 2-bromo-3-hydroxypropanoic acid using physically immobilized Candida antarctica Lipase B.

The enzyme-mediated ceramic synthesis techniques include precipitants of the metal ions that are generated by the enzymatic reactions as well as the metal ions in the solutions getting precipitated into carbonates, oxides, phosphates, or hydroxides. Enzymes can also interact directly with the metal-containing substrates for producing ceramic material [2001]and allow the synthesis of inorganic solids under mild conditions [2002] [193]

1.6 Enzymes in Medicine

1.6.1 General

Ten main applications of enzymes in medicine are [207–237]: (1) Diabetics use strips of paper fertilized with aldohexose enzyme to observe their glucose; (2) the presence of enzymes may facilitate to diagnose malady; (3) enzymes are typically used as medicines to interchange enzyme deficiencies in patients; (4) nuclease could be a potential medical care for monogenic disorder which caused by variation in a single gene; (5) proteases are accustomed to clean wounds and thus accelerate the healing method; (6) the chemical synthesis of complicated medicine is used to perform chemical conversions; (7) enzymes are accustomed to aid digestion, to supplement the natural amylase, lipase and protease produced by the pancreas; (8) enzymes such as trypsin, chymotrypsin, serratiopeptidase used to break the internal blood clots, to dissolve the hardening of walls of blood vessels, and to dissolve the wound swelling to promote healing; (9) enzymes are the preferred markers in various disease states such as myocardial infarction, jaundice, pancreatitis, cancer, and neurodegenerative disorders; and (10) enzymes such as trypsin, chymotrypsin, serratiopeptidase are used to dissolve the swelling

The unique characteristics of enzymes make them highly applicable for a number of chemical transformation reactions in pharmaceutical industries, such as group protection and deprotection, selective acylation and deacylation, selective hydrolysis, deracemization, kinetic resolution of racemic mixtures, esterification, transesterification, and many others [2005] [2005]. For example, β-glucuronidase (βGLU)—a lysosomal acid hydrolase responsible for the catalytic deconjugation of β-d-glucuronides has emerged as a viable molecular target for several therapeutic applications [207]. Due to its role in colon carcinogenesis and certain drug-induced dose-limiting toxicities, the development of potent inhibitors of βGLU in human intestinal microbiota has been aroused. For example, the aim of work [207] was to improve βG LU inhibitory potency and the development of new therapeutic agents. Enzymes in healthcare and medicine are used for treating disorders, assisting metabolism, medical device cleaning, the manufacture of medicines, regulation of digestion, detection of diseases, addressing food quality, treating heart diseases, and monitoring the health of fetus during pregnancy [208].

1.6.2 Enzymes in Cancer

Considerable support exists for the roles of metabolism in modulating the carcinogenic properties of chemicals, which are procarcinogens that require activation p[2008–213]. The metabolism and especially bioactivation of carcinogens are dominated by cytochrome **P450** enzymes (66% of bioactivations) (Fig. 1.14). The roles of these P450s were considered in issues involving enzyme induction, chemoprevention, molecular epidemiology, interindividual variations, and risk assessment. Evidence from numerous *in vitro* studies, utilizing reconstituted systems, microsomal fractions and cultured cells, implicated cytochrome P450 enzymes as being the predominant enzymes responsible for the metabolic activation of many procarcinogens.

Cytochromes P450 (CYPs) were found to be essential for the activation or inactivation of anticancer drugs. Some products, undergoing the metabolic pathway of the CYP enzyme such as hydroxylated polychlorinated biphenyls

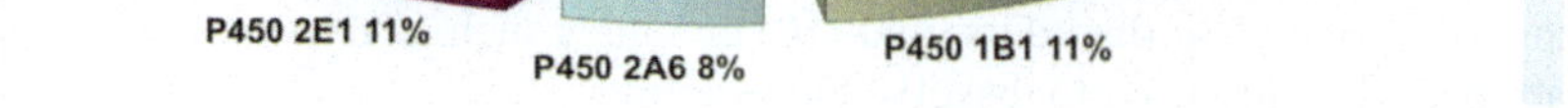

Fig. 1.14 Enzyme contributions to activation of carcinogens. **a** Fractions of activation reactions attributed to groups of enzymes. The analysis is based on 713 reactions. **b** Fractions of P450 activation reactions attributed to individual human P450 enzymes (from a total of 473 reactions considered). See text for discussion [208]. With permission from American Chemical Society

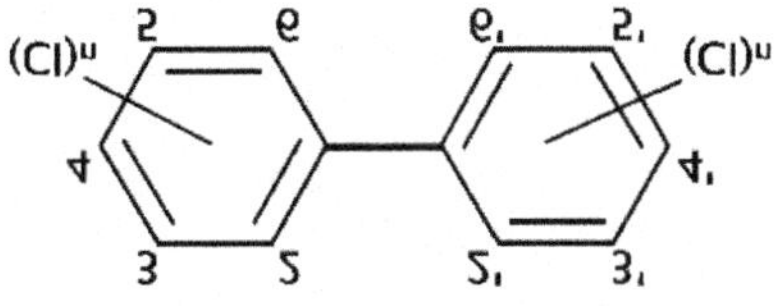

and 4-chlorobiphenyl, are toxic to humans and are considered to be potential carcinogens [211] The association between CYP expression and cancer risk, tumorigenesis, progression, metastasis and prognosis has been widely reported in basic and clinical studies. Review [209] covered studies in which five procarcinogens representing different chemical classes, benzo[*a*]pyrene, 4-aminobiphenyl (4-ABP), 2-amino-1-methyl-6-phenylimidazo[4, 5-b]pyridine, 2-amino-9*H*-pyrido[2, 3-b]indole, and 4-(methylnitrosamino)-1-(3-pyridyl)-1-butanone, have been administered to different P450 knockout mouse models. Of various human P450 enzymes determined to date, CYP1A1, 1A2, 1B1, 2A13, 2A6, 2E1, and 3A4 were reported to play critical roles in the bioactivation of carcinogenic chemicals [210]. *In vivo* studies have shown that: (1) disruption of *Cyp1b1* and *Cyp2a5* genes in mice resulted in suppression of tumor formation caused by 7,12-dimethylbenz[*a*]anthracene and 4-(methylnitrosamino)-1-(3-pyridyl)-1-butanone, respectively; (2) specific inhibitors for CYP1 and 2A enzymes are able to suppress tumor formation caused by several carcinogens in experimental animals *in vivo*; (3) inhibitors of human CYP1 and CYP2A enzymes activated carcinogenic polycyclic aromatic hydrocarbons (**PAHs**) and tobacco-related nitrosamines, respectively, in humans; and (4) the considered inhibitors include a variety of carcinogenic and/or noncarcinogenic PAHs and acetylenic PAHs, many flavonoid derivatives, derivatives of naphthalene, phenanthrene, biphenyl, and pyrene and chemopreventive organoselenium compounds, such as benzyl selenocyanate and benzyl selenocyanate; *o*-XSC, 1,2-, 1,3-, and 1,4-phenylenebis (methylene)selenocyanate.

A quantum mechanics/molecular mechanics (QM/MM) dynamic simulations were performed to explore the interaction between naphthalene and the cytochrome P450 enzyme (CYP 1B1) [211]. Computational results showed that both additions on α- and β-carbon can generate naphthalene 1,2-oxide. The activation barrier for the addition of β-carbon was found to be higher than that for the α-carbon by 2.6 kcal·mol^{-1}. It was also found that naphthalene 1,2-oxide is unstable and the O–C bond cleavage easily occurs via cellular hydronium ion, hydroxyl radical/anion; then it will convert to the potential ultimate carcinogen 1,2-naphthoquinone. Recently, metabolic therapy for cancer, particularly in amino acid deprivation by enzymes were in a focus of special attention [211]. For example, L-asparaginase was approved for the treatment of acute lymphoblastic leukemia according to the US Food and Drug Administration. Arginine deiminase and recombinant human arginase have been developed into clinical trials as potential cancer therapeutic agents for the treatment of arginine-auxotrophic tumors methionase, lysine oxidase, phenylalanine ammonia lyase. Autophagy is a catabolic process that is responsible for the degradation of dysfunctional proteins and organelles. In response to metabolism stress, autophagy could be induced by amino acid deprivation. The manipulation of autophagy in

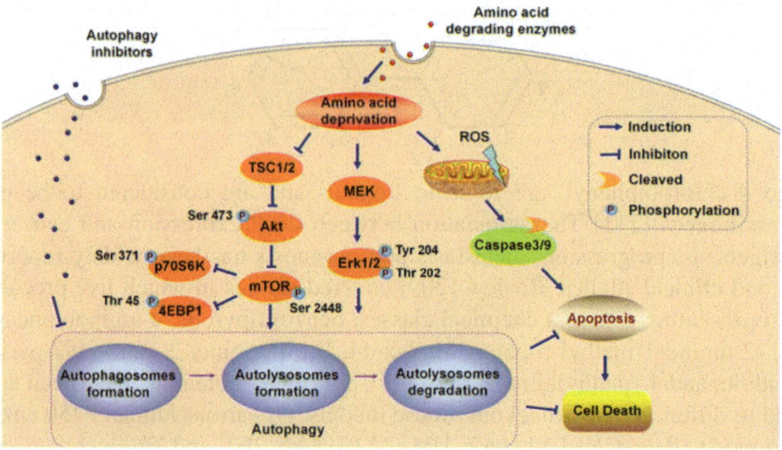

Fig. 1.15 Schematic illustration of the cytoprotective role of autophagy in the cytotoxicity induced by amino acid-degrading enzymes (L-Asparaginase, ADI, and rhArg) in cancer cells [214]. Published under the terms of the Creative Commons Attribution License CC BY

combination with amino acid-degrading enzymes is a potential therapeutic approach in preclinical studies. Figure 1.15 schematically illustrates the cytoprotective role of autophagy in the cytotoxicity induced by amino acid-degrading enzymes. In recent years, there has been increasing interest in targeting tumor amino acid metabolism as a strategy to treat cancer [218–221].

Coronavirus disease (COVID-19) is a disease caused by a virus (SARS-CoV-2) which quickly spread worldwide, resulting in the COVID-19 pandemic [222–237]

Today, COVID-19 has infected more than 300 million people and has already killed more than 5.5 million around the world (World Health Organization). A review [213, 222] discussed the main enzymes related to COVID-19, summarizing the findings on their role in viral entry mechanism, the consequences of their dysregulation, and the effects of SARS-CoV-2 mutations on them. The following important aspects were in focus of attention: (1) Initially, the SARS-CoV-2 virus came to attention through its effects on the respiratory system; (2) other organs also can be injured; (3) SARS-CoV-2 enters the human body cells through the angiotensin-converting enzyme 2 (ACE-2) on the cell surface [214] (Fig. 1.16); (4) the virus can be found in almost every tissue of body and affects the number of receptor; and (5) after invasion, the immune system reaction is responsible for the cytokine storm, which leads to inappropriate responses that cause tissue damage and hence multi-organ failure and death.

A vast amount of clinical data has been revealing COVID-19 enzymatic alteration patterns, including cardiac, pancreatic, and hepatic ones associated with organ damage. In addition, viral mutations can create new disease behaviors, and these effects are related to the activity of enzymes. The cleavage of human proteins such as MYH7 and MYH6 (cardiac myosins linked to several cardiomyopathies), FOXP3

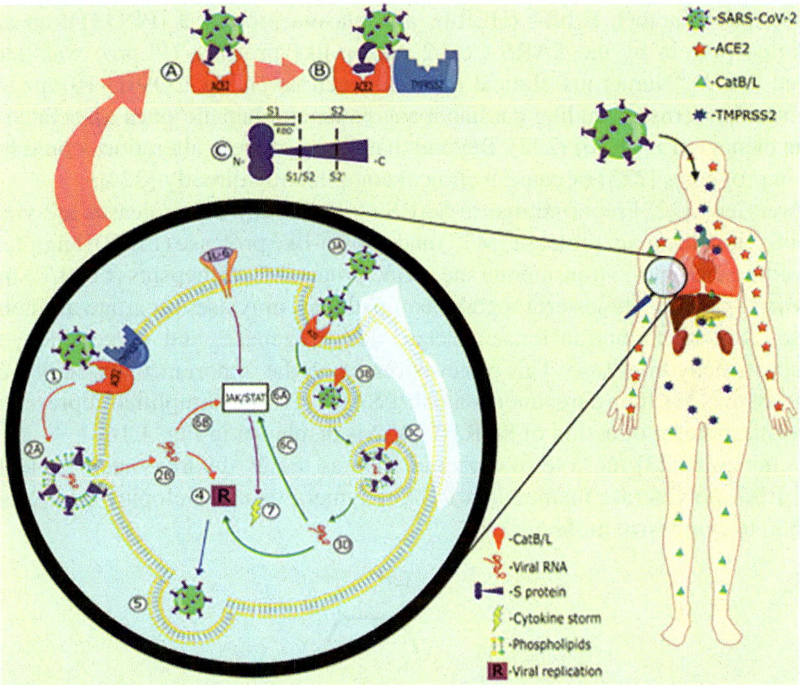

Fig. 1.16 Simplified representation of the mechanism of action of SARS-CoV-2. Initially, the severe acute respiratory syndrome-Coronavirus 2 (SARS-CoV-2) approaches the cell membrane (1) and it binds to the angiotensin-converting enzyme 2 (ACE2; colored in red) (pink arrow, step A) through the receptor binding domain (RBD) of the viral spike (S) protein. The S protein suffers a conformational change and it is cleaved by the transmembrane serine protease-2 (TMPRSS2; colored in blue) (pink arrow, step B), when this protease is expressed in the cell. The S protein has two cleavage sites, represented by the dotted lines, S1/S2, which is between the S1 and S2 domains; and S2′, which is within S2 domain; also, the RBD is housed in the S1 domain (pink arrow, step C). After binding ACE2 and being cleaved by TMPRSS2, the virus fuses with the cell membrane (2A) and the viral RNA is released in the cytosol (2B). Alternatively, the virus can enter the cell through endocytosis (3A). In the endosome, the viral particle is cleaved by cathepsins B and L (CatB/L) (3B). The viral particle fuses with the endosome membrane (3C) and the viral RNA is released in the cytosol (3D). After the release of the viral RNA, either by previous binding of the viral particle to ACE2 or by previous endosome encompassing, the virus undergoes the replication step (4), in which it subverts the cell machinery (not shown). After this step and the synthesis of the viral polyproteins (pp) pp1a and pp1ab, the viral proteases main protease (M^{pro}) and papain-like protease (PL^{pro}) play important roles in releasing themselves by autoproteolytic activity and in processing the central and the C-proximal region of pp1a/pp1ab (M^{pro}) and the N-proximal region of pp1a/pp1ab (PL^{pro}) (not shown). After replication, a new viral particle is generated and it is released from the cell. The presence of the viral RNA detected in other cells elicits the release of cytokines, such as interleukin-6 (IL-6, not discussed in this paper), in the body. IL-6 binds to its cell membrane receptor, it activates the JAK/STAT pathway (6A), and it promotes the cytokine storm (7) in the targeted cells. In parallel, the release of viral RNA in the cytosol (steps 1, 2A, and 2B, or steps 3A, 3B, and 3C) can also directly activate the JAK/STAT pathway in the same cell (6B and 6C), and then it promotes the cytokine storm (7). ACE2, catB/L, and TMPRSS2 are ubiquitously expressed in the body. TMPRSS2, transmembrane serine protease-2 [Color figure can be viewed at wileyonlinelibrary.com] [223]. Published under Creative Commons license CC BY from Wiley

(transcription factor), ErbB4 (HER4), and plasma protein S (PROS1), an antico agulation protein by the SARS-CoV-2 papain-like protease (PLpro) was demonstrated [217]. "Numerous clinical data has been revealing COVID-19 enzymatic alteration patterns, including cardiac, pancreatic, and hepatic ones associated with organ damage (Fig. 1.16) [222]. Beyond that, some of these alterations could have a role in prognosis [223] or cause a clinical complication directly [224].

Overview [222] reported that in SARS-CoV-2 following proteases are vital for the infection: the main protease (M^{pro}) and papain-like protease (PL^{pro}), angiotensin-converting enzyme-2, transmembrane serine protease-2, cathepsins (CatB/L) lipase, enzymes from the cholesterol metabolism pathway, amylase, aspartate aminotransferase, alanine aminotransferase, lactate dehydrogenase, and glyceraldehyde 3-phosphate dehydrogenase. This review stressed on the importance of detailed enzymatic studies for future treatments against SARS-CoV-2. Simplified representation of the mechanism of action of SARS-CoV-2 is displayed in Fig. 1.16.

In review, [223] the role of enzymes such as the N^4-hydroxylated nucleobase, viral RNA polymerase mammalian RNA polymerases in developing a biocatalytic method to synthesize molnupiravir,

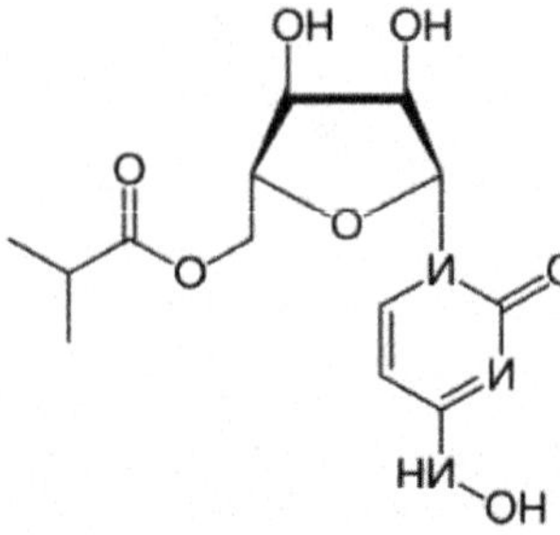

an orally available antiviral agent with clinical efficacy against COVID-19 has been evaluated. Molnupiravir is currently approved for the treatment of mild-to-moderate COVID-19 in adults by the UK's Medicines and Healthcare Products Regulatory Agency. The cytochromes P450 were associated with the metabolism of environmental toxins, drugs, and carcinogens. The role of these enzymes in COVID-19 pathogenesis and therapy, the possible changes in CYP expression and/or function in COVID-19, the effects of CYPs on the metabolism of arachidonic acid, vitamins, and steroid hormones in COVID-19; and the effects of CYPs on the metabolism of therapeutic COVID-19 drugs were evaluated [227]. The possible changes in CYP expression and/or function in COVID-19, the effects of CYPs on the metabolism of arachidonic acid, vitamins, and steroid hormones in COVID-19 the effects of CYPs on the metabolism of therapeutic COVID-19 drugs were discussed.

It was showed [237] that betacoronavirus, severe acute respiratory syndrome coronavirus 2 (SARS-CoV-2), binds to the metalloprotease angiotensin-converting enzyme 2 (ACE2)'. Middle East respiratory syndrome (MERS-CoV) utilizes dipeptidyl peptidase 4 (DPP4), a ubiquitous membrane-bound aminopeptidase. The modeling of the structure of SARS-CoV-2 spike glycoprotein predicted that it can interact with human DPP4 in addition to Angiotensin-converting enzyme 2 (ACE2). Angiotensin

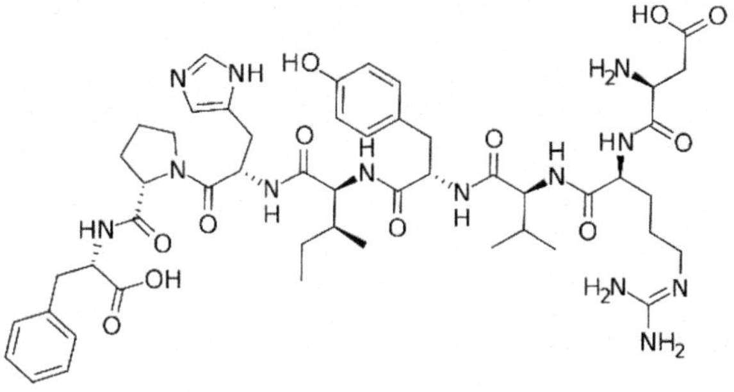

is a peptide hormone that causes vasoconstriction and an increase in blood pressure. DPP4 activity differentially regulates glucose homeostasis and inflammation via its enzymatic activity and nonenzymatic immunomodulatory effects. The dysregulation of DPP4 in COVID-19 comorbid conditions was discussed in detail.

References

1. Thatoi H, Mohapatra S, Das SK (2021) Bioprospecting of enzymes in industry, healthcare and sustainable environment. Springer, Singapore
2. Kara S, Rudroff F (2021) Enzyme cascade design and modelling. Springer
3. Wanmeng M, Zhang W, Chen Q (2021) Novel enzymes for functional carbohydrates production. From scientific research to application in health food industry. Springer Nature Singapore Pte Ltd
4. Magnani F, Marabelli C, Paradisi F (2021) Enzyme engineering: methods and protocols. methods in molecular biology. Publication City/Country New York, NY, United States
5. Morrison H (2020) Enzyme active sites and their reaction mechanisms. Elsevier Science Publishing Co Inc
6. Hussain CM, Sudarsanam P (2020) Advanced functional solid catalysts for biomass valorization. Elsevier Science Publishing Co Inc 2020
7. Bisswanger H (2019) Practical enzymology. 3rd edn. Wiley
8. Punekar NS (2018) Enzymes: catalysis, kinetics and mechanisms. Springer
9. Bisswanger H (2017) Enzyme kinetics: principles and methods. Third, enlarged and improved Edition. Wiley-VCH Verlag GmbH & Co
10. Brahmachari G, Demain A, Adrio J (2016) Biotechnology of microbial enzymes production, biocatalysis and industrial applications. Elsevier
11. Rampelotto PH (2015) Enzymes and their biotechnological applications (MDPI 2015)
12. Purich D (2009) Contemporary enzyme kinetics and mechanism. 3rd edn. Elsevier
13. Chaiyen P, Tamanoi F (2004) Flavin-dependent enzymes: mechanisms, structures and applications, handbook of proteolytic enzymes. vol 1. Elsevier
14. Likhtenshtein GI (2003) New trends in enzyme catalysis and mimicking chemical reactions. Kluwer Academic/Plenum Publishers, N.Y
15. Denisov ET, Sarkisov OM, Likhtenshtein GI (2003) Chemical kinetics. Fundamentals and recent developments. Elsevier Science

16. Likhtenshtein GI (2012) Solar energy conversion. chemical aspects. Wiley-VCH, Weinhem
17. Likhtenshtein GI (2016) Electron spin in chemistry and biology: fundamentals, methods, reactions mechanisms, magnetic phenomena, structure investigation. Springer
18. Eisenthal R, Danson M (2002) Enzyme assays. A practical approach. Oxford University Press
19. Taylor KB (2002) Enzyme kinetics and mechanisms Springer, Dordrecht; Marangoni AG (2002) Enzyme kinetics: a modern approach. Wiley-Interscience
20. Silverman RB (2002) The organic chemistry of enzyme-catalyzed reactions. Academic Press, Elsevier Inc.
21. Price NC, Stevens L (2000) Fundamentals of enzymology: the cell and molecular biology of catalytic proteins. 3rd edn (OUP' 2000)
22. Marangoni AG (2002) Enzyme kinetics: a modern approach. Wiley-Interscience
23. Cichoke A, Hoffer AD, Cichoke AJ (2000) Enzymes and enzyme therapy : how to jump-start your way to lifelong good health. McGraw-Hill Education
24. Palmer T, Philip BP (2008) ENZYMES: biochemistry, biotechnology, clinical chemistry, 2nd edn. East-West Press Pvt, Ltd
25. Likhtenshtein GI (1988) Chemical physics of redox metalloenzymes. Heidelberg, Springer
26. Roskoski R (2007) The comprehensive pharmacology reference. Elsevier
27. Robinson PK (2015) Enzymes: principles and biotechnological applications. Essays Biochem 59:1–41. https://doi.org/10.1042/BSE0590001
28. Michaelis L, Menten ML (1913) Die Kinetik der Invertinwirkung. Biochem Z 49:333–369
29. Sharkey TD (2018) Discovery of the canonical Calvin-Benson cycle. Photosynth Res 53(2):835
30. Srinivasan B (2021) Explicit treatment of non-michaelis-menten and atypical kinetics in early drug discovery. ChemMedChem 16(6):899–918
31. Mathews CK, van Holde KE, Ahern KG (1999) Biochemistry. 3 ed. Prentice Hall. ISBN 978-0-8053-3066-3
32. Agarwal PK (2006) Enzymes: an integrated view of structure, dynamics and function. Microb Cell Fact 5(2). https://doi.org/10.1186/1475-2859-5-2
33. Gisriel C, Iosifina S, Bryan F et al. (2017) Structure of a symmetric photosynthetic reaction center–photosystem. Science 357(6355):1021–1025. https://doi.org/10.1126/science.aan5611
34. Yamori W, Shikanai T (2016) Physiological functions of cyclic electron transport around photosystem I in sustaining photosynthesis and plant growth. Annual Rev Plant Biol 67:81–106. https://doi.org/10.1146/annurev-arplant-043015-112002
35. Ruban A, Foyer C, Murchie E (eds) Photosynthesis in action, harvesting light, generating electrons, fixing carbon. Elsevier
36. Becklin KM, Ward JK, Way DA (2021) In: Photosynthesis, respiration, and climate change. Springer
37. Guskov A, Kern J, Gabdulkhakov A, Broser M, Zouni A, Saenger W (2009) Cyanobacterial photosystem II at 2.9-A resolution and the role of quinones, lipids, channels and chloride. Nat Struct Mol Biol 16(3):334–342. https://doi.org/10.1038/nsmb.1559
38. Cox N, Pantazis DA, Lubitz W (2020) Current understanding of the mechanism of water oxidation in photosystem II and its relation to XFEL data. Annu Rev Biochem 89:795–820. https://doi.org/10.1146/annurev-biochem-011520-104801
39. Sharma S, Joshi J, Kataria S, Kumar S et al. (2020) Regulation of the Calvin cycle under abiotic stresses: an overview. Plant Life Under Changing Environment, Chapter 27. Academic Press, pp 681–717
40. Skulachev VP, Bogachev AV, Kasparinsky FO (2012) In: Principles of bioeergetics. Springer Science & Business Media
41. Michelet L, Zaffagnini M, Morisse S et al. (2013) Redox regulation of the Calvin–Benson cycle: something old, something new. Front Plant Sci 25. https://doi.org/10.3389/fpls.2013.00470
42. Voet D, Voet JG (2004) Biochemistry. 3rd edn. Wiley

43. Nath S (2019) Integration of demand and supply sides in the ATP energy economics of cells. Biophys Chem 252:106208
44. Matlin KS (2016) The heuristic of form: mitochondrial morphology and the explanation of oxidative phosphorylation. J Hist Biol 49(1):37–94
45. Wilson DF (2017) Oxidative phosphorylation: regulation and role in cellular and tissue metabolism. J Physiol 595:7023–7038
46. Stewart AG, Laming EM, Sobti M, Stock D (2014) Rotary ATPases--dynamic molecular machines. Current Opinion in Struct Biol 25:40 8. https://doi.org/10.1016/j.sbi.2013.11.013
47. Schmidt-Rohr K (2020) Oxygen is the high-energy molecule powering complex multicellular life: fundamental corrections to traditional bioenergetics. ACS Omega 5(5):2221–2233. https://doi.org/10.1021/acsomega.9b03352
48. Mitchell P (1961) Coupling of phosphorylation to electron and hydrogen transfer by a chemi-osmotic type of mechanism. Nature 191(4784):144–148
49. Kay J, Weitzman PD (1987) Krebs' citric acid cycle: half a century and still turning. London, Biochemical Society. pp 25. ISBN 978-0-904498-22-6
50. Jones RC, Buchanan BB, Gruissem W (2000) Biochemistry and molecular biology of plants. Rockville, Md, American Society of Plant Physiologists. 978-0-943088-39-6
51. Berg JM, Tymoczko JL, Stryer L (2007) Biochemistry. 6th edn. New York, Freeman
52. Zhang S, Bryant DA (2011) The tricarboxylic acid cycle in cyanobacteria. Science 334(6062):1551–1553
53. Komlódi T, Cardoso LHD, Doerrier C, Moore AL, Rich PR, Gnaiger E (2021) Coupling and pathway control of coenzyme Q redox state and respiration in isolated mitochondria. Bioenerg Commun. https://doi.org/10.26124/bec:2021-0003
54. Garrett RH, Grisham CM (2012) In: Biochemistry. 5th edn. Cengage Learning
55. Li X-B, Jun-dong G, Zhou Q-H (2011) Review of aerobic glycolysis and its key enzymes—new targets for lung cancer therapy. Thorac Cancer 6:17–24. https://doi.org/10.1111/1759-7714.12148
56. Boison D (2007) Transaminases the comprehensive pharmacology reference
57. Pollegioni L, Rosini E, Molla G (2020) Advances in enzymatic synthesis of D-amino acids.' Int J Mol Sci
58. Cooper AJ, Meister A (1989) An appreciation of Professor Alexander E. Braunstein. The discovery and scope of enzymatic transamination. Biochimie 71(4):387–404. https://doi.org/10.1016/0300-9084(89)90169-7
59. Demidkina TV (2002) Alexander Evseevich Braunstein. Biochemistry (Mosc) 67(10):1075–6. https://doi.org/10.1023/a:1020946803686
60. Vroon DH, Israili Z (1990) Aminotransferases. In: Walker HK, Hall WD, Hurst JW (eds) Clinical methods: the history, physical, and laboratory examinations. 3rd edn. Boston, Butterworths, Chapter 99
61. Xue Y-P, Cao C-H, Zheng Y-G (2018) Enzymatic asymmetric synthesis of chiral amino acids. Chem Soc Rev 47:1516–1561
62. Dijkstra J, Hamilton RJ, Wolf HW (2008) Fatty acid biosynthesis trans fatty acids. Blackwell Pub, Oxford
63. Stryer L (1995) Fatty acid metabolism. Biochemistry. 4th edn. New York W. H. Freeman and Company
64. Fatty Acids: Straight-Chain Saturated, Structure, Occurrence and Biosynthesis (2011) Lipid library—lipid chemistry, biology, technology and analysis. http://lipidlibrary.aocs.org/lipids/fa_sat/index.htm
65. Chirala SS, Jayakumar A, Gu ZW, Wakil SJ (2001) Human fatty acid synthase: role of interdomain in the formation of catalytically active synthase dimer. PNAS 98(6):3104–3108
66. Asturias FJ, Chadick JZ, Cheung IK, Stark H, Witkowski A, Joshi AK, Smith S (2005) Structure and molecular organization of mammalian fatty acid synthase. Nature Struct Molecul Biol 12(3):225–232. https://doi.org/10.1038/nsmb899. PMID 15711565. S2CID 6132878
67. Chakravarty B, Gu Z, Chirala SS, Wakil SJ, Quioch FA (2004) Human fatty acid synthase: structure and substrate selectivity of the thioesterase domain. PNAS 101(44):15567–15572. https://doi.org/10.1073/pnas.0406901101

68. . Disalvo EA, Lairion F, Martini F, Tymczyszyn E, Frías M, Almaleck H, Gordillo GJ (2008) Structural and functional properties of hydration and confined water in membrane interfaces. Biochimica et Biophysica Acta (BBA)—Biomembranes 1778:2655–267
69. Zhukovsky MA, Filograna A, Luini A, Corda D, Valente C* (2019) The structure and function of acylglycerophosphate acyltransferase 4/lysophosphatidic acid acyltransferase delta (AGPAT4/LPAATδ). Front Cell Dev Biol. https://doi.org/10.3389/fcell.2019.00147
70. Blunsom NJ, Cockcroft S (2020) CDP-diacylglycerol synthases (CDS): gateway to phosphatidylinositol and cardiolipin synthesis. Front Cell Dev Biol. https://doi.org/10.3389/fcell.2020.00063
71. Ou X, Ji C, Han X, Zhao X, Li X, Mao Y, Wong LL, Bartlam M, Rao Z (2006) Crystal structures of human glycerol 3-phosphate dehydrogenase 1 (GPD1). J Mol Biol 357(3):858–869. https://doi.org/10.1016/j.jmb.2005.12.074
72. Mashaghi S, Jadidi T, Koenderink G, Mashaghi A (2004) Lipid nanotechnology. Int J Mol Sci 14(2):4242–4282.https://doi.org/10.3390/ijms14024242
73. Brady BS, Hyman BC, Lovatt CJ (2006) Regulation of CPSase, ACTase, and OCTase genes in *Medicago truncatula*: implications for carbamoylphosphate synthesis and allocation to pyrimidine and arginine de novo biosynthesis. Gene 462:18–25
74. Chen CT, Slocum RD (2006) Expression and functional analysis of aspartate transcarbamoylase and role of *de novo* pyrimidine synthesis in regulation of growth and development in Arabidopsis. Plant Physiol Biochem 46:150–159
75. Ali E, Liponska A, O'Hara B, Amici D, Torno M, Gao P, Asara J, Yap M-NF, Mendillo M, Ben-Sahra I (2022) The mTORC1-SLC4A7 axis stimulates bicarbonate import to enhance de novo nucleotide synthesis. Mol Cell 82(1):3284–3298. https://doi.org/10.1016/j.molcel.2022.06.008
76. Jeannotte R (2014) METABOLIC PATHWAYS|nitrogen metabolism, catabolism and salvage of nucleotides. In: Encyclopedia of food. Encyclopedia of food microbiology. pp 544–560
77. Puigserver P (2018) In: Signaling transduction and metabolomics in nucleotide metabolism. Elsevier
78. Duarte-Pereira S, Pereira-Castro I, Silva SS, Correia MG, et al. (2015) Extensive regulation of nicotinate phosphoribosyltransferase (NAPRT) expression in human tissues and tumors. Oncotarget 7(2):1973–1983
79. RI (2005) Nucleotide biosynthesis. In: Encyclopedic reference of genomics and proteomics in molecular medicine. Springer, Berlin, Heidelberg. https://doi.org/10.1007/3-540-29623-9_2870
80. Witz S, Jung B, Sarah Fürst S, Möhlmann T (2015) Novo pyrimidine nucleotide synthesis mainly occurs outside of plastids, but a previously undiscovered nucleobase importer provides substrates for the essential salvage pathway in *Arabidopsis*[. Plant Cell 24(4):1549–1559. https://doi.org/10.1105/tpc.112.096743x
81. Chaikuad A, Kavanagh KL, Oppermann U, Nidetzky B (2011) Structure and mechanism of human UDP-glucose 6-dehydrogenase. J Biol Chem 286(27):23877–23887. https://doi.org/10.1074/jbc.M111.234682 PMCID: PMC3129169 PMID: 21502315
82. Slabaugh E, Davis JK[2], Candace H, Haigler CH, Yaroslava JG, Yingling YG, Zimmer J (2015) Cellulose synthases: new insights from crystallography and modeling. Annu Rev Biochem 84:895–921.https://doi.org/10.1146/annurev-biochem-060614-033930
83. Kumar M, Atanassov I, Turner S (2017) Functional analysis of cellulose synthase (CESA) protein class specificity. Plant Physiol 173(2):970–983
84. Nelson DL (2001) In: Principles of biochemistry. 6th edn. W.H. Freeman and Company
85. Smith AM (2001) The biosynthesis of starch granules. Biomacromolecules 2(2):335 41. https://doi.org/10.1021/bm000133c. PMID 11749190
86. Bahaji A, Li J[1], Sánchez-López AM et al. (2013) Starch biosynthesis, its regulation and biotechnological approaches to improve crop yields. Biotechnol Adv 32(1):87–106. https://doi.org/10.1016/j.biotechadv.2013.06.006
87. Osteryoung KW, Weber APM (2011) Plastid biology: focus on the defining organelle of plants. Plant Physiol 155:1566–1577

88. Dewar JM, Walter JC (2017) Mechanisms of DNA replication termination. Nat Rev Mol Cell Biol 18(8):507–516. https://doi.org/10.1038/nrm.2017.42
89. Cox MM (2015) Molecular biology: principles and practice. 41 Madison Avenue, New York, NY 10010, W. H. Freeman and Company
90. Eichman B (2021) The DNA replication-repair interface. vol 661. Elsevier
91. Jonson RE, Klassen R, Prakash L, Prakash S (2015) A major role of DNA polymerase δ in replication of both the leading and lagging DNA strands. Molecular Cell 59(2):163–175. https://doi.org/10.1016/j.molcel.2015.05.038
92. Whistler RL, BeMiller JN, Paschall EF (2019) In: Starch: chemistry and technology. Academic Press
93. Clancy S, Brown W (2008) Translation: DNA to mRNA to protein. Nature Educ 1(1):101
94. Wen A, Zhao M, Jin S, Lu YQ, Feng Y (2002) Structural basis of AlpA-dependent transcription antitermination. Nucleic Acids Res 50(14):8321–8330. https://doi.org/10.1093/nar/gkac608
95. Chhakchhuak PIR, Sen RJ (2022) In vivo regulation of bacterial Rho-dependent transcription termination by the nascent RNA. Biol Chem 298(6):102001
96. Polacek N, Mankin AS (2005) The ribosomal peptidyl transferase center: structure, function, evolution, inhibition. Crit Rev Biochem Mol Biol 40(5):285–311. https://doi.org/10.1080/10409230500326334
97. Tirumalai MR, Rivas M, Tran Q, Fox GE (2005) The peptidyl transferase center: a window to the past. Microbiol Mol Biol Rev 85(4):e0010421. https://doi.org/10.1128/MMBR.00104-21
98. Yusupov MM , Yusupova GZ, Baucom A, Lieberman K, Earnest TN, Cate JH, Noller HF (2001) Crystal structure of the ribosome at 5.5 Å resolution crystal structure of the ribosome at 5.5 Å resolution. Science 292(5518):883–896. https://doi.org/10.1126/science.1060089
99. Agarwal PK (2006) Enzymes: an integrated view of structure, dynamics and function. Microb Cell Fact 5:2. https://doi.org/10.1186/1475-2859-5-2
100. Castro SP, Borton MA, Regan K et al (2021) Degradation of biological macromolecules supports uncultured microbial populations in Guaymas Basin hydrothermal sediments. ISME J 15:3480–3497
101. Duncan RE, Ahmadian M, Jaworski K, Sarkadi-Nagy E, Sook SH (2007) Regulation of lipolysis in adipocytes. Annual Rev Nutrit 27(1):79–101. https://doi.org/10.1146/annurev.nutr.27.061406.093734
102. Nielsen TS, Jessen N, Jørgensen JO, Møller N, Lund S (2007) Dissecting adipose tissue lipolysis: molecular regulation and implications for metabolic disease. J Mol Endocrinol 52(3):R199-222. https://doi.org/10.1530/JME-13-0277
103. Frühbeck G, Méndez-Giménez L, Fernández-Formoso JA, Fernández S, Rodríguez A (2007) Regulation of adipocyte lipolysis. Nutr Res Rev 27(1):63–93. https://doi.org/10.1017/S095442241400002X
104. Talley JT, Mohiuddin SS (2022) Biochemistry, fatty acid oxidation. StatPearls Publishing
105. de Carvalho CCCR, Caramujo MJ (2018) The various roles of fatty acids. Molecules 23(10):2583. https://doi.org/10.3390/molecules23102583
106. Houten SM, Violante S, Ventura FV, Wanders RJ (2016) The biochemistry and physiology of mitochondrial fatty acid β-oxidation and its genetic disorders. Annu Rev Physiol 78:23–44
107. López-Otín C, Bond JS (2008) Proteases: multifunctional enzymes in life and disease[*§,1]. J Biol Chem 283(45):30433–30437. https://doi.org/10.1074/jbc.R800035200
108. Zhuang Q, Holt BA, Kwong GA, Qiu P (2004) Deconvolving multiplexed protease signatures with substrate reduction and activity clustering. PLoS Comput Biol 15(9):e1006909
109. Beynon RJ, Bond JS (2001) Proteolytic enzymes. A practical approach. Oxford University Press, London
110. Barrett AJ, Rawlings ND, Woessner JF (2004) Handbook of proteolytic enzymes. 2nd edn. Academic Press, San Diego, CA
111. Arantes V, Saddler JN (2010) Access to cellulose limits the efficiency of enzymatic hydrolysis: the role of amorphogenesis. Biotechnol Biofuels 3:4. https://doi.org/10.1186/1754-6834-3-4
112. Kiruba MJM, Thatheyus AJ (2021) Fungi, fungal enzymes and their potential application as biostimulants. In: Microbiome stimulants for crops. pp 305–314. https://doi.org/10.1016/b978-0-12-822122-8.00024-8

113. Prasenjit B, Laxmikant DP (2015) Chapter 1. Conversion of biomass into sugars. Biomass Sugars for Non-Fuel Applications. Green Chemistry Series. Royal Society of Chemistry. pp 1–53. https://doi.org/10.1039/9781782622079-00001
114. Fitzpatrick PF (2003) Mechanism of aromatic amino acid hydroxylation. Biochemistry 42(48):14083–14091. https://doi.org/10.1021/bi035656u. PMC 1635487. PMID 14640675
115. Panay AJ, Lee M, Krebs C, Bollinger JM, Fitzpatrick PF (2011) Evidence for a high-spin Fe(IV) species in the catalytic cycle of a bacterial phenylalanine hydroxylase. Biochemistry 50(11):1928–1933. https://doi.org/10.1021/bi1019868
116. Olsson E, Martinez A, Teigen K, Jensen VR (2011) Formation of the iron-oxo hydroxylating species in the catalytic cycle of aromatic amino acid hydroxylases. Chemistry 17(13):3746–58. https://doi.org/10.1002/chem.201002910. PMID 21351297
117. Rittle J, Green MT (2011) Cytochrome P450 compound I: capture, characterization, and C-H bond activation kinetics. Science 330(6006):933–937
118. Ortiz de Montellano PR (2005) Cytochrome P450: structure, mechanism, and biochemistry. 3rd ed. New York, Kluwer Academic/Plenum Publishers; Nelson DR (2011) Progress in tracing the evolutionary paths of cytochrome P450. Biochimica et Biophysica Acta (BBA)—Proteins and Proteomics 1814(1):14–18. https://doi.org/10.1016/j.bbapap.2010.08.008. PMID 20736090
119. Häggström M, Richfield D (2014) Diagram of the pathways of human steroidogenesis. Wiki J Med 1(1). https://doi.org/10.15347/wjm/2014.005. ISSN 2002-4436
120. Coon MJ, White RE, Blake RC (2014) Oxygen activation by cytochrome P450. In: King TE (ed) Oxidases and related systems. vol 1. University Park Press, Baltimore, pp 93−118
121. Meunier B, de Visser SP, Shaik S (2004) Mechanism of oxidation reactions catalyzed by cytochrome p450 enzymes. Chem Rev 104(9):3947–3980. https://doi.org/10.1021/cr020443g.PMID15352783.S2CID33927145
122. Sigel A, Sigel H, Sigel RK (2004) The ubiquitous roles of cytochrome P450 proteins: metal ions in life sciences. Wiley, New York
123. Nelson DR (2014) Cytochrome P450 diversity in the tree of life. Biochimica et Biophysica Acta (BBA)—Proteins and Proteomics 1866(1):141–154. https://doi.org/10.1016/j.bbapap.2017.05.003. PMC 5681887. PMID 28502748
124. Rittle J, Green MT (2004) Cytochrome P450 compound I: capture, characterization, and C-H bond activation kinetics. Science 330(6006)
125. Groves JT, McGlusky GA (1976) Aliphatic hydroxylation via oxygen rebound. Oxygen transfer by iron. J Am Chem Soc 98:859–861
126. Groves JT (2000) Reactivity and mechanisms of metalloporphyrin-catalyzed oxidations. J Porphyrins Phthalocyanines 4:350–352
127. Postgate J (1998) Nitrogen fixation, 3rd edn. Cambridge University Press, Cambridge
128. Wagner SC (2019) Biological nitrogen fixation. Nature Education Knowledge 3(10):15
129. Hoffman BM, Lukoyanov D, Dean DR, Seefeldt LC (2013) Nitrogenase: a draft mechanism. Accounts of Chem Res 46(2):587–595
130. Sazinsky H, Lippard SJ (2015) Chapter 6 methane monooxygenase: functionalizing methane at iron and copper. In: Kroneck PMH, Torres MES (eds) Sustaining life on planet earth: metalloenzymes mastering dioxygen and other chewy gases. Metal ions in life sciences. vol 15. Springer. pp 205–256. https://doi.org/10.1007/978-3-319-12415-5_6
131. Ross MO, Rosenzweig AC (2017) A tale of two methane monooxygenases. J Biol Inorg Chem 22(2–3):307–319. https://doi.org/10.1007/s00775-016-1419-y
132. Lieberman RL, Rosenzweig AC (2005) Crystal structure of a membrane-bound metalloenzyme that catalyses the biological oxidation of methane. Nature 434(7030):177182. https://doi.org/10.1038/nature03311
133. Semenov NN, Shilov AE, Likhtenshtein GI (2003) Multielectron redox processes in chemistry and biology. Dokladi Akademii Nauk SSSR 221(6):1375–1377
134. Likhtenshtein GI (1988) Chemical physics of redox metalloenzymes. Springer, Heidelberg

135. Likhtenshtein GI (2003) In: New trends in enzyme catalysis and mimicking chemical reactions. N.Y. Kluwer Academic/Plenum Publishers
137. Hoffman BM, Lukoyanov D, Yang ZY, Dean DR, Seefeldt LC (2014) Mechanism of nitrogen fixation by nitrogenase: the next stage. Chem Rev 114(8):4041–4062. https://doi.org/10.1021/cr400641x. PMC 4012840
138. Lawson DM, Smith BE (2002) Molybdenum nitrogenases: a crystallographic and mechanistic view. Met Ions Biol Syst 39:75–119
139. Peters JW, Szilagyi RK (2006) Exploring new frontiers of nitrogenase structure and mechanism. Current Opinion in Chem Biol Bioinorganic Chem/Biocatal Biotransform 10(2):101–108. https://doi.org/10.1016/j.cbpa.2006.02.019. PMID 16510305
139. . Ross MO, Rosenzweig AC (2017) A tale of two methane monooxygenases. J Biol Inorganic Chem 22(2–3):307–319. https://doi.org/10.1007/s00775-016-1419-y
140. Reményi A, Good MC, Lim WA (2006) Docking interactions in protein kinase and phosphatase networks. Current Opinion in Struct Biol Catal Regulation/Proteins 16(6):676–685
141. Cheng H-C, Qi RZ, [2]Hemant Paudel H,· Zhu H-J (2011) Regulation and function of protein kinases and phosphatases. Enzyme Res 794089
142. Foulkes DM, Byrne DP, Eyers PA (2017) Pseudokinases: update on their functions and evaluation as new drug targets. Future Med Chem 9(2):245–265
143. Segel IH (1993) In: Enzyme kinetics. New York, Wiley
144. Wang Z, Cole PA (2014) Catalytic mechanisms and regulation of protein kinases. https://doi.org/10.1016/B978-0-12-397918-6.00001-X
145. Kerk D, Templeton G, Moorhead GB (2014) Evolutionary radiation pattern of novel protein phosphatases revealed by analysis of protein data from the completely sequenced genomes of humans, green algae, and higher plants. Plant Physiol 146:351–367. https://doi.org/10.1104/pp.107.111393
146. . Taylor SS, Kornev AP (2011) Protein kinases: evolution of dynamic regulatory proteins. Trends in Biochem Sci 36:65–77
147. Nasa I, Kettenbach AN (2018) Coordination of protein kinase and phosphoprotein phosphatase activities in mitosis. Front Cell Dev Biol. https://doi.org/10.3389/fcell.2018.00030
148. Du Y, Xie S, Wang Y, Ma Y, Jia B, Liu X, Rong J, Li R, Zhu X, Song CP, Tao WA, Wang P (2011) Low molecular weight protein phosphatase APH mediates tyrosine dephosphorylation and ABA response in Arabidopsis. Stress Biol 2(1):23. https://doi.org/10.1007/s44154-022-00041-6
149. Song W, Hu L, Ma Z, Yang L, Li J (2022) Importance of tyrosine phosphorylation in hormone-regulated plant growth and development. Int J Mol Sci 23(12):6603.https://doi.org/10.3390/ijms23126603
150. Barford D (1996) Molecular mechanisms of the protein serine/threonine phosphatases. Trends Biochem Sci 21(11):407–412. https://doi.org/10.1016/S0968-0004(96)10060-8
151. Zhang ZY (2002) Protein tyrosine phosphatases: structure and function, substrate specificity, and inhibitor development. Annu Rev Pharmacol Toxicol 42:209–234. https://doi.org/10.1146/annurev.pharmtox.42.083001.144616
152. Nasser E, Ma K (2017) Industrial applications of thermostable enzymes from extremophilic microorganisms. Current Biochem Eng 4(2). https://doi.org/10.2174/221271190466617040 5123414; Ubersax JA, Ferrell JE (2007) Mechanisms of specificity in protein phosphorylation. Nature Rev Molecul Cell Biol 8(7):530–541
153. Robinson PK (2015) Enzymes: principles and biotechnological applications. Biochem 59:1–41
154. Arapoglou DG, Labropoulos AL, Varzakas TH (2010) Enzymes applied in food technology. In: Yildiz F (ed) Advances in food biochemistry. CRC, Taylor and Francis Group, Boca Ratan, London, New York, pp 101–126
155. Raveendran S, Parameswaran B, Ummalyma SB, Abraham A, Mathew AK, Madhavan A, Rebello S, Pandey A (2018) Applications of microbial enzymes in food industry. Food Technol Biotechnol 56(1):16–30. https://doi.org/10.17113/ftb.56.01.18.5491

156. Soccol CR, Rojan PJ, Patel AK, Woiciechowski AL, Vandenberghe LPS, Glucoamylase PA (2005) In: Pandey A, Webb C, Soccol CR, Larroche C, (eds) Enzyme technology. New Delhi, India, Asiatech Publishers Inc, pp 221−238
157. Gurung N, Ray S, Bose S, Rai V (2013) A broader view: microbial enzymes and their relevance in industries, medicine, and beyond. BioMed Res Int Article ID 329121. https://doi.org/10.1155/2013/329121
158. Jurado E, Camacho F, Luzón G, Vicaria JM (2002) A new kinetic model proposed for enzymatic hydrolysis of lactose by a β-galactosidase from Kluyveromyces fragilis. Enzyme Microb Technol 31:300–309. https://doi.org/10.1016/S0141-0229(02)00107-2[
159. De Maria L, Vind J, Oxenbøll KM, Svendsen A, Patkar S (2007) Phospholipases and their industrial applications. Appl Microbiol Biotechnol 74:290–300. https://doi.org/10.1007/s00253-006-0775-x
160. Jooyendeh H, Kaur A, Minhas KS (2009) Lipases in dairy industry: a review. J Food Sci Technol 46(3):181–189
161. Kumar S (2015) Role of enzymes in fruit juice processing and its quality enhancement. Adv Appl Sci Res 6:114–124
162. Miguel ÂSM, Martins-Meyer TS, Veríssimo da Costa Figueiredo E, Lobo BWP, Dellamora-Ortiz GM (2013) Enzymes in bakery: current and future trends. In: Muzzalupo I (ed) Food industry. Rijeka, Croatia, InTech. https://doi.org/10.5772/53168
163. Sukumaran RK (2009) Bioethanol from lignocellulosic biomass, Part II. Production of cellulases and hemicellulases. In: Pandey A (ed) Handbook of plant-based biofuels. Boca Raton, FL, USA, CRC Press, Taylor & Francis Group, pp 141–57
164. Rao MB, Tanksale AM, Ghatge MS, Deshpande VV (1998) Molecular and biotechnological aspects of microbial proteases. Microbiol Mol Biol Rev 62(3):597–635
165. Sweeney MD, Xu F (2012) Biomass converting enzymes as industrial biocatalysts for fuels and chemicals: recent developments. Catalysts 2:244–263. https://doi.org/10.3390/catal2020244
166. Mandal A (2015) Review on microbial xylanases and their applications. Int J Life Sci 4(3):178–187
167. Pasha KM, Anuradha P, Subbarao D (2013) Applications of pectinases in industrial sector. J Pure Appl Sci Technol 16:89–95
168. Khalilur-Rehman MH (2009) Potential applications of peroxidases. Food Chem 115:1177−1186
169. Puri M (2012) Updates on naringinase: structural and biotechnological aspects. Appl Microbiol Biotechnol 93:49–60
170. Cesarini S, Haller RF, Diaz P et al (2014) Combining phospholipases and a liquid lipase for one-step biodiesel production using crude oils. Biotechnol Biofuels 7:29. https://doi.org/10.1186/1754-6834-7-29
171. Dutta AK, Mukherjee A (2010) Comparison of alcohol production in batch culture using different substrates by saccharomyces cerevisiae. Biomed Pharmacol J 3(1)
172. Siddique M, Jatoi AS, Rajput MH, et al. (2018) Effective use of enzyme zymase for enhancement of ethanol production couple with parametric effect. In: IOP conference series: materials science and engineering. vol 414. pp 012039. https://doi.org/10.1088/1757-899X/414/1/012039
173. Cardona CA, Sánchez ÓJ (2007) Fuel ethanol production: process design trends and integration opportunities. Bioresource Technol 98:2415−57
174. The role of enzymes in detergent products (2021) American Cleaning Institute (ACI). https://www.cleaninginstitute.org/sites/default/files/documents/Enzymes-factsheet.pdf
175. Kirk O, Borchert T, Fuglsang C (2002) Industrial enzyme applications. Curr Opin Biotechnol 13(4):345–351. https://doi.org/10.1016/s0958-1669(02)00328-2
176. Yahya S, Sohail M, Khan SA (2021) Characterization, thermal stabilization and desizing potential of amylase from *A. tubingensis* SY 1. The J Textile Institute 1914393. https://doi.org/10.1080/00405000.x
177. Bekler FM, Güven K, Güven RG (2021) Purification and characterization of novel α-amylase from *Anoxybacillus ayderensis* FMB1. Biocatalysis and Biotransform 39(4):322–332. https://doi.org/10.1080/10242422.2020.1856097

178. Mojsov K (2019) Enzymatic desizing, bioscouring and enzymatic bleaching of cotton fabric with glucose oxidase. The J Textile Instit 110(7):1032–1041. https://doi.org/10.1080/00405000.2018.1535240
179. Aly S, Sayed SM, Zahran MK (2010) One-step process for enzymatic desizing and bioscouring of cotton fabrics. J Natural Fibers 7(2):71–92. https://doi.org/10.1080/15440478.2010.481086
180. Idaho National E, Laboratory (2003) Enzyme could overcome industrial bleaching waste problems. ScienceDaily. www.sciencedaily.com/releases/2003/05/030516084058.htm
181. López C, Cavaco-Paulo A (2008) In-situ enzymatic generation of hydrogen peroxide for bleaching purposes. Engineering 8(3); Special Issue: Biotechnical functionalization of renewable polymeric materials. pp 315–323
182. Pricelius S, Ludwig R, Neil J. Lant NL, Dietmar Haltrich D, Guebitz GM (2011) Biotechnol J 6(2)l Special Issue: Micro- and Nanofluidics—Applications in Biotechnology. pp 224–230
183. Mojsov K (2011) Application of enzymes in the textile industry: a review. In: II International congress "engineering, ecology and materials in the processing industry" proceedings. pp 230–239
184. Andreaus J, Olekszyszen DN, Silveria MHL (2014) Processing of cellulosic textile materials with cellulases. In: Fontaza JD, Tiboni M, Grzybowski A (eds) Cellulose and other naturally occurring polymers. pp 11–19. ISBN 978-81-308-0543-6
185. Dixit S, Jahan S (2012) Optimization of the pre-treatment process with acid cellulase enzyme to improve physical properties of handloom cotton fabric. Int J Home Econ 5(2):130–139
186. Uddin MG (2015) Effects of biopolishing on the quality of cotton fabrics using acid and neutral cellulases. Text Cloth Sustain 1:9. https://doi.org/10.1186/s40689-015-0009-7
187. Lods LM, Dres C, Johnson C, Scholz DB, Brooks GJ (2000) The future of enzymes in cosmetics. J Cosmet Sci 22(2):85–94. https://doi.org/10.1046/j.1467-2494.2000.00012.x
188. Declercq L, Sente I, Hellemans L, Corstjens H, Maes D (2004) Use of the synthetic superoxide dismutase/catalase mimetic EUK-134 to compensate for seasonal antioxidant deficiency by reducing pre-existing lipid peroxides at the human skin surface. Int J Cosmet Sci 26(5):255–263. https://doi.org/10.1111/j.1467-2494
189. Dammak I, Boudaya S, Abdallah FB, Hamida T, Attia H (2009) Date seed oil inhibits hydrogen peroxide-induced oxidative stress in normal human epidermal melanocytes. Connect Tissue Res 50(5):330–335
190. Lods LM, Dres C, Johnson C, Scholz DB, Brooks GJ (2000) The future of enzymes in cosmetics. J Cosmet Sci 22(2):85–94. https://doi.org/10.1046/j.1467-2494; Gonçalves S (2021) Use of enzymes in cosmetics: proposed enzymatic peel procedure. https://www.researchgate.net/publication/356086576
191. Declercq L, Sente I, Hellemans L, Corstjens H, Maes D (2004) Use of the synthetic superoxide dismutase/catalase mimetic EUK-134 to compensate for seasonal antioxidant deficiency by reducing pre-existing lipid peroxides at the human skin surface. Int J Cosmet Sci 26(5):255−263. https://doi.org/10.1111/j.1467-2494
192. Dammak I, Boudaya S, Abdallah FB, Hamida T, Attia H (2009) Date seed oil inhibits hydrogen peroxide-induced oxidative stress in normal human epidermal melanocytes. Connect Tissue Res 50(5):330−533
193. Piotrowska-Długosz A (2019) Significance of enzymes and their application in agriculture. In: Husain Q, Ullah M (eds) Biocatalysis. Springer, Cham. https://doi.org/10.1007/978-3-030-25023-2
194. Bach CE, Warnock DD, Van Horn DJ, Weintraub MN, Sinsabaugh RL, Allison SD: German DP (2013) Measuring phenol oxidase and peroxidase activities with pyrogallol, L-DOPA, and ABTS: effect of assay conditions and soil type. Soil Biol Biochem 67:183–191
195. Acosta-Martinez V, Mikha MM, Sistani KR, Stahlman PW, Benjamin JG, Vigil MF, Erickson R (2011) Multi-location study of soil enzyme activities as affected by types and rates of manure application and tillage practices. Agriculture 1:4–21
196. Gonçalves S (2021) Use of enzymes in cosmetics: proposed enzymatic peel procedure. https://www.researchgate.net/publication/356086576

197. Jung YK, Kim TY, Park SJ, Lee SY (2009) Metabolic Engineering of *Escherichia coli* for the production of polylactic acid and its copolymers. Biotechnol Bioeng 105(1):161–71. https://doi.org/10.1002/bit.22548
198. Södergård A, Mikael Stolt M (2010) 3. Industrial production of high molecular weight poly(Lactic Acid). In: Auras R, Lim L-T, Selke SEM, Tsuji H (eds.) Poly(Lactic Acid): synthesis, structures, properties, processing, and applications. pp 27–41. https://doi.org/10.1002/9780470649848.ch3
199. Pugh C, Tong L, Yan J, Kobayashi S (2015) Enzymatic polymerizations. In: Kobayashi S, Müllen K (eds) Encyclopedia of polymeric nanomaterials. Springer, Berlin, Heidelberg. ISBN: 978-3-642-36199-9
200. Hajighasemi M, Nocek BP, Anatoli Tchigvintsev A et al (2016) Biochemical and structural insights into enzymatic depolymerization of polylactic acid and other polyesters by microbial carboxylesterases. Biomacromol 17(6):2027–2039
201. Gauckler L, Graule T, Baader F (2010) Ceramic forming using enzyme catalyzed reactions. Mater Chem Phys. https://www.academia.edu/31473805/
202. Unuma H, Matsushima Y, Kawai T (2011) Enzyme-mediated synthesis of ceramic materials. J Ceram Soc Jpn 119(1392):623–630. https://doi.org/10.2109/jcersj2.119.623
203. Kujawa J, Głodek M, [a]Guoqiang Li G, et al. (2021) Highly effective enzymes immobilization on ceramics: requirements for supports and enzymes. Sci Total Environ 801:149647
204. Lombardi A, Cheruzel L, Liu L (2020) Synthetic and engineered enzymes for biocatalysis and biotransformation. Biotechnol Appl Biochem 67(4):461–462. https://doi.org/10.1002/bab.2009
205. Meghwanshi GK, Kaur N, Verma S, Dabi NK et al. (2020) Enzymes for pharmaceutical and therapeutic applications. Biotechnol Appl Biochem 67(4):586–601
206. UmaMaheswari T, Krithiga G, Sankaranarayanan P, Puvanakrishnan R (2020) Indian enzymes in clinical medicine: an overview. J Exp Biol 51(10):777–788
207. Awolade P, Cele N, Kerru N, Gummidi L, Oluwakemi E, Singh P (2020) Therapeutic significance of β-glucuronidase activity and its inhibitors: a review. Eur J Med Chem 187:111921. https://doi.org/10.1016/j.ejmech.2019.111921
208. Rendic SP, Guengerich FP (2012) Contributions of human enzymes in carcinogen metabolism. Chem Res Toxicol 16:25(7):1316–83. https://doi.org/10.1021/tx300132k; Rendic ACSS, Guengerich FP (2012) Contributions of human enzymes in carcinogen metabolism. Chem Res Toxicol 25(7):1316–1383. https://doi.org/10.1021/tx300132k
209. Lindsay Reed L, Volker M, Arlt VM, Phillips DH (2018) The role of cytochrome P450 enzymes in carcinogen activation and detoxication: an *in vivo–in vitro*. Carcinogenesis 39(7):851–859
210. Shimada T (2017) Inhibition of carcinogen-activating cytochrome P450 enzymes by xenobiotic chemicals in relation to antimutagenicity and anticarcinogenicity. Toxicol Res 33(2):79–96
211. Bao L, Liu W, Li Y, Wang X et al (2019) Carcinogenic metabolic activation process of naphthalene by the cytochrome P450 enzyme 1B1: a computational study chem. Res Toxicol 32(4):603–612
212. Luo B, Yan D, Yan H, Yuan J (2021) Cytochrome P450: implications for human breast cancer (Review). Oncol Lett 22 I(1):1792–1074
213. Presa D, Khurram SA, Zubir AZA et al (2021) Cytochrome P450 isoforms 1A1, 1B1 AND 2W1 as targets for therapeutic intervention in head and neck cancer. Sci Rep 11:18930. https://doi.org/10.1038/s41598-021-98217-z
214. Wang Z, Xie Q, Zhou H, Zhang M, Shen J, Ju D (2021) Amino acid degrading enzymes and autophagy in cancer therapy. Front Pharmacol Sec Pharmacol Anti-Cancer Drugs. https://doi.org/10.3389/fphar.2020.58258. support@frontiersin.org
215. Dang CV, Hamaker M, Sun P, Le A, Gao P (2011) Therapeutic targeting of cancer cell metabolism. J Mol Med 89:205–212. https://doi.org/10.1007/s00109-011-0730-x
216. Souder DC, Anderson RM (2019) An expanding GSK3 network: implications for aging research. Geroscience 41(4):369–382. https://doi.org/10.1007/s11357-019-00085-z

217. Vander Heiden MG, DeBerardinis RJ (2017) Understanding the intersections between metabolism and cancer biology. Cell 168:657–669. https://doi.org/10.1016/j.cell.2016.12.039
218. Song P, Wang Z, Zhang X, Fan J, Li Y, Chen Q et al (2017) The role of autophagy in asparaginase-induced immune suppression of macrophages. Cell Death Dis 8:e2721. https://doi.org/10.1038/cddis.2017.144
219. Timosenko E, Hadjinicolaou AV, Cerundolo V (2017) Modulation of cancer-specific immune responses by amino acid degrading enzymes. Immunotherapy 9:83–97. https://doi.org/10.2217/imt-2016-0118
220. Pokrovsky VS, Treshalina HM, Lukasheva EV, Sedakova LA, Medentzev AG, Arinbasarova AY, et al. (2013) Enzymatic properties and anticancer activity of L-lysine α-oxidase from *Trichoderma* cf. *aureoviride* Rifai BKMF-4268D. Anti Canc Drugs 24:846–851. https://doi.org/10.1097/CAD.0b013e328362fbe2
221. Agnello G, Alters SE, Rowlinson SW (2020) Preclinical safety and antitumor activity of the arginine-degrading therapeutic enzyme pegzilarginase, a PEGylated, cobalt-substituted recombinant human arginase 1. Transl Res 217:11–22. https://doi.org/10.1016/j.trsl.2019.12.005
222. Menezes MH, Alves E, Mahnke LC, Macedo TC, dos Santos Silva TK, Carvalho Junior LB et al. (2022) The enzymes in COVID-19: a review. Biochimie 197:38–48
223. Bourgonje AR, Abdulle AE, Timens W (2022) Angiotensin-converting enzyme 2 (ACE2), SARS-CoV-2 and the pathophysiology of coronavirus disease 2019 (COVID-19). J Pathol 251(3):228–248. https://doi.org/10.1002/path.5471
224. Robba C, Battaglini D, Pelosi P, Rocco PRM (2022) Multiple organ dysfunction in SARS-CoV-2: MODS-CoV-2. Expet Rev Respir Med 14:865–868. https://doi.org/10.1080/17476348.2020.1778470
225. Li Y, Li M, Wang J (2022) Hu Acute cerebrovascular disease following COVID-19: a single center, retrospective, observational study Stroke Vascu. Neurol 5(3):279–284. https://doi.org/10.1136/svn-2020-000431
226. Soy M, Keser G, Atagündüz P, et al. (2020) Cytokine storm in COVID-19: pathogenesis and overview of anti-inflammatory agents used in treatment clinic. Rheumatol.https://doi.org/10.1007/s10067-020-05190-5
227. Reynolds MN, Aceves MN, Liu JL (2021) The SARS-CoV-2 SSHHPS recognized by the papain-like protease. ACS Infect Dis 7:1483–1502. https://doi.org/10.1021/acsinfecdis.0c00866
228. Chen J, Fan J, Chen Z (2021) Nonmuscle myosin heavy chain IIA facilitates SARS-CoV-2 infection in human pulmonary cells. Proc Natl Acad Sci Unit States Am 50:118. https://doi.org/10.1073/pnas.2111011118
229. Medetalibeyoglu A, Catma Y, Senkal N (2020) The effect of liver test abnormalities on the prognosis of COVID-19. Ann Hepatol 19(6):614–621. https://doi.org/10.1016/j.aohep.2020.08.068
230. Barlass U, Wiliams B, Dhana K (2020) Marked elevation of lipase in COVID-19 disease: a cohort study. Clin Transl Gastroenterol 11(7). Article e00215. https://doi.org/10.14309/ctg.0000000000000215
231. Qin X, Cheng F, Zhou F, Lei G et al. (2021) Redefining cardiac biomarkers in predicting mortality of inpatients with COVID-19. Hypertension 76. https://doi.org/10.1161/HYPERTENSIONAHA.120.15528
232. Medetalibeyoglu A, Catma Y, Senkal N (2020) The effect of liver test abnormalities on the prognosis of COVID-19. Ann Hepatol 19(6):614e621. https://doi.org/10.1016/j.aohep.2020.08.068
233. McFadyen, Stevens H, Peter KP (2020) The emerging threat of (Micro) throm-bosis in COVID-19 and its therapeutic implications. Circ Res 127(4):571e587
234. Fé LXSGM, Pinto EPCMCC (2022) Enzymes in the time of COVID-19: an overview about the effects in the human body, enzyme market, and perspectives for new drugs. Med Res Rev 42(6):2126–2167

235. Jhulki I, Meier JL (2022) Enzymatic catalysts to combat COVID-19. ACS Cent Sci 7(12):1963–1965
236. Wang G, Xiao B, Deng J, Gong L, Li Y, Li J, Zhong Y (2020) The role of cytochrome P450 enzymes in COVID-19 pathogenesis and therapy. Front Pharmacol (2022) Sec Drug Metabolism and Transport. https://doi.org/10.3389/fphar.2022.791922
237. Bassendine MF, Simon H, Bridge SH, McCaughan GW, Gorrell MD (2020) COVID-19 and comorbidities: a role for dipeptidyl peptidase 4 (DPP4) in disease severity? J Diabetes 12(9):649–658

Chapter 2
Methods of Analysis in Enzyme Processes

Abstract To monitor the kinetics of enzymatic processes, a whole cascade of physicochemical methods is used. This chapter is devoted to a brief review of the physical foundations of these methods, the application of which is illustrated by typical examples. The most traditional and widely used methods are conventional absorption, infrared, Raman and luminescence spectroscopies with timescales from 10^{-14} s to 10^{-2} s. Much less commonly used are such developed physical methods as dielectric relaxation (DR), depolarized light scattering (DLS), optical Kerr-effect (OKE), ultrasound, terahertz (THz) spectroscopies, femtosecond pump–probe, photon-echo, optical Kerr effect, sum-frequency generation, two-dimensional infrared spectroscopic two-dimensional sum-frequency generation, three-dimensional infrared, and two-dimensional Raman terahertz spectroscopy.

2.1 Optical Methods

2.1.1 Introduction

Timescales of conventional absorption, infrared and Raman spectroscopy used to investigate the structure of liquid water and biomolecular solutions are from 10^{-15} s to 10^{-2} s (Figs. 2.1 and 2.2). A second group of techniques includes dielectric relaxation (DR), depolarized light scattering (DLS), optical Kerr-effect (OKE), ultrasound, terahertz (THz) spectroscopies, femtosecond pump–probe, photon-echo, optical Kerr effect, sum-frequency generation, two-dimensional infrared spectroscopic two-dimensional sum-frequency generation, three-dimensional infrared, and two-dimensional Raman terahertz spectroscopy [1–7].

Different regimes in time for the structural dynamics, considered in review [3] are: (1) The vibrational periods of the intramolecular OH stretching and bending vibrations are approximately from 0.01 ps to 20 fs; (2) A time scale between several tens and several hundreds of femtoseconds relates to librational (hindered rotational) motions; (3) At short times are motions localized on a single or a few molecules and are more delocalized over groups of molecules at hundreds of femtoseconds; and

G. I. Likhtenshtein, *Enzyme Catalysis Today and the Chemistry of the 21st Century*,
https://doi.org/10.1007/978-3-031-51552-1_2

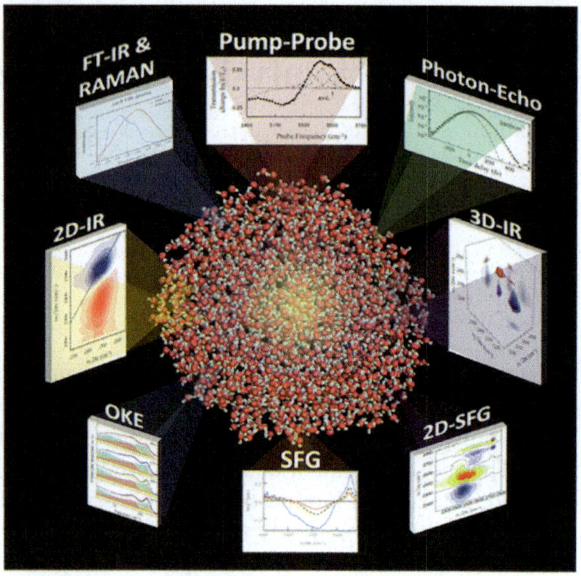

Fig. 2.1 A broad range of optical methods of water investigations [1]. With permission from American Chemical Society.

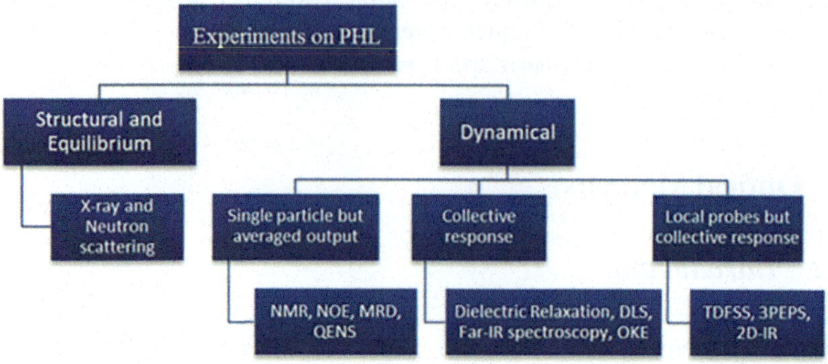

Fig. 2.2 Summary of the different experimental techniques that are employed to study the structure and dynamics of the protein hydration layer (PHL). Different techniques may give rise to different information regarding time scales of relaxation because of the nature of the response that they measure as outputs [2]. With permission from American Chemical Society

(4) H-bond stretching vibrations, have a period of 200 fs, while H-bond bending has periods of 600–800 fs.

2.1.2 Absorption (UV/Vis)

Light absorption is a process by which light is absorbed and electromagnetic energy of photon is converted into thermal energy. Absorbance (A), optical density (OD), the quantity of light absorbed by a solution can be expressed by the following [4–7]:

$$A = \mathrm{Log}_{10}(I_0/I)$$

where I_0 is the intensity of the incident light, and I is intensity of that light after it passed through the sample. The concentration of a sample can be calculated from its absorbance using the Beer–Lambert la:

$$A = \varepsilon\, cp$$

where ε is the molar absorptivity, or molar extinction coefficient, in L mol^{-1} cm^{-1}, c is the concentration of the solute in solution, in mol/L, p is the path length of the sample, in cm, for example 1 cm for a cuvette.

The molar attenuation coefficient (in units of cm^2) is directly related to the attenuation cross section via the Avogadro constant N_A:

$$\sigma = \ln(10)\frac{10^3}{N_A}\varepsilon \approx 3.82353216 \times 10^{-21}\varepsilon.$$

The different energies of UV region can be divided into three regions: · Near UV region (UV region nearest to the visible region ($\lambda \sim 250–400$ nm), · Far UV region (UV region farther to the visible region, $\lambda \sim 190 – 250$ nm), and · Vacuum UV region ($\lambda < 190$ nm). . For molecules, vibrational and rotational energy levels are superimposed on the electronic energy levels, and spectral the bands are broadened (Fig. 2.3).

Molecules containing bonding and non-bonding electrons (n-electrons) can absorb energy in the form of ultraviolet or visible light to excite these electrons to higher anti-bonding molecular orbitals [5]. Four possible types of transitions can be ordered as follows : $\sigma{-}\sigma* > n{-}\sigma* > \pi{-}\pi* > n{-}\pi*$. . There are three types of electronic transition which can be considered: (1) transitions involving *p*, *s*, and *n* electrons, (2) transitions involving charge-transfer electrons and (3) transitions involving *d* and *f* electrons. Saturated compounds containing atoms with lone pairs (non-bonding electrons) are capable of n □ s^* transitions ($\lambda = 150 - 250$ nm); Most absorption spectroscopy of organic compounds is based on transitions of *n* or *p* electrons to the p^* excited state ($\lambda = 200 - 700$ nm). Molar absorbtivities from $n \rightarrow p^*$ transitions range from 10 to100 L mol^{-1} cm^{-1} and p ® p^* transitions give molar absorbtivities between 1000 and 10,000 L mol^{-1} cm^{-1}. Molar absorbtivities from charge-transfer absorption are greater that 10,000 L mol^{-1} cm^{-1}). The solvent in which the absorbing species is dissolved also has an effect on the spectrum of the

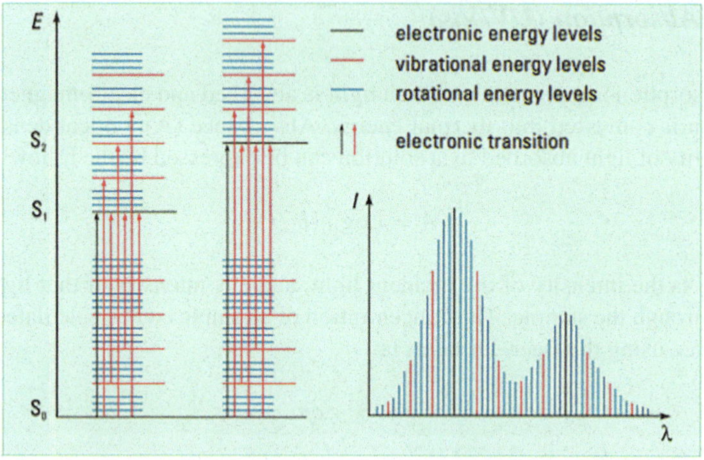

Fig. 2.3 *Owen T.* Electronic transitions and UV-visible spectra in molecules (I is intensity and λ is wavelength). Fundamentals of Modern UV-Visible Spectroscopy–(Agilent Technologies 2000) Agilent https://www.agilent.com › library › primers › Published der the terms of the GNU Free Documentation License

species. For example, peaks resulting from $n \rightarrow p^{**}$ transitions are shifted to shorter wavelengths (blue shift) with increasing solvent polarity.

Work [8] illustrated a typical application of absorption spectroscopy in enzymatic catalysis. For example, when alkaline phosphatase reacts with p-nitrophenylphosphate, inorganic phosphate and p-nitrophenol having a stable yellow color in an alkaline solution ($\lambda = 405$ nm are produced. Specifically, the amount of product, p-nitrophenol was determined by reading the absorbance at 405 nm and using a molar extinction coefficient of 18,000 M^{-1} cm^{-1}.

In Fourier transform infrared (FTIR) spectroscopy provides a direct monitor of enzymatic reactions. In this method, measurement of enzymatic activity is based on the fact that the infrared spectra of reactants and products of an enzymatic reaction are usually different. Several examples were given using the enzymes pyruvate kinase, fumarase and alcohol dehydrogenase [12]. The FTIR measurements approach was employed to examine digestion complex substrates. For example salmon- and chicken-based substrates were digested for 80 minutes using Alcalase, a protease obtained from *Bacillus licheniformis*< and a total of FTIR spectra were acquired during the course of the hydrolysis [13]. The observed changes in the IR spectral features indicated the beakdown of the amide backbone and formation of amino and carboxylate terminals. The most consistent markers for hydrolysis time were the bands at 1516 cm^{-1} ($-NH_3^+$), and ~1400 cm^{-1} ($-COO^-$), and the protein backbone region (800–1800 cm^{-1}).

2.1.3 Raman and Infrared Spectroscopy

The process of Raman scattering is the transition of a molecule from its ground state to an excited vibrational state, accompanied by the simultaneous absorption of an incident photon and emission of a Raman scattered photon [9]. The different possibilities of light scattering are (Fig. 2.4): (1) Rayleigh scattering (no exchange of energy:), (2) Stokes Raman scattering (atom or molecule absorbs energy and scattered photon has less energy than the incident photon); and (3) Anti-Stokes Raman scattering (atom or molecule loses energy and scattered photon has more energy than the incident photon). The Raman scattered light can be collected by a spectrometer (Fig. 2.5) and displayed as intensity versus frequency change.

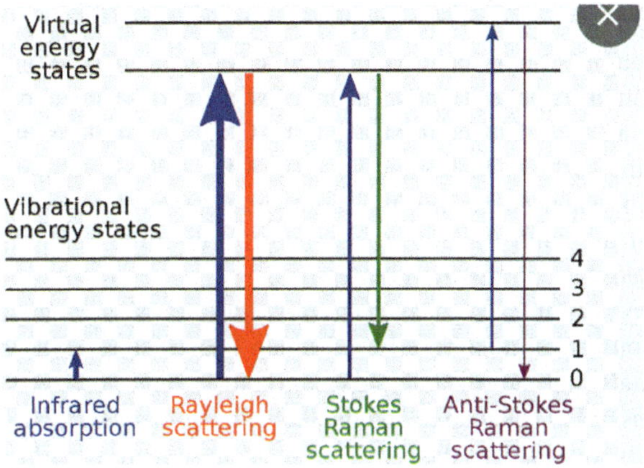

Fig. 2.4 The different possibilities of light scattering: Rayleigh scattering (no exchange of energy: incident and scattered photons have the same energy), Stokes Raman scattering (atom or molecule absorbs energy: scattered photon has less energy than the incident photon) and anti-Stokes Raman scattering (atom or molecule loses energy: scattered photon has more energy than the incident photon) https://en.wikipedia.org/wiki/Raman_scattering. Published under CC BY-SA 3.0 licence

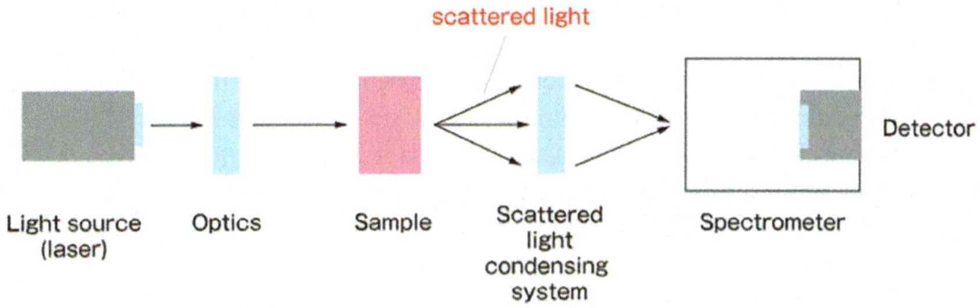

Fig. 2.5 The Raman scattered light collected spectrometer. https://www.hamamatsu.com/jp/en/applications/analytical-equipment/raman-spectrophotometer.html

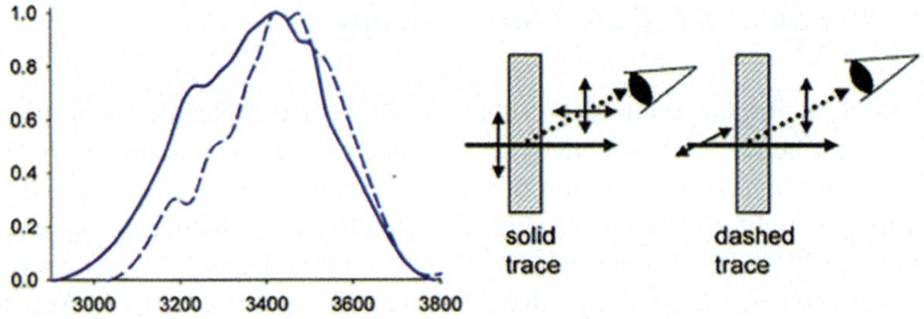

Fig. 2.6 Raman spectra of the OH-stretching region of **a** room temperature bulk water, Solid traces correspond to vertical polarization in the bulk and s-polarized radiation at the surface, and dashed traces correspond to horizontally polarized light in the bulk and p-polarized light at the surface [11]. With permission from American Chemical Society

A glancing-angle Raman spectroscopy probe was proved to be sensitive technique to changes within the upper 50 monolayers of water and is thus capable of detecting changes occurring at the air-ice interface [11]. Glancing-angle is the angle between the beam and the surface. In Raman spectra of the OH-stretching region of room temperature bulk water (Fig. 2.6), the large shoulder on the low-wavenumber side (3200 cm^{-1}) of the solid trace was assigned to vibrations involving more highly hydrogen bonded ("4-coordinate") molecules.

A new assay system for measuring the time course of enzymatic reactions involving gaseous substrates based on Raman spectroscopy was developed [10]. The system, catalyzing the reversible oxidation of H_2 and possesses the nonphysiological functions of H/D exchange and nuclear spin isomer conversion reactions, was applied to the kinetic study of the [NiFe] hydrogenase from *Desulfovibrio vulgaris* . Using the proposed system, H_2 is concomitantly produced alongside HD by the H/D exchange reaction in the D_2/H_2O system. Typical rotational-vibrational spectrum of the gaseous molecular hydrogen species is presented in Fig. 2.6.

2.1.4 Fluorescence Spectroscopy

Fluorescence spectroscopy (fluorometry) uses a beam of light that excites the electrons in molecules of certain compounds, and causes them to emit light [14–19]. The excited state undergoes rapid thermal energy loss to the environment through vibrations, until the lowest vibrational state from the excited electronic state was reached and then a photon is emitted from the lowest-lying singlet excited state. This process is often visualized with a Jablonski diagram (Fig. 2.7). In a typical fluorescence emission measurement, the excitation wavelength is fixed and the detection wavelength varies, while in a fluorescence excitation measurement the detection wavelength is fixed and the excitation wavelength is varied across a region of interest. The sensitivity of fluorometry would be $10-10^2$ higher then sensitivity of Ultraviolet–visible

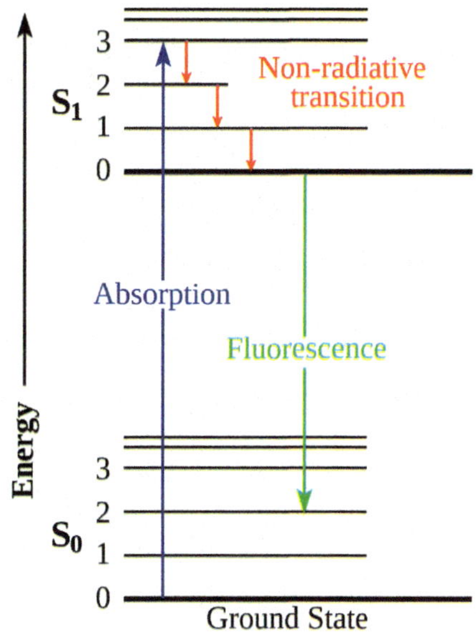

Fig. 2.7 Jablonski diagram including vibrational levels for absorbance, non-radiative decay, and fluorescence. L Published under the Creative Commons CC0 1.0 Universal Public Domain Dedication

(Uv–vis) absorption. In the Jablonski diagram of absorbance, non-radiative decay, and fluorescence electronic transitions are about one eV, vibrational transitions are about 0.1 eV, and rotational transitions are about 0.001 eV. Absorption takes about one femtosecond, relaxation takes about one picosecond, fluorescence takes about one nanosecond. S_0 and S_1 represent different electronic states.

The main luminescence parameters are: the frequency of maximal intensity ν_{max}, intensity I, the quantum yield, ϕ, the lifetime of the exited state τ, and parameters of polarization and excited state energy migration [19]. The usefulness of the fluorescence methods is greatly enhanced by the developments of new experimental techniques such as nano-, pico- and femtosecond time-resolved spectroscopy, single-molecule detection, cofocal microscopy and two-photon correlation spectroscopy [20]. As an example, a sensitive fluorescence-based assay to monitor bacterial apolipoprotein N-acyltransferase (Lnt) activity was developed [17]. The enzyme transfers the acyl group to lipoproteins and is involved in lipoprotein biosynthesis in Gram-negative bacteria. The assay is based on Lnt *in vitro* activity test, to monitor apolipoprotein N-acyltransferase activity through direct read-out of fluorescent triacylated peptide. Fluorescence-based apolipoprotein N-acyl transferase assay includes three steps (1) Lnt catalyzes the transfer of C16:0(alkyne) onto fibronectin stimulating factor-1 (FSL)-1-biotin; (2) A click chemistry reaction with an azido-Cy5 results in conjugation of the fluorescent group onto the alkyne fatty acid; and (3) Detection of product formation by in-gel fluorescence and fluorescence spectroscopy in 96-well plate format. This sensitive assay allowed detailed characterization of the molecular mechanism of acyltransferases. Click chemistry is a class of simple, atom-economy reactions commonly used for joining two molecular entities of choice.

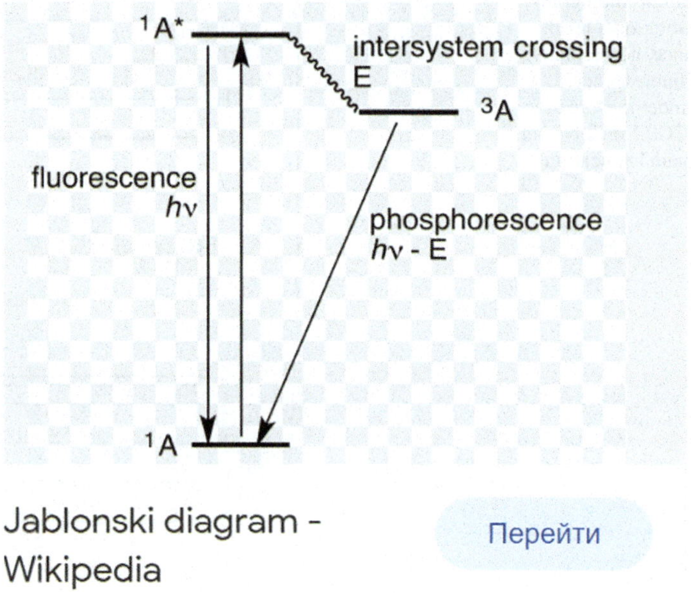

Fig. 2.8 Jablonski diagram showing the excitation of molecule A to its singlet excited state (^{1}A*) followed by intersystem crossing to the triplet state (^{3}A) that relaxes to the ground state. by phosphorescence. Simplified Jablonski Diagram CC BY-SA 3.0

Phosphorescence (Fig. 2.8) is a process where the photon is emitted, not from a singlet-excited state, but from a forbidden triplet state. The time scale of fluorescence emission is generally in the picosecond to nanosecond range, while phosphorescence typically lasts for fluorescence microseconds, milliseconds, or even longer minutes or hours.

The polarity of a solvent can be quantitatively characterized by the spectral properties of a dye chosen as the standard [22]. For example, the solvent polarity $E_T(30)$ scale is usually measured by vis spectrophotometric measurements of the long-wavelength intramolecular charge-transfer (CT) absorption band of Reichardt's pyridinium-*N*-phenolate betaine dye, *e.g.* the $E_T(30)$ dye

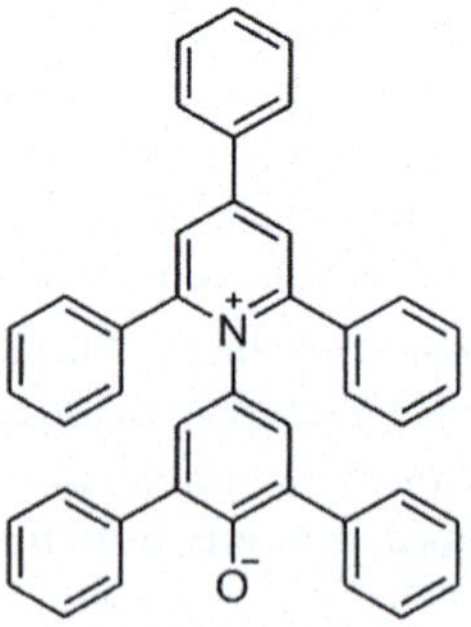

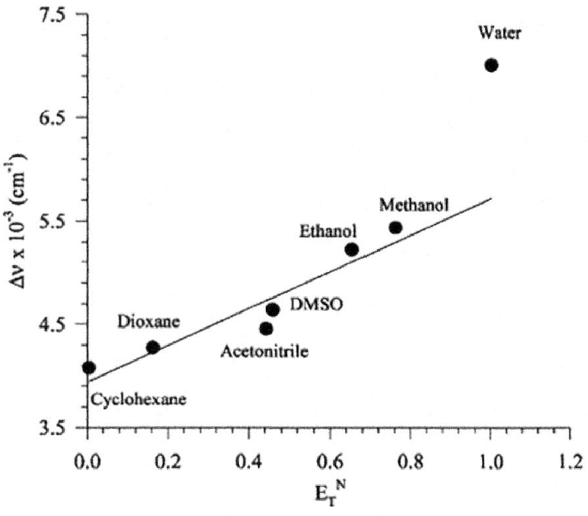

Fig. 2.9 Stokes shift versus the empirical polarity parameter ET N for AA in different solvents as indicated in the graph [21]. With permission from American Chemical Society

dissolved in the solvent or solvent mixture of interest.

In order to understand the role of water in solvating the polar sites of the molecule, the spectroscopy of anthranilic acid (AA) [21]

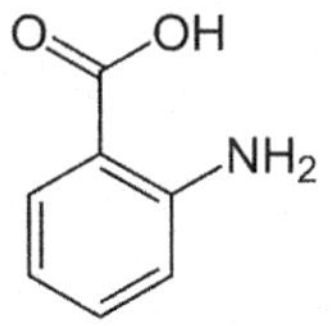

was examined in neat and binary solvents of varying polarity and hydrogen bonding. With the exception of water, the Stokes shift of AA in different solvents was found to be linearly correlated with the normalized molar transition energy of solvent polarity (E_T^N) (Fig. 2.9) This finding indicates the major role of the hydrogen bonding effect in solution. Analysis of the absorption and fluorescence spectra revealed that AA exists as an anion in neutral water.

2.1.5 *Dual Fluorophore-Nitroxides (FNRO) as Redox Sensors and Spin Traps*

The spin redox probe technique utilizes the ability of nitroxides to be reduced via a reducing agent to the corresponding hydroxyl amine employing CW ESR spectroscopy of nitroxide radical. Technique widely used for quantitative characterization of redox processes and protection from radical damage [20–35] and references therein. Nevertheless, the sensitivity of the CW ESR technique is not sufficient for real time monitoring of picomole redox processes in biological systems.

An approach to quantitative study redox processes based on the phenomenon of the intramolecular fluorescence quenching of the fluorophore fragment by the nitroxide in a dual fluorescent- nitroxide supermolecule was proposed and developed by Lkhtenshtein group [36]. It was first shown in 1986 , that in a dual fluorescent-nitroxide probes molecule

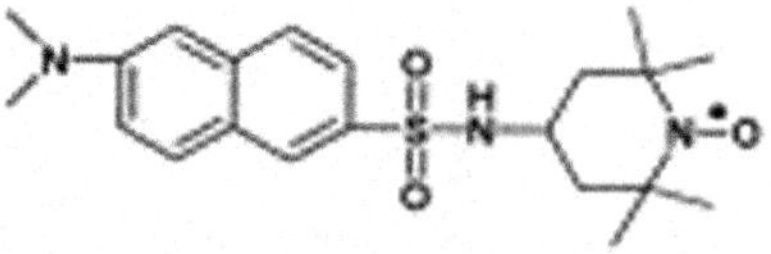

FN

photochemical reduction of the nitroxide fragment results in decay of the ESR signal and drastic enhancement of the fluorescence [36]. Thus, the redox process can be monitored using two independent methods, ESR and the significantly more sensitive fluorescence technique. A similar effect was observed after chemical reduction of dual fluorophore-nitroxide compounds by ascorbic acids and other antioxidants, as well as nitric oxide and superoxide [27–38]. Trapping of free radicals by FNRO·can be also detected following ESR signal decay and fluorescence [37, 38]. Using the abovementioned unique properties of the dual molecules, a method for the quantitative analysis of vitamin C in biological and chemical liquids has been proposed [29]. Specifically, pseudo-first-order reaction between the dual molecule FN and ascorbic acid was monitored. An increase in the fluorescent intensity and decay of the EPR signal of the dual probe FN occurred with the same rate constant k_{red}. The calibration curves of k_{red}—ascorbic acid concentration for vitamin C analysis were used for performing vitamin C analysis in human blood plasma and commercial juces

A collection of dual probes of different redox potential of the nitroxide segment and the fluorofore spectral property were synthesized, that made it possible to select reducing agents appropriate for different media (Fig. 2.10). World distribution of groups using dual fluorophore-nitroxide method of analysis after the first publication of the use of as a redox probe in 1986. is illustrated in Fig. 2.11.

The pyrene-nitronyl (PN)

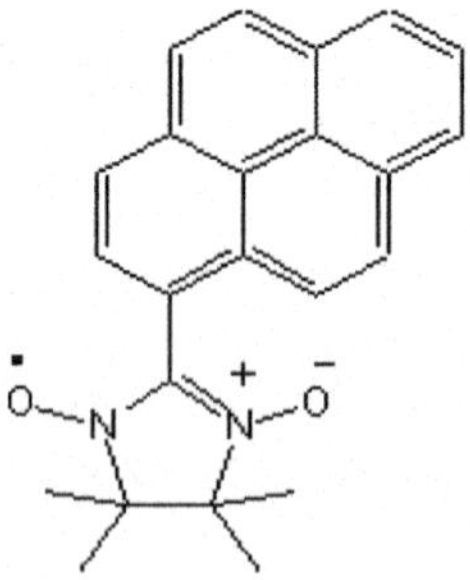

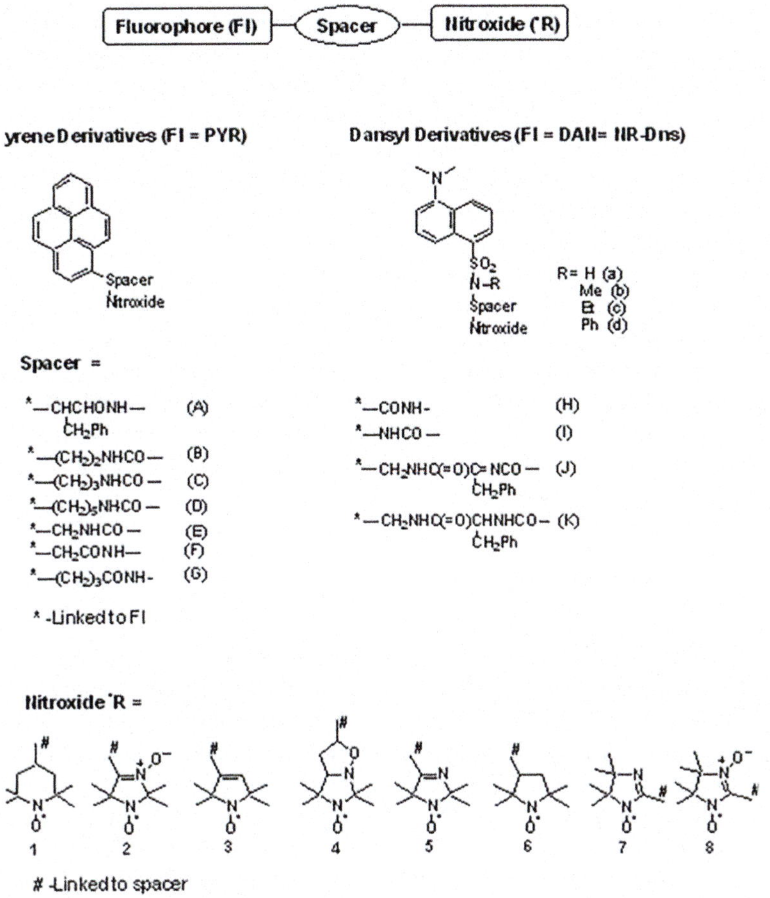

Fig. 2.10 Dual fluorescence-nitroxides. Private communication from Drs Martin VV and AL Wiess, Lipitek, Inc. International

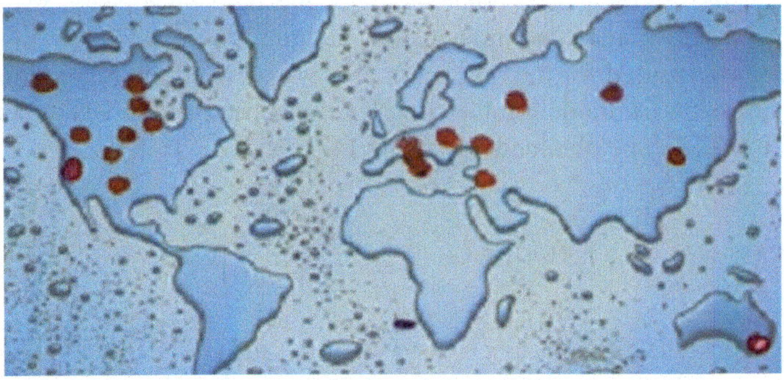

Fig. 2.11 World distribution of groups using dual fluorophore-nitroxide method of analysisThe pyrene-nitronyl (PN)

can be reduced by a variety of free radicals (e.g. superoxide) and antioxidants (e.g. ascorbic acid, quercetin, galangin, etc.) [30]. The reduction of PN to was accompanied by a drastic, about 2000 time, increase in fluorescence intensity and a corresponding decrease of the EPR signal. The developed fluorescence technique allows detection of antioxidants at the submicromolar concentration scale, while the EPR and light absorption techniques are limited to several micromolar. A new rapid and highly sensitive method for real time superoxide analysis has been developed based on a reaction of fluorophore–nitronyl with O_2^-[30] The reaction of PN with superoxide radicals, as a reduced agent, generated by the xanthine/xanthine oxidase system with slow rate ($\omega i < 2 \times 10-7$ M/min), was readily monitored by the fluorescent and ESR techniques. The dependence of rate of fluorescent increase (dI/dt) on the rate of superoxide production was used as a calibration curve for real time monitoring O_2^- in biological and chemical systems. This method has also been applied to the determination of hydrogen peroxide in submicromolar concentrations. The specificity of this approach can be enhanced by the addition of superoxide dismutase and catalase.

2.1.6 Terahertz Spectroscopy

Terahertz spectroscopy detects properties of matter with electromagnetic fields that are in the frequency range between a few hundred gigahertz and several terahertz (1 THz = 10^{12} Hz) [39–50]. Terahertz time-domain spectroscopy (THz-TDS) is a spectroscopic technique in which the properties of matter are probed with short pulses of terahertz radiation. THz spectroscopy is a powerful tool for characterizing dynamical modes {e.g., rotational, torsional, phonon, intramolecular and intermolecular). Typical laser based experiments of THz spectroscopy are: THz time-domain spectroscopy (THz- TDS), time-resolved THz spectroscopy (TRTS), THz emission spectroscopy (TES), and THz attenuated total reflection THz ATR.

The dielectric function is a measurement of the sample polarizability and could be described by using the Debye model [47]. When the applied field acts on the solution, the molecules would polarize. After the field disappears, the molecular polarization would vanish, and molecules would tend to relax back to their original state. The time required for the relaxation process is called the relaxation time τ. A general expression for modeling the complex dielectric permittivity function based on Debye model can be obtained as follows [47]:

$$\varepsilon(\omega) = \varepsilon_\infty + \sum_{n}^{j=1} \frac{\varepsilon_j - \varepsilon_{j+1}}{(1 + i\omega\tau_j)},$$

where the static dielectric constant is denoted by $\boldsymbol{\varepsilon}_\infty$ and $\boldsymbol{\varepsilon}_j$ is the intermediate steps in the dielectric constant, whereas $\boldsymbol{\tau}_j$ corresponds to the Debye relaxation time of the jth relaxation process [39]. The complex permittivity is defined as $\varepsilon^* = \varepsilon - (j\sigma/\omega)$, where s represents the electrical conductivity of the medium and particles, ω is the angular frequency of the electric field applied.

The dielectric responses of liquids have been extensively studied in the THz range and modeled using two Debye relaxation processes—a slow one (few ps) and a fast one less than femtoseconds (fs). Both processes depending on the temperature [40]. The enzymatic reaction of trypsin was controlled by fluorescent microscopy before being analyzed by sub-THz spectroscopy in the frequency range 0.06–0.11 THz [40]. Terahertz measurements were carried out using a vectorial network analyzer, which emits and detects THz radiations. The specific substrate NαBenzoyl-L-Arginine-7-Amido-4-MethylCoumarin (BA-AMC) was injected into the functionalized channels, and its hydrolysis was detected according to the reaction: BA-AMC + H_2O → Nα-Benzoyl-L-Arginine + 7-Amino-4-MethylCoumarin (AMC). Experiment showed that this activity is associated with conformational changes that occur during the conversion of substrates to products, such as protein hydrolysis into peptides after binding to a specific enzyme.

In study [50] the feasibility of employing terahertz time-domain spectroscopy (THz-TDS) for the nondestructive and label-free monitoring of protein digestion was evaluated. All experiments were carried out by employing a compact commercial THz-TDS (TR-1000; Otsuka Electronics, Osaka, Japan). The Er-doped femtosecond-pulsed laser (FemtoliteCS-20; IMRA, AISIN, Japan) in this system, which operates at a repetition rate of 50 MHz, emits a 780-nm laser beam was employed. The acid protease, pepsin, was used at its optimal pH to hydrolyze bovine serum albumin. The Debye model to analyze the dielectric properties of the solution during protein hydrolysis was explored.

2.2 Electrochemical Methods

2.2.1 Introduction

Through standard electrochemical measurements, catalytic activity, reversibility and stability, potentials of redox-active cofactors, and interfacial electron transfer rates can be measured [51–55]. The electrochemical behavior of molecules including analyte and product concentrations can be investigated, employing amperometry, potentiometry, cyclic voltammetry (CV), electrochemical impedance spectroscopy (EIS), chronoamperometry, chronocoulometry (CC) and adsorptive stripping differential pulse voltammetry (AdSDPV). The determination of analyte concentrations with selective, fast, and robust electrochemical methods is widely used for biological processes control, product analysis, environmental compliance, and medical applications.

2.2.2 Amperometry

Any analyte that can be oxidized or reduced is a candidate for amperometric detection. In this technique, a voltage (potential) is applied between two electrodes [51–57]. When an electroactive analyte is oxidized at the anode or reduced at the cathode the measured electric current changes occur. In pulsed amperometric detection (PAD), a working potential is applied for a short time (a few hundred milliseconds). Potential applications of amperometric biosensors range from the monitoring of environmental markers like pesticides, medical analytes in blood urine, glucose, lactate, and cholesterol, biomarkers for cancer or diabetes, to monitoring various industrially important substances as saccharides, alcohols, and phenols, cellulose-degradation and lignocellulosic biomass. Enzymes for application in 3rd generation amperometric biosensors are mostly heme, flavo-, or quino-enzymes, but copper enzymes are also known to interact with electrodes through direct electron transfer (DET) [51–55].

2.2.3 The Oxygen Electrode

The Clark electrode [58, 59] is an electrode that measures ambient oxygen partial pressure in a liquid amperometric oxygen measurement using a catalytic platinum surface according to the net reaction:

$$O_2 + 4\,e^- + 4\,H^+ \rightarrow 2\,H_2O$$

The Clark type electrode uses a silver/platnium-electrodes immersed in an electolyte solution (Fig. 2.12). When a polarizing voltage is applied to the cathode, electrons combine with oxygen molecules and reduce them to hydroxide ions. The electrode compartment is isolated from the reaction chamber by a thin Teflon membrane; which is permeable to molecular oxygen and allows this gas to reach the cathode.

2.2.4 Voltammetry

Voltammetry is the study of current (in amperes) as a function of applied potential (in volts) [52–54, 60]. To conduct such an experiment, one requires two electrodes. The working electrode, contacting with the an analyte, applies the desired potential and facilitate the transfer of charge to and from the analyte. A second reference electrode with a known stable reduction potential acts as the other half of the electrochemical cell. The auxiliary electrode passes all the current needed to balance the current observed at the working electrode (Fig. 2.13). The analytical data for a voltammetric experiment comes in the form of a voltammogram which plots the current produced by the analyte versus the potential of the working electrode. The

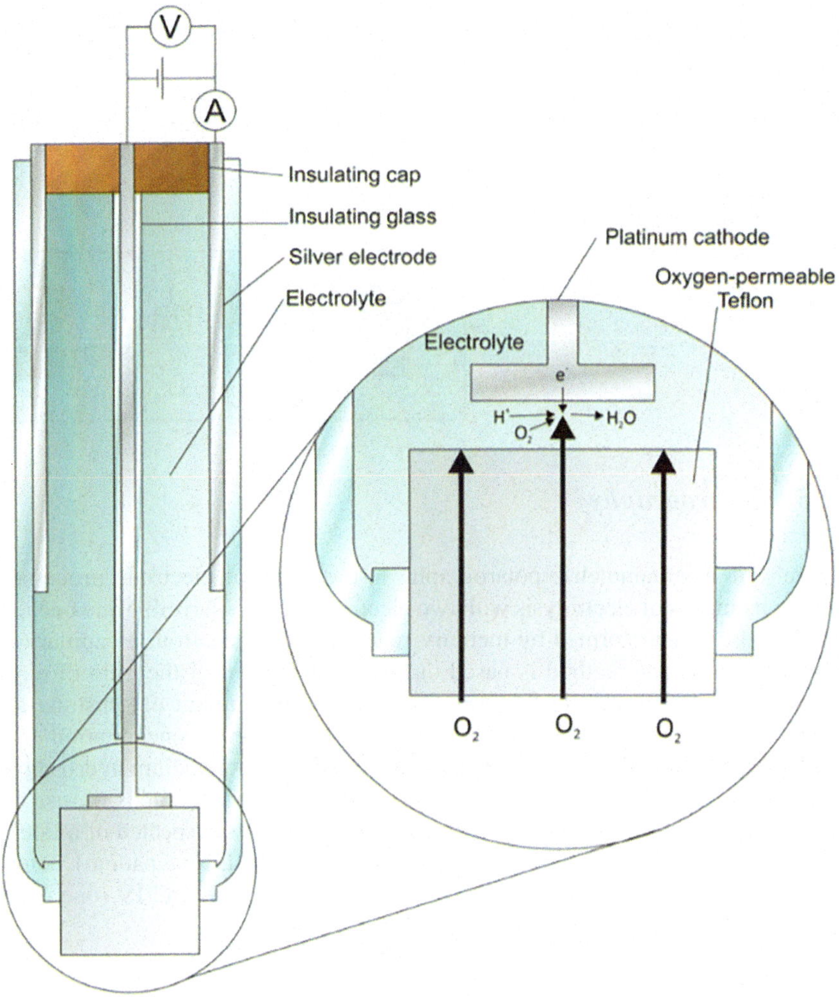

Fig. 2.12 A schematic representation of Clark's 1962 invention, the Oxygen Electrode.https://en.wikipedia.org/wiki/Clark_electrode#/media/File:Clark_Oxygen_Electrode.png. Published under CC BY-SA 4.0 licence

electrochemical (voltametric) biosensors in which voltammetric techniques such as differential pulse voltammetry (DPV) were employed for analytical purposes [60].

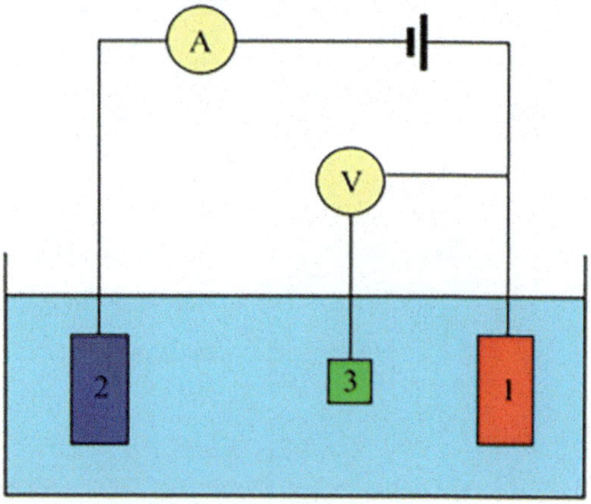

Fig. 2.13 Three-electrode setup: (1) working electrode; (2) auxiliary electrode; (3) reference electrode, A—ammeter, V—voltmeter. https://en.wikipedia.org/wiki/Voltammetry. Adam Rędzikowski Published under the Creative Commons Attribution-Share Alike 3.0 Unported license

2.2.5 Polarography

Principle of **a** voltammetric polarography is the study of electrode processes in solution by means of electrolysis with two electrodes, one polarizable and one unpolarizable, the former formed by mercury regularly dropping from a capillary tube [52, 53, 61–63]. The method is based on the measurement of the ratio of oxygen consumed by the enzyme in the presence of a limited amount of substrate as an internal standard to the total dissolved oxygen. The oxygen concentration can be measured by the addition of an excess amount of substrate or sodium hydrosulphite.

As an example, the polarographic and spectrophotometric methods were utilized for detection of mitochondrial disorders [63]. In six patients suspected of mitochondrial disease defects of complex I (two patients), complex III (one patient), complex IV (two patients) and a combination of defect of complex III and IV (one patient) were founded.

2.2.6 Cyclic Voltammetry

In cyclic voltammetry (CV) experiment, the working electrode's potential is ramped in the opposite direction to return to the initial potential [53, 64, 65]. The current at the working electrode is plotted versus the applied voltage (the working electrode's potential) to give the cyclic voltammogram trace. Cyclic voltammetry is generally used to study the electrochemical properties of an analyte in solution or of a molecule that is adsorbed onto the electrode. A standard CV experiment employs a cell fitted with three electrodes: reference electrode, working electrode, and counter (auxiliary) electrode. The homogeneous reaction between glucose oxidase and osmium bipyridine–pyridine carboxylic acid in the presence of glucose has been studied by cyclic voltammetry and digital simulation (Fig. 2.14) [66, 67].

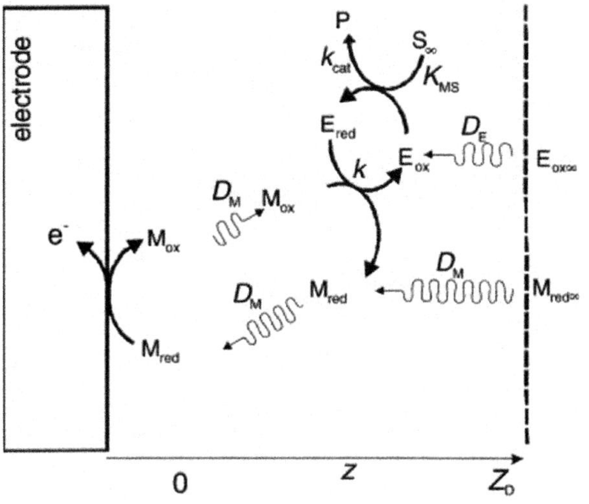

Fig. 2.14 The homogeneous system. M_{ox} and M_{red} are the oxidized and reduced forms of the mediator. E_{ox} and E_{red} are the oxidized and reduced enzyme. *S* and *P* are the substrate and product of the enzymatic reaction. D_E and D_M are the diffusion coefficients of enzyme and mediator respectively; the coefficients of the oxidized and reduced forms are assumed to be equal. Oxidized mediator Os(III) is only produced at the electrode surface by re-oxidation of Os(II) [66]. Published under the terms of the Creative Commons CC-BY license from Electrochemical Society.

A standard three-electrode electrochemical cell was employed with an operational amplifier potentiostate (TEQ-Argentina). An Ag/AgCl; 3 M KCl (0.210 V vs. NHE) reference electrode was utilized and; the auxiliary electrode was a large area platinum gauze (a thin translucent fabric) (Fig. 2.14). Measurements of the dependence of the amperometric response on enzyme, substrate and co-substrate concentrations with digital simulation of the coupled enzyme reaction diffusion allowed to extract kinetic parameters for the substrate–enzyme reaction: $K_{MS} = 10.8$ mM, $k_{cat} = 254\ s^{-1}$ and for the redox mediator–enzyme reaction, $k = 2.2 \times 10^5\ M^{-1}\ s^{-1}$. Fig. 2.15 depicts a typical set of cyclic voltammograms recorded at 5 mV s^{-1} in a solution containing 1.0 mM $[Os(bpy)_2ClPyCOOH]^+$ and 1.6 μM GOx (Fluka) in 0.1 M NaH_2PO_4/Na_2HPO_4, 0.1 M NaCl buffer solution of pH 7 in the absence (a) and in the presence of increasing concentrations of D-glucose, the enzyme–substrate, (b–e).

2.2.7 Potentiometry

Potentiometry measures the potential of a solution between two electrodes. One electrode is called the reference electrode and has a constant potential, while the other one is an indicator electrode whose potential changes with the composition of a sample. Its response is proportional to the analyte concentration by comparison of

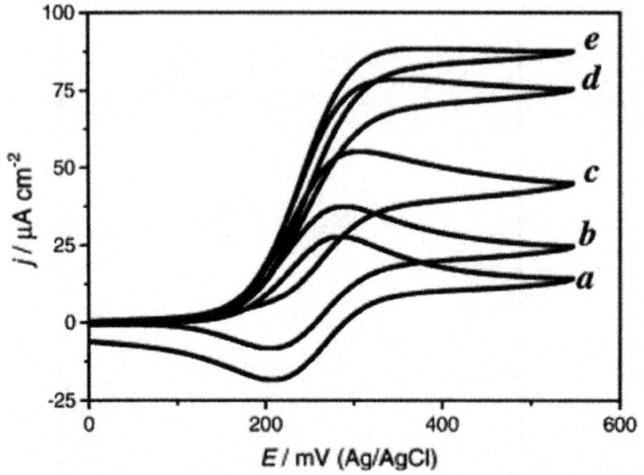

Fig. 2.15 Cyclic voltammograms recorded at different D-glucose concentrations and constant m_Σ = 1.02 mM and e_Σ = 1.6 µM in NaH_2PO_4/Na_2HPO_4 0.1 M + 0.1 M NaCl buffer solution. Glucose concentration (mM): a) 0.0; b) 1.6; c) 3.2; d) 9.9; e) 22.7. The reference electrode is Ag/AgCl in 3 M KCl, scan rate 5 mV s^{-1}. V.Flexer,[a]M.V.Ielmini,[a]E.J.Calvo, [a] Extracting kinetic parameters for homogeneous $[Os(bpy)_2ClPyCOOH]^+$ mediated enzyme reactions from cyclic voltammetry and simulations [66] Published under the terms of the Creative Commons CC-BY license from Electrochemical Society

its activity to the reference electrode. The advantage of potentiometric biosensors is their sensitivity and selectivity when a highly stable and accurate reference electrode was used [68]. Potentiometric methods of the pH measurement are based on the measurement of the activity of hydrogen ions in an acid–base system of known ratio between basic and acidic forms of the compound [69]. The pH is measured through a calibrated glass combination electrode, over a concentration range of 0.005–0.05 M and in 2–12 aqueous pH range. The pK_a is determined by a titration of the basic or acidic compound with a standardized strong acid or base. The plot of pH against the volume of the added titrant is called potentiometric titration curve.

Open circuit potential (OCP) is defined as the potential that exists in an open circuit. Figure 2.16 demonstrates the application of open circuit potentiometry (OCP) to measure glucose oxidase turnover kinetics, k_{turn} [70]. The electrode surface becomes poised by the addition of a redox pair, such as ferrocenemethanol/ ferrocenium methanol ($FcMeOH/FcMeOH^+$), which acts as the cosubstrate for the enzymatic process.

2.2.8 *Chronoamperometry*

Chronoamperometry is an electrochemical technique in which the potential of the working electrode is stepped and the resulting current from faradaic processes, occurring at the electrode, is monitored as a function of time [53, 71, 72]. Faradaic

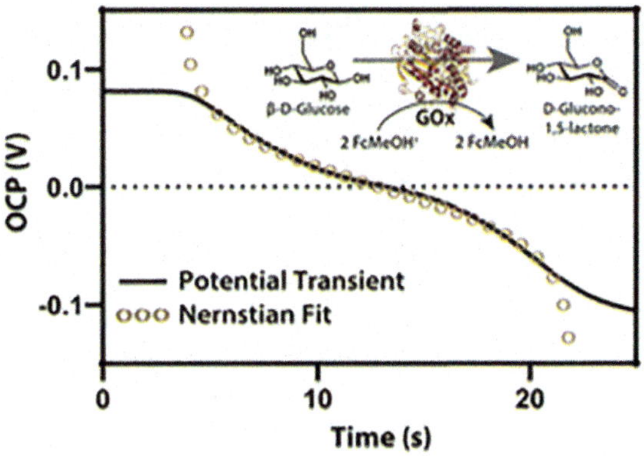

Fig. 2.16 OCP time dependence in the presence of glucose oxidase, glucose and a redox pair, ferrocenemethanol/ferrocenium methanol [70]. With permission from American Chemical Society

processes are defined as those which obey Faraday's law, that is the amount of chemical reaction occurring is directly proportional to the amount of charge passed across the electrode boundary. The functional relationship between current response and time is measured after applying a single potential step (applying a forward potential step and recording the resulting current) or a double potential step (applying a forward potential and returning the potential to the initial value during a given period of time) to the working electrode of the electrochemical system. There are two types of chronoamperometry that are commonly used, controlled-potential chronoamperometry and controlled-current chronoamperometry. Current at a given time, $i(t)$ (measured in units of amperes), is directly proportional to concentration, as shown by the Cottrell Equation:

$$\tilde{n}(t) = \frac{nFAD_o^{1/2}C_o^*}{\pi^{1/2}t^{1/2}} \tag{2.1}$$

where n is the number of electrons transferred, F is Faraday's constant (96485 C/mol), A is the working electrode area (in cm), D_o is the diffusion coefficient (in cm^2/s), C_o is the concentration (in mol/cm^3), and t is time (in s). Chronoamperometric determination of the Michaelis constant for nitrate reduction by *R. sphaeroides* NapAB is demonstrated in Fig. 2.17. The Figure illustrates the simple chronoamperometric determination of a Michaelis constant from an experiment on nitrate reduction by the respiratory nitrate reductase (NapAB) from *Rb. sphaeroides*, when the concentration of substrate was stepwise increased [73, 74], The electrode was spun at a high rate, and the change in i_{lim} against nitrate concentration was fit to the Michaelis-Menten equation in panel B

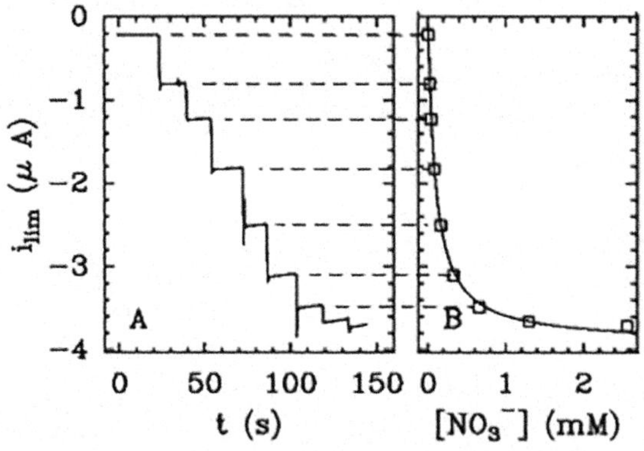

Fig. 2.17 Chronoamperometric determination of the Michaelis constant for nitrate reduction by *R. sphaeroides* NapAB. The electrode was poised at −450 mV Vs SHE, rotated at 2 krpm, and nitrate was added from stock solutions while the activity was continuously recorded as a current. In panel B, the change in current is fit to the Michaelis-Menten equation to determine Km. With permission from American Chemical Society [73]

2.2.9 *Coulometry*

Coulometry determines the amount of matter transformed during an electrolysis reaction by measuring the amount of electricity (in coulombs) consumed or produced [53, 72, 75] This method, based on the strict proportionality between the extent of a chemical change and the quantity of electricity involved (Faraday's law), can be used for determining the quantity of a substance, charge, and the amperes. There are two basic categories of coulometric techniques: potentiostatic coulometry involves holding the electric potential constant during the reaction using a potentiostat and coulometric titration or amperostatic coulometry, keeps the current \constant using an amperostat. Chronocoulometry is used to study the kinetics of chemical reactions, diffusion processes, and adsorption.

2.2.10 *Electrochemical Impedance Spectroscopy*

Electrochemical impedance spectroscopy (EIS) is characterized by an electrical circuit consists of resistances, capacitors, or constant phase elements that are connected in parallel or in a series to form an equivalent circuit Fig. 2.18 [53, 76–78], A capacitor is a device that stores electrical energy in an electric field. Capacitance is expressed as the ratio of the electric charge on each conductor to the potential difference (i.e., voltage) between them. The EIS has the ability to study intrinsic material properties or specific processes that could influence conductance, resistance, or capacitance of an electrochemical system. Thus, the EIS could be used to explore mass-transfer, charge-transfer, and diffusion processes. The excitation signal

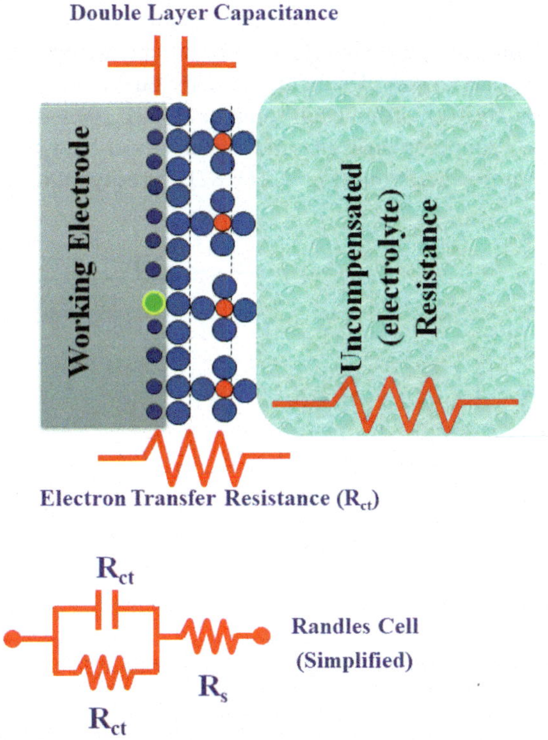

Fig. 2.18 A simple scheme to descript the EIS circuit and the redox reaction takes place at the surface of working electrodes in a conventional-electrochemical cell (i.e., three-electrode system). R_{ct} is the charge transfer resistance, R_s is electrolyte resistance, and C_{dl} is the capacitance double layer [77]. Published under Creative Commons Attribution License from MDPI the

is presented as a function of time, as shown in Equation:

$$E_t = E_0 \cdot \sin(\omega t) \tag{2.2}$$

where E_t is the potential at time t, E_0 is the amplitude of the signal, and ω is the radial frequency. The impedance of the whole system is given as

$$Z = E/I = Z_0 \exp(i\Phi) = Z_0(\cos\Phi + i \sin\Phi) \tag{2.3}$$

where Z, E, I, ω, and Φ are impedance, potential, current, frequency, and phase shift between E and I, respectively.

2.3 Enzyme Biosensors

Enzyme-based chemical biosensors are based on biological recognition and on knowledge about the target analyte [79–86]. Figure 2.19 shows schematically the parts comprising a typical biosensor: (a) bioreceptors that specifically bind to the analyte; (b) an interface architecture; (c) the transducer signal (d) computer software; and (e) an interface to the human operator. Biosensors can be applied to a large variety of samples including body fluids, food samples, cell cultures and be used to analyze environmental samples (Fig. 2.20).

Amperometric enzyme biosensors utilizing oxidoreductases were classified into three generations (Fig. 2.21) [82]. Three types of biosensors are known. 1st generation

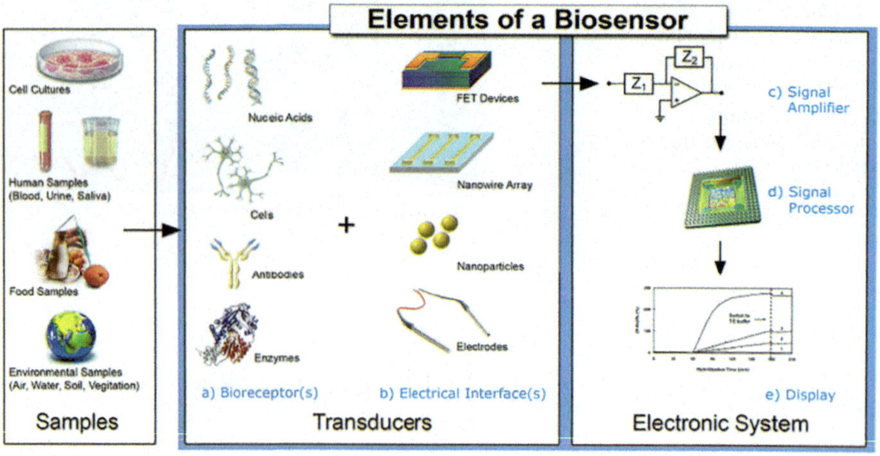

Fig. 2.19 Elements and selected components of a typical biosensor [86] Published under the Creative Commons Attribution License from MDPI

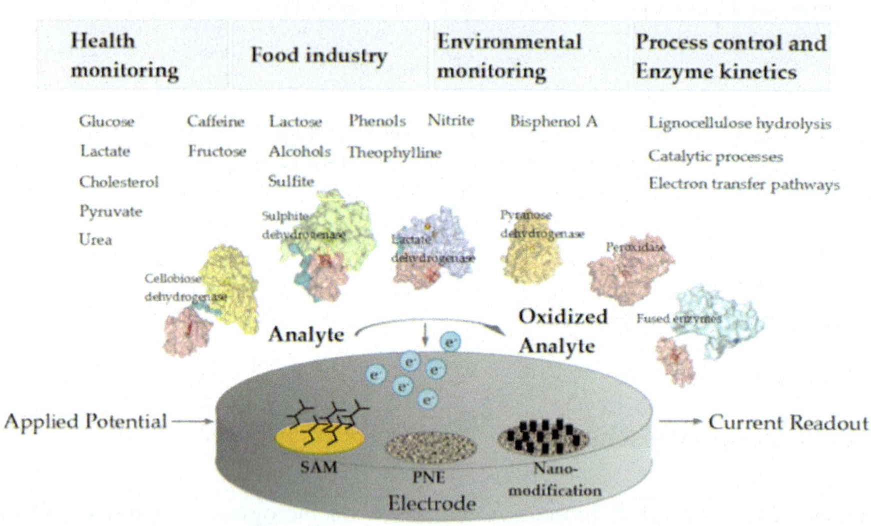

Fig. 2.20 Electrochemical Biosensors [82]. Published under the Creative Commons Attribution License from MDPI

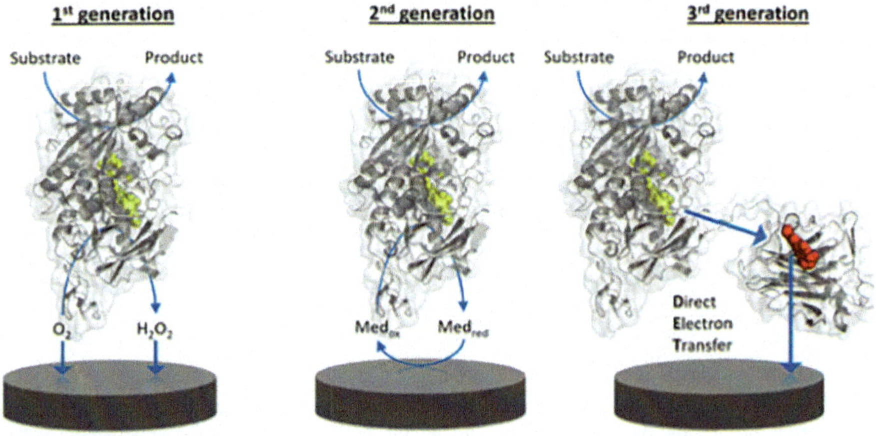

Fig. 2.21 Schematic representation of the working principle of the three generations of enzyme-based amperometric biosensors. 1st generation: based on the electrocatalytic detection of enzymatic substrate conversion or product formation. 2nd generation: based on the electrocatalytic recycling of a redox mediator. 3rd generation: based on direct electron transfer from enzyme to electrode. Yellow molecules depict the FAD cofactor of flavoenzymes and the red molecule depicts the heme cofactor in the mobile cytochrome domain of a multi-cofactor enzyme [82]. Published under the Creative Commons Attribution License from MDPI

biosensors employing oxidases based on the electrocatalytic monitoring of substrate consumption or product formation, 2nd generation biosensors employing oxidases or dehydrogenases based on the electrocatalytic recycling of suitable redox mediators, and 3rd generation biosensors employing preferentially dehydrogenases capable of direct electron transfer to bare or modified electrodes

An amperometric enzyme, biosensor incorporated device involved the amount of the current ensuing from an enzyme reaction. In amperometric biosensors, the output current of the sensor can be analyzed and used for the sensing process [85]. Figure 2.22 illustrates schematically application areas, analytes, enzymes, and the architecture of 3rd generation amperometric biosensors.

In work [80] a novel self-powered microneedle-based transdermal biosensor for pain-free high-accuracy real-time measurement of glycaemia in interstitial fluid (ISF) was designed. The glucose sensor comprises graphite electrodes having glucose oxidase entrapped into the pores. The enzyme, catalyzing the oxidation of glu cose to hydrogen peroxide and D-glucono-δ-lactone

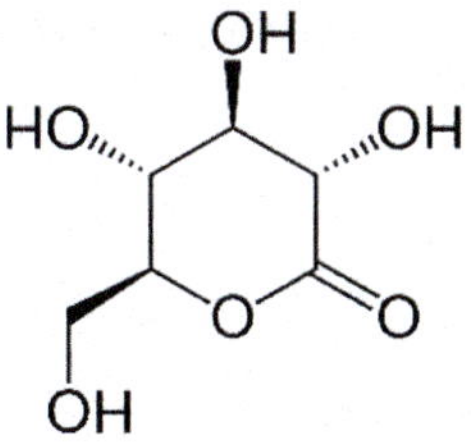

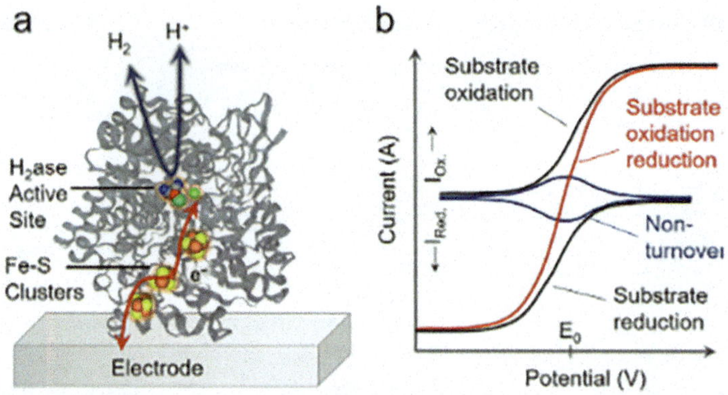

Fig. 2.22 Protein film electrochemistry. **a** Schematic representation of the electron transfer between an electrode surface and the enzyme active site, which catalyzes interconversion between oxidized and reduced substrates (given example is a hydrogenase). **b** Simulated voltammetry showing the enzyme's redox couple under non-turnover conditions and catalytic voltammetry under turnover conditions [85]. With permission from Elsevier

a coenzyme, requires flavin adenine dinucleotide (FAD). The technique allowed pump-free uptake of ISF to be carried out with high efficacy and reliability in a few seconds (uptake rate up to 1μl/s) by exploiting capillarity in the microneedles. Figure 2.23 shows the calibration curves and the experimental data depicting that the sensor can be used to detect concentrations.

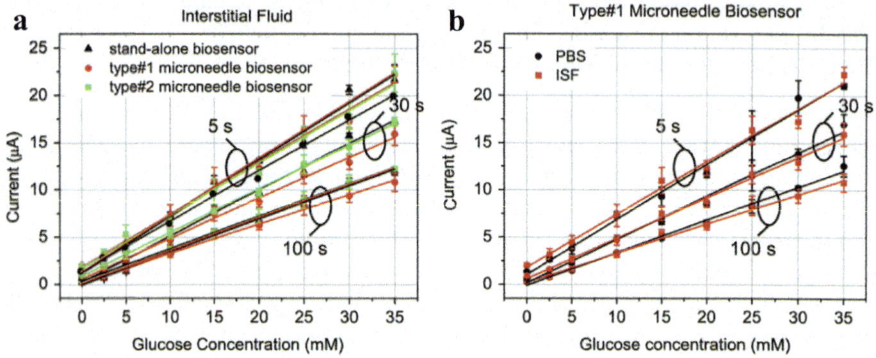

Fig. 2.23 **a** Representation of the biosensor having glucose sensor attached on the back of the microneedle; **b** Calibration curves for the concentration between 0–35 mM; **c** Calibration curves of developed biosensors and stand-alone biosensors in ISF, and **d** Calibration curves for developed biosensor for concentration 0–35 mM at 5 s, 30 s, and 100 s of sampling times [80]. With permission from Elsevier

2.4 Magnetic Resonance

2.4.1 Grounds

Electron and some nuclei possess corresponding intrinsic angular momentums that are quantized. In an external magnetic field, there is a high-energy state and a low-energy state depending on the relative orientation of the magnet to the external field, and in thermal equilibrium, the low-energy orientation is preferred [82, 86–90, 93–112]. In a constant magnetic field of strength H_0, the Electron Spin Resonance (ESR) phenomena involves the resonance absorption or dispersion of a microwave frequency (ν) (0.3–250 GHz) of electromagnetic field. The microwave absorption leads to magnetization in the exited state of the system. Accordingly, the electron magnetic resonance condition is [2.1] :

$$h\nu = g_e \beta_B \boldsymbol{H}_0, \tag{2.4}$$

where g_e is a g-factor, characterizing the value of the intrinsic electron spin moment (free electron g-value is 2.002319),), β_B is the Bohr magneton $\left(9 \cdot 27400968 \times 10^{-24} J \cdot T^{-1}\right)$. The values of g-factors and magnetic field strength H_0 dictate position of resonance frequencies in the ESR spectra. The interaction of an unpaired electron, by way of its magnetic moment, with nearby nuclear spins results in additional allowed energy states. This phenomenon leads to the superfine (hipperfine) splitting of the ESR signal into N lines in accordance with the quantum mechanical rule $N = 2I + 1$ where I is the nuclear quantum number. Three common mechanisms by which electrons and nuclei interact are: the isotropic Fermi contact interaction, dipolar interaction and dynamic nuclear spin polarization The isotropic Fermi contact interaction is the quantum mechanical magnetic interaction between an electron and an atomic nucleus when the electron is inside that nucleus. The anisotropic dipole-dipole interaction between spin electron and spin of nucleus depends on the sample orientation. Dynamic nuclear spin polarization is a mechanism for interactions between an unpaired electron and a nuclear spin. Spin polarization is the degree to which the spin is aligned with a given direction. Dynamic nuclear polarization (DNP) relates to transferring spin polarization from electrons to nuclei, resulted in aligning the nuclear spins to the extent that electron spins are aligned [10–12, 14, 105–110].

Spin relaxations times are important parameters of electron spin resonance and nuclear magnetic resonance [82, 86–94, 111–113]. The characteristic time of recovery of the induced magnetization to the ground state on account a transfer of magnetic energy to energy of media (lattice) is noticed as spin-lattice relaxation ($1/T_1$). Another type of relaxation, spin-spin or spin-phase relaxation ($1/T_2$), is related to time of return of the spin system to equilibrium in the exited magnetic state as a result of spin-spin interaction with environment. The second approach to the spin relaxations phenomena is based on properties of the electron Larmor precession, which is the precession of the magnetic moments of electrons about external magnetic field.

In effect saturation phenomenon, the population of ground and excited states, and hence the rates of both transitions are equalized, and the signal disappeared. The ability of NMR or ESR signals to become saturated in an applied microwave RF or MW field is determined by the value of the MW amplitude H_1 and the spin relaxation parameters ($T_{1e,n}$ and $T_{2e,n}$). These parameters can be measured directly be pules methods [91–93, 111–113].

2.4.2 *Nuclear Magnetic Resonance-Based Methods*

2.4.2.1 Grounds

Nuclear magnetic resonance (NMR) is a phenomenon in which nuclei in a strong constant magnetic field are perturbed by radiation of radio frequency. The nucleus bearing spin also have different energies in a non-zero magnetic field [91, 92, 114–123]. The NMR phenomena involves the resonance absorption or dispersion of a frequency range from around 20 kHz to around 300 GHz of electromagnetic field (ν). The energy difference between the two states with the nuclear quantum number $m = 1/2$ and gyromagnetic relation γ is [2.2]:

$$\Delta E = h\nu = \gamma (h/2\pi)\, H_0 \tag{2.5}$$

Chemical shift in the NMR spectra is due to shielding which is a barrier made of inner-shell electrons and it decreases the nucleus pull on the outer electrons [94, 95, 114–123]. Electrons orbiting around the nucleus generates a small magnetic field that opposes H_0, and electrons are shielding the nucleus from H_0. Due to addition of electron density, magnetic induction, or other effects chemical shift has been decreased. When an atom is deshielded, it means that nucleus chemical shift has been increased due to removal of electron density, magnetic induction, or other effects. The detected frequencies (in Hz) for ^1H, ^{13}C, ^{31}P, ^{15}N and ^{29}Si nuclei are usually referenced against position of the NMR signal tetramethylsilane (TMS). Some of the most useful information for structure determination in a one-dimensional NMR spectrum comes from J-coupling or scalar spin–spin coupling between NMR active nuclei arises from the interaction of different spin states through the chemical bonds of a molecule and results in the splitting of NMR signals. Below are some typical examples illustrating the analytical application of NMR.

The general Overhauser effect was first predicted in 1953 by Albert Overhauser [96–99, 105–108] who observed that saturation of the conduction electron spin resonance by microwave irradiation causes a polarization of nuclear spins. The nuclear Overhauser effect (NOE) is the transfer of nuclear spin polarization from one population of spin-active nuclei (e.g. ^1H, ^{13}C, ^{15}N etc.) to another via cross-relaxation induced by dipole-dipole interaction The change in the integrated intensity (positive or negative) of one NMR resonance occurs when another is saturated by irradiation with a radio frequency field. For example, in a system consisting of two spins of value 1/2, I and S, their spin components along the z-axis (parallel to the polarizing field) are 'up' or 'down' are designated by corresponding eigenstates α and β . In a system consisting of two spins of value 1/2, I and S, the four eigenstates of the system are $\alpha\alpha$, $\alpha\beta\beta\alpha$ and $\beta\beta$ take place. The transition probabilities per unit time between the states are W_{1I}, W_{1S}, $W_{0\,IS}$, and W_{2IS}.

Chemically induced spin correlated dynamic electron polarization (CIDEP) effects take place when the populations of the spin states deviate substantially from thermal equilibrium [100–104, 107–108]. Such an effect can arise in the electron spin system of reactive radicals generated as transient intermediates following the sudden initiation of chemical reactions, for example by a photochemical reaction. The main mechanisms causing spin polarization in free radicals are: the triplet mechanism (TM), the radical pair mechanism (RPM), the correlated coupled radical-pair mechanism (CRPM), the radical triplet pair mechanism (RTPM) and triplet-triplet annihilation mechanism (TTAM).

2.4.2.2 Analytical Application of NMR

Techniques for the correlation of ^{1}H and ^{15}N chemical shifts based on zero and double quantum NMR were developed [106, 114]. ^{15}N has a high spectral resolution because of its broad chemical shift range (500−900 ppm). The sensitivity of ^{15}N is 1/1000th of that of ^{1}H and 1/15th of that of ^{13}C for equal numbers of nuclei at a constant field. Proton-decoupling lead to the enhancement of the ^{15}N signal by the nuclear Overhauser effect (NOE) for ^{15}N nuclei that relax via dipolar interaction with proton(s). The sensitivity of ^{15}N detection is significantly enhanced by indirect detection through coupled ^{1}H by ^{1}H-^{15}N heteronuclear multiple-quantum coherence (HMQC) transfer magnetic resonance spectroscopy (MRS). HMQC allows the detection of low-concentration [5-^{15}N]GLN in vivo and [$^{15}N_{indole}$]serotonin *in vitro*. For example, HSQC makes it possible the detection of nmol quantities of [5-^{15}N]GLN and [^{15}N]urea in 40 μL of the extracellular fluid [107, 115–116].

Glutamine synthetase is important for the glutamine/glutamate cycle in brain. Rates of Glutamine synthetase catalyzed reaction can be inferred from an isotope chase experiment [107]. In this approach, the magnetic resonance spectroscopy (MRS) measurement shows a steady state concentration of ^{15}N-labeled glutamate while infusing a constant concentration of $^{15}NH_4Cl$. Then the infusate was switched to $^{14}NH_4Cl$ and the exponential decline in (5-^{15}N)glutamate signal was measured. In vivo ^{15}N MRS at 4 or 7 Tesla has made a contribution to kinetic studies of the individual pathways that control glutamate flux in the rat brain [108, 115–116]. In this condition, glutamine synthetase activities, measuring at several levels of hyperammonemia, showed that this enzyme becomes saturated at blood ammonia concentration > 0.9 μmol/g, and causes the elevation of brain ammonia. [^{14}N]LEU was replaced by [^{15}N]LEU and the rate of GLU synthesis from LEU was monitored in vivo. The in vivo ^{15}N spectra of the brain [5-^{15}N]GLN were acquired during isotope chase by ^{14}N (Fig. 2.24). A progressive decrease in the peak area was observed, as ^{15}N is replaced by ^{14}N while maintaining the total (i.e. $^{15}N + {}^{14}N$) concentration at a

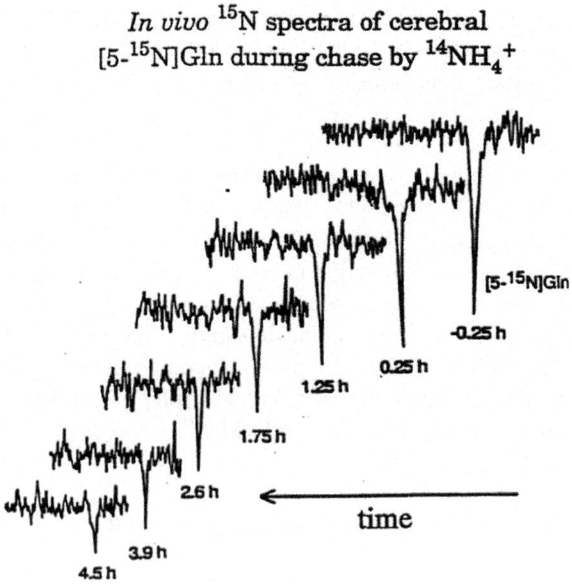

Fig. 2.24 Representative in vivo ^{15}N spectra of brain [^{15}N]GLU during [^{15}N]LEU infusion. [reproduced from [3] with permission] [108, 115–116]. With permission from Springer Nature

steady-state. These techniques was used to measure the rates of enzyme catalyzed reactions in a variety of living systems and in vivo of the ATP synthetase and the creatine kinase reactions, in particular [109, 117].

2.4.3 *Nitric Oxide Analysis by ESR*

Spin trapping is an analytical technique employed for the detection and identification of short-lived free radicals through the use of electron paramagnetic resonance (EPR) spectroscopy [186]. When the half-life of radicals is too short to detect with EPR, compounds known as spin traps are used to react covalently with the radical products and form more stable adduct that will also spectra detectable by EPR spectroscopy. The most commonly used spin traps are alpha-phenyl N-tertiary-butyl nitrone (PBN) and 5,5-dimethyl-pyrroline N-oxide (DMPO)

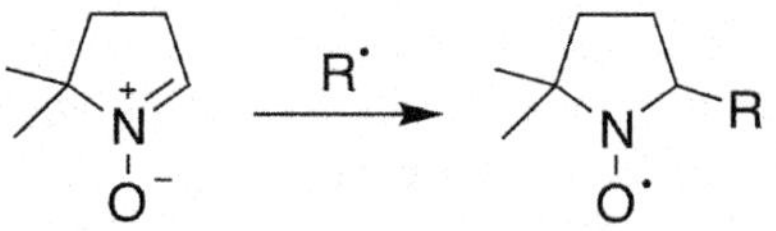

Below are some examples of the use of the method for the analysis of radicals of high chemical reactivity and diffusive mobility. A liposome-encapsulated spin-trap (LEST) method for the capture and in situ detection of NO by EPR was introduced [176, 186–190]. The method showed a linear response for $[NO] \geq 4\mu M$ and can detect $[NO] \geq 40$ nM in a 500 μL sample ($\geq$ 20 pmol) in real time over minutes to hours. LEST showed minimal abiotic NO production in the presence of nitrite and NADH. In addition, nitrate reductase-like activity was detected in cell lysates of the coccolithophore Emiliania huxleyi and was elevated in virus-infected culture.

The compound 5-Hydroxy-2,2,6,6-tetramethyl-4-(2-methylprop-1-en-yl)cyclohex-4-ene-1,3-dione [188]

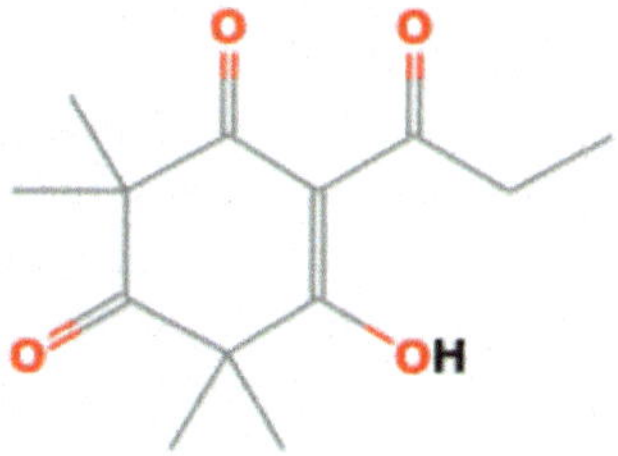

has been proved to behave as an efficient cheletropic trap for both NO and NO_2 radicals in the presence of oxygen, yielding EPR observable nitroxide and alkoxynitroxide, respectively [177, 188]. In the study [177, 188] the *in situ* reaction of the aromatic spin trap 3,5-dibromo-4-nitrosobenzene sulfonate (DBNBS)

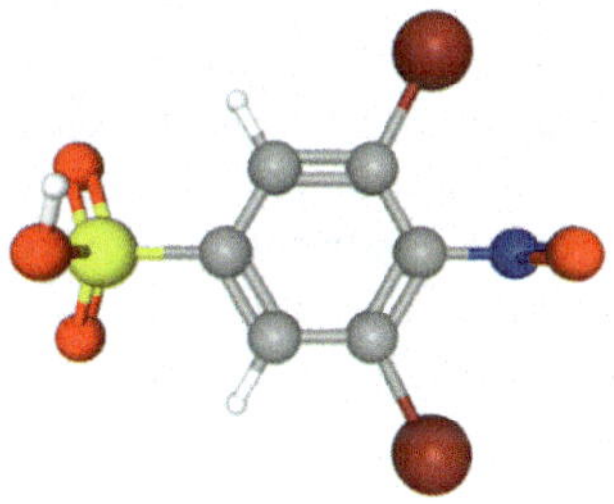

with nitric oxide (NO), generated via the reduction of nitrite under acidic conditions at room temperature, was examined [177, 187]. It was concluded that DBNBS is an efficient nitric oxide scavenger that has the potential to 3,5-dibromo-4-nitrosobenzenesulfonate (DBNBS) has been used in combination with EPR) spectrometry to trap nitric oxide [155, 189]. The reaction between DBNBS and **NO** yields a radical product which gives rise to an EPR signal consisting of three lines with a hyperfine splitting $A_N = 0.96$ mT. Then the fractions containing the radical product were identified by the presence of the three-line EPR signal, and these fractions were analyzed by negative ion fast atom bombardment–mass spectrometry (FAB–MS).The data suggested that the radical product is bis(2,6-dibromo-4-sulfophenyl)nitroxyl

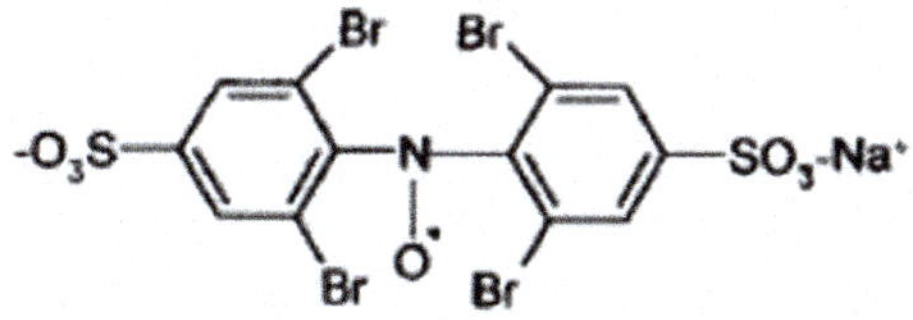

2.5 Microdialysis

The most widely used in *vivo* sampling method is microdialysis (Fig. 2.25), in which species, such as substrates and products, can diffuse through the semi-permeable membrane comprising the probe [135–141, 145–151]. Rapid sampling microdialysis (rsMD) is an enzyme-based electrochemical detection method, which is particularly suited to monitoring metabolism. In particular, the technique of microdialysis enables sampling and collecting of small-molecular-weight substances from the interstitial space. Microdialysis is a widely used method in neuroscience and permits quantification of neurotransmitters, peptides, and hormones in the behaving animal, Work [139, 145] provided a brief review of the history of microdialysis and its general application in the neurosciences including theoretical principles underlying the microdialysis process. Several important aspects were discussed in details including: (1) Methods available for estimating extracellular concentration from dialysis samples, (2) The various factors that affect the estimate of in vivo relative recovery, (3) The importance of determining in vivo relative recovery to data interpretation; and (4) Impact of tissue trauma on the interpretation of microdialysis results. The principles and applications of microdialysis in preclinical and clinical settings, microdialysis as a tool in drug research and development were discussed.

An enzyme-based electroanalysis system for real-time analysis of a clinical microdialysis sampling stream during surgery was developed [137, 147]. Free flap tissue transfer was used in reconstructive surgery after resection of tumours or in major trauma. Flap surgery is a technique in plastic and reconstructive surgery where any type of tissue is lifted from a donor site and moved to a recipient site with an intact blood supply. Rapid sampling microdialysis (rsMD) is an enzyme-based electrochemical detection method, which is particularly suited to monitoring metabolism.

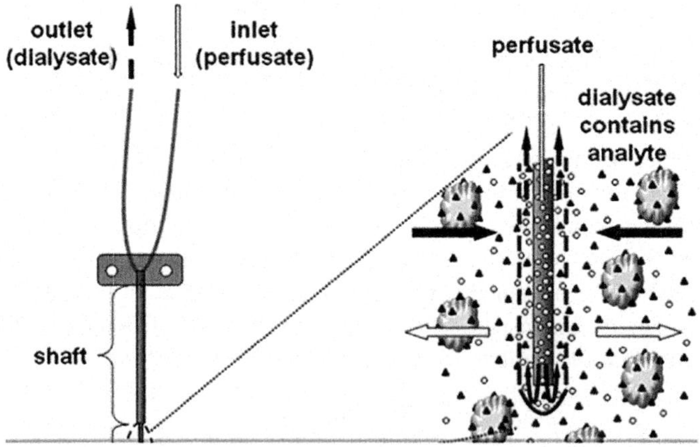

Fig. 2.25 Schematic illustration of a microdialysis probe https://en.wikipedia.org/wiki/Microdialysi

The experiment suggested that: (1) Online flow injection system can analyze a dialysate flow stream from an implanted microdialysis probe every 30 s for levels of glucose and lactate; (2) The technique can monitoring of free flap reconstructive surgery, from flap detachment to re-vascularisation and overnight in the intensive care unit; (3) The on-set of ischaemia was seen as an increase in lactate and decrease in glucose levels; and (4) Glucose levels returned to normal within 10 min of successful arterial anastomosis, whilst lactate took longer to clear.

Soil extracellular enzyme activity (EEA) represents a critical bottleneck in the release of bioavailable nutrients from organic materials. The potential of *in situ* microdialysis to directly sample extracellula soil enzymes, measuring their activity using a modified enzyme assay was explored [138, 148]. The following important results were reported: (1) The protease recovery is hindered at lower soil moisture contents; (2) The response of native protease activity after the addition of soybean litter to clay and sandy soils was measured; (3) Finding microdialysis was observed greater oil extracellular enzyme activity (EEA) in litter-amended treatments than controls in soil types; and (4) Hydrolytic enzyme activity of free and stabilized clay soil fractions were estimated.

Work [139, 149] discussed the immune response resulting from microdialysis probe implantation. The principles of microdialysis and the applications of Dexamethasone, a potent synthetic corticosteroid,

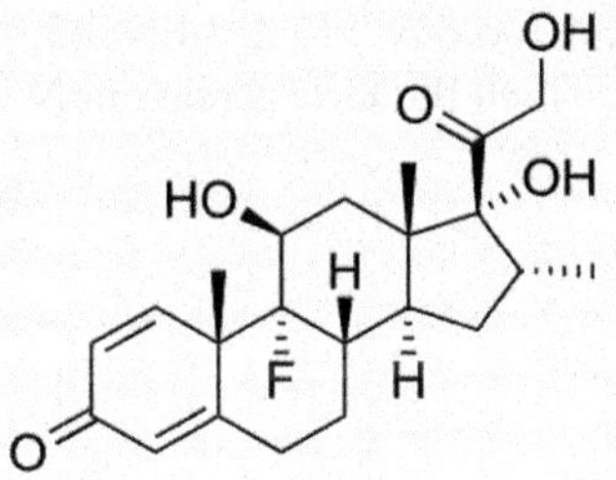

were formulated. Microdialysis in (i) neuronal devices, (ii) dopamine and fast scan cyclic voltammetry, (iii) the attenuation of microglial cells, (iv) macrophage polarization states, and (v) spreading depolarizations was briefly reviewed. Employing microdialisis, it was shown how that the addition of dexamethasone, an anti-inflammatory drug, to the perfusion fluid protects evoked dopamine responses as measured by fast-scan cyclic voltammetry next to the probes after 24 h [141, 151]. It was found that that dexamethasone stabilizes evoked dopamine responses measured at the probe outlet over a 4–24 h postimplantation interval.

2.6 Nitric Oxide Analysis

2.6.1 Introduction

Nitric oxide (NO) is a tiny molecule with enormous biological impart [129, 133, 142–145], [147–189]. Nitric oxide (NO) is essential in a significant number of vital normal and pathological physiological processes. Nowadays, a whole arsenal of chemical, physical and biological methods spanning the NO picomolar-to-micromolar concentration range in physiological milieus has been developed and widely used (Fig. 2.26) The spectroscopic and electrochemical methods described above are the most commonly employed techniques for measuring NO. Other less frequently employed approaches include mass spectrometry, X-ray photoelectron spectroscopy, quantum cascade infrared laser spectroscopy, mechatronics sensoring, and quartz crystal microbalance technique also have been developed/. Nevertheless, real-time monitoring of NO dynamics under physiological conditions of nano and less molar concentrations is still a challenging analytical problem. This has been attributed to the labile nature of NO molecule with half-life of the order of seconds, which both rapidly diffuses through the medium and readily reacts with many scavenger targets.

Because of its simplicity, availability and the intense color of an adduct, optical spectroscopy for a long time served as main practical tools for nitrite and NO analysis in chemistry and biology [129, 142–145, 147–189]. The Peter Griess diazotization reaction on which the Griess reagent N-(1-Naphthyl)ethylenediamine dihydrochloride [154]

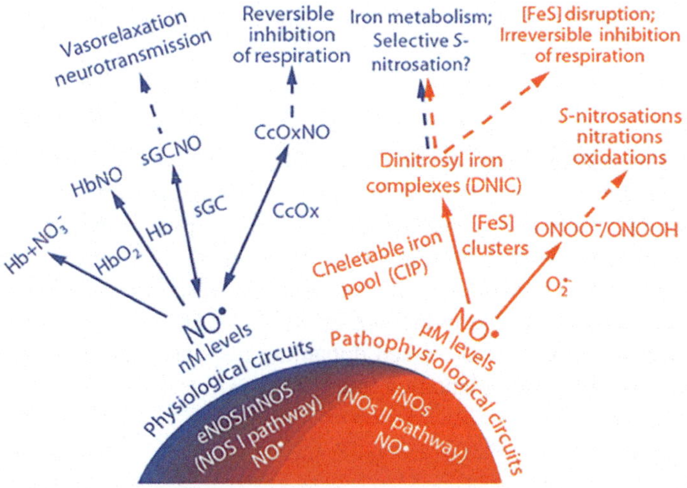

Fig. 2.26 Commonly used methods of nitric oxide analysis and action [145, 157]. With permission from American Chemical Society

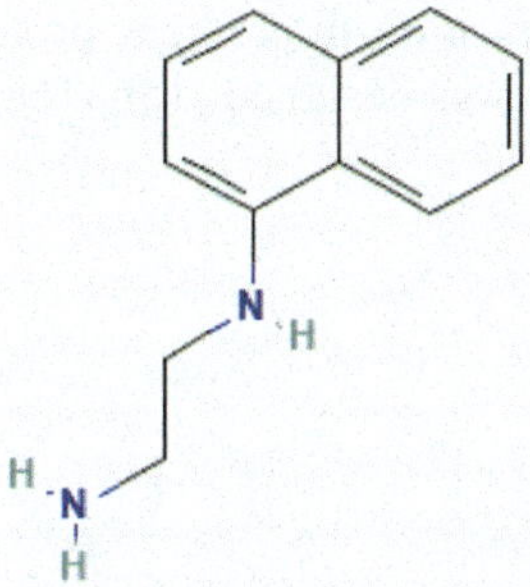

and its modification can be used for an indirect assay of nitric oxide [129, 143, 144, 152–158]. Nowadays, Sigma-Aldrich sells this compound. In transition metal proteins assay, since oxyhemoglobin (Hb) has a high affinity for NO and upon binding is converted quantitatively to methemoglobin (MetHb). spectrophotometric determination of nitric oxide using hemoglobin was employed for analysis of NO concentrations between 300 nM and 30 □M [146]m.

2.6.2 NO Luminescence Analysis

In recent years, fluorescent methods have been increasingly used due to its sensitivity from mM up to a single molecule and is widely available for routine analysis and precision investigations, and therefore, gains many advantages over other physical methods used in biology and medicine [147–150, 159–177].

Classical fluorophores such as anthracene, coumarin, are routinely employed in biology for analysis of small molecules [151–157, 159, 160]. In the last decades, a number new advance non-cytotoxic water soluble and membrane permeable probes including two-photon fluorophores, excitable at low-energy wavelengths were synthesized and investigated employing advanced spectroscopic techniques. One of the general basis of methods for NO analyses is the conversion of a fluorescence silent probe to a compound of high fluorescent quantum yield [151, 163].

A water-soluble, biocompatible, small molecular probe *o*-phenylenediamine-Phe-Phe-OH (**1**) for the highly selective and sensitive detection of NO *in vitro* and in living cells was designed [150, 162]. The compound can react with NO and turn on the fluorescence emission at 367 nm via an intramolecular charge transfer (ICT) mechanism. In PBS buffer, **1** was applied for detecting NO within the range of 012 μM with a LOD of 6 nM and was employed to sense intracellularly generated NO in living cells. Attaching a NO-responsive dihydropyridine pendant group to a fluorophore led to a probe that displays a very high sensitivity to NO concentrations down to the low nM range and a high specificity to NO while being insensitive to other oxidative oxygen/nitrogen species that often interfere with the sensing of NO [155, 167]. Nitric oxide turn-on fluorescent probe based on deamination of aromatic primary monoamines was designed [156, 168]. The stable, water-soluble, and nonfluorescent FA-OMe

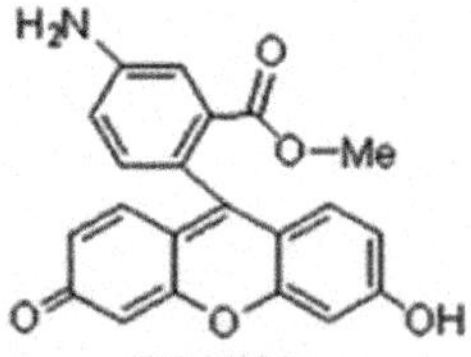

FA-OMe

can sense nitric oxide (NO) and form the intensely fluorescent product dA-FA-OMe via reductive deamination of the aromatic primary amine. The reaction was accompanied by an notable increase of the fluorescent quantum yield from 1.5 to 88.8%. The selectivity study indicated that FA-OMe had high specificity for NO oversome other reactive oxygen/nitrogen species. By connection of O^6-benzylguanine (BG) to an "*o*-phenylenediamine-locked" rhodamine spirolactam responsive to nitric oxide (NO), a novel locked" rhodamine spirolactam

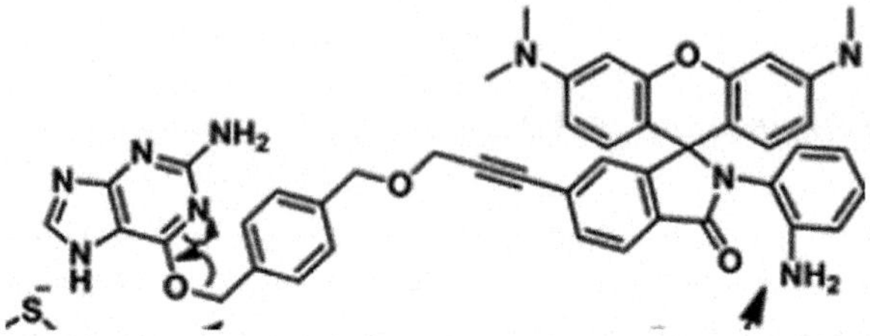

responsive to nitric oxide (NO), a novel substrate (TMR-NO-BG) of genetically encoded SNAP-tag has been constructed [157, 169]. In living cells, labeling SNAP-tag fused proteins with TMR-NO-BG will *insitu* generate corresponding probe–protein conjugates (TMR-NO-SNAP). The labeling inherited high NO sensitivity from the small-molecule parent and guaranteed the site-specificity.

Of particular interest is the analysis of nitric oxide at single cell levels. In review [159, 171], three categories of analytical techniques enabling NO detection at single cell levels, namely, fluorescence microscopy, capillary electrophoresis with laser induced fluorescence detection, and electrochemistry were utilized. The correspondent basic principles, performance, applications, figures of merits and limitations were considered. The probe, FP-NO was prepared by incorporating a recognition unit, thiosemicarbazide moiety into a coumarin fluorophore [160, 172]. The weakly fluorescent FP-NO quickly and selectively reacts with NO to form a highly fluorescent product, FP-P. with the detection limit of 47.6 nM. The feasibility of FP-NO as the fluorescence probe for imaging and flow cytometry analysis of exogenous NO in MCF-7 cells and exogenous NO production in inflamed J774A was demonstrated. Visualization of exogenous and endogenous NO production in live zebrafish was then achieved. A family of cysteine analogues tagged with fluorophores for visualizing cellular NO dynamics

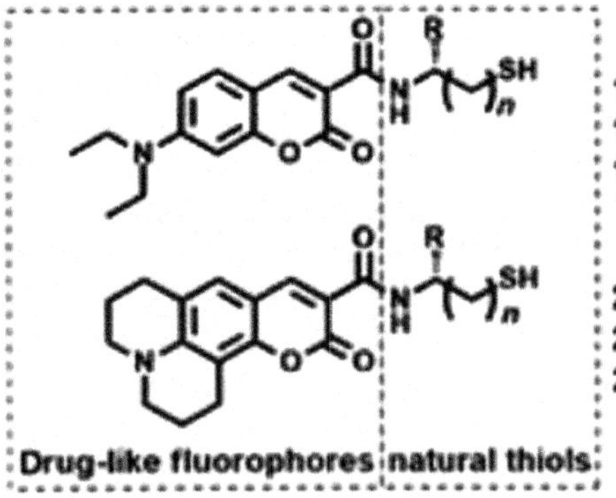

1a n = 1, R = –COOMe
1b n = 1, R = –H
1c n = 2, R = –COOMe

2a n = 1, R = –COOMe
2b n = 1, R = –H
2c n = 2, R = –COOMe

was designed [161, 173]. Biocompatible probes **1a** and **2a** were found to switch between their original fluorescent state and quenched *S*-nitrosated state in response to cellular NO dynamics. The probes were utilized to tracking NO fluctuations in activated murine macrophages in response to an anti-inflammatory agent treatment. The authors concluded that probes **1a** and **2a** will afford promising tools for studying NO pathophysiology by tracking its dynamics in health and disease.

The review [162, 174] focused on organic small-molecule fluorescent probes for selective detection and imaging of endogenous NO *in vitro* and in vivo. The progress of fluorescence imaging for real-time detection of NO in biological system was described. Due to practically unlimited variety of structures and properties, organometal-based fluorescent probes provide for new possibilities in the field [163, 175]. As an example, copper(II) complex CuRBT with a ring-closed rhodamine-containing tripodal ligand was synthesized. This highly selective fluorescent imaging agent for nitric oxide featured a 700-fold fluorescent enhancement toward NO from a dark-background with the detection limit of NO about 1 nM in aqueous solution and could be applied for monitoring intracellular NO. Nitric oxide can be detected in the gas phase by the chemiluminescence of its reaction with ozone [166]. The absolute intensity and quenching constants of the NO_2 emission in the chemiluminescent reaction between nitric oxide and ozone have been determined using a large capacity-flow reactor at low pressures. These studies revealed the following (1) The mechanism proposed is $NO + O_3 = NO^*_2(^2B_1) + O_2(1a)$ $NO + O_3 = NO_2(^2A_1) + O_2(1b)$ $NO^*_2 = NO_2 + hv(2)$ $M + NO_2 = NO_2 + M$; (2) The rate constants of the primary steps are $k_{1a} = (7.6 \pm 1.5) \times 10^{11} \exp(-4180 \pm 300/RT)$ and $k_{1b} = (4.3 \pm 1.0) \times 10^{11} \exp(-2330 \pm 150/RT)\,cm^3 mole^{-1} sec^{-1}$; and 3 The half-quenching pressures of electronically excited NO_2 produced are: $Ar =$

$13.0 \pm 1.0\mu, O_2 = 8.3 \pm 0.5\,\mu, NO = 5.2 \pm 0.3\mu, NO_2 = 3.7 \pm 0.1\mu.$. The detailed mechanism of reactions (1*a*) and (1*b*) was discussed.

A new highly sensitive chemiluminescence approach to direct NO detection in aqueous solutions using a natural nitric oxide target, soluble guanylyl cyclase (sGC) was introduced [164, 176]. The enzyme catalyzes the conversion of guanosine triphosphate to guanosine 3',5'-cyclic monophosphate and inorganic pyrophosphate. The suggested enzymatic assay based on the fact that the rate of the reaction increases by about 200 times when NO binds with sGC. The assay provided a sensor for nitric oxide luminescent detection of the above reaction which is accomplished by converting inorganic pyrophosphate into ATP with the help of ATP sulfurylase followed by light emission from ATP-dependent luciferin-luciferase reaction (Fig. 2.27). Measurement of NO generated by a nitric oxide donor (PAPA-NONOate), nitric oxide synthase, and NO gas dissolved in buffer was performed. The method allowed for the measurement of NO concentrations in the nanomolar range and NO generation rates as low as 100 pM/min. Luciferyl adenylate singlet-excited singlet state emits light upon relaxation to its ground state. For example, yellow-green light ($\lambda_{max} = 565$ nm) is produced by fireflies for reaction with the luciferin.

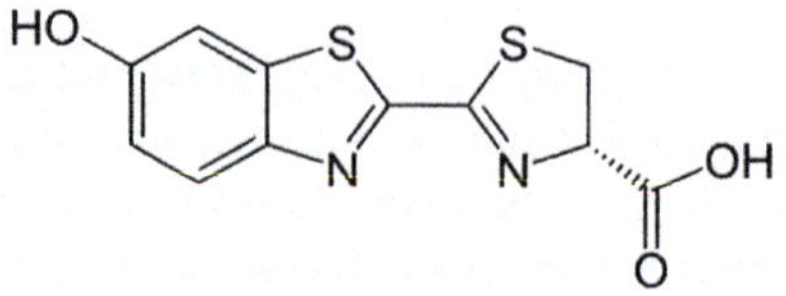

The method of NO assay developed in [165, 177] was based on the reaction catalyzed by glyceraldehyde 3-phosphate dehydrogenase (GAPDH), whose product is used as a substrate for phosphoglycerate kinase (PGK), generating adenosine 5'-triphosphate (ATP), which is a cofactor for the firefly luciferase bioluminescent reaction. The experiment indicated that the range from 10 to 100 nM of ˙NO, with limits of detection and quantitation of 4 and 15 nM, respectively can be detected.

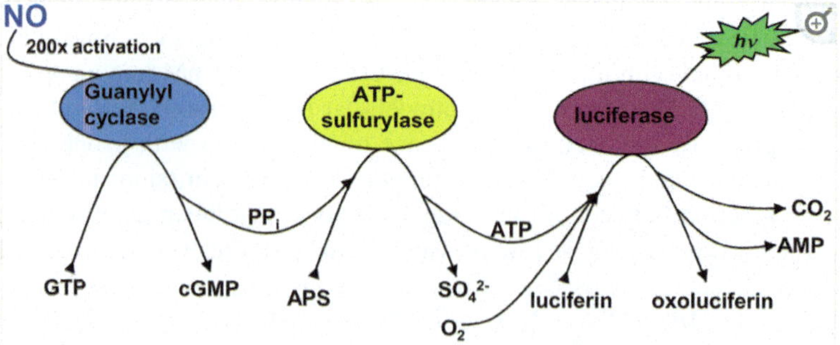

Fig. 2.27 General outline of the chemiluminescent detection of nitric oxide [164]. With permission from Elsevier

2.6.3 Nitric Oxide Analysis by Electrochemistry

- Electrochemistry studies chemical electrochemical techniques, cyclic voltammetry and amperometry, together with fluorescence, are one of the most sensitive and commonly employed analytical method for monitoring nitric oxide. Real-time monitoring, ability to high-spatial resolution, wide variation of electrodes materials and its modification to enhance selectivity and sensitivity provide specific advantages [168–174, 178–186]. A review [168–179] described the development of electrochemical sensors for NO, including the fabrication of sensors, the detection principle, calibration, detection limits, selectivity, and response time. The NO selective electrodes had the many experimental applications to measure physiological levels of NO in vitro, in vivo, and in real time. In work [169, 180], a dual amperometric/potentiometric microsensor for sensing nitric oxide (NO) and potassium ion (K^+) was developed. The dual NO/K^+ sensor was prepared based on a dual recessed electrode possessing Pt (diameter, 50 μm) and Ag (diameter, 76.2 μm) microdisks. The Pt disk surface (WE1) was modified with electroplatinization and the following coating with fluorinated xerogel; and the Ag disk surface (WE2) was oxidized to AgCl on which K^+ ion selective membrane was loaded. The dual microsensor was employed for amperometric sensing of NO (106 ± 28 pA μM^{-1}, $n = 10$, at +0.85 V applied vs Ag/AgCl) and for potentiometric sensing of K^+ (51.6 ± 1.9 mV pK^{-1}, $n = 10$), respectively, with high sensitivity and good selectivity. The sensor was applied for the simultaneous monitoring of NO and K^+ changes in a living rat brain cortex . It was indicated the dual NO/K^+ sensor's measurements well reflect membrane potential changes of neurons and associated cellular components of neurovascular coupling.

Remarkable advances of electrochemical techniques were in full extend demonstrated in [170, 178]. A sensor for real time monitoring nitric oxide based on hemin-functionalized graphene field-effect transistors with single atom thickness, and the highest carrier mobility a sub-nanomolar sensitivity was designed (Fig. 2.28). Experiments indicated that the graphene–hemin sensors can rapidly and selectively respond to NO in physiological solutions with a sub-nanomolar sensitivity, and can be used for real-time monitoring of NO released from macrophage cells and endothelial cells (ECs).

2.6.4 Nitric Oxide Analysis Based on Utilization of Dual Fluorescent-Nitroxide Probes

Although a variety of methods for NO detection have been proposed, they are not generally amenable to real-time measurement of endogenous NO *in situ*. In recent decades, a series of new fluorescent methods for the analysis of NO have been proposed and developed [27–35, 27–38,179–181, 191–193]: (1) Dual fluorescent-

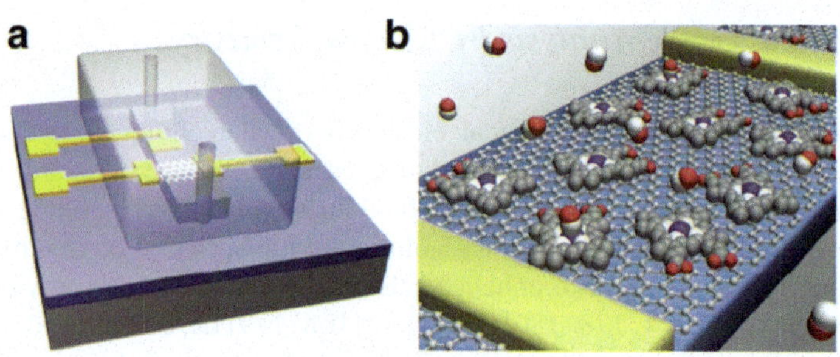

Fig. 2.28 Schematic illustration of a graphene–hemin conjugate device. **a** Schematic of a graphene sensor with integrated microfluidic sample delivery system. **b** Schematic illustration of a graphene–hemin FET that can selectively respond to NO. A typical graphene device has an active area of 0.5 × 0.5–0.5 × 2 μm^2, sizes of source and drain electrodes are 2 × 5 μm^2 (excluding the electrical leads) and the size of gate electrode is 50 × 50 μm^2. The microfluidic PDMS channel has a width of 1 mm, height of 0.5 mm and length of 1 cm [170, 178]. With permission from Springer Nature.

nitroxide probes analysis (DFNPA), (2) Fluorescence inductive-resonance method of analysis (FIRMA 1, FIRMA 2, FIRMA 3); (3) Fluorescence inductive-resonance cascade method of analysis (FIRCMA) and (4) Fluorescence spin exchange method of analysis (FSEMA). This approach of detection of nitric oxide was based on the phenomenon of the intramolecular fluorescence quenching of the fluorophore fragment by the nitroxide in a dual fluorescent- nitroxide supermolecule [36].

A nitronyl nitroxide radical covalently linked to an organic fluorophore, pyrene, was used to detect nitric oxide (NO) from from pig trachea epithelia [31]. This approach is based on the phenomenon of the intramolecular fluorescence quenching of the fluorophore fragment by the nitroxide. The pyrene-nitronyl (PN) reacts with NO to yield a pyrene-imino nitroxide radical (PI) and NO_2. The transformation of the EPR signal was accompanied by an increase in the fluorescence intensity since the imino nitroxide radical is a weaker quencher than the nitronyl. The results indicated that the fluorescence measurements enable detection of nanomolar concentrations of NO compared to a sensitivity threshold of only several micromolar for the EPR technique. The measured basal flux of *S*-nitroso compounds obtained in pig trachea was about 1.2 nmol/g × min, and NO-synthase stimulated by extracellular adenosine 5′-triphosphate produced NO flux of 0.9 nmol/g × min.

A novel assay was developed for the measurement of nitric oxide in nanoscale concentration [179–181], [27–38, 191–193]:. Suggested fluorescence inductive-resonance method of analysis (FIRMA 1).is based on phenomena of the singlet-singlet energy transfer by inductive resonance mechanism, due to the dipole-dipole interaction between excited fluorescent donor (D*) and nonexcited acceptor (A) [182]. A donor acceptor pair, in which the donor is a fluorophore (stilbene label 4-acetamido-4′-isothiocyanatostilbene- 2,2′-disulfonic acid (SITS)

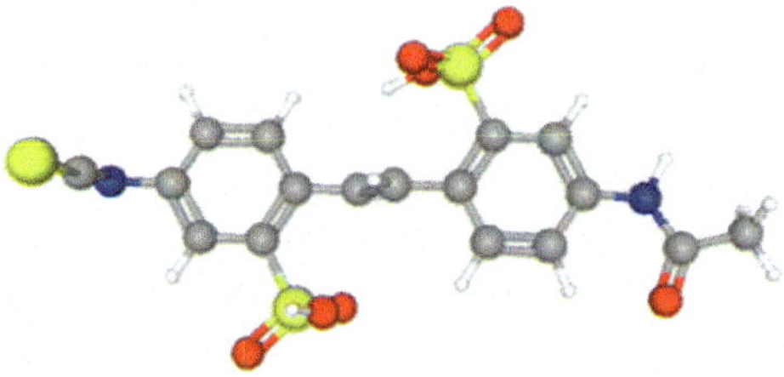

tethered to myoglobin Fig. 2.29 in exited singlet state and acceptor is a nitric oxide trap (hemin), was constructed [27, 28, 31, 33, 35, 36, 45, 179, 180]. Hemin was chosen due to its very fast reaction with NO and the high stability of the resulting complex. According to the Förster theory, efficiency of the process depends on the distance between the donor and acceptor and the value of overlap of the donor fluorescent and the acceptor spectra. Obviously, trapping NO by the hemin Fe^{2+} will lead to a change in absorption spectrum and, in consequence of this, to change of the overlap integral and the intensity of fluorescence. The rapid trapping of NO by myoglobin results in the change of the absorbance of myoglobin, which causes the change of both the absorbance–fluorescence overlap integral value and the fluorescence intensity of myoglobin–STIS. The change in emission intensity of the stilbene fragment versus an increasing concentration of NO precursors clearly demonstrated the method availability to monitor the formation of a heme–NO complex in a concentration range from 10 nM to $\mu - M$. For example, the measured NO flux in tissue sample from rabbit trachea epithelia and porcine trachea epithelia was found to be $\sim 7.9 \times 10^{-12}$ mol/s $\times$ g and $\sim 3.0 \times 10^{-12}$ mol/s $\times$ g respectively. It is necessary to stress that titration of deoxyMb–SITS at low NO concentration ($0.05\mu M$), can be carried out only by the fluorescence technique: at low concentrations in the 10–500 nM range, the fluorescence method is still highly sensitive. Another version of the FIRMA 1 is to create a supermolecule consisting of a donor (exited stilbene or excited CdSe quantum dots for example) and an acceptor (hemin) separated by a rigid bridge at the optimum distance for energy transfer (FIRMA 2). The third version (FIRMA 3) is a system consisting of acceptor molecules (hemin) surrounded at an optimal distance by fluorescent molecules Fig. 2.30 [180], [192, 193].

Figure 2.30 illustrates schematically two donor -acceptor (D-A) models, having a fluorescent chromophore (stilbene derivative for example) as a donor and hemin as an acceptor, with direct D-A energy transfer and transfer via a light harvesting antenna, correspondingly. The latter is an analog of the photosynthetic light harvesting antenna.

Fluorescence spin exchange method of analysis (FSEMA) [180, 192, 193] is based on measuring the difference in efficiency of spin exchange (electron transfer, intersystem crossing) in supermolecules, consisting of NO trap (hemin Fe^{2+}) and a fluorophore, by detecting fluorescence in absent and present NO. Binding of NO by the trap fragment should influence on efficiency of the spin exchange between excited fluorophore and the nitryl complex and, as consequence, on the fluorescence intensity. In the frame of this conception, supermolecules composed with acceptor

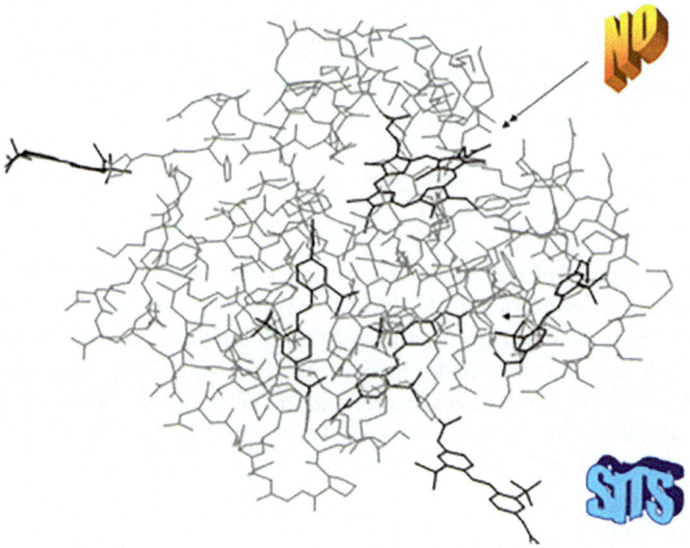

Fig. 2.29 Molecule of myoglobin with attached fluorescence label SITS (Mb(Fe2+)-SITS) [179, 191]. With permission from Elsevier

Fig. 2.30 Energy transfer in a light harvesting antenna, the asterisk is a hemin (NO trap), blue rectangulars are primary donors and orange rectangulars are antenna donors

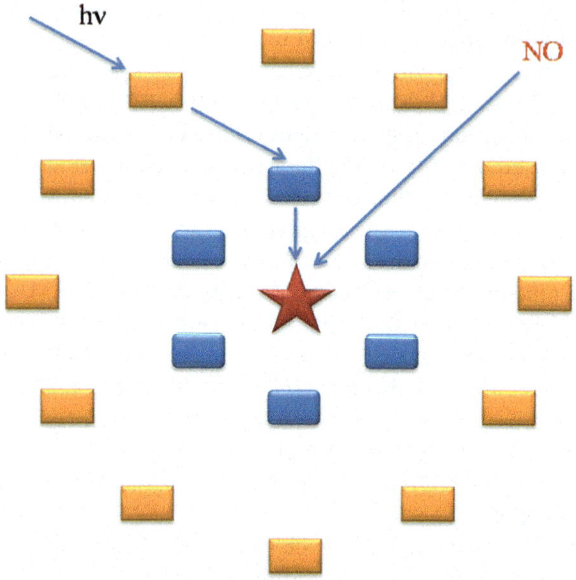

(an excited fluorophore) and donor (hemin Fe^{2+}) segments tethered via a bridge can be designed.

2.7 Calorimetry

Calorimetry is used to measure amounts of heat transferred to or from a substance. The temperature change measured by the calibrated calorimeter is used to derive the amount of heat transferred by the process under study [183–193], [195–205]. The basic principle of Isothermal titration calorimeter (ITC) is the direct measurement of the enthalpy change for a bimolecular binding interaction at a constant temperature (Fig. 2.31). This technique measures the heat released or absorbed over time, being convenient methodology to quantify the amount of reacting molecules such as the binding thermodynamics, as well as to measure the reaction rate. As an example [183, 195] this method was applied to determine k_{cat} and K_M of the enzymatic hydrolysis of urea by *Canavalia ensiformis* (jack bean) urease and calculation of intrinsic molar enthalpy (ΔHint) of the reaction was performed.

Isothermal titration calorimetry instruments measure in real time the thermal power that results when one solution (in a syringe) is titrated into another (in a sample cell), as illustrated in Fig. 2.31 [184, 196]. A pair of cells, typically coin-shaped or cylindrical with volumes on the order of 200–1,400 μL, are termed the sample

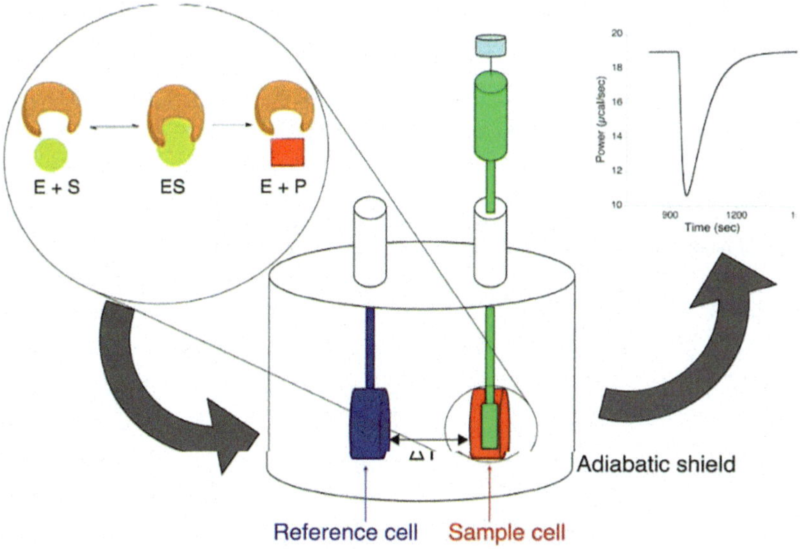

Fig. 2.31 Schematic representation of isothermal titration calorimeter to study enzymatic reactions [183, 195]. With permission from Elsevier

and reference cells and contain the analyte solution and reference buffer respectively [185, 197]. The instantaneous rate of heat production in the ITC sample cell, dQ/dt, is directly proportional to the reaction velocity ($v_0 = d[P]/dt$) and the enthalpy change of the reaction catalyzed $\left(\Delta_r H = H_{\text{product}} - H_{\text{substrate}}\right)$, according to

$$\frac{dQ}{dt} = V_{\text{cell}}\,\Delta_r H \frac{d[P]}{dt}$$

where V_{cell} is the volume of the sample cell and

$$\Delta_r H = \frac{\int_{t=0}^{\infty} \frac{dQ}{dt} dt}{n_S}$$

where n_S is the number of moles of substrate injected. In a multiple injection ITC enzyme kinetic assay, the enzyme concentration is chosen to be sufficiently low so that substrate depletion during the experiment is negligible but high enough to provide good signal [186, 198]. Thus, it is possible to calculate enzyme velocity at any point in the experiment, provided $\Delta_r H$ and V_{cell} are known. These values can be obtained after injecting a known amount of substrate into a sample cell containing sufficient enzyme to rapidly convert it entirely to product,

Exothermic and endothermic reactions give descending and ascending steps, respectively, if the raw feedback power is plotted as a function of time.

Human soluble epoxide hydrolase is a bifunctional enzyme composed of two structurally and functionally independent domains. The C-terminal domain (CTD) is responsible for the hydrolysis of numerous epoxy-fatty acids (EpFAs) [193, 205]. To probe the kinetics of EpFA hydrolysis catalyzed by hsEH CTD, an ITC single-injection method (SIM) was employed [186, 198]. The kinetic results is summarized in Table 2.1. A variant of the single-injection ITC method was applied for the detailed description of the enzyme inhibition,

Table 2.1 Mean and standard error values for the apparent enthalpy and kinetics parameters of hsEH CTD-mediated hydrolysis of the EpFAs analyzed in study [193, 205]

Substrate	$\Delta H_{\text{app}}\ [\text{kJ mol}^{-1}]$	$k_{\text{cat}}\ [s^{-1}]$	$K_M\ [\mu M]$	$K_{sp}\ [s^{-1}\mu M^{-1}]$
5(6)*EET*	−4.84 ± 0.65	0.35 ± 0.05	46.61 ± 10.98	0.01 ± 0.001
8(9)*EET*	−39.45 ± 3.88	1.28 ± 0.19	23.10 ± 2.39	0.054 ± 0.010
11(12)*EET*	−12.43 ± 0.62	4.31 ± 0.16	1.74 ± 0.20	2.53 ± 0.25
14(15)*EET*	−23.72 ± 3.46	6.64 ± 1.54	12.88 ± 1.94	0.527 ± 0.075
8(9)*EpETE*	−34.13 ± 0.93	1.61 ± 0.01	9.74 ± 0.13	0.173 ± 0.006
17(18)*EpETE*	−51.78 ± 1.73	0.64 ± 0.06	26.37 ± 4.05	0.025 ± 0.003
19(20)*EpDPE*	−23.73 ± 1.82	1.62 ± 0.08	26.60 ± 1.21	0.061 ± 0.003
12(13)*EpOME*	−5.67 ± 0.38	9.65 ± 0.19	2.98 ± 0.56	3.56 ± 0.86

2.8 Mass Spectrometry-Based Methods

Mass spectrometry (MS) is an analytical technique that is used to measure the mass-to-charge ratio of ions [2002–2006] [206–208]. The results are presented as a mass spectrum, a plot of intensity as a function of the mass-to-charge ratio. These spectra are used to determine the elemental or isotopic signature of a sample, the masses of particles and of molecules, and to elucidate the chemical identity or structure of molecules. In a typical MS procedure, a sample was ionized by bombarding with a beam of electrons to break up the sample into positively charged fragments or simply become positively charged without fragmenting. These ions are then separated according to their mass-to-charge ratio, by accelerating them using ion-spray and subjecting them to an electric or magnetic field. Mass spectrometry is a rapid, sensitive, and accurate quantitative approach for the direct monitoring of enzyme-catalyzed reactions that does not require a chromophore or radiolabeling and thus provides a viable alternative to existing analytical Review [206], was intended to know how mass spectrometry for studying enzymes activity has progressed in products of enzyme reaction detection to quantitative measures of enzyme activity and kinetics. The technique can be served as a tool for rapidly screening inhibitory compounds as an alternative to current methods of high throughput drug screening. The utility of small- and large-molecule mass spectrometry to reveal enzyme intermediates of acetyl transferases, sulfonucleotide reducatases, and phosphoenolpyruvate PEP-utilizing enzymes and illuminate mechanism was described in details in review [195, 206].

Use of mass spectrometry coupled with High Performance Liquid Chromatography (HPLC) for the kinetic analysis of enzymatic reactions in real time was reported [196, 206]. The hydrolysis of dinucleotides with bovine pancreatic ribonuclease A (RNase A) and the substrate-specific hydrolysis of lactose with beta-galactosidase were monitored employing ion-spray (pneumatically assisted electrospray) mass spectrometry. The resulting data can be used to calculate both K_M and V_{max} for system under interest. The proteolysis of intact viral capsid proteins, the alpha-glucosidase-catalyzed hydrolysis of p-nitrophenyl-alpha-glucopyranoside

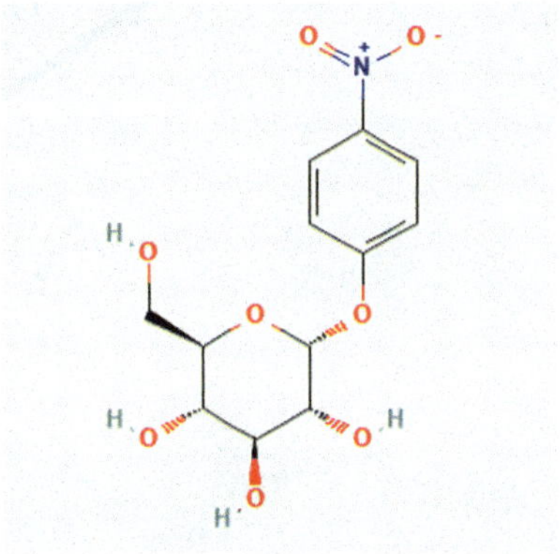

and the lipoprotein lipase-catalyzed ester hydrolysis of resorufin were examined [197, 206]. Matrix-assisted laser desorption/ionization (MALDI) and electrospray ionization (ESI) mass spectrometry were used to monitor the proteolysis of viral protein capsids and the hydrolysis of small molecules with a-glucosidase and bovine lipoprotein lipase. MALDI methodology is a several -step process: (1) The sample is mixed with a suitable matrix material and applied to a metal plate; (2) A pulsed laser irradiates the sample, triggering desorption of the sample and matrix; (3) The analyte molecules are ionized by being protonated or deprotonated in the hot plume of ablated gases; and (4) Then they can be accelerated into whichever mass spectrometer is used to analyse them. The MALDI experiments provided information about capsid (a protein shell of virus) dynamics and the stabilizing force of viral protein/RNA interactions. The kinetic of hydrolysis of p-nitrophenyla-glucopyranoside , catalyzed by a-glucosidase to produce p-nitrophenolate and glucose was examined by ESI-MS and UV-visible spectroscopy. The kinetic parameters obtained by ESI-MS are k_{cat} =256 min^{-1} and K_m = 0.535 mM. The kinetics of bovine lipoprotein lipase were also monitored with ESI-MS by determining the response of the sodium adducts of 1,2-O-dilauryl-rac-glycerol 5, and 1,2-O-dilauryl-rac-glycero-3-glutaric acid [197].

Figure 2.32 [198, 206] illustrates the principle underlying the DESI-MS method for online measurement of enzymatic reaction kinetics, namely: (1) enzyme and substrate solutions undergo mixing in a micro-Tee mixer that is connected with a short piece of silica capillary to serve as a microreactor for reaction; (2) After exiting the microreactor capillary, the reaction mixture is directly ionized by DESI for MS detection; and (3) The reaction time can be varied by changing injection flow rates of enzyme and substrate solutions.

Study [198, 207] described a new approach for online measuring enzymatic reaction kinetics using desorption electrospray ionization mass spectrometry (DESI-MS). By using DESI-MS, enzymatic reaction products in a buffered aqueous

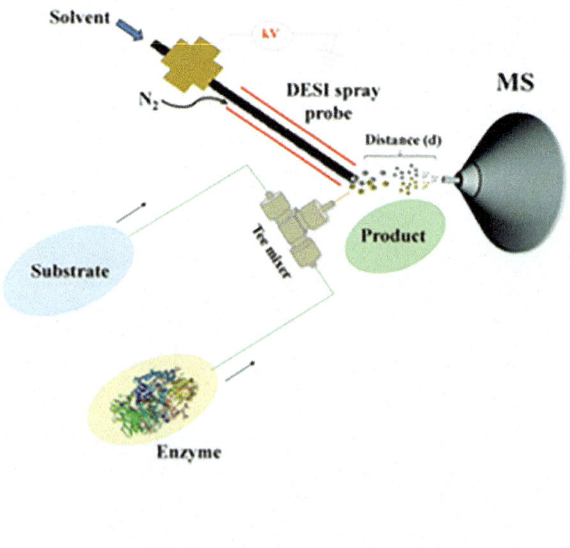

Fig. 2.32 Schematic showing the apparatus of timeresolved DESI-MS [206]. With permission from American Chemical Society

media could be directly detected [2006]. Enzymatic reactions examined in this study include hydrolysis of 2-nitrophenyl-β-D-galactopyranoside by β-galactosidase and hydrolysis of acetylcholine by acetylcholinesterase. Derived Michaelis–Menten constants K_m for these two reactions were determined to be 214 μM and 172 μM, respectively, which are in good agreement with the values of 300 μM and 230 μM reported in literature, validating the DESI-MS approach. This time-resolved DESI-MS also allowed to deter K_m and turnover number k_{cat} for trypsin digestion of angiotensin II (K_m and k_{cat} are determined to be 6.4 mM and 1.3 s^{-1}, respectively).

Peptide hydrolysis was studied *in situ* by incubating tissues with substrate with or without inhibitor. MALDI-TOF is used to detect substrate and hydrolysis product ions which can be used to create a map of enzyme activity [199, 207]. 3-Deoxy-d-manno-2-octulosonate-8-phosphate

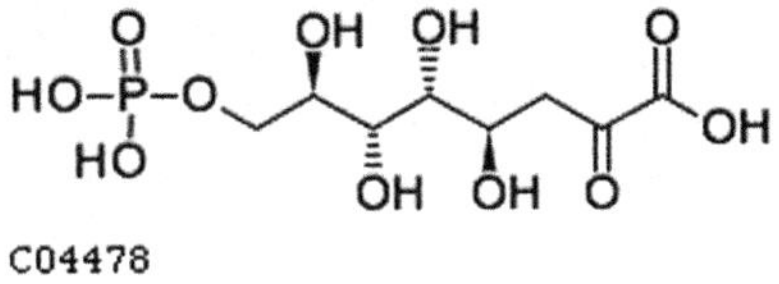

- (KDO8P) synthase catalyzes the first step of lipopolysaccharide biosynthesis, namely condensation of phosphoenolpyruvate (PEP) with arabinose 5-phosphate (A5P), to produce KDO8P [200, 208]. To detect an intermediate in the KDO8P synthase reaction, time-resolved ESI-MS TOF (Fig. 2.32) was utilized

Time-of-flight mass spectrometry (TOFMS) is a method of mass spectrometry in which an ion's mass-to-charge ratio is determined by a time of flight measurement. This time will depend on the velocity of the ion, and therefore is a measure of its mass-to-charge ratio. The principle working hypotheses of the rapid mixing, time-resolved ESI technique are: (1) A computer-controlled stepping motor or syringe pump allows the precise control of the speed at which the solutions of enzyme and substrate are mixed; (2) The enzyme and substrates come together in a mixing tee that has a short fused silica capillary column on the outlet; (3) The capillary leads directly to the needle of the electrospray unit in the mass spectrometer; and (4) There is a high voltage on the needle of electrospray unit that disperses the solution coming through the capillary into small particles for nebulization. The catalytic reaction of KDO8PS with its natural substrates, PEP

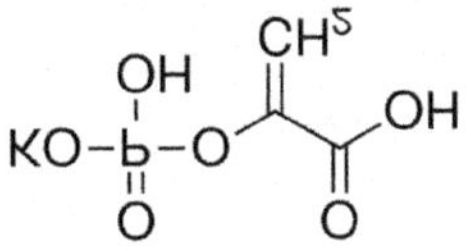

and A5P, was examined under single enzyme turnover conditions by rapidly mixing the EÆPEP solution with a limiting amount of the second substrate, A5P. The reaction was monitored over several time ranges (7–160 ms) (Fig. 2.33 and 2.34).

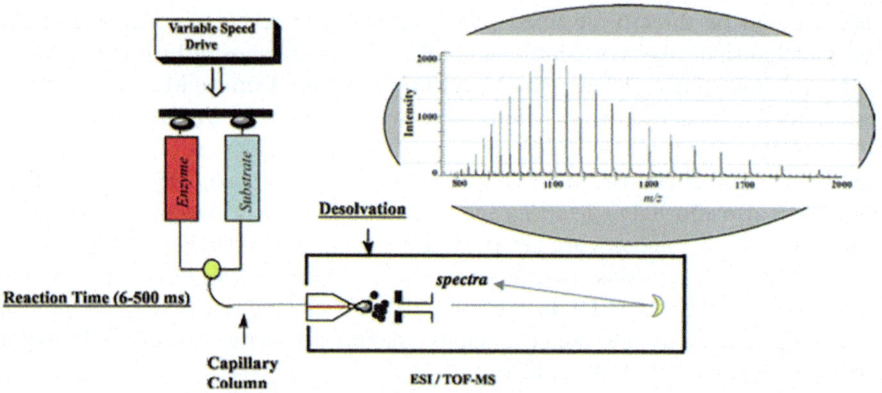

Fig. 2.33 Experimental design for time-resolved ESI-MS TOF [200, 207] With permission from Elsevier

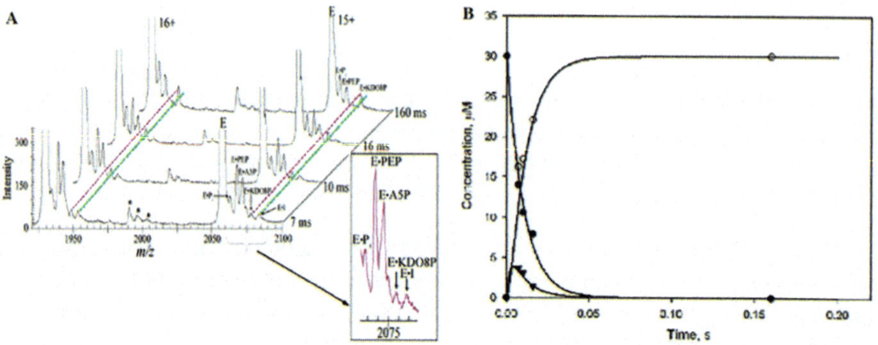

Fig. 2.34 Monitoring catalysis for KDO8P catalysis using ESI-MS TOF: **a** Detection of an enzyme intermediate. **b** Reaction kinetics using timeresolved ESI-MS illustrating that the formation and decay of the enzyme intermediate parallel the disappearance of substrate and formation of product [200, 208]. With permission from Elsevier

References

1. Perakis F, De Marco L, Shalit A, Tang F (2016) Vibrational spectroscopy and dynamics of water. Chem Rev 116:7590–7607
2. Mondal S, Mukherjees, Bagchi B (2017) Protein hydration dynamics: much ado about nothing? J Phys Chem Lett 8:4878–4882
3. Laage D, Elsaesser N, Hynes JT (2017) Water dynamics in the hydration shells of biomolecules. Chem Rev 117:10694–10725
4. Chalmers JM, Griffiths PR (eds) (2002) Handbook of vibrational spectroscopy, vol 5. Wiley, New York
5. Skoog DA, Holler FJ, Crouch SR (2007) In: Principles of instrumental analysis. 6th ed. Brooks/Cole
6. West W (2015) Absorption of electromagnetic radiation. Science. https://doi.org/10.1036/1097-8542.001600

7. Bisswanger H (2017) Enzyme kinetics: principles and methods, Third, enlarged and improved Edition. Wiley-VCH Verlag GmbH & Co.
8. Zalatan JG, Herschlag D (2006) Alkaline phosphatase mono- and di-esterase reactions: comparative transition state analysis. J Am Chem Soc 128(4):1293–1303. https://doi.org/10.1021/ja056528r
9. Mikla VI, Mikla VV (2014) In: Raman spectroscopy in medicine in medical imaging technology. Elsevier, pp 129−141
10. Kawahara-Nakagawa Y, Nishikawa K, Nakashima S, Inoue S, Ohta T (2019) New assay method based on raman spectroscopy for enzymes reacting with gaseous substrates. Protein Sci 28(3). https://doi.org/10.1002/pro.3569
11. Kahan TF, Reid JP, Donaldson DJ (2007) The first uses glancing-angle Raman spectroscopy to study the extent of hydrogen bonding in the QLL via the shape of the water OH-stretching band. J Phys Chem A 111:11006–11012
12. Kumar S, Barth A (2010) Following enzyme activity with infrared spectroscopy. Sensors (Basel) 10(4):2626–2637. https://doi.org/10.3390/s100402626
13. Böcker U, Wubshet SJ, Lunberg D, Afseth NK (2017) Fourier-transform infrared spectroscopy for characterization n of Lindberg protein chain reductions in enzymatic reactions. Analyst 142:2812–2818
14. Jabłoński A (1933) Efficiency of anti-stokes fluorescence in dyes. Nature 131:839–840. https://doi.org/10.1038/131839b0
15. Gauglitz G, Vo-Dinh T (2003) Handbook of spectroscopy. Wiley-VCH
16. Sharma A, Schulman SG (1999) Introduction to fluorescence spectroscopy. Wiley. ISBN 978-0-471-11098-9
17. Nozeret K, Boucharlat A, Agou F, Buddelmeijer N (2019) A sensitive fluorescence-based assay to monitor enzymatic activity of the essential integral membrane protein Apolipoprotein N-acyltransferase (Lnt). Sci Rep 9:15978. https://doi.org/10.1038/s41598-019-52106-8
18. Albani JR (2004) Structure and dynamics of macromolecules: absorption and fluorescence studies. Elsevier
19. Lakowicz JR (1983) Principles of fluorescence spectroscopy. Plenum Press, New York
20. Likhtenshtein GI (2021) In: Biological water: physicochemical aspect. Springer Nature
21. Abou-Zied, Al-Busaidi BY, Husband J (2014) Solvent effect on anthracitic acid spectroscopy. J Phys Chem A 118(1):103–109
22. Shakerizadeh-Shirazi F, [a]Hemmateenejad B, Mehranpour AM (2013) Determination of the empirical solvent polarity parameter $E_T(30)$ by multivariate image analysis. [b]Anal Methods 5:891−896
23. Likhtenshtein GI, Yamauchi J, Nakatsuji S, Smirnov A, Tamura R (2008) Nitroxides: application in chemistry, biomedicine, and materials science. WILEY-VCH, Weinhem
24. Kocherginsky N, Swarts HM (1995) Nitroxide spin labels. In: Reactions in biology and chemistry. CRC Press
25. Likhtenshtein GI (1993) Biophysical labeling methods in molecular biology. Cambridge University Press, New York, Cambridge
26. Khramtsov VV, Grigor'ev I.A, Foster MA, Lurie DJ, ZweierJL, Kuppusamy P (2004) Spin pH and SH probes: enhancing functionality of EPR-based techniques. Spectroscopy 18:213−225
27. Likhtenshtein GI (2009) Novel fluorescent methods for biotechnological and biomedical sensoring: assessing antioxidants, reactive radicals, NO dynamics, immunoassay, and biomembranes fluidity. Appl Biochem Biotechnol 152:135–155
28. Likhtenstein GI, Ishii K, Nakatsuji S (2007) Dual chromophore-nitroxides: novel molecular probes, photochemical and photophysical models and magnetic materials. Photochem Photobiol 83:871–881
29. Lozinsky E, Martin VV, Berezina TA, Shames A, Weis AL, Likhtenshtein GI (1999) Dual fluorophore-nitroxide probes for analysis of vitamin C in biological liquids. J Biochem Biophys Meth 38:29–42
30. Medvedeva N, Martin VV, Likhtenshten GI (2004) Dual fluorophore-nitronyl probe for investigation of superoxide dynamics and antioxidant status of biological systems. J Photochem Photobiol A: Chem 163:45–51

31. Lozinsky EM, Martina LV, Shames AI, Uzlaner N, Masarwa A, Likhtenshtein GI, Meyerstein D, Martin VV, Priel Z (2004) Detection of NO from pig trachea by a fluorescence method. Anal Biochem 326:139–145
32. Parkhomyuk-BenArye P, Strashnikova N, Likhtenshtein GI (2002) Stilbene photochrome-fluorescence-spin molecules: covalent immobilization on silica plate and applications as redox and viscosity probes. J Biochem Biophys Methods 51,(1) 4:1–15
33. Likhtenshtein GI (2010) Dual phluorophore-nitroxides as a tool for analysis of antioxidants, nitric oxide and superoxide in biological liquids and tissues. In: International conference "Recent advantages in sensoring, signals and materials" November 3–5, 2010, Faro, Portugal, pp 162–168
34. Likhtenshtein GI (2016) Electron spin in chemistry and biology: fundamentals, methods, reactions mechanisms, magnetic phenomena. Springer, Structure Investigation
35. Likhtenshtein GI (2020) Nitroxides: brief history and recent developments. Springer
36. Bystryak IM, Likhtenshtein GI, Kotelnikov AI, Hankovsky OH, Hideg K (1986) Effect of solvent molecular dynamics on the photochemical reduction of nitroxyl radicals Russ. J Phys Chem 1986(60):1679–1983
37. Blough NV, Simpson DJ (1986) Chemically mediated fluorescence yield switching in nitroxide-fluorophore adducts: optical sensors of radical/redox reactions. J Am Chem Soc 110:1915–1917
38. Braslau R, Rivera F III, Lilie E, Cottman M (2013) Urushiol detection using a profluorescent nitroxide. J Org Chem 78:238–245
39. Jepsen PU, Cooke DG, Koch M (2011) Terahertz spectroscopy and imaging—modern techniques and applications. Laser Photon Rev 5:124
40. Abbas A, Trezeibre A, Bourzgui (2009) Terahertz biomems for enzymatic catalysis monitoring. In: 5th international conference on microtechnologies in medicine and biology. Québec City, Canada. ffird-00397688f
41. Baxter JB, Guglietta GW (2011) Terahertz spectroscopy. Anal Chem 83:4342–4368
42. Popov, Ishai PB, Khamzin A, Feldman Y (2016) The mechanism of the dielectric relaxation in water. Phys Chem Chem Phys 18:13941
43. van Hoof NJJ, Ter Huurne SET, Gómez Rivas J, Halpin A (2018) Time-resolved terahertz time-domain near-field microscopy. Optics Expr 26:32118–32129
44. Mittleman D (2013) In: Sensing with terahertz radiation. Springer
45. Elton DC (2017) The origin of the Debye relaxation in liquid water and fitting the high frequency excess response. Phys Chem Chem Phys 19:1873
46. Neu J, Schmuttenmaer CA (2018) Tutorial: an introduction to terahertz time domain spectroscopy (THz-TDS). J Appl Phys 124:231101
47. Debye P (1912) Zur Theorie der spezifischen Waerme. Annalen der Physik (in German) 39(4):789–839
48. Sun Q, He Y, Pickwell-MacPherson E (2017) Recent advances in terahertz technology for biomedical applications. Quant Imaging Med Surg 7:345–355
49. Wang L (2017) Medical applications of terahertz imaging: a review of current technology and potential applications in biomedical engineering. Sensors (Basel) 21(19):6465. https://doi.org/10.3390/s21196465
50. Cao C, Serita K, Kitagishi K, Murakami H, Zhang Z-H, Tonouchi M (2020) Terahertz spectroscopy tracks proteolysis by a joint analysis of absorptance and debye model. Biophys J 119:2469–2482. https://doi.org/10.1016/j.bpj.2020.11.003
51. Bankim S, Ashwini S (2010) Simultaneous voltammetric determination of acetaminophen, aspirin and caffeine using an in situ surfactant-modified multiwalled carbon nanotube paste electrode. Electrochim Acta 55:8638–8648. https://doi.org/10.1016/j.electacta.2010.07.093
52. , Kissinger P, William R, Heineman WR (1996) Laboratory techniques in electroanalytical chemistry. Second Edition, Revised and Expanded (2 ed.) CRC 1996
53. Bard A, Larry R, Faulkner LR (2000) In: Electrochemical methods: fundamentals and applications. 2 ed. Wiley. ISBN 978-0-471-04372-0
54. Zoski C (2007) In: Handbook of electrochemistry. Elsevier 2007. ISBN 978-0-444-51958-0

55. Sanghavi BJ, Srivastava AK (2011) Adsorptive stripping differential pulse voltammetric determination of venlafaxine and desvenlafaxine employing Nafion–carbon nanotube composite glassy carbon electrode. Electrochimica Acta 56(11, 15):4188–4196
56. Schachinger F, Chang H, Scheiblbrandner S, Roland Ludwig R (2021) Amperometric biosensor s based on direct electron transfer enzymes. Molecules 26(15). https://doi.org/10.3390/molecules26154525
57. Bollella P, Gorton L (2018) Enzyme based amperometric biosensors. Current Opinion in Electrochem 10:157–173l
58. Clark Jr LC, Lyons C: Electrode systems for continuous monitoring in cardiovascular surger. Ann. N.Y. Acad. Sci., 102, 29 (1962)
59. . Clark JLC (1987) Biosensors: fundamentals and applications. In: Turner APF et al. (ed), Oxford University Press, New York, NY
60. Gupta S, Kaushal A, Kumar A, Kumar D (2017) Ultrasensitive transglutaminase based nanosensor for early detection of celiac disease in human. Int J Biol Macromol 105:905–911
61. Heineman WR, Kissinger PT, Wehmeyer KR (2021) From polarography to electrochemical biosensors: the 100-year quest for selectivity and sensitivity. J Electrochem Soc168:116504
62. Singh A, Sharma A, Ahmed A, Sundramoorthy AK, Furukawa H, Arya S, Khosla A (2021) Recent advances in electrochemical biosensors: applications, challenges, and future scope. Biosensors 11(9):336. https://doi.org/10.3390/bios11090336
63. Wenchich L, Drahota Z, Honzík T, Hansíková H, Tesarová M, Zeman J, Houstek J (2003) Polarographic evaluation of mitochondrial enzymes activity in isolated mitochondria and in permeabilized human muscle cells with inherited mitochondrial defects. Physiol Res 52(6):781–788
64. Heinze J (1984) Cyclic voltammetry-electrochemical spectroscopy. New Analytical Methods. Angewandte Chemie International Edition in English 23(11):831–847. https://doi.org/10.1002/anie.198408313
65. Elgrishi N, Rountree KJ, McCarthy BD, Rountree ES, Eisenhart TT, Dempsey JL (2017) A practical beginner's guide to cyclic voltammetry. J Chem Educ 95(2):197. https://doi.org/10.1021/acs.jchemed.7b00361
66. Flexer V, Ielmini MV, Calvo EJ (2017) Extracting kinetic parameters for homogeneous $[Os(bpy)_2ClPyCOOH]^+$ mediated enzyme reactions from cyclic voltammetry and simulations. Bioelectrochemistry 74:201–209
67. Elgrishi N, Rountree KJ, McCarthy BD, et al (2017) A practical beginner's guide to cyclic voltammetry. J Chem Educ 95(2):197
68. Koncki R (2007) Recent developments in potentiometric biosensors for biomedical analysis. Anal Chim Acta 599:7–15
69. Subirats X, Ràfols C (2015) Methods for pKa Determination (I): potentiometry, spectrophotometry, and capillary electrophoresis. In: Reference module in chemistry, molecular sciences and chemical engineering
70. Smith LA, Glasscott MW, Vannoy KJ, Dick JE (2007) Enzyme kinetics via open circuit potentiometry. Anal Chem 92(2):2266–2273
71. Fourmond V, Infossi P, Giudici-Orticoni M_T P, Christophe Léger C (2010) Two-step" chronoamperometric method for studying the anaerobic inactivation of an oxygen tolerant NiFe hydrogenase. J Am Chem Soc 132(13):4848–4857
72. Guy OJ, Walker K-AD (2016) Graphene functionalization for biosensor applications in silicon carbide biotechnology. 2nd ed. https://www.sciencedirect.com/topics/chemistry/chronoamperometry
73. Bertrand P, Frangioni B, Dementin S (2007) Effects of slow substrate binding and release in redox enzymes: theory and application to periplasmic nitrate reductase. J Phys Chem B 111:10300
74. Léger C, Bertrand P (2009) Direct electrochemistry of redox enzymes as a tool for mechanistic studies. Chem Rev 108(7):2379–2438
75. Faulkner LR, Bard AJ (2000) Basic potential step methods, electrochemical methods: fundamentals and applications, 2nd edn. Wiley, New Jersey

76. Maalouf R, Chebib H, Saïkali Y, Vittori O, Sigaud M, Jaffrezic-Renault N (2007) Amperometric and impedimetric characterization of a glu-tamate biosensor based on Nafion and a methyl viologen modified glassy carbon electrode. Biosens Bioelectron 22:2682–2688
77. Magar HS, Hassan RYA, Mulchandani A (2021) Electrochemical impedance spectroscopy (EIS): principles, construction, and biosensing applications. Sensors 21(19):6578. https://doi.org/10.3390/s21196578
78. Bahadır EB, Sezgintürk MK (2016) Artif cells. Nanomed Biotechnol 44(1):248–262
79. Kaushal A, Kumar A, Kumar D (2017) Ultrasensitive transglutaminase based nanosensor for early detection of celiac disease in human. Int J Biol Macromol 105:905–911
80. Maalouf R, Chebib H, Saïkali Y, Vittori O, Sigaud M, Jaffrezic-Renault N (2007) Amperometric and impedimetric characterization of a glu-tamate biosensor based on Nafion and a methyl viologen modifi ed glassy carbon electrode. Biosens Bioelectron 22:2682–2688
81. Koncki R (2007) Recent developments in potentiometric biosensors for biomedical analysis. Anal Chim Acta 599:7–15
82. Schachinger F, Chang H, Scheiblbrandner S, Ludwig R (2021) Amperometric biosensors based on direct electron transfer enzymes. Molecules 26(15):4525. https://doi.org/10.3390/molecules26154525]
83. Cruys-Bagger N, Ren G, Tatsumi H et al (2012) An amperometric enzyme biosensor for real-time measurements of cellobiohydrolase activity on insoluble cellulose. Biotechnol Bioeng 109:3199–3204
84. Chang H, Wohlschlager L, Csarman F et al. (2021) Real-time measurement of cellobiose and glucose formation during enzymatic biomass hydrolysis. Anal Chem 93:7732−7738
85. Khan M, Nagal V, Nakate UT et al (2021) Engineered CuO nanofibers with boosted non-enzymatic glucose sensing performance. J Electrochem Soc 168:067507
86. Grieshaber D, MacKenzie R1, Voros J, Reimhult E (2008) Electrochemical biosensors—sensor principles and architectures sensors. 8:1400−1458
87. Koncki R (2007) Recent developments in potentiometric biosensors for biomedical analysis. Anal Chim Acta 599(7–15):17765058
88. Cruys-Bagger N, Ren G, Tatsumi H, et al. (2012) An amperometric enzyme biosensor for real-time measurements of cellobiohydrolase activity on insoluble cellulose. Biotechnol Bioeng 109:3199–3204
89. Chang H, Wohlschlager L, Csarman F et al (2021) Real-time measurement of cellobiose and glucose formation during enzymatic biomass hydrolysis. Anal Chem 93:7732–7738
90. Kornienko N, Ly KH, Robinson WE et al (2019) Advancing techniques for investigating the enzyme-electrode interface. Acc Chem Res 52(5):1439–1448
91. Bollella P, Lo G (2018) Enzyme based amperometric biosensors. Current Opinion in Electrochem 10:157−173
92. Strambini LM, Longo A, Scarano, S.; Prescimone S, et al. (2015) Self-powered microneedle-based biosensors for pain-free high-accuracy measurement of glycaemia in interstitial fluid. Biosens Bioelectron 66:162–168
93. Hagen WR (2008) Biomolecular EPR Spectroscopy. CRC
94. Eaton GR, Eaton SS, Barr DP, Weber RT (2010) Quantitative EPR. Springer, Berlin
95. Jamauci J (2008) Fundamentals of electron spin resonance. In: Likhtenshtein GI, Nakatsuji S, Smirnov A, Tamura R (eds) Nitroxides: application in chemistry, biomedicine, and materials science. WILEY-VCH, Weinhem, pp 71–120
96. Likhtenshtein GI (2016) Electron spin in chemistry and biology: fundamentals, methods, reactions mechanisms, magnetic phenomena, structure investigation. Springer, Berlin
97. Hoff AJ (ed) (1989) Advanced EPR. In: Applications in biology and biochemistry. Elsevier, Amsterdam
98. Poole CP (1997) Electron spin resonance: a comprehensive treatise on experimental techniques. Dover Publication
99. Eaton GR, Salikhov KM, Eaton SS (1998) Foundations of modern EPR. World Scientific, Singapore

100. Lambert JB, Mazzola EP, Ridge CD (2019) Nuclear magnetic resonance spectroscopy: introduction to principles, applications, and experimental methods, 2nd ed. Wiley
101. Hore P (2015) Nuclear magnetic resonance. 2nd ed. Oxford Chemistry Primers
102. Savitsky A, Nalepa A,· Petrenko T, Plato M,· Möbius K, Lubitz W (2022) Hydrogen bonded complexes of neutral nitroxide radicals with 2 propanol studied by multifrequency EPR/ ENDOR. Appl Magnet Resonance 53:1239−1263
103. McConnell HM, Robertson RE (1958) Isotropic nuclear resonance shifts. J Chem Phys 29:1361–1366s
104. Levitt MY (2008) Spin dynamics: basics of nuclear magnetic resonance. 2nd ed. Wiley
105. Overhauser AW (1953) Polarization of nuclei in metals. Phys Rev 92:411–441
106. Gil RR, Navarro-Vazquez A (2017) Chapter 1 application of the nuclear overhauser effect to the structural elucidation of natural products. Modern NMR approaches to the structure elucidation of natural products: Volume 2: data acquisition and applications to compound classes. The Royal Society of Chemistry
107. Williamson MP (2009) Applications of the NOE in molecular biology. Annu Rep NMR Spectrosc 65:77–109
108. Honegger P, Steinhauser O (2020) The protein–water nuclear Overhauser effect (NOE) as an indirect microscope for molecular surface mapping of interaction patterns. Phys Chem Chem Phys 22:212
109. Kaptein R (1982) Photo-CIDNP studies of proteins. Biol Magn Res 4:145–191
110. Berliner LJ, Bagryanskaya E (2011) Chemically induced electron and nuclear polarization. In: Misra K (ed) Multifrequency electron paramagnetic resonance. Willey-VCH, pp 947−992
111. Buchachenko AL (2013) Mass-independent isotope effects. J Phys Chem B 117:2231–2238
112. Eaton GR, Salikhov KM, Eaton SS (1998) Foundations of modern EPR ☹World Scientific, Singapore
113. Salikhov KM, Molin YN, Sagdeev RZ, Buchachenko AL (1984) Spin polarization and magnetic effects in radical reactions. Elsevier, Amsterdam
114. Bax A, Griffey RH, Hawkins BL (1983) Correlation of proton and nitrogen-15 chemical shifts by multiple quantum NMR. J Magn Reson 55:301
115. Kanamori K (2017) In vivo N-15 MRS study of glutamate metabolism in the rat brain. Anal Biochem 529:179–192
116. Kanamori K, Ross B (2012) Nitrogen metabolism in vivo. In: Choi IY, Gruetter R (eds) Neural metabolism in vivo, Advances in neurobiology. vol 4. Springer, New York, pp 1005–1026
117. Alger JR, Shulman RG (1984) NMR studies of enzymic rates in vitro and in vivo by magnetization transfer. Q Rev Biophys 17:83–124
118. Brindle KM (1988) NMR methods for measuring enzyme kinetics in vivo. Prog Nucl Magn Reson Spectrosc 20:257–293
119. Rackayova V, Cudalbu C, Pouwels PJW, Braissant O (2017) Creatine in the central nervous system: from magnetic resonance spectroscopy to creatine deficiencies. Anal Biochem 529:144–157
120. Liu L, Kodibagkar VD, Yu J-X, Mason RP (2007) ^{19}F-NMR detection of lacZ gene expression via the enzymic hydrolysis of 2-fluoro-4-nitrophenyl β-D-galactopyranoside in vivo in PC3 prostate tumor xenografts in the mouse. FASEB J 2:2014–2019. https://doi.org/10.1096/fj.06-7366lsf
121. Mizukami S, Takikawa R, Sugihara F, Hori Y, Tochio H et al (2008) Paramagnetic relaxation-based 19F MRI probe to detect protease activity. J Am Chem Soc 130:794–795
122. Rogawski R, Sergeyev IV,[†§] Yinglu Zhang Y, et al. (2017) NMR signal quenching from bound biradical affinity reagents in DNP samples. J Phys Chem B 121(48):10770−10781
123. van Zijl PCM, Yadav NN (2011) Chemical exchange saturation transfer (CEST): what is in a name and what isn't? Magn Reson Med 65(4):927–948. https://doi.org/10.1002/mrm.22761
124. Sparkman OD (2000) Mass spectrometry desk reference. Pittsburgh, Global View Pub
125. Downard K (2004) Mass Spectrometry—a foundation course. Royal Society of Chemistry. https://doi.org/10.1039/9781847551306

126. Downard KM (2007) Historical account: Francis William Aston: the man behind the mass spectrograph. Eur J Mass Spectrom 13(3):177–190. https://doi.org/10.1255/ejms.878
127. Cheng† S, Wu Q, Xiao H, Chen*† H (2017) Online monitoring of enzymatic reactions using time-resolved desorption electrospray ionization mass spectrometry. Anal Chem 89(4):2338–2344
128. Greis KD (2007) Mass spectrometry for enzyme assays and inhibitor screening: an emerging application in pharmaceutical research. Mass Spectrom Rev 26:324–339
129. Kelleher NL, Hicks LM (2005) Contemporary mass spectrometry for the direct detection of enzyme intermediates. Curr Opin Chem Biol 9:424–430
130. Hsieh FY, Tong X, Wachs T, Ganem B, Henion J (1995) Kinetic monitoring of enzymatic reactions in real time by quantitative high-performance liquid chromatography-mass spectrometry. Anal Biochem 229(1):20–25. https://doi.org/10.1006/abio.1995.1373
131. Bothner R, Chavez J, Wei J, Strupp C, Phung Q, Schneemann A, Siuzdak G (2000) Monitoring enzyme catalysis with mass spectrometry. J Biol Chem 275(18):13455–13459. https://doi.org/10.1074/jbc.275.18.13455
132. Grobe N, Elased KM, Cool DR, Morris M (2012) Mass spectrometry for the molecular imaging of angiotensin metabolism in kidney. Am J Physiol 302:E1016–E1024
133. Xu L, Chang KC, Payne EM et al (2021) Mapping enzyme catalysis with metabolic biosensing. Nat Commun 12:6803. https://doi.org/10.1038/s41467-021-27185-9
134. Anderson KS (2012) Detection of novel enzyme intermediates in PEP-utilizing enzymes. Arch Biochem Biophys 1,433(1):47−58
135. Pitt JJ (2009) Principles and applications of liquid chromatography-mass spectrometry in clinical biochemistry. Clin Biochem Rev 30(1):19–34
136. Kebarle P, Verkerk UH (2009) Electrospray: from ions in solution to ions in the gas phase, what we know now. Mass Spectrom Rev 28(6):898–917
137. Bibcode:2009MSRv...28..898K. https://doi.org/10.1002/mas.20247. PMID 19551695
138. Konermann L (2013) Unraveling the mechanism of electrospray ionization. Anal Chem 85(1):2–9. https://doi.org/10.1021/ac302789c
139. Kruve A, Kaupmees K, Liigand J, Leito I (2014) Negative electrospray ionization via deprotonation: predicting the ionization efficiency. Anal Chem 86(10):4822–4830. https://doi.org/10.1021/ac404066v
140. Barrios-Collado C, Vidal-de-Miguel G, Martinez-Lozano S (2016) Numerical modeling and experimental validation of a universal secondary electrospray ionization source for mass spectrometric gas analysis in real-time. Sens Actuators, B Chem 223:217–225. https://doi.org/10.1016/j.snb.2015.09.073
141. Lomeli SH, Peng IX, Yin S, Loo RR, Loo JA (2010) New reagents for increasing ESI multiple charging of proteins and protein complexes. J Am Soc Mass Spectrom 21(1):12731. https://doi.org/10.1016/j.jasms.2009.09.014
142. Anderson KS (2005) Detection of novel enzyme intermediates in PEP-utilizing enzymes. Arch Biochem Biophys 433(1):47–58
143. de Rond T, Danielewicz M, Northen T (2015) High throughput screening of enzyme activity with mass spectrometry imaging. Curr Opin Biotechnol 31:1–9
144. Grobe N, Khalid M, Elased KM, David R, Cool DR, Morris M (2012) Mass spectrometry for the molecular imaging of angiotensin metabolism in kidney. Am J Physiol Endocrinol Metab 302(8):E1016–E1024
145. Chefer VI, Alexis C. Thompson AC Agustin Zapata A, Toni S. Shippenberg TS (2009) Overview of brain microdialysis. Curr Protoc Neurosci. Chapter: Unit7.1. https://doi.org/10.1002/0471142301.ns0701s47
146. Chaurasia CS, Müller M, Bashaw ED et al (2007) AAPS-FDA workshop white paper: microdialysis principles, application and regulatory perspectives. Pharm Res 24(5):1014–1025. https://doi.org/10.1007/s11095-006-9206-z
147. Rogers ML, Brennan PA, Leong CLS, Gowers SAN et al. (2013) Online rapid sampling microdialysis (rsMD) using enzyme-based electroanalysis for dynamic detection of ischaemia during free flap reconstructive surgery. Anal Bioanal Chem 405(11):3881–3888. https://doi.org/10.1007/s00216-013-6770-z

148. Buckley S, Allen D, Brackin R et al. (2019) [a]Sandra Jämtgård,[b]Torgny Näsholm,[b]Susanne Schmidt[a] Microdialysis as an *in situ* technique for sampling soil enzymes. Soil Biol Biochem 135:20–27
149. Jaquins-Gerstl A, Michael AC (2020) Dexamethasone-enhanced microdialysis and penetration injury front. Bioeng Biotechnol Sec Biomater. https://doi.org/10.3389/fbioe.2020.602266
150. Dreier J, Fabricius M, Ayata C et al (2016) Recording, analysis, and interpretation of spreading depolarizations in neurointensive care: review and recommendations of the COSBID research group. J Cereb Blood Flow Metab 37:1595–1625
151. Nesbitt KM, Varner EL, Jaquins-Gerstl A, Michael AC (2015) Microdialysis in the rat striatum: effects of 24 h dexamethasone retrodialysis on evoked dopamine release and penetration injury. ACS Chem Neurosci 6:163–173. https://doi.org/10.1021/cn500257x
152. Likhtenshtein GI (2018) Nitric oxide in chemistry and biology—physicochemical aspect: fundamentals and recent developments. NOVA Science Publishes
153. IgnarroLJ LJ (2010) Nitric oxide. Elsevier Inc, Second Edittion
154. Griess P (1879) Bemerkungen zu der Abhandlung der HH. Weselky und Benedikt Ueber einige Azoverbindungen. Ber Deutsch Chem Ges 12:426–428
155. Hetrick EM, Schoenfisch MH (2009) Analytical chemistry of nitric oxide. Annu Rev Anal Chem 2:409–2433
156. Sun J, Zhang X, Broderick M, Fein H (2003) Measurement of nitric oxide production in biological systems by using Griess reaction assay. Sensors 3:276–284
157. Toledo JC, Augusto O (2012) Connecting the chemical and biological properties of nitric oxide. Chem Res Toxicol 25:975–989
158. Zhang Y, Samson FE, Nelson SR, Pazdernik TL (1996) Nitric oxide detection with intracerebral microdialysis: important considerations in the application of the hemoglobin-trapping technique. J Neurosci Methods 68:165–173
159. Li H, Wan A (2015) Fluorescent probes for real-time measurement of nitric oxide in living cells. Analyst 140:7129
160. Chan J, Chang CJ (2012) Reaction-based small-molecule fluorescent probes for chemoselective bioimaging. Nat Chem 4:973–984
161. McQuade LE, Lippard SJ (2010) Fluorescent probes to investigate nitric oxide and other reactive nitrogen species in biology (truncated form: fluorescent probes of reactive nitrogen species. Curr Opin Chem Biol 14:43–49
162. Xiaomei L, Shuang L, Gaolin L (2016) Fluorescent turn-on for the highly selective detection of nitric oxide in vitro and in living cells. Analyst (Cambridge, UK) 141:2600–2605
163. Dong CH, Heo S, Chen S, Kim HM, Liu Z (2014) Fluorescent probes based on o-diamine functionality. Anal Chem 86:308–311
164. Yu-Jin Z, Wen-Jing Y, Chuan-Kui W (2016) Effect of fluorophore on sensing NO for newly synthesized PET-based two-photon fluorescent probes. Chem Phys 468:37–43
165. Yang Y, Seidlits SK, Adams MM, Lynch VM, Schmidt CE, Anslyn EV, Shear JB (2010) A highly selective low-background fluorescent imaging agent for nitric oxide. J Am Chem Soc 132:13114–13116
166. Hilderbrand SA, Lim MH, Lippard SJ (2007) Metal-based turn-on fluorescent probes for sensing nitric oxide. Acc Chem Res 40:41–51
167. Ma S, Fang DC, Ning B, Li M, He L, Gong B (2014) The rational design of a highly sensitive and selective fluorogenic probe for detecting nitric oxide. Chem Commun 50:6475–6478
168. Shiue TW, Chen YH, Wu CM, Singh G, Chen HY, Hung CH, Liaw WF, Wang YM (2012) Nitric oxide turn-on fluorescent probe based on deamination of aromatic primary monoamines. Inorg Chem 51:5400–5408
169. Chao W, Xinbo S, Zhuo H, Xiaoyu L, Yongpin X (2016) Monitoring nitric oxide and SNAP-tag. ACS Chem Biol 11:2033–2040
170. Bryan NS, Grisham MB (2007) Methods to detect nitric oxide and its metabolites in biological samples

171. Ye X, Rubakhin SS, Sweedler JV (2016) Detection of nitric oxide in single cells. Analyst 133:423–433
172. Han Q, Liu J, Meng Q et al. (2019) Turn-on fluorescence probe for nitric oxide detection and bioimaging in live cells and zebrafish. ACS Sens 4(2):309–316
173. Guo R-Y, Zhang Y-T, Chanmungkalakul S et al (2021) Bioinspired design of reversible fluorescent probes for tracking nitric oxide dynamics in live cells CC S. Chemistry 3(10):116–128
174. Chen Yi (2020) Recent developments of fluorescent probes for detection and bioimaging of nitric oxide. Nitric Oxide 98:1–19
175. Hu X, Wang J, Zhu X, Dong D, Zhang X, Wu S, Duan C (2011) A copper(II) rhodamine complex with a tripodal ligand as a highly selective fluorescence imaging agent for nitric oxide. Chem Commun 47:11507–11509
176. Woldman YY, Sun J, Zweier JL, Khramtsov VV (2009) Direct chemiluminescence detection of nitric oxide in aqueous solutions using the natural nitric oxide target soluble guanylyl cyclase. Free Radical Biol Med 47:1339–1345. PMID 19751819. https://doi.org/10.1016/J.Freeradbiomed.2009.09.007
177. Marques SM, Esteves da Silva JCG (2014) A nitric oxide quantitative assay by a glyceraldehyde 3-phosphate dehydrogenase/phosphoglycerate kinase/firefly luciferase optimized coupled bioluminescent assay. Anal Methods 6:3741–3750
178. Jiang S, Cheng R, Wang X, Xue T, Liu Y, Nel A, Huang Y, Duan X (2013) Real-time electrical detection of nitric oxide in biological systems with sub-nanomolar sensitivity. Nat Commun 4:2225
179. Davies IR, Zhang XJ (2008) Nitric oxide selective electrodes. Methods Enzymol 436:63–95
180. Moon J, Ha Y, Kim M, Sim J, Lee Y, Suh M (2016) Dual electrochemical microsensor for real-time simultaneous monitoring of nitric oxide and potassium ion changes in a rat brain during spontaneous neocortical epileptic seizure. Anal Chem 88:8942–8948
181. Jiang S, Cheng R, Wang X, Xue T, Liu Y, Nel A, Huang Y, Duan X (2013) Real-time electrical detection of nitric oxide in biological systems with sub-nanomolar sensitivity. Nat Commun 4:2225
182. Jensen GC, Zheng Z, Meyerhoff ME (2013) Amperometric nitric oxide sensors with enhanced selectivity over carbon monoxide via platinum oxide formation under alkaline conditions. Anal Chem 85:10057–10061
183. Bianco CL, Fukuto JM (2015) Examining the reaction of NO and H_2S and the possible cross-talk between the two signaling pathways. PNAS 112(34):10573–10574
184. Nyasulu FW, Mottola HA (1987) Amperometric determination of nitrogen dioxide in air samples by flow injection and reaction at a gas-liquid interface. J Automat Chem 9(1):46–49. https://doi.org/10.1155/S1463924687000105
185. Jensen GC, Zheng Z, Meyerhoff ME (2013) Amperometric nitric oxide sensors with enhanced selectivity over carbon monoxide via platinum oxide formation under alkaline conditions. Anal Chem 85(21):10057–10061. https://doi.org/10.1155/S1463924687000105
186. Blackburn BJ (1969) Detection and identification of short-lived free radicals by electron spin resonance trapping techniques (spin trapping). Photolysis of organolead, tin, and -mercury compounds. J Am ChemSoc 91:4481–4490
187. Hirsh DJ (2016) A liposome-encapsulated spin trap for the detection of nitric oxide. Free Radical Biol Med 96:199–210
188. Lauricella R, Triquigneaux M, Andre-Barres Ch, Charles L, Tuccio B (2010) 5-Hydroxy-2,2,6,6-tetramethyl-4-(2-methylprop-1-en-yl)cyclohex-4-ene-1,3-dione, a novel cheletropic trap for nitric oxide EPR detection. In: Chemical communications. vol 46. Cambridge, United Kingdom, pp 3675–3677
189. Venpin WKPF, Kennedy EM, Mackie JC, Dlugogorski BZ (2012) Mechanistic study of trapping of NO by 3,5-dibromo-4-nitrosobenzene sulfonate. Ind Eng Chem Res 51:14325–14336
190. Davies CA, Nielsen BR, Timmins G et al (2001) Characterization of the radical product oxide formed from the reaction of nitric oxide with the spin trap 3,5-Dibromo-4-nitrosobenzene sulfonate. Nitric Oxide 5:116–127

191. Chen O, Uzlaner N, Priel Z, Likhtenshtein GI (2008) Novel fluorescence method for real-time monitoring of nitric oxide dynamics in nanoscale concentration. J Biochem Biophys Methods 70(6), 24:1006–1013
192. Likhtenshtein GI (2018) Nitric oxide in chemistry and biology—physicochemical aspect: fundamental and recent developments. NOVA Science Publishes
193. Likhtenshtein GI (2009) Stilbenes: application in chemistry, life science and material science. Wiley-VCH, Weinhem
194. Förster T (1946) Energiewanderung und Fluoreszenz. Naturwissenschaften 33:166–175
195. Mazzei L, Ciurli S, Zambelli B (2014) Hot biological catalysis: isothermal titration calorimetry to characterize enzymatic. J Vis Exp 86:e51487
196. Wang Y, Wang G, Moitessier N (2020) Enzyme kinetics by isothermal titration calorimetry: allostery, inhibition, and dynamics. Front Mol Biosci 19. https://doi.org/10.3389/fmolb.2020.583826 TC
197. Malvern (2016) Microcal itc systems: understanding biomolecular interactions.Cambridge, Malvern
198. Todd MJ, Gomez J (2001) Enzyme kinetics determined using calorimetry: a general assay for enzyme activity? Anal Biochem 296:179–187. https://doi.org/10.1006/abio.2001.5218
199. Su H, Xu Y (2018) Application of ITC-based characterization of thermodynamic and kinetic association of ligands with proteins in drug design. Front Pharmacol 9:1133. https://doi.org/10.3389/fphar.2018.011332018
200. Geschwindner S, Ulander J, Johansson P (2015) Ligand binding thermodynamics in drug discovery: still a hot tip? J Med Chem 58:6321–6335. https://doi.org/10.1021/jm501511fdoi:10.1021/jm501511f
201. Di Trani JM, Moitessier N, Mittermaier AK (2018b) Complete kinetic characterization of enzyme inhibition in a single isothermal titration calorimetric experiment. Anal Chem 90:8430–8435. https://doi.org/10.1021/acs.analchem.8b00993
202. Benini S, Cianci M, Mazzei L, Ciurli S (2014) Fluoride inhibition of Sporosarcina pasteurii urease: structure and thermodynamics. J Biol Inorg Chem 19:1243–1261. https://doi.org/10.1007/s00775-014-1182-x
203. Berlemont R, Martiny AC, Martiny AC (2016) Glycoside hydrolases across environmental microbial communities. PLoS Comput Biol 12:e1005300. https://doi.org/10.1371/journal.pcbi.1005300
204. Burnouf D, Ennifar E, Guedich S et al (2012) KinITC: a new method for obtaining joint thermodynamic and kinetic data by isothermal titration calorimetry. J Am Chem Soc 134:559–565. https://doi.org/10.1021/ja209057d
205. Abis G, Pacheco-Gómez R, Bui TTT, Conte M (2019) Isothermal titration calorimetry enables rapid characterization of enzyme kinetics and inhibition for the human soluble epoxide hydrolase. Anal Chem 91:14865–14872. https://doi.org/10.1021/acs.analchem.9b01847
206. Bothner B, Chavez R, Wei J et al (2000) Monitoring enzyme catalysis with mass spectrometry. J Biol Chem 275(18):13455–13459. https://doi.org/10.1074/jbc.275.18.13455
207. Cheng S, Wu Q, Xiao H, Chen H (2017) Online monitoring of enzymatic reactions using time-resolved desorption electrospray ionization mass spectrometry. Anal Chem 89(4):2338–2344
208. Anderson KS (2005) Detection of novel enzyme intermediates in PEP-utilizing enzymes. Arch Biochem Biophys 433(1):47–58
209. Abis G, Pacheco-Gómez R, Bui, TTT, Conte M (2019) Isothermal titration calorimetry enables rapid characterization of enzyme kinetics and inhibition for the human soluble epoxide hydrolase. Anal Chem 91:14865–14872. https://doi.org/10.1021/acs.analchem.9b01847

Chapter 3
Methods of Enzyme Kinetics

Abstract The main goals of studying the kinetics of enzymatic processes are to establish their elementary mechanisms, namely, the succession and rate of intermediate of elementary stages as well as their chemical nature. To determine the parameters of the Michaelis–Menton equation, a whole arsenal of classical and modern physical methods described in Chap. 2 is used. In parallel, the following groups of methods are also successfully applied: (1) Continuous flow, temperature, pH and pressure jumps, kinetic isotope effect (KIE), artificial intermediates, and transition state analogs methods. Of particular importance are theoretical computation approaches. Foundation of these approaches, combined quantum mechanics/molecular mechanics (QM/MM) methods in computational enzymology, in particular, and examples illustrated theoretical approaches are subject of consideration in separate sections.

3.1 Introduction

In the service of enzymatic kinetics, there is a whole arsenal of classical and modern methods that make it possible to determine the parameters of the Michaelis–Menton equation, establish the nature of intermediate compounds and the detailed mechanism of the processes under study. Among the most widely used methods are flow and jump techniques, mass spectrometry, kinetic isotope effect (KEF), approaches based on the use of artificial analogues of transition state, intermediates and substrates, and mechanical dynamics and quantum mechanical computations [1–24]. The following classification of flow strategies has been accepted [8]: (1) Continuous flow involves non-stopped flow; (2) Stopped flow involves physically stopping the product zone for a period of time before online detection; (3) Quench flow involves both physically and chemically stopping the product zone and with offline detection; (4) Merging zones/bypass flow requires a manifold that can change the flow direction to selectively deliver only a certain part of the product zone to the detector; and (5) Air segmented flow employs small air segments between each solution bolus to prevent early mixing. Advantages and disadvantages of these different flow strategies were discussed. In the

G. I. Likhtenshtein, *Enzyme Catalysis Today and the Chemistry of the 21st Century*,
https://doi.org/10.1007/978-3-031-51552-1_3

case of temperature, pH and pressure jumps, the perturbation involves rapid heating, acidity and pressure which changes the value of the equilibrium constant, followed by relaxation to equilibrium at the given condition. Transition state (TS) analogues, are chemical compounds with a chemical structure that resembles the transition state of a substrate molecule in an enzyme-catalyzed chemical reaction

3.2 Artificial Intermediates

Many enzymatic reactions, including those of serine hydrolases and cysteine proteases, proceed via the formation of an unstable acyl-enzyme intermediate. Their acyl-enzyme intermediates are unstable, commonly having half-lives of minutes to hours. In some cases, acyl-enzyme complexes can be stabilized using substrate analogues or active site mutations [1]. For example, a new strategy for incorporating 2,3-diaminopropionic acid (DAP) into recombinant proteins, via expansion of the genetic code was developed [3]. Replacing catalytic cysteine or serine residues of enzymes with DAP permitted their first-step reaction with native substrates, allowing the efficient capture of acyl-enzyme complexes that are linked through a stable amide bond. A strategy for creating mechanism-based, light-activated serine or cysteine nucleophile protease and hydrolase substrate traps in complex mixtures and live mammalian cells was recently developed [4]. The research proceeds in several steps: (1) The genetically encoded, site-specific incorporation of photocaged Dap ((2*S*)-2-amino-3-{[(2-{[1-(6-nitrobenzo[*d*][1,3]dioxol-5-yl)ethyl] thio}ethoxy)carbonyl]amino}propanoic acid) (pc-Dap) into proteins expressed in *Escherichia coli*; (2) pc-Dap was converted to Dap (2,3-diaminopropionic acid) in proteins *in vitro* by illuminating purified proteins; and (3) The obtained system was incubated for up to 2 days at pH 8; 4) By incubating purified proteases or thioesterases in which the catalytic cysteine or serine was replaced with Dap; (4) Replacing the catalytic nucleophile with genetically encoded 2,3-diaminopropionic acid allowed the first-step reaction to form an acyl-enzyme intermediate in which a substrate fragment is covalently linked to the enzyme through a stable amide bond.

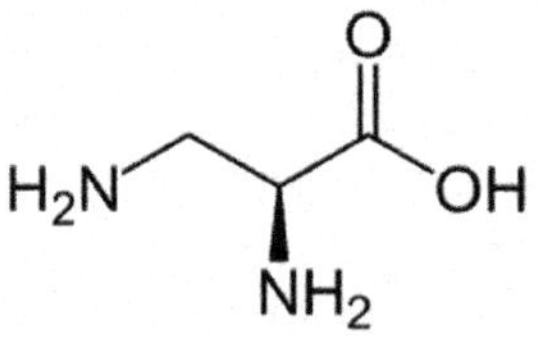

Review [2] summarized published trapping studies and their biological outcomes. It was demonstrated that in *vivo* trapping allows systematic identification of potential substrates of proteases, their adaptors, and chaperones. This approach combines *in vivo* genetic modifications of proteolytic systems, stabilized protease-substrate interactions, affinity enrichments of trapped substrates, and mass spectrometry (MS)-based identification. In *vitro* approaches, immobilized protease components were incubated with isolated cellular proteome, complement this in vivo approach.

3.3 Jump Methods [10–30]

Simple continuous flow (Figs. 3.1 and 3.2) has been applied to automate enzyme kinetic studies [6]. In the stopped flow injection technique, the flow of solution in the tubing can be stopped for a desired period of time by stopping the pump. The reaction time can be prolonged and the progress of the stopped reaction zone can be continuously monitored over time.

By using the automatic flow-based analytical systems, enzyme kinetics studies can be carried out at real-time initial rate [8]. The flow-based systems have been developed to provide rapid, low-volume, and high-precision analyses (Fig. 3.2).

Quench flow technique is based on a similar idea to the stopped flow technique [17]. The difference is that in the stopped flow technique, one reaction mixture stays in the system for continuous monitoring of change with time. More than one reaction mixture is needed to be introduced into the part of the system. As an example, the catalytic mechanism of 3-deoxy-D-manno-2-octulosonate-8-phosphate (Kdo8P) synthase from *Escherichia coli* was investigated under pre-steady state conditions using rapid chemical quench flow methods. The reaction was terminated by quenching with 0.2 N KOH at 15 °C. The results suggest the formation of acyclic bisphosphate as a reaction intermediate.

Zone trapping/merging zones) technique or Bypass zone trapping flow injection analysis (ByT-FAS) was introduced with similar principle to stopped flow technique [18]. The technique ensures that the part of the sample/reagent plug being measured had reached physical steady state and had not been affected by laminar flow and dispersion. This is done by controlling the sample size to be sufficiently large and by arranging the flow manifold in such a way that the plug of mixture contains a portion of physical steady state. Only the steady state portion is detected. Flow injection

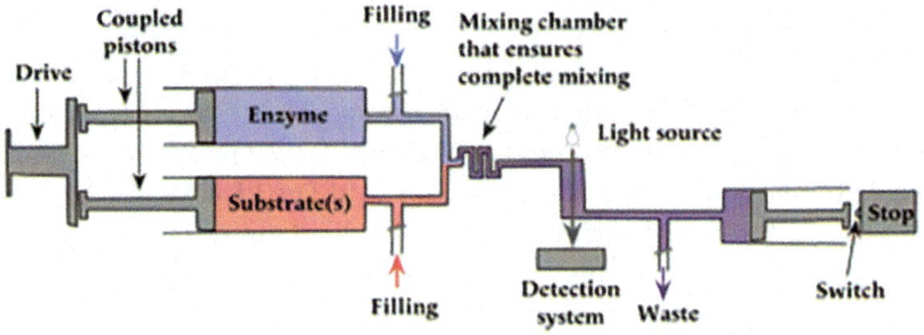

Fig. 3.1 Diagrammatic sketch of a stopped flow instrument https://en.wikipedia.org/wiki/Stopped-flow. Published under the Creative Commons Attribution-Share Alike 4.0 International

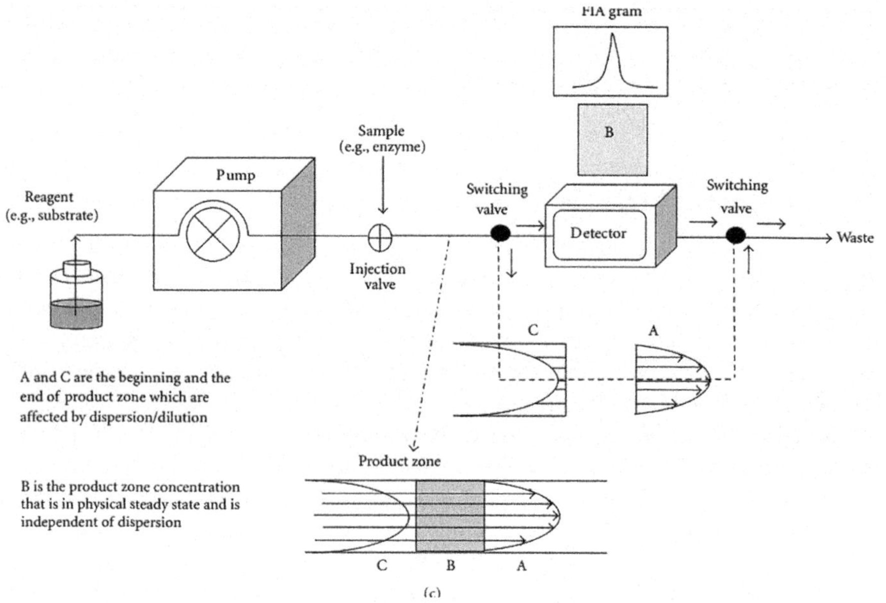

Fig. 3.2 Basic manifold diagrams of **a** continuous and stopped flow, **b** quench flow, and **c** merging zones/bypass flow [8]. Published under a Creative Commons Attribution 4.0 License from Hindawi

analysis (FIA) is accomplished by injecting a plug of sample into a flowing carrier stream [8, 9]. Flow-based analysis system (FIA) is based on dispersion of sample and reagent solutions in small tubing by flowing and merging them together. The process includes the following stages: (1) Product is formed while the merged solutions are flowing into the detector; (2) Special flow-through-cells are made for continuous flow of solution into and out of the detector; and (3) This enables flow injection analysis-based systems to couple with various types of detectors (e.g., electrochemical, mass spectrometry, chemiluminescence, visible absorption, fluorescence [16–19].

A sequential injection-Lab-at-valve (SI-LAV) segmented flow system for kinetic study of an enzyme was introduced for separation of enzyme and substrate zones and separation of the stacked zones from the carrier solution [19, 20]. As an example, the enzyme horseradish peroxidase (HRP) and tetramethylbenzidine (TMB) kinetic parameters based on initial rate was used as a model study. TMB can act as a hydrogen donor for the reduction of hydrogen peroxide to water by horseradish peroxidase. The operation of the system was the conventional batch-wise process in which a group of identical products are produced simultaneously). The SI-LAV system (Fig. 3.3) is composed of a bidirectional syringe pump, a holding coil (PTFE tubing of 350 cm in length and 0.03" ID), a 10 port selection valve, and a mixing chamber made of a small glass dropper of approximately 7.5 cm in length and 8 mm in diameter, directly connected to one of the selection ports of the multi-selection valve. The volume of solutions to be put into the chamber can be varied up to 3.5 mL. Enzyme HRP (10 mL), TMB (235 mL) and H_2O_2 were mixed in a quartz micro-cuvette 580 L of 10 mm pathlength. The cuvette was placed into the UV1800 UV-VIS spectrophotometer which was set at kinetic mode. Michaelis constant (K_m) and maximum velocity (V_{max}) were considered from the Michaelis–Menten plot and Lineweaver–Burk plot and were tabulated (Table 3.1).

Table 3.1 Michaelis constant (K_m) and maximum velocity (V_{max}) for the enzyme horseradish peroxidase (HRP) and tetramethylbenzidine (TMB) reaction [19].

In the case of temperature jump, the perturbation involves rapid heating which changes the value of the equilibrium constant, followed by relaxation to equilibrium at the new temperature [20–28]. The heating usually involves discharging of a capacitor, a device that stores electrical energy in an electric field (in the kV range) or heating by the output of a pulsed laser which emits in the near infra-red. For example, a method for generating optical control of enzyme activity based on short laser pulses as optical "switches," introducing an instantaneous change of the pH-value for activation of protein function, was developed [20, 21]. A proton release from 2-nitrobenzaldehyde was introduced to yield a pH-jump of almost 3 units on switching from inactive to active conditions for the enzyme. Following this change in pH-value, an activation

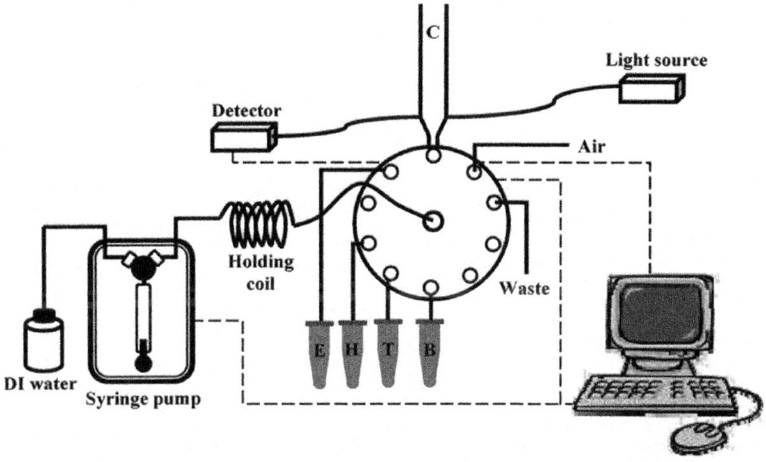

Fig. 3.3 Schematic of the SI-LAV system (*C* is mixing chamber and detection cell; *E* is enzyme HRP; *H* is H_2O_2; *T* is TMB; B is phosphate–citrate buffer) [19, 20]. With permission from Elsevier

Table 3.1 Summary of reported enzyme isotope effects on steady-state kinetic parameters (*l* and *h* Superscripts designate natural (Light) Enzyme and ^{13}C,^{15}N,^{2}H labeled enzyme (Heavy) correspondingly [49, 20]

Parameters	K_m (*mM*, *n*−3)		V_{max} (Abs s^{-1}, *n*−3)	
	SI-LAV	Batch method	SI-LAV	Batch method
H_2O_2	0.94 ± 0.07	1.02 ± 0.23	0.0125 ± 0.001	0.0150 ± 0.001
TMB	0.86 ± 0.15	0.99 ± 0.15	0.0125 ± 0.001	0.0155 ± 0.001

With permission from Elsevier

of hydrolytic activity of acid phosphatase (a nonspecific phosphomonoesterase) was occurred. The overall reaction of the enzyme was divided into two steps, namely, the pH-jump generation and the enzymatic reaction system. The kinetic data for the substrate 6-chloro-8-fluoro-4-methylumbelliferone phosphate were determined. Thus, in this study, a photoswitch the activity of a hydrolytic enzyme by applying a laser-induced pH-jump to a complex mixture of enzyme, substrate, and phototrigger was taken place (Fig. 3.4).

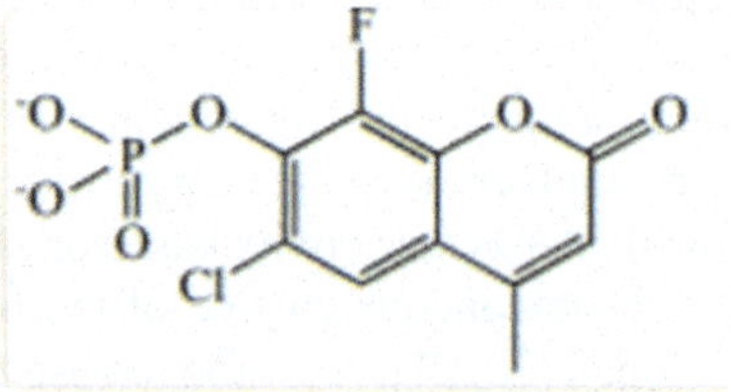

For generation of the light-induced pH-jump, 2-nitrobenzaldehyde (2-NBA) was chosen [23, 24]. Intramolecular rearrangement of 2-NBA led to fast formation of nitrosobenzoic acid, a weak acid ($pK_a \approx 3.54$) which acts as proton emitter within the nanosecond range. Excitation at 388 nm for about 250 ms resulted in to a jump in pH of approximately $2 - 3$ pH-units. A stable pH-value is given for more than 30 s after excitation. As a result of the enzymes pH profile, this laser-induced pH-jump led

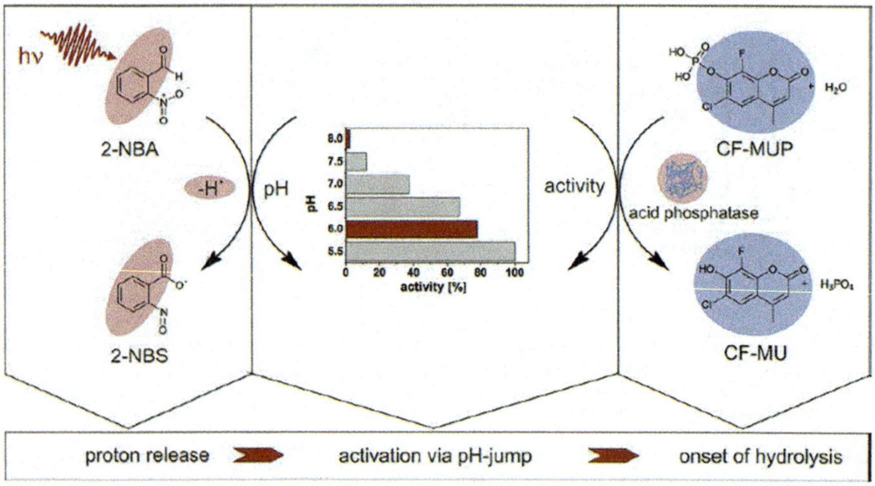

Fig. 3.4 Principle of the optical activation experiments the photoexcited intramolecular rearrangement of 2-nitrobenzaldehyde (2-NBA) to 2-nitrosobenzoic acid (2-NBS) causes a proton release and a strong change in pH. As a consequence of the enzymatic pH-profile an activation of hydrolytic activity is achieved and enzymatic conversion of 6-chloro-8-fluoro-4-methylumbelliferone phosphate (CF-MUP) to 6-chloro-8-fluoro-4-methylumbelliferone (CF-MU) is switched on [20, 21]. With permission from American Chemical Society

photoswitching of acid phosphatase activity with 6-chloro-8- fluoro methylumbelliferone phosphate (CF-MUP) as substrate. The enzymatic conversion was monitored by red-shifted to the absorption maximum of CF-MU at 370 nm. Using temperature and pressure jump methods, kinetic resolution of a conformational transition and the ATP hydrolysis step in *a Dictyostelium* myosin II mutant containing a single tryptophan residue were investigated [25, 26]. The experiment indicated that: (1) The initial binding process was accompanied by a small quench in fluorescence; (2) This is followed by a large fluorescence enhancement that appears coincident with the hydrolysis step; (3) The enhancement corresponds to the open-closed transition (k(obs) approximately 1000 s^{-1}at 20 °C; (4) The is largely in the open state; and (5) Pressure jump during steady-state ATP turnover reveals the distinct transients for the rapid open-closed transition and the slower hydrolysis step.

Temperature-jump small- and wide-angle X-ray scattering measurements on a dynamic enzyme, cyclophilin A were performed [26, 28]. It was found that cyclophilin A displays rich functional intramolecular protein dynamics on the microsecond timescale following a temperature jump. The resulting time-resolved signal to assess the kinetics of conformational changes was measured. Two relaxation processes were resolved: a fast process is related to surface loop motions, and a slower process is related to motions in the core of the protein that are critical for catalytic turnover. A new constructed temperature-jump/stopped-flow apparatus was a redesigned [27, 28]. The apparatus was capable of mixing the reactants on a submillisecond timescale concomitant with a temperature-jump as large as 60 °C (Fig. 3.5). Importantly that it was possible to study enzyme reactions that are faster than the denaturation and process can be investigated above denaturation temperatures. The temperature-jump/stopped-flow enables to investigate at physiological temperature the mechanisms of many enzymes, which was impossible until now because of their heat instability.

A schematic view of the temperature-jump/stopped-flow instrument is presented in Fig. 3.5. Temperature calibration of the heating elements (A) and determination of the dead time of the temperature-jump/stopped-flow apparatus (B) are shown in Fig. 3.6 [28].

3.4 Kinetic Isotope Effect (KIE)

A kinetic isotope effect (KIE) is the change in the reaction rate of a chemical reaction when one of the atoms in the reactants is replaced by one of its isotopes [29–60, 31–57]. Quantitatively, KIE is the ratio of rate constants for the reactions involving the light (k_L) and the heavy (k_H) isotopically substituted reactants

$$\mathrm{KIE} = \frac{k_L}{k_H} \tag{3.1}$$

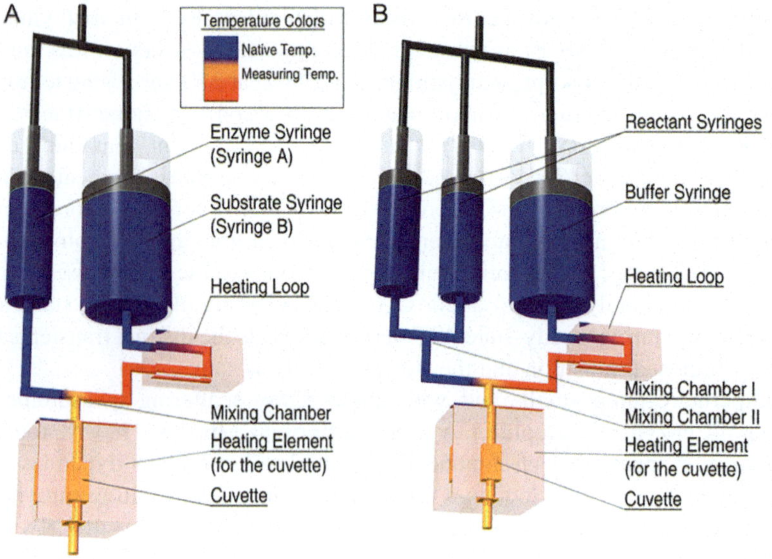

Fig. 3.5 Schematic representation of the temperature-jump/stopped-flow apparatus in single **a** and double **b** mixing experiments. **a** The heat-sensitive reactant (enzyme) is loaded into syringe **a** and the non-heat-sensitive reactant (substrate) is loaded into syringe **b**, both thermostated at 20 C (indicated by blue). Upon the stopped-flow push the non-heat-sensitive reactant flows through a heating loop in which it is heated above the reaction temperature (indicated by red). The enzyme (myosin) and the hot reactant combine in the mixing chamber, and they contribute to the new reaction temperature (indicated by orange) according to their temperatures and volumes. The process of the reaction at an elevated experimental temperature can be monitored in the cuvette using, e.g., spectroscopic signals. **b** Reactants are loaded into syringes **a** and **b**, which are both kept at native temperature (indicated by blue). Upon the first stopped-flow push, the reactants are mixed in Mixing Chamber I thermostated at native temperature (also indicated by blue) then flow into a delay line. The motor then stops for a desired delay time, al-lowing the mixture to reach equilibrium. On the next push, the solution is mixed with hot buffer (indicated by red) in Mixing Chamber II and forced into the observation cell. The process of reaching the new equilibrium upon the rapid temperature-jump perturbation can be monitored in the cuvette using, e.g., spectroscopic signals [27, 28]. With permission from Elsevier

A primary KIE is a KIE where the isotopically labeled atom is directly involved in bond cleavage or formation, secondary KIEs refers to KIEs resulting from isotopic substitution of an atom not directly involved in bond cleavage or formation. Enzymatic reactions involve multiple kinetic steps in addition to the bond cleavage/formation step (chemical step), including substrate binding, conformational changes and product release, with product release often being the rate-limiting step. In cases where these steps are isotopically insensitive, they may still mask the isotope effect on the chemical step.

Semi-classical models of KIEs are rooted in transition state theory (TST), which assumes a virtual dynamic equilibrium between the transition state (TS) and the ground state (GS). The rate constant (k) described by TST is:

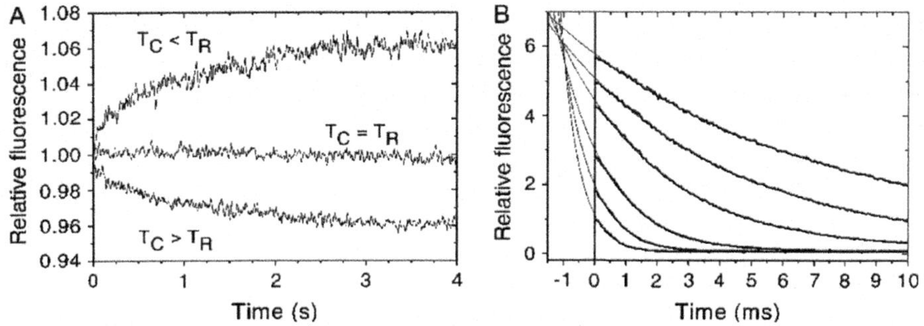

Fig. 3.6 Temperature calibration of the heating elements **a** and determination of the dead time of the temperature-jump/stopped-flow apparatus **b**. **a** Temperature-jump/stopped-flow fluorescence records on mixing 50 mM f N-acetyl-L-tryptophanamide (NATA) thermostated at 20 C with hot buffer thermostated at 58 C, 60 C, and 62 C and then pushed into a 52 C cuvette house. A fluorescent intensity increase or decrease indicates that the temperature of the reaction mixture (TR) was higher or lower than the temperature of the cuvette (TC), respectively. A constant fluorescence intensity indicates that the temperature of the reaction mixture was identical to the temperature of the cuvette. **b** Temperature-jump/stopped-flow records of the reaction of NATA with increasing concentrations of NBS at 25 C to determine mixing dead time**.** The instrument dead time was determined to be 0.9 ms [27, 28]. With permission from Elsevier

$$k = \kappa \frac{k_B T}{h} \frac{Q^{\#}}{Q} e^{-De/k_B T} \tag{3.2}$$

Here, k is the transmission coefficient (which can be smaller than unity or be larger than unity due to tunneling, T is the absolute temperature, h is Planck's constant and kB is Boltzmann's constant, Q and Q‡ are the total partition functions of the ground state (GS) and TS, respectively, and Δe is the energy difference between the TS and ground state GST. Within the framework of the Semi-Classical Models of KIEs [28, 36].

$$\text{KIE} = \left(\frac{k_L}{k_H}\right) \times \text{MMI} \times \text{TXC} \times \text{ZPT} \tag{3.3}$$

Here, k_L/k_H is the ratio of transmission coefficients, which semi-classically (no tunneling) is close to unity; MMI, the "Mass Moment of Inertia" term, refers to the isotope effect on translation and rotation; EXC refers to the isotopic variations on excited vibrational level; ZPE is the contribution arising from the isotopic difference in vibrational zero-point energies, and is the primary contributor to KIEs in semi-classical models.

According theory of solvent isotope effects, the transfer of an exchangeable proton in a rate-limiting step frequently results in an observed SKIE [33]. A variety reactions involving the transfer of hydrogen (H^+, H^-, or $H^{\cdot}$) cannot be explained by semi-classical models. For such reactions quantum mechanical tunneling mechanism was proposed [35]. Tunneling is the phenomenon where a particle's probability density spreads through a potential energy barrier without ever obtaining the energy necessary to surmount that barrier. Since a particle's ability to tunnel depends on mass, KIEs are an excellent probe for tunneling effects. A tunneling correction term QT should be added to the transition state (TST) rate constant (k_{TST}):

$$k = Q_T k_{TST} \tag{3.4}$$

Full tunneling Marcus-Levich model [36–38] is described by Eq. 3.5 and Fig. 3.7 [32, 37].

$$k = \frac{|V|^2}{\hbar}\sqrt{\frac{\pi}{\lambda k_B T}} e^{-\frac{(\Delta G+\lambda)^2}{4k_B T\lambda}} \int_0^\infty F(m, DAD) e^{-E(DAD)/k_B T} dDAD \tag{3.5}$$

In this equation, the factors in front of the integral give the rate of reaching a tunneling ready state (TRS), based on the electronic coupling between reactant and product (V = the degree of adiabaticity of the reaction), the reorganization energy (λ), and the driving force of the reaction (ΔG°). The integral measures the probability of hydrogen transfer once the system reaches a tunneling ready state (TRS). The first factor insides the integral determines the probability of tunneling as a function of mass (m) and the donor–acceptor distance (DAD). A Boltzmann factor, gives the probability of being at any given DAD.

According to Marcus-like models [36–38] (Fig. 3.7), rates are governed by three major processes: (1) Heavy atom motion that brings the system to (TRS), where the reactant and product potential surfaces are degenerate and tunneling can occur, (2) Fluctuations of the tunneling barrier, modulated by heavy atom motion, and (3) The actual process of hydrogen tunneling. Various models for the physical sources of KIEs were discussed, namely, how those models can be used to interpret data, and how the focus of traditional models has grown to a model that includes motion of the enzyme and quantum mechanical nuclear tunneling [32, 33].

In work [32, 33], two case studies of enzymes, thymidylate synthase and alcohol dehydrogenase using KIEs methods were carried out. Alcohol dehydrogenase ADH catalyzes the oxidation of an alcohol to an aldehyde using a nicotinamide cofactor (Fig. 3.8). Mechanism **of the C–**H bond cleavage by thymidylate synthase and ADH was considered in details.

Studies of ADH demonstrated how 2° KIEs can assess TS structure. The studies of TS as showed the use of temperature dependence of 1° KIEs in the investigation of both the hydride and proton transfers involved in that reaction. Combining experimental KIE measurements with QM/MM simulations has clarified the molecular mechanism, possible role of dynamics, and transition state structures for the enzyme.

Data on equilibrium binding isotope effects (BIEs) in two case studies of enzymes, thymidylate synthase and alcohol dehydrogenase, suggested the following [38, 39]: (1) How to establish the bond vibrational status of enzyme substrates in the Michaelis complex prior to the transition state; (2) How they differ from the solution state;

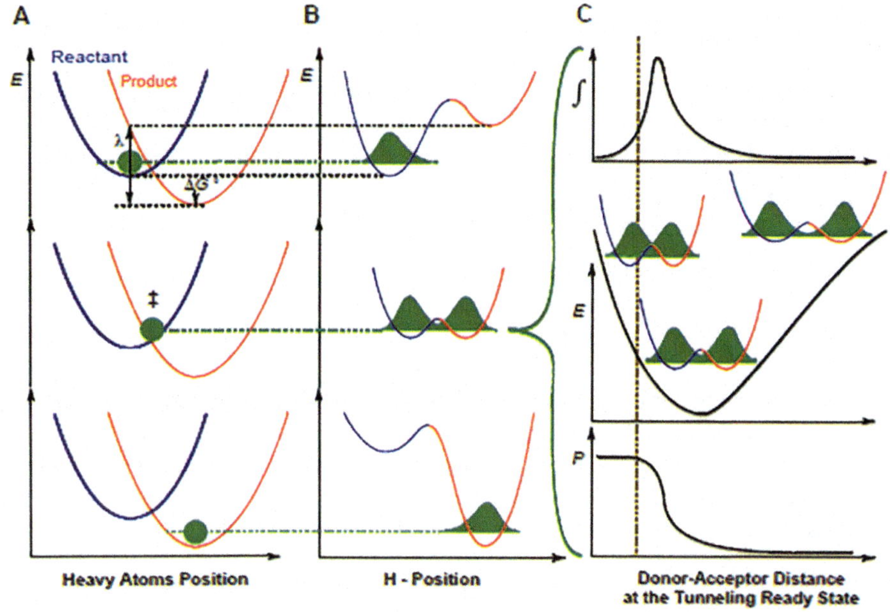

Fig. 3.7 Marcus-like models of hydrogen tunneling—a visualization of. Three slices of the potential energy surface (PES) along components of the collective reaction coordinate showing the effect of heavy atom motions on the zero point energy in the reactant (blue) and product (red) potential well. Panel **a** presents the heavy atom coordinate (also known as the Marcus parabolas), and Panel **b** shows the H-atom position, which is orthogonal to the heavy atom coordinate. In the top panels the hydrogen is localized in the reactant well, and the zero point energy of the product state is higher than that of the reactant state. Heavy atom reorganization brings the system to the tunneling ready state (TRS, middle panels **a** and **b**), where the zero point energy in the reactant and product wells are degenerate and the hydrogen can tunnel between the wells. Further heavy atom reorganization breaks the transient degeneracy and traps the hydrogen in the product state (bottom panels). The rate of reaching the TRS depends on the reorganization energy (λ) and driving force (ΔG°), which are indicated in the top panel. Panel **C** shows the effect of the donor–acceptor distance (DAD) sampling on the wave function overlap at the TRS (middle panel). Transmission probability (P) is a function of the overlap integral of the hydrogen wave functions in the reactant (blue) and product (red) wells (bottom panel **c**). The top panel **c** presents the contribution to H-transfer at each DAD as a function of the P and the population at each DAD The vertical dashed line represents the DAD under which the ZPE is greater than the barrier height. At such distances, the process of a wave function spreading from reactant well to product well is no longer "tunneling," but one can still use the particle's transmission probability analogously to the tunneling probability at longer DADs [32, 33]. Published under Creative Commons Attribution license from MDPI

(3) How BIEs provided an experimental means of interrogating enzyme-substrate interactions; and (4) How BIEs inform on the influence of enzyme-mediated atomic distortions in modulating substrate reactivity. A rapid equilibrium dialysis method used to measure BIEs for several enzyme systems was outlined. As an example, the substrate bonding environment for the human protein lysine N-methyltransferase NSD2 was investigated. Normal BIEs with

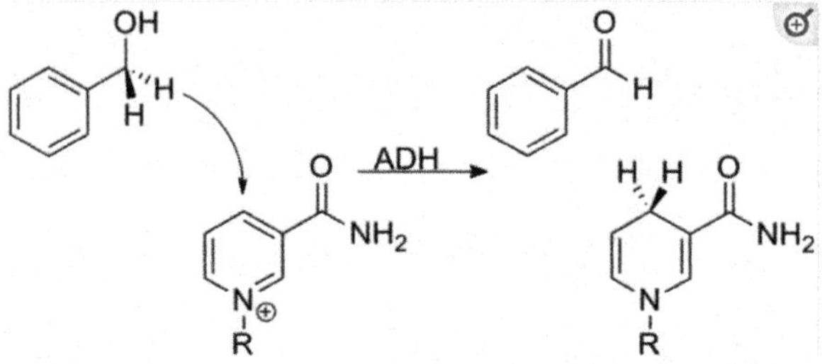

Fig. 3.8 The reaction catalyzed by ADH (R = adenine diphosphate ribosyl-) with BnOH as alternative substrate [32, 33]. Published under Creative Commons Attribution license from MDPI

$$K_{\mathrm{eq}}^{\mathrm{Light}\,/}/K_{\mathrm{eq}}^{\mathrm{Heavy}} = 1.05 \tag{3.6}$$

occurs when the bonding environment of the isotopically substituted position is less constrained in the bound state than in the free state, resulting in a decreased force constant for this bond upon enzyme binding. By contrast, inverse BIEs (<1.0) occurs when binding of the substrate bearing the heavy isotope is favored and are commonly expressed as a ratio of equilibrium constants, for example, with

$$K_{\mathrm{eq}}^{\mathrm{Light}}/K_{\mathrm{eq}}^{\mathrm{Heavy}} = 0.95 \tag{3.7}$$

or as a negative percentage value. Inverse BIEs indicates an increased force constant for the isotopically substituted bond in the bound state. Measurement of equilibrium BIEs, using the rapid equilibrium dialysis (RED) device includes: (1) determining substrate equilibration time, (2) determination of substrate stability, (3) measurement of BIEs on binary and ternary complexes association, (4) scintillation counting and spectral deconvolution, and (5) BIE interpretation and examples [38, 39].

A theory of KIE for multistep enzymatic reactions was developed by Cleland and Northrop (1999) [41, 42]. The isotope in limiting step was expressed in the experimental ratio V/K, where V and K are the reaction maximum rate and Michaelis constant. For the mechanism:

$$k_1 \quad k_3 \quad k_5 \quad k_7 \quad k_9$$

$$\mathbf{E + S \leftrightarrow ES \leftrightarrow ES* \leftrightarrow EP* \leftrightarrow EP \rightarrow E + P}$$

where ES* $\longleftrightarrow$ EP* is the isotope sensitive chemical step. The isotope effect on V/K at the substitution of a light atom X for a heaver atom Y is given by

$$\alpha_{X,Y} = \frac{V^X K^Y}{K^X V^Y} = \frac{\left[\left(\frac{k_5^X}{k_5^Y}\right) + c_f + \left(\frac{K^X}{K^Y}\right)c_r\right]}{1 + c_f + c_r} \tag{3.8}$$

where c_f and c_r are combinations of rate constants of different steps. The forward commitment

$$c_f = \left(\frac{k_5}{k_4}\right)\left(1 + \frac{k_3}{k_2}\right) \tag{3.9}$$

whereas the reverse commitment

$$c_r = \left(\frac{k_6}{k_7}\right)\left(1 + \frac{k_8}{k_9}\right) \tag{3.10}$$

When c_f and c_r are very small ($\square_x x = k_5X/k_5Y$) and experimental KIE is directly related to the limiting chemical step. If c_f is very large, the KIE will be completely suppressed.

Figure 3.9 contains the structure of the transition state analog immucillin-H bound to the enzyme with PO_4 provides the structural model for the transition state complex [42, 43].

- Time-resolved electrospray ionization mass spectrometry (TRESI-MS) method for measuring KIE was introduced [43, 44]. The approach was demonstrated first using a relatively large isotope effect (associated with yeast alcohol dehydrogenase (YADH) catalyzed oxidation of ethanol. A small $^{12}C/^{13}C$ KIE was associated specifically with pre-steady state acylation of chymotrypsin during hydrolysis of an ester substrate were measured and discussed. Paper [44, 45] provided suggestions and examples how KIE data can be used in study of enzyme reaction mechanisms. The importance of following the Standards for the Reporting of Enzymological Data committee's recommendations when reporting KIE studies was outlined. A particular focus was placed on procedures for calculating, propagating, and reporting experimental errors. A framework which integrates KIEs into kinetic and isotopic models of metabolism, accounting for their system-wide

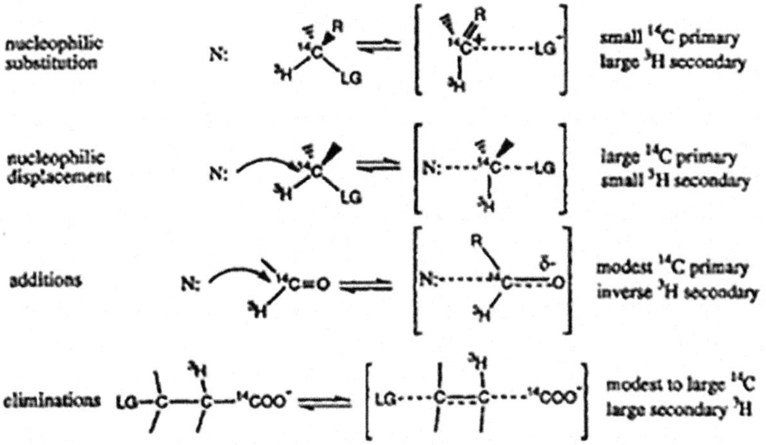

Fig. 3.9 The KIE expected for the primary $^{14}C$and a secondary ^{3}H isotopic label are indicated [42, 43] (Reproduced with permission from Elsevier)

effects on metabolite concentrations, metabolic fluxes, and isotopic patterns was developed [45, 46]. This approach was applied to assess the impact of KIEs on the central carbon metabolism of *Escherichia coli* in the context of ^{13}C-ILEs. Results showed that: (1) The impact of KIEs strongly depends on the label input and on the variable considered; (2) The framework can be used to investigate the impact of all the isotopic tracers $\left(^{2}H,^{013}C,^{015}N,^{018}O, \cdot etc.\right)$ on different isotopic datasets and metabolic systems; and (3) By allowing the integration of isotopic and metabolomics data collected under stationary and/or non-stationary conditions, it may facilitate the development of *more* accurate kinetic models with improved explicative and predictive capabilities.

- Chapters of the book [46, 47]. Focused on practical considerations, when using analytical and/or numerically integrated rate equations to model enzyme reactions. The book table contents include the following papers: Measurement of enzyme binding isotope effects by *Vern Schramm,* Chemical ligation and isotope labeling to locate dynamic effects by *Rudolf Alleman,* Measurement of heavy enzyme isotope effects by *Amnon Kohen,* KIE of metabolic flux and enzymes *Guillaume Tcherkez,* Solvent and Primary KIE on Flavin Enzymes by *Paul Fitzpatrick,* The Rapid Determination of Primary Deuterium Isotope Effects on Enzyme-Catalyzed Proton Transfer at Carbon in 50/50 HOH/DOD by *John Richard,* 18-O isotope effects on oxygenases by *Judith Klinman,* Chlorine isotope effects on dehalogenases by *Piotr Paneth,* Chlorine isotope effects (w/Paneth) by *Agnieszka Dybala-Defratyka,* Kinetic Deuterium Isotope Effects in Cytochrome P450 Reactions by *F. Peter Guengerich,* Use of Isotopes and Isotope Effects for Investigations of Diiron Oxygenase Mechanisms by *John Lipscomb,*. Characterization of substrate, co-substrate, and product isotope effects associated with enzymatic oxygenations using gas- and liquid chromatography/isotope ratio mass spectrometry by *Thomas Hofstetter.* NMR titrations to determine equilibrium isotope effects by *Charles Perrin,* NMR to characterize the nonstatistical distribution of stable isotopes in metabolism *Richard Robins,* Applications of isotope ratio mass spectrometry in sports drug testing: accounting for isotope fractionation in analysis of biological samples *Thomas Piper*, Whole molecule mass spectrometry to measure KIEs on RNA 2'O-transphosphorylation reactions by *Mike Harris,* Isotope Effects on the Concentration of Steady State Intermediates by *Hung-Wen Liu and Mark Ruszczycky, and* 19. Headspace analysis of CO_2 to measure carbon kinetic isotope effects in decarboxylation reactions by *Ron Kluger.*
- A hypothesis proposed in [49, 50] assumed that the isotopic substitution of all ^{12}C, ^{14}N, and nonexchangeable ^{1}H by ^{13}C, ^{15}N, and ^{2}H, would slow down protein picosecond to femtosecond dynamics. It was shown that in several systems the protein's fast dynamics are reduced in "heavy" enzymes in a way that reduces the probability of barrier crossing of its chemical step (Table 3.2). The observations indicated that slower protein dynamics are electrostatically altered in isotopically labeled enzymes. The paper also described the following; (1) Methods of preparation and structural characterization of "heavy" enzymes with isotopic substitution of all ^{12}C, ^{14}N, and nonexchangeable ^{1}H by ^{13}C, ^{15}N, and ^{2}H; (2) Effects of isotopic labeling on physiochemical properties of the proteins; (3) Approaches

toward measurement and interpretation of enzyme isotope effects; (4) In addition, studies of Purine nucleoside phosphorylase (PNP) HIV-1 protease, Lactate dehydrogenase (LDH) Alanine racemase (AR) and Dihydrofolate reductase (DHFR) were performed.

- Review [50, 51] surveyed inverse SKIEs in enzyme-catalyzed reactions, focusing on the combination of solvent isotope effects with other isotope effects to probe the mechanisms of amine and alcohol oxidations by enzymes. Macroscopic rate constants such as k_{cat} and k_{cat}/K_m were found to decrease when the reaction is performed in D_2O. A variety of this effect reasons including the transfer of protons were discussed. Nevertheless, under certain circumstances, these constants can increase, which can be attributed to an inverse solvent equilibrium isotope effect on a rapid equilibrium preceding the rate-limiting step(s). Interpretation of solvent isotope effects should take in account the following aspects: (1) Many protons in the enzyme are exchanged in D_2O; (2) Multiple protons in the substrate also

Table 3.2 Measurement of enzyme isotope effects. See details in [49, 50]

Enzyme	Substrate (s)	k^J_{cat}/k^h_{cat}	k^I_m/k^h_m
Chemical step is not rate limiting			
H_5PNP	Inosine	1.0 ± 0.2	1.1 ± 0.2
	Guanosine	1.0 ± 0.1	0.9 ± 0.2
E258D/L261A H_5PNP	Guanosine	1.0 ± 0.1	1.1 ± 0.2
E_cDHFR	DHF	**1.09 ± 0.02**	**0.74 ± 0.08**
	NADPH	1.12 ± 0.04	1.1 ± 0.2
	DHF	1.04 ± 0.03[b]	**1.98 ± 0.38[b]**
	NADPH		**1.3 ± 0.2[b]**
N23PP/S148A E_cDHFR	DHF	1.0 ± 0.1	1.0 ± 0.2
	NADPH		1.1 ± 0.3
TmDHFR	DHF	**1.36 ± 0.05[a]**	1.5 ± 0.5[b]
	NADPH		0.77 ± 0.3[b]
BsDHFR	DHF	**2.4 ± 0.2[a]**	0.7 ± 0.2[b]
	NADPH		1.0 ± 0.2[b]
HsDHFR	DHF	**1.44 ± 0.02**	1.07 ± 0.4
	NADPH	**1.45 ± 0.02**	1.06 ± 0.26
Chemical step is rate limiting			
HIV-1 protease	Aminobenzoyl-Thr-Ile-Nle-pNO_2-Phe-	1.19 ± 0.05	**0.62 ± 0.07**
	Gln-Arg-NH_2		
AR[c]	D-Alanine	1.32 ± 0.04	1.0 ± 0.1
	L-Alanine	1.21 ± 0.03	0.9 ± 0.1

With permission from Elsevier

can be replaced; (3) The properties of D_2O as solvent viscosity and polarity, for example, are not identical to those of H_2O and D_2O; (4) pKa values are altered in D_2O increasing by 0.3–0.7 33] [50, 51] Based on a critical analysis of numerous experimental data on isotope effects, the following conclusions were formulated [50, 51]: (1) For oxidations of both alcohols and amines by a variety of enzymes, the use of multiple isotope effects has established that proton transfer precedes hydride transfer; (2) For amine oxidation, the enzyme does not require an active site base to remove a proton from the amine; (3) For alcohol oxidation, a variety of enzymes stabilize the alkoxide sufficiently that its formation does not contribute to the rate of catalysis; (4) Methanol oxidase and galactose oxidase provide examples of how the use of alternate, typically nonphysiological, substrates can alter the relative energetics of alkoxide formation and the subsequent CH bond cleavage; and (5) the flavoprotein D-arginine dehydrogenase binds the zwitterionic form of the amino acid substrate productively, in contrast to the binding of the anionic substrate by other amine-oxidizing enzymes.

According theory of solvent isotope effects, the transfer of an exchangeable proton in a rate-limiting step frequently results in an observed SKIE [33, 34]. The transfer of solvent-sensitive protons yields primary deuterium isotope effects that occur along the reaction coordinate (Fig. 3.10). During this transfer, the bond to the donor weakens in the transition state, yielding a normal SKIE (i.e., $k_{D2O} \equiv k_{H2O}/k_{D2O} > 1$), and protons in flight in a rate-limiting step produce normal SIE contributions. Under certain circumstances, macroscopic rate constants such as k_{cat} and k_{cat}/K_m were found to increase, showing an inverse solvent kinetic isotope effect (SKIE). It was concluded that these phenomena can be attributed to an inverse solvent equilibrium isotope effect on a rapid equilibrium preceding the rate-limiting step(s). (See details in capture of Fig. 3.10).

P450 reactions (Fig. 3.11) are well-suited to the application of KIE approaches, because the reactions catalyzed by these enzymes involve the breaking of C–H bonds [47,49]. The most common P450 reactions are the conversion of a methyl, methylene, or methine group to an alcohol. Types of deuterium KIE studies used in P450 research is shown in Fig. 3.12. Many of the reported KIEs for P450 reactions are larger than the classical limit of ~ 7.

In the case of oxidation of ethanol to acetaldehyde by human P450 2E1:

$$CH_3CH_2OH \rightarrow CH_3CH(OH)_2 \rightleftarrows CH_3CHO + H_2O$$

the reaction is depicted as =

$$E \underset{k_{-1}}{\overset{k_1 S}{\rightleftarrows}} ES \rightarrow^{k2} EP \rightarrow^{k3} E + P$$

and

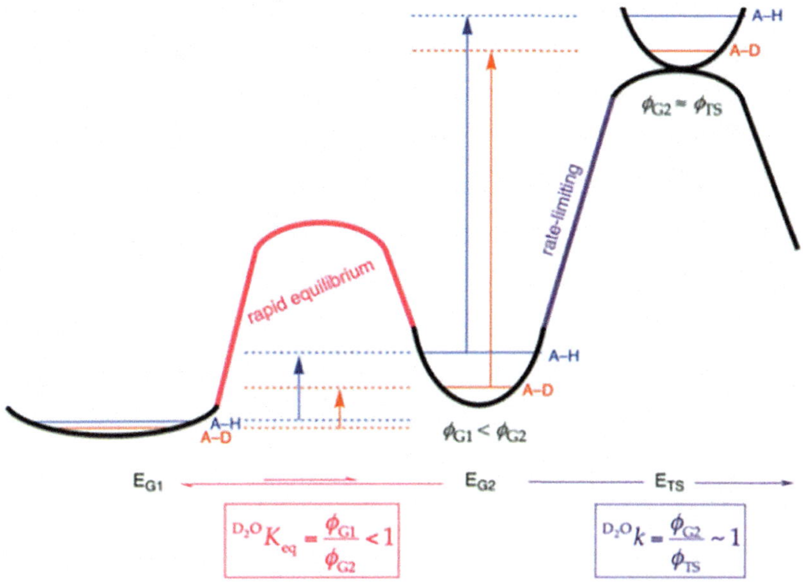

Fig. 3.10 Reaction coordinate diagram of a solvent-sensitive rapid equilibrium preceding a rate-limiting step that results in inverse solvent isotope effects. Horizontal lines represent zero-point energy levels, reflecting the mass of the hydron and the stiffness of the bond in which it is involved. When the stiffness of the bond increases, as shown in the rapid-equilibrium step (pink), the zero-point energy difference between the two isotopes increases relative to the reactant ground state (G1), leading to accumulation of deuterium in the intermediate ground state (G2); this is reflected in an increase in the fractionation factor (ϕ) for this step and an inverse solvent equilibrium isotope effect (SEIE) (i.e., $^{D2O}K_{eq} < 1$). If the stiffness of the solvent-exchangeable hydron is not altered in the transition state (TS) of a subsequent rate-limiting step (purple), then no SKIE will be contributed for this step (i.e., $^{D2O}k \sim 1$). The observed inverse solvent kinetic isotope effect (SKIE) ($^{D2O}k_{obs}$) for the overall process will be inverse (Eq. (3.2)). A = solvent-exchangeable group; E = enzyme [33, 34]. Published under Creative Commons CC BY 4.0 license from MDPI

$$D_V = \frac{D_{k_2} + k_2/k_3}{1 + k_2/k_3}$$

$$D_{V/K} = \frac{D_{k_2} + k_2/k_{-1}}{1 + k_2/k_{-1}} \tag{3.11}$$

where ^{D}V is the KIE on k_{cat} and $^{D}(V/K)$ is the KIE on the ratio of k_{cat}/K_m. The value of $^{D}(V/K)$ was found to be ~3–5 and a plot of product vs. time indicates rate-limiting step is after product formation. The ^{D}k for C-H bond breaking was found to be 3.8 and the isotopic effect on product formation measured using pre-steady-state kinetic studies was estimated to be 3.2 [52]. Thus, the presence of a significant primary deuterium KIE is evidence that hydrogen abstraction is at least partially rate-limiting in the reactions. Many of the reported KIEs for P450 reactions are larger than the classical limit of ~ 7 . No reports of serious efforts at studying tunneling have appeared.

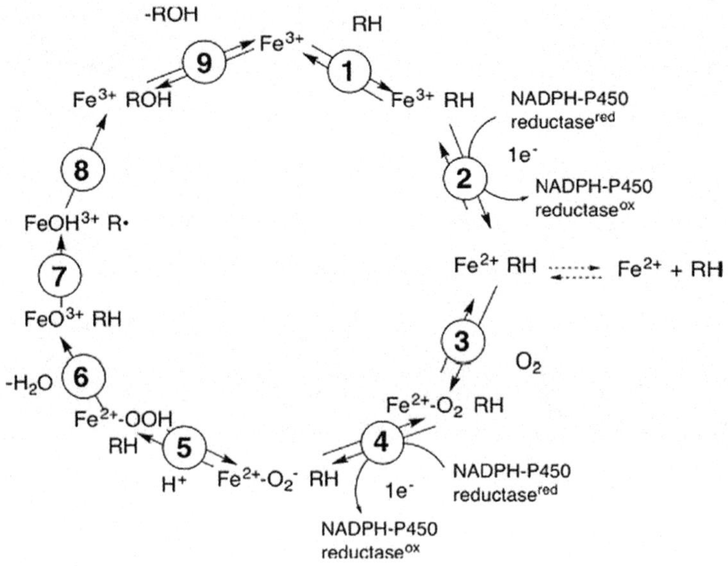

Fig. 3.11 Catalytic cycle of microsomal P450s. RH is the substrate, with a C-H bond to be broken. With the seven mitochondrial P450s and many bacterial P450s, the electrons are delivered from a ferredoxin (including adrenodoxin) instead of the diflavin NADPH-P450 reductase. In some cases, cytochrome b5 (b5) delivers an electron in step **4**, but not always. Note that the only step in which C-H bond breaking occurs [47, 49]. With permission from Elsevier

Isotope effects in the soluble methane monooxygenase system and CmlI, which catalyzes the multistep arylamine to arylnitro conversion in the biosynthetic pathway for chloramphenicol, were examined [53, 54]. The methods using isotope described in [7]. Allowed to probe the reaction mechanism of the diiron cluster-containing soluble form of methane monooxygenase (sMMO) and the N-oxygenase CmlI. The characterization of the sMMO reaction using isotopically enriched O_2 and substrates showed that overall reaction is:

$$CH_4 + O_2 + NADH + H^{+} = CH_3OH + H_2O + NAD^{+}$$

catalyzes the six-electron oxidation of an aryl-amine precursor (NH_2-CAM) to the aryl-nitro group of chloramphenicol (CAM)

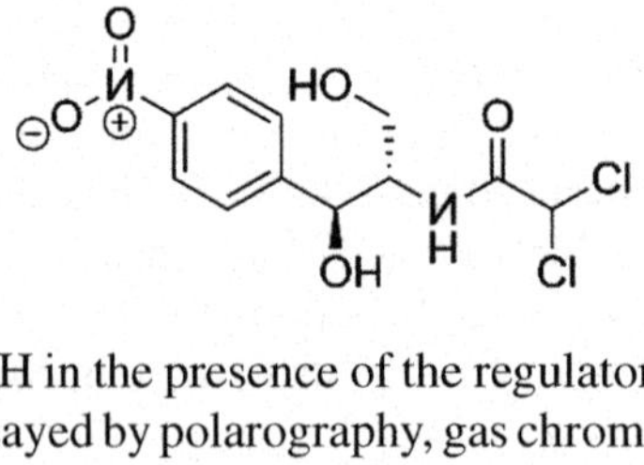

The activity of sMMOH in the presence of the regulatory protein MMOB and the reductase MMOR was assayed by polarography, gas chromatography, or colorimetric

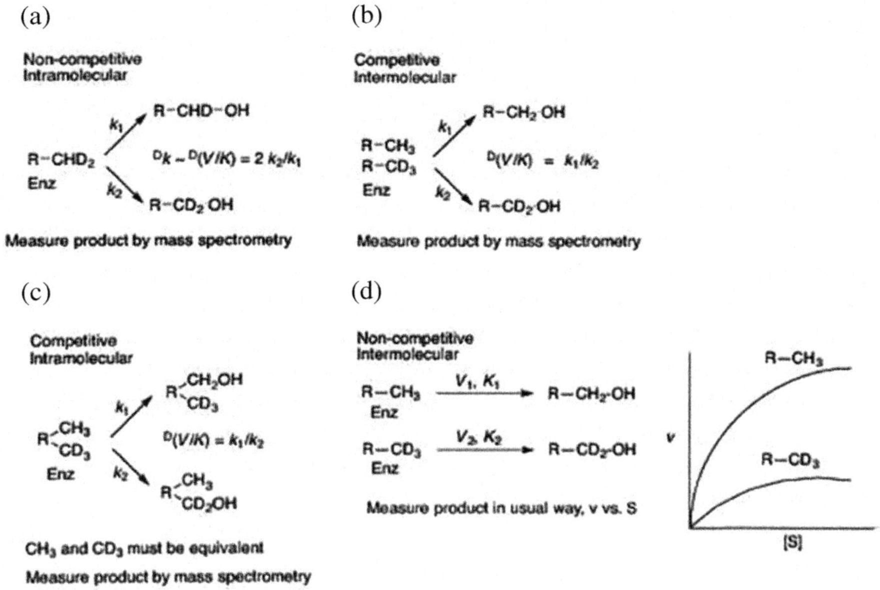

Fig. 3.12 Types of deuterium KIE studies used in P450 research. **a** Non-competitive intramolecular. The factor of 2 is a statistical correction, in that there are two deuteriums and one protium available here. The factor would be ½ if the substrate were $-CH_2D$. **b** Competitive intermolecular. **c** Competitive intramolecular. (D) Non-competitive intermolecular. In the "k_1" pathways, a C–H bond is broken; in "k_2" pathways a C–D bond is broken [47, 49]. With permission from Elsevier

NADH consumption assays. The sMMO structure was studied by time-resolved resonance, Raman Spectroscopy Mossbauer spectroscopy, and mass spectroscopy. Reaction of S-[1-^{2}H,1-^{3}H]-ethane in a multiple turnover sMMO system resulted in the formation of ethanol. The large tritium kinetic isotope effect justified that oxygen preferentially inserted in only the C–^{1}H. In work [54, 55], three aspects of chlorine kinetic isotope effects (Cl-KIEs) on enzymatic dehalogenations were considered: (1) Methodology for experimental measurement of Cl-KIEs, employing fast-atom bombardment-isotope ratio mass spectrometer FAB-IRMS; (2) The consequences of the reaction complexity; and (3) theoretical evaluation of the experimental results. Dechlorination of 1,2-Dichloroethane (DCE) by the DhlA Dehalogenase was investigated using isotopologues $^{35}\mathrm{Cl}-(\mathrm{CH_2})_2-^{35}\mathrm{Cl}_2^{35}\mathrm{Cl}-(\mathrm{CH_2})_2-^{37}\mathrm{Cl}$, and $^{37}\mathrm{Cl}-(\mathrm{CH_2})_2-^{37}\mathrm{Cl}$ presented in the ratio of about 9:6:1, respectively [55, 57]. The Cl-KIE for DCE dechlorination by DhlA was found to be 1.0068. The theoretical prediction of the intrinsic isotope effect indicated that in both cases dehalogenation is a rate-determining step. For benzyl alcohol, an enzyme isotope effect larger than unity was observed, whereas the enzyme isotope effect was close to unity for isopropanol.

The origin of a substrate (benzyl alcohol or isopropanol) preference in the alcohol dehydrogenase (ADH) from *Geobacillus* (BsADH) was examined by enzyme isotope labeling [32, 33]. ADH catalyzes the oxidation of an alcohol to an aldehyde using a nicotinamide cofactor. The main benefits of using ADH as a model are that one can study both the forward (alcohol to aldehyde) and reverse reactions. Employing both experimental and computational methods, it was shown that: (1) The observed KIEs were equal to the intrinsic KIEs at the physiological temperatures (20–30 °C); (2) An elevated or reduced temperatures minor commitments appear, partially masking intrinsic KIEs; (3) The reaction required a small activation energy (about 4.0 kcal/mol); (4) The intrinsic KIEs are temperature independent, indicating that the hydride tunnels from a narrow distribution of DADs is unaffected by temperature; (5) For benzyl alcohol, an enzyme isotope effect is larger than unity; and (6) The enzyme isotope effect is close to unity for isopropanol.

The enzyme kinetic isotope effects for reactions catalyzed by Alcohol Dehydrogenase (BsADH) and the effect of dynamic coupling in BsADH) catalysis for different substrates were examined [55, 56]. The natural substrates of BsADH served as hydride acceptors under anaerobic conditions. The several important results of the work were reported, namely: (1) The hydride transfer from benzyl alcohol to NAD^+ during catalysis by BsADH is rate-limiting under steady state conditions at pH 7.0; (2) Substrate KIEs (KIE=$k_{cat}{}^{H}/k_{cat}{}^{D}$) was determined by measuring the steady-state rate constant (k_{cat}) using protonated and deuterated substrates; (3) The resulting substrate KIE (KIE=$k_{cat}{}^{H}/k_{cat}{}^{D}$) is highly temperature-dependent, increasing sharply at lower temperatures but remaining constant above 40 °C; and (4) The substrate KIE measured with protonated and deuterated isopropanol is temperature-independent for the examined temperature range (20–50 °C), with an average value of 2.42 ± 0.42.

Dihydrofolate Reductase (R67 DHFR) catalyzes the reduction of dihydrofolate to tetrahydrofolate (THF) [57, 60].To investigate the correlation between the KIEs, their temperature dependence, DAD distribution, and active site preorganization and dynamics the KIEs for the enzyme were computed. Initial Michaelis complexes were constructed using the X-ray crystal structures of plasmid-encoded DHFR with bound folate and the oxidized cofactor $NADP^+$. The potential energy surface was described by a hybrid QM/MM Hamiltonian, where the catalytically active QM region was treated with a modified AM1 semiempirical Hamiltonian denoted AM1-specific reaction parameter (AM1-SRP). By employing mass-perturbation PI simulations (PIFEP/ UM with SQCP), nuclear quantum effects for both light (H) and heavy (D) isotopes in the hydride transfer reaction were measured. As a result, the classical mechanics PMF profiles at 278, 288, 298, 308, and 318 K for the catalyzed hydride transfer from NADPH to H3folate+ (N5-protonated DHF) in the R67 DHFR were obtained (Fig. 3.13). Studies of enzymes such as DHFR, glycerol-3-phosphate-dehydrogenase, triosephosphate isomerase, paraoxonase, and organophosphate have shown that penetration of solvent molecules to the active site is detrimental to catalytic activity. Employing mass-perturbation PI simulations (PIFEP/ UM with SQCP), revealed nuclear quantum effects for both light (H) and heavy (D) isotopes in the hydride transfer reaction. The quantum corrected ratios H/D KIEs are shown in Fig. 3.13 [57, 60].

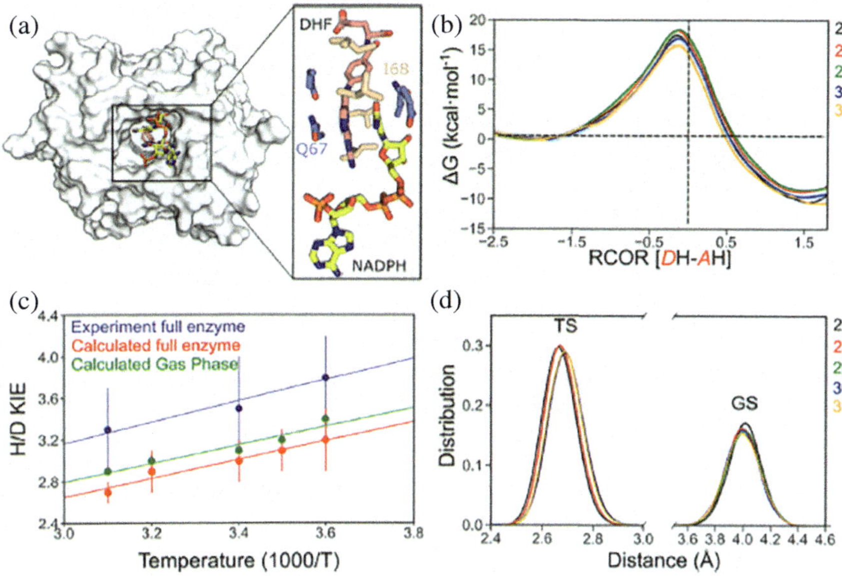

Fig. 3.13 **a** Schematic representation of R67 DHFR shown as surface and ligands shown as sticks in the active site pore. A subset of the active site is shown in the enlarged version of the ligands and some active site residues. **b** Classical free-energy profile for the hydride transfer reaction catalyzed by R67 DHFR at different temperatures. The error bars are shown as highlighted regions around the solid lines. **c** Computed H/D**KIEs** for the R67 catalyzed hydride transfer reaction. Red and green points represent calculated KIEs in the enzyme and gas phase (with AM1-SRP parameterization), respectively, while blue points are experimental KIEs. **d** Distribution of distances calculated between donor and acceptor atoms from ground state (GS) and TS trajectories. Colors: black, red, green, blue, and orange represent energies at temperatures 278, 288, 298, 308, and 318 K, respectively. The color-temperature notation also applies to **b** [57, 60]. Published under Creative Commons public license, CC-BY from American Chemical Society

3.5 Method of Transition State Analogs [59–89]

In 1946 Pauling introduced idea that lowering of the activation energy in enzyme catalysis, as compare with a chemical reaction, stems from the enzyme's affinity for the transition state exceeding its affinity for in the chemical process [61, 65]. Tight-binding transition state analogues evolved enzymatic geometry to form the transition state (Figs. 3.14 and 3.15). Knowledge of the TS can provide information to design stable TS analogs which can be effective enzyme inhibitors. Such an approach has three important aspects: (1) hinting to chemists about the plausible structure of specific inhibitors for synthesis, (2) using these inhibitors to test working hypothesizes about TS structures, and (3) using these inhibitors for the regulation of enzymatic processes in vitro and in vivo..The transition states for chemical reactions are proposed to have lifetimes near 10^{-13} s, the time for a single bond vibration. Transition state analogues convert the short-lived transition state to a stable thermodynamic state. Analogues of the transition state can bind millions of times more

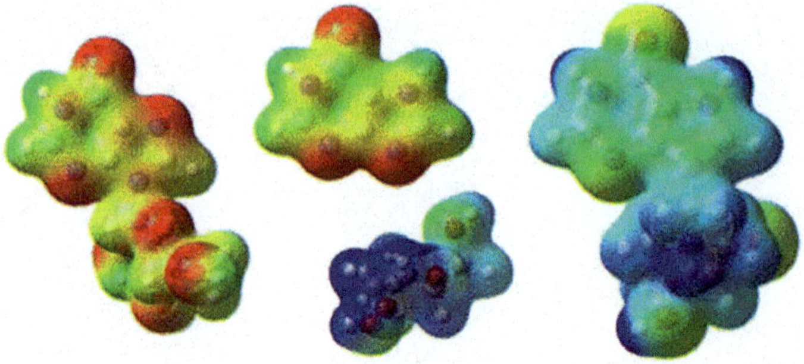

Fig. 3.14 Schematic characterization of principle of the transition state analogs [58, 61]. With permission from American Chemical Society

tightly than substrates and have been used for generating catalytic antibodies that catalyze chemical reactions. If an antibody could bind the transition state tightly like an enzyme, and it should also catalyze the reaction just like an enzyme (Chap. 13). Antigens stable mimics, the transition-state analogs, which have long half-lives and could be synthesized chemically, may function as an antigen to generate the antibody.

The full TS analogue design (TSAD) must include the following steps [62, 66]: (1) Chemical or biochemical synthesis of substrate molecules containing isotopic labels at specific positions; (2) Determination of experimental conditions wherein intrinsic isotope effects, that is, KIEs reflecting only the chemical step –can be measured; (3) Measurement of KIEs for each position labeled in step 1; and (4) Iterative computation of theoretical KIEs from quantum mechanical calculations with model structures until the KIEs match those from experiment in step 3. The following are typical examples of the transition state analogs method application.

Purine nucleoside phosphorylase (PNP) is an enzyme comprising the purine salvage pathway. PNP catalyzes the reversible phosphorolysis of purine ribo- or 2′-deoxyribonucleosides to the purine and ribose- or 2-deoxyribose 1-phosphate [62]. Structures of inosine, the PNP TS and the proposed analogue obtained by interpretation of KIEs for inosine arsenolysis used computer modeling and started with the crystal structure of inosine as the ground state model are presented. The state of the art of combined quantum mechanical/molecular mechanical calculations on enzyme systems and carry out an analysis of the role of preorganization effects on enzyme catalysis were reviewed [63, 66]. The calculation indicated that the free energy cost of aligning the functional groups for catalysis is responsible for a substantial fraction of the $10^{11} - 10^{13}$ faster rates of the enzyme-catalyzed reactions for trypsin and catechol *O*-methyltransferase compared to analogous reference reactions in aqueous solution. Figure 3.15 shows Schematic diagram of a representative enzyme-catalyzed and reference solution reaction.

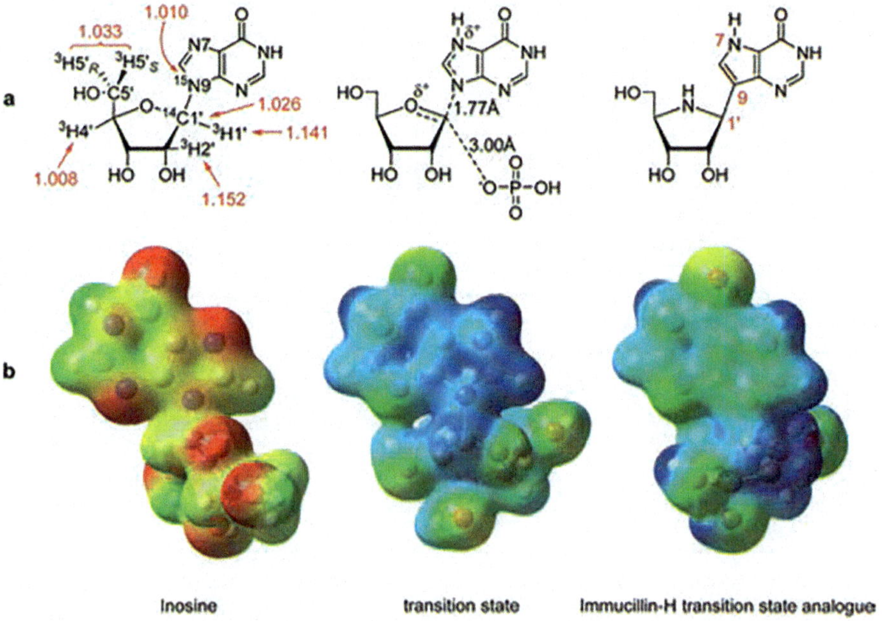

Fig. 3.15 Schematic diagram of a representative enzyme-catalyzed and reference solution reaction in which the preorganization free energy cost is a total of 10 kcal/mol in solution (5 kcal/mol to form ES and a further 5 kcal/mol to form ES(NAC)) and 1.4 kcal/mol in the enzyme. The value from the enzyme is an average of the two values (0.4 and 2.4 kcal/mol) noted in the text. The sum of the quantum mechanical and free energy perturbation difference between the correctly oriented noncovalent complex, ES(NAC), and the transition state, ETS, is 15 kcal/mol in the enzyme and 20 kcal/mol in solution. E = enzyme, S = substrate, ES = noncovalent enzyme − substrate complex, ES(NAC) = noncovalent enzyme − substrate complex [63, 66]. With permission from American Chemical Society

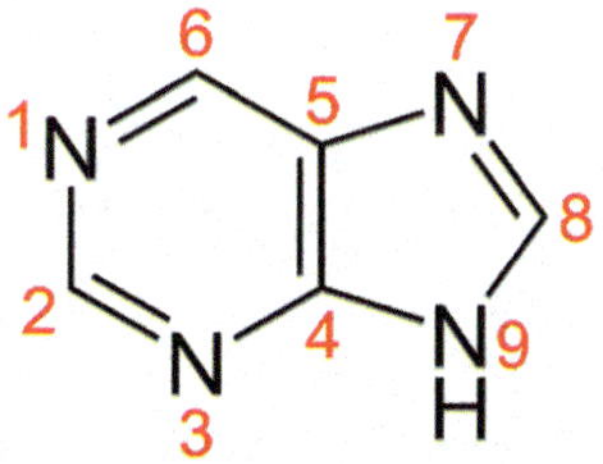

Extrapolating from the acid- and base-catalyzed solvolysis of purine nucleosides, purine nucleoside phosphorylase (PNP) increased the reaction rate by approximately 10^{13} [58, 64]. Steady-state, pre-steady-state, and T-jump studies of PNPs together with structural and computational studies demonstrated a similarity of transition state in the purine nucleoside phosphorylase (PNP) reaction and a transition state analog (Fig. 3.16). It was suggested that the protein architecture remote from the catalytic site can become dynamic when transition state analogues are bound, resulting in

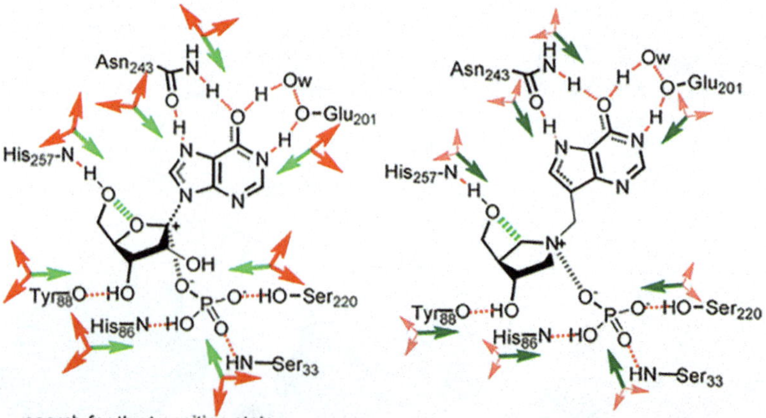

Fig. 3.16 Comparison suggested transition state in the purine nucleoside phosphorylase (PNP) reaction and a transition state analog [58, 64]. With permission from American Chemical Society

favorable or unfavorable entropic contributions to binding. Figure 3.17 shows the distinct time scales for purine nucleoside phosphorylase (PNP), catalysis and transition state analogue binding. Substrate and the pretransition state conformational changes are on the ms to ns time scale.

The enzymes responsible for DNA methylation are the DNA methyltransferases (DNMTs) [64]. DNMTs add a methyl group on the carbon-5 (C5) position of the deoxycytidine at the CpG site in the DNA by using S-adenosyl-L-methionine as methyl donor.

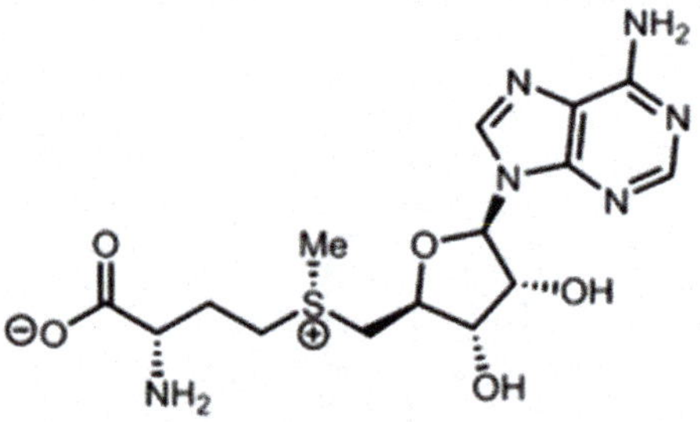

A protected bromomethylcytosine derivative was used to obtain transition state analogues of the DNMTs bearing a 5- methylcytosine scaffold linked to adenosine via an N-alkyl linker. The approach led to seven 5-methylcytosine-adenosine compounds that were inactive against hDNMT1, hDNMT3Acat, TRDMT1, and other RNA human and viral methyltransferases. Two of these compounds showed an inhibitory activity against human protein arginine methyltransferases (PRMT4) in the micromolar range.

Sialidases (SA) are a family of exoglycosidases hydrolyzing sialic acid, a negatively charged a-keto monosaccharide with nine carbon backbone atoms displayed

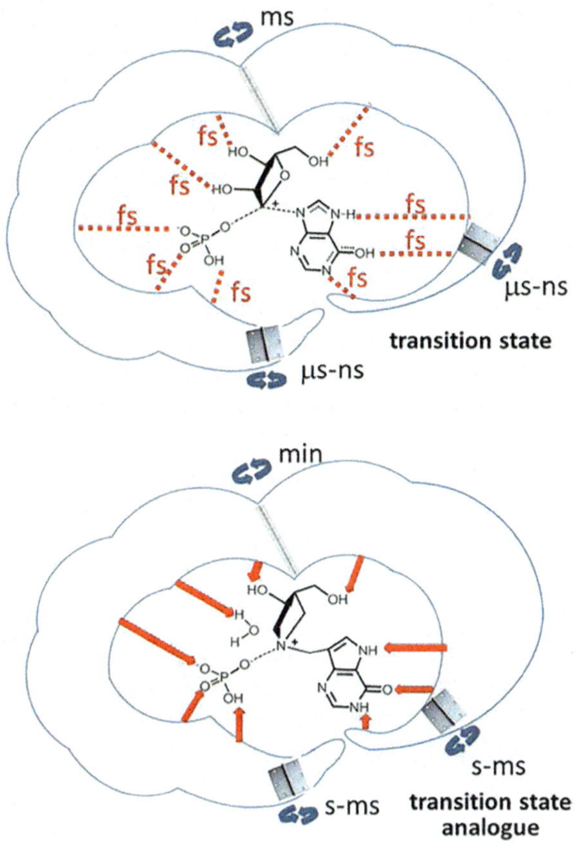

Fig. 3.17 Cartoons of the distinct time scales for purine nucleoside phosphorylase (PNP), catalysis and transition state analogue binding. Substrate and the pretransition state conformational changes are on the ms to ns time scale (upper panel). Within these times, the local catalytic site contacts (red dashed lines) vary their distance on the fs time scale. Optimized fs interactions locate the transition state which exists only for 10 fs. The fs excursions are from a relaxed form of the enzyme. When a transition state analogue is bound (bottom panel), each of the interactions is stabilized (red arrows), the protein condenses around the analogue, and the motions [58, 64]. With permission from American Chemical Society

at the outermost end of glycans, glycoproteins, or glycolipids. In work [65], multi- and polyvalent compounds were synthesized, based on the transition-state analogue 2-deoxy-2,3-didehydro-*N*-acetylneuraminic(**DANA**).

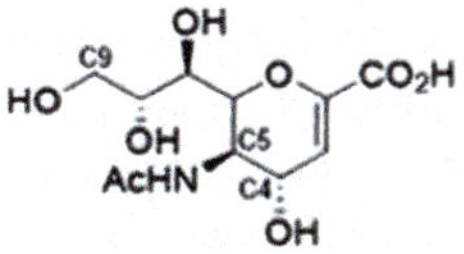

DANA

The following results were reported: (1) The most potent SA inhibitors were obtained by making structural modifications at C-4, C-5, and C-9 positions of DANA, a pan-inhibitor of bacterial SA with Ki in the mm range; (2) Poly-DANA inhibited the catalytic activity of SA from *S. pneumoniae* (NanA) and the symbiotic microorganism *B. thetaiotaomicron* (BtSA) at the picomolar and low nanomolar level: (3)

Each DANA grafted to the polymer surpasses the inhibitory potential of the monovalent analogue by more than four orders of magnitude; (4) The inhibitory capacity and binding interaction of the compounds on the full-length BtSA and NanA and on the truncated NanA catalytic domain (NanA-CAT) was established; and (5) Structural information on ligand binding was obtained from a NanA-CAT co-crystal structure, molecular dynamic simulation on DANANanA complexes, and dynamic light scattering (DLS) experiments. The authors concluded that polymers of the sialyl cation transition-state analogue DANA have a high level of inhibition for a bacterial SA of the GH33 family.

Metal fluoride complexes mimic the transferring phosphoryl group in many enzyme-catalyzed reactions [66, 69]. The trifluoromagnesate transition state analog (TSA) was employed to study a Zika virus NS3h helicase. The enzyme uses energy from ATP hydrolysis to reorganize ssRNA, leading to completion of virus replication. The octahedral trifluoromagnesate core of a transition state complex displays two conformations for a catalytically important loop, demonstrating how ATP hydrolysis can be coupled with RNA translocation. This structure was further probed by quantum mechanical calculations of the catalytic core (200 atoms), confirming the structural data interpretation and the concerted mechanism of ATP hydrolysis by this class of helicase.

A spectroscopic examination of two of six examinated galactonoamidines with inhibition constants and efficacy in the low nanomolar concentration range (K_i = 6–11 nM, IC_{50} = 12–36 nM) were found to be putative transition state analogs for the hydrolysis of *β*-galactosides by *β*-galactosidase [67, 71]. For example, **p-**methylbenzyl galactonoamidine

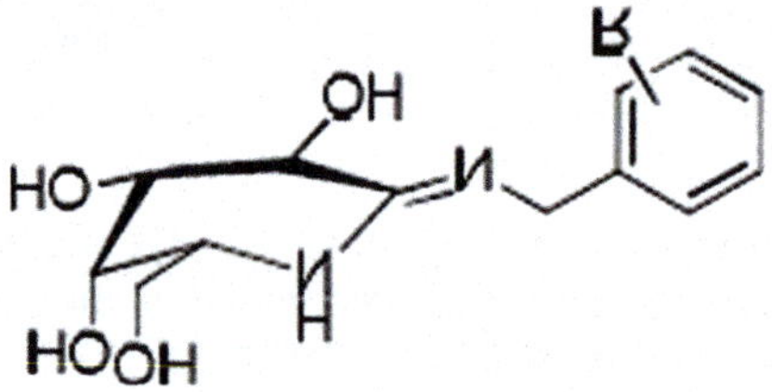

can be characterized as a putative transition state analog. The experimental results elaborated using docking and molecular dynamics studies indicated that the putative transition state analogs (TSAs): (1) have a similar orientation in the active site; (2) show conserved positioning of the glycon ; (3) display a large number of H-bond interactions toward the catalytically active amino acid residues via their glycon a branched glucose polymer; and (4) exhibit hydrophobic interactions at the outer rim of the active site with small changes of the position and orientation of their respective aglycons.

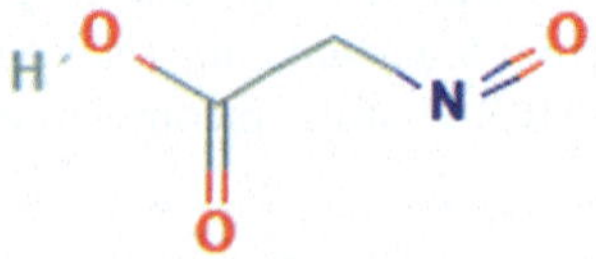

3.6 Theoretical Methods

3.6.1 *Foundation of Theoretical Approaches*

To understand enzyme catalysis mechanism at the molecular level, it is essential to figure out structural or conformational change of enzyme and substrate during catalysis, nature of transition states, the bottlenecks to control reaction kinetics, and the decrease of corresponding free energy barriers due to the catalytic effect of enzymes. A number of computational simulation tools, in particular the combined quantum mechanics and molecular mechanics (QM/MM) method were introduced [90–168]. These methods allow to understand enzyme function based on three-dimensional structures and dynamics of enzyme-substrate complexes.

Theoretical methods of investigation structure of enzyme reaction are based on the Schrodinger equation for relatively small quantum particles such as electrons and nucleus and heavy species motion of which can be described be the Newton lows. The Schrodinger equation gives the positions of the nuclei and the number of electrons in order to yield information such as electron densities, energies and other properties of the system [123, 130]. *Ab initio* quantum chemistry methods were designed to solve the electronic Schrödinger equation [124, 130]. The methods aim is to calculate the many electron function which is the solution the electronic Schrödinger equation in the Born–Oppenheimer approximation. This approximation is the assumption that the electronic motion and the nuclear motion in molecules can be separated and a molecular wave function can be considered in terms of electron positions and nuclear positions. For a complete description of quantum phenomena, the quantum nature of both electrons and nuclei has to be taken into account.

To overcome sophistical problems of the Schrödinger equation solving for molecules, a number of approximations were proposed. The simplest type of *ab initio* electronic structure calculation is the Hartree–Fock (HF) scheme, in which Coulombic electron-electron repulsion effect (mean field) is included in the calculation. [125, 126, 130, 131]. In the Hartree-Fock method, the many electron function is a linear combination of many simpler electron functions with the dominant function being the Hartree–Fock function. The method scales nominally as N (N being a relative measure of the system size, a number of the spin orbitals). A molecular orbital is a mathematical function describing the location and wave-like behavior of an electron in **a** molecule. In the case where the particles are fermions, for example electron with spin 1/2, the Hartree–Fock method can be approximated by a single Slater determinant (SD) [128, 130]. Slater determinant is an expression that describes the wave function of a multi-electronic system. The SD satisfies anti-symmetry requirements, and consequently the Pauli principle, by changing sign upon exchange of two electrons. The Slater determinant arises from the consideration of a wave function for a collection of electrons, each with a wave function known as the spin-orbital where denotes the position and spin of a single electron.

$$\Psi(x_1, x_2) = \chi(\mathbf{x})^1 \cdot \chi(\mathbf{x})^2 \tag{3.12}$$

Molecular dynamics (MD) simulation methods analyze the physical movements of atoms and molecules [129, 133]. The force of attraction tends to pull the particles closer together. To overcome the force of attraction and move apart, the particles need a corresponding kinetic energy. In the most common version of the methods, the trajectories of atoms and molecules are determined by numerically solving Newton's equations of motion for a system of interacting particles. Forces between the particles and their potential energies are calculated using interatomic potentials or molecular mechanics force fields. In fact, Molecular mechanics uses classical mechanics to model molecular systems [130, 131, 136, 137]. The Born–Oppenheimer approximation is assumed valid and the potential energy of all systems is calculated as a function of the nuclear coordinates using force fields. This model involves the development of a simple molecular mechanics energy equation representing the sum of various energy interaction terms comprised of bonds, angles, torsions of bonded, as well as, nonbonded atoms. Molecular mechanics methods are widely used to give accurate structures and energies for molecules.

A force field method [135, 141–143, 149, 150] is used to estimate the forces between atoms within molecules and also between molecules. The force of attraction tends to pull the particles closer together. To overcome the force of attraction and move apart, the particles need a corresponding kinetic energy. The force field refers to the functional form and parameter sets used to calculate the potential energy of a system of atoms or coarse-grained particles in molecular mechanics, molecular dynamics, or Monte Carlo simulations. Coarse-grained modeling aims at simulating the behavior of complex systems using their coarse-grained (simplified) representation. Monte Carlo simulations [144, 151] are used to model the probability of different outcomes in a process that cannot be predicted due to the intervention of random variables. It is a technique used to understand the impact of risk and uncertainty in prediction and forecasting models. Monte Carlo methods are a broad class of computational algorithms that rely on repeated random sampling to obtain numerical results. The underlying concept is to use randomness to solve problems that might be deterministic in principle.

The variational method [145, 152] is approximation to the lowest energy eigenstate or ground state, and some excited states. This allows calculating approximate wavefunctions such as molecular orbitals. A variational principle is one that enables a problem to be solved using calculus of variations, which concerns finding functions that optimize the values of quantities that depend on those functions.

Density functional theory (DFT) is a computational quantum mechanical modeling method used to investigate the electronic and nuclear structure ((principally the ground state) of many-body systems, in particular atoms, molecules, and the condensed phases. Using density functional theory (DFT) theory [148] the properties of a many electron system can be determined by employing functionals, i.e., functions of another function. In DFT the functional is the electron density which is a function of space and time. In DFT the key variable is the electron density $n(\mathbf{r})$, normalized. For a given ground-state density $n_0(\mathbf{r})$ it is possible, in principle, to calculate the corresponding ground-state wavefunction $\Psi_0(\mathbf{r}_1, \ldots, \mathbf{r}_N)$ [154].

Umbrella sampling is a technique in computational physics and chemistry, used to improve sampling of a system where ergodicity is hindered by the form of the system's energy landscape. In mathematics, ergodicity expresses the idea that a point of a moving system, either a dynamical system or a stochastic process, will eventually visit all parts of the space that the system moves in, in a uniform and random sense [148, 150]. In umbrella sampling approach, the standard Boltzmann weighting for Monte Carlo sampling is replaced by a potential chosen to cancel the influence of the energy barrier present.

3.6.2 Combined Quantum Mechanics/Molecular Mechanics (QM/MM) Methods in Computational Enzymology

Quantum Mechanics (QM) methods are restricted to systems of up to a few hundred atoms. However, the size and conformational complexity of biopolymers calls for methods capable of treating up to several 100,000 atoms and allowing for simulations over time scales of tens of nanoseconds. The combined quantum mechanics and molecular mechanics (QM/MM) method which allow to understand enzyme function based on three-dimensional structures and dynamics of enzyme-substrate complexes obeyed these requirements [132–140, 132–151]. The principle of QM/MM approaches is to treat a small part of the system with a quantum mechanical (QM) method (i.e., a method that describes the electronic structure of molecules) and the rest of the system with a molecular mechanical method (i.e., a method that describes interactions between atoms using a simple potential energy function, a force field (Fig. 3.18) [152, 157]. Thus, the hybrid QM/MM(approach is a molecular simulation method that combines the strengths of ab initio QM calculations (accuracy) and molecular mechanics MM (speed) approaches.

Five general aspects are important in planning a QM/MM calculation on an enzyme [51, 156]: (1) Choice of the QM method; (2) Choice of MM forcefield, (including MM parameters required for the QM region); (3) Partitioning of the system into QM and MM regions, (including treatment of covalent bonds at the boundary);

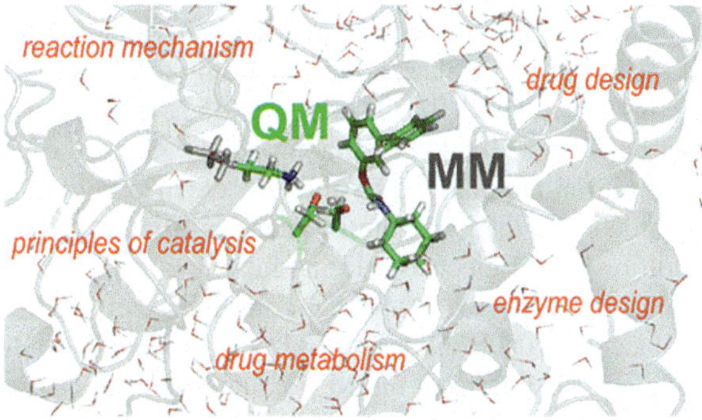

Fig. 3.18 Schematic illustration of the QM MM approach [149, 156]. With permission from American Chemical Society

and (4) type of simulation (e.g., a molecular dynamics simulation, or calculation of potential energy); and (5) construction (and testing) of an accurate molecular model of the enzyme complex. The equation for QM/MM methods consists of 3 terms. In the additive approach

$$E_{\text{total}} = E_{\text{QM(QM)}} + E_{\text{MM(MM)}} + E_{\text{QM-MM}}. \tag{3.13}$$

the energy of the QM region according to the QM method, $E_{\text{MM(MM)}}$ is the energy of the MM region according to the MM method, and $E_{\text{QM-MM}}$ is the interaction energy between the two regions). In the subtractive approach

$$E_{\text{total}} = E_{\text{MM total}} + E_{\text{QM(QM)}} - E_{\text{MM(QM)}}, \tag{3.14}$$

with E_{MM}, total the energy of the total system and $E_{\text{MM(QM)}}$ the energy of the QM region as calculated by the MM method. Specifically, in a subtractive approach, the total QM/MM energy of the system is calculated stepwise [136, 138]: The total energy of the system (QM and MM regions) is evaluated using the MM methodology; (2) The incorporation of the isolated QM energy; (3) The MM energy of the QM subsystem is calculated and subtracted; and (4) Last step corrects the calculations by including the interaction within the QM subsystem. Both approaches allowed electrostatic embedding, which takes into account the electrostatic influence of the MM region on the QM region, i.e., polarization of the QM region by the atomic charges of the MM region.

Electrostatic interactions between the QM and MM regions can be classified as either mechanical embedding, electrostatic embedding or polarized embedding (https://www.ks.uiuc.edu/Research/namd/2.13/ug/node82.html). In the mechanical embedding, the QM region is embedded into the MM region and the interaction between both regions is treated at the MM level and the QM–MM electrostatic interaction is treated on the same footing as the MM–MM electrostatics. The charge model of the MM method, taking in account rigid atomic point charges and bond dipoles, can be applied to the QM region as well. In the electrostatic embedding, the MM charges surrounding the QM region are considered in the Hamiltonian as one-electron operators, and, the electron density of the QM region can be modified by the external field created by the MM charges [151, 153]. Electrostatic embedding also accounts for the interaction of the polarizable QM density with rigid MM charges [151, 153]. In the polarizable embedding the MM region is allowed to be polarized by the interaction with the QM density [152, 154]. These polarized embedding schemes can be further divided into approaches where the polarizable charge model in the MM region is polarized by the QM electric field and the polarizable MM model is included into the QM Hamiltonian and therefore allows for mutual polarization. In addition to the electrostatic interactions, there are also van der Waals and bonded contributions to the QM–MM coupling term [153, 155]. Boundary-atom schemes replace the MM frontier atom M1 by a special boundary atom that participates as an ordinary MM atom in the MM calculation but also carries QM features to saturate the free valency of Q1 [154, 157].

The approach of using a frozen hybrid orbital, that do not participate in the Local Self-Consistent Field **(LSCF),** to saturate the dangling bond at the QM–MM boundary was introduced by Warshel and Levitt [133, 139]. Different schemes based on **the** idea of placing a set of suitably oriented localized orbitals on one of the frontier atoms have been elaborated [125, 146, 157]. LSCF links the quantum mechanical portion of the molecule to the classical portion using a strictly localized bond orbital extracted from a small model molecule for each bond. The LSCF allows quantum electronic computations on a fragment of a large molecule depicted by a classical force field [158, 167]. The bond separating the quantum subsystem from the classical one was assumed to be represented by localized orbitals introduced as data. The molecular orbitals of this subsystem are developed on a basis of functions, orthogonal to the localized orbitals, which derive from the atomic orbitals of the usual basis set by a simple linear transformation. The algorithm of the LSCF method was obtained by a simple modification of the standard Hartree-Fock or DFT codes.

Configuration Interaction Singles (CIS) is a method in QM/MM applications due to its computational expedience that allows for the treatment of several hundred atoms [159, 167]. Configuration interaction (CI) is a post-Hartree–Fock linear variational method for solving the nonrelativistic Schrödinger equation within the Born–Oppenheimer approximation for a quantum chemical multi-electron system. The method allows one to simultaneously solve for a large number of excited states and to optimize the geometry of any selected state. Both spin singlet and spin triplet states can be generated. The techniques of investigation of very large systems proposed in [160, 167] can be used to investigate very large systems. The techniques can be subdivided into two main categories: fragmentation and embedding strategies. The LSCF was enabled to treat chemically relevant regions of large biological molecules through methods of quantum chemistry, while describing the remaining parts of the systems by means of frozen extremely localized molecular orbitals (ELMOs) (Fig. 3.19). ELMOs are molecular orbitals strictly localized on small molecular units and, due to their extreme localization, they can be transferred from molecule to molecule and considered as plausible electronic building blocks [160, 167].

Achievements and challenges in this area have been summarized and discussed in detail in a series of comprehensive reviews [134, 147, 161–164].

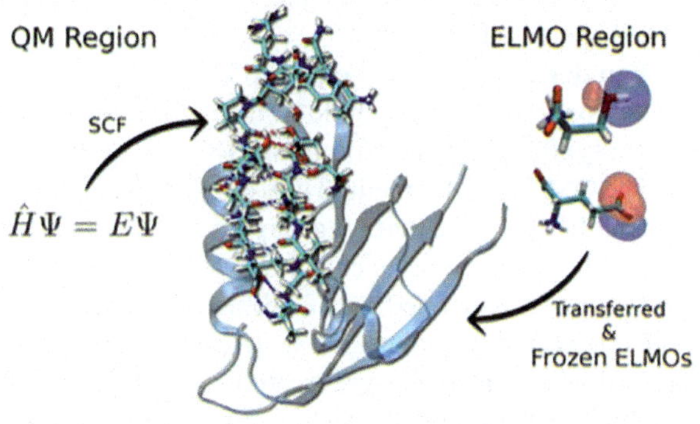

Fig 3.19 Transient frozen extremely localized molecular orbitals (ELMOs) [160, 167] With permission from American Chemical Society Sent

For example, QM/MM methods and their application to enzyme-catalyzed reactions to investigate fundamental and practical problems in enzymology were reviewed. Special emphasis on very general link-atom methods and various ways to treat the charge near the boundary were introduced. Review [162, 169] presented the general methodological aspects of the QM/MM approach, which is used within optimization and simulation techniques. Review [149, 156] discussed how modeling of reactions using QM/MM methods can provide detailed insight into enzyme mechanisms and highlighted applications and can contribute to the practical development and application of enzymology. The current state of the art of QM/MM) approaches in ground and electronically excited states and their applications to biological problems published by 2015 was reviewed [134, 140]. The review [163, 170] summarized the advances that have been made in QM–MM 29 apply approaches to UV–visible, Raman, IR, NMR, electron paramagnetic resonance, and Mössbauer spectroscopies, Recent review [147, 154] discussed methods for the incorporation of quantum mechanical effects into enzyme kinetics simulations in which the enzyme is an explicit part of the model. The following three aspects were emphasized: (1) Use of quantum mechanical electronic structure methods such as molecular orbital theory and density functional theory, usually in conjunction with molecular mechanics; (2) Treating vibrational motions quantum mechanically, either in an instantaneous harmonic approximation, or by path integrals, or by a three-dimensional wave function coupled to classical nuclear motion; and (3) Incorporation of multidimensional tunneling approximations into reaction rate calculations.

3.6.3 Examples Illustrating Theoretical Approaches

Citrate synthase catalyzes the first reaction in the citric acid cycle the conversion of oxaloacetate into citrate. QM/MM modeling has helped elucidate the enzyme mechanism [149, 156]. The reaction features proton abstraction from acetyl-coenzyme A (acetyl-CoA), condensation of acetyl-CoA with oxaloacetate to form citryl-CoA, and hydrolysis of this stable intermediate to form citrate and CoA. This finding indicated that the enolate was energetically favored in the enzyme active site, with conventional hydrogen bonds from His274 and a conserved water molecule providing stabilization of this species (Fig. 3.20).

Van der Kamp MW, Perruccio F, Mulholland, AJ: (2008)) High-level QM/MM modeling predicts an arginine as the acid in the condensation reaction catalyzed by citrate synthase. Chem. Commun., 1874 – 1876.

The QM/MM/MD method was employed to investigate the mechanism of hydride transfer between the methyl group on the Nmethylated lysyl residue and the middle ring of the flavin cofactor catalyzed by Lysine Specific Demethylase [132, 138]. The QM region was defined as a truncated FAD with a structure formed by three rings and a truncated lysine residue with three methyl groups. All the other protein atoms and solvent molecules were treated as MM region. The hydride transfer process, a reaction coordinate that is the difference between the breaking C–H bond length (r1)

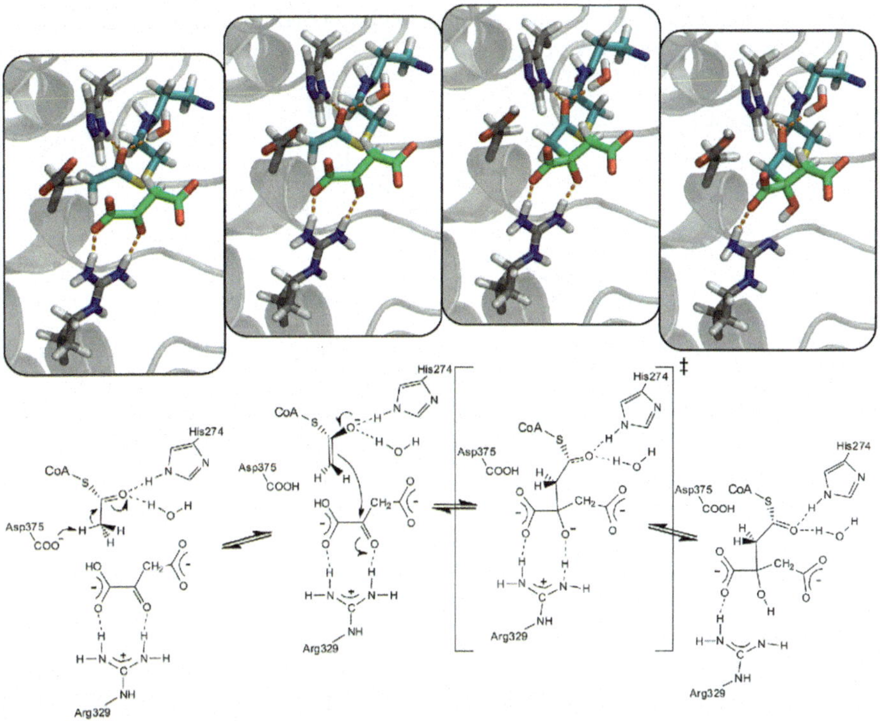

Fig. 3.20 Reaction mechanism for carbon – carbon bond formation in citrate synthase as suggested by high-level QM/MM calculations. Proton transfer to Asp375 leads to an acetyl-CoA enolate intermediate (light blue), which subsequently attacks the carbonyl carbon of oxaloacetate (light green). To arrive at the citryl-CoA intermediate, an unusual proton transfer from Arg329 appears to be required. Structures are obtained by B3LYP/ 6–31 + G(d)-CHARMM27 optimization [149, 156]. With permission from American chemical Society

and the forming N–H bond length (r2) was monitored. Integrating the QM/MM/MD simulation and the umbrella sampling method allowed to calculate the free energy profile, reaction barrier in the proposed hydride transfer mechanism and potential of mean force (PMF). The transition state with the maximum PMF was found near the reaction coordinate with $s = 0.02$ Å, and the corresponding free energy barrier is about 10 kcal/mol. The Suv4-20h2 enzyme acts on histone 4 lysine 20 (H4K20) catalyzing its methylation. The QM/MM MD simulations were applied to the Suv4-20h2-catalyzed reaction for understanding of its preference with monomethylated H4K20 peptide as the substrate [164, 171]. The computation showed that: (1) the free energy barrier for monomethylation is 23.9 kcal/mol), (2) the free energy barrier for the di-methylation (17.9 kcal/mol) is much lower than those of the monomethylation and trimethylation, and (3) the enzyme can catalyze the di-methylation reaction using H4K20me1 substrate much more efficiently compared to the monomethylation and trimethylation.

The enzyme para-hydroxybenzoate hydroxylase (PHBH) catalyzes the transformation of p-hydroxyben [116, 122].

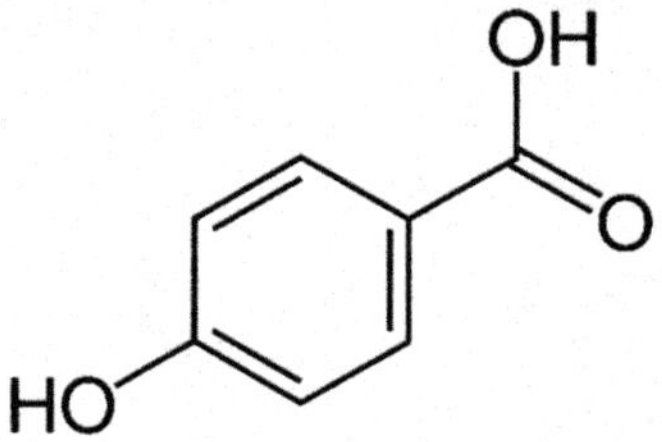

The aromatic hydroxylation of *p*-hydroxybenzoate by the reactive C4a-hydroperoxyflavin cofactor intermediate in PHBH has been studied by a combined ab initio quantum mechanics and molecular mechanics (QM/MM) method. The calculation showed the following results: (1) The calculation started from a model of the C4a-hydroperoxyflavin intermediate in the PHBH reaction cycle, built on the basis of the crystal structure of the enzyme–substrate complex; (2) A pathway for the hydroxylation step was calculated by imposing a reaction coordinate involving cleavage of the peroxide oxygen–oxygen bond and bond formation between the C3 atom of the substrate and the distal oxygen of the peroxide moiety of the cofactor; (3) A QM/MM potential was used in system in which the QM region (49 atoms) was treated at the ab initio HF level with the 3-21G(d) or 6-31G(d) basis sets; (4) Inclusion of electron correlation, applied as B3LYP/6-311+G(d,p) and LMP2/6-31+G(d) single point energy corrections to the *ab initio* QM/MM structures, was shown to be essential to obtain barriers in agreement with the experimental rate constant; (5) Electrophilic aromatic substitution was suggested as the mechanism of this rate-limiting step in the PHBH catalyzed reaction cycle; (6) The polarization of the QM region by the enzyme has been investigated; (7) A potentially important catalytic interaction between the reacting OH group in the transition state (formally OH^{+}) and the backbone carbonyl of the Pro293 residue was identified; and (8) This interaction is calculated to lower the barrier by a catalytically significant 2–3 kcal/mol, corresponding to a 100-fold rate enhancement.

The cytochrome P450 enzymes catalyze many types of oxidation reactions, often with high regio- and stereoselectivity, including hydroxylation, epoxidation, and heteroatom oxidation. A comprehensive review [167, 171] covers the extensive theoretical work on these enzymes until 2004, in particular DFT model studies and early QM/MM studies. A work [151, 158] focused on QM/MM investigations of cytochrome P450cam that catalyzes the stereospecific hydroxylation of the non-activated C5 Hexo bond in camphor under physiological conditions. All the intermediates in the catalytic cycle of cytochrome P450cam have been characterized through DFT/MM calculations. The assignment of the spin states of these intermediates was performed. The computation indicated that their relative energies are strongly dependent on the admixture of Hartree–Fock exchange in the chosen density functional which are tuned by the interactions with the protein environment. In addition, the

DFT/MM studies of the P450cam reactions have been reported for the protonation steps and the hydroxylation steps [168, 169].

References

1. Doerr A (2019) A trap for intermediate enzyme complexes. Nat Methods 16:217. https://doi.org/10.1038/s41592-019-0339-5
2. Rei Liao JY, van Wijk KJ (2019) Discovery of AAA+ protease substrates through trapping approaches. Trends Biochem Sci 44:528–545
3. Huguenin-Dezot N, Alonzo DA, Heberlig GW et al. (2019) Trapping biosynthetic acyl-enzyme intermediates with encoded 2,3-diaminopropionic acid. Nature 565(7737):112–117. https://doi.org/10.1038/s41586-018-0781-z
4. Tang S, Beattie AT, Kafkova L etal. (2022) Mechanism-based traps enable protease and hydrolase substrate discovery. Nature 602:701–707. https://doi.org/10.1038/s41586-022-04414-91819
5. Isfandbod M, Martínez-Pañeda E (2021) A mechanism-based multi-trap phase field model for hydrogen assisted fracture. Int J Plasticity 144:103044
6. Tang S, Beattie AT, Kafkova L et al. (2022) Mechanism-based traps enable protease and hydrolase substrate discovery. Nature 602:701–707. https://doi.org/10.1038/s41586-022-04414-918,19.7
7. Priestley ND, Floss HG, Froland W, Lipscomb JD et al (1992) Cryptic stereospecificity of of methane monooxygenase. J Am Chem Soc 114:7561
8. Hartwell K, Grudpan K (2012) Flow-based systems for rapid and high-precision enzyme kinetics studies. J Anal Methods in Chem 450716:10. https://doi.org/10.1155/2012/450716x
9. Holliday GL, Mitchell, JM, Thornton JM (2009) Understanding the functional roles of amino acid residues I JBOn enzyme catalysis. J Molecul Biol 390(3):560–577
10. Kauler B (1999) Flow design for embedded systems. 2nd ed. R&D Books/Miller Freeman. ISBN 978-0-87930-555-0
11. Mervartova K, Polasek M, Mart J, Calatayud MJ (2007) Recent applications of flow-injection and sequential-injection analysis techniques to chemiluminescence determination of pharmaceuticals. J Pharm Biomed Anal 45(3):367–381
12. Kolev S, McKelvie I (2008) Advances in flow injection analysis and related techniques. Elsevier, New York, NY, USA
13. Vahl K, Kahlert H, Bottcher D et al (2008) A potential high-throughput method for the determination of lipase activity by potentiometric flow injection titrations. Anal Chim Acta 610(1):44–49
14. Ogonczyk D, Koncki R (2007) Potentiometric flow-injection' system for determination of alkaline phosphatase in human serum. Anal Chim Acta 600(1–2):194–198
15. Hungerford JM, Christian GD, Ruzicka J, Giddings JC (1985) Reaction rate measurement by flow injection analysis using the gradient stopped-flow method. Anal Chem 57(9):1794–1798
16. Faller B, Cadene V, Bieth JG (1993) Demonstration of a two- 'step reaction mechanism for the inhibition of heparin-bound neutrophil elastase by α1-proteinase inhibitor. Biochemistry 32(35):9230–9235
17. Liang PH, Kohen A, Baasov T, Anderson KS (1997) Catalytic mechanism of Kdo8P synthase. Pre-steady-state kinetic analysis using rapid chemical quench flow methods. Bioorganic and Med Chem Lett 7(19):2463–2468
18. Vida ACF, Sasaki MK et al. (2011) Zone trapping/merging zones in flow analysis: a novel approach for rapid assays involving relatively slow chemical reactions. Talanta 85(1):259–263
19. Gross JW, Frey PA (2002) Rapid mix-quench MALDI-TOF mass spectrometry for analysis of enzymatic systems. Methods Enzymol 354:27–49

20. Ponhong K, Hartwell SK, Grudpan K (2011) Sequential injection lab-at-valve (SI-LAV) segmented flow system for kinetic study of an enzyme. Talanta 85(1):804–808. https://doi.org/10.1016/j.talanta.2011.03.086
21. Kohse S, Neubauer A, Pazidis A, Lochbrunner S, Kragl U (2013) Photoswitching of enzyme activity by laser-induced pH-jump. J Am Chem Soc 135(25):9407–9411
22. Nunes RMD, Pineiro M, Arnaut LG (2009) Photoacid for extremely long-lived and reversible pH-jumps. J Am Chem Soc 131:9456–9462
23. Tian N, Sun W, Feng Y (2019) Chloromethyl-modified Ru(II) complexes enabling large pH jumps at low concentrations through photoinduced hydrolysis. Chem Sci 10:9949–9953. https://pubs.rsc.org/en/content/articlehtml/2019/sc/c9sc03957k-fn1
24. Donten ML, Hamm P, VandeVondele J (2011) A consistent picture of the proton release mechanism of oNBA in water by ultrafast spectroscopy and ab initio molecular dynamics. J Phys Chem B 115:1075–1083
25. Urbanke C, Wray J (2001) A fluorescence temperature-jump study of conformational transitions in myosin subfragment. Biochem J 358(Pt 1):165–173. https://doi.org/10.1042/0264-6021:3580165
26. Malnasi A, Pearson DS, Kovács M et al (2001) Kinetic resolution of a conformational transition and the ATP hydrolysis step using relaxation methods with a dictyostelium myosin II mutant containing a single tryptophan residue. Biochemistry 40(42):12727–12737. https://doi.org/10.1021/bi010963q
27. Thompson MC, Barad BA, Wolff AM et al (2019) Temperature-jump solution X-ray scattering reveals distinct motions in a dynamic enzyme. Nat Chem 11:1058–1066. https://doi.org/10.1038/s41557-019-0329-3
28. Kintses B, Simon Z, Gyimesi M (2006) Enzyme kinetics above denaturation temperature: a temperature-jump/stopped-flow apparatus. Biophysical J. https://doi.org/10.1529/BIOPHYSJ.106.092833
29. Świderek K, Ruiz-Pernía J, Vicent Moliner V, IñakiTuñón: Heavy enzymes — experimental and computational insights in enzyme dynamics. Current Opinion in Chem Biol 21:11–18
30. Swain CG, Stiver EC, Reuwer JF, Schaad LJ (1958) Use of hydrogen effects to identify attacking nucleophile in enolization of ketons. J Am Chem Soc 80:5885–5893
31. Northrop DB, Cho Y-K (2000) Effect of high pressure on solvent isotope effects of yeast alcohol dehydrogenase. Biophys J 79:1621–1628
32. Harris ME, Vernon Anderson V (2017) Measurement and analysis of kinetic isotope effects: 596. Methods in Enzymology. Elsevier Science Publishing Co Inc
33. Roston D, Islam Z, Kohen A (2013) Isotope effects as probes for enzyme catalyzed hydrogen-transfer reactions molecules. 18:5543–5567. https://doi.org/10.3390/molecules18055543
34. Fernandez PL, Murkin AS (2020) Inverse solvent isotope effects in enzyme-catalyzed reactions. Molecules 25(8):1933. https://doi.org/10.3390/molecules25081933
35. Schowen BK, Schowen RL (2005) Solvent isotope effects on enzyme systems. In: Methods Enzymol; Purich DL (ed) Academic Press, Cambridge, MA, USA, vol 87. pp 551–606 (1982); Quinn D (2005) Theory and practice of solvent isotope effects. In: Kohen A, Limbach H-H (eds) Isotope effects in chemistry and biology. CRC Press, Boca Raton, FL, USA, pp 995–1018
36. Bigeleisen J (1949) The relative reaction velocities of isotopic molecules. J Chem Phys 17:675–678
37. Marcus RA, Sutin N (1985) Electron transfers in chemistry and biology. Biochim Biophys Acta 811:265–322
38. Levich VG, Dogonadze R, German E, Kuznetsov AM, Kharkats YuI (1970) Theory of homogeneous reactions involving proton transfer. Electrochem Acta 15:353–368
39. Stratton CF, Poulin MD, Schramm VL (2017) Binding isotope effects for interrogating enzyme-substrate interactions. Methods Enzymol 596:1–21. https://doi.org/10.1016/bs.mie.2017.07.019
40. Swiderek K, Paneth P (2017) Binding isotope effects. Chem Rev 113:7851–7879
41. Anderson VE (2017) Quantifying energetic contributions to ground state destabilization. Arch Biochem Biophys 433:27–33

42. Cleland WW, Northrop DB (1999) Energetics of substrate binding, catalysis, and product release. In: Schramm VL, Purich DL (eds) Methods in enzymology **308**, Enzyme Kinetics and Mechanism, Part E. Academic Press, San Diego, pp 3−27
43. Schramm VL (2001) Atomic motion in enzymatic reaction coordinates. Curr Opin Struct Biol 11:657–665
44. Liuni P, Olkhov-Mitsel E, Orellana A (2013) DW: measuring kinetic isotope effects in enzyme reactions using time-resolved electrospray mass spectrometry. Anal Chem 85(7):3758–3764
45. Francis K, Kohen A (2013) Standards for the reporting of kinetic isotope effects in enzymology. Perspect Sci 1(1–6):110–120
46. Millard P, Portais J-C, Mendes P (2015) Impact of kinetic isotope effects in isotopic studies of metabolic systems. BMC Syst Biol 9(64)
47. Brookes JC (2017) Quantum effects in biology: golden rule in enzymes, olfaction, photosynthesis and magnetodetection. Proc Math Phys Eng Sci 473(2201). https://doi.org/10.1098/rspa.2016.0822
48. Hay S (2013) Extracting kinetic isotope effects from a global analysis of reaction progress curves. Methods Enzymol 596:85–111
49. Guengerich FP (2017) Kinetic deuterium isotope effects in cytochrome P450 reactions methods enzymol. 596:217–238
50. Kholodar SA, Ghosh A, Kohen A (2017) Measurement of enzyme isotope effects. Methods Enzymol 596:43–83
51. Fitzpatrick PF (2015) Combining solvent isotope effects with substrate isotope effects in mechanistic studies of alcohol and amine oxidation by enzymes*. Biochim Biophys Acta 1854(11):1746–1755
52. Ortiz de Montellano PR (2007) Substrate oxidation. In: Ortiz de Montellano PR (ed) Cytochrome P450: structure, mechanism, and biochemistry. New York, Springer, pp 111–176
53. Quinn DM (2006) Theory and practice of solvent isotope effects. In: Kohen A, Limbach HH (eds) Isotope effects in chemistry and biology. Taylor and Francis, Place Published, pp 995–1018
54. Northrop DB (1982) Deuterium and tritium kinetic isotope effects on initial rates. Methods Enzymol 87:607–625
55. Banerjee R, Komor AJ, Lipscomb JD (2017) Use of isotopes and isotope effects for investigations of diiron oxygenase mechanisms. Methods Enzymol 596:239–290. https://doi.org/10.1016/bs.mie.2017.07.016
56. Szatkowski L, Manna RN, Grzybkowska A (2017) Measurement and prediction of chlorine kinetic isotope effects in enzymatic. Systems Methods Enzymol 596:179−215
57. Westaway KC, Koerner T, Fang Y-R, Rudzinski J, Paneth P (1998) A new method of determining chlorine kinetic isoope effects. Anal Chem 70:3548–3552
58. Behiry EM, Ruiz-Pernia JJ, Luk L et al. (2018) Substitution of promiscuous alcohol dehydrogenase reveals the origin of substrate preference in the transition state. Ang Chem 130(12):3182–3185
59. Zoi I, Javier Suarez J, Antoniou D (2016) Modulating enzyme catalysis through mutations designed to alter rapid protein dynamics. J Am Chem Soc 138(10):3403–3409
60. Mhashal AR, Major DT (2021) Temperature-dependent kinetic isotope effects in R67 dihydrofolate reductase from path-integral simulations. J Phys Chem B 125(5):1369–1377
61. Schramm VL* (2013) Transition states, analogues, and drug development. ACS Chem iol 8(1):71–81
62. Schramm VL (2015) Transition states and transition state analogue interactions with enzymes. Acc Chem Res 48(4):1032–1039
63. Silverman B (2004) The organic chemistry of drug design and drug action. San Diego, CA, Elsevier Academic Press. ISBN 0-12-643732-7
64. Schramm VL (2011) Enzymatic transition states, transition-state analogs, dynamics, thermodynamics, and lifetimes. Annu Rev Biochem 80(1):703–732. https://doi.org/10.1146/annurev-biochem-061809 100742. PMC 5502542

65. Pauling L (1948) Nature of forces between large molecules of biological interest. Nature 161:707–709. https://doi.org/10.1038/161707a0
66. Evans GB, Schramm VI (2018) Tyler PC[a] The transition to magic bullets—transition state analogue drug design. Med Chem Commun 9:1983–1993
67. Kollman P, Kuhn B, Peräkylä M (2002) Computational studies of enzyme-catalyzed reactions: where are we in predicting mechanisms and in understanding the nature of enzyme catalysis? J Phys Chem B 106(7):1537–1542. https://doi.org/10.1021/jp012017p
68. Halby L, Marechal N, Pechalrieu1 D, et al. (2018) Hijacking DNA methyltransferase transition state analogues to produce chemical scaffolds for PRMT inhibitors. Hijacking DNA methyltransferase transition state analogues to produce chemical scaffolds for PRMT inhibitors. Phil Trans R Soc B373:20170072
69. Assailly C, Clarisse Bridot C, Saumonneau A et al (2021) Polyvalent transition-state analogues of sialyl substrates strongly inhibit bacterial sialidases. Chemistry 27(9):3142–3150
70. Ge M, Molt Jr RW, Jenkins HT, Blackburn GM, Jin Y, Antson A (2020) Unprecedented Octahedral Trifluoromagnesate MgF3(Wat)–transition state analog reveals the molecular mechanism of ATP hydrolysis by Zika virus helicase. https://chemrxiv.org/engage/chemrxiv/article-details/60c748724c8919ced7ad2f17
71. Pickens JB, Logan G, Mills LG, Wang F, Susanne Striegler S (2018) Evaluating hydrophobic galactonoamidines as transition state analogs for enzymatic β-galactoside hydrolysis. Bioorganic Chem 77:144–151
72. Keinan E (ed) In: Catalytic antibodies. Wiley & Sons
73. Abe A, Kamiya M (2021) A versatile toolbox for investigating biological processes based on quinone methide chemistry: from self-immolative linkers to self-immobilizing agents. Bioorg Med Chem 44:116281. https://doi.org/10.1016/j.bmc.2021.116281
74. Motherwell WB, Bingham M, Six Y (2001) Recent progress in the design and synthesis of artificial enzymes. Tetrahedron 57(22):4663–4686. https://doi.org/10.1016/S0040-4020(01)00288-5
75. Agarwal PK (2005) Role of protein dynamics in reaction rate enhancement by enzymes. J Am Chem Soc 127(43):1524856. https://doi.org/10.1021/ja055251s. PMID 16248667
76. Xu Y, Janda KD (2004) Catalytic antibodies: hapten design strategies and screening methods. Bioorg Med Chem 12(20):5247−5268. 10.1016/j
77. Barrera GJ, Portillo R, Mijares A et al (2009) Immunoglobulin A with protease activity secreted in human milk activates PAR-2 receptors, of intestinal epithelial cells HT-29, and promotes beta-defensin-2 expression. Immunol Lett 123(1):52–59
78. Bar-Even A, Noor E, Savir Y, Liebermeister W, Davidi D, Tawfik DS, Milo R (2011) The moderately efficient enzyme: evolutionary and physicochemical trends shaping enzyme parameters. Biochemistry 50:4402–4410. https://doi.org/10.1021/bi2002289
79. Dubowchik M, Walker MA (1999) Receptor-mediated and enzyme-dependent targeting of cytotoxic anticancer drugs Gene. Pharmacol Ther 83(2):67–123
80. Padiolleau-Lefèvre S, Naya RB, Shahsavarian MA et al (2014) Catalytic antibodies and their applications in biotechnology: state of the art. Biotechnol Lett 36:1369–1379. https://doi.org/10.1007/s10529-014-1503-8
81. Schultz P, Lerner R (2002) Antibody catalysis: completing the circle. Nature 418:485. https://doi.org/10.1038/418485a10.1016/j.bmc.2021.116281
82. Bolon DN, Voigt CA, Mayo SL (2002) De novo design of biocatalysts Curr Opin Chem Biol 6(2):125–129. https://doi.org/10.1016/s1367-5931(02)00303-4
83. Nicholas K (2004) Catalytic metalloantibodies: biology in service of chemistry. southeastern.edu. Retrieved 11 Nov 2019
84. Hanson CV, Nishiyama Y, Sudhir P (2005) Catalytic antibodies and their applications. Current Opinion in Biotechnol 16(6):631−636
85. Padiolleau-Lefèvre S, Naya RB, Shahsavarian MA, Alain Friboulet A (2014) Catalytic antibodies and their applications in biotechnology: state of the art. Biotechnol Lett 36:1369–1379. https://doi.org/10.1007/s10529-014-1503-8

86. Yin J, Andryski SE (2003) Structural evidence for substrate strain in antibody catalysis. PNAS 100:856–861
87. Paul S, Volle DJ, Beach CM, Johnson DR, Powell MJ, Massey RJ (1989) Catalytic hydrolysis of vasoactive intestinal peptide by human autoantibody. Science 244:1158–1162. https://doi.org/10.1126/science.2727702S
88. Bowen A, Wear M, Casadevall A (2017) Antibody-mediated catalysis in infection and immunity. Infect Immun 85(9)
89. Holmdah Rl, Nimmerjahn F, Ludwig RJ (2019) Editorial: Autoantibodies. Front Immunol 2019 Sec. Immunological Tolerance and Regulation. https://doi.org/10.3389/fimmu.2019.0048
90. Witte M, Zillikens D, Schmidt E (2018) Diagnosis of autoimmune blistering diseases. Front Med 5:296. https://doi.org/10.3389/fmed.2018.00296
91. Yajima K, Nakamura A, Sugahara A, Takai T (2003) FcgammaRIIB deficiency with Fas mutation is sufficient for the development of systemic autoimmune disease. Eur J Immunol 33:1020–1029. https://doi.org/10.1002/eji.200323794
92. Zong S, Hoffmann C, Mané-Damas M, Molenaar P, Losen M, Martinez-Martinez P (2017) Neuronal surface autoantibodies in neuropsychiatric disorders: are there implications for depression. Front Immunol 8:752. https://doi.org/10.3389/fimmu.2017.00752
93. Kenyon GL (2000) Enzyme, enzyme mechanisms, proteins and aspects of NO chemistry. In: Dale Poulter C (ed) (University of Utah)
94. Tipton CM, Fucile CF, Darce J, Chida A, Ichikawa T, Gregoretti I et al (2015) Diversity, cellular origin and autoreactivity of antibody-secreting cell population expansions in acute systemic lupus erythematosus. Nat Immunol 16:755–765. https://doi.org/10.1038/ni.3175
95. McQuarrie DA (2007) In: Quantum chemistry. 2nd ed. University Science Books. ISBN 978-1891389504
96. Pauling L (1939) In: The nature of the chemical bond and the structure of molecules and crystals: an introduction to modern structural chemistry. 1st ed. Cornell University Press
97. Shaik SS, Hiberty PC (2007) A chemist's guide to (Valence Bond Theory Wiley-Interscience). ISBN 978-0470037355
98. Chapela GA, Scriven LE, Davis HT (1989) Molecular dynamics for discontinuous potential. IV. Lennard-Jonesium. The J Chem Phys 91(7):4307–4313. Bibcode:1989JChPh..91.4307C. https://doi.org/10.1063/1.456811
99. Allinger NL, Burkert U (1997) Molecular mechanics. An American Chemical Society Publication. ISBN 978-0-8412-0885-8
100. Mechanistic Enzymology: Bridging Structure and Function Chapter 6. pp 121−137.https://doi.org/10.1021/bk-2020-1357.ch006
101. Warshel A, Levitt M (1976) Theoretical studies of enzymic reactions: dielectric, electrostatic and steric stabilization of the carbonium ion in the reaction of lysozyme. J Mol Biol 103(2):227–249. https://doi.org/10.1016/0022-2836(76)90311-9
102. Brunk E, Rothlisberger U (2015) Mixed quantum mechanical/molecular mechanical molecular dynamics simulations of biological systems in ground and electronically excited states. Chem Rev 115(12):6217–6263. https://doi.org/10.1021/cr500628b
103. Mieres-Perez J, Sanchez-Garcia E (2009) Quantum mechanics/molecular mechanics multiscale modeling of biomolecules. In: Advances in physical organic chemistry, Angew Chem Int Ed 48:1198–1229 2020
104. Papamichael EM, Gkini OA (2019) Enzyme kinetics and modeling of enzymatic systems. In: Advances in enzyme technology.https://doi.org/10.1007/978-0-387-35141-4_3
105. Sekharan S, Morokuma K (2011) Description of ONIOM (QM/MM) in annual reports in computational chemistry
106. Senn HM, Thiel W (2009) QM/MM methods for biomolecular systems. Angewandte Chemie 48(7):1198–229.https://doi.org/10.1002/anie.200802019
107. Ranaghan KE, Mulholland AJ (2010) xInvestigations of enzyme-catalysed reactions with combined quantum mechanics/molecular mechanics (QM/MM) methods. Int Rev Phys Chem 29:65−133

108. Vreven T, Byun KS, Komaromi I (2006) (Combining quantum mechanics methods with molecular mechanics methods in ONIOM. J Chem Theory Comput 2:815–826
109. Vijayakumar S, Hirsch M, Uritsky N, Shokhen M, Albeck A (2006) EMBM—a new enzyme mechanism-based method for rational design of chemical sites of covalent inhibitors. J Chem Inf Model 50(12):2256–2265
110. Swanson DJ, Creed AS (2014) Sharpening the focus of force field analysis. J Change Managem 14(1):28–47. https://doi.org/10.1080/14697017.2013.788052. S2CID 144716750
111. Mick J, Hailat E, Russo V, Rushaidat K, Schwiebert L, Potoff J (2013) GPU-accelerated gibbs ensemble Monte Carlo simulations of Lennard-Jonesium. Comput Phys Commun 184(12):2662–2669
112. Jurečka P, Černý J, Hobza P, Salahub DR (2006) Density functional theory augmented with an empirical dispersion term. Interaction energies and geometries of 80 noncovalent complexes compared with ab initio quantum mechanics calculations. J Comput Chem 28:555–569. https://doi.org/10.1002/jcc.20570.PMID17186489
113. Yu T, Guo H (2020) Understanding enzyme catalysis mechanism using QM/MM simulation methods. In: Mechanistic enzymology: bridging structure and function Chapter 6. pp 121−137. https://doi.org/10.1021/bk-2020-1357.ch006
114. van der Kamp MW, Mulholland AJ (2006) Combined quantum mechanics/molecular mechanics (QM/MM) methods in computational enzymology. Biochemistry 52(16):2708–2728
115. Macetti G, Genoni A (2019) Quantum mechanics/extremely localized molecular orbital method: a fully quantum mechanical embedding approach for macromolecules. J Phys Chem A 123(43):9420–9428
116. Morzan UN, Alonso de Armiño DJ, Foglia NO, Ramírez F, González Lebrero MC, Scherlis DA, Estrin DA (2018) Spectroscopy in complex environments from QM-MM simulations. Chem Rev 118(7):4071−4113. https://doi.org/10.1021/acs.chemrev.8b00026
117. Macetti G, Genoni A (2019) Quantum mechanics/extremely localized molecular orbital method: a fully quantum mechanical embedding approach for macromolecules. J Phys Chem A. https://doi.org/10.1021/acs.jpca.9b08882.hal-02324013
118. Gao J, Truhlar DJ (2020) Quantum mechanics/molecular mechanics multiscale modeling of biomolecules. Adv Phys Org Chem 54:144–182
119. Qian P, Guo HB, Wang L, Guo H (2017) QM/MM investigation of substrate and product specificities of Suv4-20h2: how does this enzyme generate dimethylated H4K20 from monomethylated substrate? J Chem Theory Comput 13(6):2977–2986
120. Del Rizzo PA, Trievel RC (2014) Molecular basis for substrate recognition by lysine methyltransferases and demethylases. Biochim Biophys Acta Gene Regul Mech 1839(12):1404–1415
121. Cortopassi WA, Kumar K, Duarte F, Pimentel AS, Paton RS (2016) Mechanisms of histone lysine-modifying enzymes: a computational perspective on the role of the protein environment. J Mol Graph Model 67:69–84
122. Ridder L, Mulholland AJ, Rietjens IM, Vervoort J (1999) Combined quantum mechanical and molecular mechanical reaction pathway calculation for aromatic hydroxylation by p-hydroxybenzoate-3-hydroxylase. Mol Graph Model ч 17(3–4):163–175
123. Guimares CRW, Udier-Blagovic M, Tubert-Brohman I, Jorgensen WL (2005) Contributions of conformational compression and preferential transition state stabilization to the rate enhancement by chorismate mutase. J Chem Theory Comput 1:617–625
124. Shaik S, Kumar D, de Visser SP, Altun A, Thiel W (2005) Theoretical perspective on the structure and mechanism of cytochrome P450 enzymes. Chem Rev 105:2279–2328. https://doi.org/10.1021/cr030722j
125. Senn HM , Walter Thiel W (2009) QM/MM methods for biomolecular systems. Angew Chem Int Ed Eng 48(7):1198−1229
126. Wang D, Zheng J, Shaik S, Thiel W (2008) J Phys Chem B 112:5126–5138
127. Dubey KD, Wangm B, Shaik S (2016) Molecular dynamics and QM/MM calculations predict the substrate-induced gating of cytochrome P450 BM3 and the regio- and stereoselectivity of fatty acid hydroxylation. J Am Chem Socx 138(3):837–845

128. Lonsdale R, Harvey JN, Mulholland AJ (2012) A practical guide to modelling enzyme-catalysed reactions. Chem Soc Rev 41:3025–3038
129. Lonsdale R,[†] Kerensa T, Houghton KT et al. (2013) Quantum mechanics/molecular mechanics modeling of regioselectivity of drug metabolism in cytochrome P450 2C9. J Am Chem Soc 135(21):8001–8015
130. Levine IN (1991) Quantum chemistry. Englewood Cliffs, New Jersey, Prentice Hall, pp 455–544. ISBN 978-0-205-12770-2
131. Hartree DR (1930) The wave mechanics of an atom with a non-coulomb central field. Math Proc Camb Philos Soc 24(1):111
132. Fock VA (1930) Näherungsmethode zur Lösung des quantenmechanischen Mehrkörperproblems. Z Phys 61(1):126–148
133. Fischer CF (1987) Charlotte "General Hartree-Fock program." Comput Phys Commun 43(3):355–365
134. Slater JC (1928) The self consistent field and the structure of atoms. Phys Rev 32(3):339–348
135. Frenkel D, Smit B (2002) In: Understanding molecular simulation: from algorithms to applications. 2nd edn. Academic Press, San Diego
136. Kollman PA, Massova I, Reyes C et al (2000) Calculating structures and free energies of complex molecules: combining molecular mechanics and continuum models. Acc Chem Res 33(12):889–897
137. Vanommeslaeghe K, Guvench O, MacKerell AD (2014) Molecular mechanics. Curr Pharm Des 20(20):3281–3292.https://doi.org/10.2174/13816128113199990600
138. Yu T, Guo H (2020) Understanding enzyme catalysis mechanism using QM/MM simulation methods in mechanistic enzymology: bridging structure and function Chapter 6 121–137. https://doi.org/10.1021/bk-2020-1357.ch006
139. Warshel A, Levitt M (1976) Theoretical studies of enzymic reactions: dielectric, electrostatic and steric stabilization of the carbonium ion in the reaction of lysozyme. J Molecul Biol 103(2):227–249. https://doi.org/10.1016/0022-2836(76)90311-9. PMID 985660
140. Brunk E, Rothlisberger U (2015) Mixed quantum mechanical/molecular mechanical molecular dynamics simulations of biological systems in ground and electronically excited states. Chem Rev 115(12):6217–6263. https://doi.org/10.1021/cr500628b. PMID 25880693
141. Mieres-Perez J, Sanchez-Garcia E (2020) Quantum mechanics/molecular mechanics multiscale modeling of biomolecules. In: Advances in physical organic chemistry. Angew Chem Int Ed 48:1198–1229
142. Papamichael EM, Gkini OA (2019) Enzyme kinetics and modeling of enzymatic systems. In: Advances in enzyme technology.https://doi.org/10.1007/978-0-387-35141-35243
143. Sekharan S, Morokuma K (2011) Description of ONIOM (QM/MM) in annual reports in computational chemistry. pp 132
144. Ranaghan KE, Mulholland AJ (2010) Investigations of enzyme-catalysed reactions with combined quantum mechanics/molecular mechanics (QM/MM) methods. Int Rev Phys Chem 29:65–133
145. Chung LW, Sameera WMC, Ramozzi R (2015) The ONIOM method and its applications. Chem Rev 115(12):5678–5796
146. Senn HM, Thiel W (2007) QM/MM methods for biological systems. Atomistic Approaches. Mod Biol From Quantum Chem Mol Simul 268:173−290
147. Vreven T, Byun KS, Komaromi I et al (2006) Combining quantum mechanics methods with molecular mechanics methods in ONIOM. J Chem Theory Comput 2:815–826
148. Senn HM, Thiel W (2009) QM/MM methods for biomolecular systems. Angewandte Chemie 48(7):1198–1229. https://doi.org/10.1002/anie.200802019. PMID 1917332819
149. Heinz H, Lin TJ, Mishra RK, Emami FS (2013) Thermodynamically consistent force fields for the assembly of inorganic, organic, and biological nanostructures: the interface force field. Langmuir 29(6):1754–1765
150. Leach A (2001-01-30) In: Molecular modelling: principles and applications. 2nd ed. Harlow, Prentice Hall

151. Kroese DP, Brereton T, Taimre T, Botev ZI (2014) Why the Monte Carlo method is so important today. WIREs Comput Stat 6(6):386–392. https://doi.org/10.1002/wics.1314.S2CID1852 1840
152. Nesbet RK (2003) Variational principles and methods in theoretical physics and chemistry. Cambridge U.P, New York
153. Medvedev MG, Bushmarinov IS, Jianwei S, Perdew JP, Lyssenko KA (2017) Density functional theory is straying from the path toward the exact functional. Science 355(6320):49–52
154. Gao J, Truhlar DG (2020) Quantum mechanics/molecular mechanics multiscale modeling of biomolecules. Adv Phys Org Chem 54:144−182
155. Torrie GM, Valleau, JP (1977) Nonphysical sampling distributions in Monte Carlo free-energy estimation: umbrella sampling. J Comput Phys 23(2):187–199. Bibcode: 1977 JCoPh..23..187T. https://doi.org/10.1016/0021-9991(77)90121-8
156. van der Kamp MW*, Mulholland AJ* (2013) Combined quantum mechanics/molecular mechanics (QM/MM) methods in computational enzymology. Biochemistry 52(16):2708–2728
157. Bonfrate S, Nicolas Ferré N, Miquel Huix-Rotllant M (2023) An efficient electrostatic embedding QM/MM method using periodic boundary conditions based on particle-mesh Ewald sums and electrostatic potential fitted charge operators. J Chem Phys 158:021101. https://doi.org/10.1063/5.0133646
158. Zinovjev K (2023) Electrostatic embedding of machine learning potentials. J Chem Theory Comput 19(6):1888–1897
159. Huang Z, Silva A, Singh A (2022) Polarized embedding for signed networks. https://arxiv.org/abs/2110.09899
160. He1 Z, Mokhberian N, Câmara A (2021) Detecting polarized topics using partisanship-aware contextualized topic embedding's. https://aclanthology.org/2021.findings-emnlp
161. Riccardi D, Li G, Cui Q (2004) Importance of van der Waals interactions in QM/MM simulation. J Phys Chem B108(20):6467–6478. https://doi.org/10.1021/jp037992q
162. Riccardi D, Li G, Cui Q (2019) Importance of van der Waals interactions in QM/MM simulations. J Phys Chem B 108(20):6467–6478. https://doi.org/10.1021/jp037992q
163. Murphy RB, Philipp DM (2000) Friesner RAr: frozen orbital QM/MM methods for density functional theory. Chem Phys Lett 321(1–2):113–120. https://doi.org/10.1016/S0009-2614(00)00289-X
164. Ferr N, Assfeld X, Rivai JL (2002) Local self-consistent field (LSCF). J Comput Chem 23:610–624
165. Assfeld X, Rivail J-L (1996) Quantum chemical computations on parts of large molecules: the ab initio local self consistent field method. Chem Phys Lettr 263(1–2, 6):100–106
166. Senn HM, Thiel W (2009) QM/MM methods for biomolecular systems. Angew Chem Int Ed 48:1198–1229 9
167. Macetti G, Genoni A (2019) Quantum mechanics/extremely localized molecular orbital method: a fully quantum mechanical embedding approach for macromolecules. J Phys Chem A. https://doi.org/10.1021/acs.jpca.9b08882.hal-02324013, https://hal.univ-lorraine.fr/hal-02324013/document
168. Lin H, Truhlar DG (2007) QM/MM: what have we learned, where are we, and where do we go from here? Theoret Chem Acc 117(2):185–199. https://doi.org/10.1007/s00214-006-0143-z
169. Boulanger E, Harvey JN (2018) QM/MM methods for free energies and photochemistry. Current Opinion in Struct Biol 49:72−76
170. Morzan UN, Alonso de Armiño DJ, Foglia NO et al. (2018) Spectroscopy in complex environments from QM-MM simulations. Chem Rev 118 (7):4071−4113. https://doi.org/10.1021/acs.chemrev.8b00026. PMID 29561145
171. Ridder L, Harvey JN, Rietjens MCM et al. (2003) *Ab Initio* QM/MM modeling of the hydroxylation step in *p*-hydroxybenzoate hydroxylase. J Phys Chem B 107(9):2118–2126. https://doi.org/10.1021/jp026213n
172. Martí S, Moliner V, Tuñón I, Williams IH (2003) [c]QM/MM calculations of kinetic isotope effects in the chorismate mutase active site. Org Biomol Chem 1:483−487

173. Shaik S, Kumar D, Sde Visser SP, Altun A, Thiel W (2005) Theoretical perspective on the structure and mechanism of cytochrome P450 Enzymes†Chem. Rev 2005(105):2279–2328
174. Zurek J, Foloppe N, Harvey JN, Mulholland AJ (2006) QM/MM study of mechanisms for compound I formation in the catalytic cycle of cytochrome P450cam. Org Biomol Chem 4:3931–3937

Chapter 4
Elementary Mechanisms of Enzyme Reactions (Part 1)

Abstract The outstanding catalytic, regulatory, and self-preserving properties of enzymes have been developed over the course of several million years of molecular evolution. Proximity and Oriental Effects are an important part of these mechanisms due to the pre-binding of the substrate by unique orientation relative to the catalytic groups. For example, cyclization of aliphatic molecule provides decrease of entropy of $DS_0 = -3.6 - 4.7$ eu. Difference in rates is 3–4 orders of magnitude (intramolecular versus intermolecular). In a concerted reaction, when a substrate is simultaneously attracted by different active donor and acceptor reagents such as acid and basic groups, nucleophile and electrophile, or reducing and oxidizing agents a significant decrease in the activation energy is expected. General theoretical considerations in favor of synchronous or sequential mechanisms and specific cases are discussed in a separate section. In the case of effective concerted mechanism, the decrease in the synchronization probability with increasing number of atoms or groups, participating in an elementary stage, can be compensated for by an appreciable decrease in the activation energy due to the inclusion of nuclei of donor and acceptor groups in the process. This consideration has led us to the formulation of the principle of optimum motion (POM). In 1975, the author of this monograph has suggested that rapid electron transfer in photosynthetic reaction centers in the forward direction and significantly slower transfer in the reverse direction may account for the cascade structure of RC which provides tunneling (long-distance) mechanism of the photoseparated charges. The discovery and experimental confirmation of Long-Range Electron Transfer (LRET) between donor (D) and acceptor (A) centers in model and biological systems were described. Data on Photosystem I and photosystem II which are the two multi-protein complexes that contain the pigments necessary to harvest photons and use light energy to catalyze the primary photosynthetic endergonic reactions producing high energy compounds were briefly reviewed. Magnetic isotope effects arise when a chemical reaction involves spin-selective processes proceeded via the radical pair mechanism was described.

G. I. Likhtenshtein, *Enzyme Catalysis Today and the Chemistry of the 21st Century*,
https://doi.org/10.1007/978-3-031-51552-1_4

4.1 Proximity and Oriental Effects

To accelerate chemical reactions, enzymes make use of a variety of molecular strategies, often with a remarkable effectiveness [1–27]. Strategy, called the proximity and oriental effects, is to organize the substrates within the active site of the enzyme such that the reactants are much closer together and oriented than they would be in the reaction corresponding solution (Fig. 4.1) [1–7]. These effects of proximity are analogous to an effective increase in concentration of the reagents and endows the reaction with an intramolecular character with a rate increase. Difference in rates can be 3–4 orders of magnitude (intramolecular versus intermolecular (Fig. 4.1). According to Jencks Circe effect a large entropy penalty for the uncatalyzed reaction is eliminated by the enzyme because of tight binding of the substrate [3]. Thus, the catalytic efficiency of enzymes depends on their ability to juxtapose the reacting atoms and to steer their orbitals along a path (orientation effect) (Fig. 4.1). Below the effects mentioned above are illustrated with corresponding examples.

Benson [4, 5] introduced three rules for estimation of decrease of standard entropy (DS_0 in eu) at frozen of mobility molecules in gas face at 25 °C (the entropy unit "e.u." is equal to one calorie per kelvin per mole, or 4.184 joules per kelvin per mole): (1) cyclization of aliphatic molecule decrease of entropy of $DS_0 = -3.6 - 4.7$ eu; (2) introduction of double bond in cycles of 4, 5, and 6 bonds lead to decrease of entropy, $DS_0 = -3.2$ eu, $DS_0 = -5.4$ eu and $DS_0 = -0.6$ eu, respectively; (3) entropies of free rotation of small radicals (fragments) such as CH_3 and OH are

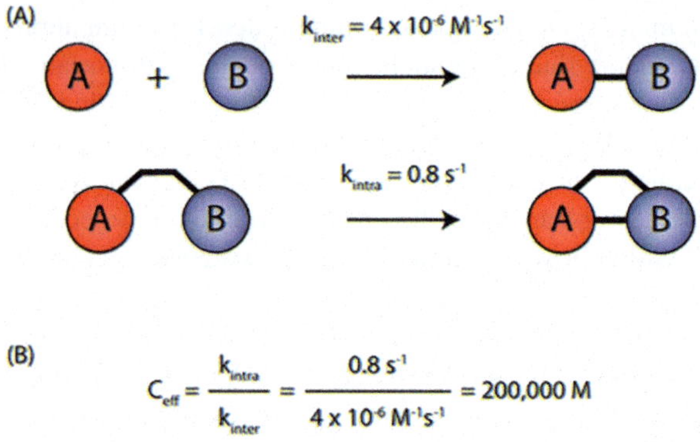

Fig. 4.1 Effective concentration measures the rate enhancement of an intramolecular reaction relative to a corresponding intermolecular reaction **A** Shown are two hypothetical reactions between reactants A and B. The top row shows an intermolecular reaction while the bottom row shows an otherwise equivalent intramolecular reaction. The rate constants of both reactions, k_{inter} and k_{intra}, are indicated. **B** Effective concentration (Ceff) equals the ratio of the intramolecular rate constant over the intermolecular rate constant. In this example Ceff equals 200,000 M, meaning that the reactants in the intermolecular reaction would have to be present at 200,000 M in order to match the rate of the intramolecular reaction when its reactants are present at 1 M. https://www.labxchange.org/library/pathway/lx-pathway:f1580683-16ef-442d-a278-8c5d21813bb2/items/lx-pb:f1580683-16ef-442d-a278-8c5d21813bb2

$DS_0 = -5{,}8$ and 4.6 eu, respectively; (4) for more voluminous fragments DS_0 are in a range from − 8.3–11eu; and (5) having rotation barriers of 2–5 kcal/mol can decrease of DS_0 for 0.5–2 eu. In real systems, the entropy change can be governed by the environment viscosity.

In a typical example of the oriental effect, as a result of alkyl substitution in both aromatic ring and side chain, the rate constant for acid-catalyzed lactonization of hydrocoumaric acid in compound IV

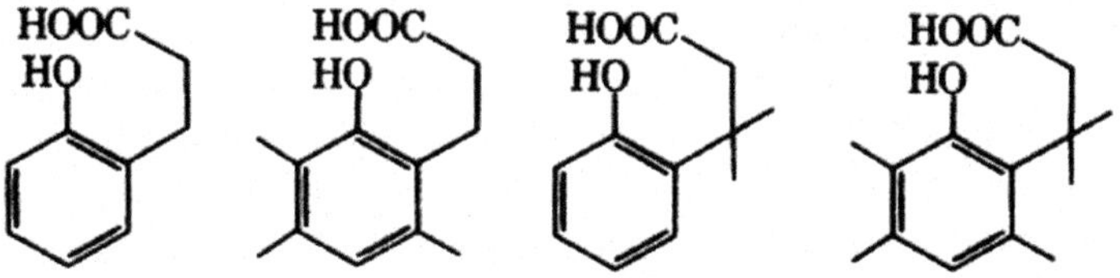

I II III IV

increased by factors as high as 10^{11} and, in comparison with the bimolecular esterification of phenol and acetic acid, by almost 10^{16} [5, 6]. The effect was attributed to a unique interlocking of methyl groups, which produces a severe conformational restriction of the side chain and increases greatly the population of the most productive conformer. To test the sensitivity of a chemical reaction to the orientation of the reacting atoms, the velocities of acid-catalyzed esterification and γ-lactonizations were measured [6, 7]. Variation in orientation of the attacking oxygen atom relative to the carbon atom of the carboxylic acid was achieved by using bicyclic ring systems to limit the conformational mobility of the γ-hydroxy acids

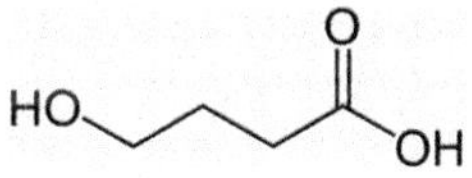

Factors as high as 2×10^4 were observed for the acceleration of a reaction due to this "orientation factor" even after corrections for proximity and torsional strain have been made. The enzymatic reaction of cytidine deaminase is an example of proximity, oriental, and concert effects in enzyme reactions [2]. Computer simulations of the cytidine deaminase reaction and its temperature dependence, using density functional theory calculations and an empirical valence bond description of the reaction, showed that: (1) Substrate binding is associated with a significant entropy loss; (2) the latter closely matches the activation entropy penalty for the uncatalyzed reaction in water; (3) the activation entropy for the rate-limiting catalytic step in the enzyme is close to zero; and (4) the entropic activation parameters for the hydrolytic deamination of cytidine to uracil catalyzed by cytidine deaminase are $T\Delta S^{\ddagger} = 8.6$ kcal/mol showing that the entropic penalty has totally vanished in the enzyme reaction.

4.2 Concerted Chemical Processes

4.2.1 Theoretical Consideration

A concerted (synchronous) reaction is a chemical reaction in which all bond breaking and bond making occur in a single step. In order to explain the high efficiency of many chemical and enzymatic processes, the concept of energetically favorable, concerted mechanisms is widely used [7–14]. In a concerted reaction, a substrate is simultaneously attracted by different active donor and acceptor reagents such as acid and basic groups, nucleophile and electrophile, or reducing and oxidizing agents. It may, however, be presumed that certain kinetic statistic limitation exists on realization of reactions that are accompanied by a change in the configuration of a large number of nuclei [8–14].

A concerted reaction can be modeled as a result of the simultaneous elementary transition (taking approximately 10^{13} s) of a system of n independent oscillators, with the mean displacement of nuclei w_0, from the ground state to the activated state in which this displacement exceeds for each nucleus a certain critical value ($w_{\underline{cr}}$) [8]. The theory gives the following expression for the synchronization factor a_{syn}, which is the ratio of preexponential factors synchronous and regular processes:

$$\alpha_{syn} = \frac{n}{2^{n-1}} \left(\frac{n\mathrm{RT}}{\pi E_{syn}} \right)^{n-1/2}. \tag{4.1}$$

Value of rate constant can be given as

$$k_{syn} = a_{syn}(k_B T/h) \cdot \exp \cdot \left(\mathrm{DS}^{\#}/R - E_{syn}/\mathrm{RT} \right). \tag{4.2}$$

Analysis of Eq. 4.1 and Fig. 4.2 provides a clear idea of the scale of the synchronization factor and dependence of this factor on the number of n and the energy activation (Fig. 4.2) [9]. For example, at moderate energy activation 20–30 kJ/mol, the incorporation of each new nucleus into the transition state can lead to about tenfold decrease in the rate of the process. The model of concerted processes discussed above is only a crude approximation of the motion of a complex system of nuclei along the reaction coordinate. However, such an approximation apparently permits one to estimate a scale of the reaction of interest and to choose from among the possible reaction mechanisms.

The competition between stepwise and concerted (CPET) pathways in certain chemical reactions was discussed on the bases of experimental kinetics and theoretical calculation [9–28].

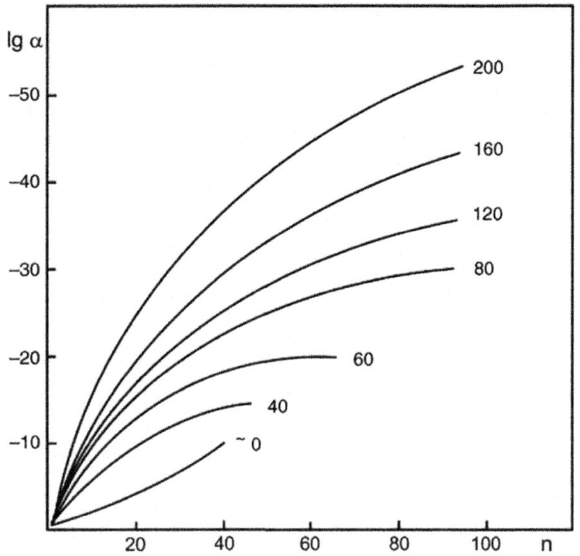

Fig. 4.2 Theoretical dependence of the synchronization factor (a_{syn}) on the number of degree of freedom (n) of nuclei involved in a concerted reaction. The curves have been constructed in accordance with Eqs. 4.11 and 4.12. Energy activation of the synchronous reaction is given in kJ/mol [9]. Published by Kluwer Academic Publisher

4.2.2 Simple Molecules Reactions

The concerted and the stepwise mechanisms of the Diels–Alder reaction between butadiene and ethylene, involving a [4π + 2π] cycloaddition between a diene, and an olefin (dienophile)

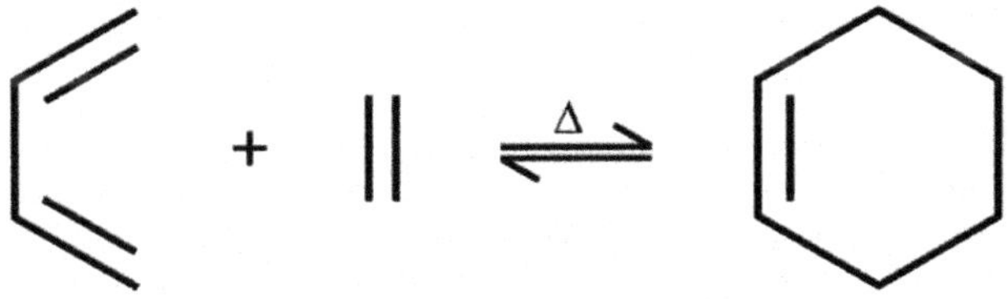

were studied by *ab initio* MO methods [14, 15]. For the concerted mechanism, the energy barrier height of the transition state was found to be 2.7 kcal/mol lower than that of the formation of biradical intermediate with a transition state in the stepwise mechanism. The CiLC–IRC analysis indicated the nature of the aromaticity for the transition state of the concerted mechanism. It was shown that in the case of cyclopropanes substituted with donor–acceptor (DA) groups, a synergistic effect occurs due to the "pushpull effect" of substituents [15, 16]. The concerted and the stepwise mechanisms of the Diels–Alder reactions of butadiene with silaethylene and disilene were studied by *ab initio* MO methods [16, 17]. For this reaction, the activation energy barrier of the concerted-type transition state constrained with Cs symmetry was found to be about 9 kcal/mol higher than that of the stepwise transition stat. For the reaction

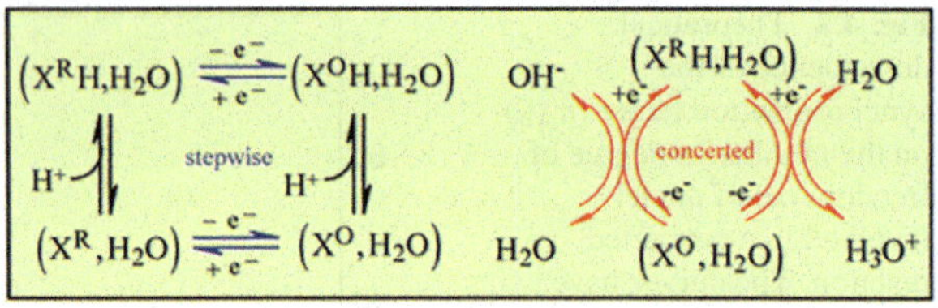

Fig. 4.3 Alternative mechanisms in proton-coupled electron transfer reactions in water [19]. With permission from American Chemical Society

of butadiene and silaethylene, an asymmetric concerted two stepwise processes can occur. Allenyl has the structure R_2C=C=CR—where *R* is H or some alkyl group and enyne contain C–C double bond and C≡C triple bond. The C^2–C^6cyclization of enyne-allenes was studied by a combination of kinetic isotope effects, theoretical calculations, and dynamics trajectories [18]. For the cyclization of allenol acetate, the isotope effect ($k_{CH3}/k_{CD3} = 1.43$) was interpreted in terms of a highly asynchronous transition state near the concerted stepwise boundary. This conclusion was supported by density functional theory calculations that locate a highly asynchronous transition structure for the concerted reaction. The stepwise mechanism and the asynchronous concerted mechanism start out geometrically similar, and the two pathways appear to have merged.

The competition between stepwise and concerted (CPET) pathways in proton-coupled electron transfer reactions in water was discussed on thermodynamic and kinetic bases (Fig. 4.3) It was found that in the case where water is the proton acceptor, the CPET pathway may compete favorably with the stepwise pathway.

4.2.3 *Enzymatical Reaction*

The NAD^+-dependent protein deacetylation catalyzed by Sir2 enzymes was characterized employing method on-the-fly Born–Oppenheimer molecular dynamics simulations with the B3LYP/6-31G(d) QM/MM potential and the umbrella method [20]. The results indicated that the nicotinamide cleavage reaction employs a highly dissociative and concerted displacement mechanism because (1) the cleavage of the glycosidic bond is facilitated by the nucleophilic participation of the acetyl-lysine, (2) the dissociative transition state has a significant oxocarbenium ion character, and (3) during this step of the reaction, the Sir2Tm enzyme strongly stabilizes the covalent *O*-alkylamidate intermediate. In addition, functional roles of key residues and motifs have been elucidated. Thymidylate synthase (TSase) catalyzes the intracellular *de novo* formation of thymidylate (a DNA building block). Two mechanisms have been proposed for the rate-limiting hydride transfer step in TSase catalysis: a stepwise mechanism in which the hydride transfer precedes the cleavage of the covalent bond between the enzymatic cysteine and the product and a mechanism where both happen concertedly (Figs. 4.4 and 4.5) [20, 21]. These two possibilities were tested using secondary kinetic isotope effect (KIE), mutagenesis study, and primary KIEs. In order to distinguish between the different mechanisms and to

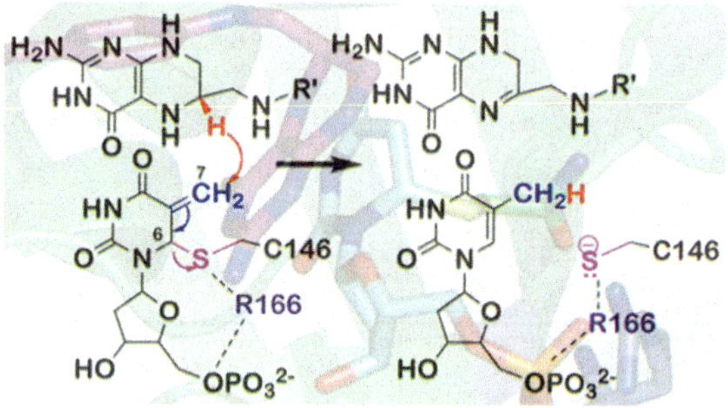

Fig. 4.4 Proposed a stepwise mechanism of thymidylate synthase (Tsase) catalysis [20, 21]. Published under Creative Commons public use license CC–BY from American Chemical Society

examine the role of an arginine R (166) in the rate-limiting hydride transfer step, the following types of experimental investigations were carried out: (1) secondary (2°) kinetic isotope effect (KIE) measurements on C6 of dUMP, (2) mutation of R166; and (3) an examination of primary (1°) KIEs on the hydride transfer. Because the irreversible hydride transfer precedes the C6–S cleavage, in the stepwise mechanism (Fig. 4.4) no 2° KIE on k_{cat}/K_M for the conversion of compound D to dTMP was expected. For the concerted hydride transfer mechanism, the sp^3–sp^2 rehybridization of C6 from compound D to dTMP was predicted. Involving 13 heavy atoms in the thymidylate synthase (Tsase) catalysis, according to Eqs. 4.1 and 4.2, with suggested energy activation $E_{syn} \approx 5$ and 10 kcalmol^{-1}, would "cost" a decrease of the process probability by synchronization factor $a_{syn} \approx 10^{-6}$ and 10^{-8}, respectively, and the estimated reaction constant rate can be $k \approx 10^2$ s^{-1} and 10^{-4} s^{-1}, respectively, Thus, this consideration allows the concerted mechanism only if energy activation would be markedly less than 10 kcalmol-1 which agrees with experimental kinetic data [20, 21].

Malic enzyme catalyzes the oxidative decarboxylation of Lmalate (1) to pyruvate (2) (Fig. 4.6) and this reaction could proceed via a stepwise ($1 \to 3 \to 4$) or concerted mechanism [23]. By measuring the isotope effect induced on V/K when C4 of malate is labeled with ^{13}C, it was found that this effect was decreased from 1.031 to 1.025, when ^{13}C] malate was simultaneously labeled with a deuterium at C2. It was concluded that dehydrogenation of the C2 hydroxyl takes place prior to the elimination of CO_2 in a stepwise mechanism.

Proline racemase catalyzes the racemization of proline

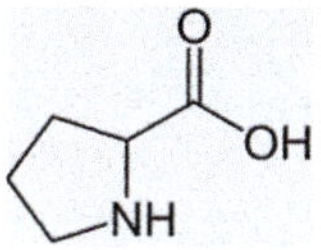

() via two proton transfers at opposite faces of the proline αcarbon (Hydron transfers) [24], It was tested whether these Hydron transfers take place in one versus two steps by essentially measuring the solvent deuterium isotope effect on the net rate constant

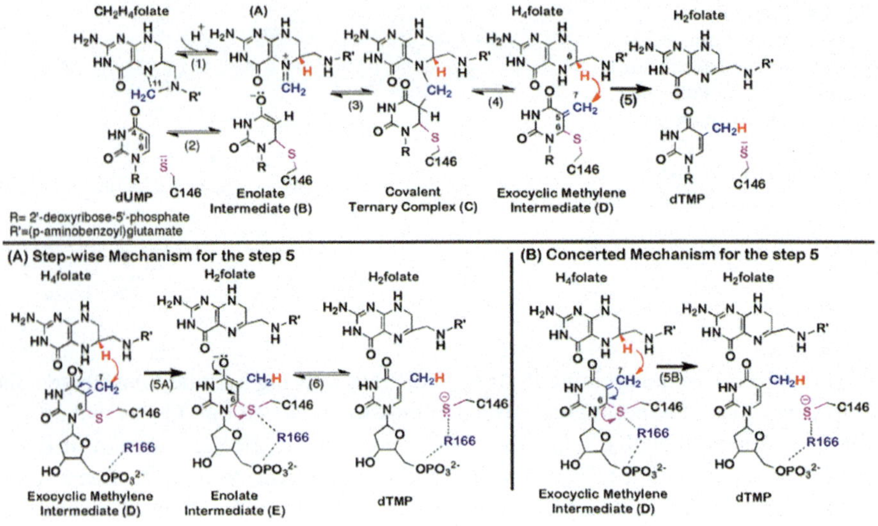

Fig. 4.5 (Top panel) Proposed mechanism for Tsase. The reaction is initiated by Michael addition of the enzymatic cysteine (C146) at C6 of dUMP (step 2) that leads to the formation of a covalent Tsase-dUMP enolate intermediate **B**, followed by Mannich condensation (step 3) to form a covalent ternary complex of Tsase-dUMP-CH2H4folate **C**. Following the Hoffman elimination (step 4), the methylene group forms an enzyme-bound exocyclic methylene intermediate **D**. Finally, a hydride is transferred from C6 of H4folate to C7 of compound D and C146 departs from the product dTMP (step 5). (Bottom panels) Proposed mechanisms for step 5. **A**. A stepwise mechanism that proceeds through the formation of a covalent Tsase-dTMP enolate intermediate (compound E, similar to the compound B in the top panel).8,9 **B** A concerted mechanism where the hydride transfer takes place concomitantly with the C–S bond cleavage [21]. Published under Creative Commons public use license CC–BY from American Chemical Society

for the Michaelis complex. Experimentally, it was found that: (1) The discrimination against solvent deuterium in the product L-proline is the same, whether D-[2-1H]proline or D-[2-2H]proline is the substrate' (2) No significant difference in the solvent deuterium isotope effect regardless of whether the substrate was initially deuterated at the α-carbon; (3) The substrate and product "on-off" steps are faster than the racemization step; and (4) The racemization reaction proceeds either in a concerted manner or in a stepwise fashion involving enzyme catalytic groups (e.g., thiols)

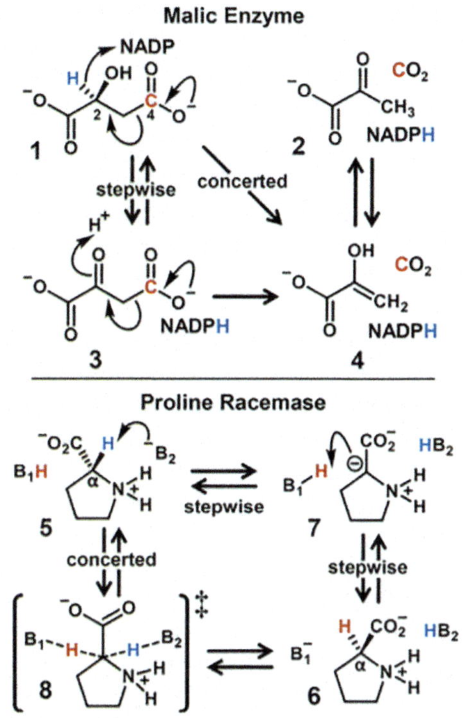

Fig. 4.6 Mechanisms proposed for malic enzyme and proline racemase. The positions of the fractionating and reweighting labels are colored red and blue, respectively [22, 23]. With permission from American Chemical Society

4.3 Principle of "Optimum Motion" and Mechanism of Enzymatic Reactions

According to the principle of least motion (PLM), when multiple species with different nuclear structures could theoretically form as products of a given chemical reaction, the least amount of change in nuclear structure or the smallest change in nuclear positions is expected [24, 25].

In the case of effective concerted mechanism, the decrease in the synchronization probability a_{syn} (Eqs. 4.1 and 4.2, Fig. 4.2) with increasing n can be compensated for by an appreciable decrease in the activation energy due to the inclusion of nuclei of donor and acceptor groups in the process. This consideration has led us to the formulation of the principle of optimum motion (POM) [9–12]. According to this principle, the number of nuclei whose configuration is changed in the elementary act of a chemical reaction (including electron transfer) must be sufficiently large to provide favorable energetic of the step and, at the same time, sufficiently small for the maintenance of a high value of the synchronization probability during motion along the reaction pathway to the reaction products., The formation of pretransition states can be preceded by a number of elementary steps or a concerted single step optimally combining enthalpy, entropy, and synchronization factors. POM allowed

Table 4.1 Enthalpies of activation for enzyme-catalyzed (k_{cat}) and nonenzymatic (k_{non}) reactions [26]. With permission from American Chemical Society

Reaction	$\Delta H^{\ddagger}(k_{cat})$	$\Delta H^{\ddagger}(k_{non})$
Yeast OMP decarboxylase	11	44.4
Urease	9.9	32.4
Bacterial α-glucosidase	10.5	29.7
Staphylococcal nuclease	10.8	25.9
Chymotrypsin	8.6	24.4
Chorismate mutase	12.7	20.7

discussion of alternative mechanism mechanisms of transformation of the pretransition to the transition state in a number enzyme reactions [9, 25-29]. Beneath some chosen enzymatic systems will be analyzed. Typical values of enthalpies of activation for enzyme-catalyzed (k_{cat}) and nonenzymatic (k_{non}) reactions are presented in Table 4.1 It should be noted that the experimental values of the activation energies and entropies of enzymatic reactions may include corresponding parameters of preceding conformational transitions and water reorganization which can be the cause of the enthalpy–entropy compensation effect (Chap. 8)

In the formate dehydrogenase reaction, as follows from Fig. 4.7 [25, 26], the transition from substrate to the transition state includes a change in the position of 12 heavy atoms, namely, three formate atoms, eight NAD^+ atoms, and two shifts of amino acid fragments of the enzyme. According to Eqs. 4.1 and 4.2 and Fig. 4.3, and taking into account a typical entropy activation in monomolecular step of enzymatical processes (about 8 J.K^{-1}. mol^{-1} [27]), we have the values of $k \approx 10\,s^{-1}$ and $k \approx 10^{-3}$ s^{-1} f. or suggested energies activation of 4 kcalmol^{-1} and 8 kcalmol^{-1}, respectively. Thus, for the given reaction, the concert mechanism is possible only if the activation energy is significantly lower than 8 kcalmol^{-1}.

Lactate dehydrogenase (LDH) accelerates the oxidation of lactate by NAD $^+$ to pyruvate and NADH by hydride transfer from the C-4 carbon of NAD. Figure 4.8 shows the active site of LDH with the bound substrate pyruvate and the NAD cofactor as derived from X-ray crystallographic data. This reaction is also a good example for the analysis of weather the process is stepwise or concerted [28]. As it is seen from Fig. 4.8, two atoms of substrate, six atoms of the nicotine ring (conversion of NADH to NAD+), all in all 8 heavy atoms, should change their position at the formation of the transition state. Equations 4.1 and 4.2 and Fig. 4.2 with experimental energy activation $E_{syn} \approx 10$ kcal mol^{-1} [26, 28]. gives $a_{syn} \approx 10^{-7}$ and the value of rate constant $k \approx 10^{-3}$ s^{-1}. Accordingly, the concerted mechanism reaction has very low probability with above-mentioned value of the experimental energy activation, which definitely include the parameters of previous conformational processes.

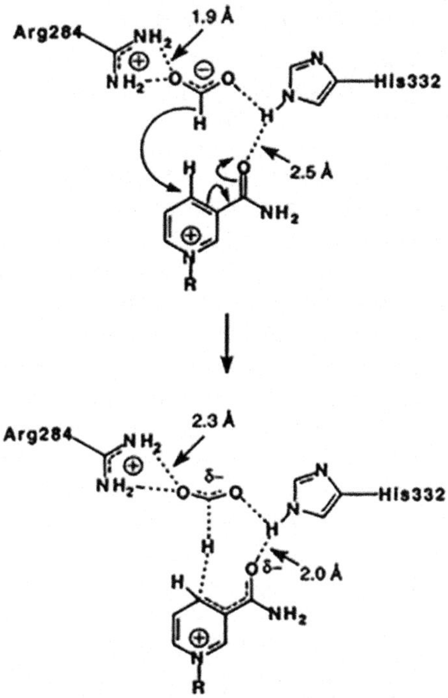

Fig. 4.7 Suggested concert mechanism of formate dehydrogenase reaction [25, 26]. With permission from American Chemical Society

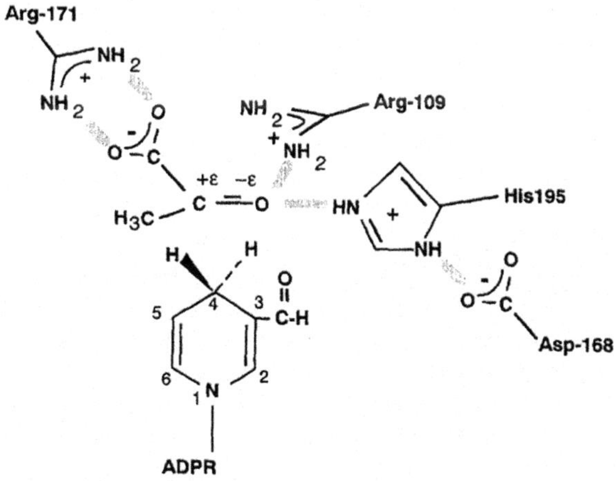

Fig. 4.8 Active site of lactate dehydrogenase (LDH) [27, 28]. With permission of Elsevier

4.4 Electron Transfer

4.4.1 First Grounds

Electron transfer (ET) is one of the most ubiquitous and fundamental phenomena in chemistry, physics, and biology [29, 30–70]. As a light microscopic particle, an electron easily tunnels through a potential barrier. The Franck–Condon principle is applicable not only to the vertical radiative processes but also to nonradiative horizontal electron transfer. The Franck–Condon Principle describes states that when a molecule is undergoing an electronic transition, such as ionization, the nuclear configuration of the molecule experiences no significant change. The decisive step in the field was taken by Marcus and his colleagues [30] and Hash [31, 32]. These authors articulated the need for readjustment of the coordination shells of reactants in self-exchange reactions and of the surrounding solvent to the electron transfer. According to the fundamental Marcus two-state model (Fig. 4.9) [31], the distortion of the reactants, products, and solvent from their equilibrium configuration can be described by identical parabolas in reaction coordinate space for the initial and final states shifted related to each other according to the driving force of the value of the process (DF_0).

Within the adiabatic regime (assuming strong electronic coupling, where the resonance integral $V > 200$ cm^{-1}), the Eyring theory of the transition state predicts that the value of the electron transfer rate constant is:

$$k_{ET} = \frac{k_B T}{h} \exp - \left[\frac{(\lambda + \Delta G_0)^2}{4\lambda k_B T} \right] \tag{4.3}$$

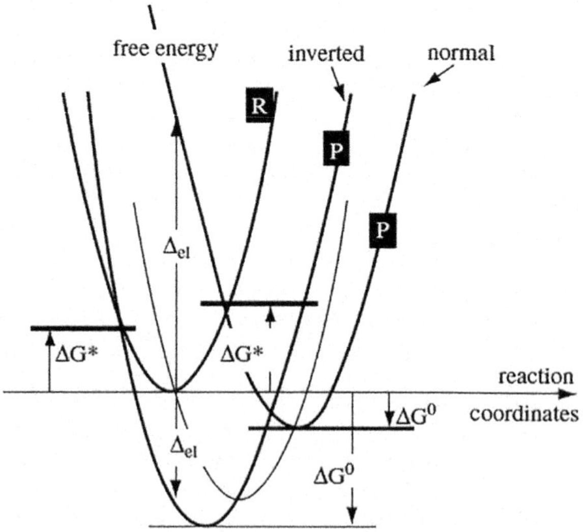

Fig. 4.9 The Marcus profile of potential energy surfaces versus reaction coordinates for reactants environment (curve *R*) and products + environment (curve *P)* in the normal regime, the inverted regime and at the inversion point (thin line).standard Gibbs free energy □G_0 [32, 33]. Published under the Creative Commons license (CC–BY-4.0) from IOPscience

where λ is the reorganization energy defined as the energy for electron excitation without distortion of the nuclear frame. The following relationships based on log $k_{ET} - \Delta G_O$ graphs were predicted from Eq. 4.3: (1) $\lambda > \Delta G_O$, log k increases if ΔG_0 decreases (normal Marcus region), (2) $\lambda = \Delta G_0$, the reaction becomes barrierless, and (3) $\lambda < \Delta G_0$, log k decreases with increasing driving force.

The theory of nonadiabatic electron transfer when the resonance integral V less than 200 cm^{-1} was developed by Levich et al. [34, 35]. These authors, suggesting harmonic one-dimensional potential surface, proposed a formula for nonadiabatic ET energy:

$$k_{\mathrm{ET}} = \frac{2\pi V^2}{h\sqrt{4\pi\lambda k_{\mathrm{B}}T}} \exp\left[-\frac{(\lambda + \Delta G_0)^2}{4\lambda k_{\mathrm{B}}T}\right] \tag{4.4}$$

4.4.2 Long-Range Electron Transfer

Long-Range Electron Transfer (LRET) between donor (D) and acceptor (A) centers] can occur by three mechanisms: (1) direct transfer that involves direct overlap between electron orbitals of the donor and acceptor that can be described in the frame of the Marcus–Levich theory [29, 33], (2) consecutive electron jumps via chemical intermediates with a fixed structure, and (3) superexchange via intermediate orbitals. Various aspects of the LRET) and coupling of electron transport with proton transport are discussed in the papers [9, 36–50]

In the superexchange process [41], the nonadiabatic ET rate constant is strongly dependent on electronic coupling between the donor state D and the acceptor state connected by a bridge (V_{AB}):

$$V_{\mathrm{AB}} = \frac{\sum V_{\mathrm{A}\alpha} V_{\alpha B}}{\Delta E_\alpha} \tag{4.5}$$

where V_{Aa} and V_{aB} are the couplings between bridge orbitals and acceptor and donor orbitals, respectively, and $\mathrm{DE_a}$ is the energy of the bridge orbitals relative to the energy of the donor orbital. The summation over includes both occupied and unoccupied orbitals of the bridge.

According to the approach of Onuchic and coworkers [46], for a pathway between bridged donor and acceptor groups, the coupling element can be written as

$$V_{\mathrm{AB}} = V_0 \prod_i^N \varepsilon_{\mathrm{i}} \tag{4.5}$$

where V_0 is the coupling between the donor and the first bond of the pathway and e_i is a decay factor associated with the decay of electron density from one bond to another.

The basic idea underlying our approach to LRET is an analogy between superexchange in electron transfer and such electron exchange processes as triplet–triplet energy transfer (TTET) and spin exchange (SE) [47–49]. The spin exchange and TT phenomena may be considered an idealized model of ET without or with only a slight replacement of the nuclear frame. The proposed approach is based on the following considerations: (1) The experimental dependence of exchange parameters J_{TT} and J_{SE} on the distance between the exchangeable centers (Eq. 4.7) may be used for evaluating such dependences for the resonance integral $V_{ET}{}^2$ in the ET equation; (2) All three integrals $V_{ET}{}^2$, J_{SE}, and J_{TT} are related to the overlap integral (S_i), which quantitatively characterizes the degree of overlap of orbitals involved in these processes

$$V_{ET}^2, J_{SE}, J_{TT} \propto S_i^n \propto \exp(-\beta_i R_i) \tag{4.7}$$

where R_i is the distance between the interacting centers and the distance decay constant *bi* is a coefficient that characterizes the degree of the integrals decay; (3) In the first approximation, a number of orbitals involved in an exchange process $n = 2$ for the ET and SE processes with the overlap of two orbitals and $n = 4$ for the TT process in which four orbitals overlap (ground and triplet states of the donor and ground and triplet states of the acceptor) therefore, $b_{SE} \approx 1/2\ b_{TT}$. As it is seen in Fig. 4.10, experimental data on the dependence of k_{TT} and J_{SE} on the distance between the centers (DR) lie on two curves, which are approximated by the following equation [9, 47];

$$k_{TT}, J_{SE} \propto \exp(-\beta \Delta R) \tag{4.8}$$

For systems in which the centers are separated by a nonconductive medium (molecules or groups with saturated chemical bond), b_{TT} equals 2.6 Å^{-1}, while For systems in which the radical centers are linked by conducting conjugated bonds, b_{SE} is 0.3 Å^{-1}. The value of b_{ET} and $b_{SE} \approx 1/2\ b_{TT}$ (2.6 Å^{-1}) may be estimated as 1.3 Å^{-1}.

The semiempirical approach for the quantitative estimation of the effect of bridging the group on LRET was applied to the analysis of experimental data on electron transfer in biological systems [47, 48]. As an example, experimental rate constants of ET (k_{ET}) in reaction centers of purple bacteria and PSI were plotted against the edge–edge distance I between the donor and acceptor centers [47] (Fig. 4.11). The dependence of log k_{ET} on R with $b_{ET} \approx 1.4$ Å^{-1} is similar to $b_{ET} \approx 1.3$ Å^{-1}, predicted for long-distance superexchange orbital overlap through "nonconducting" media by the shortest pathway.

An examination of the empirical data on the exchange integral values (J_{ET}) for the spin–spin interactions in systems with known structure, that is, biradicals, transition metal complexes with paramagnetic ligands, and monocrystals of nitroxide radicals

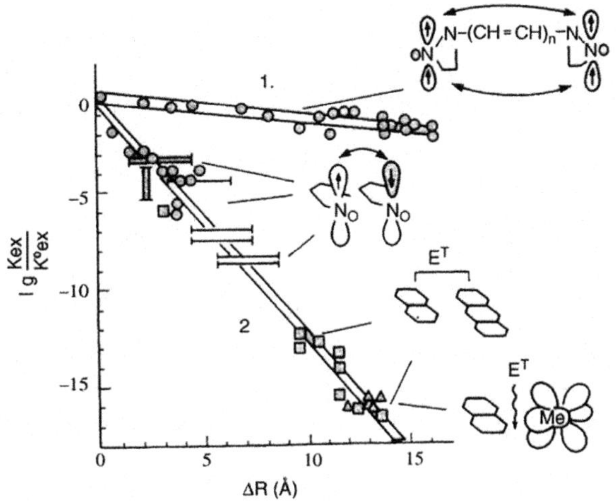

Fig. 4.10 Dependence of the logarithm of relative parameters of the exchange interaction on the distance between the interacting centers (DR). k_{TT} is the rate constant of triplet–triplet electron transfer, and J_{SE} is the spin exchange integral. Index 0 is related to van der Waals contact [47]. With permission from Elsevier

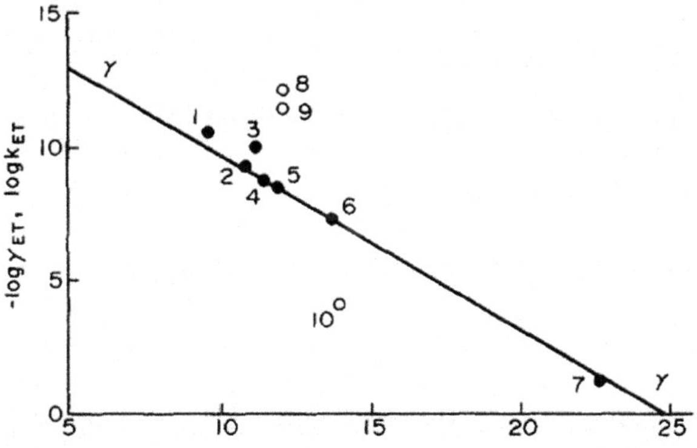

Fig. 4.11 Dependence of the experimental rate constants of ET (*kFr*) in photosynthetic RCs of bacterial and plant PSI 1 1 on the edge-edge distance between the donor and acceptor centers. The straight line is related to the dependence of the attenuation parameter χl ($\beta_{ET} \approx$ (β_{SE}) for SE and ET processes in homogeneous "nonconducting" media filled circles correspond to a "regular" dependence, open circles to a "deviation" [47]. With permission from Elsevier

were performed [47, 48]. Values of the attenuation parameter g_x for the exchange interaction through a given group were estimated by an examination of the empirical data on the J_{ET}) for the spin–spin interactions in systems with known structure [9, 47]. By our definition, the g_x is

Table 4.2 Values of the attenuation parameter of individual groups (γ_X), van der Waals contact (γ_v), and hydrogen bond (χ_{hb}) for spin exchange in biradicals and paramagnetic complexes of transition metals with nitroxide ligands (see details in text) [47] With permission from Elsevier

Group, *X*	γ_x	Group, *X*	γ_x
C_6H_4	6.00 ± 0.03		
C=C	1.7	–NH–Co–	55[a]
C=O	8.4 ± 0.4	γ_v	50
C			
NH	6.5	γ_{hb}	10
O	5	H	12
S =O	2.1	SO_2	2.2
	3.5	RP=O	2.40 ± 0.03[b]

$$\gamma_X = \frac{J_{RYZP}}{J_{RYXZP}} \tag{4.9}$$

where *R* is a nitroxide or organic radical, *P* is a paramagnetic complex or radical, and *X*, *Y*, and *Z* are chemical groups in the bridge between *R* and *P*. Table 4.2 shows the results of the calculation parameter c_X from empirical data using Eq. (4.9) These values and $b_{ET} \approx 1.3$ Å^{-1} may be used for the analysis of alternative electron transfer pathways in biological systems.

Several experimental and theoretical investigations have demonstrated that media states can affect long-range electron tunneling [50–60]. Numerous data convincingly proved that the transition of proteins can occur at a distance of at least 25 A° (Fig. 4.12) [50–52] Some theoretical aspects of the tunneling in various environments and in protein, in particular, were discussed in [46–60] First direct evidence for long-range electron tunneling through proteins was emerged by Gray and Winkler group in 1982 work on cytochrome *c* modified with a $Ru^{III}(NH_3)_5$ moiety coordinated to His33 on the protein surface (Ru(His33)-Fe-cyt *c*) [50]. In a kinetics study, flash photochemical electron injection generated transient Ru^{II}(His33)-Fe^{III}-cyt *c* that relaxed to the Ru^{III}–Fe^{II} thermodynamic product with a time constant of 30 ms ($-\Delta G°$ = 0.2 eV). This finding was explained as a result of the tunneling long-distance electron transfer between redox centers separated by 15Å in the protein. A technique utilizing flash photolysis to generate this key intermediate, thereby enabling to measure ET rates from the iron (I1)-heme to different ruthenium(II1) histidine-48 acceptors in sperm whale myoglobin (Mb) was developed [57]. In this system, the closest donor–acceptor edge-edge distance was found to be also13 Å./

Insight into the physical mechanisms of short—and long-range biomolecular electron transfer was gained [51, 58]. The blue copper protein from *Pseudomonas aeruginosa*, azurin, immobilized at gold electrodes through hydrophobic interaction with alkanethiol self-assembled monolayers (SAMs) of the general type [-S-$(CH_2)_n$-CH_3] (n = 4, 10, and 15) was employed [51, 58]. Fast scan cyclic voltammetry and a Marcus equation analysis were used to determine unimolecular standard rate constants and reorganization free energies for variable *n*, temperature (2–55 °C), and pressure (5–150 MPa) conditions. These experiments revealed that electron exchange at distance over *ca.* 12 Å is a direct ET process. All the ET data was described by using a single

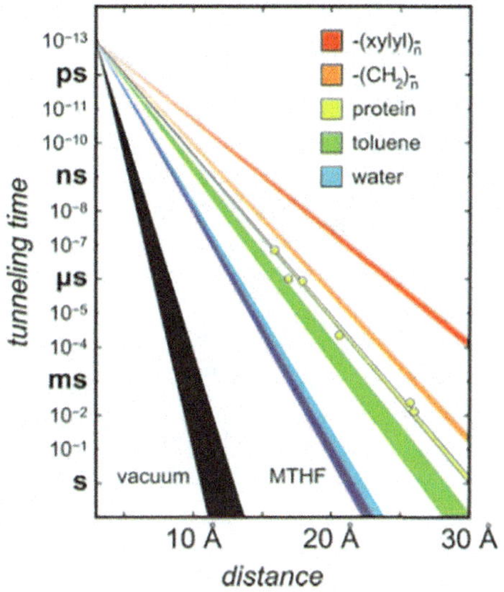

Fig. 4.12 Timetable for activationless electron tunneling through various media: vacuum (black, $\beta = 2.9–4.0$ Å^{-1}), (MTHF) glass (violet, $\beta = 1.57–1.67$ Å^{-1}), aqueous glass (cyan, $\beta = 1.55–1.65$ Å^{-1}), and toluene glass (green, $\beta = 1.18–1.28$ Å^{-1}). Investigations of ET rates in D-(bridge)-A complexes have produced exponential distance dependences: *xylyl* bridges, $\beta = 0.76$ Å^{-1} (red) (12); alkane bridges, $\beta = 1.0$ Å^{-1} (orange) (10); and—strand bridges in ruthenium-modified azurin, $\beta = 1.1$ Å^{-1} (yellow) [51]. With permission from PNAS

reorganization energy (0.3 eV), however, the values for the enthalpies and volumes of activation were found to vary with *n*.

Figure 4.13 shows that the logarithm of the optimal rates spanning 12 orders of magnitude for intraprotein electron transfer reactions as a function of the edge-to-edge distance generates a linear relationship (Fig. 4.13) [60]. It appears that the protein medium acts as a relatively uniform barrier to electron tunneling with an average/*b* value of 1.7 Å. The rates of high driving-force ET reactions have been measured for 30 Ru(iamine)-labeled metalloproteins (Fig. 4.14) [51, 52]. Two important parameters were extracted from the data: a rate constant for ET at close contact (k_{ET}); and an exponential distance decay constant β ($k_{ET} = k_{ET}\ e^{-\beta(r-r_0)}$) that describes the efficiency of the long-range coupling.

The electrochemical tunneling spectroscopy (ECTS) (Fig. 4.15) and molecular dynamics simulations were used to show that the current between cytochrome c (Cc) and the mitochondrial complex III (CIII), or cytochrome bc1 decays along more than 10 nm in the solution [55]. Human cytochrome c (hCc) forms a natural complex with the cytochrome c1 (Cc1) subunit of cytochrome bc1. ECTS experiments were performed at electrochemical potentials, in which pCc1 on the sample electrode is reduced ($U_S = -200$ mV) and hCc on the ECTS probe is oxidized ($U_P = 600$ mV), resulting in a constant positive bias $U_{bias} = U_P – U_S = 800$ mV. The distance decay values β were directly measured from the current at constant electrochemical potential. It was found that b and the calculated local barrier for electron transfer

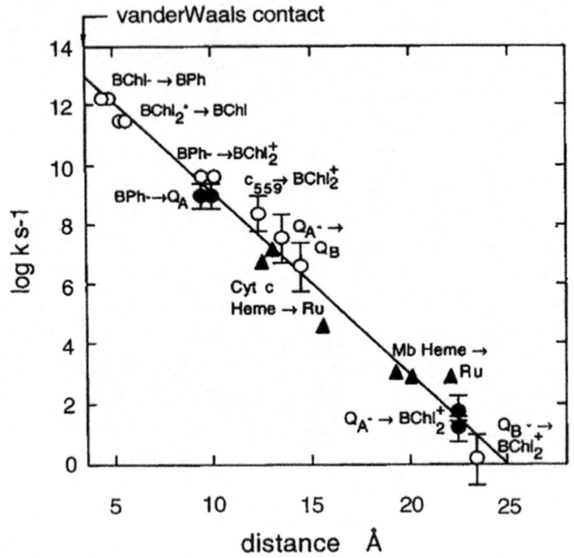

Fig. 4.13 Free energy optimized rate versus edge-to-edge distance relationship for intraprotein electron transfer. Photosynthetic reaction center rates are shown as circles and excited heme-ruthenium electron transfers in modified myoglobin and cytochromes c are shown as triangles. Filled symbols are for reactions with extensive free energy dependence, while open circles are for reactions with a smaller experimental free energy dependence and more uncertain optimal rates, with uncertainties indicated by error bars. The fit line represents a simple exponential decay with a/3 of 1.4, ~ —I. Distance, derived from crystal structures, is defined as center of edge atom of donor to center of edge atom of acceptor; thus the vertical line at 3.6, ~ represents van der Waals contact [60]. With permission from Elsevier

can be regulated by the electrochemical potential applied to the proteins and are distributed around $\beta = 9 \pm 2 \text{ nm}^{-1}$ (Fig. 4.15) In addition, molecular dynamics simulations revealed a reduced ionic density and extended electric field in the volume confined between the proteins.

4.4.3 Photosynthetic Light Energy Conversion

4.4.3.1 General

Photosystem I and photosystem II are the two multi-protein complexes that contain the pigments necessary to harvest photons and use light energy to catalyze the primary photosynthetic endergonic reactions producing high energy compounds. In the Nature, the primary photochemical and photophysical processes occur in the photosynthetic reaction centers (RCs) of microbes, plant, and algae [62–92]. The processes in the RCs donor–acceptor pair D–A lead to charge photoseparation,

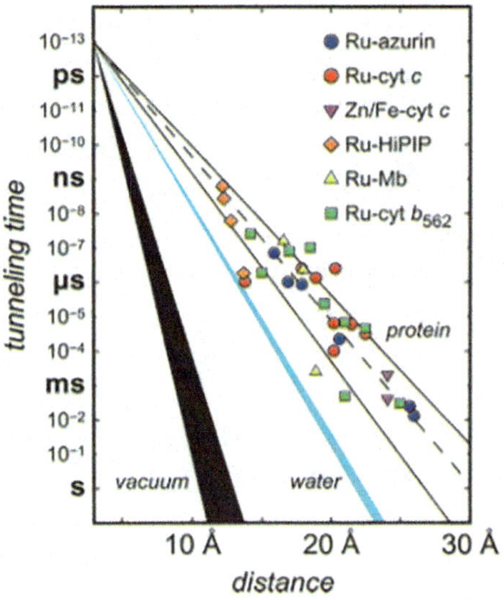

Fig. 4.14 Tunneling timetable for intraprotein ET in Ru-modified azurin (blue circles), cyt c (red circles), myoglobin (yellow triangles), cyt b562 (green squares), HiPIP (orange diamonds), and for interprotein ET Fe: Zn-cyt c crystals (fuchsia triangles). The solid lines illustrate the tunneling-pathway predictions for coupling along β-strands ($\beta = 1.0$ Å^{-1}) and—α-helices $\beta = 1.3$ Å^{-1}); the dashed line illustrates a 1.1-Å^{-1} distance decay. Distance decay for electron tunneling through water is shown as a cyan wedge. Estimated distance dependence for tunneling through vacuum is shown as the black wedge [51]. With permissions from PNAS

which is an appearance of the pair of two charges (D^+A^-), where (D^+) the cation-radical is a strong oxidant and A^- is the anion-radical is a strong reducing agent. The most important problems are to prevent fast recombination of D^+ and A^- centers of high chemical reactivity and to provide relatively long lifetime for these centers, involving it in subsequent chemical reactions eventually results in the formation of stable compounds such as ATP and NADPH.

In 1975 the author of this monograph has suggested that rapid electron transfer in reaction centers in the forward direction and significantly slower transfer in the reverse direction may account for the cascade structure of RC which provides tunneling (long-distance) mechanism. [61]. A key of this concept was an assumption, which RC brings about positioning of donor and acceptor pigments in the cascade at optimum distances. A basic idea underlying above-mentioned mechanism included the following assumptions: (1) The primary donor (D) and several acceptor (A_i) centers compose a cascade in an ordered structure, in which all these centers are placed at an optimum distance (6–10 Å) from each other and are separated by a nonconducting protein medium; (2) such a separation slows down the forward electron transfer between adjacent D–A_1 and A_i–A_{i+1} pairs as compared to electron transfer in a system with close contacts between the centers; (3) the direct transfer can be sufficiently fast, if the optimum distances do not exceed 6–10 Å; (4) the recombination of each D^+–A_i^- pair becomes slower and slower as A_i moves away

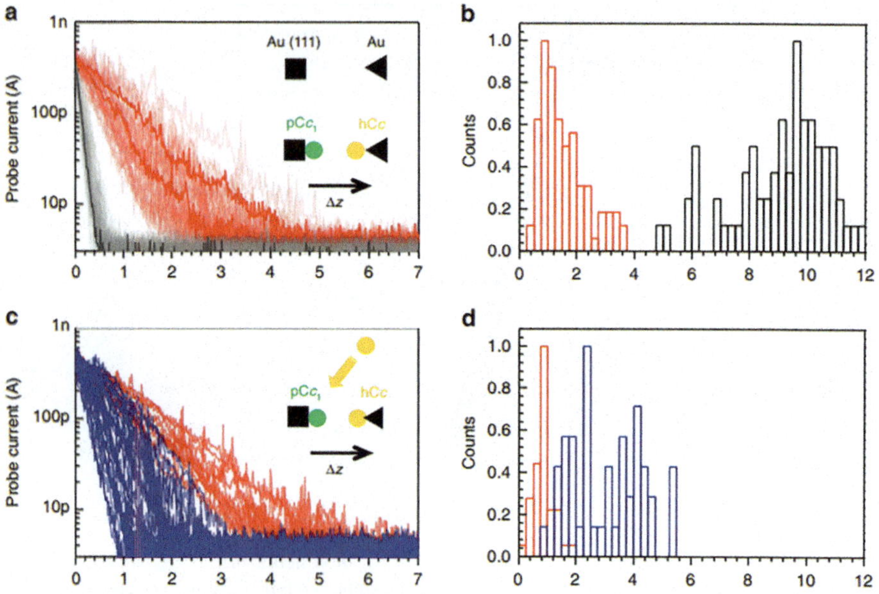

Fig. 4.15 Current–distance electrochemical tunneling spectroscopy of pCc1–hCc. **a** Ensemble of semilogarithmic current–distance (*I-z*) plots obtained for pCc1–hCc proteins (red) and for bare gold (gray) during probe retraction, showing a more gradual current decay for pCc1–hCc. Selected representative traces are depicted in bold. Sample and probe electrodes are represented by a square and a triangle, respectively. **b** Histograms of distance decay factors (β) quantified from individual curves in a. **c** Inhibition experiment conducted to test the specificity of pCc1–hCc. Semilogarithmic I-z plots obtained for pCc1–hCc during probe retraction before (red), and after (blue) the addition of WT hCc. The measured probe current is reduced more abruptly upon WT hCc addition. **d** Histograms of β from individual curves in c. **e** Selected semilogarithmic I-z plots obtained for pCc1–hCc obtained in approach experiments after retracting the probe 100 nm. **f** Histogram of β obtained from individual curves in e. All the experiments were performed at Us = − 200 mV and at constant bias of 800 mV in 50 mM sodium phosphate buffer, pH = 6.5. Initial current set point 0.4 nA [55]. Published under the terms of the Creative Commons CC BY license, from Springer Nature

from the donor; and (5) in contrast, in the system of tightly packed centers the recombination rate is expected to be very fast. The cascade tunneling hypothesis has been supported in subsequent combined physical experiments reported in [62–68].

In 1984, after having analyzed the X-ray diffraction pattern from the reaction center crystals, Nobel Prize winners Johann Deisenhofer, Robert Huber, and Hartmut Michel presented the three-dimensional structure of the reaction center, the first high-resolution structure of a membrane protein] https://www.nobelprize.org/prizes/chemistry/1988/press-release/. The structure convincingly confirmed the cascading arrangement of key pigments at optimal distances from each other as was predicted *i* in 1975 [61]. Later, the cascade structure was also confirmed in X-ray studies of the reaction centers of Photosystem I and Photosystem II at 2.5 Å resolution [70], Recently the first structure of a homodimeric type I RC, an X-ray crystal structure at 2.2 Å resolution was reported [71] Various aspects of the structure and mechanism of action of photosynthesis reaction centers have been described and discussed in a number of recent reviews and original papers [62–92].

4.4.3.2 Light Harvesting Antenna

Photosynthesis in photosynthetic reaction centers begins when light is absorbed by an antenna pigments which can be a (bacterio) chlorophyll, carotenoid, or open chain tetrapyrrole depending on the type of organism [78]. The light harvesting antenna (LHA) is "designed" by the Nature to increase the efficiency of light absorption in the photosynthetic systems of bacteria and plants. The accumulation of light energy and its delivery to chlorophyll of reaction center is carried out by the mechanism of transfer of energy from the pigment (mostly chlorophyll) in the excited singlet state to the singlet level of the neighboring pigment in the ground state. According to Forster Resonance Energy Transfer (E) (FRET) theory (https://en.wikipedia.org/wiki/F%C3%) the efficiency of singlet -singlet transfer between an excited fluorophore (donor, D) and a quenching chromophore (acceptor, A)

$$E = \frac{k_{\mathrm{ET}}}{k_f + k_{\mathrm{ET}} + \sum k_i} \tag{4.10}$$

where k_{ET} is the rate of energy transfer, k_f is the radiative decay rate, and the k_i is the rate constants of any other de-excitation pathway and

$$E = \frac{1}{1 + (r/R_0)^6} \tag{4.11}$$

where r is the distance (R) between the centers and R_0 is the Förster distance (critical radius) of this pair of donor and acceptor, which is the distance at which the energy transfer efficiency is 50%.

$$R_0^6 = \frac{9Q_0(\ln 10)\kappa^2 J}{128\pi^5 n^4 N_A} \tag{4.12}$$

where Q_0 is the fluorescence quantum yield of the donor in the absence of the acceptor, κ^2 is the dipole orientation factor, n is the refractive index of the medium, N_A is Avogadro's number, J is the overlap integral (J) expresses the degree of spectral overlap between the donor emission and the acceptor absorption

$$J = \int f_{\mathrm{D}}(\lambda)\varepsilon_{\mathrm{A}}(\lambda)\lambda^4 \mathrm{d}\lambda \tag{4.13}$$

Therefore, the efficient energy transfer in the light harvesting antenna should follow a number of requirements: (1) The light must have the range from < 1000 nm (the near IR), to > 340 nm (the near UV); (2) for energy transfer, the molecules must

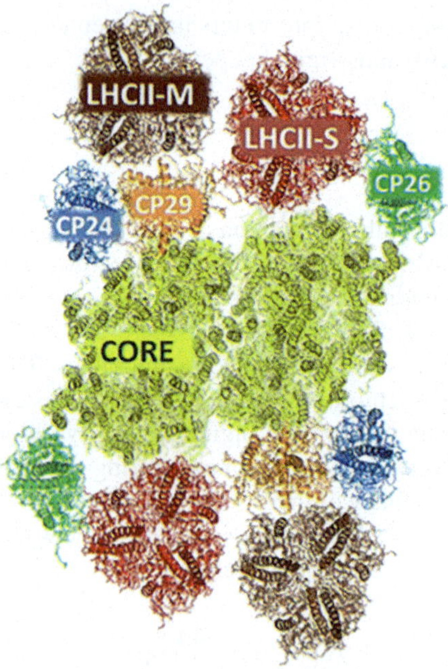

Fig. 4.16 Model of the PSII supercomplex C2S2M2 from higher plants. Top-view for the stromal side on a C2S2M2 supercomplex from *A. thaliana*. Based on the projection map of C2S2M2 at 12 Å resolution, a model of the 3D structure of the supercomplex was reconstructed (using the crystal structures of PSII core and LHCII). For the minor antenna complexes, the structure of a monomer of LHCII was used while the pigment composition/occupancy was assigned based on the results of mutation analysis experiments on in vitro reconstituted complexes [72]. Published under the terms of the Creative Commons CC BY license, from Springer Nature

be the optimal distance apart, appropriately aligned and have overlapping spectra for fluorescence and absorption; and (3) the system should be organized with a thermodynamic gradient from antenna to reaction center (an energy "funnel").

An example would be light harvesting antenna (LHA) as a model of a plant PSII supercomplex [72]. In a supercomplex of a plant PSII, the reaction center is enclosed by an overall elliptically shaped LH1 consisting of 15 αβ-apoproteins, each accommodating two BChl *a* molecules. The LHA main core is surrounded by outer light harvesting complexes (Lhcs), together forming the supercomplexes (Fig. 4.16) which are complemented by some "extra" Lhcs.

4.4.3.3 Bacterial Photosynthetic Reaction Center

The reaction center of photosynthetic bacteria (RCPB) is composed of four protein subunit. The structure showed the precise arrangement in the L and M subunits of the photochemically active groups—two chlorophyll molecules forming a dimer, serving in exited states as the primary electron donor, two monomeric chlorophylls, two pheophytin molecules, one quinone molecule, called Q_A and a second quinone molecule, Q_B, and one iron ion (Fe) (Fig. 4.17) [78, 79]). The arrangement of electron transfer cofactors in the wild-type *Rba. sphaeroides* RC in two pseudo-symmetrical membrane-spanning wires is shown in the Figure. The ET chain in heliobacterial RC (HbRC) contains three chemically distinct BChls being on the P-side P800, a BChl g′ dimer that serves as the primary electron donor, and ending on the N-side with F_X, a [4Fe-4S] cluster, that serves as the terminal acceptor [71]. Two different models for charge separation can be proposed: one in which Acc serves as the initial acceptor (HbRC*→ $P800^+$ Acc^- →$P800^+$ A_0^-→$P800^+$ FX^-) and one in which Acc serves as the initial donor (HbRC*→ Acc^+ A_0^-→$P800^+$ A_0^-→$P800^+$ FX^-). A_0 is the primary acceptor in the HbRC the primary acceptor in the HbRC.

The basis of asymmetric charge separation and the blockage in photocurrent generation by the AM260W mutation is explored in (Fig. 4.18) (see details in the Capture).

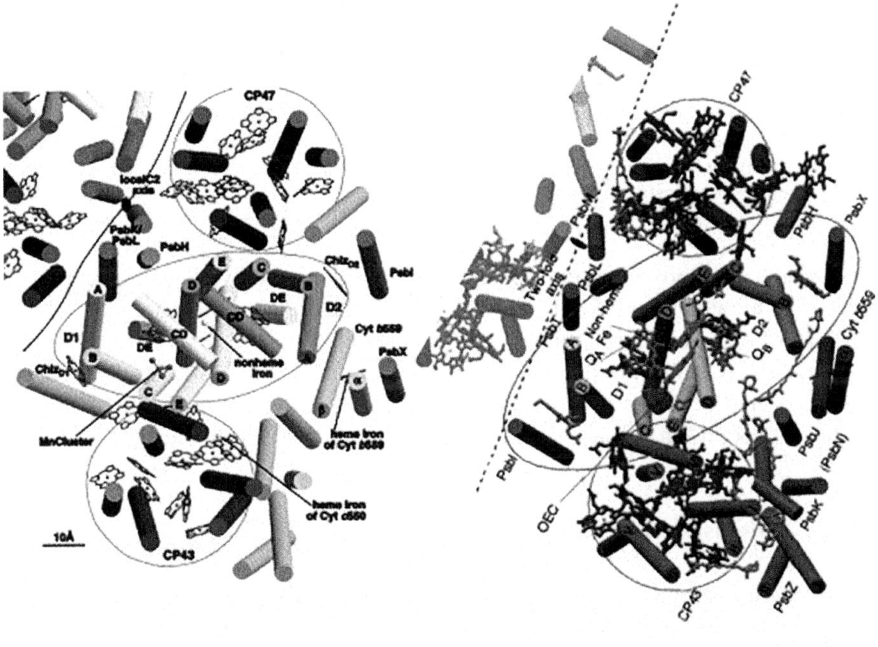

Fig. 4.17 **a** Two wires of cofactors connect the *P*-side (positive terminus) and *Q*-side (negative terminus) of the wild-type RC. Photoexcitation of the *P* dimer triggers electron flow along the *A*-wire (red arrows) forming, sequentially, $P^+B_A^-$, $P^+H_A^-$, and $P^+Q_A^-$ radical pairs. Electrons exit the complex via the dissociable Q_B ubiquinone. **b** Overlay of the X-ray crystal structures of the wild-type and AM260W mutant RCs shows that replacement of Ala M260 by Trp prevents incorporation of the Q_A ubiquinone due to steric overlap (top), breaking the *A*-wire (bottom). **c** Similarly, replacement of Ala M149 by Trp prevents incorporation of the H_B BPhe (top), breaking the *B*-wire (bottom). Key for all panels: nitrogen—blue, oxygen—red, cofactor carbons—yellow, green, pink or cyan, protein carbons—orange (wild-type RC) or yellow (mutant RC), magnesium—magenta spheres, iron—brown sphere. [79]. Published under the terms of the Creative Commons CC BY license from Springer Nature

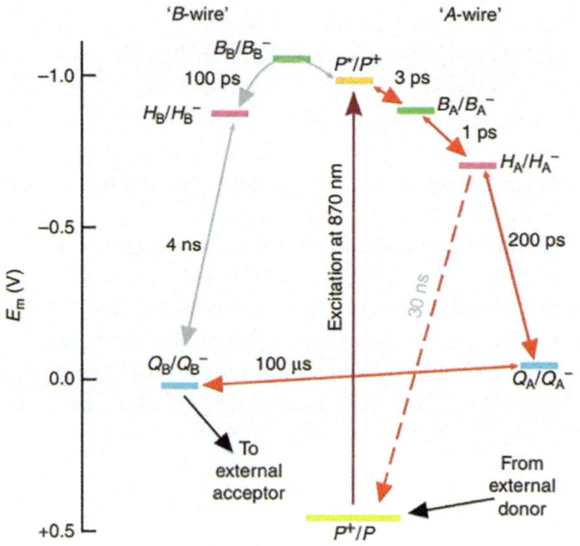

Fig. 4.18 Redox basis for asymmetric photochemical charge separation. Plot shows the mid-point potentials of redox couples involved in A-wire photochemical charge separation (right) and equivalent states involving B-wire cofactors (left) [79]. Published under the terms of the Creative Commons CC BY license, from Springer Nature

This arrangement provided relevant redox potentials on the two cofactor wires. The process is initiated by the formation of the first singlet electronic state of a "special pair" (*P*) of bacteriochlorophylls (BChls) located close to one side of the membrane (*P*-side). A dissociable ubiquinone (Q_B) on the opposite side (*Q*-side) is then reduced in a four-step process involving a "redox wire" comprising a monomeric BChl (B_A), a bacteriopheophytin (BPhe—H_A) and a nondissociable ubiquinone (Q_A).

4.4.3.4 Photosystems I and II

Photosystem I and photosystem II are the two multi-protein complexes that contain the pigments necessary to harvest photons and use light energy to catalyze the primary photosynthetic endergonic reactions producing high energy compounds [62–95] (Fig. 4.19). In oxygenic photosynthetic organisms, plant and green bacteria, the reaction centers of two systems Photosystem I (PS I) and Photosystem II (PS II) convert the absorbed light energy into energy of stable products. Ferredoxin in reduced state and dioxygen PS I from plants and cyanobacteria mediates light-induced electron transfer from plastocyanin to ferredoxin (flavodoxin) at the stromal membrane side, while PS II is a photoenzyme that catalyzes oxidation of the water in a water-splitting Mn-containing system. Subsequent absorption of four light quanta by PS I and PS II results in evaluation of dioxygen from two water molecules.

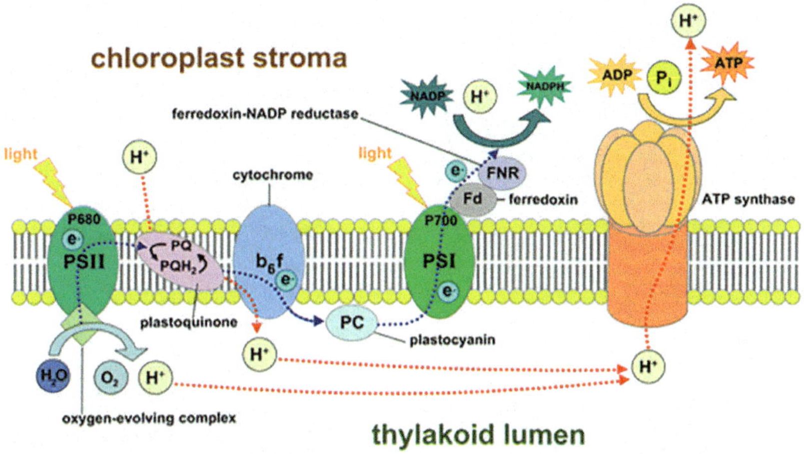

Fig. 4.19 Light-dependent reactions of photosynthesis at the thylakoid membrane https://en.wikipedia.org/wiki/Photosystem#/media/File:Thylakoid_membrane_3.svgю Published under CC BY-SA 4.0 license

Plants, algae, and some bacteria use two photosystems, PSI with chlorophyll P700 and PSII with chlorophyll P_{680}. The following principle steps in this unique physicochemical system for converted light energy are: (1) Using light energy, PSII acts first to channel an electron through a series of acceptors that drive a proton pump to eventually generate ATP; (2) the electron passes on to PSI; (3) second photon of light captured by P700 provides the required energy to channel the electron to ferredoxin, generating reducing power in the form of NADPH; and (4) after the photoexcitation of the primary donor, a dimer of chlorophyll a $(Chla)_2$, P700, an electron is transferred via a chlorophyll (A_0) to a phylloquinone (A_1) and then to the iron sulfur clusters, F_x, F_A, and F_B. The primary donor, P_{680}, of PS II is the most oxidizing of any RC discovered thus far with an E_m of $\sim$ 1,200 mV [80]. The crystal structure of the complete photosystem I (PSI) from higher plants to 4.4 A° and 3.3 A° resolution was determined [70].

PS I from *S. elongates* contains nine protein subunits featuring transmembrane □-helices and three stromal subunits. The organic cofactors are arranged in two branches along the pseudo-C2 axis [81]. The distances between adjacent donor and acceptor centers of the system vary from 8.2 to 14.9 Å for the "right" brunch (A) and from 8.6 to 22 Å for the "left" brunch (B) (Fig. 4.20). Therefore, "nonconducting" zones are similar to those in the bacterial RC separate the centers. The crystal structure of the complete photosystem I (PSI) from higher plants to 4.4 A° and 3.3 A° resolution was determined [81]. Figure 4.20 shows a schematic picture of the structure of the trimeric Photosystem I based on the structure of the complex at 2.5 Å resolution as well as spectroscopic and biochemical data [85].

\ Structures of Photosynthetic Reaction Centers of Photosystems I and II are presented in Fig. 4.21

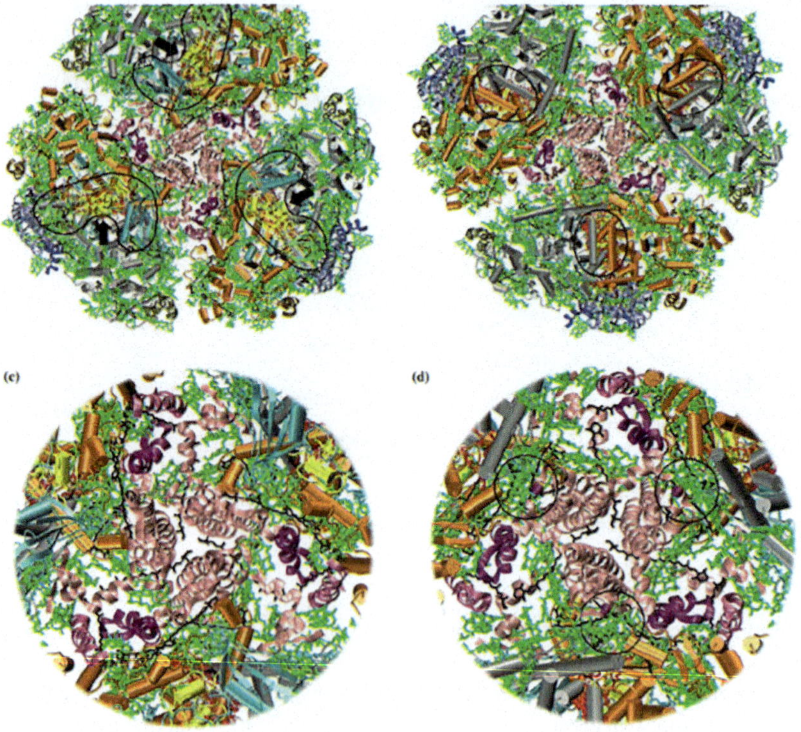
(c)
(d)

◀**Fig. 4.20** Trimeric Photosystem I at 2.5 A resolution. Aschematic picture of the structure of the trimeric Photosystem I as viewed from the stromal side (**a**, **c**) and lumenal side (**b**, **d**) onto the membrane plane. One trimeric complex consists of 3 · 12 proteins (shown in a backbone representation) to which 3 · 127 cofactors are noncovalently bound. The molecular weight of the whole trimeric Photosystem I complex exceeds 1000,000 Da. The center of one monomeric unit is built by the large subunits PsaA (silver) and PsaB (orange), which are shown in a cartoon representation with alpha helices shown as columns and beta sheets as arrows. The small membrane intrinsic subunits are depicted in a ribbon representation: PsaF is shown in ice blue, PsaI in purple, PsaJ in blue, PsaK in tan, PsaL in pink, PsaM in mauve and PsaX in ochre. The membrane extrinsic stromal proteins are depicted in: yellow (PsaC), cyan (PsaD), and lime (PsaE). The 96 chlorophylls per monomer are depicted in green, except the six chlorophylls of the electron transfer chain, ETC, which are shown in red. Further pigments are: two phylloquinones of the ETC (orange) and the three FeS clusters (yellow/S and red/Fe). The carotenoids can be seen in black, and the four lipids that have been identified in the structure are cyan. The colors are the same in all panels. a View of PS I from the stromal side. The stromal hump that contains the three extrinsic proteins PsaC, PsaD, and PsaE is outlined in black. It provides the docking site for ferredoxin (indicated by black arrows). PsaC forms the center of the stromal hump; PsaD is located close to the trimerization domain, and PsaE is located on the distal side of the hump; b depicts the PS I trimer as it can be seen from the lumenal side of the membrane. The black oval marks the potential docking site for the soluble electron donor proteins plastocyanin or cytochrome c6. The chlorophylls of the primary electron donor P700 can be partially seen below the prominent surface helices that provide the main interaction site for the soluble electron carriers; c shows a magnification of the trimerization site from the stromal side. The main contacts in the middle of the trimer (the trimerization domain) are provided by the three PsaL subunits (pink) of the respective monomers. Other proteins, such as PsaD (cyan), PsaI (purple), or the membrane intrinsic PsaM (mauve), are forming further contacts between the monomers, either directly or indirectly by interaction with bridging cofactors or interaction with PsaL. The trimerization domain contains several chlorophylls that are organized in a "circle" around PsaL and can serve as sites for excitation energy transfer between different monomers; d depicts a lumenal view of the trimerization domain. The three calcium ions are especially remarkable. Calcium has been shown to be important for the trimerization of PS I. These three calcium ions engage in hydrogen bonds with PsaL and PsaA of one monomer and PsaL of the other monomer (as depicted within the area of the black circles [85] with permission fr. m Springer Nature

Structure and action mechanism of Oxygen Evolution Complex of Photosystem II in particular is described in Section *X*

4.5 Magnetic Isotope Effects

Magnetic isotope effects arise when a chemical reaction involves spin-selective processes proceeded via the radical pair mechanism [93–99]. The result is that some isotopes react preferentially, depending on their nuclear spin quantum number In contrast to classical isotope effect (CIE), which is governed by chemical energy of the starting and transition states of reactant molecules, this new isotope effect is controlled by magnetic interactions. The first mass-independent isotope effect which chemically discriminates isotopes by their nuclear spins and nuclear magnetic moments rather than by their masses was detected in 1976 by Buchachenko et al.

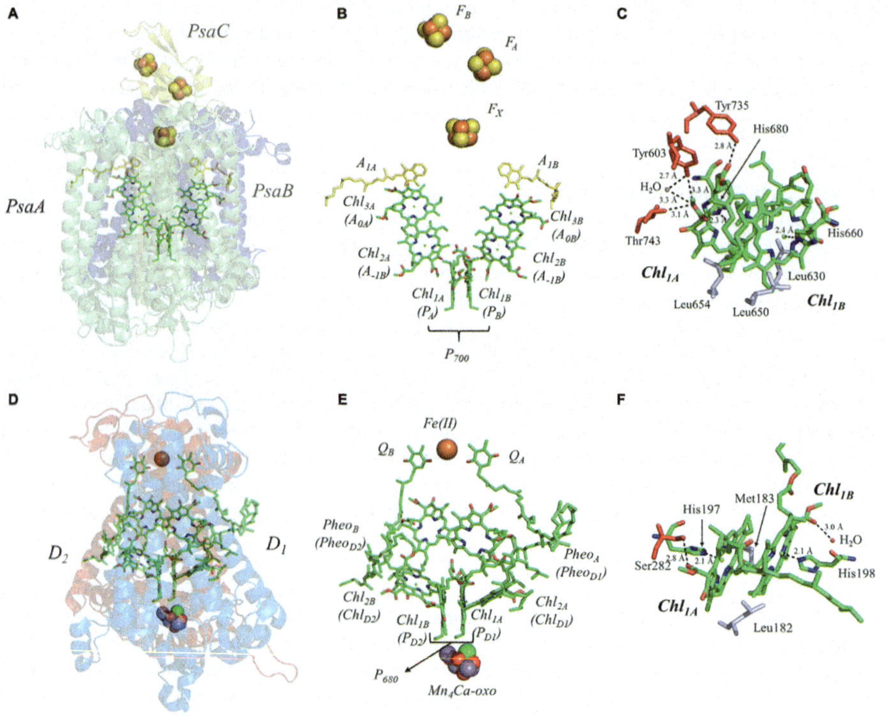

Fig. 4.21 **A** The core subunits (PsaA, green; PsaB, blue; and PsaC, yellow) and cofactors of PS I as observed by X-ray crystallography (PDB ID: 1jb0), **B** the cofactors that participate in the primary electron transfer pathway of PS I and **C** the binding pocket of the primary donor, P700. **D** The core subunits (D1, red; D2, blue) and cofactors of PS II as observed by X-ray crystallography (PDB ID: 3wu2), **E** the cofactors that participate in the primary electron transfer pathway of PS II and **F** the binding pocket of the primary donor, P680. Commonly used labels are in parentheses. Part (C, F) emphasize the prominent protein-matrix effects of the primary donor of PS I and PS II, respectively, where the residues that are hydrogen bonded to the primary donors are shown in red, and nearby nonpolar or π-stacked residues are colored in gray [87]. Published under Creative Commons CC–BY license from Frontiers

[93–97] and by Sagdeev and Molin group [98]. This new phenomenon was independently confirmed within by the groups of Buchachenko, Turro and Pines, and others [94–97, 99]. In particular, on the basis of observed magnetic isotope and magnetic field effects in enzymatic ATP synthesis catalyzed by magnesium ion, a new, ion-radical mechanism of ATP synthesis was formulated [95–99], This mechanism was suggested for enzymatic adenosine triphosphate (ATP) and deoxynucleic acid (DNA) synthesis by using catalyzing metal ions with magnetic nuclei (^{25}Mg, ^{43}Ca, ^{67}Zn) and supported by magnetic field effects on these reactions [99].

References

1. Page MI, Jencks WP (1971) Entropic contributions to rate accelerations in enzymic and intramolecular reactions and the chelate effect. Proc Nat Acad Sci 68(8):1678–1683
2. Kazemi M, Flynn A, Hammes GG (2011) Flexibility diversity and cooperativity pillars of enzyme catalysis. Biochemistry 50(48):10422–10430
3. Jencks WP (1975) Binding energy, specificity, and enzymic catalysis: The Circe effect. Adv Enzymol Related Areas Mol Biol 1(43):219–410
4. Benson SW Thermochemical kinetics: methods for the estimation of thermochemical data and rate parameters 2ND edition https://www.amazon.com]/
5. Milstien S, Cohens LA (1970) Rate acceleration by stereopopulation control: models for enzyme action. Proceed Nat Acad Sci 67:31143–31147
6. Storm DR (1970) A source for the special catalytic power of enzymes: orbital steering. PNAS 66(2):445–452
7. Alexandrov IV (1976) Dependence of the rate of the concerted chemical reactions on the number of degree of freedom involved in the transition. Teor Exper Khim Theor Experim Chem 12:299–306
8. Likhtenshtein GI (2003) New trends in enzyme catalysis and mimicking chemical reactions Kluwer academic/ plenum publishers, Dordrecht Bordwell
9. Likhtenshtein GI (1977) Kinet Catal 28:878–882
10. Likhtenshtein GI (1977) Kinet Catal 28:1255–1260
11. Likhtenshtein GI (1988) Chemical physics of redox metalloenzymes. Springer, Heidelberg
12. Bernasconi CF (1992) The principle of non-perfect synchronization. Chem Res 25:9–16
13. Denisov ET, Sarkisov OM, Likhtenshtein GI (2003). Chemical kinetics: fundamentals and recent developments, Elsevier Science
14. Sakai S (2000) Theoretical analysis of concerted and stepwise mechanisms of diels−alder reaction between butadiene and ethylene. J Phys Chem A 104(5):922–927
15. Kyriacou P Novel reactions of donor-acceptor cyclopropanes, and diels—alder approach. Towards Fargesine, and Fumimycin https://ir.lib.uwo.ca/cgi/viewcontent.cgi?article=5664&context=etd the
16. Wakayama H, Sakai S (2007) Theoretical analysis of concerted and stepwise mechanisms of Diels–Alder reactions of butadiene with silaethylene and disilene. J Phys Chem A 111(51):13575–13582
17. Bekele T, Christian CF, Lipton MA, Singleton DA (2005) Concerted' transition state, stepwise mechanism dynamics effects in C2–C6 Enyne Allene Cyclizations. J Am Chem Soc 127(25):9216–9223
18. Costentin C, Robert M, Savéant JM (2007) Concerted proton−electron transfer reactions in water are the driving force and rate constant depending on pH when water acts as proton donor or acceptor? J Am Chem Soc 129(18):5870–5879
19. Po H, Wang S, Zhang Y (2008) Highly dissociative and concerted mechanism for the nicotinamide cleavage reaction in Sir2Tm enzyme suggested by Ab Initio QM/MM molecular dynamics simulations. J Am Chem Soc 130(49):16721–16728
20. Islam Z, Strutzenberg TS, Gurevic I, Kohen A (2014) Concerted versus stepwise mechanism in thymidylate synthase. J Am Chem Soc 136(28):9850–9853
21. Carreras CW, Santi DV (1995) The catalytic mechanism and structure of thymidylate synthase. Annu Rev Biochem 64:721
22. Ruszczycky MW, Liu HW (2021) Distinguishing concerted versus stepwise mechanisms using isotope effects on isotope effects. Biochemistry 60(46):3416–3418
23. Belasco JG, Albery WJ, Knowles JR (1986) Energetics of proline racemase: double fractionation experiment, a test for concertedness and for transition-state dominance. Biochemistry 25(9):2552–2558
24. Hine J (1977) The principle of least nuclear motion. Adv Phys Org Chem 15:1–6
25. Bruice TC, Benkovic SJ (2000) Chemical basis for enzyme catalysis. Biochemistry 39:6267–6274

26. Villà J, Štrajbl M, Glennon TM et al (2000) How important are entropic contributions to enzyme catalysis? Proc Natl Acad Sci USA 97(22):11899–11904
27. Deng H, Callender R (1999) Raman spectroscopic studies of the structure, energies, and bond distortions of substrates bound to enzymes. In: Schramm VL, Purich DL (eds) Methods in enzymology 308, enzyme kinetics and mechanism, Part E. Academic Press, San Diego, pp 176–215
28. Vocadlo DJ, Davies GJ, Laine R, Withers SG (2001) Catalysis by hen egg-white lysozyme proceeds via a covalent intermediate. Nature 412:835–838
29. Marcus RA (1956) On the theory of oxidation-reduction reactions involving electron transfer. J Chem Phys 24:966–978
30. Marcus RA (2020) Reflections on electron transfer theory cite as. J Chem Phys 153:210401
31. Hush NS (1958) Adiabatic rate processes at electrodes. Energy-Charge Relat J Chem Phys 28(5):962–972
32. Aubry S (2007) A nonadiabatic theory for electron transfer and application to ultrafast catalytic reactions. J Phys Condensed Matter 19(25):255204
33. Levich VG, Dogonadze R (1959) Quantum mechanical theory of electron transfer in polar media. Dokl Acad Nauk 78:2148–2153
34. Likhtenshtein GI (2012), Solar energy conversion. Chemical aspects, First Edition. (Wiley-VCH)
35. Dogonadze RR, Kuznetsov AM, Zakaraya MG, Ulstrup J (1979) A quantum theory of low-temperature chemical and biological rate processes in condensed media, Tunn Biol Syst Proc Symp, 145–171
36. Likhtenshtein GI (1976) Spin labeling method in molecular biology. Wiley Interscience, NY
37. Cukier RI (2002) A theory that connects proton-coupled electron-transfer and hydrogen-atom transfer reactions. J Phys Chem B 106(7):1746–1757
38. Hammes-Schiffer S, Soudackov AV (2008) Proton-coupled electron transfer in solution, proteins, and electrochemistry. J Phys Chem B 112(45):14108–14123
39. Bramwell ST, Giblin SR, Calder S, Aldus R, Prabhakaran D, Fennell T (2009) Measurement of the charge and current of magnetic monopoles in spin ice. Nature 461(7266):956–959
40. Kokkanen AA, Kuznetsov AM, Medvedev IG (2008) Approximate method for calculation of electron transition probability for simple outer-sphere electrochemical reactions. Russ J Electrochem 44(4):397–407
41. McConnell HM (1961) Intramolecular charge transfer in aromatic free radicals. J Chem Phys 35:508–515
42. Stuchebrukov AA, Marcus RA (1995) Theoretical study of electron transfer in ferricytochromes. J Phys Chem 99(19):7581–7590
43. Scourotis SS, Beratan DN (1995) Theories of structure–function relationships for bridge-mediated electron transfer reaction, in Advances in Chemical Physics, vol 107, Part 1 (eds Jortner J, Bixon M), (John Wiley & Sons, Inc. NY), 377–452
44. Zamaraev KI, Molin YN, Salikhov KM (1981) Spin exchange, in theory and physicochemical application. Springer, Heidelberg
45. Cook WR, Coalson RD, Evans DG (2009) Effectiveness of perturbation theory approaches for computing non-condon electrontransfer dynamics in condensed phases. J Phys Chem B 113(33):11437–11447
46. Tanimura Y, Leite VBP, Onuchic JN (2002) The energy landscape for solvent dynamics in electron transfer reactions: a minimalist model. J Chem Phys Melville Amer Inst Physics 11(5):2172–2179
47. Likhtenshtein GI (1996) Role of orbital and dynamic factors in electron transfer in reaction centers of photosynthetic systems. J Phochem Photobiol A Chem 96:79–92
48. Likhtenshtein GI (1974) Spin Labeling Method in Molecular Biology. Nauka, Moscow
49. Denisov ET, Sarkisov OM, Likhtenshtein GI (2003) Chemical kinetics. Fundamentals and recent developments. (Elsevier Science)

50. Winkler JR, Nocera DG, Yocom KM, Bordignon E, Gray HB (1982) Electron-transfer kinetics of pentaammineruthenium(III)(histidine-33)-ferricytochrom e c. Measurement of the rate of intramolecular electron transfer between redox centers separated by 15 Å in a protein. J Am Chem Soc 104:5798–5800
51. Gray HB, Winkler JR (2005) Long-range electron transfer. PNAS 103(10):3534–3539
52. Winkler JR, Gray HB (2014) Long-range electron tunneling. J Am Chem Soc 136:2930–2939
53. Newton MD (1997) Medium reorganization and electronic coupling in long-range electron transfer. J Electroanal Chem 438(1–2):3–10
54. Voityuk AA (2011) Long-range electron transfer in biomolecules. Tunn Hopping J Phys Chem B 115(42):12202
55. Lagunas A, Alejandra Guerra-Castellano A, Nin-Hill AB et al (2018) Long distance electron transfer through the aqueous solution between redox partner proteins. Nat Commun 9:5157. https://doi.org/10.1038/s41467-018-07499
56. Page CC, Moser CC, Dutton PL (2003) Mechanism for electron transfer within and between proteins. Curr Opin Chem Biol 7(5):551–556
57. Davis WB, Ratner MA, Wasielewski MR (2001) Conformational gating of long distance electron transfer through wire-like bridges in donor–bridge–acceptor molecules. J Am Chem Soc 123(32):7877–7886
58. Khoshtariya DE, Dolidze TD, Shushanyan M, van Eldik R (2010) Fundamental signatures of short—and long-range electron transfer for the blue copper protein Azurin at Au/SAM junctions. PNAS 107(7):2757–2762
59. Winlker JJ, Dmochowski IJ, Dawson JH, Winkler JR, Gray HB (1999) Substrates for rapid delivery of electrons and holes to buried active sites in proteins. Angew Chem Int Ed 38:90–92
60. Moser CC, Dutton PL (1992) Engineering protein structure for electron transfer function in photosynthetic reaction centers. Biochim et Biophys Acta Bioenerg 1101(2):171–176
61. Likhtenshtein GI, Syrtsova LA, Samuilov VD, Frolov EN, Borisov AU (1975) Bogatyrenko VR) XII International botanic congress: theses of reports. Nauka, Leningrad, p 429
62. Kulikov AV, Bogatyrenko VR, Melnikov AV, Syrtzova LA, Likhtenshtein GI (1979) Biofisika 24(2):178–183
63. Likhtenshtein GI, Kotel'nikov AI, Kulikov AV, Syrtsova LA, Bogatyrenko VR, Mel'nikov AI, Frolov EN, Berg AI (1979) Some peculiarities of the electronic transfer in redox enzymes. Intern J Quant Chem 16(3):419–435
64. Likhtenshtein GI (1979) Multinuclear redox metalloenzymes. Nauka, Moscow
65. Likhtenshtein GI, Kulikov AV, Kotelnikov AI, Bogatyrenko VR (1982) Photobiochem Photobiol 3(2):178–182
66. Adachi T, Kitazumi Y, Shirai O, Kano K (2020) Direct electron transfer-type bioelectrocatalysis of redox enzymes at nanostructured electrodes. Catalysts 10(2):236
67. Jeuken LJC, Jones AK et al (2002) Electron-transfer mechanisms through biological redox chains in multicenter enzymes. J Am Chem Soc 124(20):5702–5713
68. Likhtenshtein GI (2008) Pure Appl Chem 80(10):2125–2139
69. Likhtenshtein GI (1988) J Mol Catal 48(1):129–138
70. Jordan P, Fromme WHT, Klukas O, Saenger W, Krau N (2001) Three-dimentional structure of cyanobacterial photosystem I at 2.5 Å resolution. Nature 411:909–917
71. Gisriel C, Sarrou I, Ferlez B (2017) Structure of a symmetric photosynthetic reaction center–photosystem. Science 357(6355):1021–1025
72. Cherepanov DA, Ivan V, Shelaev IV, Fedor E (2021) Gostev FE: Primary charge separation within the structurally symmetric tetrameric Chl2APAPBChl2B chlorophyll exciplex in photosystem I. J Photochem Photobiol B 217:112154
73. Duan H-G, Prokhorenko VI, Wientjes E, Croce R, Thorwart M, Miller RJD (2017) Primary charge separation in the photosystem II reaction center revealed by a global analysis of the two-dimensional electronic spectra. Sci Rep 7:12347
74. Kojima R, Yamamoto H, Azai C, Uragami C, Hashimoto H, Kosumi D, Oh-oka H (2020) Energy transfer and primary charge separation upon selective femtosecond excitation at 810 nm in the reaction center complex from Heliobacterium modesticaldum. J Photochem Photobiol A Chem 1(401):112758. https://doi.org/10.1016/j.jphotochem.2020.112758

75. Pierella Karlusich JJ, Carrillo N (2017) Evolution of the acceptor side of photosystem I: ferredoxin, flavodoxin, and ferredoxin-NADP+ oxidoreductase. Photosynth Res 134:235–250
76. Sirohiwal A, Neese F, Pantazis DA (2020) Protein matrix control of reaction center excitation in photosystem II. J Am Chem Soc 142:18174–18190
77. Song Y, Sechrist R, Nguyen HH et al (2021) Excitonic structure and charge separation in the heliobacterial reaction center probed by multispectral multidimensional spectroscopy. Nat Commun 12:2801
78. Chauvet A, Josephine Sarrou J, Lin S et al (2013) Temporal and spectral characterization of the photosynthetic reaction center from Heliobacterium modesticaldum. Photosynth Res 116:1–9
79. Kamran M, Friebe VM, Delgado JD et al (2015) Demonstration of asymmetric electron conduction in Pseudosymmetrical photosynthetic reaction Centre proteins in an electrical circuit. Nat Commun 6(1):6530
80. Rappaport F, Guergova-Kuras M, Nixon PJ, Diner BA, Lavergne J (2002) Kinetics and pathways of charge recombination in photosystem II. Biochemistry 41:8518–8527
81. Amunts A, Toporik H, Borovikova A, Nelson N (2010) Structure determination and improved model of plant photosystem. J Biol Chem 285(5):3478–3486
82. Caspy I (2018) Nelson N: structure of the plant photosystem I. Biochem Soc Trans 46(2):285–294
83. Mazor Y, Borovikova A, Nelson N (2015) The structure of plant photosystem I super-complex at 2.8 Å resolution. Elife 15(4):e07433
84. Kubota-Kawai H, Mutoh R, Shinmura K, Sétif P, Nowaczyk MM, Rögner M, Ikegami T, Tanaka H, Kurisu G (2018) X-ray structure of an asymmetrical trimeric ferredoxin–photosystem I complex. Nat Plants 4(4):218–224
85. Grotjohann I, Fromme P (2005) Structure of cyanobacterial photosystem I. Photosynth Res 85(1):51–72. https://doi.org/10.1007/s11120-005-1440-1444
86. Lubitz W (2006) EPR studies of the primary electron donor P700 in photosystem I. In: Golbeck JH (ed) Photosystem I: the light-driven plastocyanin: ferredoxin oxidoreductase. Advances in Photosynthesis and Respiration, (Publishers: Springer, Dordrecht) 24, 245–269
87. Gorka M, Baldansuren A, Malnati A, Gruszecki E, Golbeck JH, Lakshmi KV (2021) Shedding light on primary donors in photosynthetic reaction centers. Front Microbiol 12:735666. https://doi.org/10.3389/fmicb.2021.735666
88. Yano J, Kern J, Sauer K, Latimer MJ, Pushkar Y, Biesiadka J et al (2006) Where water is oxidized to dioxygen: structure of the photosynthetic Mn4Ca cluster. Science 314(5800):821–825
89. Dismukes GC (2001) Splitting water. Science 292(5516):447–448
90. Shen JR (2015) The structure of photosystem II and the mechanism of water oxidation in photosynthesis. Ann Rev Plant Biol 66(23):48
91. Wang J, Gisriel CJ, Reiss K, Huang HL, Armstrong WH, Brudvig GW, Batista VS (2021) Heterogeneous composition of oxygen-evolving complexes in crystal structures of dark-adapted photosystem II. Biochemistry 60(45):3374–3384
92. Lubitz W, Chrysina M (2019) CoxN: water oxidation in photosystem II. Photosynth Res 142(1):105–125
93. Buchachenko A, Lawler RG (2017) New possibilities for magnetic control of chemical and biochemical reactions. Acc Chem Res 50(4):877–884. https://doi.org/10.1021/acs.accounts.6b00608
94. Brocklehurst B (1997) Magnetic isotope effects in biology: a marker for radical pair reactions and electromagnetic field effects. Int J Radiat Biol 72(5):587–596. https://doi.org/10.1080/095530097143086
95. Buchachenko AL, Bukhvostov AA, Ermakov KV, Kuznetsov DA (2020) A specific role of magnetic isotopes in biological and ecological systems. Phys Biophys Beyond Prog Biophys Mol Biol 155:1–19
96. Buchachenko A (2015) Magneto-biology and medicine. Nova Biomedical, New York
97. Buchachenko AL, Kuznetsov DA (2006) Magnesium magnetic isotope effect: a key to the mechanochemistry of phosphorylating enzymes as molecular machines. Mol Biol 40(1):9–15

98. Sagdeev RZ, Leshina TV, Kamkha M, Belchenko O, Molin YN, Rezvukhin AA (1977) Magnetic isotopic effect in the triplet sensitized photolysis of dibenzoyl peroxide. Chem Phys Lett 48:89–94
99. Turro NJ, Kraeutler B (1980) Magnetic field and magnetic isotope effects in organic photochemical reactions. A novel probe of reaction mechanisms and a method for enrichment of magnetic isotopes. Acc Chem Res 13:369–377

Chapter 5
Enzymes Mechanisms. (Part 2) Radical Processes

5.1 Cytochromes P450

5.1.1 General Properties and Action Mechanism

Cytochromes P450 (CYPs) are a superfamily of enzymes containing heme as a cofactor that functions as monooxygenases with more than 11,500 isozymes. CP450 (CYPs) are powerful enzyme hydroxylation systems of organic substrates, i.e., steroids, hydrocarbons, organic acids, alcohols, and amines, which operatives in animal and plant tissues and bacteria [1–15]. Historically, these enzymes received their name from "pigment 450" due to the unusual position of the Soret band in UV-Vis absorption spectra of the reduced CO-saturated state.

Cytochromes contain protocheme IX as a cofactor. These enzymes catalyze oxidation processes according to the following general scheme:

$$DH + RH + O_2 \rightarrow ROH + H_2O + D$$

where RH is the substrate and DH is the reducing agent (e.g., NADH) with the use of the appropriate reductase, as well as adrenodoxin, rubredoxin, and cytochrome b_5. Dioxygen, being a weak one-electron oxidant, is activated after the reduction in the enzyme heme coordination sphere.

Though hydroxylation of an organic substrate with dioxygen and biological reducing agents is thermodynamically favorable, serious energetic problems along the reaction pathway are expected. Dioxygen is a very weak one-electron reductant ($E_0 = -0.3$ eV) [6, 7]. and the energy gain in formation of O_2-H bond is small. Therefore, dioxygen cannot abstract a hydrogen atom directly from such inert molecules as saturated hydrocarbons in ambient conditions and is activated after the reduction in the enzyme heme coordination sphere. Midpoint reduction potentials for native reducing agents (ferredoxin, flavin cofactors in NADPH-cytochrome P450 oxidoreductase) are determined as $E_0 = -0.240 \div -0.420$ eV. Such potential is

G. I. Likhtenshtein, *Enzyme Catalysis Today and the Chemistry of the 21st Century*,
https://doi.org/10.1007/978-3-031-51552-1_5

sufficient to reduce the cytochrome P450 heme iron Fe^{3+} to Fe^{+2} ($E_0 = -0.410$ eV), and dioxygen to H_2O_2 by two-electron mechanism (E_0 for the transition $O_2 + H^+ +$ 2e at pH 7 is equal 0. 282 eV). H_2O_2 is essentially a stronger two-electron oxidizing reagent than those dioxygen in a one-electron process. Oxidation of a hydrocarbon by H_2O_2 with the formation of a hydroxylated product and water is a highly exothermic process. For instance, in the reaction $H_2O_2 = H_2O + O$, the estimated loss of enthalpy is $DH_0 = 35{,}7$ kcal/mole, whereas the oxygen atom insertion across the R-H group of hydrocarbons gains about 190 kcal/mole. The next thermodynamically allowed step in the cytochrome P450 reaction is a full or partial electron transfer from Fe^{+2} to H_20_2 (the Fenton-like process) with the formation of a strong oxidant, whose center can be an analog of an OH radical with $E_0 = 2.7$ eV. The appearance of such an active radical would immediately destroy chemical groups in the vicinity of an active site. Nevertheless, the Nature solved this problem by oxygen rebound mechanism for C−H hydroxylation catalyzed by cytochrome P450s.

Information about the structure and activity of cytochromes obtained in different periods was summarized in reviews [4, 7–15].

A comprehensive review [10] summarized data on the mechanism of the cytochrome P450 catalyzed hydroxylation reaction available up to 2011. Separate sections have been devoted to the key intermediates of the process, namely the ferric resting and ferrous states, the oxygenated complex, the peroxo states of P450, and the Compound I and Compound II Ferryl-oxo Species and have been discussed in detail.

Review [2] summed up the important results in this area presented in related articles published up to 2018

5.1.2 *Crystal Structure*

Cinnamate 4-hydroxylase (C4H; CYP73A), a cytochrome P450 monooxygenase uses NADPH-cytochrome P450 reductase as a donor of electrons and hydroxylates cinnamic acid to form 4-coumaric acid in phenylpropanoid metabolism. In order to understand the structure and function of this class of plant P450 enzymes, the enzyme CYP73A33 was characterized [15]. The 1.7 Å resolution crystal structure of CYP73A33, along with the results of the steady-state kinetic analysis and the absorption spectroscopy titration were reported (Fig. 5.1). Obtained data suggest the presence of a putative allosteric substrate-binding site in a hydrophobic pocket on the enzyme surface. In addition, comparing the newly resolved structure CYP73A33 with those of well-investigated cytochromes P450 from mammals and bacteria enabled to identify residues of critical functional importance and revealed a unique sequence signature that is potentially responsible for substrate specificity and catalytic selectivity.

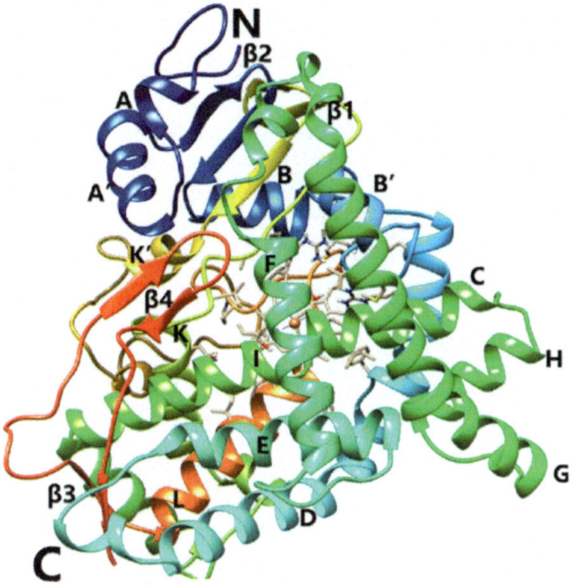

Fig. 5.1 Ribbon diagram representing the crystal structure of SbC4H1. Secondary structure elements have been numbered sequentially as A–L and β1–β4. "N" and "C" refer to the N- and C-terminal regions, respectively. Color compositions indicate different secondary structural elements in SbC4H1. The heme is represented by a stick model. Molecular graphics images were produced using the chimera package (UCSF0) [15]. Published under the CCC's Annual Academic Copyright License from Oxford University Press

Figure 5.2 illustrates the size and shape of the substrate-binding pocket in P450 isozymes based on the crystal structure extracts of protein databank (pdb) files for P450cam (2CPP pdb), (P450BM3) and (P4502D6) [16] and references therein. Specifically, P450BM3 binds long chain fatty acids and hydroxylates these at one of the carbon atoms in the terminus. P4502D6 has a more spherical substrate-binding pocket, and the P450s have a substrate-binding pocket with a size that is dependent on the P450 isozyme. P450cam, a camphor hydroxylating P450 isozyme, where atom C5 of the substrate is regioselectively hydroxylated, has a binding pocket that tightly binds camphor in a specific orientation that enables this reaction.

- CYP199A2, a cytochrome P450 enzyme from *Rhodopseudomonas palustris*, oxidatively demethylates 4-methoxybenzoic acid to 4-hydroxybenzoic acid. 4-Ethylbenzoic acid converts to a mixture of predominantly 4-(1-hydroxyethyl)-benzoic acid and 4-vinylbenzoic acid [17]. The crystal structure of CYP199A2 has been determined at 2.0 Å resolution. The experiment revealed the following: (1) The channel in the enzyme is hydrophobic except for the basic residue Arg246 at the entrance, which may play a role in the specificity of this enzyme for charged benzoates over neutral phenols and benzenes; (2) the substrate-binding pocket is hydrophobic; (3) computer docking of 4-ethylbenzoic acid into the active site suggested that the substrate carboxylate oxygens interact with Ser97 and Ser247; and (4) the beta-methyl group is located over the heme iron by Phe185, the side chain of which is only 6.35 Å above the iron in the native structure.

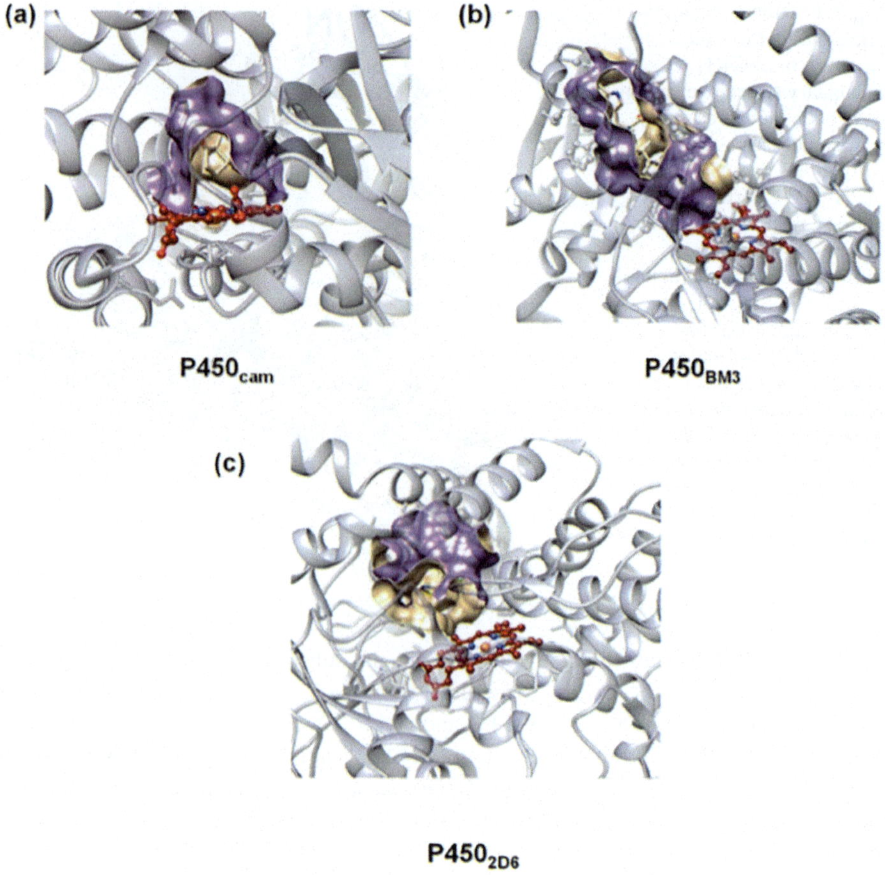

Fig. 5.2 Active site structures of **a** P450cam; **b** P450BM3; and **c** P4502D6 with the substrate-binding pocket shape highlighted in purple and the protein backbone is in light blue and the heme is red [16]. With permission from MDPI

Cinnamate 4-hydroxylase (C4H; CYP73A) forms 4-coumaric acid in phenylpropanoid metabolism. In order to understand the structure and function of class of plant P450 enzymes, the C4H; CYP73A using NADPH-cytochrome P450 reductase as a donor of electrons the enzyme were characterized [15]. The obtained structural information on the 1.7 Å resolution crystal structure of cinnamate 4-hydroxylase (CYP73A3, along with the results of the steady-state kinetic analysis and the absorption spectroscopy titration, displays a high degree of similarity of the structural and functional features of C4H to those of other P450 proteins. The data also suggest the presence of a putative allosteric substrate-binding site in a hydrophobic pocket on the enzyme surface

5.1.3 Mechanism and Active Cite Structure

As is depicted in Figs. 5.3 and 5.4, the catalytic cycle of the cytochromes P450 includes eight distinct states, with three different formal oxidation states of iron, namely (Fe^{2+}, $Fe^{3}$and Fe^{4+}). [10, 14] Active intermediates in heme monooxygenase reactions were revealed by cryoreduction/annealing and EPR/ENDOR studies [7]. Review [7, 8] described the use of cryoreduction/annealing EPR/ENDOR techniques for determining the active oxidizing species in reactions catalyzed by heme monooxygenases. The three candidate heme states are: ferric peroxo, ferric hydroperoxo, and compound I intermediates. Include cytochromes P450, nitric oxide synthase and heme oxygenase was also discussed. Key steps of the catalytic cycle of cytochrome P450 are: (1) The substrate-free form is an aquo/hydroxo entity, dioxygen (superoxide) in the oxygenated complex, peroxo-dianion, or protonated hydroperoxide compound 0 (Cpd 0) in {5a, 5b}; (2) after cleavage of the dioxygen bond and release of water, the higher valent "oxo" state compound I(Cpd I) contains a single oxygen atom {6}; and (3) following oxygen atom transfer to substrate, there are potential transient complexes with the product molecule prior.

The general catalytic cycle for oxygen reduction is shown in Fig. 5.4, which starts with the binding of O_2 to a ferrous heme to afford a ferric-superoxo adduct, followed by reduction to ferric−hydroperoxo, and high-valent, iron−oxo (ferryl) intermediates via O−O bond scission.

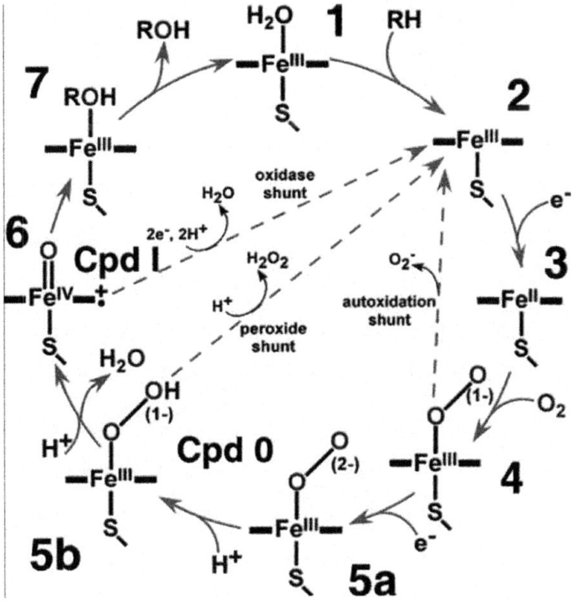

Fig. 5.3 Catalytic cycle of cytochrome P450 showing the intermediates involved in the oxygenase mechanism [10]. With permission from Elsevier

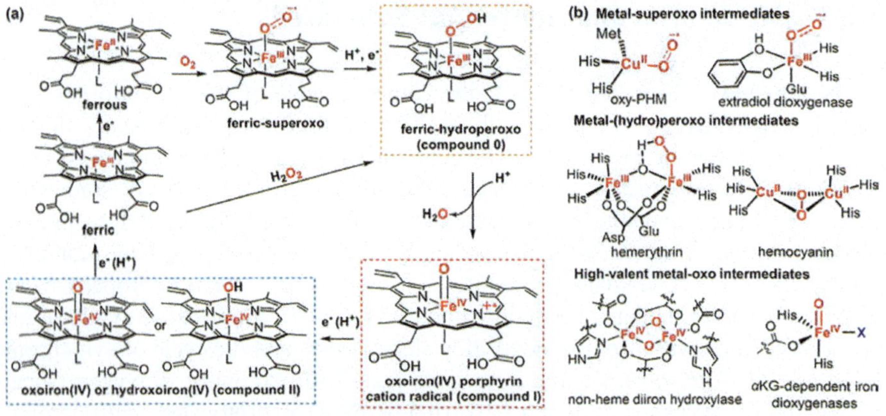

Fig. 5.4 **a** Interaction of dioxygen with heme and heme—oxygen intermediates in heme proteins. **b** Representative metal—oxygen intermediates in metalloproteins (PHM: peptidylglycine α-hydroxylating monooxygenase) [14]. Published under Creative Commons public use license CC-BY from American Chemical Society

The oxygen rebound mechanism (Fig. 5.5), first proposed by Groves in the 1970 [18] 5297 invoked the oxoiron(IV) porphyrin cation radical, compound I, as the active oxygen species. Compound I was proposed to activate aliphatic C–H substrates via hydrogen atom transfer (HAT) to the ferryl oxygen. Subsequently, the incipient substrate radical recombines with the Fe(IV)–OH of compound II (oxygen rebound) to afford hydroxylation products. The main catalytically active intermediate in cytochrome P450s enzymatic cycle is the Compound I (P)Fe(IV)=O, ferryl-oxo complex possessing a π-cation radical localized on porphyrin ring.

Figure 5.6 gives the orbital shapes and Cpd I and Cpd II occupation numbers.

A whole arsenal of physicochemical methods is widely used to study the structure and mechanism of action of cytochromes P450. Spectroscopic studies of the cytochrome P450 reaction mechanisms were the focus of the review [2]. Classical spectroscopic methods, such as electronic absorption (UV-Vis), electron paramagnetic resonance (EPR), nuclear magnetic resonance (NMR), electron nuclear double resonance (ENDOR), Mössbauer, X-ray absorption (XAS), resonance Raman (rR) in the combination of with novel approaches, such as cryoreduction and nanodisc technology have been proved to be powerful tools in providing detailed mechanistic insights into the biophysics and biochemistry. Below are examples of the use of physicochemical methods to study the details of the structure and mechanism of action of cytochromes P450.

Replica exchange molecular dynamics simulations of cytochrome P450 BM-3 in complex with *N*-palmitoylglycine (NPG) were employed to study of conformational change occurs in the active site of the complex where the terminal atoms of the ligand move from a site distant from the heme iron [21]. The population of the proximal state was found to increase with temperature in agreement with UV–vis

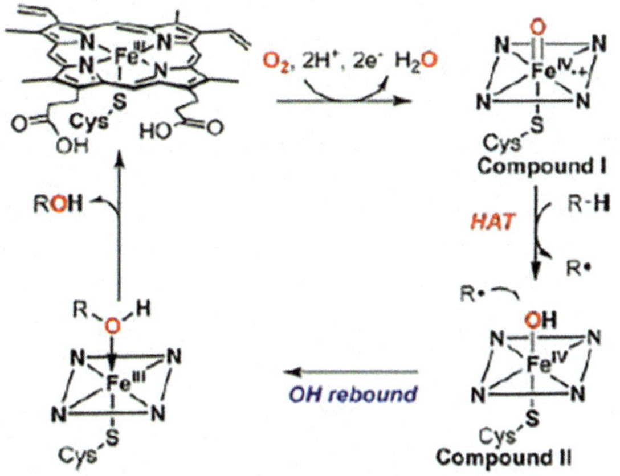

Fig. 5.5 Oxygen rebound mechanism for C–H hydroxylation catalyzed by cytochrome P450s. **a** Interaction of dioxygen with heme and heme—oxygen intermediates in heme proteins. **b** Representative metal—oxygen intermediates in metalloproteins (PHM: peptidylglycine α-hydroxylating monooxygenase) [14]. Published under Creative Commons public use license CC-BY from American Chemical Society

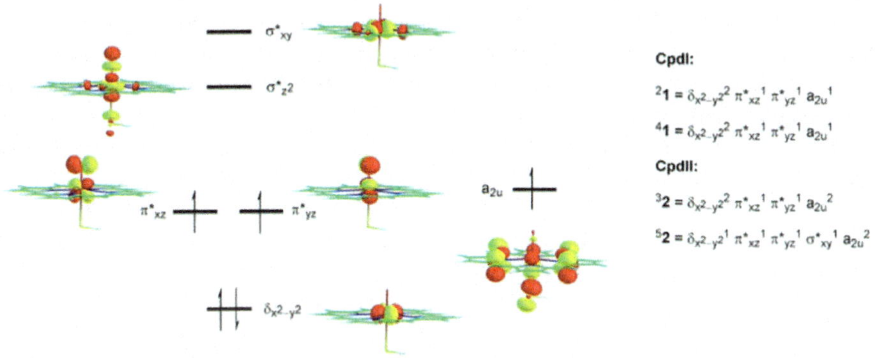

Fig. 5.6 Orbital diagram of Cpd I and Cpd II with occupation in the relevant spin states [16]. With permission from MDPI

absorbance and NMR measurements. The conformations were characterized by X-ray crystallography and computer modeling. The observed increase in the population of proximal conformations indicated that the proximal state is stabilized by conformational entropy. The temperature weighted histogram (T-WHAM) method was used to characterize the transition state regions of the conformational ensemble.

Heme protein cryoreduction is a process of formation of freeze-trap dioxygen adducts of heme proteins within an aqueous buffer containing glycerol (Fig. 5.7) [2]. The process using the 60Co nuclide as source of γ-rays (~ 4 Mrad) included the following steps: (1) Upon subsequent irradiation with γ-rays at 77 K or lower temperature, free-electrons and organic radicals are produced; (2) the electrons, migrate

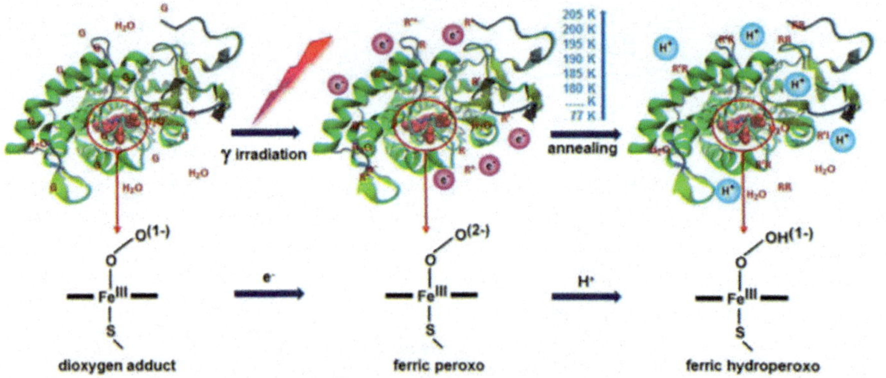

Fig. 5.7 Schematic representation of heme protein cryoreduction; glycerol (G) [2]. With permission from Elsevier

to the Fe-O-O fragment, while other movements, including proton transport, are generally restricted at 77 K and the freeze-trapped Fe-O-O– fragment is generated; (3) the sample can be then annealed to permit proton transfer, thereby generating a trapped hydroperoxo-heme fragment (Fe-O-O-H); (4) subsequent annealing to higher temperatures allowed delivery of another proton to facilitate O-O bond cleavage and formation of Compound I or Compound II species; and (5) cryoradiolyticaly generated intermediates can then be studied by EPR, ENDOR, Mössbauer, UV-Vis, and resonance Raman spectroscopies.

It was also shown that substrate binding to the six-coordinate low-spin (LS) ferric enzyme displaced the water cluster coordinated as the sixth ligand of the heme iron and changes the spin state to the five-coordinate high spin (HS). The 5cHS ferric substrate-bound form exhibited more positive reduction potential, rising by 80–130 mV, resulting in much easier reduction to a ferrous state.

Work [23] extended the characterization of bacterial reductase from *Bacillus megaterium* (3A4-BMR) chimeras using differential scanning calorimetry to evaluate stabilizing role of BMR. A chimeric molecule is an engineered construct in which two or more components are linked to form a novel biological agent. The "molecular Lego" approach to CYP19A1 (aromatase) and the data showed that the activity of the chimeras is very low (0.003 min^{-1}) for all the constructs tested with a different linker loop length: ARO-BMR, ARO-BMR-3GLY, and ARO-BMR-5GLY. The "Lego" approach is based on pulling the proteins apart and rebuilding them, segment by segment. In contrast, the fusion to BMR showed a remarkable effect on thermal stability studied by differential scanning calorimetry as indicated by the increase in T_{onset} by 10 °C and the presence of a cooperative unfolding process driven by the BMR protein domain.

Pressure-induced transitions in the heme domain of cytochrome P450BM3 (P450BMP) were studied versus the concentration of palmitic acid [24]. The experiments indicated that: (1) An increase in hydrostatic pressure causes a high- to low-spin shift and subsequent P450 to P420 transition. which was associated with important conformational and hydration changes of the protein; (2) the interactions of P450 2B4 (2B4) and P450BMP with their substrates (benzphetamine and palmitic acid) were accompanied by larger changes in the partial volume of the proteins (+ 267 and + 248 mL/mol, respectively) than the interactions of P450cam with camphor (+ 106 mL/mol); (3) the reaction volumes of the low- to high-spin transitions of the substrate-free cytochromes (20–23 mL/mol) were consistent with the displacement of one water molecule; and (4) the volume changes in the high- to low-spin transition of the substrate-bound P450cam, 2B4, and P450BMP were found to be – 90, – 49, and – 16 mL/mol correspondingly.

Authors of work [25] used high hydrostatic pressure in combination with spin labeling, ESR, and fluorescence spectroscopy as a tool for exploring the conformational landscape of human cytochrome P450 3A4 (CYP3A4). Points of incorporation of the environment-sensitive EPR label (MTSSL) and fluorescent probe (MDCC) are indicated in Fig. 5.8. Site-directed incorporation of a luminescence resonance energy transfer donor–acceptor pair allowed to identify a pressure-dependent equilibrium between two states of the enzyme. Increase in pressure increased the spatial separation between the two distantly located fluorophores. The transition is characterized by volume change ($\Delta V°$) and P1/2 values of – 36.8 ± 5.0 mL/mol and 1.45 ± 0.33 kbar, respectively, which corresponds to a $K_{eq}°$ of 0.13 ± 0.06, This pressure-promoted displacement of the equilibrium is eliminated by the addition of testosterone, an allosteric activator. Using site-directed spin labeling demonstrated that: (1) The testosterone-sensitive transition is revealed by pressure-induced changes in the electron paramagnetic resonance spectra of a nitroxide side chain placed at position 85 or 409 of the enzyme; (2) an effector-dependent conformational equilibrium between open and closed states of CYP3A4 occurs; and (3) the process involves a pronounced change at the interface between the region of α-helices A/A' and the meander loop of the enzyme, where residues 85 and 409 are located. Points of incorporation of the environment-sensitive EPR label (MTSSL) and Fluorescent probe (MDCC). The cartoon representation of the structure of CYP3A4 is based on the coordinates of the enzyme complex with peripherally bound progesterone (PDB ID: 1W0F). Side chains of the naturally present cysteine residues C64, C98, and C468 are shown in white spheres. Gray spheres show the residues in the regions of the meander and C-terminal loops, respectively, which were chosen for label incorporation on the basis of our studies of high-pressure adaptation in CYP261 enzymes. Residues A121 and S495 are shown in black spheres Points of incorporation of the environment-sensitive EPR label (MTSSL) and fluorescent probe (MDCC). The cartoon representation of the structure of CYP3A4 is based on the coordinates of the enzyme complex with peripherally bound progesterone (PDB ID: 1W0F). Side chains of the naturally present cysteine residues C64, C98, and C468 are shown in white spheres. Gray spheres show the residues in the regions of the meander and C-terminal loops, respectively, which were chosen for label incorporation on the basis

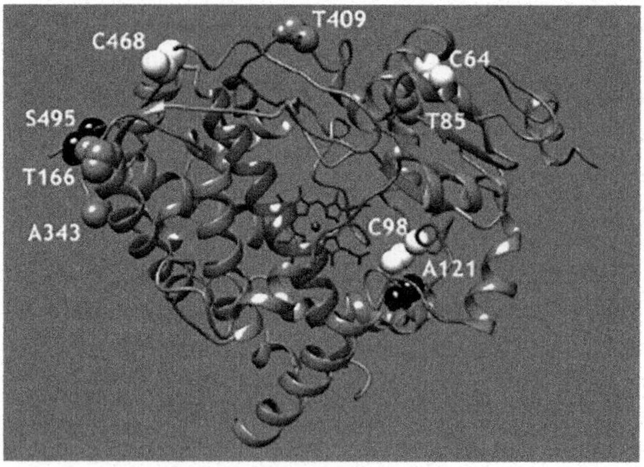

Fig. 5.8 Points of incorporation of the environment-sensitive EPR label (MTSSL) and fluorescent probe (MDCC). The cartoon representation of the structure of CYP3A4 is based on the coordinates of the enzyme complex with peripherally bound progesterone (PDB ID: 1W0F). Side chains of the naturally present cysteine residues C64, C98, and C468 are shown in white spheres. Gray spheres show the residues in the regions of the meander and C-terminal loops, respectively, which were chosen for label incorporation on the basis of our studies of high-pressure adaptation in CYP261 enzymes. Residues A121 and S495 are shown in black spheres. Pressure-Perturbation Study of CYP3A4 [25]. With permission from Elsevier

of our studies of high-pressure adaptation in CYP261 enzymes. Residues A121 and S495 are shown in black spheres

An approach to study protein-protein interactions, using Fourier transform infrared spectroscopy in combination with titration experiments and principal component analysis (FTIR-TPCA), was developed [26]. The FTIR-TPCA results were obtained for the interaction between cytochrome P450 and the redox partner protein in two P450 systems, *the Pseudomonas putida* P450cam (CYP101) with putidaredoxin (P450cam-Pdx), and the *Bacillus megaterium* P450BM-3 (CYP102) heme domain with the FMN domain (P450BMP-FMND). The structural changes that occur upon redox partner complex formation involved the following: (1) An increase in beta-sheets and alpha-helix content; (2) a decrease in the population of random coil/3(10)-helix structure; (3) a redistribution of turn structures within the interacting proteins; and (4) changes in the protonation states or hydrogen-bonding of amino acid carboxylic side chains.

Many substrates and inhibitors of Cytochrome P450 (CYP) monoxygenses alter the spin state of the ferric heme by displacing the heme's axial water ligand in the resting enzyme to yield a five-coordinate iron complex, or they replace the axial water to yield a nitrogen-ligated six-coordinate iron complex [27]. Pulsed and continuous wave (CW) EPR demonstrated that water-bridged complexes are common among a range of nitrogenous drugs or drug fragments that bind to CYP3A4 or CYP2C9. Principal component analysis revealed a distinct clustering of CW EPR spectral parameters for water-bridged complexes. CW EPR indicated heterogeneous mixtures of ligated states, including multiple directly-coordinated complexes and water-bridged complexes.

Crystallographic and kinetics studies have shown that the F429H mutation of cytochrome P450 2B4 introduced an H-bond between His429 and the proximal thiolate ligand, Cys436. The mutation did not alter the protein fold but sharply decreases the enzymatic activity and stabilizes the oxyferrous P450 2B4 complex [28]. To characterize the influence of hydrogen bond on the states of the catalytic cycle of the radiolytic cryoreduction combined with EPR and (electron nuclear double resonance (ENDOR) spectroscopy to study were employed. Analysis experimental data revealed a number important findings: (1) The addition of an H-bond to the axial Cys436 thiolate significantly changes the EPR signals of both low-spin and high-spin heme iron(III) and the hyperfine couplings of the heme-pyrrole ^{14}N; (2) change of H ENDOR spectra of the water ligand in the six-coordinate low-spin ferriheme state indicated that the H-bond introduced between His and the proximal cysteine decreases the extent of S $\rightarrow$ Fe electron donation and weakens the Fe(III)-S bond; (3) the added H-bond changes the primary product of cryoreduction of the Fe(II) enzyme, which is trapped in the conformation of the parent Fe(II) state; (4) in the wild-type (WT) enzyme, the added electron localizes on the porphyrin, generating an $S = (3)/2$ state with the anion radical exchange-coupled to the Fe(II); (5) the additional H-bond has little effect on g values and ^{1}H-^{14}N hyperfine couplings of the cryogenerated, ferric hydroperoxo intermediate but noticeably slows its decay during cryo-annealing; (6) in both the WT and the mutant enzyme, this decay showed a significant solvent kinetic isotope effect, indicating that the decay reflected a proton-assisted conversion to Compound I (Cpd I); (7) Cpd I is formed during the annealing of the cryogenerated hydroperoxy intermediate; and (8) there is the active hydroxylating species in both WT P450 2B4 and the F429H mutant; and (8) the added H-bond of the mutation diminished the reactivity of Cpd I.

The cytochrome P450 enzyme CYP119, its compound II derivative, and its nitrosyl complex were studied by iron K-edge X-ray absorption spectroscopy employing Advanced Photon Source of Argonne National Laboratory at temperatures below 140 (K) [29]. The compound II derivative was prepared by reaction of the resting enzyme with peroxynitrite and had a lifetime of $\approx$10 s at 23 °C. The CYP119 nitrosyl complex was obtained by reaction of the enzyme with nitrogen monoxide gas or with a nitrosyl donor and was stable at 23 °C for hours. The X-ray absorption near-edge structure spectra displayed shifts in edge and pre-edge energies consistent with increasing effective positive charge on iron in the series native CYP119 < CYP119 nitrosyl complex < CYP119 compound II derivative. The obtained structural features for the compound II derivative were an iron-oxygen bond length of 1.82 Å and an iron-sulfur bond length of 2.24 Å, both of which indicate an iron-oxygen single bond in a ferryl-hydroxide, $Fe^{IV}OH$, moiety.

Pulsed EPR methods were employed for the study of drug binding to heme-thiolate enzymes such as cytochrome P450 and nitric oxide synthase [30]. Pulsed Hyperfine Sublevel Correlation Spectroscopy (HYSCORE) and ENDOR methods to measure ^{1}H of axial ligands of the heme group were described. The experiment illustrated water serving as the axial ligand in the drug-free enzyme and ligands coordinating directly to the heme or through one or more bridging water molecules. It has been shown by absorbance spectroscopy that the CYP3A4 substrate testosterone (TST) induced spin state changes that are characteristic of a low-spin to high-spin conversion [31]. The thermodynamic landscape of testosterone binding to cytochrome P450 3A4 including ligand binding and spin state equilibria was established. EPR was employed to investigate the equilibrium binding of TST to CYP3A4 at [CYP3A4] > K(D), allowing for characterization of the singly occupied state (i.e., CYP3A4.TST). The observance spectroscopy was used to examine equilibrium binding, which was found to be positively cooperative with respect to ligand-dependent spin state conversion. Thus, an analysis of the EPR and absorbance spectral results suggested that the binding of the second TST induces a shift to the high-spin state and that the second TST binding causes displacement of the bound water.

Solution magnetic circular dichroism (MCD) spectra of P450terp were compared with spectra of other P450 enzymes (e.g., P450cam, P450BMP, P450BM-3holo, and P450BM1), as well as with spectra of chloroperoxidase and NO synthase [32]. The following results were reported: (1) Upon substrate-binding, the MCD spectra of ferric P450terp and all other thiolate-ligated heme systems showed a strong Soret band; (2) the band position is unique relative to the typical Soret MCD pattern(s) of catalases or other 5-coordinate ferric heme systems; (3) this intense negative MCD feature can be diagnostic for cysteinate-linked ferric hemes; (4) in the case of ferrous P450s, the intensity of the Soret-region MCD varied between substrate-bound and substrate-free enzymes; (5) a finding was the clear spectral shifts of the Soret MCD band between the substrate-bound and substrate-free forms of ferrous-CO-P450terp; and (6) the band positions for both types of P450terp are red-shifted from known bands of ferrous-CO-P50cam.*x*

The nanodisc approach utilizes an encircling amphipathic membrane scaffold protein (MSP) which can solubilize a circular fragment of phospholipid [9, 33–36]. Nanodiscs provide a powerful tool for a broad spectrum of biochemical and biophysical studies of membrane proteins including P450 [9]. Nanodiscs are composed of phospholipids and an encircling amphipathic helical belt protein, termed a membrane scaffold protein (MSP) (Fig. 5.9) which can solubilize a circular fragment of phospholipid. The initial MSP sequences were based on the Human ApoA1 protein component of high density lipoprotein particles. Other proteins and peptides of similar amphipathic structure can also form discoidal bilayers. These include synthetic amphipathic peptides, apolipophorin, and other apolipoproteins. Empty nanodiscs and nanodiscs with membrane proteins (MP) can be assembled from a detergent solubilized mixture of all components by gradual removal of detergent via adsorption on hydrophobic beads or by dialysis (Fig. 5.10).

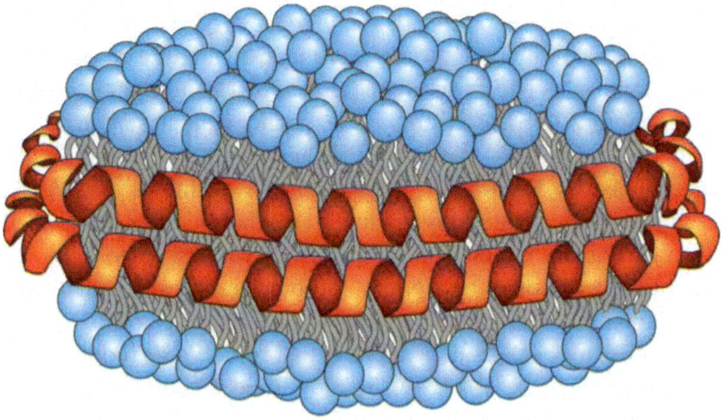

Fig. 5.9 Nanodiscs are discoidal lipid bilayer stabilized by encircling amphipathic helical scaffold proteins termed MSPs [9]. With permission from American Chemiical Society

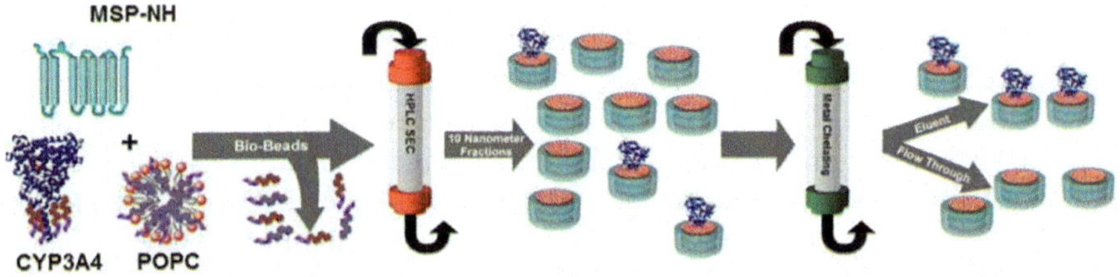

Fig. 5.10 Schematic descriptions of the self-assembly process where detergent solubilized target, lipid, and membrane scaffold protein yield nanodiscs upon removal of detergent [9]. With permission from American Chemiical Society

The demonstration of the nanodiscs approach was reported when microsomal membranes from Sf9 cells overexpressing cytochrome P450 CYP6B1 from black swallowtail butterfly *(Papilio polyxenes)* were solubilized in cholate and used for nanodisc assembly [34, 36]. Direct assembly of nanodiscs from the native membranes, described for the pig kidney a^+, K^+ ATPase, provided an active target solubilized in a lipid environment [35]. Refolding of membrane proteins in nanodiscs from an unfolded state via solubilization in SDS detergent was demonstrated by successful reconstitution of the homotetramer KcsA channel and bacteriorhodopsin.

Review [8] provided a general discourse on nanodisc and its application to solution NMR. This method has been widely used to investigate the structure of nanodiscs and membrane proteins in nanodiscs. Various NMR methods have been used to study empty nanodiscs. Solid state NMR (ss-NMR) revealed the detailed structure of the encircling MSP, details of lipid-MSP interactions and some of the physiologically important interactions of lipid head-groups with cations in solution. The application of nanodiscs for mechanistic studies of steroidogenic human cytochromes P450 using cryo-radiolysis was reported [9]. Enzymes CYP17A1 or CYP19A1 incorporated in nanodiscs at ambient conditions were investigated by stopped flow mixing. Oxygenated cytochromes P450 was characterized by low-temperature spectroscopic methods, including UV-VIS absorption and resonance Raman scattering. The binding of cytochrome b_5 to CYP2B4 assembled in nanodiscs using amphipathic 22-mer peptides was probed by NMR [36]. Reconstitution of the P450-cytb5 complex in peptide-based nanodiscs, containing no detergents, has been demonstrated and characterized by size exclusion chromatography and NMR spectroscopy. In addition, NMR experiments were used to identify the binding interface of the P450-cytb5 complex in the nanodisc.

The interactions of NADPH-dependent cytochrome P450 reductase (CPR) with full-length and truncated CYP2J2 were monitored by the kinetics of NADPH supported reduction in the first electron transfer step [33]. The binding of CPR to human cytochrome P450 CYP3A4 incorporated in POPC nanodiscs was studied by size exclusion chromatography with monitoring of catalytic activity by the rate of NADPH reduction. It was found that CPR shuttles electrons from NADPH through the FAD and FMN-coenzymes into the iron of the prosthetic heme group of the CYP. In the course of these electron transfer reactions, CPR undergoes large conformational changes open" and "compact" conformation detected by NMR and small-angle X-ray scattering (SAXS) (Fig. 5.11) [38]. In the open conformation, the solvent exposed FMN coenzyme favors efficient transfer of electrons from FMN to the heme iron, while the compact conformation serves to facilitate FAD to FMN electron transfer. It was suggested that the rotating movement of the FMN-binding domain as mediated by the flexible hinge could be a key mechanism to enable tight binding of CPR to CYPs in aNADPH-heme electron transfer compatible conformation. A rotation would allow CPR to remain bound to the CYP and shuttle electron transfer between the FADcoenzyme to theFMNcoenzyme and from the reduced FMN to the heme.

Several works use computer methods to solve cytochrome problems that cannot be solved by experimental methods alone. For example, a simulation method, random expulsion molecular dynamics was introduced. [39]. Substrate-exit channels from the buried active site were found by imposing an artificial randomly oriented force on the substrate, in addition to the standard molecular dynamics force field [39]. The computation revealed that the pathways obtained by the random expulsion molecular dynamics method for simulations of the substrate-bound structure of cytochrome P450BM-3 and application to complexes of cytochrome P450cam with different substrates and with product matched with thermal motion pathways obtained by an analysis of crystallographic B-factors. In contrast to large backbone motions observed in cytochrome P450BM-3 for the exit of palmitoleic acid, passage of camphor through cytochrome P450cam only requires small backbone motions in conjunction with side chain rotations. A series of density functional theory (DFT) calculations to predict the ^{57}Fe Mössbauer quadrupole splittings (DeltaEQ) and isomer shifts (deltaFe) for an Fe(IV)=O model compound ($[Fe(O)(TMC)(NCCH3)](OTf)_2$, TMC = 1,4,8,11-tetramethyl-1,4,8,11-tetraazacyclotetradecane, OTf = $CF_3SO_3^-$) as well as a cytochrome P450 reaction intermediate (P450-RI) was carried out [40]. The DFT

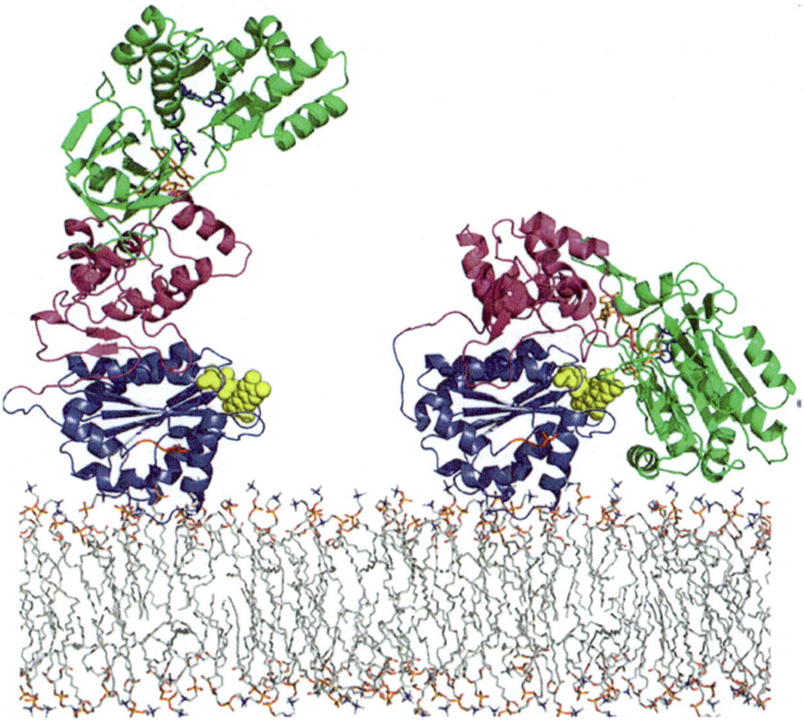

Fig. 5.11 Model of CPR in open and closed conformations in a lipid bilayer. In this model, the FMN-binding domain is positioned at the membrane surface, and the open and closed conformations are illustrated as predicted to be oriented in vivo. The open structure (left) is compatible with FMN-heme electron transfer. The closed structure (right) is optimal for FAD-FMN electron transfer. The FMN domain is shown in blue, linker domain in purple, FAD and NADPH-binding domain in green, FMN coenzyme as yellow spheres, FAD coenzyme as orange sticks and NADPH as blue sticks [38].

predictions on the model compound were in agreement with experiment. The same DFT methods permitted good predictions of both DeltaEQ and deltaFe when using a geometry optimized structure having a porphyrin dianion, a protonated cysteine, and $S = 1$.

In work [41], crystallography and molecular simulation to study the effect of single mutations on binding and regioselective metabolism of mefenamic acid by the heme domain of the protein M11 mutants were combined. A multistep approach was used that combines molecular docking, molecular dynamics (MD) simulation, and binding free-energy calculations to address protein flexibility. In this way, several important findings were obtained, namely: (1) Preferred binding modes that are consistent with oxidation at the experimentally observed sites of metabolism (SOMs) were identified; (2) binding modes in line with the preferred SOMs of mefenamic acid by M11 and its mutants by including protein flexibility and dynamics in free-energy computation were ranked; and (3) structural insights into the change in regioselectivity of mefenamic acid hydroxylation due to single active site mutations were demonstrated.

Interactions between membrane-bound CYP 1A1, an antitumor drug target, and cytochrome P450 (CYP) heme monooxygenases (CPR) by a multiresolution computational approach were investigated [42]. The computation indicated that CPR may affect ligand passage to the CYP active site. It was also shown that in the complexes, the interactions of the CPR FMN domain with the proximal side of CYP 1A1 are supplemented by more transient interactions of the CPR NADP domain with the distal side of CYP 1A1. Several arrangements of CPR around CYP 1A1 that are compatible with ET were identified

5.2 Nitric Oxide Synthase

5.2.1 General

Nitric oxide (NO) is a tiny molecule with enormous biological impact. NO mediates a large number of physiologic and pathophysiologic processes including vascular relaxation, inhibition of platelet aggregation, regulation of endothelial cell adhesivity, preservation of the normal vessel wall structure, etc. [43–64] and references therein. The family of nitric oxide synthases (NOS) has significant importance in various physiological mechanisms and is also involved in many pathological processes. In physiological conditions, besides nitric oxide (NO), nNOS also produces hydrogen peroxide (H_2O_2) and superoxide ($O\cdot{-}2O_2\cdot^-$). For example, in mammals, NO acts as an intercellular messenger in a variety of signaling reactions, as well as a cytotoxin in the innate immune response.

Three isomers of the enzyme are intensively investigated: neuronal (nNOS), endothelial (eNOS), and cytokine-inducable (iNOS). NO is generated in an enzymatic process of oxidation of L-arginine (Arg) by dioxygen catalyzed nitric oxide synthase (NOS) in the presence of NADPH:

$$\begin{aligned} 2\,\mathrm{L} - \underline{\mathrm{arginine}} + 3\underline{\mathrm{NADPH}} + 3\,\mathrm{H}^+ \\ + 4\,\mathrm{O}_2 2\,\underline{\mathrm{citrulline}} + 2\,\underline{\mathrm{nitric}}\ \underline{\mathrm{oxide}} \\ + 4\,\mathrm{H}_2\mathrm{O} + 3\underline{\mathrm{NADP}} \end{aligned}$$

The process involves stepwise oxidation of Arg to N-hydroxyl-L-arg,

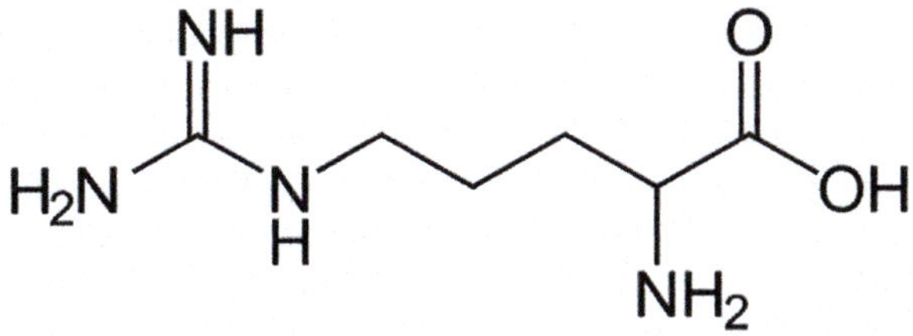

which is converted to cirulline and nitric oxide (Fig. 5.12).

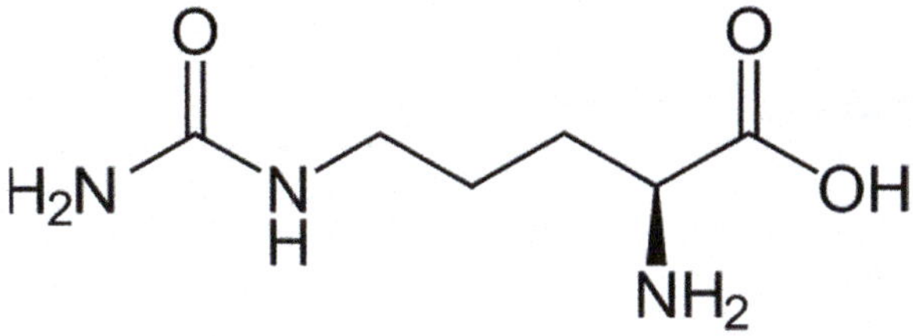

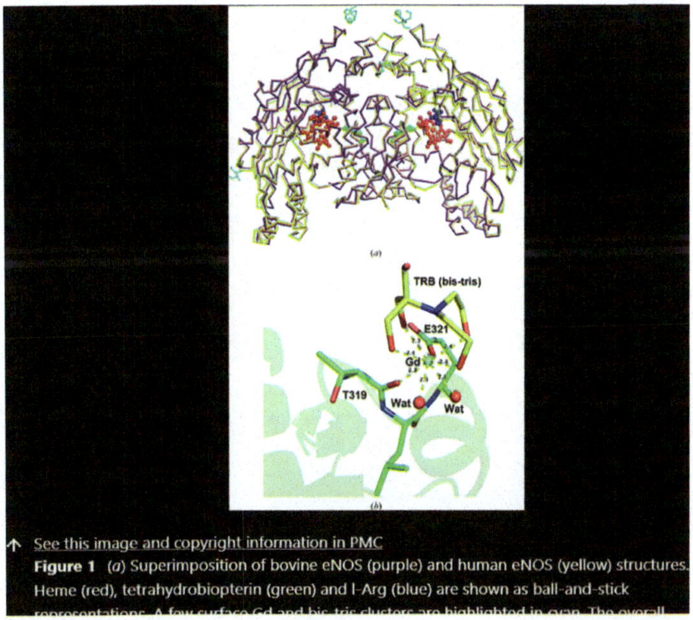

Fig. 5.12 **a** Superposition of bovin eNOS (purle) and 1-Arg (blue) and human eNOS (yellow) structures [53, 54]. With permission from Wiley

In the reactions, the hydroxylase domain of NOS containing heme, the cofactor tetrahydrobiopterin (H_4B), the reductase domain, containing FMN, FAD and NADPH, calmodulin (CaM), and the Arg binding site involved [43–53]. The reductase domain provides electrons for the active site reduction. Calmodulin lies between these two domains and promotes electron transfer from NADH to heme. The H_4B cofactor is indispensable for coupled catalysis [44]. Active intermediates in heme mono reactions of NOC were revealed by EPR/ENDOR studies using cryoreduction/annealing technique [61]

5.2.2 *Structure of NOS and It Active Cite*

Structures of bovin (eNOS), human eNOS and **human nNOSox** dimer are depicted in Figs. 5.12 and 5.13.

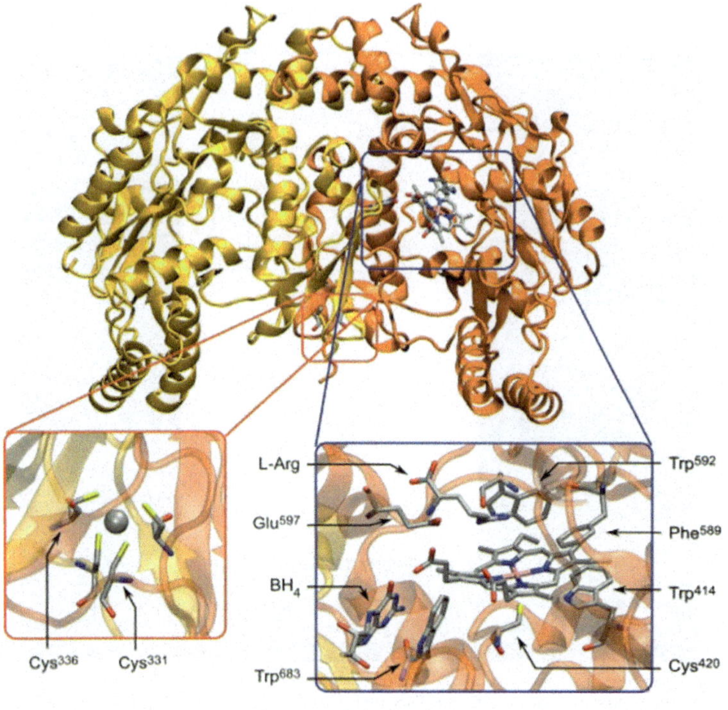

Fig. 5.13 Structure of the human nNOSox dimer (PDB: 4D1N) [43]. Published under CC BY licence

To provide a detailed understanding of the domain orientations and interactions in the iNOS holoenzyme complex, homology models and regulation of interdomain interactions by calmodulin in inducible nitric oxide synthase were proposed using previously solved crystal structures of the murine iNOS domains and CaM [55]. The higher-order domain architecture of iNOS is presented in Fig. 5.14 [61, 62]. For group I, two copies of the reductase homology model were docked manually into the corresponding terminal densities. In group II, one of the FMN subdomains is released from its interaction with the FAD/NADPH subdomain. Group III structures show conformations whereby one of the reductase domains and CaM have folded over to directly interact with the oxidase domain. In these conformations, CaM remains in the same position as in group II

Mammalian NO synthases domain organization is presented in Fig. 5.15. The structures and higher-order domain organization of all three NOS holoenzymes were examined [61]. Three NOS are composed of an N-terminal oxidase domain and a C-terminal reductase domain. Calmodulin (CaM) activates NO synthesis by binding to the helical region connecting these two domains. The structures of NOS with and without CaM bound are similar, consisting of a dimerized oxidase domain flanked by two separated reductase domains.

The crystal structure of the rat nNOS reductase provided insight into the structural features and organization of the enzymes [55]. The structure revealed the extensive network of hydrogen bonds between the three cofactors and the protonation state of the FMN, which appeared as a semiquinone. In addition, inspection of the FMN/FAD domains interface indicated important hydrophobic contacts and salt bridges, especially between the Glu816 and Arg1229 residues, both conserved in cNOSs. Arg1400, present in both nNOS and eNOS (Arg1165), Fig. 5.16 schematically represents NOS reactivity.

The NOS cofactor tetrahydrobiopterin (BH4) is indispensable for coupled catalysis [44]. The rates couple/uncoupled catalysis and the monomer/dimer equilibrium were determined by measuring NADPH oxidation spectrophotometrically at 340 nm. Citrulline synthesis, as the formation of [^{3}H]-citrulline from [^{3}H]-Arg, and by FPLC and low-temperature polyacrylamide gel electrophoresis was performed. In the case of uncoupling at the heme and at the flavins, the absence of Arg and BH4 results in NOS monomerization, with the monomers incapable to sustain NO synthesis but still able to support uncoupled catalysis. Omission of Ca^{2+}/CaM can completely block heme reduction.

Cryoreduction/annealing- EPR/ENDOR experiments with the arginine-bound oxy NOS from *Geobacillus stearothermophilus* confirmed that iron(IV)oxo porphyrin π-cation radical (Fe(IV)=O P^{+}), Compound I (6) (Fig. 5.17), is the reactive heme species in the first stage I of NO production, the hydroxylation of L-arginine to N-hydroxy–L-arginine (NOHA) [60, 61]. The experiment revealed that: (1) Cryoreduction of ferrous oxy gsNOS complex with Arg (or N-Me-Arg) at 77K generates intermediate ferric hydroperoxo intermediate (FeIII-OOH 5); (2) the conversion of 5 to 6 shows a large solvent kinetic isotope effect, sKIE $\cong$8; (3) subsequent annealing of 5 at 145K generates the primary product state of reaction; (4) EPR/^{15}N ENDOR spectroscopy shows NOHA coordinated to the ferric heme; (5) in the second stage

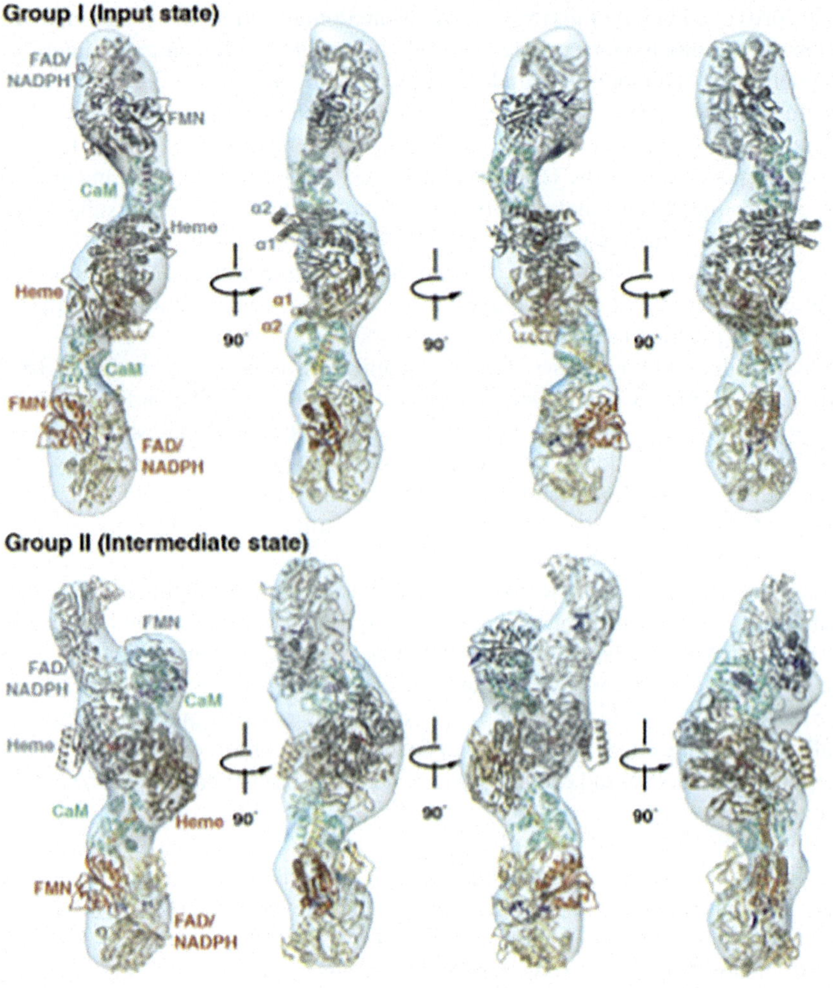

Fig. 5.14 Higher-order domain architecture of iNOS. Crystal structures and homology models of the individual domains and subdomains are shown as ribbon structures and were fit into a representative EM map from the input (group I), the intermediate (group II), and the output (group III) state. The ribbon structures shown in shades of brown belong to one monomer, and those in shades of gray belong to the second monomer. CaM is shown in green, the heme cofactors are shown in red, and FAD and FMN are shown in navy blue. The progression of electron transfer can be observed by tracing the movement of the FMN subdomain in the gray monomer. In group I, the FMN subdomain is bound within the FAD/NADPH subdomain. In group II, CaM has docked to the oxidase domain, and the FMN subdomain has begun to pull away from the rest of the reductase domain. By group III, the FMN subdomain has also docked to the oxidase domain allowing electron transfer from FMN to the heme cofactor in the oxidase domain of the opposite (brown) monomer. The 2D class average used to create the 3D reconstruction, and the corresponding forward projection of the proposed models for each of the structures are shown in Fig. S3. (Scale bar: 10 nm.) [61, 62]. With permission from PNAS

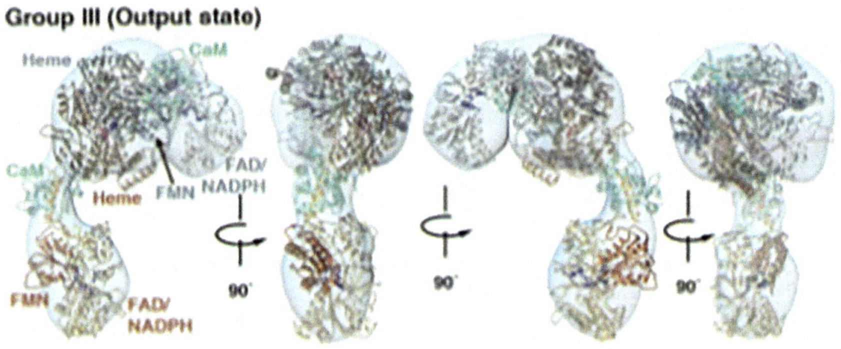

Fig. 5.14 (continued)

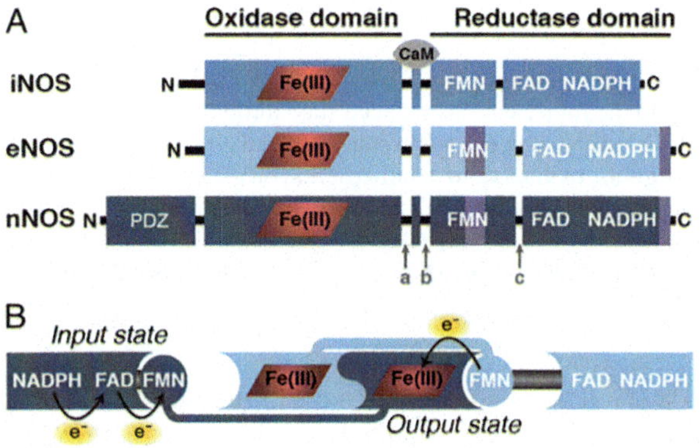

Fig. 5.15 Mammalian NO synthases. **A** NOS domain organization. Each NOS monomer is composed of two domains: the oxidase domain and the reductase domain. The heme cofactor (red) binds within the oxidase domain. The reductase domain is divided further into the FAD/NADPH-binding and the FMN-binding subdomains. CaM binds to a small helical segment that connects the oxidase and reductase domain. Three linker regions connect these domains: linker-a is between the oxidase and the CaM-binding helix; linker-b is between the CaM-binding helix and the FMN subdomain; and linker-c is between the FMN subdomain and the FAD/NADPH subdomain. Although largely similar, each NOS isoform possesses a few unique features. iNOS is the smallest (260 kDa for the homodimer without CaM bound) and lacks the autoinhibitory helix within the FMN subdomain and the C-terminal extension (both shown in purple) present in eNOS and nNOS. eNOS has a molecular weight of 266 kDa for the homodimer, and nNOS is the largest at 321 kDa for the homodimer as a result of an N-terminal extension of ~250 aa containing a PDZ domain. **B** NO synthase electron transfer pathway. NOS is shown as its native homodimer. The monomer on the left shows the input state, and the monomer on the right shows the output state. In the input state, the FMN subdomain interacts with the FAD/NADPH subdomain allowing electron transfer between the flavin cofactors. In the output state, the FMN subdomain interacts with the heme domain allowing electron transfer between FMN and heme [61, 62]. With permission from PNAS

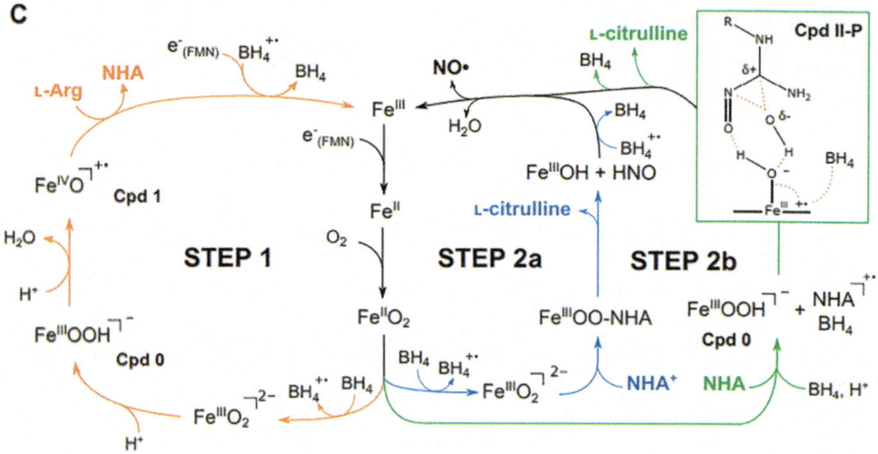

Fig. 5.16 Schematic representation of NOS reactivity. **a** Electron and proton transfers in the reductase domain leading to the four-electrons reduced state through two electrons reduced to the quasi-equilibrium state. **b** Three different oxidation states of the flavin moiety (present in FAD and FMN). **c** Mechanism of L-arginine oxidation: first half-reaction on the left (step1, orange path), second half-reaction on the right, with NHA protonated as reviewed by Santolini et al. (step2a, blue path), and NHA neutral as proposed by Shamovsky (step2b, green path) involving formation of the Cpd II-P intermediate with extensive spin delocalization depicted by red dashed lines [43]. With permission from Elsevier

of NOS catalysis (NOHA is converted to citrulline and NO. The suggested enzymatic cycle of mammalian NOSs under physiological conditions included several steps (Fig. 5.18): (1) Tetrahydrobiopterin cofactor reduces the ferrous oxy NOS/ NOHA complex to (4) FeIII-O_2^{2-})/ FeIII-OOH. 4/5, forming the BH4 radical; (2) the resulting enzyme intermediate goes on to form Fe(III)NOS, citrulline, and NO- / HNO; (3) the latter is oxidized by the BH4 radical to the final product, NO; (4) final product of reaction is expected to be the Fe(II)NO complex, formed by reaction of the Fe(III) heme with NO- /HNO; and (5) cryoreduction at 77K of ternary complexes of both oxy gsNOS and oxy eNOS with bound NOHA yields ferric peroxo intermediate (4)(FeIII-O_2 $^{2-}$); and (6) annealing of the 4/NOHA complex for gsNOS gave evidences for 5 as the active oxidant in the second half of nitric oxide synthase reaction.

To understand how heme and (6R)-5,6,7,8-tetrahydro-l-biopterin (H4B) participate in nitric oxide synthesis ferrous-dioxy heme (Fe(II)O_2), formation and disappearance, H4B radical formation, and Arg hydroxylation during a single catalytic turnover by the inducible nitric oxide synthase oxygenase domain (iNOSoxy) were examined employing rapid kinetic technique [47]. Rapid kinetic studies link tetrahydrobiopterin radical formation to heme-dioxy reduction and arginine hydroxylation in inducible nitric oxide synthase. The suggested scheme of the process and obtained

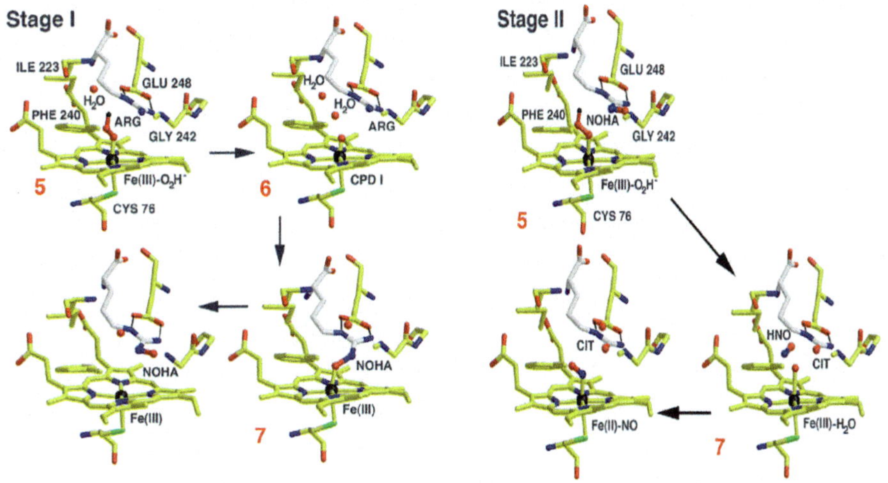

Fig. 5.17 Catalytically active states and primary products of reaction for Stages I and II of the gsNOS catalytic cycle as revealed by cryo-annealing EPR/ENDOR studies. These representations are based on crystallographic structures of bacterial NOS active centers with substrates and heme ligands bound, (PDB codes: 1M7V, 1M7Z, 2FC1, 2FBZ), but are purely figurative, and not meant to capture the yet uncharacterized conformational changes that likely accompany these states [60, 61]. With permission from Elsevier

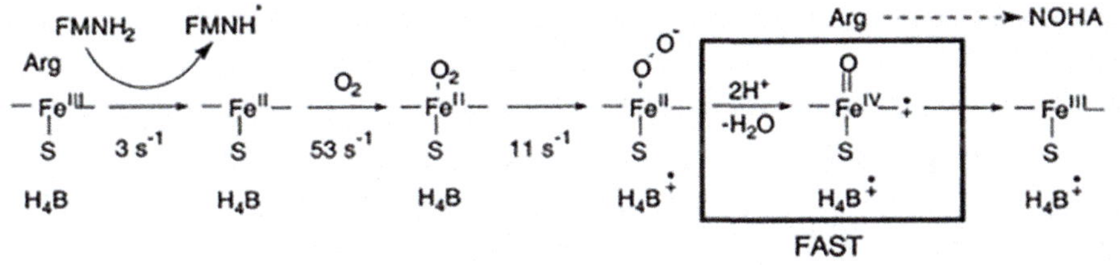

Fig. 5.18 Kinetics of Arg hydroxylation. In NADPH-driven reaction, shown step in the transfer of an electron from the reductase flavins ($FMNH_2$) to the ferric heme. Oxygen then binds to form the $Fe^{II}O_2$ species, which receives an electron from H_4B to form a heme-peroxo intermediate. Subsequent steps that lead to Arg hydroxylation are rapid and generate ferric iNOSoxy containing NOHA and the H_4B radical [64] With permission from Elsevier

rate constants are indicated in Fig. 5.19. The experiment t showed hat: (1) A ferrous-dioxy intermediate in iNOS forms for 53 s^{-1} and then is transformed to the [S-Fe(IV)=O] state; (2) the rate of the [S-Fe(IV)=O] decay is equal to the rate of H_4B radical formation and the rate of Arg hydroxylation; (3) the rate of ferrous-dioxy decay (12 s^{-1})) was equivalent to the rate of H4B radical formation (11 s^{-1}) and the rate of Arg hydroxylation (9 s^{-1}). *oxidation*, The authors have come to the reasonable conclusion that H_4B transfers electrons to the active site to form a center with strong oxidizing power. Figure 5.18 contains a mechanism for Arg hydroxylation that incorporates in the kinetic results.

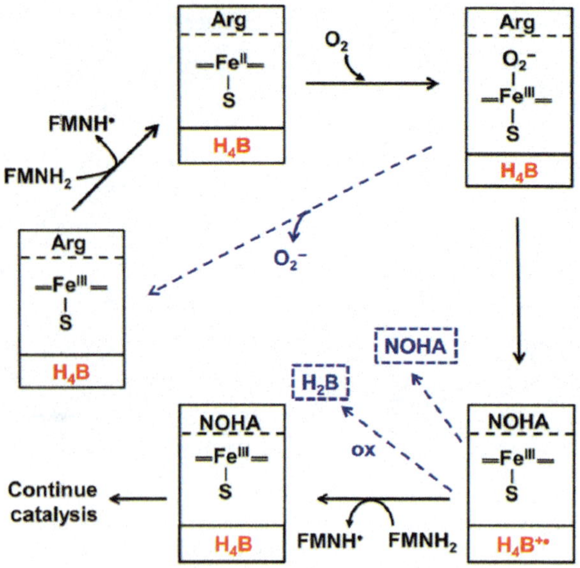

Fig. 5.19 Electron transfer steps and the possible uncoupling events that occur during Arg hydroxylation to NOHA by NOS. The solid boxes indicate detected NOSheme species that form in the reaction, and contain the substrate species and H4Bredox species (red) that are bound within the enzyme at each step. The dashed blue arrows indicate three uncoupling events that compete kinetically with the productive steps: Decay of the heme-dioxy species to form superoxide and ferric heme, NOHA dissociation from the enzyme, and oxidation of the H4B radical to inactive H2B [65]. With permission from Wiley

It was suggested that NOS flavoprotein domain might reduce the H4B radical by utilizing the NOS heme porphyrin as a conduit to deliver the electron [64, 65]. To test model, in which NOS flavoprotein domain might reduce the H4B radical by utilizing the NOS heme porphyrin as a conduit to deliver the electron (Fig. 5.19), a single catalytic turnover and stop-freeze methods, along with electron paramagnetic resonance spectroscopy were employed. This approach allowed to measure the rate and extent of reduction of the 5-methyl-H4B radical formed in neuronal NOS (nNOS) as a consequence of its catalyzing Arg hydroxylation. The experiments showed the rates and extents of nNOS heme reduction correlated with the rates and extents of 5-methyl-H4B radical reduction supporting a model where the heme porphyrin transfers an electron from the NOS flavoprotein to the H4B radical formed during catalysis.

A 10-step kinetic model, in which the ferric heme-NO enzyme forms as the immediate product of catalysis, and then partitions between NO dissociation versus reduction to a ferrous heme-NO complex, was introduced [56, 57]. Computer simulations of the model described both pre-steady and steady-state features of nNOS catalysis, including NADPH consumption and NO production, buildup of a heme-NO complex, changes between pre-steady and steady-state rates, and the change in enzyme K(m,O(2)) in the presence or absence of NO synthesis. In the frame of model, catalytic turnover is regulated by the following two distinct features: (1) partitioning of an immediate Fe(III) heme-NO product between a futile and productive regenerating pathway and (2) the relative rate of the futile regenerating pathway.

5.3 Methane Mono Oxygenase

5.3.1 General

In microorganisms utilizing methane, a methane-hydroxylating system, methane monooxigenase, (MMO) which catalyzed the reaction has been detected [65–92, 93, 94–97]:

$$CH_4 + O_2 + NAD(P)H + H^+ -> CH_3OH + NAD(P)^+ + H_2O$$

There are two distinct types of MMO, a soluble, cytoplasmic MMO (sMMO) and a membrane-bound, particulate MMO (pMMO) (Fig. 5.21). Both oxidize methane at metal centers within a complex, multisubunit scaffold, but the structures, active sites, and chemical mechanisms are different. The active site in sMMO contains a diiron center bridged by an oxygen atom (Fe-O-Fe), whereas the active site in pMMO utilizes copper. sMMO consist of three protein components for activity: a hydroxylase (MMOH, 251 kDa), a reductase (MMOR, 38.5 kDa), and a regulatory protein (MMOB, 15.9 kDa). MMOH is responsible for methane oxidation. The pMMO hydroxylase consists of the pmoB, pmoA, and pmoC subunits. Two mechanisms, general for both enzymes, suggested for the reaction between an active intermediate compound and the alkane: radical and nonradical has been discussed. The radical process starts with abstraction of the hydrogen atom from the substrate to form the free alkyl radical and proceeds via bound-radical mechanism. The nonradical mechanism implies a concerted incorporation of oxygen atom to the C–H bond of alkane pathway, occurring via a four-center transition state and leading to a “hydrido-alkyl-Q” compound.

5.3.2 Soluble Methane Monooxigenase

To date more than 28 crystal structures of MMOH are known. The first crystal structure of MMOH with a resolution of 1.7 Å was determined in 1993 [74, 76]. According to suggested model, the diiron active site located in each α subunit in the α2β2γ2 dimer within a four-helix bundle of which surface-exposed helices. Fe1 is coordinated by His147, Glu114, and a water molecule, and Fe2 is coordinated by His246, Glu243, and Glu209. The two iron ions are bridged by residue Glu144 and two hydroxides. Upon reduction, the Fe–Fe distance increases from ~3.1 Å to ~3.3

Å. For *M. capsulatus* (Bath) sMMO, the turnover number for methane is in the range of 0.2–1.0 s^{-1} at 45 °C. The K_m value for methane is 0.94 μM and the K_m value for O_2 is ~16 μM. For *M. trichosporium* OB3b sMMO, turnover number of 3.7 s^{-1} and the Km for methane is ~12 μM.47. at 30 °C and has been reported. Overall architecture of MMOs is displayed in Fig. 5.20.

For the last three decades, the enzymes have been intensively investigated with the entire arsenal of biochemical, kinetic and physical methods [65–83]. Study [75] demonstrated the value of X-ray free electron lasers (XFEL) and serial femto second crystallography (SFX) methods for investigating the structures of metalloenzymes with radiation sensitive metal active sites. Microcrystals of the sMMOH:MMOB complex from *Methylosinus trichosporium* OB3b were exposed to X-XFEL pulses. Importantly, the ≤ 35 fs duration of exposure of an individual crystal yields diffraction data before photoreduction-induced structural changes. 1.95 Å resolution crystal structures for the fully oxidized and fully reduced states of the sMMOH:MMOB complex were obtained. The results showed how the diiron cluster and the active

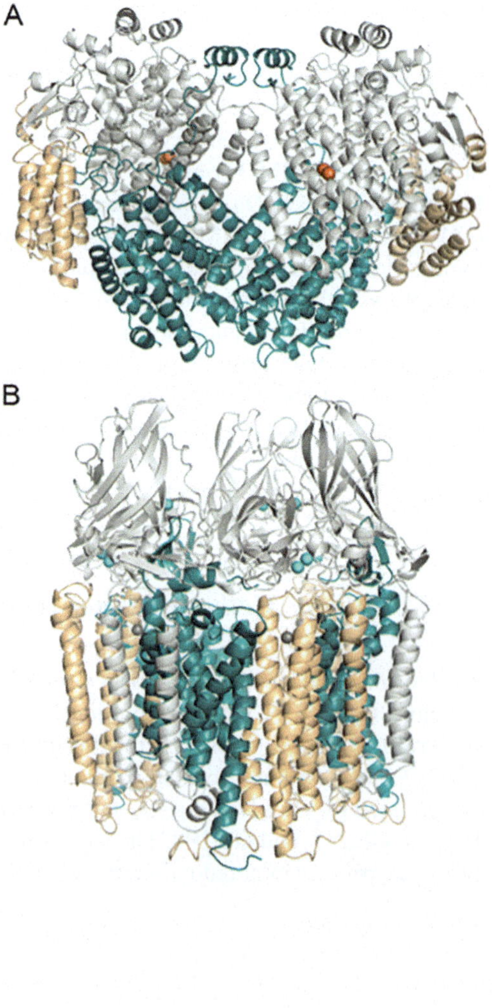

Fig. 5.20 Overall architecture of MMOs. **A** The sMMO hydroxylase (MMOH, PDB accession code 1MTY) with α subunits shown in gray, β subunits shown in teal, and γ subunits shown in wheat. Each α2β2γ2 dimer contains two diiron active sites (orange spheres). **B** The pMMO trimer (PDB accession code 3RGB) with pmoB subunits shown in gray, pmoA subunits shown in teal, and pmoC subunits shown in wheat. Copper ions are shown as cyan spheres and zinc ions are shown as gray spheres [71] With permission from American Cemical Society

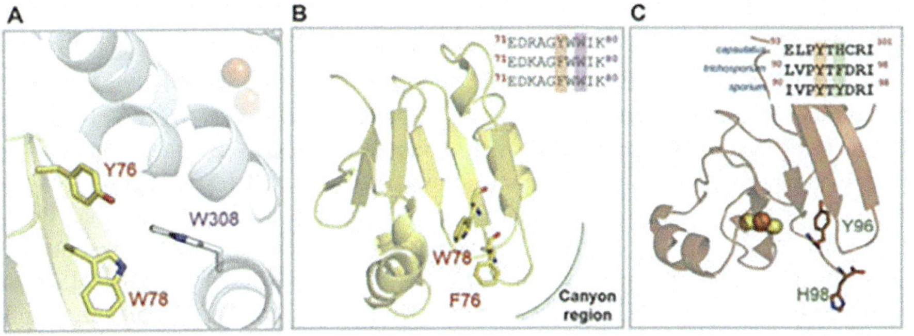

Fig. 5.21 Multicomponent interactions between hydroxylase and MMOB or MMOR **for** the catalytic reactions. **A** Hydrophobic interactions are generated between MMOH (gray) and MMOB (yellow) from *Methylococcus capsulatus* (Bath) (PDB: 4GAM). The orange balls present diiron in active sites. **B** Core region of MMOB from *Methylosinus trichosporium* OB3b has Phe76 and Trp78 that generate hydrophobic interactions (PDB: 2MOB). **C** Ferredoxin domain of MMOR (MMOR-Fd) is a putative binding site to the surface of MMOH α-subunit for electron transfers from M. capsulatus (PDB: 1JQ4). [85]. Published under the terms of Creative Commons Attribution (CC BY) license

site environment are reorganized by the regulatory protein component in order to enhance the steps of oxygen activation and methane oxidation.

Lower-resolution structures of the sMMOH:MMOB complex from the sMMO of *Methylococcus capsulatus* Bath revealed a series of cavities through sMMOH postulated to serve as the O_2 conduit [76]. High-resolution X-ray crystal structures of the diferric and diferrous states of both sMMOH and the sMMOH:MMOB complex using the components from *Methylosinus trichosporium* OB3b indicated a narrow molecular tunnel, termed the W308-tunne. This tunnel was found to be sized appropriately for O_2 and traverses the sMMOH-MMOB interface. In the MMOB V41R variant of diferrous sMMOH:MMOB complex, the rate constant for O_2 binding decreases by 25000-fold. This finding indicates the tunnel serving as the primary route for the transfer of O_2 into the active site. In addition, the crystal structures showed that formation of the diferrous sMMOH:MMOB complex restricts access through the chain of cavities while opening the W308-tunnel..

In comprehensive review [73, 74], the following active site intermediates and steps leading to methane oxidation were described: (1) The resting state MMOH is reduced to $MMOH_{red}$ by MMOR; (2) molecular oxygen interacts with $MMOH_{red}$ to irreversibly bind O2O2 at the active site to form **O** (at this point, O2O2 is not directly bound to the diiron cluster); (3) **O** undergoes ligand substitution to yield an antiferromagnetically coupled diiron(III)-peroxo species (**P***); (4) in a redox reaction the bridging peroxide is reduced and Fe(IV) is oxidized; (5) transfer of electrons yields antiferromagnetically coupled diiron(II)-peroxo species (**P**); (6) subsequently oxygen activation occurs via O-O bond cleavage within **P** to intermediate **Q,** which decays to catalyze methane oxidation; (7) species **R** forms with a methyl radical species and hydroxyl radical ($\cdot CH_3$) bound to the diiron cluster; (8) the radicals recombine with oxygen inserted into a C–H bond to form species **T**; and (9) in a rate limiting step, species **T** decays into $MMOH_{ox}$ and releases methanol (CH_3OH).

Recently, X-ray crystal structures were reported for the sMMOH complex with two double variants within the core region of MMOB, DBL1 (N107G/S110A), and DBL2 (S109A/T111A), as well as two variants in the MMOB N-terminal region, H33A and H5A [77] The experiment showed that: (1) DBL1 causes a 150-fold decrease in the formation rate constant of the reaction cycle intermediate **P;** (2) DBL2 accelerates the reaction of the dinuclear Fe(IV) intermediate **Q** with substrates larger than methane by three- to fourfold; (3) H33A slows **P** formation; (4) H5A AA slows both formation of **Q** and its reactions with substrates; (4) complexation with DBL1 or H33A alters the position of sMMOH residue R245, which is part of a conserved hydrogen-bonding network encompassing the active site diiron cluster where **P** is formed; (5) electron paramagnetic resonance spectra of sMMOH:DBL1 and sMMOH:H33A complexes differ markedly from that of sMMOH:MMOB; (6) In the sMMOH:DBL2 complex, the position of M247 in sMMOH is altered such that it enlarges a molecular tunnel associated with substrate entry into the active site; and (7) The H5A variant causes only subtle structural changes despite its kinetic effects, emphasizing the precise alignment of sMMOH and MMOB required for efficient catalysis. The kinetics of intermediate O conversion was monitored through the application of rapid freeze quench (RFQ) techniques [79]. In two chromophorically and electronically distinct intermediates (P* and P) the binding and activation of O_2 by the iron atoms occurs through a metal-to-ligand transfer of electrons.

Geometric and electronic structures of a peroxo intermediate in the catalytic cycle of methane monooxygenase hydroxylase (MMOH) were examined utilizing spin density functional theory using ^{57}Fe Mössbauer spectral parameters as a reference [81, 82]. Computed isomer shifts ($\delta_{Fe} = +0.68, +0.66$ mm s^{-1}) and quadrupole splittings ($\Delta E_Q = -1.49, -1.48$ mm s^{-1}) for the predicted structure were in agreement with experimental values of a peroxo MMOH intermediate. It was concluded that in the active site's electronic structure, inducing a ferromagnetic to antiferromagnetic transition of the diiron core, weakening the O–O bond in preparation for cleavage in subsequent steps of the catalytic cycle take place. The diiron clusters exhibited the EPR spectra characterized by a signal at g = 16 arising from ferromagnetic coupling of the two high spin S = 2 irons to yield an S = 4 species [73, 87]. The EXAFS of the methane C–H bond breaking diiron(IV) intermediate Q of sMMO was revisited using high-energy resolution fluorescence-detected extended X-ray absorption fine structure (HERFD-EXAFS) technique [82, 83]. The presented data showed a Fe-Fe distance as long as 3.4 Å, which close to that observed in open-core synthetic model complexes.

In study [83, 84] it was found that the two tryptophan residues in MMOB and the single tryptophan residue in MMOR are converted to 5-fluorotryptophan (5FW) by expression in media containing 5-fluoroindole. The significant N-terminal region of MMOB was ^{19}F-labeled by reaction of the K15C variant with 3-bromo-1,1,1-trifluoroacetone (BTFA). The approach showed directly that MMOR and MMOB competitively bind to sMMOH with similar K_D value. These findings suggested a new model for regulation of the dynamic equilibration of MMOR and MMOB with sMMOH which allowed a transient formation of key reactive complexes that irreversibly pull the reaction cycle forward.

Studies on the interaction of MMOH-MMOB from *Methylococcus capsulatus* Bath (Figs. 5.12, 5.22) were carry out by absorbance spectroscopy demonstrated that Tyr76 and Trp78 induce hydrophobic interactions through π-π stacking and suggested that Tyr93 and Tyr95 could be key residues for electron transfer [84, 85]. The studies revealed that: (1) Interaction of MMOB and MMOD form a canyon region to induce allosteric effects in diiron active sites; (2) the regulatory component generates hydrogen bonds and hydrophobic packing on the surface of MMOH (Fig. 5.22); (3) MMOB and MMOD bind a canyon region to induce allosteric effects in diiron active sites; (4) the indole moiety from Trp308 on MMOH generates π-stacking hydrophobic interactions with MMOB, including Tyr76 and Trp78; (5) the superimposed structure of MMOR-Fd has an aromatic residue, Tyr96, in this hydrophobic interaction area; (6) His98 is positioned in the solvent-accessible area to be putative binding residues to MMOH; and (7) Tyr76 and Trp78 induce hydrophobic interactions through π-π stacking and suggested that Tyr93 and Tyr95 could be key residues for electron transfer

5.3.3 Particulate MMO

Various aspects of the structure and mechanism of action of pMMO (Fig. 5.22) are widely discussed in the corresponding bibliography [85–92, 93, 94–97]. The first crystal structure of pMMO was that of the *M. capsulatus* (Bath) enzyme, determined in 2.8 Å resolution [85, 86].

The pMMO hydroxylase from *M. capsulatus* (Bath) consists of the pmoB, pmoA, and pmoC subunits (Fig. 5.25) [85]. The subunits are arranged as a 300 kDa $\alpha3\beta3\gamma3$ trimer which composes of PmoA (24 kDa) and PmoC (22 kDa), and PmoB (42 kDa). A primarily periplasmic subunit comprising two cupredoxin-like β-barrels joined by two transmembrane helices All structures **contain a copper** binding site at the N-terminus of the pmoB subunit with ligands His33, which is the amino terminal residue of pmoB, His137, and His139. The Cu···Cu distance in the dinuclear model refines to, 2.6 A, consisting with the 2.57-A° Cu–metal interaction determined by EXAFS.

Early data, namely EPR, Extended X-ray absorption fine structure (EXAFS) and X-ray structural analysis indicated that *Methylocystis sp.* strain M pMMO contains a mixture of Cu-O/N and Cu-Cu ligand environments with a Cu-Cu interaction at 2.52–2.68 Å. Lawton et al. (2014) Proteins Struct Funct Bioinf 82:2263–2267Sarah Sirajuddin and Amy C. Rosenzweig*Enzymatic Oxidation of Methane Biochemistry. 2015 April 14; 54(14): 2283–2294.

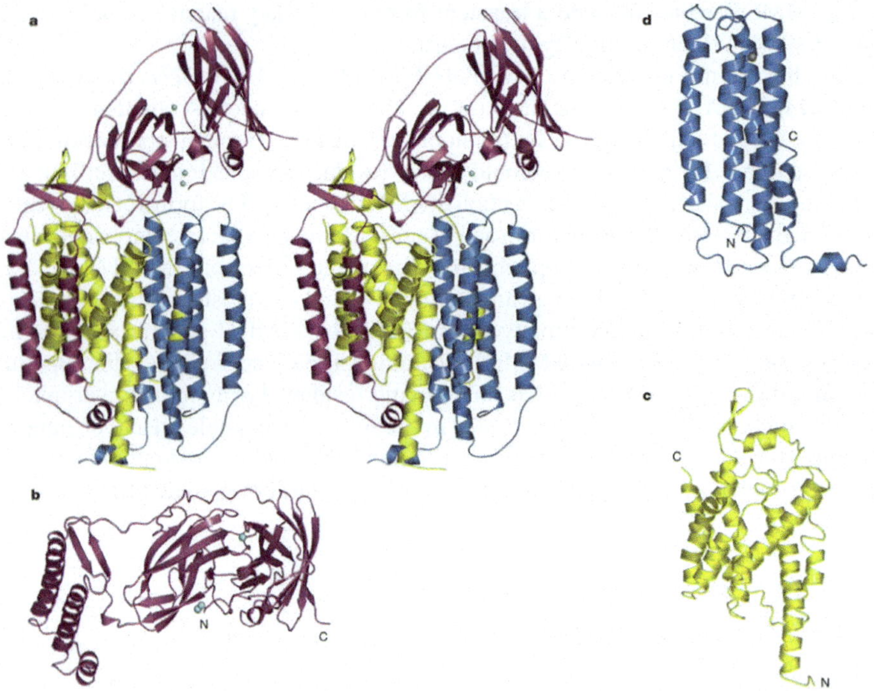

Fig. 5.22 pMMO subunits. **a** Stereoview of a single protomer with pmoB shown in magenta, pmoA shown in yellow, and pmoC shown in blue. **Copper ions** are shown as cyan spheres, and a zinc ion is shown as a gray sphere. **b** The pmoB subunit viewed looking down the membrane normal. The N-terminal b-barrel is in the middle. **c** The pmoA subunit. **d** The pmoC subunit [86] With permission from Springer Nature

On the basis of multiple pMMO crystal structures, two copper centers have emerged as potential active sites, one located in PmoB (Cu_B) and one located in PmoC (Cu_C) (Fig. 5.23) [96, 99]. The CuB site includes three histidine ligands, His29, His133, and His135, using type *II Methylocystis* (Mc.) species and metal-binding site in the PmoC subunit whose ligand set, which includes Asp129, His133, and His146. The one-electron reduction of the oxygen molecule is thermodynamically forbidden while the enzyme reduction requires the transfer of two electrons with the formation of a derivative of hydrogen peroxide. It seems impossible to transfer two electrons from one copper ion and much more likely that two electrons are concerted transferred to O_2, for example, one from nearby Cu^{2+}c, and the second from $Cu^{2+}{}_B$, according to the mechanism of long-distance transfer.

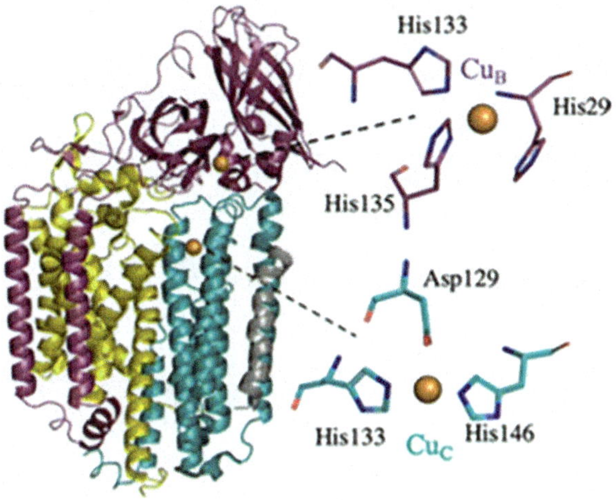

Fig. 5.23 Single protomer from the crystal structure of Mc. sp. str. Rockwell pMMO (PDB accession code 4PHZ). PmoA is light gold, PmoB is purple, PmoC is teal, and an unidentified helix is gray. The two copper ions are shown as spheres and to the right are expanded to show the modeled coordinating residues [93] With permission from American Cemical Society

EPR and ENDOR studies have established the presence of two monocopper sites. Report [97] showed that the type II *Methylocystis* species strain Rockwell pMMO, like the type I pMMOs, contains two monocopper sites and that its Cu_B site has a coordination environment identical to that of type I enzymes. EPR/ENDOR measurements showed the coordination environment of the spectroscopically observed site, provisionally denoted Cu_C, in both types of pMMO, indicating the assignment of this site observed by EPR to the crystallographically observed metal-binding site in the PmoC subunit. The crystal structure of Mc. sp. str. Rockwell pMMO is presented in Fig. 5.23.

According to density functional theory calculations [97], the formation of a dicopper–oxygen species of particulate methane monooxygenase (pMMO) is more favorable than that one-cooper species. One copper ion is coordinated by the side chain nitrogens of His137 and His139 while the other is coordinated by the side chain of His33, and by an oxygen from the side chain of Glu35. In this pathway a μ-η 2 :η 2—peroxo- or μ-η 1 :η 2 -peroxo-CuICuII species is converted to a reactive di(μ-oxo) CuIICuIII or (μ-oxo)(μ-hydroxo)CuIICuIII species capable of methane oxidation. Upon reduction and treatment of spmoB with O_2 or H_2O_2 or pMMO with H_2O_2, an absorbance feature at 345 nm is generated evidencing for oxygen binding. The calculations also suggested that the H-atom transfer or proton-coupled electron transfer from the tyrosine residue can induce the O–O bond dissociation of the μ-η^2:η^2-peroxo$Cu^{II}Cu^{II}$ species to form the reactive (μ-oxo)(μ-hydroxo)$Cu^{II}Cu^{III}$ species. Figure 5.23 illustrates possible O_2 activation intermediates for pMMO [89, 93]. (Fig. 5.24).

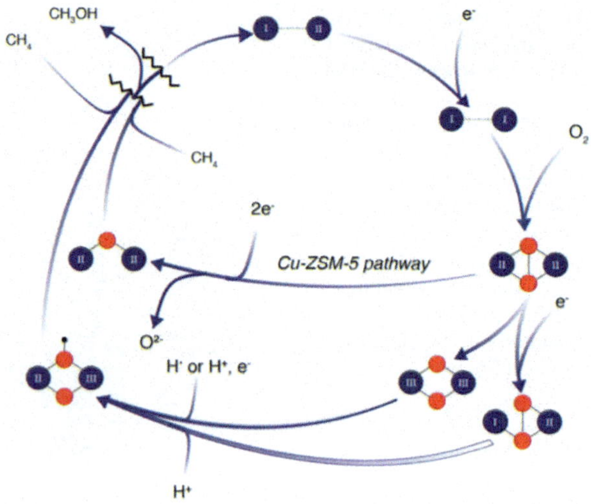

Fig. 5.24 Possible O_2 activation intermediates for pMMO. Copper ions with oxidation states designated are shown in dark blue, oxygen atoms in red, and hydrogen atoms in black [92, 93] With permission from Springer Nature

Data on Extended X-ray absorption fine structure (EXAFS) and X-ray structural analysis indicated that *Methylocystis sp.* strain M pMMO contains a mixture of Cu–O/N and Cu-Cu ligand environments with a Cu-Cu interaction at 2.52–2.68 Å. EPR, ^{1}H ENDOR, DEER, native top-down mass spectrometry (nTDMS) and the crystallography analysis of the type II Mc. sp. str. Rockwell pMMO provided new insights into the pMMO copper centers [91, 95]. Experiments revealed the following: (1) The CuB sites in pMMOs from type I and type II methanotrophs have identical molecular and electronic structures: each is a mononuclear site, a single Cu(II) ion) with four nitrogen ligands, comprising three histidyl imidazoles and one terminal amine; (2) the ^{15}N ENDOR and DEER data indicated the presence of two histidine ligands; (3) Cu_C exhibits properties that are completely consistent with coordination by the three protein-derived ligands of the crystallographically observed Cu site: the two histidyl imidazoles of His133 and His146, along with the carboxylate of Asp129; (3) the ^{1}H ENDOR data showed that the enzyme in detergent exhibits signals from nearby axial water molecule at a distance of ~ 3 Å; and (4) quantum refinement of the original *Methylococcus capsulatus* (Bath) pMMO crystal structure and native top-down mass spectrometry (nTDMS) indicated the presence of a single copper ion localized to the amino terminus of PmoB [92, 95]. Recent crystallographic data, advanced electron paramagnetic resonance (EPR), high-energy resolution fluorescence detected (HERFD), and partial fluorescence yield (PFY) EXAFS spectra of *Methylococcus* (*M.*) *capsulatus* (Bath) pMMO revealed only **monocopper site's** [95, 98]. It is important that open Cu_C coordination sites provides space for dioxygen and methane to bind, suggesting a role of Cu_C as the site of methane oxidation. Density functional theory (DFT) and quantum mechanics/molecular mechanics (QM/ MM) studies suggested that a monocopper CuIII-oxo species or a di(μoxo)CuIICuIII species could oxidize methane and that formation of the dicopper species is more favorable [94, 97]. In later studies, a (μ-oxo)(μ-hydroxo)CuIICuIII center was proposed to be the active species. Formation of this (μ-oxo)(μ-hydroxo)CuIICuIII species can be facilitated by proton transfer involving residues Tyr374 and Glu35 (*M. capsulatus* (Bath) numbering.

The electronic, structural, and reactivity properties of a possible dinuclear copper species (μ-oxo)(μ-hydroxo)$Cu^{II}Cu^{III}$ of particulate methane monooxygenase (pMMO) obtained using density functional theory calculations were discussed with respect to the C–H bond activation of methane [90, 92]. The calculation proposed that: (1) The tyrosine residue in the second coordination sphere of the dicopper site donates an H atom to the μ-η^2:η^2-peroxo$Cu^{II}Cu^{II}$ species; (2) the resultant (μ-oxo)(μ-hydroxo)$Cu^{II}Cu^{III}$ species can hydroxylate methane; (3) the H-atom transfer or proton-coupled electron transfer from the tyrosine residue can induce the O–O bond dissociation of the μ-η^2:η^2-peroxo$Cu^{II}Cu^{II}$ species; (4) the reactive (μ-oxo)(μ-hydroxo)$Cu^{II}Cu^{III}$ species for the conversion of methane to methanol at the dicopper site of pMMO are formed; and (5) the rate-determining step for the methane hydroxylation is the C–H cleavage, which is in good agreement with experimental KIE values.

Reported in [89, 91] spectroscopic data indicated that *Methylocystis* sp. strain M pMMO contains a mixture of Cu(I) and Cu(II), of which the latter exhibits two distinct type 2 Cu(II) electron paramagnetic resonance (EPR) signals. Extended X-ray absorption fine structure (EXAFS) data showed a mixture of Cu-O/N and Cu-Cu ligand environments with a Cu-Cu interaction at 2.52-2.64 Å, n consent with the crystal structure of Methylocystis sp. strain M pMMO which was determined to 2.68 Å resolution

Data on nanodiscs coupled with native top-down mass spectrometry (nTDMS) to determine the copper stoichiometry in each pMMO indicated the presence of a mononuclear copper center in both PmoB and PmoC. pMMO-nanodisc complexes with a higher stoichiometry of copper-bound PmoC exhibited increased activity [101, 103]. This finding suggested that the PmoC copper site plays a role in methane oxidation activity.

Multiple nanodisc-embedded pMMO structures reconstituting pMMO in nanodiscs and restoring methane oxidation activity, determined by cryo–electron microscopy to 2.14- to 2.46-angstrom resolution revealed the structure of pMMO in a lipid environment [98, 101]. The resulting model includes stabilizing lipids, regions of the PmoA and PmoC subunits, and a copper-binding site in the PmoC subunit with an adjacent hydrophobic cavity. Structural characterization of *M. capsulatus* (Bath) pMMO in the native lipid nanodisc showed: (1) Methane oxidation activity of cryo-EM sample; (2) Cryo-EM map indicating the trimer and the encircling nanodisc; (3) the map for lipids on the periphery and in the inner pore of the enzyme; (4) copper sites in the PmoB subunit of *M. capsulatus* (Bath) pMMO, regions of pMMO newly observed in the cryo-EM structures; and (5) active site architecture in the PmoC subunit of pMMO.

The hydroxylase component of methane monooxygenase contains a binuclear iron cluster in which the iron appears to be oxo or R-oxo bridged. Massbauer and EPR studies have demonstrated antiferromagnetic coupling for the [Fe(III)Fe(III)]

and [Fe(II)Fe(III)] states of the cluster. In the [Fe(II)Fe(II)] form the cluster exhibits an intense X-band EPR signal with zero crossing near g = 16, which originates from an electronic system with integer spin.

References

1. Coon MJ, White RE, Blake RC (1981) Oxygen activation by cytochrome P450, in King TE (ed) Oxidases and related systems (University Park Press, Baltimore) 93–118
2. Mak PJ, IG Denisov IG (2018) Spectroscopic studies of the cytochrome P450 reaction mechanisms Biochim Biophys Acta Proteins Proteom 1866 1 178
3. Ortiz de Montellano PR (2015) (ed), Cytochrome P450: structure, mechanism, and biochemistry. Springer International Publishing, New York
4. Manikandan P, Nagini S (2018) Cytochrome P450 structure, function and clinical significance: a review Curr Drug Targets 19 1 38 54
5. M Parvez M, Qhanya LB, Mthakathi MT (2016) Molecular evolutionary dynamics of cytochrome P450 monooxygenases across kingdoms: special focus on mycobacterial P450s Sci Rep 6 33099
6. Poulos TL (2014) Heme enzyme structure and function Chem Rev 114 3919 3962
7. Kobayashi K, Iwamoto T, Honda K (1994) Spectral intermediate in the reaction of ferrous cytochrome P450 with superoxide anion Biochem Biophys Res Commun 201 1348 1355
8. Davydov R, Hoffman BM (2011) Active intermediates in heme monooxygenase reactions as revealed by cryoreduction/annealing EPR/ENDOR Studies Arch Biochem Biophys 507 1 36 43
9. Denisov IG, Sligar SG (2017) Nanodiscs in membrane biochemistry and biophysics Chem Rev 117 4669 4713
10. Luthra A, Denisov IG, Sliga SG (2011) Spectroscopic features of cytochrome P450 reaction intermediates Arch Biochem Biophys 507 1 26 35
11. Cafiso DS (2014) Identifying and quantitating conformational exchange in membrane proteins using site-directed spin labeling Acc Chem Res 47 10 3102 3109
12. Jensen K, Møller BL, Laursen T, Jensen K (2011) Conformational changes of the NADPH-dependent cytochrome P450 reductase in the course of electron transfer to cytochromes P450 Biochim Biophys Acta 1 132 138
13. Denisov IG, Thomas M, Makris TM, Stephen G, Sligar SG (2005) Ilme Schlichting* Structure and chemistry of cytochrome P450 Chem Rev 105 6 2253 2278
14. Huang X, Groves JT (2017) Oxygen activation and radical transformations in Heme proteins and Metalloporphyrins Chem Rev 118 5 2491 2553
15. Zhang B, Lewis KM, Abril A, Davydov DR, Vermerris W, Sattler SE, Kang C (2020) Structure and function of the Cytochrome P450 monooxygenase Cinnamate 4-hydroxylase from *Sorghum bicolor* Plant Physiol 183 3 957 973
16. Postils V, Saint-André M, Timmins A, Li XX, Wang Y, Luis JM, Solà M, Visser De SP (2018) Quantum mechanics/molecular mechanics studies on the relative reactivities of compound I and II in cytochrome P450 enzymes Int J Mol Sci 19 7 1974
17. Xu F, Forward I, Bartlam M, Rao Z, Wong LL (2008) Crystal structure of CYP199A2, a para-substituted benzoic acid oxidizing cytochrome P450 from Rhodopseudomonas palustris J Mol Biol 383 3 561 574
18. Groves JT, Vanderpuy M (1976) Stereospecific aliphatic hydroxylation by iron-hydrogen peroxide—evidence for a stepwise process J Am Chem Soc 98 5290 5297
19. Poulos TL, Meharenna YT 2007 Structures of P450 proteins and their molecular phylogeny Met Ions Life Sci 30 3 57 96
20. Fisher MT, Sligar SG (1987) Temperature jump relaxation kinetics of the P450cam spin equilibrium Biochemistry 26 4797 4803

21. Ravindranathan KP, Gallicchio E, Friesner RA, McDermott AE, Levy RM (2005) Conformational equilibrium of cytochrome P450 BM-3 complexed with Palmitoylglycine: a replica exchange molecular dynamics study J Am Chem Soc 2006 128 5786 5791
22. Tian WD, Wells AV, Champion PM, Di Primo C, Gerber N, Sligar SG (1995) Investigations of heme ligation and ligand switching in cytochromes P450 and P420 J Biol Chem 270 8673 8679
23. G Catucci A Ciaramella G Nardo Di C Zhang S Castrignanò G Gilardi 2022 Molecular lego of human cytochrome P450: the key role of heme domain flexibility for the activity of the chimeric proteins Int J Mol Sci 23 7 3618
24. Davydov DR, Hui Bon Hoa G, JA Peterson JA (1999) Dynamics of protein-bound water in the heme domain of P450BM3 studied by high-pressure spectroscopy: comparison with P450cam and P450 2B4 Biochemistry 38 2 751 761
25. Davydov DR, Yang Z, Davydova N, Halpert JR, Hubbell WL 2016 Conformational mobility in cytochrome P450 3A4 explored by pressure-perturbation EPR spectroscopy Biophys J 110 7 1485 1498
26. Karyakin A, Motiejunas D, Wade RC, Jung C (2007) FTIR studies of the redox partner interaction in cytochrome P450: the Pdx-P450cam couple Biochim Biophys Acta 1770 3 420 431
27. Lockart MM, Rodriguez CA, William M, Atkins WM, Bowman MK (2018) CW EPR parameters reveal cytochrome P450 ligand binding modes J Inorg Biochem 183 157 164
28. Davydov R, Im S, Shanmugam M, Gunderson WA Pearl NM Hoffman BM, Waskell L (2016) Role of the proximal cysteine hydrogen bonding interaction in cytochrome P450 2B4 studied by cryoreduction, electron paramagnetic resonance, and electron–nuclear double resonance spectroscopy Biochemistry 55 6 869 883
29. Newcomb M, Halgrimson JA, Horner JH (2008) X-ray absorption spectroscopic characterization of a cytochrome P450 compound II derivative PNAS 105 24 8179 8184
30. Cruce AA Lockart M, Bowman MK (2015) Pulsed EPR in the study of drug binding in cytochrome P450 and NOS Methods Enzymol 563 311 340
31. Roberts AG, Campbell AP, Atkins WM (2005) The thermodynamic landscape of testosterone binding to cytochrome P450 3A4: ligand binding and spin state equilibria Biochemistry 44 4 1353 1566
32. Andersson LA , Johnson AK Peterson JA (1997) Active site analysis of P450 enzymes: comparative magnetic circular dichroism spectroscopy Arch Biochem Biophys 345 1 79 87
33. Luthra A, Gregory M, Grinkova YV, Denisov IG, Sligar SG (2013) Nanodiscs in the studies of membrane-bound cytochrome P450 enzymes. Cytochrome P450 Protoc. 115-27
34. Civjan NR, Bayburt TH, Schuler MA ,Sligar SG, 2003 Direct solubilization of heterologously expressed membrane proteins by incorporation into nanoscale lipid bilayers Biotechniques 35 556 558
35. Gregersen JL, Fedosova NU, Nissen P, Boesen T (2016) Reconstitution of Na(+), K(+)-ATPase in Nanodiscs Methods Mol Biol 2016 1377 403 409
36. Zhang M, R Huang R, Ackermann R, Im SC, Waskell L, A Schwendeman A, A Ramamoorthy A (2016) Reconstitution of the Cytb5–CytP450 complex in nanodiscs f or structural studies using NMR spectroscopy Angewandte Chemie Int Ed 55 14 4497 4499
37. Laursen T, Jensen K, Møller BL (2011) Conformational changes of the NADPH-dependent cytochrome P450 reductase in the course of electron transfer to cytochromes P450 Biochimica et Biophys Acta 1814 132 138
38. Ellis J, Gutierre A, Barsukov IL (2009) Domain motion in cytochrome P450 reductase: conformational equilibria revealed by NMR and small-angle X-ray scattering J Biol Chem 284 36628 36637
39. Lüdemann SK, Lounnas V, Wade RC (2000) How do substrates enter and products exit the buried active site of cytochrome P450cam?: 1. Random expulsion molecular dynamics investigation of ligand access channels and mechanisms J Mol Biol 303 5 797 811
40. Zhang Y , Oldfield E (2004) Cytochrome P450: an investigation of the Mössbauer spectra of a reaction intermediate and an Fe (IV) O model system J Am Chem Soc 126 14 4470 4471

41. Capoferri L, Leth R, Haar ter E, Mohanty AK, Grootenhuis PD, E Vottero E, Commandeur JN, Vermeulen NP, Jørgensen FS, Olsen L, Geerke (2016) Insights into regioselective metabolism of mefenamic acid by cytochrome P450 BM3 mutants through calculations Proteins 84 3 383 396
42. Mukherjee G, Nandekar PP, Wade RC (2021) An electron transfer competent structural ensemble of membrane-bound cytochrome P450 1A1 and cytochrome P450 oxidoreductase Commun Biol. 24 1 55
43. Bignon E, S Rizza S, G Filomen G, E Papaleo E (2019) Use of computational biochemistry for elucidating molecular mechanisms of nitric oxide synthase Comput Struct Biotechnol J 17 415 429
44. Reiß K, A Kollau A, Mayer B, Gorren ACF (2019) Site and mechanism of uncoupling of nitric-oxide synthase: Uncoupling by monomerization and other misconceptions Nitric Oxide 89 14 21
45. Carvajal A, Shaik S (2009) First half-reaction mechanism of nitric oxide synthase: the role of proton and oxygen coupled electron transfer in the reaction by quantum mechanics/molecular mechanics J Phys Chem B 113 336
46. Cho KB, Gauld JW (2005) Second half-reaction of nitric oxide synthase: computational insights into the initial step and key proposed intermediate J Phys Chem B 109 49 23706 23714
47. Wei C-C, Wang Z-Q, Wang Q , Meade A, Hemann C, Hille R, Stuehr DJ (2001) Rapid kinetic studies link tetrahydrobiopterin radical formation to heme-dioxy redyction and arginine hydroxylation in inducible nitric-oxide synthase J Biol Chem 276 315 319
48. Liu Y, Q Feng Q (2012) NOing the heart: role of nitric oxide synthase-3 in heart development Differentiation 84 1 54 61
49. Förstermann U, Sessa W (2012)Nitric oxide synthases: regulation and function Eur Heart J 33 7 829 837
50. Loeb E, K Asmar K, Trabado El S (2022) Nitric oxide synthase activity in major depressive episodes before and after antidepressant treatment: results of a large case-control treatment study Psychol Med 52 1 80 89
51. Kopincová J A Púzserová A, I Bernátová I, (2011) Biochemical aspects of nitric oxide synthase feedback regulation by nitric oxide Interdisc Toxicol 4 2 63 68
52. Cinelli MA, Do HT, Miley JP,Silverman RB (2019) Inducible nitric oxide synthase: regulation, structure, and inhibition Med Res Rev 40 1 158 189
53. Santolini J (2011) The molecular mechanism of mammalian NO-synthases: a story of electrons and protons J Inorg Biochem 5 127 141
54. Li H, Jama J, C Plaza C, Pineda SH, G Chreifi G, Q Jing Q, Cinelli MA, Silverman RB, Poulos TL (2014) Structures of human constitutive nitric oxide synthases Acta Crystallogr Sect D Biol Crystallogr 70 10 2667 2674
55. Garcin ED, Bruns CM, Lloyd S J (2004) Structural basis for isozyme-specific regulation of electron transfer in nitric-oxide synthase Biol Chem 279 36 37918 37927
56. Xia C, Misra I, Iyanagi T, Kim JJ (2009) Regulation of interdomain interactions by calmodulin in inducible nitric-oxide synthase J Biol Chem 284 44 30708 30717
57. Santolini J, S Adak S, Curran CM, Stuehr DJ (2001) A kinetic simulation model that describes catalysis and regulation in nitric-oxide synthase J Biol Chem 276 2 1233 1243
58. Fischmann TO, A Hruza A, Niu XD, Fossetta JD, Lunn CA, E Dolphin E (1999) Structural characterization of nitric oxide syntha,se isoforms reveals striking active-site conservation Nat Struct Biol 16 233 242
59. Tejero J, L Hannibal L, A Mustovich A, Stuehr DJ (2010) Surface charges and regulation of FMN to heme electron transfer in nitric-oxide synthase J Biol Chem 285 27232 27240
60. R Gachhui DK Ghosh C Wu J Parkinson BR Crane DJ Stuehr 1997 Mutagenesis of acidic residues in the oxygenase domain of inducible nitric-oxide synthase identifies a glutamate involved in arginine binding Biochemistry 36 17 5097 5103 https://doi.org/10.1021/bi970331x

61. R Davydov BM Hoffman 2011 Active intermediates in heme mono reactions as revealed by cryoreduction/annealing EPR/ENDOR studies Arch Biochem Biophys 507 1 36 43
62. Campbell MG, Brian C, Smithc BC, Clinton S, Pottera CS, et al ((2014)) Molecular architecture of mammalian nitric oxide synthases. PNAS E3614–E3623
63. Billiar TR, Cirino G, Fulton D, Motterlini R, Papapetropoulos A, Szabo C (2019) Nitric oxide synthases in the IUPHAR/BPS guide to pharmacology database. IUPHAR/BPS Guide to Pharmacology. (4), https://doi.org/10.2218/gtopdb/F253/2019.4
64. ZQ Wang Q Wang AL Meade C Hemann R Hille DJ Stuehr 2001 Rapid kinetic studies link tetrahydrobiopterin radical formation to heme-dioxy redyction and arginine hydroxylation in inducible nitric-oxide synthase J Biol Chem 276 315 319
65. S Ramasamy MM Haque M Gangoda DJ Stuehr 2016 Tetrahydrobiopterin redox cycling in nitric-oxide synthase: evidence supports a through-heme electron delivery the FEBS Journal 283 4491 4501
66. VS Belova RI Gvozdev YUR Malashenko AP Sadkov VV Yurchenko 1976 The isotope effect in the enzymatic oxidation of methane Biochem Moscow 41 1903 1904
67. BJ Wallar JD Lipscomb 1996 Dioxygen activation by enzymes containing binuclear non-heme iron clusters Chem Rev 96 2625 2657
68. RN Austin H-K Chang GJ Zylstra JT Groves 2000 The non-heme diiron alkane monooxygenase of *Pseudomonas oleovorans* (AlkB) hydroxylates via a substrate radical intermediate J Am Chem Soc 122 11747 11748
69. V Guallar BF Gherman SJ Lippard RA Friesner 2002 Quantum chemical studies of methane monooxygenase: comparison with P450 Curr Opin Chem Biol 6 236 242
70. Stahl SS, Fransicsco WA, Merkx M, Klinman JP, Lippard SJ (2001) Oxygen kinetic isotope effects in soluble methane monooxygenase J Biol Chem 276 4549 4553
71. S Sirajuddin AC Rosenzweig 2015 Enzymatic oxidation of methane Biochemistry 54 14 2283 2294
72. R Banerjee KK Meier E Münck JD Lipscomb 2013 Intermediate P from soluble methane monooxygenase contains a diferrous cluster Biochemistry 52 4331 4342
73. CE Tinberg SJ Lippard 2009 Revisiting the mechanism of dioxygen activation in soluble methane monooxygenase from M. capsulatus (Bath): evidence for a multi-Step, proton-dependent reaction pathway Biochemistry 48 12145 12158
74. R Banerjee JC Jones JD Lipscomb 2019 Soluble methane monooxygenase Ann Rev Biochem 88 409 431
75. AC Rosenzweig CA Frederick SJ Lippard P Nordlund 1993 Crystal structure of a bacterial non-haem iron hydroxylase that catalyses the biological oxidation of methane Nature 366 537 543
76. V Srinivas R Banerjee H Lebrette 2020 High-resolution XFEL structure of the soluble methane monooxygenase hydroxylase complex with its regulatory component at ambient temperature in two oxidation states J Am Chem Soc 142 33 14249 14266
77. JC Jones R Banerjee K Shi H Aihara JD Lipscomb 2020 Structural studies of the *Methylosinus trichosporium* OB3b soluble methane monooxygenase hydroxylase and regulatory component complex reveal a transient substrate tunnel Biochemistry 59 32 2946 2961
78. JC Jones R Banerjee MM Semonis K Shi H Aihara JD Lipscomb 2022 X-ray crystal structures of methane monooxygenase hydroxylase complexes with variants of its regulatory component: correlations with altered reaction cycle dynamics Biochemistry 61 1 21 33
79. BJ Brazeau RN Austin C Tarr JT Groves JD Lipscomb 2001 Intermediate Q from soluble methane monooxygenase hydroxylates the mechanistic substrate probe norcarane: evidence for a stepwise reaction J Am Chem Soc 123 11831 37118
80. SK Lee JC Nesheim JD Lipscomb 1993 Transient intermediates of the methane monooxygenase catalytic cycle J Biol Chem 268 21569 21577
81. Y Liu JC Nesheim SK Lee JD Lipscomb 1995 Gating effects of component B on oxygen activation by the methane monooxygenase hydroxylase component J Biol Chem 270 24662 24665

82. T Chachiyo JH Rodriguez 2012 Structure, electronic configuration, and Mössbauer spectral parameters of an antiferromagnetic Fe_2-peroxo intermediate of methane monooxygenase Dalton Trans 41 995 1003
83. GE Cutsail R Banerjee A Ang Zhou 2018 High-resolution EXAFS provides evidence for a longer Fe Fe distance in the Q intermediate of methane monooxygenase J Am Chem Soc 140 48 16807 16820
84. JC Jones R Banerjee K Shi MM Semonis H Aihara WC Pomerantz JD Lipscomb 2021 Soluble methane monooxygenase component interactions monitored by 19F NMR Biochemistry 60 25 1995 2010
85. C Lee Y Hwang HG Kang SJ Lee 2022 Electron transfer to hydroxylase through component interactions in soluble methane monooxygenase J Microbiol Biotechnol 32 3 287
86. RL Lieberman AC Rosenzweig 2005 **Crystal structure** of a membrane-bound metalloenzyme that catalyses the biological oxidation of methane Nature 2005 434 177 182
87. MP Hendrich E Münck BG Fox JD Lipscomb 1990 Integer-spin EPR studies of the fully reduced methane monooxygenase hydroxylase component J Am Chem Soc 112 5861 5865
88. Y Shiota G Juhász K Yoshizawa 2013 Role of tyrosine residue in methane activation at the dicopper site of particulate methane monooxygenase: a density functional theory study Inorgan Chem 52 14 7907 7917
89. JL Blazyk GT Gassner J Lippard 2005 Intermolecular electron-transfer reactions in soluble methane monooxygenase: a role for hysteresis in protein function J Am Chem Soc 127 49 17364 17376 https://doi.org/10.1021/ja0554054
90. T Chachiyo JH Rodriguez 2012 Structure, electronic configuration, and Mössbauer spectral parameters of an antiferromagnetic Fe_2-peroxo intermediate of methane monooxygenase Dalton Trans 41 3 995 1003
91. SM Smith S Rawat J Telser BM Hoffman TL Stemmler AC Rosenzweig 2011 Crystal structure and characterization of particulate methane monooxygenase from Methylocystis species strain M Biochemistry 50 10231 10240
92. Y Shiota G Juhasz K Yoshizawa 2013 Role of tyrosine residue in methane activation at the dicopper site of particulate methane monooxygenase: a density functional theory study Inorg Chem 52 7907 7917
93. RJ Jodts MO Ross CW Koo PE Doan AC Rosenzweig BM Hoffman 2021 Coordination of the copper centers in particulate methane monooxygenase: Comparison between methanotrophs and characterization of the CuC site by EPR and ENDOR spectroscopies J Am Chem Soc 143 37 15358 15368
94. MA Culpepper GE Cutsail III BM Hoffman AC Rosenzweig 2012 Evidence for oxygen binding at the active site of particulate methane monooxygenase J Am Chem Soc 134 18 7640 7643
95. MO Ross AC Rosenzweig 2017 A tale of two methane monooxygenases J Biol Inorg Chem 22 2–3 307 319
96. LL Cao O Caldararu AC Rosenzweig U Ryde 2018 Quantum refinement does not support dinuclear copper sites in crystal structures of particulate methane monooxygenase Angew Chem Int Ed 57 162 166
97. K Yoshizawa Y Shiota 2006 Conversion of methane to methanol at the mononuclear and dinuclear copper sites of particulate methane monooxygenase (pMMO): a DFT and QM/MM study J Am Chem Soc 128 9873 9881
98. GE Cutsail MO Ross C Amy AC Rosenzweig S De Beer 2021 Towards a unified understanding of the copper sites in particulate methane monooxygenase: an X-ray absorption spectroscopic investigation Chem Sci 12 6194 6209
99. CE Elwell NL Gagnon BD Neisen 2017 Copper-oxygen complexes revisited: structures, spectroscopy, and reactivity Chem Rev 117 2059 2107
100. RJ Jodts MO Ross CW Koo PE Doan AC Rosenzweig BM Hoffman 2021 Coordination of the copper centers in particulate methane monooxygenase: comparison between methanotrophs and characterization of the CuC site by EPR and ENDOR spectroscopies J Am Chem Soc 143 37 15358 15368

101. CW Koo FJ Tucci Y He AC Rosenzweig 2022 Recovery of particulate methane monooxygenase structure and activity in a lipid bilayer Science 375 6586 1287 1291
102. MO Ross F MacMillan J Wang 2019 Particulate methane monooxygenase contains only mononuclear copper centers Science 364 566 570
103. W Wang SJ Lippard 2014 Diiron oxidation state control of substrate access to the active site of soluble methane monooxygenase mediated by the regulatory component J Am Chem Soc 136 2244 3224
104. SY Ro LF Schachner CW Koo 2019 Native top-down mass spectrometry provides insights into the copper centers of membrane-bound methane monooxygenase Nat Commun 10 2675 https://doi.org/10.1038/s41467-019-10590-6
105. BJ Brazeau JD Lipscomb 2000 Kinetics and activation thermodynamics of methane monooxygenase compound Q formation and reaction with substrates Biochemistry 39 44 13503 13515

Chapter 6
Multielectron Redox Mechanisms

Abstract Two electron mechanism may involve the direct transport of two electrons from a mononuclear transition complex to a substrate. The probability of two electron processes, however, increases sharply if two electrons are transferred together with a proton as a hydride or the change in electric charge is compensated by the simultaneous shift of the electron cloud from neighboring groups. Hydrogenases are a diverse group of metalloenzymes that catalyze the conversion of dihydrogen into protons and electrons and the reverse reaction, the generation of dihydrogen. The [NiFe] hydrogenase and the [FeFe] hydrogenase contain sulfur bridged bimetallic centers. The concept of four-electron mechanism of N_2 reduction at mild conditions was first introduced by Likhtenshtein and Shilov in 1970. It was suggested that the four-electron process can occur in a multinuclear cluster of transition metals. The multielectron nature of the energetically favorable processes in clusters does not evidently impose any new additional restrictions. The need for a synchronous four-electronic elementary act with the formation of a hydrazine-like intermediate is dictated by thermodynamics. The central enzyme of biological nitrogen fixation catalyzes in the nitrogen-fixing bacteria the reduction of molecular nitrogen to ammonia by a reductant with the assistance of ATP hydrolysis: The nitrogenase active has iron-sulfur F cluster (F_4S_4), P-clusters (F_8S_7) and FeMo cofactor (F_7S_9Mo). After the accumulation of four and six reduction equivalents in, the reduction of u occurs first to the hydrozine derivative and then to ammonia, respectively, the reduction of u occurs first to the hydrozine derivative and then to ammonia, respectively. The role of ATP in the nitrogenase reaction was discussed. Subsequent absorption of four light quanta by Photosystem I (PS)I and photosystem II (PS II) results in evaluation of dioxygen from a two-water molecule. Photosynthetic water oxidation occurs at the oxygen-evolving complex (OEC) of (PSII) which contains a Mn_4CaO_5 inorganic cluster ligated by oxides, waters and amino acid residues. In 1976, Semenov, Shilov, and Likhtenstein suggested that in plant photosynthesis, the splitting of a water molecule with the release of molecular oxygen can occur only according the thermodynamically allowed four-electron mechanism.

G. I. Likhtenshtein, *Enzyme Catalysis Today and the Chemistry of the 21st Century*,
https://doi.org/10.1007/978-3-031-51552-1_6

6.1 Two Electron Processes. Hydride Transfer in Hydrogenases

There are a considerable number of reactions in which the products contain two electrons more than the starting compounds. These reactions can proceed as two successive one-electron transitions or by a single two electron process. Two electron mechanism may involve the direct transport of two electrons from a mononuclear transition complex to a substrate. Such a transport may alter sharply the electrostatic states of the systems and obviously require a substantial rearrangement of the nuclear configuration of ligands and polar solvent molecules. For instance, the estimation of the synchronization factor for an octahedral complex with low-molecular saturated ligands showed a very low value of $a_{syn} = 10^{-7}$–10^{-8} and, therefore, a very low rate of reaction. In addition, a significant reorganization energy of the process is expected. The probability of two electron processes, however, increases sharply if two electrons are transferred together with a proton as a hydride or the change in electric charge is compensated by the simultaneous shift of the electron cloud from neighboring groups. Involvement of bi- and, especially, polynuclear transition metal complexes as well as synchronous protons transfer, may essentially decrease the environment reorganization and, therefore, provide a high rate for the two-and four-electron mechanisms.

Hydrogenases are found in a wide range of archaea, prokaryotes and eukaryotes as soluble and membrane-bound enzymes [1–6]. Hydrogenases are a diverse group of metalloenzymes that catalyze the conversion of dihydrogen into protons and electrons and the reverse reaction, the generation of dihydrogen:

$$H_2\left(\rightleftharpoons H^+ + H^-\right) \rightleftharpoons 2H^+ + 2e^-$$

The reaction takes place at a specialized metal center that increases the acidity of H_2 and leads to a heterolytic splitting of the molecule which is strongly accelerated by the presence of a nearby base. Polarized H_2 involves in the respective coupling of H^+ and H^-. The heterolytic mechanism has been proven by H/D isotope exchange experiments. Hydrogenases are sub-classified into three different types based on the active site metal content: iron-iron ([FeFe]) hydrogenase, nickel-iron hydrogenase [NiFe], and iron hydrogenase [Fe]. In the hydrogenases, hydrogen uptake is coupled to the reduction of electron acceptors such as oxygen, nitrate, sulfate, carbon dioxide, and fumarate. On the other hand, proton reduction is coupled to the oxidation of electron donors such as ferredoxin. Both low-molecular weight compounds and proteins such as cytochrome c_3, and cytochrome c_6 can act as physiological electron donors or acceptors for hydrogenases.

The [NiFe] hydrogenase from DvMF26 and the [FeFe] hydrogenase contain sulfur bridged bimetallic centers, typically with an open coordination site on one metal (Fig. 6.1). [Fe] hydrogenases with a Fe cofactor (Fe-guanylylpyridinol, Fe-GP) catalyze the reversible reduction of the substrate methenyltetrahydromethanopterin

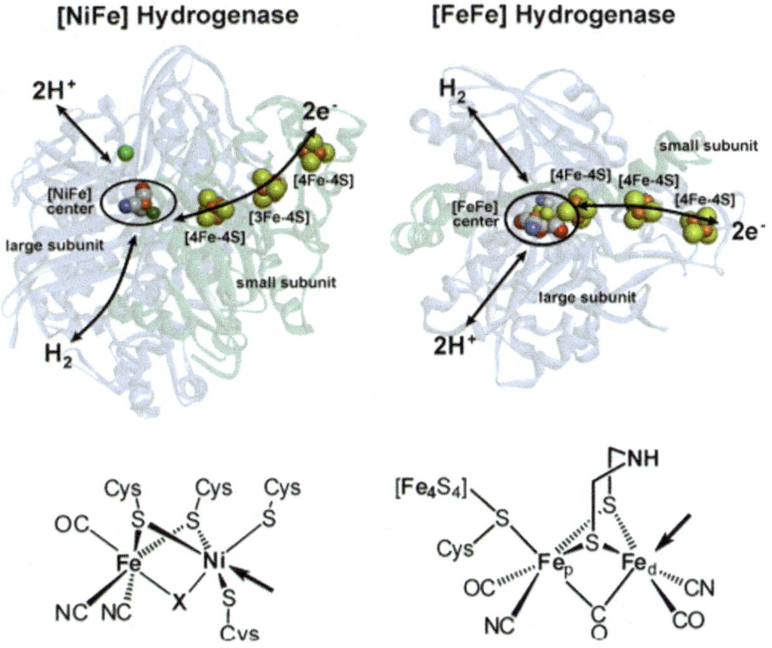

Fig. 6.1 Structures of the [NiFe] hydrogenase from DvMF26 and of the [FeFe] hydrogenase from Dd. 27. Schematically indicated are the ET chain (via iron – sulfur centers), as well as pathways for the dihydrogen and the H + transfer. At the bottom, the chemical structures of the active sites of the two types of hydrogenases are given; the arrows indicate the open metal coordination site 6 [2]. With permission from American Chemical Society

(methenylH4MPT +) with H_2 to methylene-H4MPT and a proton by transferring a hydride ion to the proR position of the C14a carbon of methylene-H4MPT.

The basic catalytic cycle of hydrogenases bearing a [NiFe] active site contains three states (Ni-SIa, Ni-C, and Ni-R) and runs in several steps: (1) In a first step the H_2 is polarized at the four-coordinated Ni and heterolytically split; (2) The resulting Ni-R states carry the H^- in the bridge between Ni(II) and Fe(II); (3) The H^+ is attached to a sulfur of the terminal Cys546; (4) Release of a proton and an electron leads to the paramagnetic Ni-C state, carrying a hydride bridge; and (5) Loss of another electron and a proton (from the hydride bridge) then leads back to Ni-Sia and completes the cycle [2].

[FeFe] hydrogenases are highly active catalysts for hydrogen conversion [3]. Their active site has two components: a [4Fe-4S] electron relay covalently attached to the H_2 binding site and a diiron cluster ligated by CO, CN^-, and 2-azapropane-1, 3-dithiolate (ADT) ligands. Reduction of the [4Fe-4S] site was proposed to be coupled with protonation of one of its cysteine ligands. In [FeFe]-hydrogenases, catalysis takes place at a di-iron center, which contains a bridging dithiolate ligand, three CO ligands and two CN (−) ligands (Fig. 6.2) [4]. The enzyme can be in two active states, H_{ox} and H_{red}. H_{ox} has one unpaired electron and a Lewis acidic Fe (I) center

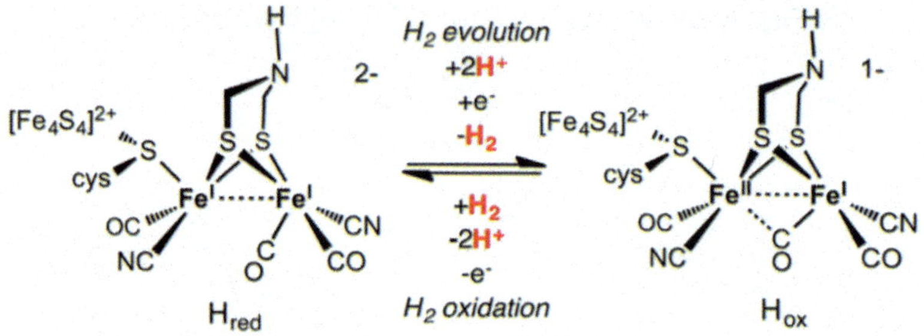

Fig. 6.2 Active site of [FeFe]-H_2ase in two catalytically significant states [5]. With permission from American Chemical Society

poised to bind H_2. H_{red} is the reduced state with all electrons are paired (an $S = 0$ in ground state). Protonation of H_{red} or H_2 activation at H_{ox} (concomitant with electron transfer) can give a presumed catalytic intermediate such as a diferrous hydride state Fe(II)Fe(II)-H^-. A catalytically relevant super-reduced state has also been identified in the enzyme from *C. reinhardtii*, featuring a reduced $[4Fe–4S]^+$ cluster and possibly an Fe–H bond.

It was assumed that in the [FeFe] hydrogenase from CpI, the hydride can be formed after the two electron shift in H_{red} (Fig. 6.3) [2]. DFT and experimental studies on model systems have indicated that, only a terminal hydride has sufficient reactivity to recombine with a proton and form molecular hydrogen.

[FeFe]-hydrogenase from *Clostridium acetobutylicum* (CaHydA) was adsorbed to a negatively charged, self-assembled monolayer (SAM). Electrochemical and single-molecule imaging experiments (scanning tunneling microscopy (EC-STM) and cyclic voltammetry) were carried out on a catalytically active hydrogenase preparation [6]. The single enzyme distribution on the Au-SAM surface coupled with macroscopic electrochemical measurements was observed. Employing the same electrode allowed the evaluation of a turnover frequency (TOF) as a function of potential for single [FeFe]-hydrogenase molecules.

6.2 Multielectron Processes. Four-Electron Reduction and Oxidation

6.2.1 General

The concept of four-electron mechanism was first introduced by Likhtenshtein and Shilov in [7] and then was developed and applied to such "heavy" enzymatic and chemical reactions as the reduction of molecular nitrogen and the water splitting under mild conditions [8–17]. It was suggested that the four-electron process can occur in a multinuclear cluster of transition metals. The multielectron nature of the

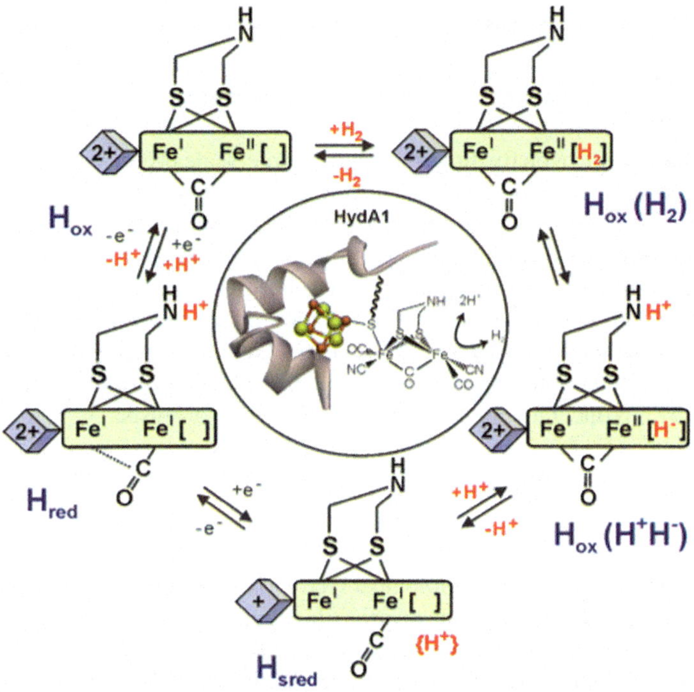

Fig. 6.3 Proposed catalytic mechanism of [FeFe] hydrogenase including the Hsred state. (H +) in Hsred indicates that the proton is believed to be associated with the H-cluster and possibly bound to a nearby amino acid residue [2]. With permission from American Chemical Society

energetically favorable processes in clusters does not evidently impose any new, additional restrictions on its rate: (1) Within the clusters coordination sphere, the multiorbital overlap is effective and, therefore, the resonance integral V is high; (2) The electron transfer from (or to) the orbitals of the metal to substrate orbital can be accompanied by the simultaneous shift of electron clouds to the reverse direction and, therefore, may prevent significant changing of local charges and does not violate markedly the reaction complex nuclear frame; and (3) The strong delocalization electron in clusters and polynuclear complexes reduces to minimum the reorganization of the nuclear system during electron transitions and, therefore, provides low energy activation and a relatively high value for the synchronization factor. The data presented in this section showed that a particularly significant contribution has been recently made to investigation of structure and action mechanism of the reduction of molecular nitrogen and the water splitting [8–17].

6.2.2 Nitrogen Fixation

6.2.2.1 Structure and Action Mechanism of Nitrogenase

Microbiological nitrogen fixation is the global large-scale process of the atmospheric nitrogen reduction to ammonia with the yield approximately 200,000 million tons per annum. The process occurs in anaerobic and aerobic bacteria such as *Azotobacter vinelandii, Clostridium pasteurianum, Klebsiella pneumonia, etc.*, and *Rizobium* from the root nodules of legumes. The Bulen and LeCompt [18] and Mortenson et al. [19, 20] groups first reported the isolation of the first partially purified dinitrogen fixing complex (nitrogenase). The central enzyme of biological nitrogen fixation catalyzes in the nitrogen-fixing bacteria the reduction of molecular nitrogen to ammonia by a reductant with the assistance of ATP hydrolysis:

$$\begin{aligned} N_2 + 8H^+ + \text{reducing agent} + 16\text{MgATP} &\rightarrow 2NH_3 \\ &+ \text{oxidant} + H_2 + 16\text{MgADP} + 16\text{Pi}^{\text{i}} \end{aligned}$$

The reducing agent can be biological (ferredoxin) and non-biological ($S_2O_4{}^{2-}$) compounds or a photochemical electron donation systems [21, 22]. Apart from N_2 and H^+, nitrogenase catalyzes reduction of many substrates (C_2H_2, C_2H_4, NO, HCN, HN_3, cyclopropene, etc.), which are also inhibitors of nitrogen reductions. The biological reduction of N_2 to NH_3 catalyzed by Mo-dependent nitrogenase requires ATP-driven electron transfer from the Fe protein to the catalytic MoFe protein. Each ET is coupled to the hydrolysis of two ATP molecules. Developments in this important field have been reviewed in the last decades [8–20, 23–47].

Crystallographic data have been first produced for the nitrogenase complex of FeP (A2) and FeMoP (A1) from *Azotobacter vinelandii* [30, 32] and for the corresponding complex of Cp2 and Cp1 from *Clostridium pasteurianum* [20, 22]. The solutions of a number of high-resolution X-ray structures of the nitrogenase component proteins and use of an arsenal biochemical and physical methods [31–59] have provided insights into the nature of the nitrogenase active site including iron-sulfur F-and P-clusters and FeMo cofactor (Figs. 6.4 and 6.5) [24]. The cluster structure of the iron-sulfur complex in the active center of nitrogenase was proven for the first time by Likhtenshtein group in the early 1970's by the method of nitroxide spin labels [12–16, 31]. Later, such structure was confirmed by ESR and Mössbauer spectroscopy [32].

The Lowe–Thorneley (LT) kinetic model for nitrogenase function was constructed on the base of the kinetic studies, which involved steady-state, stopped-flow, and freeze–quench kinetics measurements (Fig. 6.6) [33]. The model describes the kinetics of transformations among catalytic intermediates (denoted En) where n is the number of steps of electrons/ protons delivery to MoFe protein. Note that, according to the scheme, dinitrogen is attached to the active center of the enzyme only after the delivering n of four electrons and four protons.

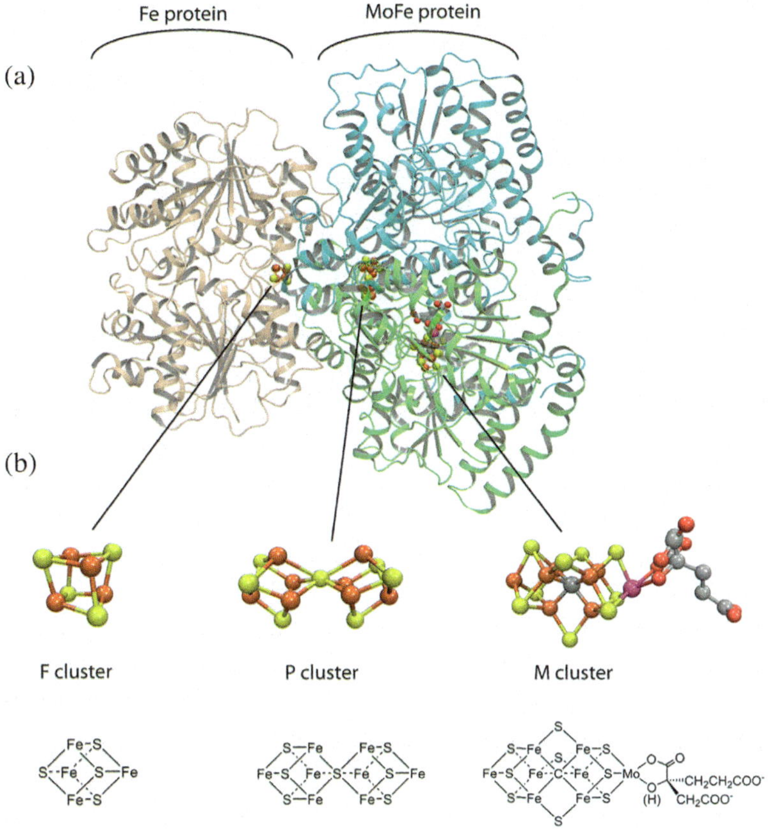

Fig. 6.4 Molybdenum nitrogenase. **A** One catalytic half of the Fe protein: MoFe protein complex with the Fe protein homodimer shown in tan, the MoFe protein α subunit in green, and the β subunit in cyan. **B** Space filling and stick models for the 4Fe − 4S cluster (F), P-cluster (P), and FeMo-co (M) [24]. Made with Pymol and ChemDraw using PDB:2AFK [24]. Published under Creative Commons public use license CC-BY from American Chemical Society

The calculations of the binding energies of the intermediates along the reaction path for hydrogenation of N_2 (Fig. 6.7) [34, 43] were based on density function theory, with plane wave expansion of the Kohn-Sham wave functions and a generalized approximation for the exchange correlation term. Two different clusters to mimic the central part of the FeMoco have included the effect of the surrounding by invoking a proton donor in the vicinity of the cofactor. The authors have come to the important conclusions that only the fourth electron/proton transfer will make the reaction irreversible.

The chemical consequences of a central ligand in the nitrogenase FeMo cofactor were investigated using density functional calculations [46]. Several possible routes for insertion of the central nitrogen ligand were considered, which allowed to conclude c that all routes involve barriers and intermediate states, are inaccessible

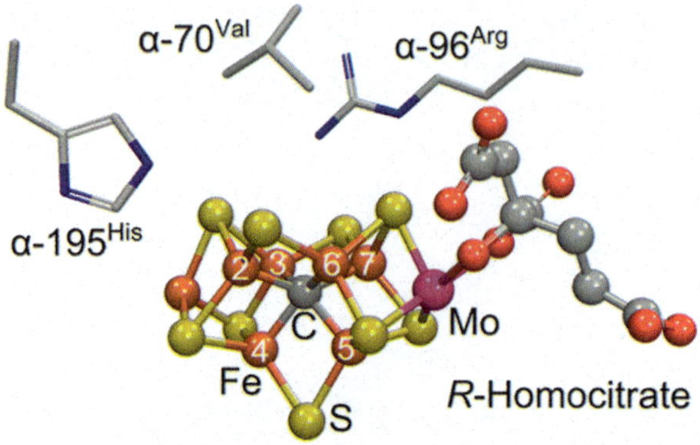

Fig. 6.5 FeMo cofactor and the side chains of selected amino acid residues of the MoFe protein. Numbering of iron atoms is according to the structure PDB coordinate 2AFK. Iron is shown in rust, molybdenum in magenta, nitrogen in blue, sulfur in yellow, carbon in gray, and oxygen in red [24]. Published under Creative Commons public use license CC–BY from American Chemical Society

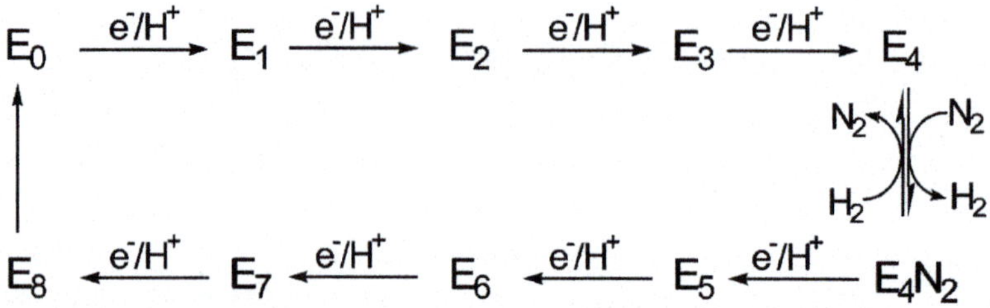

Fig. 6.6 Simplified LT kinetic scheme that highlights correlated electron/proton delivery in eight steps. Although in the full LT scheme N2 binds at either the E3 or E4 levels, the pathway through E3 is de-emphasized here. LT also denotes the protons bound to FeMoco (e.g., E1H1) [24]. Published under Creative Commons public use license CC-BY from American Chemical Society

at ambient conditions. It was also investigated how the FeMoco with the central ligand can interact with N_2 and reduce it. The possible conversion of bidentate into monodentate homocitrate on this three electron-reduced species in FeMoco has been proposed to free up a binding site for N_2 on the molybdenum atom [35]. Experiment indicated that α-Lys^{426} can facilitates chelate ring opening and subsequent orientation of the monodentate homocitrate by forming a specific hydrogen bond to the homocitrate -$CH_2CH_2CO_2^-$ carboxylate group and that mutation of α-Lys can selectively perturb N_2 reduction without affecting acetylene reduction.

A comprehensive theoretical investigation of the binding of N_2 to the Fe(7)MoS(9)N(homocitrate)(cysteine)(histidine) active site (FeMo-co) of the nitrogenase (Fig. 6.8) was carried out [36]. The model used to calculate FeMo-co (Fig. 6.8), in which 442His is truncated to imidazole, 275Cys was truncated to SCH3, and homocitrate is truncated to glycolate, and -OCH_2COO^- was considered. Ninety-four local energy minima were located for N_2 coordinated in η2-N_2 (side) and η2-N_2

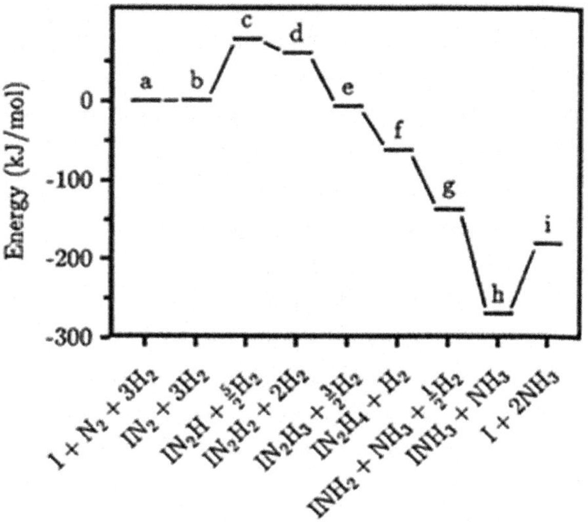

Fig. 6.7 The calculated binding energies of the intermediates along the reaction path for hydrogenation of N_2 on model I are plotted [34]. With permission from American Chemical Society

(end) modes at the endo and exo coordination positions of Fe2 and Fe6. The computation showed that: (1) Barriers to association of N_2 depend mainly on the location of the hydrogenation and the location of N_2 coordination; (2) Dissociation barriers depend primarily on whether N_2 is eta(2)- and eta(1)-coordinated, and secondarily on the location of the hydrogenation, (3) Increase negative charge on FeMo-co increases the barriers;, (4) The first transfer of H to N_2 is likely to have a relatively large activation barrier; and 6) The small barriers for dissociation of η2-N_2 suggest that this is unlikely to be a competitive coordination mode. From data on mutation to affect the N_2 activity of nitrogenase, and from the N_2coordination profiles, it was concluded that endo- η2-N_2 coordination at Fe6 is most probable.

Based on the density functional theory, calculation for ammonia production at the FeMo cofactor of the nitrogenase was performed [37]. It was concluded that: (1) During the catalytic conversion, nitrogen bridges two Fe atoms of the central cage, replacing a sulfur bridge present before dinitrogen binds to the cofactor; (2) Transformation from *cis*- to *trans*-diazene occurs; (3) Exothermic cleavage of the dinitrogen bond takes place, while the Fe atoms are bridged by a single nitrogen atom, and (4) The dissociation of the second ammonia from the cofactor is facilitated by the closing of the sulfur bridge following an intramolecular proton transfer. It was found [38] that turnover of a doubly-substituted MoFe protein (α-70Ala/α-195Gln) with hydrazine generates a new state with a rhombic $S = 1/2$ EPR signal ($> 60\%$ of the total FeMo cofactor signal). Pulsed ENDOR spectroscopy using uniformly ^{15}N-labeled hydrazine revealed that the trapped intermediate was a hydrazine-derived species bound to FeMo cofactor with a single type of ^{15}N signal. H-ENDOR spectroscopy showed that strongly coupled proton(s) were associated with the bound N, consistent with a FeMo-cofactor-NyHx species. In addition, the single observed N bound could result from end-on binding of hydrazine or from a single N atom fragment of hydrazine after N-N bond cleavage. Suggested distal (D) and alternating (A)

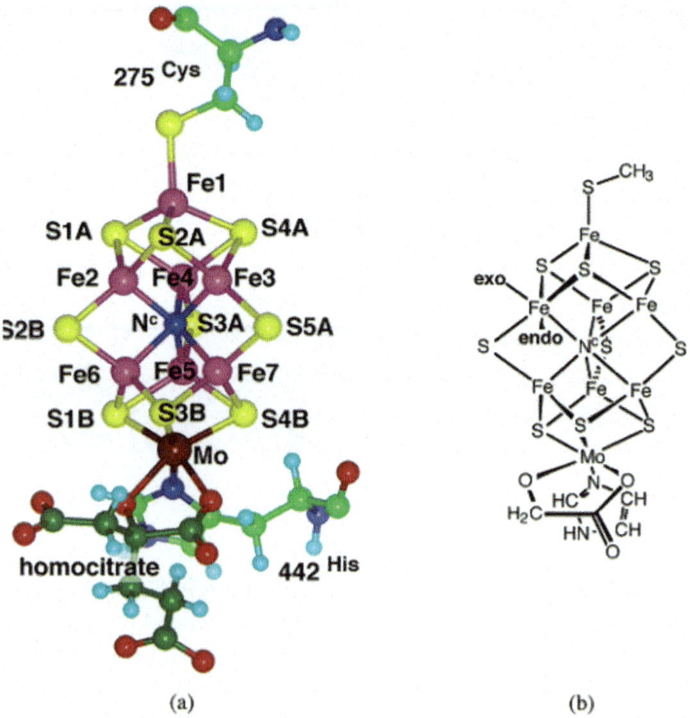

Fig. 6.8 **a** The structure of FeMo-co, connected to the protein via R-275Cys and R-442His (*Azotobacter Vinelandii*), with atom labeling for crystal structure 1M1N. The C atoms of homocitrate are dark green. **b** The exo and endo coordination positions at one Fe atom, marked on the model used in density functional calculations [36]. With permission from American Chemical Society

pathways for N_2 hydrogenation with the Janus E_4 state is presented in Fig. 6.9. As can be seen in the Figure, characterization of catalytic intermediates formed during the reduction of N_2 could distinguish between the D and A pathways.

In early 1970th, it was suggested Semenov et al. [9, 40] that four-electron reduction N_2 can occur at the stage of formation of the binuclear transition metal complex with this molecule whose structure is close to that of the hydrazine derivative.

M=N—N=M or 

Below is an estimate of the energy of possible mechanisms for the reduction of molecular nitrogen by molecular hydrogen, the redox potential of which is close to the potential of reducing agents in the nitrogenase reaction. The following known data on the average binding energies between atoms D (in bracket) in kcal/mol were used for the calculation: $D^{N\text{-}N}$ (34), $D^{N\equiv N}$ (100), $D^{N\equiv N}$ (22), $D_{H\text{-}H}$ (91), $D_{N\text{-}H}$ (93.5) and calculated energies of first and second broken bonds: $E_1 = 125$ kcal/mol and

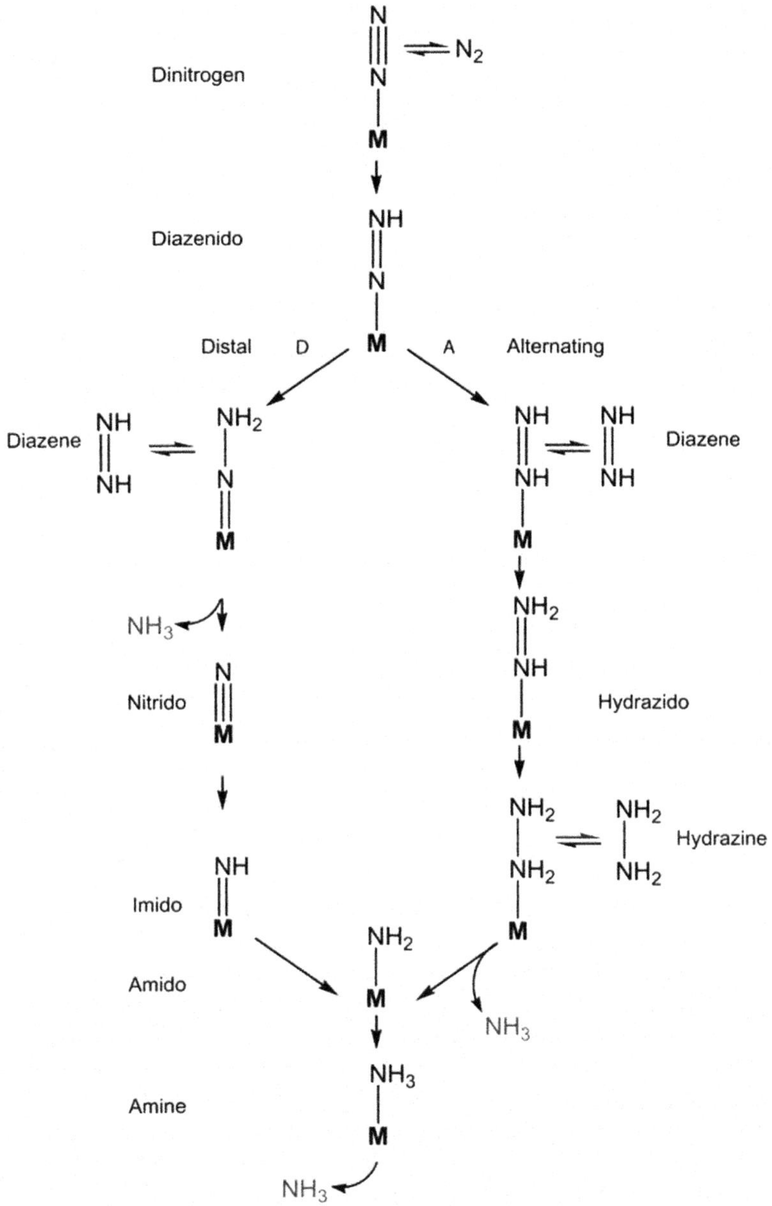

Fig. 6.9 Comparison of distal (D) and alternating (A) pathways for N_2 hydrogenation, highlighting the stages that best distinguish them, most especially noting the different stages at which NH3(1) is released [39]. With permission from American Chemical Society

$E_2 = 66$ kcal/mol. The following are probable mechanisms of the N_2 reduction by hydrogen atom and molecule:

(1) One electron reduction N_2 by hydrogen atom (Mechanism I)

$$N_2 + H = N_2H$$

$$EI = -E1 + D_{NH} + B = -125 + 93.5 + 30 = -1.5\text{kcal/mol}$$

B is the conjugation energy (tree electron bond energy) in N_2H estimated in [7, 12] as 30 kcal/mol

(2) One electron reduction N_2 by hydrogen molecule (Mechanism II)

$$\begin{aligned} N_2 + 1/2 \cdot H_2 = -N_2H\ \ EII - &= -E1 - 1/2D_{HH} + D_{NH} + B \\ &= -125 - 45.5 + 93.5 + 30 = -47\text{kcal/mol} \end{aligned}$$

(3) Two electron reduction N_2 by hydrogen molecule

$$\begin{aligned} N_2 + H_2 = HN_2H\ E_{EII} - &= -E1 - DHH' + 2D_{NH} \\ &= -125 - 91 + 2 \times 93.5 = -26\text{kcal/mol} \end{aligned}$$

(4) Four-electron reduction N_2 by hydrogen molecule

$$\begin{aligned} N_2 + 2H_2 = H_2N_2H_2\ E_{IV} - &= -E1 - E2 - 2D_{HH} + 4D_{NH} \\ &= -125 - 2 \times 91 - 66 + 4 \times 93.5 = 1\text{kcal/mol} \end{aligned}$$

According to the Eqs. 6.3 and 6.4, the mechanism IV for the formation of hydrazine is 25 kcal/mol more energetically favorable than the mechanism III occurring the formation of diazene. Therefore, the formation of hydrazine by reaction with molecular hydrogen is likely to be 10^{18} times as compared with the formation of diazene. Consideration of the data on the standard enthalpies, at 298.15 K, of diazene ($\Delta H_0 = 41.7$ kcal/mol and hydrazine ($\Delta H_0 = 23$ kcal/mol) [42]. also leads to a similar conclusion. The energy difference in this case is 18.7 kcal/mol in favor of the formation of hydrazine. It can be assumed with a high probability that the significant thermodynamic preference of the four-electron process over the two electron one will be retained in the reduction reaction of N_2 in the active center of FeMOco.

The above data on the energy of one-electron and two electron mechanisms for the reduction of molecular nitrogen indicate high energy barriers in the reactions, which thermodynamically excludes the formation of single or double bonds of N_2 with a single atom of transition metal complex in FeMOco. In addition to the energy considerations presented above, below are the arguments in favor of four electronic mechanism for the reduction of molecular nitrogen in the active center of nitrogenase with a hydrazine derivative as the first intermediate product.

1. Only the fourth electron/proton transfer will make the reaction irreversible [43].
2. In FeMoco, N_2 is placed, bridging four iron atoms inside the cavity between the two incomplete F_4S_3 cubanes [44].
3. During the catalytic conversion, nitrogen bridges two Fe atoms of the central cage, replacing a sulfur bridge present before dinitrogen binds to the cofactor [36].
4. Pulsed ENDOR spectroscopy using uniformly ^{15}N-labeled hydrazine revealed that the trapped intermediate was a hydrazine-derived species bound to FeMo cofactor with a single type of ^{15}N signal [45].
5. Hydrazine was revealed in experiments on the nitrogen reaction acid quenching [33].
6. Diazene decomposes readily even at low temperatures [47].
7. The multi-orbital binding of nitrogen in a polynuclear complex (binuclear complex in the simplest case), makes it possible to donate electrons from d-orbitals of metals to anti-binding π^* orbitals of N_2 and to accept electrons from binding π-orbitals of N_2 to empty d-orbitals of the metal.
8. The presence in FeMoco a large, unique in biology cluster of transition metals in the reduced state makes it possible not only simultaneously breaking of two bonds in N_2 and the formation of four new ones with metal atoms, but also compensation for changes in local charges due to the displacement of the general electron cloud.
9. The principal product of N_2 reduction catalyzed by a model chemical cluster is hydrazine [10, 48].

It should be noted that due to the high complexity of FeMoco and the proximity theoretical methods, the results of calculations cannot be considered final.

6.2.2.2 Role of ATP Hydrolysis

Though reduction of N_2 by biological reducing agents to NH_3 is thermodynamically favorable ($\Delta Go = -7.8$ kcal/mole, 1 atm, 298 K), the catalytic process in mild conditions faces a serious thermodynamic problem. The values of standard redox potential (E_0) of plausible intermediate products N_2-H or N_2H_2, were estimated to be approximately equal to − 3.2 and − 1.2 eV, respectively. In fact, these reactions are thermodynamically forbidden under mild conditions. The mechanism implicated in the formation of hydrazine ($E_0 = -0,75$ eV), as the first intermediate, is thermodynamically more preferable, though requiring additional energy consumption provided most likely for account of energy of ATP hydrolysis (Fig. 6.10) [12, 16, 28].

Available experimental structural and kinetics data and energetic considerations indicate two plausible roles of ATP in the nitrogenase reduction: (a) the triggering of electron transfer from iron protein to iron-molybdenum protein [23] and the strengthening reducing power of the enzyme catalytic redox centers [12, 49, 50]. The energetic profile of the nitrogenase reaction, presented in Fig. 6.10, indicates

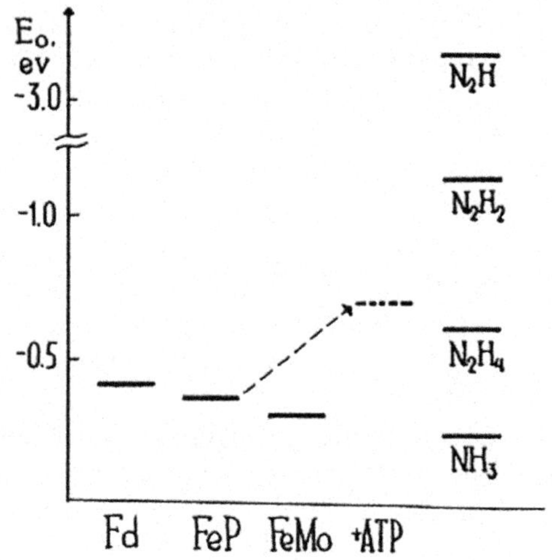

Fig. 6.10 The energy profile of a nitrogenase reaction. E_0 is the standard redox potential of the reactants, intermediates and products of the reaction; Fd = ferredoxin; FeP = Fe protein; FeMo = FeMo protein. The arrow indicates the increase of the reduction potential upon ATP hydrolysis [16]

that formation of reaction plausible intermediates is thermodynamically forbidden without utilization energy of ATP hydrolysis [12, 16, 49, 50, 53, 54].

A chemical mechanism of coupling of ATP hydrolysis reactions in nitrogenase was proposed in 1976 [9]. Specifically, the role of ATP hydrolysis can consist of a forced non-equilibrium protonation of a protein group, which increases the iron- sulfur cluster positive charge. This increase contributes in turn to the transfer of an extra electron and, after removal of the proton to the consequent formation of a super-reduced form of the cluster. Thus, ATP hydrolysis energy may be consumed for producing a center with high reducing power. This conclusion was supported in work [51]. The electrostatic and thermodynamic analysis showed that the ATP hydrolysis energy in nitrogenase can be utilized to increase the driving force of the electron-transfer reaction in the protein–protein complex, while MgATP binding induces conformational changes and protein–protein association. The induced redox-potential change was estimated *found* to be equal to or larger than one-third of an electronvolt, which is roughly the energy associated with the hydrolysis of one MgATP molecule.

The isolated Fe protein can bind MgATP and MgADP at a stoichiometry of two nucleotides per dimer [52, 55]. An ATP analog, ADP·AlF_4, forms a stable and non-active complex with A2 component of nitrogenase (Av1-Av2), in which it is located between two domains at a distance about 15 Å from the [Fe_4S_4] cluster. This value was predicted in early 1970th by estimation of the distance between the ATPase and substrate-binding sites in nitrogenase by the NMR-^{1}H method [53, 54]. The cluster and ADP·AlF_4 are separated by a region, which includes Asp125, Glu128, Asp129, and Cys132 (Fig. 6.11). Toward an understanding of the allosteric regulation of electron transfer in nitrogenase, an analysis of the mechanical coupling in the ATP- and ADP-bound was performed using a graph-based method [55]. The method was employed for investigating the mechanical coupling between distant sites in proteins and on the base crystal structures of the ADP-bound nitrogenase complex (Fig. 6.11)

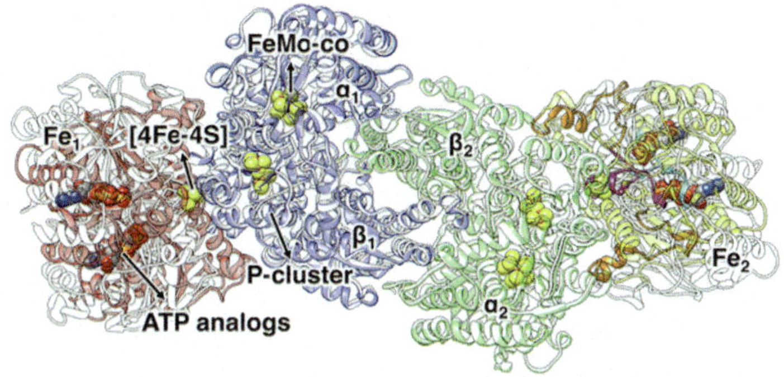

Fig. 6.11 Ribbon representation of the crystal structure of the ATP-bound nitrogenase complex (PDB ID: 2AFK). Fe1, red; Fe2, yellow; α1β1 unit of MoFe, blue; α2β2 unit of MoFe, green. The ADP-bound complex (PDB ID: 2AFI [I]) shown in white, is aligned to the ATP-bound complex on the MoFe protein. In the ATP-bound complex, the P-loop (cyan), switch I region (orange) and switch II region (magenta) are highlighted in Fe2, and the ATP analogues (AMPPCP), [4Fe-4S] clusters, P-clusters and FeMo-co are shown in spheres [55]. Publisher under the terms of the

[55] Graph theory was used to find shortest path in road or a network. Mechanically coupled regions of the nitrogenase complex as inferred computationally were validated against hydrogen/deuterium (H/D) exchange experiments. Specifically, the P-loop in the Fe protein, residues adjacent to β-188Ser in the MoFe protein (in particular β-189Phe), as well as residues at the Fe protein/MoFe protein interface, were found to be strongly coupled.

First-order rate constants for each key step in the reaction sequence, including direct measurement of the ATP hydrolysis rate constant were measured [56]. Experiments performed at with a Hi-Tech SF61 SF UV-visible spectrophotometer, was used to monitor oxidation of Fe Protein. Pre-steady state ATP hydrolysis assays were carried out at 25 °C on a rapid chemical quench-flow instrument (KinTek Corp.). The time course of Pi release was determined in a SF fluorometer (Auto SF-120, KinTek Corp.) using the coumarin (N-[2-(1-maleimidyl)ethyl]-7-(diethylamino) coumarin-3-carboxamide) labeled phosphate binding protein assay. The dissociation of the Fe protein from the MoFe protein was determined by monitoring the release of the Fe protein from the MoFe protein utilizing the SF spectrophotometer. The experiments suggested the following sequential events in the nitrogenase cycle: (1) conformationally gated ET, (2) ATP hydrolysis, (3) Pi release, (4) Fe protein dissociation from the MoFe protein, (5) the free energy of ATP binding and protein–protein association control the ET events, and (6) ATP hydrolysis and Pi release cause dissociation of the reduced MoFe protein from the $Fe^{ox}(ADP)_2$ protein. Figure 6.12 presents the key Fe protein reactions formulated as a thermodynamic cycle.

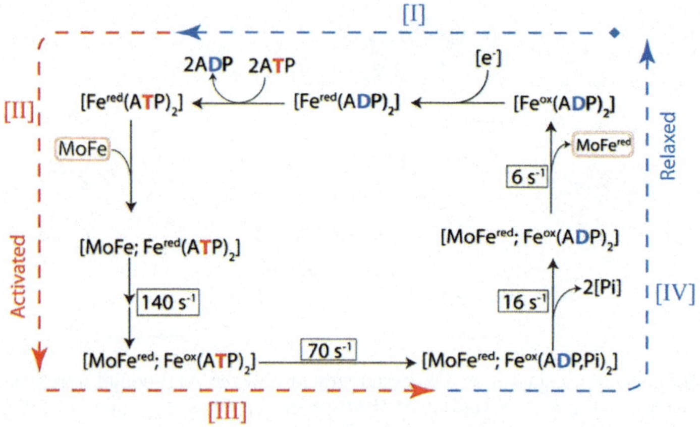

Fig. 6.12 Thermodynamic Fe–protein cycle. Drawn with a focus on Fe protein electron transport to MoFe protein, it can be viewed as starting at upper right. MoFered represents reduced MoFe protein. Dashed arrows and notation that surround the cycle refer to analogy with ABC exporter thermodynamic cycle [56]. With permission from PNAS

- Thus, the triple role of ATP in the nitrogenase reaction was assumed, namely: (1) The binding of this molecule to the [Fe_4S_4] cluster causes the triggering of electron transfer from iron protein to iron-molybdenum protein; (2) The hydrolysis of ATP provides allosterically the energy for the formation of the super-reduced state, what is needed to reduce N_2; and (3) The dissociation of the nitrogenase complex occurs, which prevents the energetically favorable reverse transition of the electron. Changes in the midpoint potentials of the nitrogenase metal centers as a result of iron protein-molybdenum-iron protein complex formation was reported [59]. The midpoint potential of the Fe protein [4Fe-4S]2 + /1 + cluster couple was observed to change by – 200 mV from – 420 mV in the uncomplexed L127Delta Fe protein to – 620 mV in the L127Delta Fe protein-MoFe protein complex. The midpoint potential of the two electron oxidized couple of the P-clusters (P2 + / N) of the MoFe protein was observed to shift by – 80 mV upon protein–protein complex formation.

6.3 Photosynthetic Water Oxidation Complex

Cyanobacteria, green algae, and higher plants provide the major part of molecular O_2 of Earth atmosphere via water oxidation of oxygenic photosynthesis. PS II is a photoenzyme that catalyzes oxidation of the water in a water splitting (Fig. 6.13) [60–89]. Subsequent absorption of four light quanta by PS I and PS II results in evaluation of dioxygen from a two-water molecule [61]. The overall process occurs by the following scheme:

$$2H_2O + 4[-\text{ferredox in (oxidized)} + 4hv \rightarrow O_2 + 4[\text{ferredox in (reduced)}] + 4H^+$$

Photosynthetic water oxidation occurs at the water oxidation system (WOS) or oxygen-evolving complex (OEC) of Photosystem II (PSII) (6.13). The OEC contains a Mn_4CaO_5 inorganic cluster ligated by oxides, waters and amino acid residues. The OEC is embedded in Photosystem II (PS II), a membrane pigment–protein complex, where the primary charge separation by absorbed sunlight energy and successive electron transfer occur through several vectorially arranged pigment molecules.

The key step of the process is the water splitting under absorption of light quanta of relative low energy. The oxidation of two water molecules to produce a dioxygen proceeds in a stepwise manner is described in the four S-state cycle with the following intermediate states (The Kok Scheme [61]):

$$S_0 \rightarrow S_1 \rightarrow S_2 \rightarrow S_3 \rightarrow S_4$$

In 1976, Semenov, Shilov, and Likhtenstein suggested that in plant photosynthesis, the splitting of a water molecule with the release of molecular oxygen can occur only according to the four-electron mechanism [40]. Indeed, one and two electron oxidation of water requires an oxidant with a standard potential of + 2.7 V and 1.35 V, respectively. The formation of an oxygen molecule requires a potential of only + 0.8 V [62]. In such a case, the PS II photosystem must contain a molecular device that switches four one-electron processes initiated by light into a four-electron process. In this regard, the following mechanism of water splitting in a four-nuclear magnesium complex was proposed [40]:

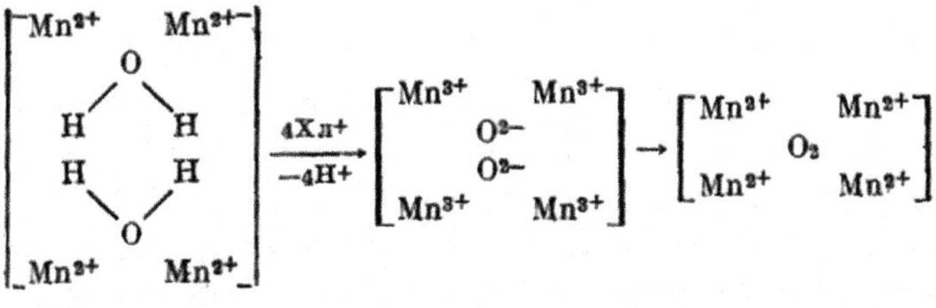

Subsequent studies confirmed the participation of the four nuclear manganese complex in the process of water photooxidation and also, to a large extent, the four electronic oxidation mechanism. In the PSII the chlorophyll cationic radical $P680^+$ is reduced by a redox-active tyrosine, known as Tyr[subz] (Tyr[sup161] of D1 subunit), to generate a neutral tyrosine radical Tyr[subz], which acts as a strong oxidant for the water oxidation process. The radical cation $P680^+$[sup þ] has a very high oxidizing potential, estimated to be 1.2 V [62]. which is sufficient for the water-splitting reaction by the four-electron mechanism with O_2 evolution.

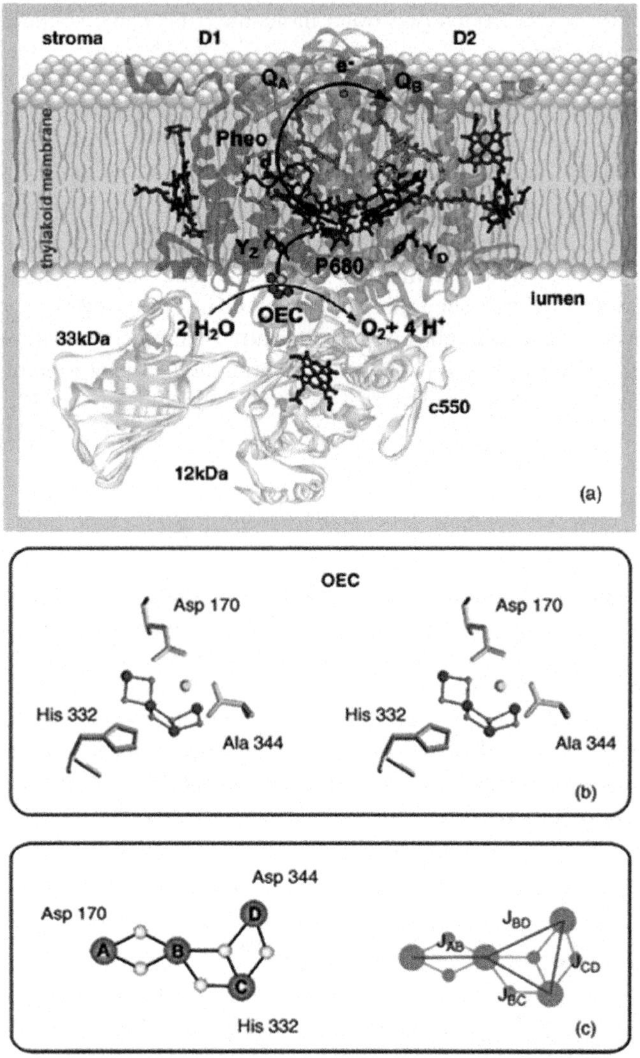

Fig. 6.13 Panel a shows a schematic view of the photosystem II (PS II) complex in the thylakoid membrane that is based on the 3.0 A For clarity, the inner antenna proteins CP43 and CP47, which are involved in harvesting the light energy, and the cytochrome b559 subunits are not shown. The other core proteins of PS II are shown with corresponding labels. The cofactors of the D1, D2, cyt c550, and cyt b559 proteins are placed on top of the proteins. Panel b presents a stereoview of the Mn_4O_xCa cluster together with selected ligands that are referred to in this chapter. The structure shown was derived by Yano et al. on the basis of polarized EXAFS spectroscopy on PS II single crystals. The view is also approximately along the membrane plane and from a similar angle. Panel c displays a schematic top view (approximately along the membrane normal) of the Mn4O5 core of the Mn_4O_xCa cluster using the same color code as that above. This schematic representation allows a clearer visualization of the bridging motifs between the four Mn ions, which are labeled A, B, C, and D. On the right-hand side of panel c, the general coupling scheme is shown that is employed in this study to derive the electronic structure of the S0 and S2. Reproduced with permission from American Chemical Society [60]

The structure and mechanism of the Mn-Ca cluster are important in understanding the mechanism of water oxidation and for the design of efficient artificial water-oxidizing catalysts. Research in this area has been the subject of numerous studies, including the use of advanced physical methods and was comprehensively reviewed [10, 12, 16, 28, 40, 60–74]. In this sections, we will focus mainly on the water-splitting process that appears to be one of the most enigmatic reactions in chemistry and photochemistry.

A possible cycle of the water splitting shown in Fig. 6.14 [87] and references therein, includes the following stages: (1) Water binds as an OH− (O5) in the S0 state, being deprotonated during the subsequent oxidation step to S1; (2) In the S3 state the second substrate water enters the reaction as OH^- (O6) to Mn4 (or Mn1); (3) A further oxidation and deprotonation step leads to S4, formation of which is triggered by proton release; (4) A proton is transferred from W1 (H_2O) to Asp61, as in the preceding S2–S3 transition; and (5) Mutants of PSII lacking Asp61 show a pH dependency of the lag phase during the S3–S0 transition which has been attributed to a deprotonation of the Mn4CaO_5 [87].

Simplified S-state Kok scheme for oxygen evolution along with proposed five oxidation states of Mn in the various intermediate S-states, structure of the OEC and kinetics of electron transfer using data available by 2014 was presented in [63]. The geometry of the Mn_4CaO_5 cluster in the OEC has been revealed in the 1.9 Å crystal structure. X-ray diffraction (XRD) studies, detailed extended X-ray absorption fine structure (EXAFS) studies, using oriented membranes, single crystals, and range-extended methods, have indicated that the Mn4CaOx cluster in the dark stable S1 state consisted of three di-μ-oxo and one mono-μ-oxo-bridged Mn−Mn interactions. Kok et al. [63].

(Paper [64]) introduced structural models of the oxygen-evolving complex of photosystem II (PSII) in the dark-stable S_1 state, as well as in the reduced S_0 and oxidized S_2 states, with complete ligation of the metal−oxo cluster by amino acid residues, water, hydroxide, and chloride. The models developed according to state-of-the-art quantum mechanics/molecular mechanics (QM/MM) hybrid methods, being applied in conjunction with the X-ray crystal structure of PSII from the cyanobacterium *Thermosynechococcus elongatus* were consistent with available data on X-ray diffraction models and extended X-ray absorption fine structure. The models suggested the following: (1) Manganese and calcium ions are ligated consistently with standard coordination chemistry assumptions, supported by biochemical and spectroscopic data; (2) The calcium-bound chloride ligand is found to be bound in a position consistent with pulsed electron paramagnetic resonance data obtained from acetate-substituted PSII; (3) The ligation of protein ligands includes monodentate coordination of D1-D342, CP43-E354, and D1-D170 to Mn(1), Mn(3), and Mn(4), respectively; and (4) η^2 coordination of D1-E333 to both Mn(3) and Mn(2); and ligation of D1-E189 and D1-H332 to Mn(2) take place.

The mechanism of water splitting in photosystem II (PSII) as described by chemical models of the oxygen-evolving complex in the S_0-S_4 states was examined [65]. Molecular models of the OEC Mn_3CaO_4Mn catalytic cluster were constructed using state-of-the-art quantum mechanics/molecular mechanics (QM/MM) hybrid

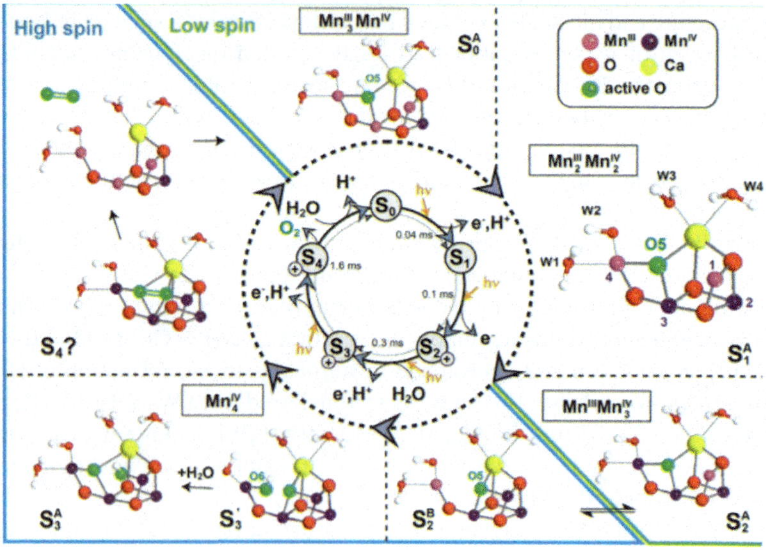

Fig. 6.14 Model for the water oxidation cycle in PS II based on spectroscopic and theoretical work detailing the structures of the Mn cluster in the different S states, the water binding events and the O–O bond formation and O2 release in S4. The boxes show the determined oxidation states of the Mn ions in the respective S state. A color code is used for the assignment of the MnIII (light purple) and MnIV (dark purple) ions when the OEC is passing through the S states. Note that the oxygens of the waters and μ-oxo bridges are given in red except for the proposed substrate oxygens O5 and O6 (green). The S2 state exists in two conformations (closed/open cubanes), see text. S3 may also exist in an open (S3 A) and a closed (S3 B) cubane form (not shown). For S3 a state S3′ is shown in which manganese oxidation has occurred but the 2nd water has not been introduced. This state is present in native sample in a minority of PS II centers, but it is stabilized upon specific chemical treatment (methanol addition or Ca^{2+} to Sr^{2+} exchange). A switching of the preferred total spin ground state configuration of the cluster is thought to take place in S2 from low to high spin and between S4 and S0 back from high to low spin; this is indicated by the diagonal line dividing the green and blue boxes. This switching of the spin state may be necessary for the formation of triplet dioxygen $^3O_2$ in the final step of the cycle [87]. With permission from Springer Nature

methods, in conjunction with the X-ray diffraction (XRD) structure of PSII from the cyanobacterium *Thermosynechococcus elongatus.* Obtained structures of the S_3, S_4, and S_0 states include an additional μ-oxo bridge between Mn(3) and Mn(4) being essential for the deprotonation of substrate water molecules. A detailed mechanism suggested the following: (1) Substrate water molecules coordinated as terminal ligands to Mn(4) and calcium; (2) The formation of an oxyl radical by deprotonation of the substrate water molecule ligated to Mn(4); (3) Accumulation of four oxidizing equivalents nucleophilic attack by a substrate water molecule initially coordinated to calcium and activated by two basic species, including CP43-R357 and the μ-oxo bridge between Mn(3) and Mn(4). The authors concluded that the reaction is concerned with water ligand exchange, swapping the activated water by a water molecule in the second coordination shell of calcium.

Electronic structure of the Mn_4O_xCa cluster in the S_0 and S_2 states of the oxygen-evolving complex of photosystem II was investigated using advance pulse ^{55}Mn-ENDOR and EPR spectroscopy [83]. ^{55}Mn hyperfine coupling constants of the S_0 and S_2 states of the OEC were interpreted on the basis of Y-shaped spin-coupling schemes with up to four nonzero exchange coupling constants J. This analysis established: (1) The oxidation states of the manganese ions in S_0 and S_2; (2) Shortening of 2.85 A Mn–Mn distance in S_0 to 2.75 A in S_1; and (3) Deprotonation of hydroxo bridge between MnA and MnB, that is, between the outer Mn; and its neighboring Mn of the ı3-oxo bridged moiety of the cluster. These results were summarized in a molecular model for EOS (Fig 6.13). Using 9.4 GHz (X-band) hyperfine sublevel-correlation (HYSCORE) and 34 GHz (Q-band) electron spin–echo electron nuclear double resonance (ESE-ENDOR) spectroscopies, the binding mode of ^{13}C-labeled methanol was revealed. These results were compared to analogous experiments on a mixed-valence Mn(III)Mn(IV) complex (2-OH-3,5-Cl_2-salpn)$_2$Mn(III)Mn(IV) (salpn = *N*,*N*′-bis(3,5-dichlorosalicylidene)-1,3-diamino-2-hydroxypropane) in which methanol ligates to the Mn(III) ion [85].

Intense femtosecond X-ray pulses produced at the Linac Coherent Light Source (LCLS) were used for investigation of the dark state (S_1) and the first illuminated state (S_2) of Mn_4CaO_5 cluster, employing simultaneous X-ray diffraction (XRD) and X-ray emission spectroscopy (XES) of microcrystals of photosystem II (PS II) at room temperature [66]. Employed ultrashort X-ray pulses (UXP) were the pulses at 30 femtosecond or shorter with wave lengths occurring at interatomic distances. The inherent X-ray's abilities to capture the change in position of atoms, or molecules during transient processes were used. The UXP method was probed for the overall protein structure and the electronic structure of the Mn_4CaO_5 cluster. XRD data were presented from both the dark state (S_1) and the first illuminated state (S_2). Structure details deduced from diffraction of micron-sized crystals of PS II using sub 50 fs X-ray pulses at room temperature at LCLS were displayed in a figure which includes the following features: (1) $2mF_o$-DF_c electron density map for the PS II complex in the dark S_1 state obtained at LCLS contouring at 1.2 σ shown for a radius of 5 Å around the protein; (2) Detail of the same map in the area of the Mn_4CaO_5 cluster in the dark S_1 state, with mesh contoured at 1.0 σ and 4.0 σ; and (3) Positions of selected residues in D1 fragment ant atoms Mn and Ca.

The progress in investigation of the catalytic intermediate states of the Mn_4CaO_5 cluster using an X-ray free-electron laser to determine simultaneously the light-induced protein dynamics via crystallography and the chemistry that occurs at the catalytic center was summarized [67]. An experimental set-up for the simultaneous collection of both synchrotron radiation (SR)-based X-ray diffraction (XRD) and X-ray emission spectroscopy (XES) using the coherent X-ray imaging (CXI) instrument at the Linac Coherent Light Source (LCLS) was designed. A set of tools for the study of PSII at XFELs, including a new liquid jet based on electrofocusing, an energy dispersive von Hamos X-ray emission spectrometer for the hard X-ray range and a high-throughput soft X-ray spectrometer based on a reflection zone plate was developed. The diffraction data obtained showed that the structure of PSII at

RT were found to be isomorphous to the structure obtained using cryogenic XRD at Synchrotron radiation (SR) sources.

A "radiation-damage-free" structure of PSII from *Thermosynechococcus vulcanus* in the S_1 state at a resolution of 1.95 Å s using femtosecond X-ray pulses of the SPring-8 ångström compact free-electron laser (SACLA) and hundreds of large, highly isomorphous PSII crystals was examined [68]. The valences of each manganese atom was assigned as Mn1D(III), Mn2C(IV), Mn3B(IV) and Mn4A(III), based on the average Mn–ligand distances and analysis of the Jahn–Teller axis on Mn(III). Importantly, one of the oxo-bridged oxygens, O5, has significantly longer distances to Mn than do the other oxo-oxygen atoms, suggesting that O5 is a hydroxide and therefore may serve as one of the substrate oxygen atoms. Time-resolved X-ray emission spectroscopy (XES) of the Mn_4Ca cluster, in which a core electron was excited by an incident X-ray photon was utilized [69]. The excited state decays by emitting an X-ray photon to fill the core hole and the process of oxygen formation in real time was observed. These experiments revealed that: (1) The oxygen evolution step, initiated by three sequential laser flashes, is accompanied by rapid (within 50 μs) changes to the Mn Kβ XES spectrum; (2) No oxidation of the Mn4Ca core above the all-MnIV state was detected to precede O–O bond formation; (3) O–O bond formation occurs prior to the transfer of the final (fourth) electron from the Mn_4Ca cluster to the oxidized tyrosine TyrZ residue; and (4) The kinetic limitations associated with O–O bond formation exist. The model also suggests an evolutionary adaptation to avoid releasing harmful peroxide species.

The recent structural studies, including a high-resolution, radiation-damage-free structure of the water-oxidizing complex, and structures of intermediate S-states revealed by femtosecond X-ray free electron lasers were taking in account in comprehensive review [70]. The structure and functions of the biologically active site and its implications for the development of inorganic catalysts for solar fuels production were discussed in detail. On the base of the combination of serial femtosecond X-ray crystallography and Kβ X-ray emission spectroscopy (XES), a multimodal approach for simultaneously probing the overall protein structure and the electronic state of the Mn_4CaO_5 cluster throughout the catalytic (Kok) cycle was developed. Data on use X-ray free-electron lasers (XFELs) which generate X-ray by electrons flying through a periodic magnetic field, beam focus, and dose on the Mn $K\beta_{1,3}$ XES spectra in PS II over a wide range of pulse parameters utilizing both microcrystal and solution PS II samples were reported. The experiments were performed at LCLS using a nominal beam energy of 9.5 keV, a pulse duration of 35 fs, and beam focus of 2.5–10 lm corresponding to pulse intensities ranging from 6.6 10^{15} to 5.7 10^{17} W/cm^2 and doses ranging from 1.3 10^6 to 1.1 10^8 Gray. The findings showed beam intensities ranging from $\sim 5 \times 10^{15}$ to 5×10^{17} W/cm^2 at a pulse length of ~35 fs. The spectral effects are compared to those observed between S-states in the Kok cycle.

To monitor apparent volume changes of the PSII protein, membrane particles were investigated employing photothermal beam deflection (PBD) [72]. The PBD technique relied on the detection with microsecond resolution of diffractive index changes upon excitation by a nanosecond laser flash, using an infrared laser providing a continuous-wave probe beam, was applied to study PSII protein in aqueous suspension. A Q-switched, frequency-doubled Nd:YAG laser was served as excitation light source (pump flashes) to step the PSII through the S-state cycle to induce the measured PBD signals. Driven by nanosecond laser flashes, the PSII was synchronously stepped through its water oxidation cycle involving four (semi)stable states (S_0, S_1, S_2, and S_3) and minimally three additional transiently formed intermediates. This technique allowed for determination of the associated enthalpy changes (ΔH) from the heat release into the medium and of apparent volume *DV*. The associated volume changes involved the following: (1) Initial formation of charged groups resulting in contraction assignable to electrostriction; (2) Volume contraction caused by reduced metal–ligand distances upon manganese oxidation; (3) Charge-compensating proton removal resulting in volume expansion due to electrostriction reversal., *S_2, and S_3*).The basic reaction scheme of the WOC was extended to include further intermediates, which reflect the alternating charge state of the Mn_4CaO_n complex due to ET and PT events, resulting in the following sequence [72]:

$$S_1^n \rightarrow S_2^+ \rightarrow S_2^n \rightarrow S_3^+ \rightarrow S_3^n \Rightarrow S_0^+ \rightarrow S_0^n \rightarrow S_1^+ \rightarrow S_1^n$$

where the indices "*n*" or "+" denote a neutral or positively charged WOC in comparison to the S_1 state. The bold arrows indicate the transitions occurring first after light-flash excitation. According to suggested scheme, O_2 is released during the $S_3{}^n \Rightarrow S_0{}^n$ transition. Each $Si^n \rightarrow Si+$ transition reflects the oxidation of the WOC by ET to redox-active tyrosine residue (Tyr161) YZ + , whereas each $Si+ \rightarrow Si^n$ transition involves charge-compensating proton removal from the WOC. On the basis of the photothermal beam deflection (PBD) results, a completed reaction scheme of PSII volume changes in the S-state cycle due to the proton transfer (PT) and electron transfer ET events was suggested [72]. Suggested reaction scheme for the O_2-evolving transition of the WOC was displayed in a Fig. 6.15. As a conclusion, the following processes were detected: (1) Seven transitions attributed, according to the H/D-exchange, temperature, and pH effects on their time constants, to ET or proton transfer (PT) events; (2) The oxidizing the Mn_4CaO_n cluster in the $S_2 \rightarrow S_3$ and $S_0 \rightarrow S_1$; (3) A biphasic PT prior to the O_2-evolving reaction; and (4) The reoxidation of the primary quinone acceptor ($Q_A{}^-$) at the PSII acceptor.

PS II WOS in visible-light-driven water-splitting action was monitored by using in *situ,* aqueous-environment surface-enhanced Raman scattering (SERS) [73]. SERS is a surface-sensitive technique that enhances Ramans scattering by molecules adsorbed on rough metal surfaces or by nanostructures such as plasmonic-magnetic silica nanotubes The enhancement factor can be as much as 10^{10}–10^{11}. The electromagnetic theory proposes the excitation of localized surface plasmons, while the chemical theory proposes the formation of charge-transfer complexes. In work [73], experimental SERS supported by density functional theory and isotopologue comparisons was used as a tool for study the WOS dynamic structure in the course of catalytic action, employing a Ag nanoparticle (NP) film as the SERS substrate. The snapshots in the dynamic SERS were found to correspond

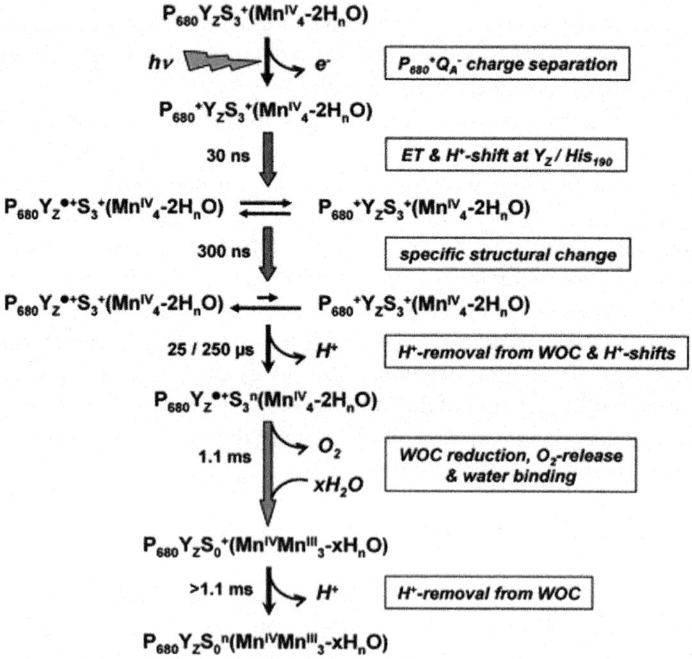

Fig. 6.15 Suggested reaction scheme for the O_2-evolving transition of the WOC [72] With permission from American Chemical Society

to intermediate states S_0–S_3 of the water-splitting cycle, from which the protonation configuration of the S_1 state and likely structures for S_2 and S_3 intermediates were determined. The following structural mechanism of the S_0–S_3 transitions in the water-splitting catalytic cycle and structures for the S_0–S_3 intermediate states of the WOS 's catalytic cycle were assigned: (1) Intermediate structures are based on the protonation state for the S_1 state; (2) The isomer of the S_2 state (open cubane) l involved in the catalytic cycle was chosen on the basis of its structural similarity to the S_1 and S_3 geometries; and (3) An original configuration was chosen for S_3 on the basis of its prevalence in experimental SERS spectra and its consistency with a structure obtained by water insertion in the $S_2 \rightarrow S_3$ transition.

The yield of *in vitro* WOS photoassembly was found to be higher in deuterated water, D_2O, compared with H_2O, when chloride is limiting [74]. The kinetic studies, and the analysis the protonation states of the OEC protein environment using *classical* Multiconformer Continuum Electrostatics showed the following: (1) A significant lag phase in H_2O at limiting chloride concentrations with an apparent H/D solvent isotope effect of 0.14 ± 0.05; (2) The growth phase of WOS photoassembly with an H/D solvent isotope effect of 1.5 ± 0.2; (3) Protons lost from amino acid that will serve as OEC ligands as metals are bound.; (4) Chloride and D_2O increase the proton affinities of key amino acid residues which tune the binding affinity of $Mn^{2+/3+}$ and facilitate the deprotonation of water to form a proposed μ-hydroxo bridged

$Mn^{2+}Mn^{3+}$ intermediate; and (5) Optimal photoassembly was observed after 20 min at 50 μE·m^{-2}·s^{-1}. In addition, in proposed mechanism of WOS photoassembly the following steps were taking in consideration: (1) Mn^{2+} first binds to the HAS in the apo-OEC protein environment; (2) The proton affinities of carboxylic acid groups in the pocket are relatively high, and electrostatic repulsion is minimized; (3) The first Mn^{2+} is oxidized to Mn^{3+}; (4) A second Mn^{2+} binds to the now vacated HAS. and is trapped in place by the formation of a μ-hydroxo bridge with the Mn^{3+} ion (the HAS represents the location where Mn^{2+} binds during photoassembly; (5) Formation of this ligand requires proton release and an efficient Brønsted base located nearby; and (6) the second Mn^{2+} is oxidized to Mn^{3+}, forming a $Mn^{3+}Mn^{3+}$ dimer which is highly stable and templates the remaining photoassembly process, while remaining steps are kinetically rapid.

Electron magnetic resonance has been proved to be a powerful method for studying the electron structure of paramagnetic complexes [75–79]. Below are some typical examples of the application of the method for the study of WOS. The experiment revealed the appearance of various EPR signals initiated by flashes of light. Specifically, it was found that a Mn EPR signal from *t* S_0 state oscillates with a period of four, indicating that it originates from fully active PSII centers [80]. The signal characteristics are consistent with the S_0 state EPR signal arising from a coupled $Mn^{II}-Mn^{III}$ intermediate. The new signal is more stable than the S_2 state signal and its decay in tens of minutes is indicative of it originating from the S_0 state. The effect of illumination at 5 K of photosystem II in different S-states was also investigated with EPR spectroscopy and showed that [81]: (1) Two split radical EPR signals around g ≈ 2.0 were observed from samples given 0 and 3 flashes, respectively; (1) The signal from the 0-flash sample, with a width of ~80 G *was* oscillated with the S_1 state in the sample; (2) The signal from the 3-flash sample, with a symmetric shape of ~160 G width from peak to trough, varied with the concentration of the S_0 state; (3) Both signals were formed within 10 ms after a flash, and decayed with half-times at 5 K of 3–4 min; (3) The signal in the S_0 state, was likely originated from magnetic interaction between $Y_Z\cdot$ and the S_2 state and was assign to $Y_Z\cdot$ in magnetic interaction with the S_0 state, $Y_Z\cdot S_0$; (4) The other signal was attributed to the magnetic interaction between $Y_Z\cdot$ and the S_1 state $Y_Z\cdot S_1$; (5) Y_Z can be oxidized at 5 K in the S_0 and S_1 states; and (6) Oxidation of Y_Z involves deprotonation of the tyrosine.

In study [82], the S_2 state of PSII-enriched membrane fragments was examined at Q-band (frequency 34 GHz) ESR spectroscopy. Light-induced signals at g = 3.1 and g = 4.6 were observed. The observation of the signal at $g = 3.1$ was agreed with a previous study at 15.5 GHz, in which the X-band $g = 4.1$ signal was proposed to arise from the middle Kramer's doublet of a near rhombic S = 5/2 system. Zero-field splitting values of D = 0.455 cm (−1) and E/D = 0.25 were used to simulate the spectra. Q-band EPR spectroscopy was applied to study the S_3 state of the Mn cluster in untreated and in 4% v/v methanol treated PSII membranes [84]. Prominent EPR signals were observed in both sets of samples. A simultaneous analysis of the spectra at both frequency bands indicated that the spin associated with the signals of S_3 is $S = 3$. The $S_1 \rightarrow S_2$ redox transition of the WOS has been studied employing

EPR and Fourier transform infrared (FTIR) spectroscopy [62]. In the S_2 state, two isomers of the WOS were observed by EPR: a $S_T = 1/2$ form and a $S_T = 5/2$ form. Using structural information on the WOS and its surroundings, in conjunction with spectroscopic information available on the $S_1 \rightarrow S_2$ transition for a variety of site-directed mutations, Ca^{2+} and Cl^- substitutions, and small molecule inhibitors, it was proposed that the hydrogen-bonding network encompassing D1–D61 and the OEC-bound waters play an important role in stabilizing one spin isomer. The S_2 state of WOS was studied using EPR spectroscopy at X-band frequencies [86]. It was found that a single manganese center oxidation occurs on $S_1 \rightarrow S_2$ state transition and S_2 state has multiline EPR including g4 signal. The observed multiline signal was arised from a ground state spin ½ center. The g4.1 signal generated at ≈140 K NIR illumination was proposed to arise from a spin 5/2 center with rhombic distortion. These data supported an X-band CW-EPR-generated g4.1 signal as originating from a near rhombic spin 5/2 of the S_2 state of the PSII manganese cluster.

Energetic and kinetic parameters of intermediates formed in the course of the reaction cycle (*S*-state cycle) of photosynthetic water oxidation were obtained [88]. Specifically, time-resolved measurements of prompt chlorophyll fluorescence and time-resolved delayed fluorescence (DF) measurements were employed to estimate rate constants, activation energies, free energy differences, and to discriminate between the enthalpic and the entropic contributions. A joint-fit simulation approach was used to analyze decays of the DF of PSII extending from 10 μs to 10 ms. Nine kinetic parameters of four exponential components were determined for the reaction intermediates in the *S*-state transitions in buffers with different pH in H_2O and in D_2O. Then, the resulting fit parameters were used to calculate the time-constant of O_2 formation, the total Gibbs free energy of the proton removal steps, and a mean time-constant for all proton removal steps.

The progress in investigation of the catalytic intermediate states of the Mn_4CaO_5 cluster was reported [67]. An X-ray free-electron laser was utilized to determine simultaneously the light-induced protein dynamics via crystallography and the local chemistry that occurs at the catalytic center. An experimental setup for the simultaneous collection of both synchrotron radiation (SR)-based X-ray diffraction (XRD) and X-ray emission spectroscopy (XES) using the coherent X-ray imaging (CXI) instrument at LCLS was designed. The diffraction data were obtained by a new liquid jet based on electrofocusing. In the frame of this technique, an energy dispersive von Hamos X-ray emission spectrometer for the hard X-ray range, and a high-throughput soft X-ray spectrometry based on a reflection zone plate showed that the structure of PSII at RT is isomorphous to the structure obtained using cryogenic XRD at SR sources. The RT Mn Kβ XES data obtained from high-valent Mn model compounds and from solutions and microcrystals of PSII, using the femtosecond pulses at LCLS, matched with data collected under non-damaging conditions at low temperatures using SR sources.

References

1. Díez-Pérez I, Brown KA, King PW et al (2012) Catalytic turnover of [FeFe]-hydrogenase based on single-molecule imaging. J Am Chem Soc 134(3):1577–1582
2. Lubitz W, Ogata H, Rüdiger O, Reijerse E (2014) Hydrogenases. Chem Rev 2014(114):4081–4148
3. Sanchez MLK, Wiley S, Reijerse E, Lubitz W, Birrell JA, Dyer RB (2022) Time-resolved infrared spectroscopy reveals the pH-independence of the first electron transfer step in the [FeFe] hydrogenase catalytic cycle. J Phys Chem Lett 13(25):5986–5990
4. Berggren G, Adamska A, Lambertz C, Simmons TR (2013) Biomimetic assembly and activiation of [FeFe]-hydrogenases. Nature 499(7456):66–69
5. Rauchfuss TB (2015) Diiron Azadithiolates as models for the [FeFe]-hydrogenase active site and paradigm for the role of the second coordination sphere. Acc Chem Res 48(7):2107–2116
6. Madden C, Vaughn MD, Díez-Pérez I (2012) Catalytic turnover of [FeFe]-hydrogenase based on single-molecule imaging. J Am Chem Soc 134(3):1577–1582
7. Liechtenstein GI, Shilov AE (1970) On the thermodynamic and kinetic features of the reductive fixation of molecular nitrogen. Zh Fiz Khim 44:849
8. Likhtenshtein GI, Therneley RNF (1995) Nitrogenase catalysis, in Tikhanovitch IA, Provorov N, Romanov VI, Newton WE (eds), Nitrogen fixation: fundamentals and applications (Kluwer Academic Publishers, Dodrecht)
9. Shilov AE, Likhtenshtein GI (1971) Biological fixation of molecular nitrogen and its chemical modelling. Russ Biol Bull 4:518–537
10. Shilov AE (1997) Metal complexes in biomimetical reactions. N_2 Fixation, activation and oxidation of alkanes, Chemical Models of Photosyntheses, (CRC Press, Boca Raton, New-York)
11. Likhtenshtein GI (1990) The use of nitroxides in the solution of some problems of chemical biophysics. Pure Appl Chem 62:281–288
12. Likhtenshtein GI (1988) Chemical physics of redox metalloenzyme catalysis. Springer-Verlag, Berlin
13. Likhtenshtein GI (1986) On mechanisms of redox reactions with participation of metalloenzymes and their models, in Shilov AE (eds), Fundamental research in homogeneous catalysis, (Gordon and Breach, New York), 325–334
14. Likhtenshtein GI (1985) The role of multielectron and synchronous processes in enzyme catalysis. Proc. 16th FEBS Cong. Part A (VNU Science Press) 9–15
15. Likhtenshtein GI (1980) Structure and mechanism of action of nitrogenase active center. In: Newton WE, Orme-Johnson WH (eds) Nitrogen fixation, V 1. Baltimore, Free-living Systems and Chemical Models (University Park Press, pp 195–210
16. Likhtenshtein GI (2003) New trends in enzyme catalysis and mimicking chemical reactions. N.Y. Kluwer Academic/ Plenum Publishers
17. Likhtenshtein GI (1979) Multinuclear redox metalloenzymes. Nauka, Moscow
18. Likhtenshtein GI (1974) Spin labeling method in molecular biology. Nauka, Moscow
19. Thorneley RNF, Lowe DJ (1985) Kinetics and mechanism of the nitrogenase enzyme system. In: Spiro TG, editor. Molybdenum Enzymes. (Wiley-Interscience; New York). 7, 89–116
20. Bulen WA, LeComt JR (1966) The nitrogenase system from Azotobacter: two-enzyme requirement for N2 reduction, ATP-dependent H2 evolution and ATP hydrolysis. Proc Natl Acad Sci 56:979–986
21. Mortenson LE, Morris JA, Jeng DY (1967) Purification, metal composition, and properties of mollibdoferredoxin and azoferredoxin, two components of thenitrogen-fixing system of Clostridium pasterianum. Biochim Biophys Acta 141:516–538
22. Bolen JT, Cambasso N., Muchmore SW, Morgan TV, Mortenson LE (1993) Structure and Environment of metal clusters of the nitrogenase molybdenum iron protein from Clostridium pasterianum, in Stiefel E, Coucouvanis D, Newton WE (eds), Molibdenum Enzymes, Cofactors, and Model Systems (Am. Chem. Soc., Wahington, DC)

23. Syrtsova LA, Druzhinin SY, Rubtsova ET, Shkondina NI (1998) New possibilities for studying mechanism of nitrogenase reaction with photodonors of electron, Curr. Plant Sci. Biotechnol. Agric. 31 (Biological Nitrogen Fixation for the 21st Century), 49–50
24. Syrtsova LA, Nadtochenko VA, Denisov NN, Timofeeva EA, Shkondina NI, Gak V (2000) Yu: x Kinetics of elementary steps of electron transfer in nitrogenase in the presence of a photodonor. Biochem Mosc 65:1145–1152
25. Rees DC, Howard JB (2000) Nitrogenase: standing at crossroads. Curr Opin Chem Biol 4:559–566
26. Hoffman BM, Lukoyanov D, Yang ZY, Dean DR, Seefeldt LC (2014) Mechanism of nitrogen fixation by nitrogenase: the next stage. Chem Rev 114(8):4041–4062
27. Jasniewski AJ, Lee CC, Ribbe MW, Hu Y (2020) Reactivity, mechanism, and assembly of the alternative nitrogenases. Electron Transfer Nitrogenase Chem Rev 120(12):5107–5157
28. Rutledge HL, Tezcan FA (2020) Electron transfer in nitrogenase. Chem Rev 120(12):5158–5193
29. Jasper J, Ramos JV, Trncik C, Jahn D, Einsle O, Layer G, Moser J (2020) Chimeric interaction of nitrogenase-like reductases with the MoFe protein of nitrogenase. J Chembiochem 21(12):1733–1741
30. Bolen JT, Cambasso N, Muchmore, SW, Morgan TV, Mortenson LE (1993) Structure and environment of metal clusters of the nitrogenase molybdenum iron proten from Clostridium pasterianum, In: Stiefel EI, Coucouvanis D, Newton WE (eds), Molibdenum Enzymes, Cofactors, and Model Systems, Am. Chem. Soc., Wahington, DC
31. Einsle O, Rees DC (2020) Structural enzymology of nitrogenase enzymes. Chem Rev 120(12):4969–5004
32. Kim J, Rees DC (1992) Crystollographic structure and functional implication of the nitrogenase molybdenum-iron protein from *Azotobacter vinelandii.* Nature 360:556–560
33. Syrtsova LA, Levchenko LA, Frolov EN, Likhtenshtein GI, Pisarscaya TN (1971) Vorob'ev, LV, Gromoglasova VA: Structure and function of the nitrogenase components from *Azotobacter vinelandii.* Mol Biol Moscow 5:726–734
34. Münck E, Rhodes H, Orme-Johnson WH, Davis LC, Brill WJ, Shah VK (1975) Nitrogenase VIII Mossbauer and ESR spectroscopy. Biochim Biophys Acta 400:32–53
35. Thorneley, RNF, Lowe, DJ: In Molybdenum Enzymes; Spiro,T G, Ed.; (Wiley, New Yok, 1985) pp 221−284
36. Rod TH, Norskov JK (2000) Modeling the nitrogenase cofactor. J Am Chem Soc 122:12751–12763
37. Durrant MC, Francis A, Lowe DJ, Newton WE, Fisher K (2006) Evidence for a dynamic role for homocitrate during nitrogen fixation: the effect of substitution at the α-Lys426 position in MoFe-protein of *Azotobacter vinelandii.* Biochem J 397(2):261–270
38. Dance IJ (2007) The mechanistically significant coordination chemistry of dinitrogen at FeMo-co, the catalytic site of nitrogenase. Am Chem Soc 129:1076
39. Kastner J, Blochl PE (2007) Ammonia production at the FeMo cofactor of nitrogenase: results from density functional theory. J Am Chem Soc 129:2998
40. Seefeldt LC, Hoffman BM, Dean DR (2009) Mechanism of Mo-dependent nitrogenase. Ann Rev Biochem 78(1):701–722
41. Hoffman BM, Lukoyanov D, Dean DR, Seefeldt LC (2013) Nitrogenase: a draft mechanism. Acc Chem Res 46:587
42. Semenov AE, Shilov G, Likhtenshtein GI (1975) Multielectron redox processes in chemistry and biology Doklady Academy Nauk. SSSR 221:1374–1736
43. Likhtenshtein GI (2016) Electron spin in chemistry and biology: fundamentals, methods, reactions mechanisms, Magnetic Phenomena, Structure Investigation. Springer
44. Feller D, Bross DH, Ruscic B (2017) Enthalpy of formation of N2H4 (hydrazine) revisited. J Phys Chem 121(32):6187–6198
45. Siegbahn PEM, Westerberg J, Svensson M, Crabtree RH (1998) Nitrogen fixation by nitrogenases: a quantum chemistry study. J Phys Chem B 102:1615–1623

46. Kastner J, Bloch PE (2007) Ammonia production at the FeMo cofactor of nitrogenase: results from density functional theory. J Am Chem Soc 129:2998
47. Seefeldt LC, Hoffman BM, Dean DR (2009) Mechanism of Mo-dependent nitrogenase. Annu Rev Biochem 78:701
48. Hinnemann B, Nørskov JK (2004) Chemical activity of the nitrogenase FeMo cofactor with a central nitrogen ligand: density functional study. J Am Chem Soc 126(12):3920–3927
49. Wiberg E, Holleman AF (2001) Diimine N_2H_2. Inorganic chemistry, Elsevier, 628
50. Shilov AE, Denisov NT, Efimov NO, Shuvalov V, Shuvalova NI, Shilova AK (1971) (New nitrogenase model for reduction of molecular nitrogen in protonic media. Nature 231:460–461
51. Likhtenshtein GI, Shilov AE (1976) On the chemical mechanism of coupling of ATP hydrolysis reactions Dokl. Acad Nauk SSSR 229:1127–1130
52. Syrtsova LA, Timofeeva EA (2001) Electron transfer coupled with ATP hydrolysis in nitrogenase. Russ Chem Bull 50:1789–1794
53. Kurnikov IV, Charnley AK, Beratan DN (2001) From ATP to electron transfer: electrostatic s and free-energy transduction in nitrogenase. J Phys Chem B 105(23):5359–5367
54. Chiu H-J, Peter JW, Lanzillota WN et al (2001) MgATP-bound and nucleotide-free structure of a nitrogenase protein complex between the 127 D-Fe-protein and the MoFe-protein. Biochemistry 40:641–650
55. Syrtzova LA, Likhtenshtei GI, Nazarova II, Pisarkaya TN, Nazarov VV (1972) Study on the ATP-ase active center of nitrogenase by NMR method Doklad. Akad Nauk USSR 206:367–370
56. Syrtsova LA, Likhtenstein GI, Pisarkaya TN et al (1974) (l974) Estimation of the distance between the ATPase and substrate-binding sites in nitrogenase by the NMR-^{1}H method. Mol Biol Moscow 8:656–662
57. Huang Q, Tokmina-Lukaszewska M, Johnson LE, Kallas H, Ginovska B, Peters JW, Seefeldt LC, Bothner B, Raugei S (2021) Mechanical coupling in the nitrogenase complex. PLOS Comput Biol 17(3):e1008719
58. Simon Duval S, Karamatullah Danyal K, Shaw S et al (2013) Electron transfer precedes ATP hydrolysis during nitrogenase catalysis. Proc Natl Acad Sci USA 110(41):16414–16419
59. Linton KJ, Higgins CF (2007) Structure and function of ABC transporters: the ATP switch provides flexible control. Pflügers Arch Eur J Physiol 453(5):555–567
60. Locher KP (2009) Review. Structure and mechanism of ATP-binding cassette transporters. Philos Trans R Soc Lond B Biol Sci 364(1514):239–245
61. Lanzilotta WN, Seefeldt LC (1997) Changes in the midpoint potentials of the nitrogenase metal centers as a result of iron protein-molybdenum-iron protein complex formation. Biochemistry 36(42):12976–12983
62. Kulik LV, Epel B, Lubitz W, Messinger J (2007) Electronic structure of the Mn_4O_xCa cluster in the S0 and S2 states of the oxygen-evolving complex of photosystem II based on pulse ^{55}Mn-ENDOR and EPR spectroscopy. J Am Chem Soc 129(1):13421–13435
63. Kok B, Forbush B, McGloin MM (1970) Cooperation of charges in photosynthetic O2 evolution-I. A linear four step mechanism. Photochem Photobiol 11(6):457–475
64. Rappaport F, Guergova-Kuras M, Nixon PJ, Diner BA, Lavergne J (2002) Kinetics and pathways of charge recombination in photosystem II (2002). Biochemistry 41(26):8518–8527
65. Yano J, Yachandra V (2014) Mn_4Ca cluster in photosynthesis: where and how water is oxidized to dioxygen. Chem Rev 114(8):4175–4420
66. Sproviero EM, Gascón JA, JMcEvoy JP, Brudvig GW, Batista VS (2006) QM/MM models of the O_2-evolving complex of photosystem II, J Chem Theory Comput. 1119–1134
67. Sproviero EM, Gascón JA, McEvoy JP, Brudvig GW, Batista VS (2008) Quantum mechanics/ molecular mechanics study of the catalytic cycle of water splitting in photosystem II. J Am Chem Soc 130(11):3428–3442
68. Kern J, Alonso-Mori R, Tran R et al (2013) Simultaneous femtosecond X-ray spectroscopy and diffraction of photosystem II at room temperature. Science 340:491–495
69. Tran R, Kern J, Hattne J (2014) The Mn4Ca photosynthetic water-oxidation catalyst studied by simultaneous X-ray spectroscopy and crystallography using an X-ray free-electron laser. Philos Trans R Soc B 369:20130324

70. Suga M, Akita F, Hirata K et al (2015) Native structure of photosystem II at 1.95 A° resolution viewed by femtosecond X-ray pulses. Nature 517:99–103
71. Davis M, Sullivan BT, Palenik MC et al (2018) Rapid evolution of the photosystem II electronic structure during water splitting. Phys Rev X 8:041014
72. Najafpour MM, Zaharieva I, Zand Z, Hosseini SM, Kouzmanova M, Hołyńska M, Tranca I, Larkum AW, Shen JR, Allakhverdiev SI (2020) Water-oxidizing complex in photosystem II: its structure and relation to manganese-oxide based catalysts. Coord Chem Rev 1(409):213183
73. Fransson T, Alonso-Mori R, Chatterjee R et al (2021) Effects of x-ray free-electron laser pulse intensity on the Mn Kβ1,3 x-ray emission spectrum in photosystem II—A case study for metalloprotein crystals and solutions. Struct Dyn 8:064302
74. Klauss A, Haumann M, Dau H (2015) Seven steps of alternating electron and proton transfer in photosystem II water oxidation traced by time-resolved photothermal beam deflection at improved sensitivity. J Phys Chem B 119:2677–2689
75. Wilson AJ, Jain PK (2018) Structural dynamics of the oxygen-evolving complex of photosystem II in water-splitting action. J Am Chem Soc 140(17):5853–5859
76. Vinyard DJ, Badshah SL, Riggio MR, Kaur D, Fanguy AR, Gunner MR (2019) Photosystem II oxygen-evolving complex photoassembly displays an inverse H/D solvent isotope effect under chloride-limiting conditions. Proceed Nat Acad Sci 116(38):18917–18922
77. Hagen WR (2008) Biomolecular EPR Spectroscopy (CRC)
78. Eaton GR, Eaton SS, Barr DP, Weber RT (2010) Quantitative EPR, Springer, Berlin
79. Savitsky A, Mobius K (2009) High-field EPR spectroscopy on proteins and their model systems, Wiley-VCH Verlag GmbH
80. Jamauci J (2008) Fundamentals of electron spin resonance, In: Nitroxides: application in chemistry, Biomedicine, and Materials Science, (ed) by Likhtenshtein GI, Yamauchi J, Nakatsuji S, Smirnov A, Tamura R (WILEY-VCH, Weinhem), 71–120
81. Likhtenshtein GI (2016) Electron spin in chemistry and biology: fundamentals, methods, reactions mechanisms, magnetic phenomena, structure investigation. Springer, Switzerland
82. Åhrling KA, Peterson S, Styring S (1997) An oscillating manganese electron paramagnetic resonance signal from the S_0 state of the oxygen evolving complex in photosystem II. Biochemistry 36(43):13148–13152
83. Zhang C (2003) Styring S: formation of split electron paramagnetic resonance signals in photosystem II suggests that $Tyrosine_z$ can be Photooxidized at 5 K in the S_0 and S_1 states of the oxygen-evolving complex. Biochemistry 42(26):8066–8807
84. Haddy A, Lakshmi KV, Gary W, Brudvig GW, Harry A, Frank HA (2004) Q-band EPR of the S2 state of photosystem II confirms an S = 5/2 origin of the X-band g = 41 signal. Biophys J 87(4):2885–2896
85. Sanakis Y, Sarrou J, Zahariou, G, Petrouleas (2008) Q-band electron paramagnetic resonance studies of the S_3 State of the OEC of Photosystem II. In: Allen, JF, Gantt E, Golbeck JH, Osmond B (eds) Photosynthesis. Energy from the Sun. Springer, Dordrecht
86. Pokhrel R, Brudvig GW (2014) Oxygen-evolving complex of photosystem II: correlating structure with spectroscopy. Phys Chem Chem Phys 16:11812–11821
87. Oyala PH, Stich TA, Stull JA et al (2014) Pulse electron paramagnetic resonance studies of the interaction of methanol with the S2 State of the Mn4O5Ca cluster of photosystem II. Biochemistry 53(50):7914–7928
88. Baituti B (2021) Sebusi odisitse a computational study of the S_2 state in the oxygen-evolving complex of photosystem II by electron paramagnetic resonance spectroscopy. Molecules 26(9):2699
89. Lubitz W, Chrysina M, Cox N (2019) Water oxidation in photosystem II. Photosynth Res 142(1):105–125

Chapter 7
Molecular Dynamics of Proteins and Their Functional Activity

Abstract This chapter presents a brief overview of physical and computer methods for studying the mechanisms and frequencies of the molecular dynamics of proteins and enzymes, illustrated with typical examples. The list of methods used includes the following: spin and Mössbauer labels and probes, fluorescence quenching, fluorescence dynamic Stokes shift, Förster resonance energy transfer (FRET), nuclear magnetic resonance, differential scanning calorimetry, small-wide-angle X-ray scattering, femtosecond time-resolved X-ray solution scattering, and neutron scattering. Protein dynamics occur at time scales from millisecond to subfemtosecond. The range of time-scales involved in substrate turnover step of enzyme catalyzed reactions and internal protein dynamics are similar. The protein motions necessary for catalysis are an intrinsic property of the enzyme and may limit the overall turnover rate. Recent works on selected protein systems have addressed the role of dynamics in enzyme evolution. In some cases, connections among hydration layer dynamics, solvation shell structure, protein surface structure and their function has been revealed. Particular attention was paid to works in which a relationship is established between specific stages of the enzymatic process and specific molecular dynamic modes.

7.1 Introduction

Enzymes must be ordered to allow the stabilization of transition states in their active sites, yet dynamic enough to adopt alternative conformations suited to each steps in their catalytic cycles (principle of dynamic adaptation (Sect. 7.2.1). Increasing evidence indicated that internal protein motions can play a role in promoting protein function, in particular [1–37]. The thermodynamic fluctuations of the solvent, surrounding the protein, can have an impact on internal protein motions and, therefore, on enzyme function. The role of such motions has been investigated by using a whole arsenal of classical and modern physical methods including utilization isotopically substituted enzymes and its mutants through a combination of experimental and computational analyses. Rational design and directed evolution have proved to be

G. I. Likhtenshtein, *Enzyme Catalysis Today and the Chemistry of the 21st Century*,
https://doi.org/10.1007/978-3-031-51552-1_7

successful approaches to increase catalytic efficiencies of both natural and artificial enzymes.

Proteins undergo a broad range of motions under physiological conditions, spanning from local to global changes in conformations (Figs. 7.1 and 7.2) [1–91]. Among them, highly collective "soft" modes, i.e., driving the dynamics in the range of correlate cooperative motions of entire domains/subunits modes. Moreover, the thermodynamic fluctuations of the solvent, surrounding the protein, have an impact on internal protein motions and, therefore, on enzyme function. The contribution of protein flexibility on the ps–ns and μs–ms timescales to enzymatic catalysis has been focus of several computational and experimental studies [28–33]. A brief characterization of physical methods used to investigate frequency $\tau_c^{-1} = 10^{-1}$–$10^{14}\ s^{-1}$ is presented in Table 7.1 [12]. τ_c^{-1} is the correlation frequency; k_{tran} is the rate constant of collisions; τ_{tran}^{-1} is the translational jump time.

Loops are among the most structurally dynamic features of proteins [36, 37]. They always link more ordered secondary structures (i.e., α-helices and β-strands), and can range from well-defined turns to random-coil-like structures. Loops can be important for a range of functions, including interactions with ligands, other parts of the protein, or other macromolecular binding partners. Loop motions can also play direct roles in chemical catalysis through positioning catalytically important amino acid residues [36, 37]. Availability of supercomputers and development of elegant molecular methods of dynamics simulation have been made a basis for the employment of explosive methods and a wide range of successful applications [28, 39] The computer simulation produces individual particle motions as a function of

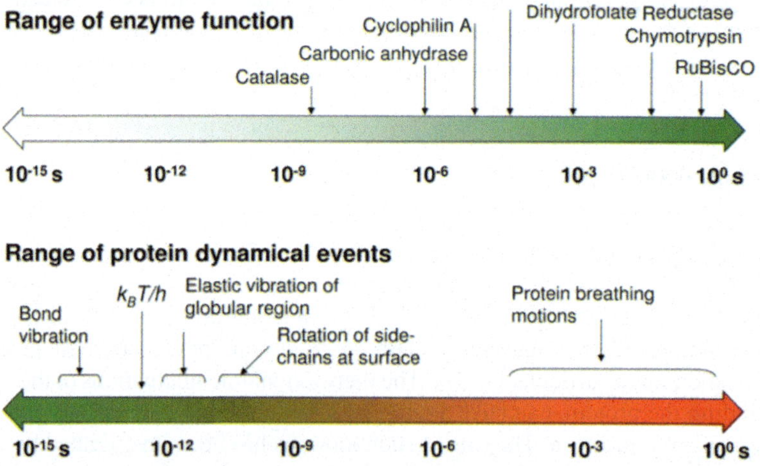

Fig. 7.1 The range of time-scales involved in substrate turnover step of enzyme-catalyzed reactions and internal protein dynamics are similar. Note the universal frequency factor ($k_B T/h$), which is commonly used in transition state theory; k_B is the Boltzmann's constant, T represents the ambient temperature and h is the Planck's constant [11]. Published under Creative Commons CC-BY from Springer Nature

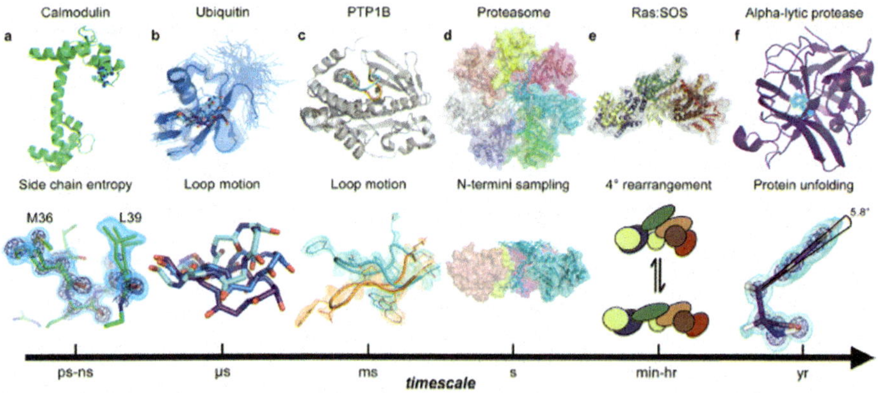

Fig. 7.2 Protein dynamics occur at different time scales. **a** Motions on the picosecond to nanosecond scale involve small changes in backbone or side chain torsion angles. Calcium bound calmodulin (1exr, upper) exhibits conformational heterogeneity on the interface of the peptide binding site. The residual conformational entropy of binding depends on side chains sampling alternative conformations as exemplified by Met36 and Leu39 (lower). Electron density contoured to 2.5 $e^-/Å^3$ in a dark blue mesh and 0.8 $e^-/Å^3$ in cyan volume representation. The lessons from calmodulin likely apply to enzymes where the loss of conformational entropy associated with the rigidification of active-site loops or side chains can specifically weaken binding to substrate or product complexes and promote flux through the catalytic cycle. **b** A model of ubiquitin (2k39) derived from RDC data reporting on motions up to microseconds is shown as cartoon, with the other models in the ensemble shown as transparent ribbons (upper). The dynamic β1β2 loop moves between alternative loop conformations, represented as sticks (upper and lower). The population of the up (cyan), mid (blue), and down (purple) β1β2 conformations can be a critical determinant of binding preferences for protein–protein interactions. The rates of transition between these states discriminate between induced fit and conformational selection mechanisms, which can influence catalytic mechanisms and inhibitor discovery. **c** For enzymes, loop motions on the millisecond time scale are often rate limiting for catalytic cycles, with essential roles for governing ligand flux and repositioning key catalytic residues for catalysis. The WPD loop of protein tyrosine phosphatase 1B (PTP1B) moves between the "closed" (1sug, orange) and "open" (1t49, cyan) form on the millisecond time scale, forming the catalytically competent closed active site conformation. Further molecular detail of the two conformations are shown in the lower panel with electron density contoured to 0.3 $e^-/Å^3$. **d** The archaeal proteasome, a ~700-kDa complex, controls active site access through the dynamic exchange of the N-terminus to block or reveal the central pore on the time scale of seconds. The structure of the proteasome is shown as a homoheptamer with each subunit in a different color (upper). In the lower panel, the ensemble of structures of the N-terminus of one of the seven subunits is shown in blue (2ku1). **e** Many enzymes enter long-lived states, with distinct catalytic activities, through stochastic fluctuations. Quaternary structure reconstruction of two RAS molecules (yellow) and son of sevenless (SOS, gray surface) complex. The Cdc25, REM, DH, histone, and PH domains of SOS are colored blue, green, orange, brown, and red, respectively [1xd2 (RAS) and 3ksy (SOS)]. This complex exchanges between long-lived states with distinct catalytic rates. The structural basis of this exchange is currently unknown but likely involves rearrangements of protein–protein interfaces shown schematically in equilibrium. **f** The folded crystal structure (1ssx) of α-lytic protease (upper) is a kinetically trapped structure. After folding catalyzed by a proline domain, the kinetic barrier to unfolding makes this protein stable on the scale of years. The benzoyl moiety of Phe228 deviates by 6° from planarity. Removing this distortion can change the unfolding barrier from over a year to less than 2 weeks. Electron density contoured to 4.75 $e^-/Å^3$ (dark blue mesh, lower) and 0.5 $e^-/Å^3$ (cyan volume representation) [35]. Published under Creative Commons public use license CC-BY from American Chemical Society

Table 7.1 A brief characterization of physical methods used to investigate protein dynamics [34]

Method	Parameters measured	Information obtained[a]
IR and Raman spectroscopy	Spectra	$\tau_c^{-1} = 10^{12}$–10^{14} s^{-1}
Fluorescence	Depolarization Radiation autocorrelation Relaxational shift Quenching by reagents	$\tau_c^{-1} = 10^{8}$–10^{9} s^{-1} $\tau_c^{-1} = 10^{8}$–10^{9} s^{-1} $\tau_c^{-1} = 10^{8}$–10^{9} s^{-1} $k_{tran} = 10^{8}$–10^{10} M^{-1} s^{-1}
Phosphorescence	Depolarization Relaxational shift Quenching by the exchange mechanism	$\tau_c^{-1} = 10^{-1}$–10^{6} s^{-1} $\tau_c^{-1} = 10^{-1}$–10^{6} s^{-1} $\tau_c^{-1} = 10^{8}$–10^{9} s^{-1}
Rayleigh scattering	Doppler shift of frequency Line broadening	$\tau_c^{-1} = 10^{9}$–10^{10} s^{-1} $\tau_c^{-1} = 10$–10^{7} s^{-1}
Newton non-elastic scattering		$\tau_c^{-1} = 10^{10}$–10^{13} s^{-1}
Spin labeling	Shape of absorption lines Saturation transfer parameters Quenching by paramagnetics	$\tau_c^{-1} = 10^{7}$–10^{10} s^{-1} $\tau_c^{-1} = 10^{4}$–10^{6} s^{-1} $k_{tran} = 10^{6}$–10^{10} M^{-1} s^{-1}
Recombination-Kinetic method	Kinetics of radical recombination	$k_{tran} = 10$–10^{8} M^{-1} s^{-1}
Mossbauer labels	Recoil-less absorption Diffusion line broadening	$\tau_c^{-1} = 10^{7}$ s^{-1} $\tau_c^{-1} = 10^{7}$ s^{-1}
Dielectric measurement	Absorption and dispersion of electromagnetic radiation	$\tau_c^{-1} = 10^{3}$–10^{10} s^{-1}
Ultrasonic methods	Sound absorption and dispersion	$\tau_c^{-1} = 10^{5}$–10^{9} s^{-1}
NMR	Line splitting in high-resolution spectra Frequency and temperature dependencies T_1 and T_2 Relaxation in the magnetic field gradient Narrowing of broad line	$\tau_c^{-1} = 10$–10^{3} s^{-1} $\tau_c^{-1} = 10^{4}$–10^{10} s^{-1} $\tau_c^{-1} = 10^{7}$–10^{10} s^{-1} $\tau_c^{-1} = 10$–10^{4} s^{-1}

τ_c^{-1} is the correlation frequency; k_{tran} is the rate constant of collisions; τ_{tran}^{-1} is the translational jump time. Reproduced with permission from Servier

time followed by the examination of specific contributions to the process. According to a 300 ps molecular dynamics simulation of myoglobin, different conformational minima are accessed every tenth of a picosecond.

A critical review of the idea that dynamical effects contribute to catalysis considering all reasonable definitions of what does and does not qualify as a dynamical effect was presented [28]. The authors claimed that: (1) no dynamical effect has ever been experimentally shown to contribute to catalysis; (2) the existence of non-negligible dynamical contributions to catalysis is not supported by consistent theoretical studies; and (3) despite a substantial body of experimental effort, there has not yet been any study that consistently established a connection between an enzyme's conformational dynamics and a significant increase in the catalytic contribution of the chemical step. Nevertheless, a direct relationship between the nanosecond dynamics of enzyme proteins and their functional activity was already demonstrated in early studies (Sect. 7.2.1). Obviously, in in the solid state of proteins, a complex multi-stage chemical reaction is impossible. The problem is that it is needed to find out directly what specific dynamic modes is associated with given chemical step. Indeed, "the dynamical proposal is not a semantic issue but a well-defined scientific hypothesis with well-defined conclusions" [28]. Although protein dynamics has been recognized as important due to the inherent flexibility of biological macromolecules, it is often difficult to distinguish which conformational changes are directly related to function. The fundamental relationship between the dynamics of proteins and surrounding water molecules and the functional activity of proteins appears to be a challenging problem and a subject of hot debate.

7.2 Early Studies

7.2.1 Hypothesizes and First Experiments

The present conception of the intramolecular dynamics of proteins is based on hypothesizes put forward in the 1950s and 1960s, namely: (1) Assumption that substrate-enzyme interaction was accompanied by a certain structural distortion of both substrate and enzyme [40]; (2) The theory of "complementarity" of the free energy of the chemical reaction in the enzyme-active center and the enzyme conformational free energy [41]; (3) Hypothesis regarding the structural fluctuation of the protein macromolecule, which is visible in the ability of inner peptide groups to exchange hydrogen atoms for water protons [42]; and (4) The conformational structure of the enzyme can be induced to fit the structure of the substrate during their interaction [43]. At a later stage, the concepts of protein dynamics were supplemented by the principle of dynamic adaptation of the enzyme conformational structure to the substrate configuration in consecutive chemical reactions of the enzyme [5–14, 45]. Such an adaptation promotes both the first step of precise orientation and the subsequent chemical steps, without allowing cleavage of the contacts needed for the chemical mechanism. This general property of the protein may be defined as dynamic adaptability [1, 2, 12, 44].

In 1950, while analyzing data on the binding of organic molecules to bovine serum albumin (BSA), Karush came to a conclusion on the flexibility of the protein binding site structure [46]. For example, it was later discovered that the binding of steroid molecules in the binding site of albumins causes a change in the conformation of the protein globule [47]. The change was observed in the rotational mobility of the spin labels located at a distance greater than 16 Å from the binding site, into which steroid molecules were introduced. Another example is a transglobular conformational transition at the binding of specific inhibitors to a lysozyme molecule labeled with a nitroxide spin label at histidine-16, located 16 Å away from the active center [48]. Addition of a corresponded amount of inhibitors NAG and NAG-NAG induced distinct changes in the label ESR spectra.

The first direct experimental evidence of the nanosecond intramolecular mobility of a protein matrix at ambient temperatures was obtained in 1970th by Likhtenshtein group using spin and Mössbauer labels and probes [6, 44, 45, 56–58]. To illustrate, a hydrophobic aromatic derivative of a nitroxide radical was embedded in the human serum albumin binding site and the mobility of the spin probe was traced by ESR spectroscopy. The apparent correlation frequency $\upsilon_c = 10^8\ s^{-1}$ of the probe motion was found to be essentially faster than macromolecular tumbling and was not found to be dependent on viscosity. Therefore, this finding was attributed to the local mobility of the label. Employing the phenomenon of fluorescence quenching of the buried tryptophane residues Lackovicz and Weber [65] explained the quenching effect by dynamic opening and closing of the gap that prevents the quencher from reaching tryptophan in the closed state. The conclusion about the nanosecond time scale of protein dynamics was confirmed in a series of early studies including the following: (1) Neutron scattering within the molecular dynamics of solid-state lysozyme showed a significant effect of glycerol and water on dynamic transition temperatures (T_d) [49]; (2) Solvent dependence of dynamic transition in protein solution was supported by neutron scattering and computer simulation methods [50]; (3) Importance of solvent mobility for protein dynamics at temperatures above 180 K was demonstrated by molecular dynamics simulation [51]; and (4) This conclusion was confirmed by low-temperature experiments with the use of the fluorescence dynamic Stokes shift, neutron scattering and molecular dynamics simulation in other proteins, superoxide dismutase, lysozyme, elastase, bacteriorhodopsin, and references therein [52–55].

Nuclear magnetic resonance can yield reach information on the dynamics of various parts of the protein measuring relaxation times such as T_1 and T_2 to determine order parameters, correlation times, and chemical exchange rates [55, 66]. NMR relaxation studies can provide detailed information pertaining to the internal dynamics in proteins on a time scale from milliseconds to picoseconds. The ^{15}N, ^{13}C and ^{2}H spin–lattice relaxation rate ($1/T_1$) and heteronuclear NOE's are sensitive to high frequency motion (10^8–$10^{12}\ s^{-1}$), while the spin–spin relaxation rate ($1/T_2$) is a function of much slower processes. For example, the NMR relaxation technique was used to investigate the backbone dynamics of staphylococcal nuclease complexed with a ligand and Ca^{2+} and labeled with ^{15}N [55]. Information on internal motions was extracted from experimental data using the model-free approach [67].

7.2.2 *Early Mechanisms of Protein Dynamics*

At any conditions, protein macromolecules exhibit high-frequency low-amplitude harmonic nuclear vibrations with $\upsilon_c = 10^{13}$–10^{14} s^{-1} and amplitude $A = 0.01$–0.05 Å. This type of motion, directly detected by the methods of IR, Raman, Mössbauer, NMR and ESR spectroscopy, takes place in all proteins, at all temperatures and degrees of humidity, and apparently is not directly related to their functions and stability. The comparative analysis of available data [44, 45, 58–60] revealed an apparent discrepancy between the physical labeling approach and certain other physical methods. Thus, the measurements of the temperature dependence of the heat capacity (c_p) of proteins, α-chymotrypsin, lysozyme, myoglobin, collagen, at various temperature and degrees of hydration indicated only a monotonic increase of c_p and did not detect pronounced phase transitions [68, 69]. Monotonic character of the temperature dependences of such physical parameter of proteins as circular dichroism was also confirmed. However, spin, fluorescence and Mössbauer labeling, H–D exchange, non-elastic neutron scattering and absorption spectra of heme in heme proteins detected sharp transitions within a temperature range before thermal denaturation [44, 45, 56–64, 70, 73].

The above-mentioned discrepancy was readily explained by the following proposal [44, 59]: (1) The suggested type of nanosecond intermolecular mobility consists of movement of the relatively large and rigid parts of the protein macromolecules (domains); (2) Such hinge-like oscillation of the tightly packed polypeptide blocks does not give a measurable contribution to the overall heat capacity and the helicity degree of polypeptide chains; and (3) The hinge-like motion of domains strongly affects the mobility of the Mössbauer labels firmly bound to the protein blocks, the mobility of spin and fluorescence labels and native chromophores located in cavities between the blocks. At a later stage, the hinge-bending or domain motion mechanism proposed was confirmed by independent experimental investigation and theoretical calculations [59] and references within). Depending on protein specifics, the correlation time of domain motion can vary in temporal regions from milliseconds to nanoseconds. Experimental results from studies of Arrhenius dependence of different characteristics of lysozyme are presented in Fig. 7.3 [34, 59, 70]. The discontinuities on the curves indicate local conformational transitions and apparently likely due to the appearance of a more open conformation of the protein. As can be seen from Fig. 7.3, these methods reveal conformational transitions at a temperature of about 30 °C, whereas the temperature dependence of the partial heat capacity decreases monotonically in this temperature region. Against the background of the general contribution of high-frequency dynamics in heat capacity from a large number of protein atoms, the contribution of domain motions is negligible.

In the search for a correlation between functional activity of α-chymotrypsin the protein dynamics, the enzyme was modified by a spin label attached to the methionine-92 group in the region of the active site and by a Mössbauer label in the enzyme globule periphery [61]. The protein tryptophane group was also served as an intrinsic fluorescence label. The experiment showed that agitation of the label

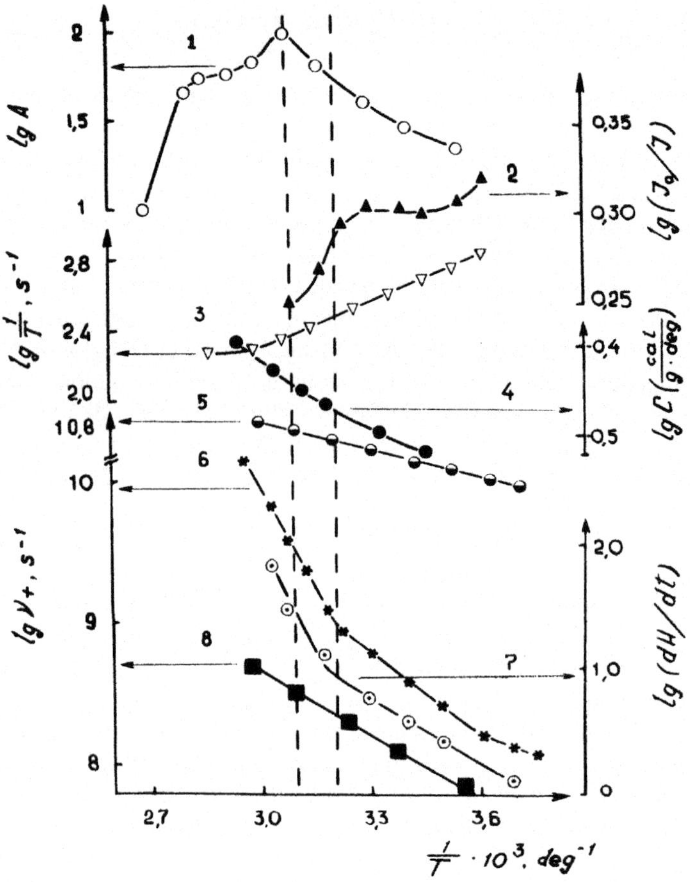

Fig. 7.3 Arrhenius dependence of different characteristics of lysozyme: (1) enzyme activity; (2) relative fluorescence intensity; (3) water proton relaxation in the presence of the sample spin labeled by His-15; (4) partial heat capacity; (5) H–D exchange; (6, 7) spin labels rotation; and (8) the lysozyme globule rotation [34]. With permission from Elsevier

mobility in the nanosecond region at relative humidity about $P_{rel} = 0.6$ correlates with a sharp increase in the hydrolysis degree of the cyanomoyl α-cymotrypsin covalent compound [72]. This result highlighted the significance of nanosecond dynamics in the chemical step of the enzyme reaction.

Figure 7.4 shows that a dramatic increase in the rate constant of electron transfer (k_{ET}) in myoglobin modified covalently by donor–acceptor groups occurs within the same temperature interval (200–220 K) as the animation of the nanosecond dynamics of Mössbauer atom located in porpherin. A similar effect was observed for Long Distance Electron Transfer (LDET) between a heme group and an excited

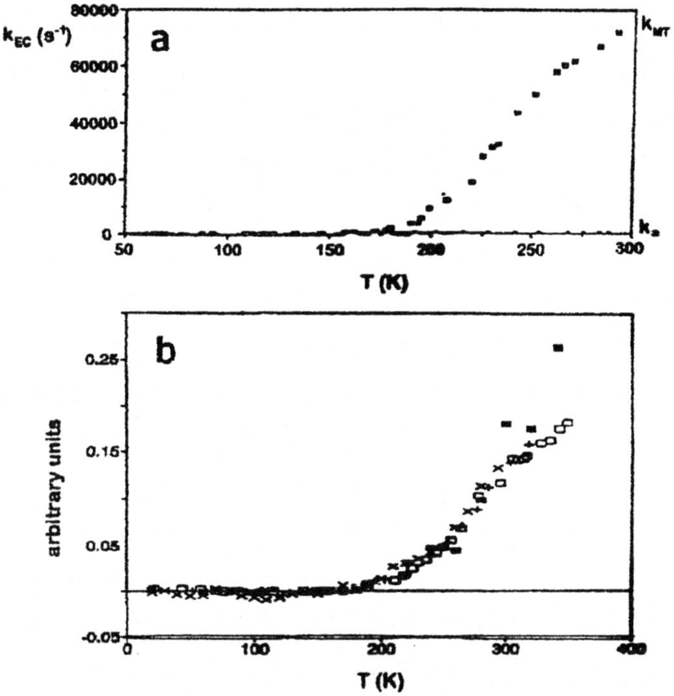

Fig. 7.4 Temperature dependence of the rate constant of electron transfer (k_{ET}) in myoglobin modified covalently by donor–acceptor groups (**a**) and the deviation of various dynamic quantities from normal harmonic behavior obtained by molecular dynamic simulation, inelastic neutron scattering, Mössbauer spectroscopy and spectral broadening analysis (**b**) [34]. Reproduced with permission from Elsevier

Zn–porphyrin of modified hemoglobin [34, 74]. The above examples clearly highlighted the significance of proteins nanosecond dynamics in the chemical step of some enzyme reactions.

7.3 Further Developments

7.3.1 Proteins Dynamics Computations

Significant progress in the study of the molecular dynamics of proteins and enzymes for two last decades is reflected in a series of comprehensive reviews and original theoretical and experimental works [1–4, 11–123]. Among the spectrum of methods for studying the molecular dynamics of such complex structures as proteins, a special place is occupied by methods based on simulation [4, 27, 28, 39, 53, 54, 84, 120]. As a typical example, results of Mixed quantum/classical molecular dynamics simulations of the hydride transfer reaction catalyzed by dihydrofolate reductase, an enzyme that reduces dihydrofolic acid

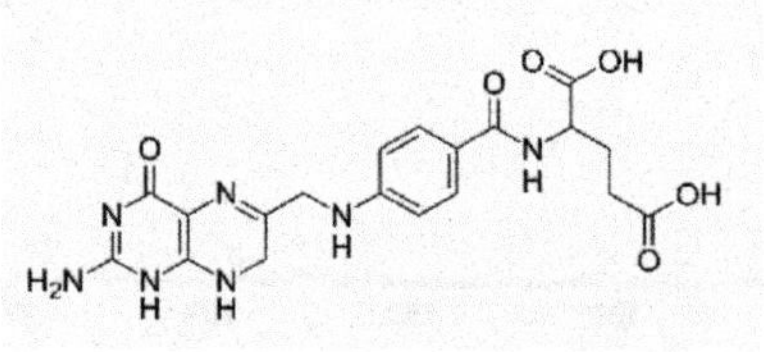

to tetrahydrofolic acid, using NADPH as electron donor, were presented in [84]. The nuclear quantum effects such as zero point energy, hydrogen tunneling, and the motion of the entire solvated enzyme, were taking in account during the generation of the free energy profiles and the real-time dynamical trajectories. Data on the tunneling transmission coefficient was found to be 0.80 for hydrogen and 0.85 for deuterium, indicated the significance of dynamical barrier recrossings. The computation implied that motions of residues both in the active site and distal to the active site can impact the free energy of activation and the degree of barrier recrossing. In addition, it was shown that a network of coupled promoting motions, extending throughout the protein, involves motions spanning femtosecond to millisecond time scales. A study of the *Thermus thermophilus* chorismate mutase (TtCM) was carried out by using self-consistent-charge density-functional tight binding)/molecular mechanics, umbrella sampling, and the weighted histogram analysis method [120]. The computation led to the following results: (1) The computed free energies of activation for the reactions in water and TtCM are comparable to the experimental values; (2) The free energies for formation of near attack conformer are 8.06 and 0.05 kcal/mol in water and TtCM, respectively; (3) The near attack conformer stabilization contributes $\approx 90\%$ to the proficiency of the enzymatic reaction compared with the reaction in water; (4) The transition state (TS) structures and partial atom charges are almost the same in the enzymatic and water reactions; (5) The difference in the electrostatic interactions of Arg-89 with O13 in the enzyme–substrate complex and enzyme–TS complex provides *of* 1.92 kcal/mol total TS stabilization; and (6) Differences in electrostatic interactions between components at the active site in the enzyme–substrate complex and enzyme–TS complex are significant, such that TS stabilization is of minor importance and the enzymatic catalysis is through an entropic advantage.

Experimental and computational works relating to enzyme dynamics and function, including the role of solvent were reviewed [75]. The evidences suggested that: (1) Fast motions on the 100 ps timescale, and any motions coupled to these, are not required for enzyme function; (2) Proteins where the function is electron transfer, proton tunneling, or ligand binding may have different dynamical dependencies from those for enzyme; and (3) Enzymes with large turnover numbers may have different dynamical dependencies from those that turn over more slowly. Quantized molecular dynamics simulations to characterize the role of enzyme vibrations in facilitating dihydrofolate reductase (DHFR) hydride transfer were carry out [87]. All simulations were performed using a modified version of the Gromacs-4.0.7 molecular dynamics package Ring Polymer Molecular Dynamics (RPMD). The RPMD equations of motion were used to simulate the dynamics of DHFR. To quantify and distinguish between statistical and dynamical correlations in the enzyme motion, sampling the full ensemble of reactive trajectories was performed. The computation led to a number important results including the following: (1) Statistical correlations give rise to nanometer-scale coupling between distal protein residues and the intrinsic reaction; (2) Dynamical correlations vanish at distances beyond 4–6 Å from the transferring hybrid; (3) Nonequilibrium dynamical coupling between protein residues and the hydride tunneling reaction exists; and (4) The calculated time-dependence of the transmission coefficient confirmed that reactive trajectories commit to the reactant or product basins within 25 fs.

Authors of work [86] characterized enzyme conformational dynamics (ECD) in cyclophilin **A** (CypA), peptidyl-prolyl cis–trans isomerase (Fig. 7.5), using normal and accelerated, atomistic dynamics simulations by employing a combination of normal MD (nMD) and accelerated MD (aMD). This approach provides atomistic detail with extended timescale (10^7–10^{11} s^{-1}). The simulation showed that Kramers' rate theory can be used to present a rationale of how enzyme conformational dynamics (ECDs) affects the reaction dynamics and catalytic rates. According to the theory, in the high friction regime the reaction rate constant is

$$k = \frac{\omega_o \omega_b D_{\text{eff}}}{2\pi k_B T} \exp\left(\frac{-\Delta G^{\#}}{k_B T}\right)$$

where k is the rate of escape from the trans well with curvature ω_o over the free energy barrier $\Delta G^{\#}$ with curvature ω_b, k_B is the Boltzmann constant and T is the temperature. D_{eff} is the effective diffusion coefficient on a one-dimensional free energy profile and assumed to be independent of the reaction coordinate. To probe the effects of ECD in catalysis, several lowered torsional energy barriers around the –Ala–Pro–ω bond of a substrate analogue, Ace–Ala–Ala–Pro–Phe–Nme were considered. The simulation indicated that substrate dynamics involved in the chemical step, is coupled to the dynamics of the surrounding medium which can either be the solvent or the active site of the enzyme.

An unified view of the role of protein dynamics in dihydrofolate reductase (DHFR) catalysis was suggested [89]. DHFR transfers a hydride from NADPH and a proton from the solvent to C-6 and N-5 of 7,8-dihydrofolate (DHF) to generate 5,6,7,8-tetrahydrofolate (THF). Computational studies indicated that side chain rotations of Ile14 and Ile94 of EcDHFR bring the reactants closer toward the reaction-ready active site configuration. Heavy enzymes were produced in media containing appropriate labeled ingredients such as $^{15}NH_4Cl$, U-$_{13}$C,2 H-glucose, and ^2H_2O. The an enzyme kinetic isotope effect (KIE) was assessed by comparing the reaction rate constants for the light enzyme, k_{LE}, and heavy enzyme produced in media containing appropriate labeled ingredients such as $^{15}NH_4Cl$, U-^{13}C,2 H-glucose, and ^2H_2O, k_{HE}. The typically > tenfold difference between k_{cat} (rate constant of fast chemical step) and k_H

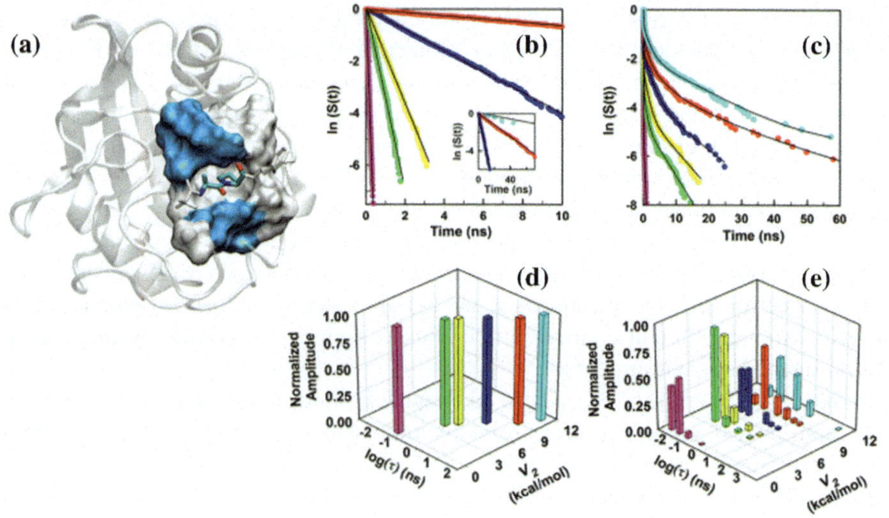

Fig. 7.5 Structure of cyclophilin A (CypA) and the influence of its dynamics on the kinetics of prolyl isomerization in its substrate. **a** CypA (gray) with its substrate Ace–Ala–Ala–Pro–Phe–Nme in the binding pocket. The binding cavity (VDW surface) is defined by ten residues (nonpolar (white): Phe60, Met61, Ala101, Ala103, Phe113, Leu122, and polar (cyan): Arg55, Gln63, Asn102, and His126) that are within 4 Å of the substrate's Ala-Pro motif. Decay of probability of survival in the trans well as a function of time obtained from nMD simulations of **b** RO and **c** RC with V2 set to 11.0 (cyan), 9.0 (red), 7.0 (blue), 5.0 (yellow), 4.0 (green), and 0.0 (magenta) kcal/mol. Continuous lines are fits to single exponential in (B) and multiexponential functions in (C). (Inset to **b**) Plots of survival probability functions on an extended timescale when V2 ¼ 7.0, 9.0 and 11.0 kcal/mol. Parameters from exponential fits of isomerization kinetics, i.e., the amplitudes and the time constants, τ, are plotted for **d** RO and **e** RC with the same color code as in (**b**) and (**c**) [86]. With permission from PNAS

(the single turnover rate constant that reports predominantly on hydride transfer were obtained.

The study [78] was concentrated on the effect of protein chain flexibility on the properties of hydration water, using short peptides as a model. 18 linear peptides with the sequence (XXGG) $\times$ 5, where X represents one of the common amino acids, other than glycine and proline. Two versions of each peptide were studied using molecular dynamics (MD) simulations: one was flexible (only the dynamics of the N and C terminals), and the other was constrained. The computation found that: (1) Restricting the peptide dynamics can slow down the translational motions of water molecules to a distance of at least 12–13 Å; (2) For the peptides, water molecules were selected based on the distance between the water oxygen atoms and the peptide heavy atoms, for different cutoff distances, i.e., 3.5 Å, 5 Å; (3) For bulk, water molecules were selected in a layer beyond a sphere with a radius of 50 Å; (4) Analysis of the "slow" water molecules (residence time 100 ps) together with a thermodynamic analysis of water within 4.5 Å of the peptide revealed significant differences between the hydration properties of the peptides. The solvation of flexible and restrained versions of each constructed peptide is illustrated in Fig. 7.6. Following the fit, water molecules with short (t_1 = 1 ps), intermediate (t_2 = 4–9 ps) and longer (t_3 = 10–30 ps) residence times can be distinguished. The calculations showed that t_{avg} is 4.8 and 7.2 ps within 3.5 and 5 Å of the bulk water molecules, respectively, and water molecules with short (t_1 = 1 ps), intermediate (t_2 = 4–9 ps) and longer (t_3 = 10–30 ps) residence times can be distinguished.

The relationship between measures of fast side-chain motion and the underlying conformational entropy was evaluated using dynamic proxy approach [18]. A dynamic proxy class is a class that implements a list of interfaces specified at runtime such that a method invocation through one of the interfaces on an instance of the class will be encoded and dispatched to another object through a uniform interface. The dynamical proxy showed that the contribution of conformational entropy can range from highly favorable to highly unfavorable and demonstrates the potential of key thermodynamic variable to modulate protein–ligand interactions. The role of solvent entropy and the loss in rotational-translational entropy that occurs

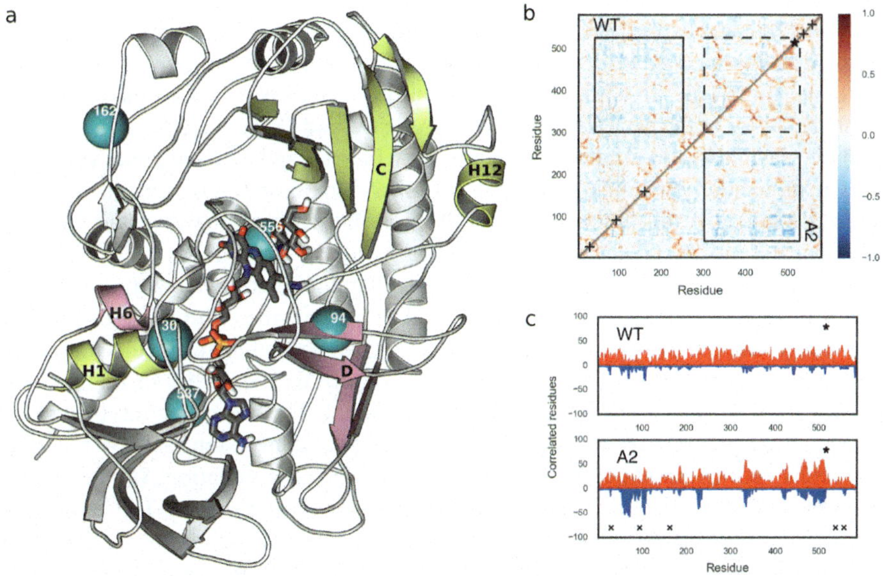

Fig. 7.6 **a** A2 GOx crystal structure with glucose docked into the active site. The secondary structures involved in the anticorrelated motions are shown in yellow and pink, and the positions of mutations are designated by cyan spheres. His516, FAD, and glucose are shown by sticks and colored by atom type). **b** Dynamic cross-correlation maps (DCCMs) of the WT GOx (top half) and the A2 mutant (bottom half). The most discriminating regions are indicated by black rectangles: solid lines for anticorrelated and dashed lines for correlated motions. DCCMs for all GOx variants are given in a figure. **c** Per-residue count of the correlated and anticorrelated motions (based on a cutoff of ± 0.3) in the WT and A2 GOx. Correlated and anticorrelated motions are shown in red and blue, respectively. The positions of the mutations are represented with × and that of His516 with ★. The count plots for all variants are given in a figure in the Supporting Information [79]. With permission from American Chemical Society

upon formation of high-affinity complexes was also refined. Enzyme lipase LipA (BSLA), was selected as a model protein for thermostabilization designs and the circular dichroism measurements, showed six combinatorial designs with improved stability at temperatures from 5.8 °C to 13.6 °C [76]. Molecular dynamic simulations revealed that: (1) The mutations alter the populations of conformational states and the increased ensembles of inactive conformations leading to a reduction in activity; (2) The mutations responsible for the reduced enzyme catalytic activity involved a short dynamic correlation path to disturbing the equilibrium conformation of active sites; (3) By removing N82V, which had a close dynamic correlation to the active sites in mutant D3, the redesigned mutant RD3 had an increased activity of 57.6%; and (4) The mutations responsible for the reduced enzyme catalytic activity involved a short dynamic correlation path to disturbing the equilibrium conformation of active sites. In study [88] attention has been paid to the role of ligand-gated conformational changes in triosephosphate isomerase and orotidine 5′-monophosphate decarboxylase. Specifically, combined conventional and enhanced molecular dynamics simulations focused on the $(\beta\alpha)8$ Barrel Enzymes Histidine and Tryptophan showed that μs timescale motion of a single loop dominates the transition between catalytically inactive and active conformations. Isomerase enzymes HisA and TrpF, and the bifunctional isomerase PriA multiple long loops undergo conformational transitions on slower timescale. In addition, the simulations provided details of the conformational behavior of the catalytic loops in HisA, PriA, and TrpF, and elucidated the role of their plasticity in facilitating bifunctionality in PriA and evolved HisA variants.

The molecular dynamics simulations of glucose oxidase (GO) performed in [79] allowed to quantify this floppiness, revealing that His516 exists in two states: catalytic and noncatalytic. A2 GOx crystal structure with glucose docked into the active site is displayed in Fig. 7.6. The simulation indicated that the relative populations of these two sub-states are almost identical in the wild-type enzyme, with His516 shuffling between them. In the glucose oxidase mutants, on the other hand, the mutations enrich the catalytic His516 conformation and reduce the flexibility of this residue, leading to an enhancement in their catalytic efficiency. This study stresses the benefit of active site preorganization with respect to enzyme conversion rates by reducing molecular reorientation needs. To test the influence of the mutations on the protein dynamics, MD simulations of several GOx variants were performed. The simulation showed the root-mean-square deviation of the protein backbone motion, indicates that the WT and GOx mutants are stable during the 100 ns trajectories. A slight decrease in residue flexibility, especially in the active site region was revealed by a detailed analysis of the enzyme dynamics related to the laboratory evolution of GOx.

7.3.2 *Miscellaneous Dynamic Properties of Proteins Studied by Physical Methods*

The activity and dynamics of a simple, single subunit enzyme, the xylanase from *Thermotoga maritima* strain FJ SS3B.1 have been measured, from − 70 to + 10 °C [112]. Data on Neutron scattering indicated no evidence for anharmonic local diffusive motions or jumps between alternative conformations in picosecond-timescale at temperatures below – 50°, whereas these motions are strongly activated at higher temperatures. The activity follows Arrhenius behavior over the whole of the temperature range was investigated. Therefore, a temperature range exists over which the enzyme rate-limiting step is independent of fast anharmonic dynamics. Adenylate kinase (AK) involves a large-amplitude rearrangement of the enzyme's lid domain. Using method of high-resolution single-molecule Förster resonance energy transfer (FRET), it was possible to follow AK's domain movements on its catalytic time scale [108]. Maximum entropy-based methods showed that AK is capable of dynamically sampling two distinct states, which correlate well with those observed by X-ray crystallography. The experiment indicated that the equilibrium favors the closed, active-site-forming configurations even in the absence of substrates. In addition, the experiments indicated that interaction with substrates restricts the spatial extent of conformational fluctuations and shifts the enzyme's conformational equilibrium toward the closed form by increasing the closing rate of the lid. Thus, integrating these microscopic dynamics into macroscopic kinetics allowed to model lid opening-coupled product release as the enzyme's rate-limiting step.

In work [116] studies capable of examining μs relaxations in response to very rapid ns jumps in temperature of approximately 7 °C were implemented. T-jump induced fluorescence ($\lambda_{ex} = 295$ nm and $\lambda_{em} = 355$ nm) time was measured as a function of final temperature. A thermophilic alcohol dehydrogenase dynamics on the microsecond (μs) time scale were investigated in two variants of the enzyme, W87F and W87F:H43A. A biexponential fitting of the data yielded two kinetically important processes: an initial, temperature independent loss of fluorescence (τ_1) with an average relaxation lifetime of 0.77 μs, and a slower microsecond relaxation that returns the system to its initial intensity (τ_2). The present results suggested that motions can be critical for global conformational sampling in proteins. Frequency fluctuations of the azide anion bound to the active site of formate dehydrogenase measured via 2D IR spectroscopy were examined [96]. These measurements revealed an underdamped oscillatory component to the frequency-frequency correlation function when the azide anion is bound to the NAD^+ ternary complex. This oscillation disappears when the reduced cofactor is added and was suggested to be relevant to donor–acceptor distance sampling of the catalyzed hydride transfer. IR photon echo measurement of the formate dehydrogenase-NAD^+-azide ternary complex was carried out [107]. The data showed complete spectral diffusion on the femtosecond-picosecond time scale. This finding indicated that for transition-state analog complex samples the distribution of azide vibrational frequencies occurs within a few picoseconds.

Temperature-jump small- and wide-angle X-ray scattering measurements on enzyme cyclophilin (CypA) demonstrated that these experiments are able to capture functional intramolecular protein dynamics on the microsecond timescale [99]. The resulting time-resolved signal was used to assess the kinetics of conformational changes. The experiment indicated that the enzyme displays rich dynamics following a temperature jump including intramolecular protein dynamics on the microsecond timescale [114]. Overview of temperature-jump SAXS/WAXS experiments is displayed in Fig. 7.7. Two relaxation processes were resolved: a fast process is related to surface loop motions, and a slower process is related to motions in the core of the protein that are critical for catalytic turnover. Specific mutants in the "loops" or the "core" regions of CypA indicated that the two processes are independent. Using a nonlinear least-squares curve-fitting algorithm, rate constants of $98 \times 10^4\ s^{-1}$ for the fast process (k_1) and $2.5 \times 10^4\ s^{-1}$ for the slow process (k_2) at 26.7 °C (299.7 K) were measured. In addition, the thermodynamic properties of transition states were calculated: for the fast process $\Delta H = 37$ KJ mol^{-1} and $\Delta S = 32$ mol^{-1} K^{-1} and for the slow process $\Delta H = 27$ KJ mol^{-1} and $\Delta S = -34$ mol^{-1} K^{-1}.

NMR spectroscopy being applied to the study of conformational dynamic processes in enzymes was utilized for investigation of the bacterial RNase H dynamics [110]. Among the results of the work, the following ones can be noted: (1) Spin relaxation measurements and molecular dynamics (MD) simulations of the enzyme showed that one of three loop regions that interact with substrates, interconverts between two conformations; (2) NMR spin relaxation and fluorescence spectroscopy have characterized a conformational transition in AlkB between an open state, in which the side chains of methionine residues in the active site are disordered, and a closed state, in which these residues are ordered; (3) Closure of active site loop 6 in triosephosphate isomerase is critical for forming the Michaelis complex; (4) Reopening of the loop after the reaction is rate limiting; (5) NMR spin relaxation and MD simulations of triosephosphate isomerase in complex with glycerol 3-phosphate

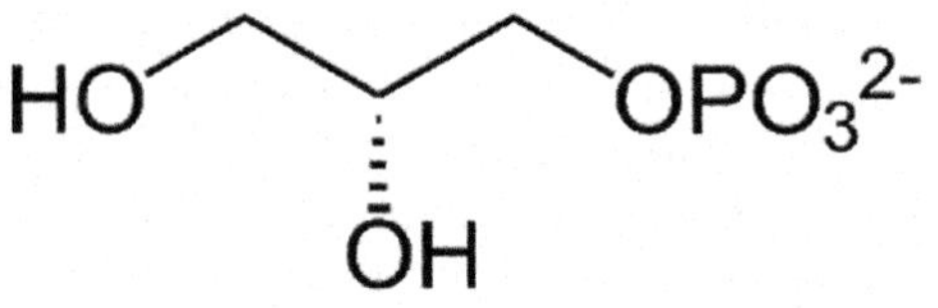

demonstrated that closure of loop is a correlated rigid-body motion; (6) The MD simulations also indicate that motions of Gly173 in the most flexible region of loop 6 contribute to opening of the active site loop for product release; and (7) The enzyme undergoes conformational exchange with similar values of $k_{ex} = \sim 4 \times 10^4\ s^{-1}$.

It was found that tetrameric hemoglobin binds ligands cooperatively by undergoing a ligand-induced T → R quaternary structure transition [121]. A structure–function relationship that has served as a model system for understanding allostery in proteins was examined. Time-resolved small- and wide-angle X-ray scattering (SAXS/WAXS) patterns of HbCO following a temperature jump provided evidence

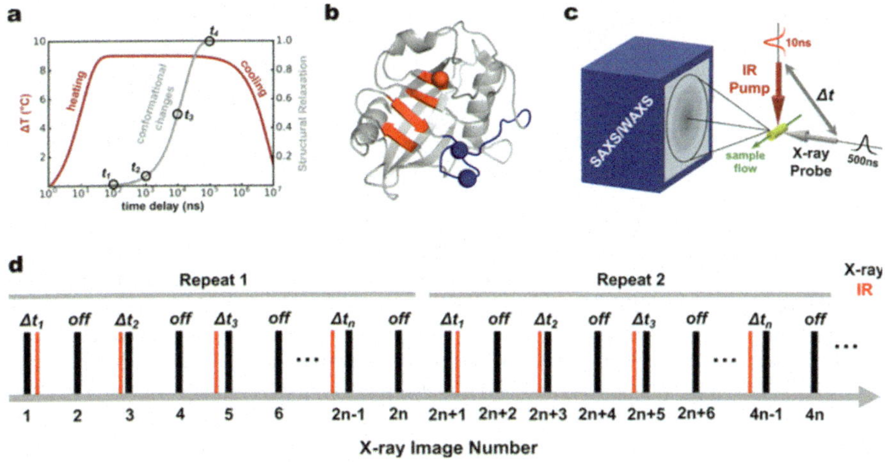

Fig. 7.7 Overview of temperature-jump SAXS/WAXS experiments. **a** During a temperature-jump experiment, an infrared (IR) laser pulse, several nanoseconds in duration, vibrationally excites the water O–H stretch and rapidly heats an aqueous solution of protein molecules (red curve). Heating is fast, but kinetic barriers to protein motions cause a lag in the structural relaxation to a new thermal equilibrium of conformational states (gray curve). **b** Ribbon model depicting a single CypA molecule, with the "core" dynamic residues (linked to catalysis) colored red and the "loop" region adjacent to the active site (determines substrate specificity) colored blue. Sites of key mutations (S99T) are identified by spheres at the corresponding Cα positions (red: S99T; blue: D66N/R69H). **c** Schematic of the temperature-jump SAXS/WAXS instrumentation, with key features highlighted. Liquid sample flows through the interaction region, where it interacts with mutually perpendicular IR pump and X-ray probe beams. Both pump and probe sources are pulsed, with a defined time delay between arrivals at the sample. SAXS/WAXS patterns are recorded on a single detector panel. **d** Data collection sequence used for the temperature-jump experiments. For each pump–probe time delay, a pump–probe measurement ("laser on") was performed, followed by a measurement with no application of the pump laser ("laser off"). On–off pairs with increasing pump–probe time delays were measured in succession until all of the desired delay times were acquired, then the sequence was repeated as many as 50 times to improve the signal-to-noise ratio of the data. Note that the first measurement within each repeat is a control measurement, wherein the probe pulse arrives at the sample before the pump pulse (negative time delay) [99]. With permission from Springer Nature

for more than one R state and unveil the time scale for interconversion between them. The hemoprotein myoglobin is a model system for the study of protein dynamics. Time-resolved serial femtosecond crystallography utilizing at an X-ray free-electron laser was used to resolve the ultrafast structural changes in the carbonmonoxy myoglobin complex upon photolysis of the Fe–CO bond [122]. The experiments indicated that structural changes appear throughout the protein within 500 femtoseconds, with the C, F, and H helices moving away from the heme cofactor and the E and A helices moving toward itю. So the protein globule behaves like a molecular sponge. Together with the observed oscillations of residues contacting the heme, hybrid quantum mechanics/molecular mechanics simulations predicted that an immediate collective response of the protein occurs upon ligand dissociation, as a result of heme vibrational modes coupling to global modes of the protein. Light

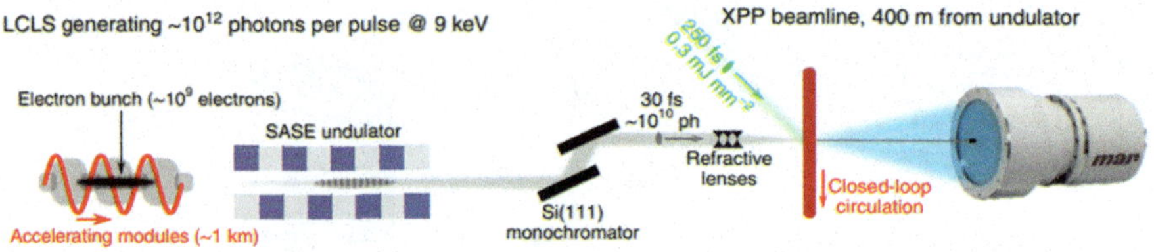

Fig. 7.8 The sample (a 2.35-mM MbCO solution) was circulated in a closed loop through a fused-silica capillary (0.3 mm diameter) positioned at the overlap between the X-ray pulses (9 keV, 30 fs) produced by the LCLS (~ 10^{10} photons per pulse at 9 keV are transmitted by a silicon monochromator) and the photolysis pulses (538 nm, 250 fs) produced by the optical laser system (~ 0.3 mJ mm^{-2}). By monitoring the pattern of the X-rays scattered by the sample at different time delays between the optical and X-ray pulses, it was possible to track the structural changes occurring in the sample after photolysis [123]. Published under the terms of the Creative Commons CC-BY license with permission from Springer Nature

absorption by MbCO in solution can trigger protein conformational changes -the light-induced structural rearrangement, propagating through the protein as a quake-like intramolecular motion (Fig. 7.8) [123]. To decipher the exact timescale of the ultrafast structural dynamics in Mb, femtosecond time-resolved X-ray solution scattering experiments on Mb at the LCLS X-FEL have been performed. Laser-induced time-resolved difference scattering curves obtained by photoexciting a solution of horse MbCO with 250 fs pulses at 538 nm were presented in a figure. The experiment showed that an ultrafast increase of myoglobin radius of gyration occurs within picosecond. The process is followed by a delayed protein expansion and equilibrium undergoes with damped with a ~ 3.6-picosecond time period oscillations undergo.

Figure 7.9 illustrates how the collective mode involves large protein regions from the active site to the protein surface and indicates an area of protein quake. Such coupling with the mode at ~ 4 ps was already suggested on the basis of transient grating investigations [125, 126].

- Ultrafast structural dynamics with different spatial and temporal scales were investigated during photodissociation of carbon monoxide from iron(II)-heme in bovine myoglobin after the first 3 ps following laser excitation [124]. Simultaneous X-ray transient absorption (XTA) spectroscopy and X-ray transient solution scattering (XSS) at an X-ray free-electron laser source with a time resolution of 80 fs were employed. In order to extract the reaction coordinates along different directions of the CO departure, XTA data were collected with parallel and perpendicular relative polarizations of the laser pump and X-ray probe pulse. Time-dependent density-functional theory calculations were used to model the time evolution of the iron K-edge X-ray absorption near edge structure (XANES). Time-resolved small-angle X-ray scattering data during the same process indicated that the local CO dissociation causes a protein quake propagating on different spatial and temporal scales. Thus, these experiments made it possible to establish interplays of electron and nuclear motions along CO dissociation trajectory in myoglobin.

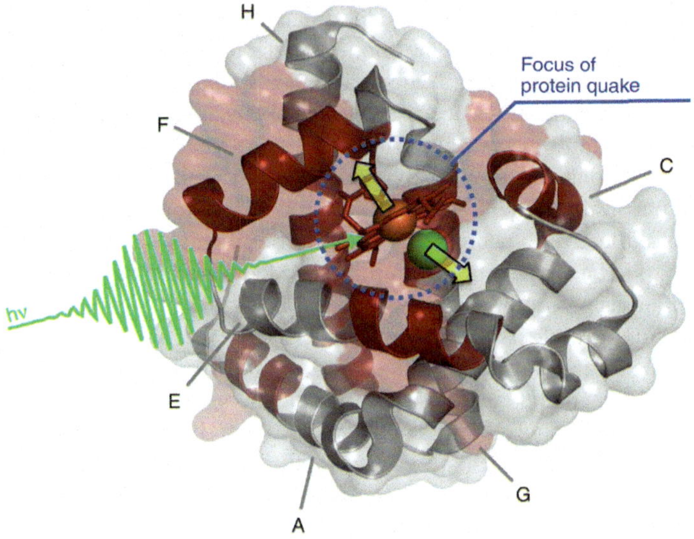

Fig. 7.9 Cartoon representation of the Mb three-dimensional structure; the segments of the polypeptide chain contributing to the collective mode having a ~ 4 ps period (which is activated by ligand release) are highlighted in red This vibrational mode involves the heme, the helix to which the heme is covalently bound (F), and parts of the A, C, E, G and H helices. Note that the regions involved participate to define the heme pocket and extend to the protein surface [123]. Published under the terms of the Creative Commons CC-BY license with permission from Springer Nature

7.3.3 Protein Dynamics and Their Functional Activity and Evolution

NMR studies of cyclophilin A (CypA), rotamase A, prolyl cis–trans isomerase or peptidyl-prolyl isomerase A (PPIA) have revealed a link between internal protein dynamics and substrate isomerization step [1, 33]. The studies, basing on ^{15}N spin relaxation investigations of small peptide substrate as well as two-dimensional (2D) ^{1}H–^{15}N heteronuclear exchange studies of the N-terminal of capsid protein (CA^N) from HIV-1 showed that: (1) Conformational fluctuations within the active site of CypA were detected on the time scale of the reaction (hundreds of microseconds); (2) The rates of conformational dynamics correlated with the substrate isomerization step; (3) Several active-site and surface loop regions showed motions only in the presence of substrate; and (4) The residues Arg55, Lys82, Leu98, Ser99, Ala101, Asn102, Ala103, and Gly109 involve in this process.. The authors proposed a reaction mechanism for PPIase activity of CypA, where the isomerization step takes place with a rate constant of about 9000 s^{-1} and motions of the protein coincide with the rate of substrate turnover step. CypA residue Arg55 was suggested to be a major contributor to catalysis, for which the observed changes in backbone conformation are likely to be coupled with motions of the catalytically essential side chain. The dynamics of the cyclophilin A (CypA) in its substrate-free state and during catalysis were characterized with NMR relaxation experiments [33]. The experiments indicated that: (1) The characteristic enzyme motions detected during catalysis are already present in the free enzyme with frequencies corresponding to the catalytic turnover rates; (2) The protein motions necessary for catalysis are an intrinsic property of the enzyme and may limit the overall turnover rate; (3) Motion is localized not only to the active site but also to a wider dynamic network; and (4) The collective nature of motions occur also in mutant forms of CypA. The authors proposed that the pre-existence of collective dynamics in enzymes before catalysis is a common feature of biocatalysts. The enzyme investigations have led to the discovery of a network of protein vibrations promoting catalysis [11].

The peptidyl-prolyl *cis/trans* isomerase (PPIase) activity of (cyclophilin A) CypA has been investigated in detail to establish the link between protein dynamics and enzymatic catalysis, both by biochemical experiments and theoretical methods. Temperature-jump small- and wide-angle X-ray scattering measurements on cyclophilin A which are able to capture functional intramolecular protein dynamics on the microsecond timescale were performed. The authors of the review [35] focused on the role of mutations in the study of possible links between the dynamics of molecular enzymes and their catalytic activity. Structure of CypA and the influence of its dynamics on the kinetics of prolyl isomerization in its substrate is displayed in Fig. 7.10. The following main results were reported: (1) In dihydrofolate reductase (DHFR) and CypA mutations, that alter the ability of the active site to sample productive higher energy states on the millisecond time scale, reduced the rate of hydride transfer; (2) For CypA, a network of side-chains that exchange between two conformations is critical for catalysis exists; (3) Mutations that lock the network in one state also reduce catalytic activity; and (4) In computationally designed DHFR and CypA directed evolution experiments increase catalytic rates. The connection between stochastic protein dynamics and function for the CypA in wild-type form (WT), and three variants (ST, STCS, and STCSIV) that feature several mutations distal from the active site was theoretically examined [91]. Eighty independent 200 ns MD trajectories of the apo WT, ST, STCS, and STCSIV proteins (20 each) were generated using Gromacs 5.052. Series of umbrella sampling US simulations of the "in" and "out" conformers were performed to compute free energy profiles along ω procedure. Markov state models (MSM) analysis was carried out with the software package pyemma version 2.3.2. The simulation indicated that: (1) In WT free form the enzyme rapidly interconverts between a catalytically active Phe113 "in" form and a catalytically inactive Phe113 "out" form (Fig. 7.10a) with the interconversion rate between in and out forms ($7 \times 10^7\ s^{-1}$); (2) The latter is faster than the substrate binding rate as suggested by NMR experiments (ca. $2 \times 10^4\ s^{-1}$), based on k_{on} rate ca. $2 \times 10^7\ s^{-1}\ M^{-1}$; and (3) For the mutants, the interconversion rates between catalytically active and inactive forms are within the μs^{-1} timescale (Fig. 7.10b). It was concluded that the minor active site conformation is catalytically impaired, and decreased catalytic activity of the mutants is caused by changes in Phe113 motions on a ns-μs timescale.

The application of NMR probes to the investigation of motions in enzymes at the picosecond to nanosecond as well as microsecond to millisecond time scales was

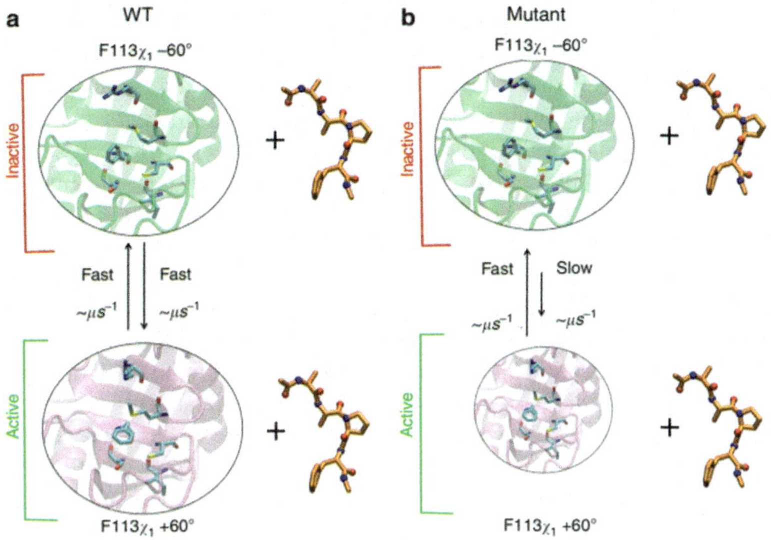

Fig. 7.10 Proposed mechanism for allosteric inhibition of cyclophilin *A* function. **a** Catalysis in WT, with the favored route being the "in" conformations, which are similarly populated to the "out" conformations. **b** Catalysis in the mutants still occurs through the "in" conformations, which have a lower population than the "out" conformations [91]. Published under the terms of the Creative Commons CC-BY license with permission from Springer Nature

described [105]. Studies in enzyme dynamics including conformational selection in RNase A, reaction coordinate compression in *Esch coli* dihydrofolate reductase, substrate binding and drug resistance in HIV protease were carried out. Protein dynamics during turnover in cyclophilin A, mesophilic and thermophilic enzymes, and **protein dynamics** in larger enzymes were in the center of interest of the author's work. The role of adenylate kinase (AK) protein component at atomic resolution was directly investigated utilizing NMR and differential scanning calorimetry (DSC) [106]. This approach allowed to establish a connection between pico- to nano-second backbone fluctuations and the slower motions associated with catalytic activity. NMR relaxation data (R1, R2, and ^{1}H–^{15}N heteronuclear Overhauser enhancement (NOE) were acquired using standard pulse sequences on Varian Innova 500 and 600 MHz spectrometers. Fast-ModelFree10 and ModelFree 4.1512 were employed to optimize the fit of both internal dynamics and global tumbling parameters. The NMR module in CHARMM27 allowed to calculate the auto-correlation function of the second Legendre polynomial of every N–H bond vector. Differential scanning calorimetry (DSC) was performed with a VP-DSC microcalorimeter (MicroCal). The experiments indicated that the correspondence of ps–ns flexibility in the hinges provides the link between local, fast-timescale dynamics and slower, global lid-opening/-closing. It was concluded that pico- to nano-second timescale fluctuations in hinge regions of AK facilitate the large-scale slower lid motions that produce a catalytically competent state.

Work [79] presented a synergistic combination of experimental and computational methods, indicating how mutations, even located at distant from the active site, positively affect glucose oxidase catalysis. Determination of the crystal structures of glucose oxidase mutants containing molecular oxygen in the active site and the molecular dynamics simulations allowed: (1) quantifying the enzyme floppiness, (2) revealing that His516 exists in two states: catalytic and noncatalytic; and (3) Understanding that in the glucose oxidase mutants the mutations enrich the catalytic His516 conformation and reduce the flexibility of this residue, leading to an enhancement in their catalytic efficiency. The authors stressed the benefit of active site preorganization with respect to enzyme conversion rates by reducing molecular reorientation needs. Using Förster resonance energy transfer (FRET) response from a Trp-NADH donor–acceptor pair in T-jump studies of a thermophilic alcohol dehydrogenase, microsecond protein motions that can be directly related to active site chemistry were detected [115]. Two distinctive transients were observed: a slow, kinetic process lacking a temperature break and a faster transient that is only detectable above 30 °C. The latter establishes a link between enzyme activity and microsecond protein motions near the cofactor binding site.

Work [91] investigates the connection between stochastic protein dynamics and function for the enzyme cyclophilin A (CypA) in wild-type form, and three variants that feature several mutations distal from the active site.. The present study uses a variety of extensive equilibrium and biased molecular dynamics (MD) simulations to clarify the link between catalytic activity and rates of molecular motions of CypA in wild-type and the three mutant variants. The simulation indicated that the free enzyme rapidly equilibrates between catalytically active and inactive forms before substrate binding (Fig. 7.10a) and for the mutants, the equilibrium is shifted toward the catalytically inactive form (Fig. 7.10b).

Recent works on some selected protein systems have addressed the role of dynamics in enzyme evolution [92–94]. The role of protein dynamics in the evolution of new enzyme function obtained during the laboratory evolution of a phosphotriesterase (PTEs) from *Pseudomonas diminuta* to an arylesterase was examined [93]. Adaptive laboratory evolution (ALE) is an innovative approach for the generation of evolved microbial strains with desired characteristics, by implementing the rules of natural selection as presented in the Darwinian Theory. Analysis of the structures and dynamics of nine protein variants along this trajectory, and three rationally designed variants, revealed cycles of structural destabilization and repair. Evolutionary pressure can "freeze out" unproductive motions and sampling of distinct conformations. This data set also demonstrated how the destabilizing effects from many mutations can be repaired in subsequent generations. Role of unique conformations that can catalyze either phosphotriester or arylester hydrolysis in bifunctional intermediates was discussed in detail.

Review [93, 94] examined various aspects of the evolutionary relation between protein dynamics and the chemical step in dihydrofolate reductases and alcohol dehydrogenases. The tool kit of kinetics, kinetic isotope effects, molecular biology, biophysics, and bioinformatics provided means to link evolutionary changes in structure-dynamics function to the vibrational and conformational states of each protein. The findings suggested that protein dynamics across an entire enzyme can play a role in adaptation to differing physiological conditions. Author of paper [24] focused on **a** thermophilic alcohol dehydrogenase for which the chemical step of hydride transfer is rate determining. Kinetic isotope effects, time-dependent hydrogen/deuterium exchange, fluorescent lifetimes and Stokes shifts were employed to probe the properties of the chemical and conformational coordinates. A long-range network of protein motions in alcohol dehydrogenase oxidoreductase-like protein was observed. The motion extended from a dimer interface to the substrate binding domain across to the cofactor binding domain, over a distance of ca. 30 Å, were defined.

Two second-shell mutations that partially restore its catalytic activity were designed and examined [92]. To identify mutations that rescue the catalytic activity of mutant S99T CypA the effect of the mutations was investigated employing NMR, a room-temperature X-ray crystallography and multiconformer modeling-[92]. CypmA mutants were generated employing site-directed mutagenesis on the CypA S99T plasmid utilizing the Phusion High-Fidelity DNA Polymerase and the primer sequences. Enzyme activity. k_{cat}/K_M for the enzyme-catalyzed cis-to-trans isomerization of succinyl-AAPF-p-nitroanilide

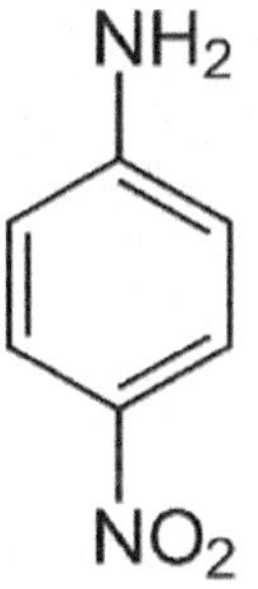

and succinyl-AFPF-p-nitroanilide was measured at 10 °C using the standard chymotrypsin-coupled assay. NMR experiments were recorded on an Agilent DD2 600 MHz four-channel spectrometer equipped with a triple-resonance cryogenically cooled probe-head or a Varian Unity Inova 500 MHz spectrometer. Relaxation dispersion experiments on CypA were carry out on a 600 MHz spectrometer at different temperatures using chemical exchange saturation transfer (CEST). ^{15}N-CEST experiments on CypA were recorded on a 500 MHz spectrometer at 10 °C for two (single- and triple-mutants) or three (doublemutant) different ^{15}N B1 field strengths. The following results were reported: (1) The changes in an enzyme's energy landscape upon directed evolution, designed second-shell mutation, S99T, in CypA were evaluated; (2) Mutations that rescue the catalytic activity of S99T CypA was identified; (3)The X-ray measurement directly revealed increased conformational exchange between two catalytically essential sub-states (See details in capture of Fig. 7.11). In addition, three effects: inverting the equilibrium between two states that are essential

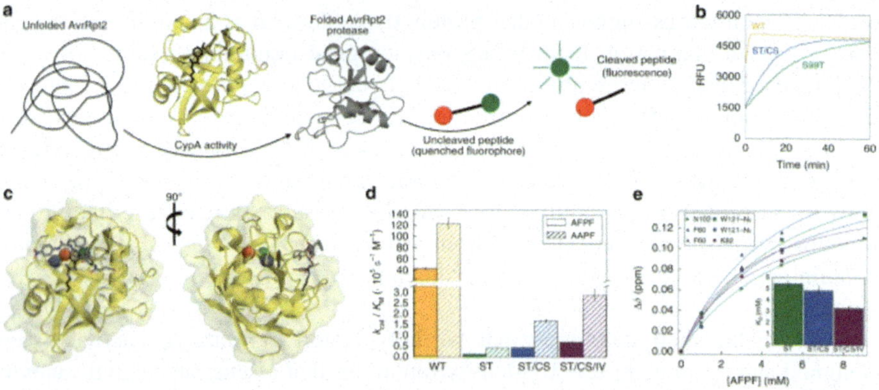

Fig. 7.11 Directed evolution selects rescue mutations for catalysis. **a** Scheme of the assay used in directed evolution to identify CypA mutations with increased catalytic activity: CypA activity for folding of AvrRpt2 protease measured by AvrRpt2-mediated cleavage of the peptide Abz-IEAPAFGGWy-NH2 (y = 3-nitro-Tyrosine). **b** Assay of directed evolution performed on cell lysate in 96-well plates to identify rescue mutations for S99T with increased CypA activity. Kinetics of peptide cleavage is shown for wild-type (yellow), S99T (green), and S99T/C115S (blue) CypA. **c, d** The severely catalytically compromised S99T mutant (green) is rescued by second-shell mutations (C115S, blue and I97V, red). **c** Sites of mutations are plotted onto the crystal structure (1RMH68) of CypA bound to Suc-AAPF-pNA (gray sticks) and the active-site residues Arg55 and Phe113 are shown in black stick representation. The overlay of NMR spectra shows that the overall structure of all CypA forms is very similar, with perturbations observed for residues close to the mutation site or in the dynamic network **d** kcat/KM values for wild-type, S99T, S99T/C115S, and S99T/C115S/I97V CypA measured by protease coupled hydrolysis of Suc-AFPF-pNA and Suc-AAPF-pNA peptides. Error bars indicate the standard deviation obtained from triplicate measurements on at least three different enzyme concentrations. **e** KD values for the three mutant forms of CypA for Suc-AFPF-pNA measured by NMR chemical shift analysis from peptide titrations (see also). Error bars denote the standard errors in the fitted parameters obtained from the global fit [92]. Published under a Creative Commons Attribution 4.0 International License from Springer Nature

for catalysis, decreasing their interconversion rate, and causing a parallel reduction in catalysis were observed.

7.3.4 *Role Solvation Dynamics*

Solvation dynamics at the active site of an enzyme, glutaminyl-tRNA synthetase (GlnRS), was studied using a fluorescence probe, acrylodan 1-(6-dimethylaminonaphthalen- 2-yl)prop-2,

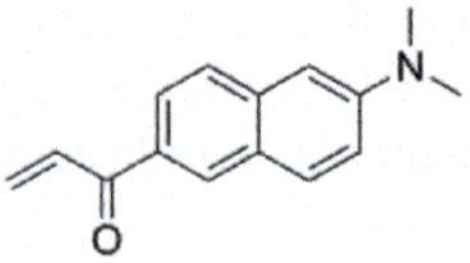

which site-specifically attached at cysteine residue C229, near the active site [104]. The picosecond time-dependent fluorescence Stokes shift indicated slow solvation dynamics at the active site of the enzyme, in the absence of substrate. The solvation dynamics becomes still slower when the substrate (glutamine or tRNA(Gln)) binds to the enzyme. Thus, the slow dynamics is indeed associated with the active site. A detailed study of functionally relevant RNase A dynamics in the wild-type enzyme and a D121A mutant form by NMR spin-relaxation techniques showed that in the wild-type enzyme, the dynamic properties were conserved in the apo, enzyme–substrate, and enzyme-product complexes [111]. Mutation of Asp-121 to Ala (mutant D121A) disrupted the timing of active site dynamics in the product release step, and resulted in 90% loss of catalytic activity. Through characterization of the solvent isotope effect on protein dynamics, determinants of the rate limitation to enzyme catalysis were examined. A global conformational change in RNase A, was found to be sensitive to solvent deuterium content. The authors concluded that Asp-121 plays a significant role in coordinating the millisecond dynamic events in RNase.

Two methods of small-/wide-angle X-ray scattering (SWAXS) experiments on ubiquitin [a small (8.6 kDa) regulatory protein Ub], hen egg white lysozyme (HEWL), and myoglobin (Mb) by which is possible calculate SWAXS patterns were developed [90]. The suggested first approach used all-atom explicit-solvent molecular dynamics (MD) simulations, while the second method involved prediction of the hydration density around a protein employing new HyPred solvation model. A solvent model is a computational method which enable simulations and thermodynamic calculations applicable to reactions and processes which take place in solution. All-atom explicit-solvent MD simulations have been performed at 4 °C for 50 ns for Ub (Protein Data Bank (PDB) code 1UBQ, HEWL (PDB 6LYZ), and Mb (PDB 1WLA), employing NAMD and the CHARMM 27 all-atom force field. HyPred proceeded by calculating a set of proximal radial distribution functions (pRDFs), for ~ 300 different categories of atoms in proteins using all-atom, explicit-solvent MD simulations. The SWAXS patterns obtained from the HyPred model agreed to both experimental data and the patterns predicted by the MD simulations.

The enzyme Candida Antarctica lipase B (CALB) was served as a model for understanding connections among hydration layer dynamics, solvation shell structure, and protein surface structure [19]. The starting coordinates for simulations of CALB enzyme were taken from the X-ray crystallographic structure (resolution: 1.55 Å). MD simulations were performed using GROMACS software package. AMBER03 force field48 was used for the CALB enzyme, and the SPC/E water model49 was employed to represent water. The protein force field GROMOS53a650 was used for the CALB enzyme with SPC/E water. The structure and dynamics of water molecules in the hydration layer were characterized for regions of the CALB surface, divided around each α-helix, β-sheet, and loop structure. The regional hydration dynamics within the 4 conformational stages of the enzyme (open, closed, crystal like, and transitioning) were examined. A color-coded map of CALB–water hydrogen bond lifetimes (HBLTs) for α-helices, β-sheets and connector regions were shown in a figure.

Results indicated that: (1) Regional differences in the structure of the biomolecular hydration layer were found to be concomitant with variations in dynamics; (2) Regions of higher density exhibit faster water dynamics; (3) Effects of protein surface topology and hydrophobicity on water structure and dynamics were evaluated using multiregression analysis; (4) Concave and hydrophobic protein surfaces favor a less dense and more tetrahedral solvation layer, akin to a more ice-like structure, with slower dynamics); (5) Pairwise entropies of local hydration layers were calculated from regional radial distribution functions, scale logarithmically with local hydration dynamics; and (6) The Rosenfeld relationship (relationship between entropy and diffusion) describes the heterogeneous structure–dynamics of the hydration layer around the enzyme Candida Antarctica lipase B (CALB). According to Rosenfeld [95], the self-diffusivity D_0 of a bulk fluid is correlated with the excess entropy in terms of an exponential relation

$$D_R \equiv D_0 \frac{\rho^{1/3}}{(k_B T/m)^{1/2}} \approx a \exp(b S_{ex}),$$

where

$$S_{ex} = \frac{S - S_{id}}{N k_R}$$

is the reduced (dimensionless) excess entropy per molecule, *a* and *b* are the constants which depend on the system.

Dispersion measurements was used to probe the tryptophan indoles in the enzyme dihydrofolate reductase (DHFR), an enzyme in which μs-ms motions are closely coupled to enzyme function [113]. Experimental chemical shifts and density functional theory based chemical shift were compared. The experiment and theory revealed that: (1) During structural changes on the μs-ms timescale, large chemical shift changes for the NH proton on the indole ring take place; (2) Relatively smaller chemical shift changes are for the ring nitrogen atom; (3) During the structural change the tryptophan indole NHs remains bound to water, but the geometry of the protein-bound water networks changes; and (4) Relaxation dispersion measurements can indirectly probe water dynamics on the μs ms timescale. As an illustration of the approach, Fig. 7.12 indicates the conserved active site waters in the closed ground state conformation of WT *E. coli* DHFR bound to folate: $NADP^+$. These findings emphasize the role of the water molecules in binding and coupling of the substrate to the protein and in transmitting information between functionally important regions of the protein. In addition, these finding provide evidence that internal protein motions can be coupled through the mediation of hydrogen-bonded water bound in the protein structure.

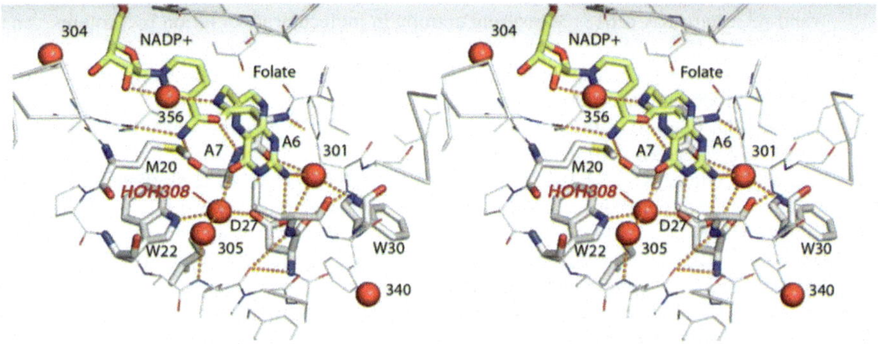

Fig. 7.12 Stereo image of the water-mediated hydrogen bond coupling between the ligands and the protein mediated by water molecules in the active site of WT *E. coli* DHFR. Hydrogen bonds are shown as orange dashed lines [113]. With permission from American Chemical Society

References

1. Likhtenshtein GI (1988) Chemical physics of redox metalloenzyme catalysis. Springer-Verlag, Berlin
2. Likhtenshtein GI (2003) New trends in enzyme catalysis and mimicking chemical reactions. Kluwer Academic/Plenum Publishers, New York
3. Denisov ET, Sarkisov OM, Likhtenshtein GI (2003) Chemical kinetics: fundamentals and recent developments. Elsevier, Amsterdam
4. McCammon JA, Harvey SC (1988) Dynamics of proteins and nucleic acids. Cambridge University Press, Cambridge
5. Bagchi B (2013) Water in biological and chemical processes: from structure and dynamics to function. Cambridge University Press, Cambridge
6. Lemay-St-Denis C, Doucet N, Pelletier JN (2022) Integrating dynamics into enzyme engineering. Protein Eng Des Select 35:gzac015
7. Klinman J, Hammes-Schiffer S (2013) Dynamics in Enzyme Catalysis. Springer, Berlin
8. Likhtenshtein GI, Yamauchi J, Nakatsuji S, Smirnov A, Tamura R (2008) Nitroxides: application in chemistry, biomedicine, and materials science. Wiley-VCH, Weinhem
9. Likhtenshtein GI (1976) Spin labeling method in molecular biology. Wiley Interscience, New York, NY
10. Likhtenshtein GI (2021) Biological water: physicochemical aspects. Springer, Berlin
11. Agarwal PK (2006) Enzymes: an integrated view of structure, dynamics and function. Microb Cell Fact 5:2. https://doi.org/10.1186/1475-2859-5-2
12. Luk LYP, Loveridge EL, Allemann RL (2015) Protein motions and dynamic effects in enzyme catalysis. Phys Chem Chem Phys 17:30817–30827
13. Campbell E, Kaltenbach M, Correy G (2016) The role of protein dynamics in the evolution of new enzyme function. Nat Chem Biol 12:944–950
14. Shina CL, Kamerlin WA (2010) At the dawn of the 21st century: is dynamics the missing link for understanding enzyme catalysis? Proteins 78(6):1339–1375
15. Zhao X, Gentile K, Mohajerani F, Sen A (2018) Powering motion with enzymes. Accounts Chem Res 51(10):2373–2381
16. Wapeesittipan P, Mey AS, Walkinshaw MD, Michel J (2019) Allosteric effects in cyclophilin mutants may be explained by changes in nano-microsecond time scale motions. Commun Chem 2(1):41
17. Conti Nibali V, D'Angelo G, Paciaroni A, Tobias DJ, Tarek M (2014) On the coupling between the collective dynamics of proteins and their hydration water. J Phys Chem Lett 5:1181–1186

18. Wand AJ, Sharp KA (2018) Measuring entropy in molecular recognition by proteins. Annu Rev Biophys 47:41–61
19. Dahanayake JN, Mitchell-Koch KR (2018) Entropy connects water structure and dynamics in protein hydration layer. Phys Chem Chem Phys 20:14765–14777
20. Leuchter JD, Levy Y (2017) On the coupling between the dynamics of protein and water. Phys Chem Chem Phys 19:8243–8257
21. Leidner F, Yilmaz KN, Paulsen J, Muller YA, Schiffer CA (2018) Hydration structure and dynamics of inhibitor-bound HIV-1 protease. J Chem Theory Comput 14:2784–2796
22. Kim JG, Muniyappan S, Oang KY, Kim TW (2016) Cooperative protein structural dynamics of homodimeric hemoglobin linked to water cluster at subunit interface revealed by time-resolved X-ray solution scattering. Struct Dynam 3:023610
23. Otten R, Lin Liu L, Kenner LR (2018) Rescue of conformational dynamics in enzyme catalysis by directed evolution. Nat Commun 9:1314
24. Klinman JP (2015) Dynamically achieved active site precision in enzyme catalysis. Acc Chem Res 48(2):449–456
25. Turton D, Senn HM, Harwood T (2014) Terahertz underdamped vibrational motion governs protein-ligand binding in solution. Nat Commun 5:3999
26. Kim KH, Oang KY, Kim J, Lee JH, Kim Y, Ihee H (2011) Direct observation of myoglobin structural dynamics from 100 picoseconds to 1 microsecond with picosecond X-ray solution scattering. Chem Commun 47(1):289–291
27. McGeagh JD, Ranaghan KE, Mulholland AJ (2011) Protein dynamics and enzyme catalysis: insights from simulations. Biochim Biophys Acta (BBA) Proteins Proteom 1814(8), 1077–1092
28. Warshel A, Bora RP (2016) Perspective: defining and quantifying the role of dynamics in enzyme catalysis. J Chem Phys 144:180901
29. Kohen A (2015) Role of dynamics in enzyme catalysis: substantial versus semantic controversies. Acc Chem Res 48:466–473
30. Delgado M, Görlich S, Longbotham JE (2017) Convergence of theory and experiment on the role of preorganization, quantum tunneling, and enzyme motions into flavoenzyme catalyzed hydride transfer. ACS Catal 7:3190–3198
31. Doshi U, Holliday MJ, Eisenmesser EZ, Hamelberg D (2016) Dynamical network of residue-residue contacts reveals coupled allosteric effects in recognition, catalysis, and mutation. Proc Natl Acad Sci USA 113:4735–4740
32. Holliday MJ, Camilloni C, Armstrong G (2017) Networks of dynamic allostery regulate enzyme function. Structure 25:276–286
33. Eisenmesser EZ, Millet O, Labeikovsky W (2005) Intrinsic dynamics of an enzyme underlies catalysis. Nature 438:117–121. https://doi.org/10.1038/nature04105
34. Likhtenshtein GI, Febbrario F, Nucci R (2000) Intramolecular dynamics and conformational transitions in proteins studied by biophysical labeling methods. Common and specific features of proteins from thermophylic microorganisms. Spectrochem Acta Part A Biomol Spectrosc 56:2011–2031. https://doi.org/10.1016/s1386-1425(00)00290-0
35. Bhabha G, Biel JT, Fraser JS (2015) Keep on moving: Discovering and perturbing the conformational dynamics of enzymes. Acc Chem Res 48(2):423–430
36. Papaleo E, Saladino G, Lambrughi M et al (2016) The role of protein loops and linkers in conformational dynamics and allostery. Chem Rev 116:6391–6423
37. Levantino M, Yorke BA, Monteiro DC, Cammarata M, Pearson AR (2015) Using synchrotrons and XFELs for time-resolved X-ray crystallography and solution scattering experiments on biomolecules. Curr Opin Struct Biol 35:41–48
38. Petrovic D, Risso VA, Kamerlin SCL, Sanchez-Ruiz JM (2018) Conformational dynamics and enzyme evolution. J R Soc Interf 15(144):201803301–2018033018
39. Karplus M, Petsko GA (1990) Molecular dynamics simulations in biology. Nature 347:631–639
40. Lumry R, Eyring H (1954) Conformational changes in proteins. J Phys Chem 58:110–120

41. Lumry R, Biltonen R (1969) Thermochemical and kinetical aspects of protein conformation. In: Timashev SN, Fasman CD (eds) Structure and stability of biological molecules, vol 2. New York, Biological macromolecules series (Dekker, pp 146–306
42. Linderstrom-Lang KH, Schellmann JA (1959) Protein structure and enzyme activity. In: Boyer PD, Lardy M, Myrback K (eds) The enzyme, vol 1. Academic Press, New York, pp 443–510
43. Koshland DJ (1959) Mechanism of transfer of enzymes. In: Boyer PD, Lardy M, Myrback M (eds) The enzyme, vol 1. Academic Press, New York, pp 305–346
44. Likhtenshtein GI (1976) Water and protein dynamics. In: Alfsen A, Bertran AJ (eds) L'eau et les Systemes Biologoque. CNR, Paris, pp 43–45
45. Likhtenshtein GI (1990) The use of nitroxides in the solution of some problems of chemical biophysics. Pure Appl Chem 62:281–288
46. Karush J (1950) Heterogeneity of the binding site of bovin serum albumin. J Am Chem Soc 72:2705–2713
47. Sergeev PV, Maga IV, Lexina VP, Stepanyants AU, Likhtenshtein GI (1976) Investigation of the interaction of steroids with human serum albumin by the spin-echo-paramagnetic probe technique. Mol Biol (Moscow) 10:1166–1174
48. Likhtenshtein GI, Akhmedov YD, Ivanov LV, Krinitskaya LA, Kokhanov YV (1974) Investigation of the lysozyme macromolecule by a spin-labeling method. Mol Biol 8:40–48
49. Tsai AM, Neumann DA, Bell LN (2000) Molecular dynamics of solid-state lysozyme as affected by glycerol and water: a neutron scattering study. Biophys J 79:2728–2732
50. Reat V, Dunn R, Ferrand M, Finney JL, Daniel RM, Smith JC (2000) Solvent dependence of dynamic transition in protein solution. Proc Natl Acad Sci USA 97:9961–9966
51. Vitkup D, Ringe D, Petsko GA, Karplus M (2000) Solvent mobility and the protein 'glass' transition. Nat Struct Biol 7:34–35
52. Vincent M, Gilles A-M, Li de la Sierra I et al (2000) Nanosecond fluorescence dynamics Stockes shift of tryptophan in a protein matrix. J Phys Chem B 104:11286–11295
53. Diaz N, Suarez D, Merz KM Jr (2001) Molecular dynamics simulation of the mononuclear zinc-β-lactomase from Bacciluscereus complexed with benzylpenicillin and quantum chemical study of the reaction mechanism. J Am Chem Soc 123:9867–9879
54. Tarek M, Martina GJ, Tobias D (2000) Amplitude and frequency of protein dynamics: analysis of discrepancies between neutron scattering and molecular dynamics simulations. J Am Chem Soc 122:10451–10459
55. Kay E (1998) Protein dynamics from NMR. Biochem Cell Biol 76:145–152
56. Frolov EN, Mokrushin AN, Likhtenshtein GI, Trukhtanov VA, Goldansky VI (1973) Investigation of dynamic structure of proteins using gamma-resonance labels. Dokl Akad Nauk SSSR 212:165–168
57. Frolov EN, Belongova OV, Likhtenshtein GI (1977) Investigation of the mobility of spin and Mössbauer labels bound to macromolecules. In: Burshtein EA (ed) Equilibrium dynamics of the native structure of protein. Izdatelstvo Akademii Nauk SSSR, Pushchino, pp 99–142
58. Likhtenshtein GI (1979) Study of protein dynamics by spin-labeling, Mösbauer spectroscopy, and NMR. In: Losche A (ed) Special collogue amper on dynamic processes in molecular systems. Karl-Marx University, Leipzig, pp 100–107
59. Likhtenshtein GI (1993) Biophysical labeling methods in molecular biology. Cambridge University Press, Cambridge N. Y.
60. Likhtenshtein GI (1985) Amorphous-crystalline model of intermolecular dynamics in global proteins. Biofizika 30:30–37
61. Belonogova OV, Likhtenshtein GI, Krinichnyi VI (1997) Electron transfer in donor-acceptor pairs in modified α-chymotrypsin. Effect of macroviscosity and local polarity. Photochem Photobiol A Chem 107:35–42. Likhtenshtein, GI (1988) Structure and dynamics of metalloenzymes studied by biophysical labeling methods. J Mol Catal 48:129–138
62. Likhtenshtein GI (1986) Water and dynamics of proteins and membranes. Stud Biophys 111:89–100
63. Likhtenshtein GI, Kulikov AV, Kotelnikov AI, Levchenko LA (1986) Methods of physical labels - a combined approach to the study of microstructure and dynamics of biological systems. J Biochem Biophys Meth 12:1–28

64. Likhtenshtein GI (1985) Amorphous-crystaline model of intermolecular dynamics in global proteins. Biofizika 30:30–37
65. Lakowicz J, Weber G (1973) Quenching of protein fluorescence by oxygen. Detection of structural fluctuations in proteins on the nanosecond time scale. Biochemistry 12, 4171–4179
66. Wütrich K (1986) NMR of proteins and nucleic acids. Wiley-Interscience Publication, New York
67. Lipardi G, Szabo A (1982) Model-free approach to the inter-retation of nuclear magnetic resonance relaxation in macromolecules. 1. Theory and range of validity. J Am Chem Soc 104:4546–4569
68. Rialdi G, Battistel E (1993) Thermodynamic properties of immobilized enzymes. Calorim Anal Therm 24:401–404
69. Privalov PL, Gill JG (1998) Stability of protein structure and hydrophobic interaction. Adv Prot Chem 39:191–234
70. Alfimova EY, Likhtenshtein GI (1979) Investigation of structure and conformational change of proteins by the inductive resonance energy transfer. Adv Mol Relax Interact Proc 14:47–89
71. Frauenfelder H, McMahon BH (2001) The energy landscape. In: Springer Series in Chemical Physics 67 (Single Molecule Spectroscopy), pp 257–276
72. Roslyakov BY, Khurgin YI (1972) Study of cyanamoyl—a chymotrypsin hydrolysis in the solid state. Biokhimiya 37:493–447 (1972)
73. Krinichnyi VI, Grinberg OY, Bogatirenko VR, Likhtenshtein GI, Lebedev YS (1985) Study of microenvironment effect on magnetic resonance parameters of spin-labeled human serum albumin in 2-mm range. Biofizika 30:203–216
74. Peterson-Kennedym SE, Mc Courty JL, Hoffman BM (1984) J Am Chem Soc 106(1984):5010
75. Daniel RM, Dunn RV, Finney JL, Smith JC (2003) The role of dynamics in enzyme activity. Annu Rev Biophys Biomol Struct 32:69–92
76. Dong F, Zhang M, Rui M, Lu F (2022) Insights of conformational dynamics on catalytic activity in the computational stability design of *Bacillus subtilis* LipA. Arch Biochem Biophys 722:109196
77. Kurkcuoglu Z, Bakan A, Kocaman D, Bahar I, Doruker P (2012) Coupling between catalytic loop motions and enzyme global dynamics. PLoS Comput Biol 8(9):e1002705
78. Gavrilov Y, Leuchter JD, Levy Y (2017) On the coupling between the dynamics of protein and water. Phys Chem Chem Phys 19:8243
79. Petrović D, Frank D, Kamerlin SCL, Hoffmann K, Strodel B (2017) Shuffling active site substate populations affects catalytic activity: the case of glucose oxidase. ACS Catal 7:6188–6197. https://doi.org/10.1021/acscatal.7b01575
80. Romero-Rivera A, Garcia-Borràs M, Osuna S (2017) Role of conformational dynamics in the evolution of retro-aldolase activity. ACS Catal 7:8524–8532. https://doi.org/10.1021/acscatal.7b02954
81. Risso VA, Gavira JA, Mejia-Carmona DF, Gaucher EA, Sanchez-Ruiz JM (2013) Hyperstability and substrate promiscuity in laboratory resurrections of Precambrian β-lactamases. J Am Chem Soc 135:2899–2902
82. Hammes GG, Benkovic SJ, Hammes-Schiffer S (2011) Flexibility, diversity, and cooperativity: pillars of enzyme catalysis. Biochemistry 50:10422–10430
83. Ma H, Szeler K, Kamerlin SCL, Widersten M (2016) Linking coupled motions and entropic effects to the catalytic activity of 2-deoxyribose-5-phosphate aldolase (DERA). Chem Sci 7:1415–1421. https://doi.org/10.1039/C5SC03666F
84. Agarwal PK, Billeter SR, Hammes-Schiffer S (2002) Nuclear quantum effects and enzyme dynamics in dihydrofolate reductase catalysis. J Phys Chem B 106(12):3283–3293
85. Kurkcuoglu Z, Bakan A, Kocaman D et al (2012) Coupling between catalytic loop motions and enzyme global dynamics. PLoS Comput Biol 8(9):e1002705
86. Doshi U, McGowan LC, Lauren C, Ladani ST et al (2012) Resolving the complex role of enzyme conformational dynamics in catalytic function. Proc Natl Acad Sci USA 109(15):5699–5704

87. Boekelheide N, Salomón-Ferrer R, Miller TF (2011) III Dynamics and dissipation in enzyme catalysis. PNAS 108(39):16159–16163
88. Romero-Rivera A, Corbella M, Parracino A, Patrick WM, Kamerlin SCL (2022) Complex loop dynamics underpin activity, specificity, and evolvability in the (βα)8 barrel enzymes of histidine and tryptophan biosynthesis. JACS 2(4):943–960
89. Luk LYP, Loveridge EJ, Allemann RK (2015) Protein motions and dynamic effects in enzyme catalysis. Phys Chem Chem Phys 17:30817
90. Virtanen JJ, Makowski L, Sosnick TR, Freed KF (2011) Modeling the hydration layer around proteins: applications to small- and wide-angle X-ray scattering. Biophys J 101:2061–2069
91. Wapeesittipan P, Mey ASJS, Walkinshaw MD, Michel J (2019) Allosteric effects in cyclophilin mutants may be explained by changes in nano-microsecond time scale motions. Commun Chem 2:41
92. Otten R, Liu Lin, Kenner LR et al (2018) Rescue of conformational dynamics in enzyme catalysis by directed evolution. Nat Commun 9:1314
93. Campbell E, Kaltenbach M, Correy GJ (2016) The role of protein dynamics in the evolution of new enzyme function. Nat Chem Biol 12:944–950
94. Klinman JP, Kohen AK (2014) Evolutionary aspects of enzyme dynamics. J Biol Chem 289(44):30205–30212
95. Banerjee A, Nandi MK, Bhattacharyya SM (2017) Validity of the Rosenfeld relationship: a comparative study of the network forming NTW model and other simple liquids. J Chem Sci 129:793–800
96. Pagano P, Guo Q, Kohen A, Cheatum CM (2016) Oscillatory enzyme dynamics revealed by two-dimensional infrared spectroscopy. J Phys Chem Lett 7(13):2507–2511
97. Palmer AG III (2015) Enzyme dynamics from NMR spectroscopy. Acc Chem Res 48(20):457–465
98. Agarwal PK (2018) A biophysical perspective on enzyme catalysis. Biochemistry 58(6):438–449
99. Thompson MC, Barad BA, Fraser JS (2019) Temperature-jump solution X-ray scattering reveals distinct motions in a dynamic enzyme. Nat Chem 11:1058–1066
100. Bahar I, Lezon TR, Yang LW, Eran E (2010) Global dynamics of proteins: bridging between structure and function. Annu Rev Biophys 39:23–42
101. Ma B, Nussinov R (2010) Enzyme dynamics point to stepwise conformational selection in catalysis. Curr Opin Chem Biol 14(5):652–659
102. Gao S, Klinman JR (2022) Functional roles of enzyme dynamics in accelerating active site chemistry: emerging techniques and changing concepts. Curr Opin Struct Biol 75:102434
103. Biel JT, Thompson MC, Cunningham CN (2017) Flexibility and design: conformational heterogeneity along the evolutionary trajectory of a redesigned ubiquitin. Structure 25:739-749.e3
104. Guha S, Sahu K, Roy D (2005) Slow solvation dynamics at the active site of an enzyme: implications for catalysis. Biochemistry 44(25):8940–8947
105. Boehr DD, Dyson HJ, Wright PE (2006) An NMR perspective on enzyme dynamics. Chem Rev 106(8):3055–3079
106. Henzler-Wildman KA, Ming L, Vu T (2007) A hierarchy of timescales in protein dynamics is linked to enzyme catalysis. Nature 450(7171):913–916
107. Bandaria JN, Dutta S, Hill SE et al (2008) Fast enzyme dynamics at the active site of formate dehydrogenase. J Am Chem Soc 130(1):22–23
108. Hanson JA, Duderstadt K, Watkins LP et al (2007) Illuminating the mechanistic roles of enzyme conformational dynamics. Proc Natl Acad Sci USA 104(46):18055–18060
109. Dong F, Zhang M, Ma R, Lu C, Xu F (2022) Insights of conformational dynamics on catalytic activity in the computational stability design of *Bacillus subtilis* LipA. Arch Biochem Biophys 722:109196
110. Palmer AG III (2015) Enzyme dynamics from NMR spectroscopy. Acc Chem Res 48(2):457–465

111. Kovrigin EL, Loria JP (2006) Enzyme dynamics along the reaction coordinate: critical role of a conserved residue. Biochemistry 45(8):2636–2647
112. Dunn RV, Reat V, Finney J (2000) Enzyme activity and dynamics: xylanase activity in the absence of fast anharmonic dynamics. Biochem J 346(2):355–358
113. Fenwick RB, Oyen DH, Dyson J, Wright PE (2018) How dynamics of tryptophan-water networks in proteins. J Am Chem Soc 140(2):675–682
114. Thompson MC, Barad BA, Wolff AM (2019) Temperature-jump solution X-ray scattering reveals distinct motions in a dynamic enzyme. Nat Chem 11:1058–1066
115. Vaughn MB, Zhang J, Spiro TG et al (2018) Activity-related microsecond dynamics revealed by temperature-jump Förster resonance energy transfer measurements on thermophilic alcohol dehydrogenase. J Am Chem Soc 140:900–903
116. Meadows CW, Balakrishnan G, Kier BL, Spiro TG, Klinman JP (2015) Temperature-jump fluorescence provides evidence for fully reversible microsecond dynamics in a thermophilic alcohol dehydrogenase. J Am Chem Soc 137(32):10060–10063
117. Boehr DD, D'Amico RN, O'Rourke KF (2018) Engineered control of enzyme structural dynamics and function. Protein Sci 27(4):825–838
118. Schwartz SD (2013) Protein dynamics and the enzymatic reaction coordinate. Top Curr Chem 337:189–208
119. Klinman JP (2015) Dynamically achieved active site precision in enzyme catalysis. Acc Chem Res 48:449–456
120. Zhang X, Bruice TC (2005) The proficiency of a thermophilic chorismate mutase enzyme is solely through an entropic advantage in the enzyme reaction. Proc Natl Acad Sci USA 102:18356–18360
121. Cho HS, Schotte F, Stadnytskyi V, DiChiara A, Henning R, Anfinrud F (2018) Dynamics of quaternary structure transitions in R-state carbonmonoxyhemoglobin are unveiled in time-resolved X-ray scattering patterns following a temperature jump. J Phys Chem B 122:11488–11496
122. Barends TRM, Foucar L, Ardevol A (2015) Direct observation of ultrafast collective motions in CO myoglobin upon ligand dissociation. Science 350:445–450
123. Levantino M, Schirò G, Lemke HT, Cottone C (2015) Ultrafast myoglobin structural dynamics observed with an X-ray free-electron laser. Nat Commun 6:6772
124. Shelby ML, Wildman A, Hayes D et al (2021) Interplays of electron and nuclear motions along CO dissociation trajectory in myoglobin revealed by ultrafast X-rays and quantum dynamics calculations. Proc Natl Acad Sci USA 118(14):e2018966118
125. Kim KH, Oang KY, Kim J, Lee JH, Kim Y, Ihee H (2011) Direct observation of myoglobin structural dynamics from 100 picoseconds to 1 microsecond with picosecond X-ray solution scattering. Chem Commun (Camb) 47(1):289–291
126. Frauenfelder H, Sligar SG, Wolynes PG (1991) The energy landscapes and motions of proteins. Science 254:1598–1603

Chapter 8
Enthalpy–Entropy Relationships in Enzyme Reactions

Abstract Enthalpy–entropy compensation (EEC) referred to the behavior of a series of closely related enzyme reactions exhibits a linear relationship between one of the following kinetic or thermodynamic parameters and isokinetic or thermodynamic temperatures. The enthalpy–entropy compensation effect in enzymatic catalysis was first described by Likhtenshtein in 1966. In this chapter, separate sections are devoted to statistical artifacts at apparent enthalpy–entropy compensation, theoretical consideration of EEC, EES for small molecules, EES for peptides and proteins, and miscellaneous examples of EEC in enzyme catalysis. The following compensation pairs of energy–entropy activation were considered: hydrolysis catalyzed by carboxypepsidase, transformation fumarate to malate catalyzed be fumarase, α-amylase reaction of starch hydrolysis, reactions catalyzed by unvertase, urease, lipase as well as enthalpy–entropy compensation at binding various inhibitors with catalase, and at inhibition bacterial luminescence. The predominant contribution of water reorganization to the thermodynamic and kinetic parameters of enzymatic reactions was stressed.

8.1 Introduction

Many kinetic and thermodynamical studies concerning homologous reaction series reported the existence of an activation enthalpy–entropy linear correlation (compensation plot. The temperature at which all the members of the series have the same rate constant called isokinetic temperature [1–24]. The enthalpy–entropy compensations (EEC) have been reported in many chemical, physical, biological and food processes and can give information about the reaction mechanism and whether the reaction is controlled by enthalpy or entropy [1–46]. Compensations phenomena also have been expanded for the thermodynamic enthalpy and entropy phenomena.

Constable [25] found a straight line relation between the logarithm of the frequency factor (corresponding approximately to the entropy of activation $\Delta S^{\#}$) and the energy of activation (corresponding approximately to the enthalpy of activation $\Delta H^{\#}$). The next principle step in the area was made by Leffler [1] and Exner [2].The enthalpy–entropy compensation effect in enzymatic catalysis was first described by

G. I. Likhtenshtein, *Enzyme Catalysis Today and the Chemistry of the 21st Century*,
https://doi.org/10.1007/978-3-031-51552-1_8

Likhnenshtein in 1966 [3, 4]. The following compensation pairs of energy–entropy activation were considered: hydrolysis catalyzed by carboxypepsidase, transformation fumarate to malate catalyzed by fumarase, α-amylase reaction of starch hydrolysis, reactions catalyzed by unvertase, urease, lipase as well as enthalpy–entropy compensation at binding various inhibitors with catalase, and at inhibition bacterial luminescence. The energy E_a and entropy ($\Delta S^{\#}$) activation of enzymatic reactions were found to be sensitive to the structure of substrate, temperature regions, pH, solvent composition, etc. Later compensation phenomena in the enzyme catalysis were confirmed by Lumry and Rajender [5, 6]. Likhtenshtein and Lumry independently explained this effect by the cooperative rearrangement ("melting") of the water ensemble accompanying the elementary chemical act of an enzyme reaction [4–6, 14–24] and the thermodynamic parameters of the water reorganization can be included in the experimental values of the energy and entropy of the activation. The fact that compensation effects can be associated with conformational rearrangements and reorganization of cooperative ensembles of water molecules in proteins was indicated by data on protein denaturation, which showed the decisive role of moisture on the enthalpy and entropy parameters of protein denaturation in the enormous range of energy activation (E_a) values (from 0 to 120 kcal/mol) and $\Delta S^{\#}$ (from 10 to 400 eu) [21] and references therein. Nanosecond intermolecular mobility of proteins consists of movement of the relatively large and rigid parts of the protein (domains) [15–19] (Sect. 8.4). Such hinge-like oscillation of the tightly packed polypeptide blocks does not give a measurable contribution to the helicity degree of polypeptide chains, the overall heat capacity and entropy. The point is that a change in the structure of a substrate or an inhibitor upon binding to an enzyme or a shift in protein domains upon a change in protein conformation should not significantly affect the entropy and enthalpy of the processes. Subsequent experimental and theoretical studies, on the one hand, confirmed the predominant contribution of water reorganization to the thermodynamic and kinetic parameters of enzymatic reactions and revealed many important aspects of the process, on the other hand. Recently various aspects of the enthalpy–entropy compensation effect (EECE) in kinetics and thermodynamics in molecular biological processes were widely discussed [47–65].

8.2 Early Theoretic Considerations

For numerous enzyme reactions, the following linear dependences are satisfied:

$$\Delta H = \Delta H^\circ + \beta \Delta S \tag{8.1}$$

where ΔH° is constant regardless of substituents, ΔH and ΔS is different for each substituent. Inserting the Gibbs free-energy equation and combining like terms produces the following equation:

$$\Delta G = \Delta H^\circ - (T - \beta_c)\Delta S \tag{8.2}$$

β_c has the dimension of temperature and is referred to as the isothermodynamic temperature because at $\beta_c = T$, $\Delta G = \Delta H°$, that is, all reactants in the series having different ΔS ran with the same rate. For enzyme reactions, the experimental values of ΔH and ΔS can be apparent parameters, which jointly reflect the entire complexity of the process (presence of intermediate microsteps, involving accompanying conformational transitions and solvent molecules reorganization). In particular, following the assumption about involving water molecules reorganization in an enzymatic reaction, the contribution of the corresponding parameters of the reorganization of a cooperative ensemble of n molecules of water, to the experimental thermodynamic and kinetic parameters of enzymatic reactions is expected [4]. Equations (8.3) and (8.4) can be written as a first approximation:

$$\Delta H = \Delta H_r + n\Delta H_0 \tag{8.3}$$

and

$$\Delta S = \Delta S_r + n\Delta S_o \tag{8.4}$$

where ΔH_r and ΔS_r are attributed to enthalpy and entropy of the chemical step, respectively; n is the number water molecules in the ensemble; ΔH_0 and ΔS_o are the standard enthalpy and entropy of the rearrangement of on elementary units. Assuming that the reorganization of water makes a decisive contribution and taking into account the Eqs. (8.1–8.4), then we obtain

$$\beta = \Delta H_0/\Delta S_o = T_m \tag{8.5}$$

where T_m is the "melting" temperature of the cooperative assemble which is expected to be close to average temperature of the experiment T_{av}. In general case, the chemical reaction and the accompanying water rearrangement cannot be regarded as independent events. If the quantities of ΔH and ΔS can be represented, to a first approximation, as linear function of n with the coefficient being, respectively, equal to a and b, then the equilibrium (or isokinetic) temperature, under the condition of the predominant contribution of the process of reorganization of the "melting" of the water ensemble, will be given by

$$\beta = (a/b)\,(\Delta H_0/\Delta S_o) = (a/b)T_m \approx (a/b)\,T_{av}. \tag{8.6}$$

Thus, depending on the mechanism of the influence of hydration on the enthalpy and entropy parameters of the chemical reaction, the value of β will be greater or less than the average temperature of the experiment. Equation (8.3) allows to roughly estimate number of water molecules involving in reaction in the enzyme active site. For example, taking energy activation of a chemical step of an enzyme reaction $E_r =$ 4–5 kcal mol^{-1}, the molar enthalpy of fusion for ice $\Delta H_o = 1.400$ kcal mol^{-1} (https://en.wikipedia.org/wiki/Enthalpy) and typical experimental values $E_a = 25$–35 kcal mol^{-1}, Eq. (8.3) gives reasonable number of the water molecules in the ensemble in vicinity of the active cite $n = 15$–20. These values are in agreement with the corresponding experimental data on water in proteins (Chap. 7). A similar analysis can be applied to the binding of substrates and inhibitors to proteins.

The values of enthalpy and entropy can be temperature dependent, that is,

$$\Delta H = \Delta H(\mathrm{d}H/\mathrm{d}T) \text{ and } \Delta S = \Delta S\,(\mathrm{d}\Delta S/\mathrm{d}T). \tag{8.7}$$

Once, having lived these functions in a series and limiting ourselves to the first two terms of the expansion, we get:

$$\Delta H = \Delta H^{\circ} + \alpha(\mathrm{d}H/\mathrm{d}T) \text{ and } \Delta S = \Delta S^{\circ} + \beta(\mathrm{d}\Delta S/\mathrm{d}T). \tag{8.8}$$

If the values of ΔH° and ΔS° refer to the chemical stage of the process and can be considered constant and small in the given series, then $\alpha(\mathrm{d}H/\mathrm{d}T)$ and $\beta(\mathrm{d}\Delta S/\mathrm{d}T)$ will determine the experimental values of the corresponding parameters. In this case

$$\Delta H/\Delta S = [\alpha(\mathrm{d}\Delta H/\mathrm{d}T)]/[\beta(\mathrm{d}\Delta S/\mathrm{d}T)] \tag{8.9}$$

8.3 Statistical Artifacts at Apparent EEC

In a quantitative approach [40], a statistical test to compare the harmonic mean experimental temperature T_{hm} with the interval

$$\left[T_{\mathrm{comp}} - t\sigma, T_{\mathrm{comp}} + t\sigma\right] \tag{8.10}$$

was proposed. In Eq. (8.10) σ is the standard error in T_{comp} and the Student's t-value depends on the chosen confidence level and the number of samples, i.e. the number of ΔH, and ΔS data pairs. For a 95 % confidence level t is close to 2 for large data sets. The authors concluded that if T_{hm} falls within experimental temperature interval, the EEC is probably of statistical nature. The general statistical mechanical basis for solvent induced isokinetic/isoequilibrium entropy–enthalpy compensation in chemical reactions and adsorption which can be used to control of rate processes and binding constants in diverse applications, was established in [38]. When the temperature rate-constant data are fitted according to an Arrhenius plot, the experimental errors associated with the pre-exponential factor and the activation energy are mutually interdependent. This result has been qualitatively illustrated in Fig. 8.1. Thus, there is a statistic compensation that consists of the propagation of experimental errors, and this effect has to be considered before concluding kinetic and/or thermodynamic compensation.

A program for statistically correct calculations of the enthalpy—entropy relationship was introduced [41]. The program is applicable generally for linear regression when several regression lines are constrained by a common point of intersection.

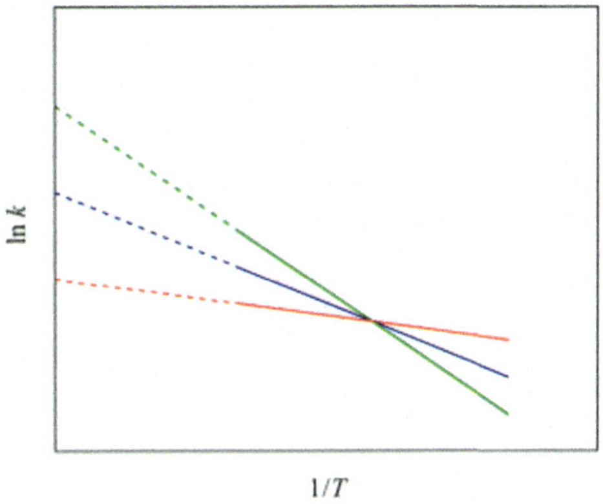

Fig. 8.1 Qualitative representation illustrating how experimental errors affect the Arrhenius plots. The blue straight line corresponds to the real behavior of the process. The green line simulates the situation found when the errors associated with the higher temperatures are positive and those associated with the lower temperatures negative, whereas the red line shows the opposite case. The intercrossing point marks the position of the apparent isokinetic temperature [38]. With permission from American Chemical Society

The difference between a correct and incorrect statistical treatment was illustrated by several examples.

A general statistical mechanical Gaussian density of states model of a complex system was analyzed to explore whether and under what conditions extra-thermodynamic compensation can occur [42]. This model suggested that the most likely behavior to be seen is linear *S–H* compensation over a rather limited range of perturbations with a compensation temperature $\beta = \mathrm{d}\Delta H/\mathrm{d}\Delta S$ within 20% of the experimental temperature. Using suggested probability distribution, a plot of ΔS versus ΔE was generated for a systematic series of perturbations. The resulting plot was found to be ellipsoidal direction (Fig. 8.2).

Three different topics were covered theoretically and illustrated by examples [36]: (1) The effect of systematic errors in experimental kinetic data which created apparent compensation effects; (2) Analysis of kinetic data when the Arrhenius parameters depend on another parameter; and (3) The effect of analyzing the temperature dependence of an overall rate of reaction, rather than a rate constant. It was concluded that a mathematical origin of the effect should be carefully considered before any physicochemical conclusions are drawn. An account of the role of ΔG along with ΔH and ΔS in the understanding of such EEC correlations using authentic literature data was presented and discussed [34]. It was stressed that planned studies on similar systems with a wide range of ΔG values are required for realistic evaluation of the EEC. As a result of study [39], a computer program has been developed to evaluate the probability that experimental errors might lead to a linear compensation plot parting from an initial randomly scattered set of activation parameters (*p*-test). The application of the *p*-test to a reaction series is illustrated in Fig. 8.3. Use of this program to kinetic data for 100 homologous reaction series allowed concluding that most

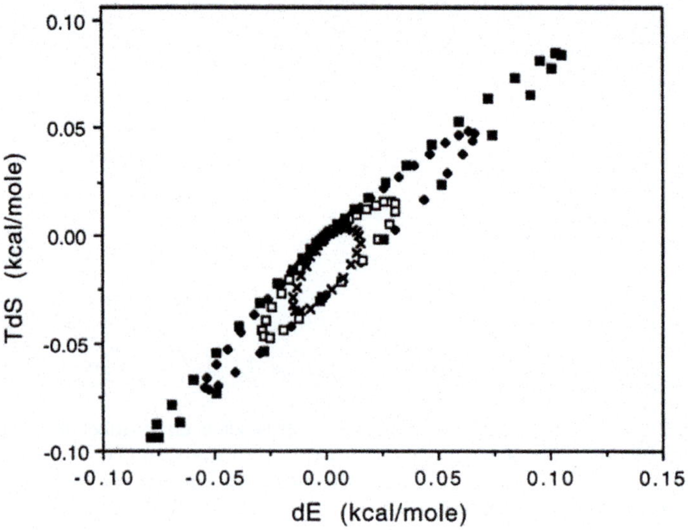

Fig. 8.2 Entropy (TS) versus energy (*E*) plot for the Gaussian density of states model at 298 K, with energy level spreads (SD, σ) of 0.5 (×), 1 (□), 2 (◆), and 3 (■) kcal mol^{-1}, respectively [42]. With permission from Wiley

of the reported compensation plots can hardly be explained by the accumulation of experimental errors.

A new method, called Combined K-CQF developed in [44] made it possible both to quantify the degree of coalescence of experimental Van 't Hoff lines and to verify whether or not the Enthalpy–Entropy Compensation is of a statistical origin at a desired confidence level. The new method requires only a standard least square fit

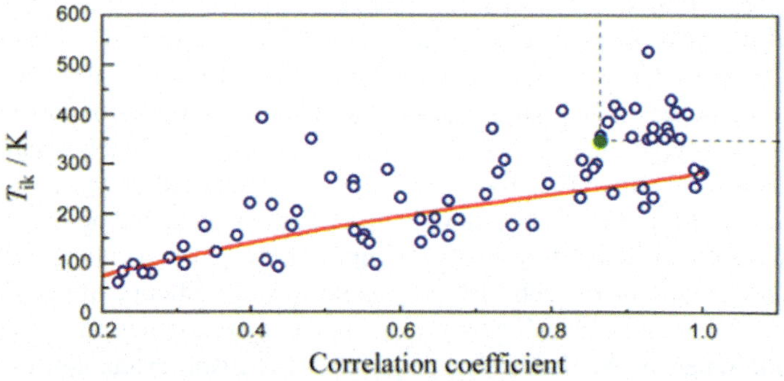

Fig. 8.3 Isokinetic temperature as a function of the correlation coefficient for the application of the *p*-test to a reaction series. Red lines: maximum probability curves. Blue points: simulation data. The dashed lines show the limits of the quadrants corresponding to potentially valid simulations. See details in [39]. With permission from American Chemical Society

of the enthalpy ΔH versus entropy ΔS plot to determine the two essential dimensionless parameters K and CQF. The parameter K indicates the position (in inverse temperature) of the coalescence region of Van 't Hoff plots and CQF is a quantitative measure of the smallest spread of the Van 't Hoff plots. The position of the (K, CQF) couple with respect to universal confidence contours determined from a large number of simulations of random Van 't Hoff plots indicated whether or not the ΔH–ΔS compensation is a statistical artifact. In work [45] on the base of isothermal titration calorimetry (ITC) measurements, statistical analysis of the selected experimental thermodynamic data was performed. Data on a variety of 171 protein–peptide, protein–carbohydrate, and protein–nucleotide interactions in addition to natural and synthetic inhibitors involving 32 proteins were analyzed. $\Delta\Delta H$ versus $\Delta\Delta G$ plots revealed a wide variation in the extent of compensation for different ligand modifications. Figure 8.4 illustrates a strong diagonal distribution in a ΔH versus $T\Delta S$ plot with a Pearson's correlation coefficient of 0.97. This coefficient is a measure the strength and direction of the relationship between two variables.

The slope of the $\Delta\Delta H$ versus $T\Delta\Delta S$ plot for data drawn from the 32 proteins was found to be close to unity for both the experimental data (1.01) and for the constrained random model (1.04) for same number of data points (Fig. 8.5).

Work [13] reviewed how to deal with kinetic and thermodynamic compensations physically, mathematically and statistically. Statistical analysis of data from 32 diverse proteins showed a significant and widespread tendency to compensation. $\Delta\Delta H$ versus $\Delta\Delta G$ plots revealed a wide variation in the extent of compensation for different ligand modification.

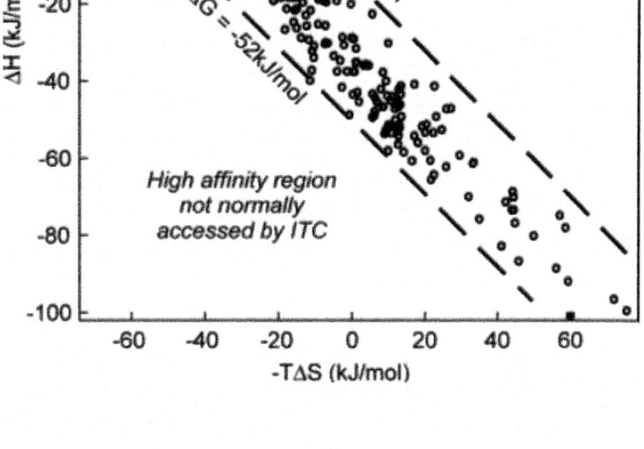

Fig. 8.4 Thermodynamic parameters derived from ITC experiments for 171 protein–ligand interactions. The limitations of experimental procedures constrain measurements to the affinity window bounded by the dashed lines [45]. With permission from Wiley

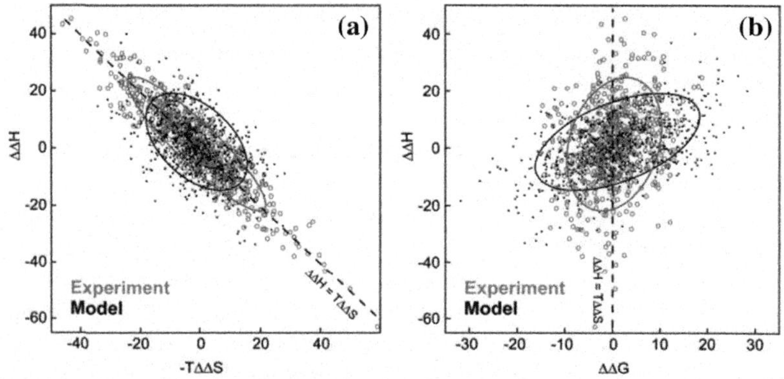

Fig. 8.5 The experimental relative thermodynamics of all pairs of ligands binding each protein have features consistent with the presence of enthalpy–entropy compensation. **a** $\Delta\Delta H$ versus $T\Delta\Delta S$ plot of all experimental differences for pairs of ligands binding to each protein (open circles) superimposed on a sample drawn from a constrained random model (points). The ellipses surround 75% of the experimental data ordered by difference from the mean (gray ¼ experimental data, black model). The experimental distribution is narrower than would be expected in the absence of compensation. **b** An alternative $\Delta\Delta H$ versus $\Delta\Delta G$ plot of the same data. The experimental data are closer in average behavior to that expected for an underlying compensation mechanism, but not equivalent to the fully compensated model [45]. With permission from Wiley

8.4 General Theoretical Considerations

Theory Gaussian distribution of internal vibration energy levels to explain changes in internal enthalpy ($\Delta\Delta H = 0.8$ kcal mol^{-1}) and entropy ($\Delta T\Delta S = 0.8$ kcal mol^{-1}) for a series of perturbations of different energy levels was developed [42]. The general form of its probability density function is

$$f(x) = \frac{1}{\sigma\sqrt{2\pi}} e^{-\frac{1}{2}\left(\frac{x-\mu}{\sigma}\right)^2}$$

The parameter μ is the mean of expectation of the distribution while the parameter is its standard deviation. Concerning the problem considered here, $x = \Delta H$, ΔS and μ is the correspondent mean or expectation of the distribution. Data on occupancy probabilities [$P(U)$] for a system with a Gaussian distribution of energy levels ($T =$ 298 K; $\sigma = 1.5$ kcal mol^{-1}, correlated changes in internal energy (U) and entropy (TS) for a series of perturbations of different energy levels, and H/S compensation for a hydrogen bond modeled by a Morse potential were presented in a separate figure and discussed in details.

According to [46], the nature of EEC can be connected with the behavior of phase volume of systems, **Ω**. A phase space is a space in which all possible states of a system are represented, with each possible state corresponding to one unique point in the phase space. The role of other thermodynamic parameters, which describe series in manifestation of this or that types of the ΔH compensation, was discussed in light of molecular statistical mechanics. Specifically, it was shown that the nature of the $\Delta H - \Delta S$ compensation can be associated with the behavior of the phase volume of systems, Ω, namely, the $\Delta H - \Delta S$ compensation is realized in the case if only in a series of systems the value of Ω is constant. The $\Delta H - (\Delta S + R\Delta \ln \Omega)$ compensation take place in the case of variable phase volume in a series of systems, that is, $\delta\Delta H/(\delta\Delta S + \mathrm{R}\delta\Delta \ln \Omega) = \mathrm{Ti}$. In different series of systems, the following types of compensation were revealed:

$$\Delta H = T_i \Delta S - T_i R\Delta \ln \Omega = \beta \Delta S + \alpha$$

$$\Delta H = T_i(\Delta S + R\Delta \ln \Omega) - T_i R\Delta \ln \sqrt{M}$$

$$\Delta H = T_i(\Delta S + R\Delta \ln \Omega) - T_i R\Delta \ln Mn$$

M is some constant characterizing a state. These three types of compensation effect received analytical argumentation and have experimental confirmation.

8.5 Miscellaneous Examples

The study on single-ion hydration thermodynamics from clusters to bulk solution [47] was based on the cluster-pair-based approximation (CPA) technique. The resulting hydration thermodynamic properties of ions were obtained by the simulation methodology and revealed enthalpy–entropy compensation relationship. Value $T\Delta_{\mathrm{aq}}S^{\mathrm{con}}/\Delta\mathrm{aq}H^{\mathrm{con}} = 0.09$ indicated a predominance of the enthalpy factor over the entropy factor during hydration of ionic species (Fig. 8.6).

Hydrophobicity is the association of nonpolar groups or molecules in an aqueous environment which arises from the tendency of water to exclude nonpolar molecules [48]. When two nonpolar surfaces associate with each other, ordered molecules of water solvating each surface are released to the bulk yields a favorable change in entropy and unfavorable change in enthalpy and an unfavorable change in enthalpy leading to enthalpy/entropy (H/S) compensation. This effect has been illustrated by the example thermodynamic properties of the solvation process thermodynamic properties of the amino acids [49]. Solvation theory simulations of amino acids surrounded by water molecules were performed to calculate thermodynamic properties of the solvation process of these amino acids on a grid around the solute. The results for the free energy, the total enthalpy, and the overall entropy for the three different water models TIP3P, TIP4P, and TIP5P were tabulated (Table 8.1).The results showed that: (1) The free energy directly correlates with the hydrophilicity/hydrophobicity of the given molecule; (2) For hydrophobic amino acids the thermodynamic values change in the following ranges: $\Delta\Delta F = 9$ kcal mol^{-1}, $\Delta\Delta H = 13$ kcal mol^{-1}, $\Delta(-T\Delta S) = 8$ kcal K^{-1} mol^{-1}, and $\Delta(-T\Delta S)/\Delta\Delta H = 0.61$; (3) For hydrophilic amino acids these values are $\Delta\Delta \mathrm{F} = 9$ kcal mol^{-1}, $\Delta\Delta H = 13$ kcal mol^{-1}, $\Delta(-T\Delta S) = 6$ kcal K^{-1}mol^{-1}, and $\Delta(-T\Delta S)/\Delta\Delta H = 6/13 = 0.45$. Thus, in both series, the enthalpic factor of the

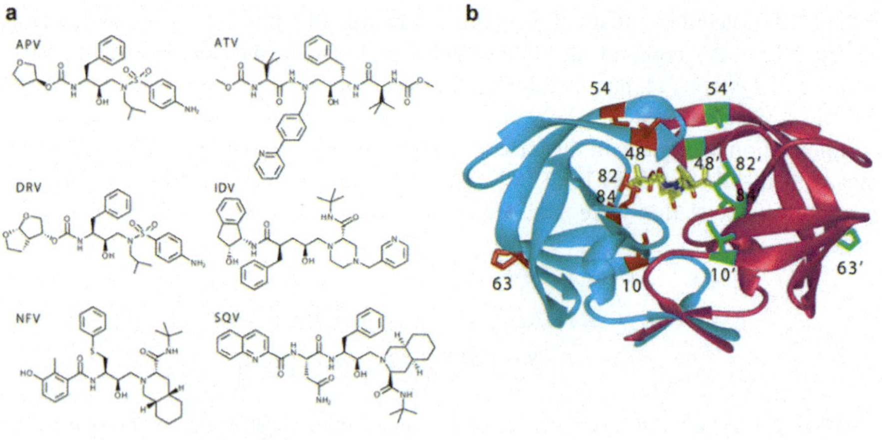

Fig. 8.6 Structure of inhibitors and HIV-1 protease. **a** Chemical structures of inhibitors. **b** Overview of the mutation sites of Flap+ and Act mutants mapped on an HIV-1 protease dimer. The monomers are distinguished in cyan and magenta, while the inhibitor ATV is shown in yellow stick model. The mutation sites of Flap+ (L10I/G48V/I54V/V82A) and Act (V82T/I84V) along with the site of the natural polymorphism L63P are highlighted in red and green stick models [61]. With permission from American Chemical Society

influence of the structure prevails over the entropy factor and, to a greater extent, for charged amino acids.

A formalism for treating the solvent reorganization that accompanies chemical reactions and physical processes in liquid solutions was developed [31]. Suggested equations were separated into a conventional chemical equation and an environmental equation. It was shown that ΔH_{env} and ΔS_{env} can be substantial, especially when the solvation involves hydrogen bonding. Therefore, environmental ΔH_{env} and ΔS_{env} can be part of, ΔH° and ΔS° for the overall process. When $\Delta H_{env} \gg \Delta G^{\circ}$, the plot of ΔH° versus ΔS° is nearly a straight line, with a slope close to the experimental temperature. Data on compensation phenomena between the enthalpy and entropy changes of the reduction reaction for all classes of electron transport metalloproteins, namely cytochromes, iron-sulfur, and blue copper proteins were presented and discussed [52]. Following Grunwald's approach for the interpretation of *H*/*S* compensation for solution reactions, it was concluded that reduction-induced solvent reorganization effects involving the hydration shell of the molecule dominate the reduction thermodynamics.

A combined terahertz and femtosecond infrared spectroscopic study of water dynamics around different ions (specifically magnesium, lithium, sodium, and cesium cations, as well as sulfate, chloride, iodide, and perchlorate anions) reveals that the effect of ions and counterions on water can be strongly interdependent and nonadditive, and in certain cases extends well beyond the first solvation shell of water molecules directly surrounding the ion [8]. To know how changes in the organization of water within a binding pocket can alter the thermodynamics of protein-ligand association, mutants of human carbonic anhydrase (HCAII) were examined [53]. The

Table 8.1 Thermodynamic quantities: free energy, enthalpy, and entropy for all 20 amino acids calculated with the amber force field and three water models [49]

Amino acid	TIP3P		
	ΔH	$-T\Delta S$	ΔG
ASP(−)	− 80.5 ± 0.3	26.4 ± 0.0	− 54.2 ± 0.3
GLU(−)	− 78.3 ± 1.9	27.7 ± 0.2	− 50.6 ± 1.7
LYS(+)	− 66.1 ± 2.5	22.8 ± 0.7	− 43.4 ± 1.9
ARG(+)	− 64.7 ± 2.1	24.1 ± 0.5	− 40.7 ± 1.6
HIS(+)	− 57.7 ± 1.6	21.5 ± 0.3	− 36.2 ± 1.3
GLN	− 35.8 ± 1.1	20.6 ± 0.4	− 15.2 ± 0.7
HIS	− 36.0 ± 1.0	21.1 ± 0.2	− 14.9 ± 0.8
ASN	− 33.8 ± 4.0	19.1 ± 0.9	− 14.7 ± 3.2
SER	− 28.0 ± 1.2	15.6 ± 0.1	− 12.4 ± 1.1
TYR	− 32.4 ± 2.1	20.2 ± 0.3	− 12.2 ± 1.9
TRP	− 33.8 ± 0.5	22.3 ± 0.2	− 11.5 ± 0.4
THR	− 27.1 ± 1.1	16.7 ± 0.4	− 10.3 ± 0.9
MET	− 27.6 ± 0.6	18.1 ± 0.1	− 9.5 ± 0.5
PHE	− 29.1 ± 0.1	19.9 ± 0.2	− 9.2 ± 0.1
CYS	− 23.9 ± 2.0	15.3 ± 0.4	− 8.5 ± 1.7
GLY	− 22.4 ± 0.4	14.0 ± 0.1	− 8.4 ± 0.4
ALA	− 22.5 ± 0.8	14.8 ± 0.1	− 7.7 ± 0.7
ILE	− 25.4 ± 0.5	18.1 ± 0.1	− 7.3 ± 0.6
VAL	− 24.5 ± 0.8	17.2 ± 0.3	− 7.3 ± 0.5
LEU	− 24.9 ± 0.7	17.8 ± 0.3	− 7.1 ± 0.5
PRO	− 22.9 ± 0.8	16.0 ± 0.2	− 6.9 ± 0.6

[a] Values are given in kcal mol^{-1}
With permission from Elsevier

association of a protein and ligand in buffered aqueous solution can be considered as the sum of nine processes [53]. Five of them are: (1) The formation of protein–ligand contacts; (2) The rearrangement of water initially solvating the protein; (3) The rearrangement of water initially solvating the ligand; (4) The formation of a hydration structure around the protein–ligand complex; and (5) Changes in interactions associated with buffer ions can be accompanied by the reorganization of a large number of water molecules and therefore can show significant compensation. For the other four processes, one should not expect a wide range of compensatory effects: (1) Changes in the conformation of the protein); (2) Changes in the conformation of the ligand; (3) changes in the dynamics of the protein; and (4) Changes in the dynamics of the ligand. In addition, the influence of differences in ligand structure—properties of human carbonic anhydrase were examined. The ligands 1,3-thiazole-2-sulfonamide (TA) and benzo[d]thiazole-2-sulfonamide (BTA) differed by a benzene ring but not in binding geometry, revealed the thermodynamic contribution of the benzene ring to binding (Fig. 8.7).

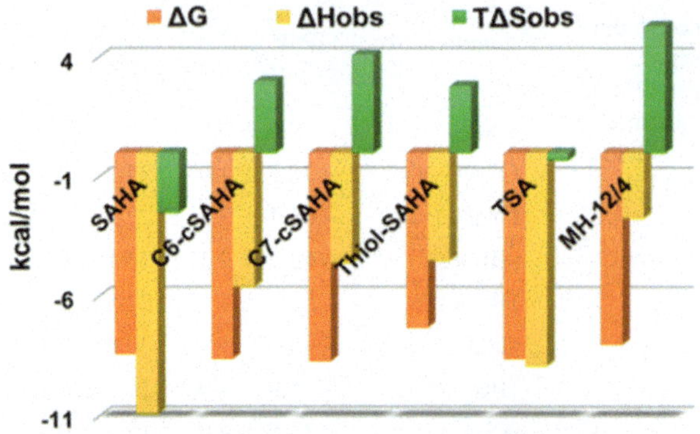

Fig. 8.7 Thermodynamic signatures for the binding of the different SAHA analogues to HDAC8. Despite the marked differences in the binding thermodynamic signatures of these ligands, their binding free energies are nearly the same, highlighting an enthalpy–entropy compensation effect [59]. Published under an ACS Author Choice License

To examine how changes in the organization of water within a binding pocket can alter the thermodynamics of protein-ligand association, mutants of human carbonic anhydrase (HCAII) were examined [54]. Results from calorimetric, crystallographic, and theoretical analyses suggested that most mutations strengthen networks of water-mediated hydrogen bonds. In addition, reduce binding affinity by increasing the enthalpic cost and, to a lesser extent, the entropic benefit of rearranging those networks during binding took place.

In work [55], experimental evidence of enthalpy/entropy (H/S) compensation was presented in a figure and discussed in detail in text. The authors concluded that within the confines of the binding pocket of HIV-1 protease is *HCAII*, binding events associated with enthalpically favorable rearrangements of water. HIV-1 protease is a (retropepsin), an enzyme involved with peptide bond hydrolysis in retroviruses. To show that EEC in HIV-1 protease is influenced by the molecular conformational changes (CCs) of the entire system comprising the solute and by solvent reorganization, free energy perturbation formalism was used. Perturbation theory comprises methods for finding an approximate solution to a problem, by starting from the exact solution of a related, simpler problem. The internal changes of enthalpy and the entropy due to CCs upon modifying the interactions compensate each other. Using molecular simulations, it was shown that, for solvation of six different HIV-1 protease inhibitors, CCs in the solute cause EEC as large as 10–30 kcal mol^{-1}.

Enthalpy–entropy compensation during the association of two molecules was studied by minimizing model systems with molecular mechanics (MM) or quantum mechanics (QM) [56]. Translational, rotational, and vibrational contributions to the enthalpy and entropy with standard statistical-mechanics methods were considered using the rigid-rotor harmonic-oscillator approximation. EEC was observed, owing to the loss of translational and rotational entropy. Linear relations between $T\Delta S$ and ΔH with slopes of 0.1–1.7 (but $\sim$ 0.01 for ionic and covalent interactions) were observed within homologous series. It was predicted that for a real binding reaction in water solution, both enthalpy–entropy compensation and anti-compensation can be also observed.

To estimate the contribution from water to the free energy of binding of a series of structurally homologous heterocyclic aromatic sulfonamides to human carbonic anhydrase (HCA), the WaterMap method (*Schrödinger Inc.*) was utilized [57]. The method allowed to predict the positions, enthalpies, and entropies of the water molecules in the binding pockets of the complexes of HCA–**T**, HCA–**BT**, HCA–**F**, and HCA–**BF, were F, T and BF:**

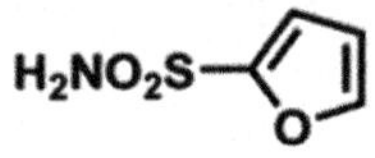

F

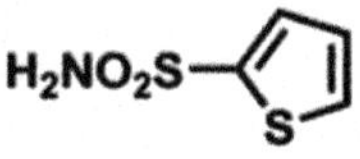

T

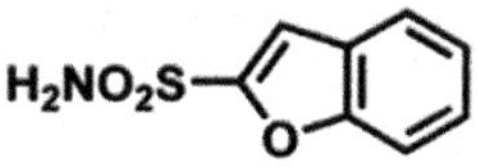

BF

Thermodynamic parameters calculated by WaterMap for benzo-extension were consistent with those determined experimentally by isothermal titration calorimeters (ITC) and the calculation indicated that the gain in binding free energy comes from enthalpic stabilization. Average values for converting F to BF and T to BT were $\Delta\Delta G^{\circ}_{\mathrm{WM;benzo}} = -2.8$ kcal mol^{-1}, $\Delta\Delta H_{\mathrm{WM;benzo}} = -3.0$ kcal mol^{-1}, and $-T\Delta\Delta S_{\mathrm{WM;benzo}} = 0.2$ kcal mol^{-1}.

EEC from a new standpoint was reviewed [50]. The authors came to conclusion that large fluctuations in enthalpy and entropy observed are too great to be a result of only conformational changes and must result, to a considerable degree, from variations in the amounts **of** water immobilized or released on forming complexes. A structural link between the amount of bound water and the protein flexibility and substantial changes in solvation-an intrinsically compensatory process was assumed. Experimental, theoretical, and molecular dynamics (MD) simulations showed that changes in water properties can be observed at distances of more than 10 Å from a biomolecule [50]. The authors concentrated on the effect of protein chain flexibility and on the properties of hydration water, using short peptides as a model of 18 linear peptides with the sequence (XXGG) $\times$ 5, where X represents one of the common amino acids, other than glycine and proline were constructed. It was demonstrated: (1) How restricting the chain flexibility can affect the structural, dynamic, and thermodynamic properties of hydration water; (2) How restricting the peptide dynamics can slow down the translational motions of water molecules to a distance of at least 12–13 Å; and (3) How analysis of the "slow" water molecules (residence time $\geq$ 100 ps) together with a thermodynamic analysis of water within 4.5 Å of the peptide can be performed.

In the study [61], the crystal structures and thermodynamics were compared for the binding of inhibitors APV, atazanavir (ATV), DRV, indinavir (IDV), nelfinavir (NFV), and saquinavir (SQV) to the wild-type (WT) HIV-1 protease and two multi-drug-resistant (MDR) variants (Fig. 8.6). A variant (Flap+) has a combination of flap and active site mutations and exhibits extremely large entropy–enthalpy compensation compared to wild-type protease, 5–15 kcal mol^{-1}, while losing only 1–3 kcal mol^{-1} in total binding free energy for any of six FDA approved inhibitors.

Detailed thermodynamic studies of the binding of structurally similar ligands, which differed with respect to the "cap," "linker," and "metal-binding" regions of the suberoylanilide hydroxamic acid (SAHA) pharmacophore

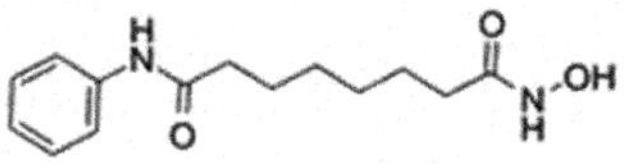

to human histone deacetylase HDAC8, were carried out (Fig. 8.7) [59]. The presented experimental data led to the following conclusions: (1) The enthalpic ($\Delta H°$) and entropic ($T\Delta S°$) contributions for the binding of the different SAHA analogues with varying "cap", "linker", and "metal-binding" regions to HDAC8 are markedly different; (2) Their standard free energy changes ($\Delta G°$) are identical among hydroxamate-containing ligands; (3) The enthalpy–entropy compensation effect accounts for the changes in the temperature dependence in the parameters of binding of SAHA analogues to HDAC8; (4) Within the given series of analogues the enthalpic ($\Delta H°$) changes from 10.95 kcal mol^{-1} to 2.72 kcal mol^{-1}, while the entropic ($T\Delta S°$) changes from – 2.52 to 5.3 kcal mol^{-1} (Fig. 8.8); and (5) The main contribution to these parameters is made by the process of reorganization of weakly bound water.

Ligands of human carbonic anhydrase (HCA) and a series of benzothiazole sulfonamide ligands with different patterns of fluorination were used to define enthalpy/entropy (*H*/*S*) compensation in this system [60]. The data analysis revealed that differences in the structure and thermodynamical properties of the waters surrounding the bound ligands are an important contributor to the observed *H*/*S* compensation. Results supported the hypothesis that the molecules of water filling the active site of a protein, and surrounding the ligand, are as important as the contact interactions between the protein and the ligand for biomolecules recognition, and in determining the thermodynamics of binding. In the study [61] changes were observed in the thermodynamics of inhibitors bounding to a drug-resistant variant (Flap+ protease) compared to wild-type HIV-1 protease (Fig. 8.8). The co-crystal structures of Flap+ protease with four of the inhibitors were determined and compared. Taking into account that the change in the enthalpy of one water molecule in the "melting" of the **water cluster** in the process of binding the substrate $\Delta H_0 = 1.4$ kcal mol^{-1}, it can be concluded that the cluster can include up to 10 water molecules.

Experimental data on entropy–enthalpy compensation plots with a high correlation coefficient for calcium binding to proteins, protein unfolding, hydrogen exchange in cytochrome c, and alkane solvation are indicated in Fig. 8.9 [42]. The analysis of the data presented in the figure led to the following values of the relative thermodynamic parameters in the systems indicated in the caption to the figure.

(A) $\delta T\Delta S = 35$ kcal mol^{-1}, $\delta\Delta H = 37$ kcal mol^{-1}, $\delta T\Delta S/\delta\Delta H = 0.95$.
(B) $\delta T\Delta S = 2.2$ kcal mol^{-1}, $\delta\Delta H = 2.2$ kcal mol^{-1} $\delta T\Delta S/\delta\Delta H = 1.0$.
(C) $\delta T\Delta \mathrm{S} = 42$ kcal mol^{-1}, $\delta\Delta H = 44$ kcal mol^{-1}, $\delta T\Delta S/\delta\Delta H = 1.0$.
(D) $\delta T\Delta S = 7$ kcal mol^{-1}, $\delta\Delta H = 14$ kcal mol^{-1}, $\delta TDS/\ \delta DH = 0.5$.

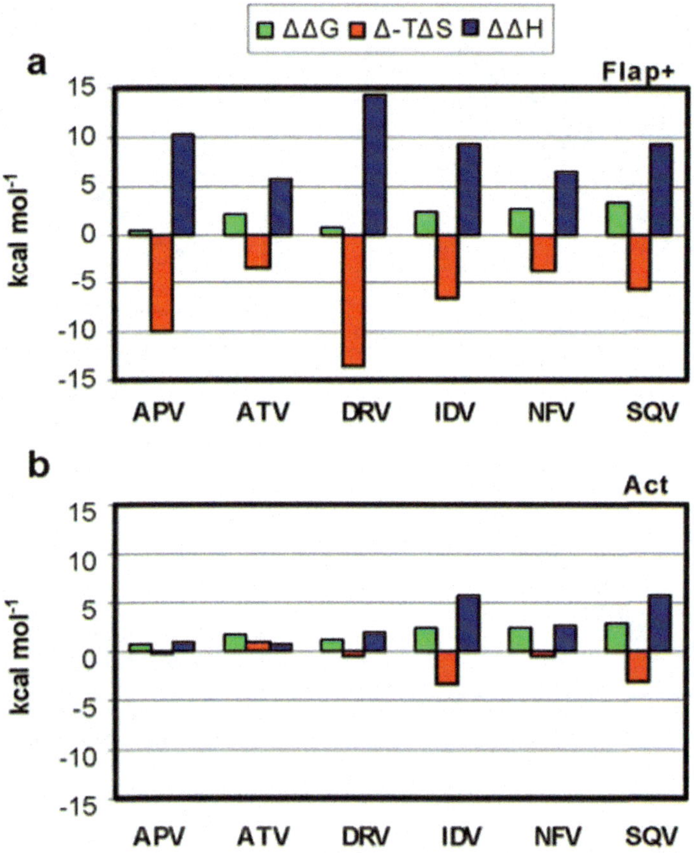

Fig. 8.8 Thermodynamics of inhibitor binding. Differences in binding energetics between **a** WT and Flap+ and **b** WT and r Act protease Act variants. The differences in ΔG, TΔS, and ΔH are shown in green, red, and blue, respectively [61]. With permission from American Chemical Society

Thus, the data on proteins show an almost complete mutual compensation of the enthalpy and entropy factors, while the enthalpy factor prevails in the solvation of alkane.

The results obtained in the work [12] by titrations of human nucleoside phosphorylase (PNP) using isothermal calorimetry (indicated)в the following: (1) Binding of a structurally rigid first-generation Immucillin ($K_d = 56$ pM) is driven by large negative enthalpy values ($\Delta H = -21.2$ kcal mol^{-1}) with a substantial entropic ($-T\Delta S$) penalty; (2) Flexible inhibitors bind with high affinity because of reduced entropic penalties; (3) Entropic penalties arise from conformational freezing of the PNP·inhibitor complex with the entropy term dominated by protein dynamics (Fig. 8.10); (4) The conformationally flexible Immucillins reduce the system entropic penalty; (5) Disrupting the ribosyl 5′-hydroxyl interaction of transition state analogues with PNP causes favorable entropy of binding; (6) Tight binding of the 17 Immucillins

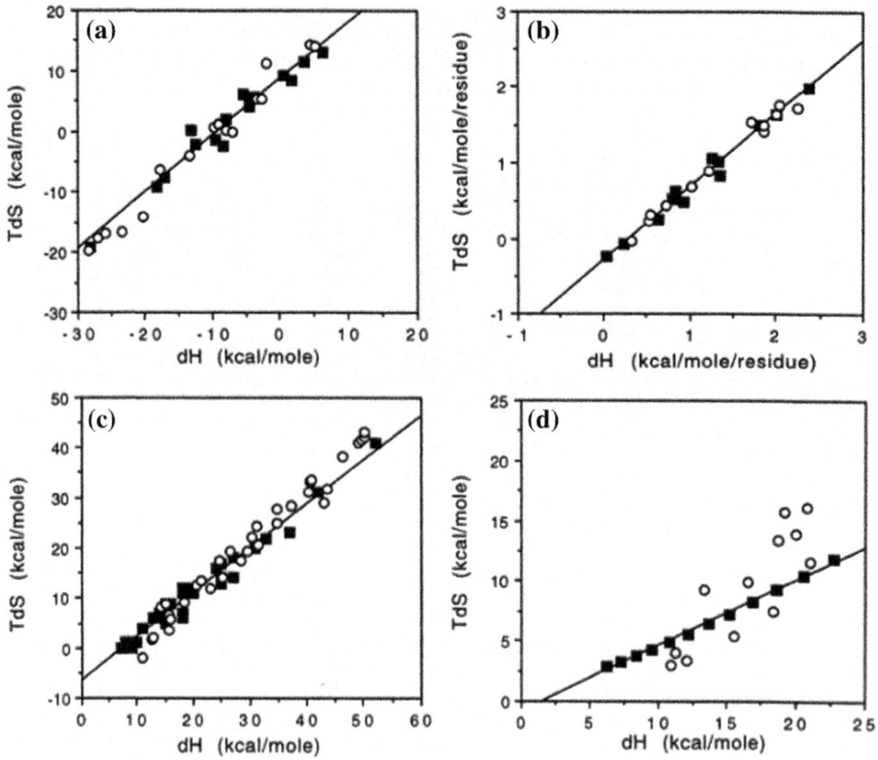

Fig. 8.9 Entropy (expressed as TS at 298 K)–enthalpy compensation plots for **a** calcium binding to proteins, **b** protein unfolding, **c** hydrogen exchange in cytochrome c, and **d**) alkane solvation. Using randomly generated ΔH values with experimental ΔG's (○) level spreads (SD, σ) of 0.5 (×), 1 (□), 2 (•), and 3 (×) kcal mol^{-1}, respectively [42]. With permission from Wiley

is characterized by large enthalpic contributions; and emphasizing their similarity to the transition state; and (7) Via introduction of flexibility into the inhibitor structure, the enthalpy−entropy compensation pattern is altered to permit tighter binding.

Thermolysin is metalloproteinase catalyzing the hydrolysis of peptide bonds containing hydrophobic amino acids. The enzyme was used to determine how water near the surface of a protein–ligand complex contributes to the thermodynamics of binding, as measured by isothermal titration calorimetry [62]. In this study, the binding modes of nine congeneric phosphonamidate-type inhibitors with systematically varied, size-increasing hydrophobic substituents were analyzed on the base of high-resolution crystal structures data profiles. Overall, $\Delta\Delta G$ spreads over 7.0 kJ mol, $\Delta\Delta H$ varies by 15.8 kJ mol^{-1}, and $-T\Delta\Delta S$ by 12.1 kJ mol^{-1} with $T\Delta\Delta S/\Delta\Delta H = 0.77$. Thus, in this series the enthalpy component somewhat prevails over the corresponding entropy component. In addition, ligands that stabilized networks of water on the surface of the protein–ligand complex had more favorable enthalpies of binding, and less favorable entropies of binding, than ligands that destabilized those

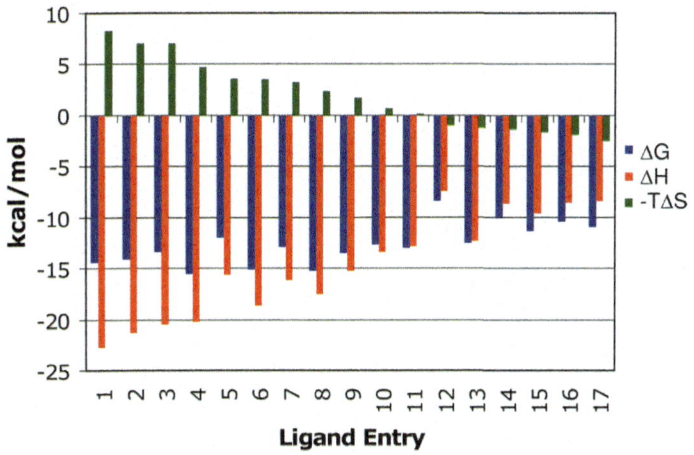

Fig. 8.10 Thermodynamic signatures of a variety of Immucillins binding to the first subunit of human PNP at 300 K: ΔG (*blue*), ΔH (*red*), and $-T\Delta S$ (*green*), exhibiting EEC compensation. **b** Plot of the data showing that as the enthalpy change $\Delta\Delta H$, becomes more positive by about 16 kcal mol^{-1}, the entropy factor $-T\Delta S$ compensates by becoming more negative by about 11 kcal mol^{-1} [12]. With permission from American Chemical Society

networks The following examples of enthalpy/entropy compensation resulting from the entropic cost of enthalpically favorable contacts were described and analyzed [53]: (1) Two inhibitors of HIV-1 protease that differ by a sulfonyl group and, thus, in their ability to form a sulfonyl–amide bond with the protein backbone; (2) A comparison of B-factors between the two protein–ligand complexes shows that the additional hydrogen bond rigidifies the protein–ligand complex; (3) Peptide mimics with different cycloalkyl moieties; (4) The ligands bind trypsin with similar free energies (G∘b) but large differences in enthalpy and entropy of binding (H∘b and $-TS$∘b, respectively); (5) Molecular dynamics simulations suggest that favorable entropies correspond to ligands that are more "mobile" in the binding pocket; (6) A zinc finger peptide on which pairs of charged residues at positions 3 and 10; (7) Enthalpies and entropies of interaction between charged residues (H∘b and $-TS$∘b, respectively; each calculated from thermodynamic cycles of cation binding to incrementally varied pairs of residues) suggest strong compensation; free energies of interaction (G∘b) vary little between pairs. Computational analysis of the most enthalpically stable pair indicated that it is also the most rigid. Abbreviations: *B*, B-factor; *G*, Gibbs free energy; *H*, enthalpy; *TS*, entropic component of free energy.

To show that EEC in HIV-1 protease is influenced by the molecular conformational changes (CCs) of the entire system comprising the solute and solvent reorganization, free energy perturbation formalism was used [55]. The internal changes of enthalpy and the entropy due to CCs upon modifying the interactions cancel each other. Using molecular simulations, it was shown that, for solvation of six different HIV-1 protease inhibitors, CCs in the solute cause EEC as large as 10–30 kcal mol^{-1}.

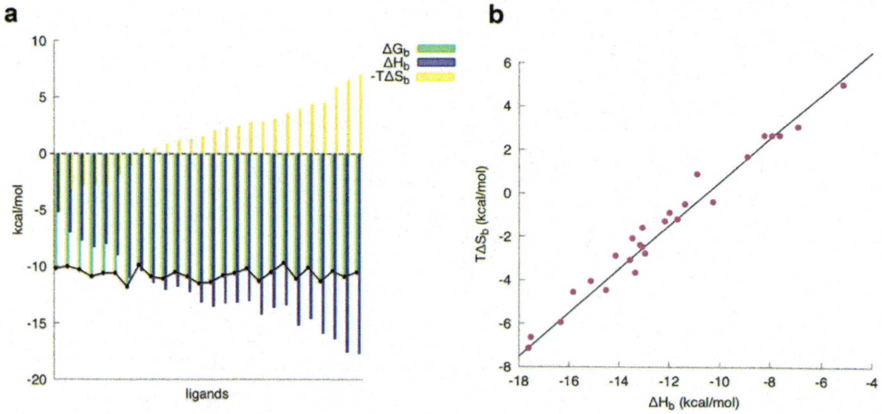

Fig. 8.11 **a** Binding enthalpy (ΔH_b, blue bars), entropy ($-T\Delta S_b$, yellow bars), and free energy (ΔG_b, green bars and black curve) measured experimentally by ITC for a series of ligands targeting HMG-CoA reductase. Despite the large variations observed in both enthalpic and entropic terms, the binding affinity remains nearly constant across the whole data set due to *H*/*S* compensation. **b** Plot of enthalpy (ΔH_b) versus entropy ($T\Delta S_b$) terms showing the linear relationship between them [64]. Published by American Chemical Society

Mini-review [64] provided a short survey of the enthalpy–entropy compensation phenomenon in the simulation of biomacromolecular recognition from the point of view of practical computational chemistry. Ligand binding found particular emphasis. A series of examples to illustrate different approaches to interpret and predict compensation phenomena at an atomistic level was presented. A typical example of ligands binding enthalpy, entropy, and free energy for a series of ligands targeting 3-hydroxy-3-methyl-glutaryl-coenzyme A reductase, (HMG-CoA reductase) is shown in Fig. 8.11. It follows from the figure that in this series there are almost equal enthalpy and entropy contributions, and the main contribution is made by the process of reorganization of the **water ensemble**.

EEC from a new standpoint was reviewed [29]. Assuming EEC to be a physical correlation a linear relationship between entropy *S* and enthalpy *H*, an equation $H + a{*}S + b$, was proposed with the slope $a = T_{comp}$ being the compensation temperature and the intercept *b* as an energy-related quantity. For example, the DNA ligands DAP (2,7-diazapyrene), MDAP (*N*-methyl-2,7-diazapyrene monocation), and DMDAP (*N*,*N*-dimethyl-2,7-diazapyrene dication) upon binding to DNA showed the fitting line $H + a{*}S + b$ and plot of ¢*H* versus −*T*¢*S* (with $T = 298$ K). Specifically, for the diazapyrene homological row $T_{comp} = 282$ K and $b = -6.78$ kcal mol^{-1}, and for the piperazinylcarbonyloxyalkyl ligands, $T_{comp} = 361$ K and $b = -7.63$ kcal mol^{-1}. The authors stressed that changes of structured water at hydrophobic surfaces, conformational transitions upon ligand-biopolymer binding, and so forth can underlie physically valid EEC.

Analysis of the experimental binding thermodynamics for approximately 100 protein–ligand complexes provided insights into the factors governing ligand affinity

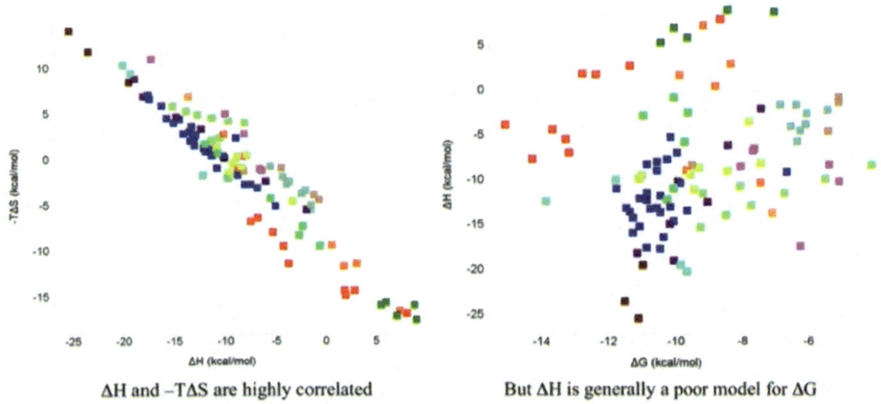

Fig. 8.12 Plot of enthalpy (ΔH) and entropy ($-T\Delta S$) versus free energy (ΔG) [65]. Published by American Chemical Society

[65]. These data indicated correlation between enthalpy and $-T\Delta S$ observed for this diverse data set (Fig. 8.12). Nevertheless, the interaction enthalpies in the figure show no correlation with free energy. The trends observed for ligand efficiencies with respect to molecular size are a consequence of enthalpic, not entropic, effects.

The interaction of pairs of oppositely charged residues on zinc finger **peptides** by measuring the binding of cations to those pairs was studied using isothermal titration calorimetry **(ITC)** [66]. For example, analysis of the binding **of** Cu^{2+} to Ser3Ser10, Ser3Asp10, Lys3Ser10, and Lys3Asp10 was used to study the binding of Lys3 and Asp10 to one another. ITC measurements revealed a range of over 3 kcal mol^{-1} in relative binding enthalpies. Double mutant cycle analysis indicated a range of interaction enthalpies ranging from − 3.1 to − 3.4 kcal mol^{-1} for the Arg-Asp pair to − 0.8 kcal mol^{-1} for the Lys-Glu pair. The magnitudes of the entropy–enthalpy compensation effects were consistent with the formation of a structurally rigid Arg-Asp contact ion pair.

References

1. Leffler JE (1955) The enthalpy–entropy relationship and its implications for organic chemistry. J Org Chem 20:1202–1231
2. Exner O (1973) The enthalpy–entropy relationship. Prog Phys Org Chem 10:411–482
3. Likhtenshtein GI (1966) Regularities in the energy and entropy properties of enzymatic processes. Biofizika 11:24–32
4. Likhtenshtein GI (1970) On the nature of compensation phenomena in liquid phase reactions. Russ J Phys Chem 44:1980–1990
5. Lumry R, Rajender S (1970) Enthalpy–entropy compensation phenomenon in water solutions of proteins and small molecules: ubiquitous property of water. Biopolymers 9:1125–1227

6. Lumry R, Gregory B (1986) Free energy management in protein reactions: concepts, complications, and compensation. In: Welch GR (ed) The fluctuating enzymes. Wiley, New York
7. Horinek D, Mamatkulov SI, Netz RR (2009) Rational design of ion force fields based on thermodynamic solvation properties. J Chem Phys 130:124507
8. Tielrooij K-J, Garcia-Araez N, Bonn M, Bakker HJ (2010) Cooperativity in ion hydration. Science 328(5981):1006–1009
9. Dragan AI, Read CM, Crane-Robinson C (2017) Enthalpy–entropy compensation: the role of solvation. Eur Biophys J 46:301–308
10. Cornish-Bowden A (2017) Enthalpy–entropy compensation and the isokinetic temperature in enzyme catalysis. J Biosci 42:665–670
11. Schauperl M, Podewitz M, Ortner TF, Waibl F, Thoeny A, Loerting T, Liedl KR (2017) Balance between hydration enthalpy and entropy is important for ice binding surfaces in antifreeze proteins. Sci Rep 7:11901
12. Edwards JM, Mason JM, Clinch K et al (2009) Altered enthalpy–entropy compensation in picomolar transition state analogues of human purine nucleoside phosphorylase. Biochemistry 48(23):5226–5238
13. Garvín A, Ibarz R, Ibarz A (2017) Kinetic and thermodynamic compensation: a current and practical review for foods. Food Res Int 2017:132–153
14. Likhtenshtein GI (1988) Chemical physics of redox metalloenzymes. Springer-Verlag, Heidelberg
15. Likhtenshtein GI (2003) New trends in enzyme catalysis and mimicking chemical reactions. Kluwer Academic/Plenum Publishers, New York
16. Denisov ET, Sarkisov OM, Likhtenshtein GI (2003) Chemical kinetics. Fundamentals and recent developments. Elsevier Science, Amsterdam
17. Likhtenshtein GI (1930) Biophysical labeling methods in molecular biology. Cambridge University Press, Cambridge, NY
18. Likhtenshtein GI (1976) Study of protein dynamics by spin-labeling, Mösbauer spectroscopy, and NMR. In: Losche A (ed) Special Collogue Amper on dynamic processes in molecular systems. Karl-Marx University, Leipzig, pp 100–107
19. Likhtenshtein GI (1976) Water and protein dynamics. In: Alfsen A, Bertran AJ (eds) L'eau et les Systemes Biologoque. CNR, Paris, pp 45–43. Likhtenshtein GI (1974) Spin labeling method in molecular biology. Nauka, Moscow
20. Likhtenshtein GI, Febbrario F, Nucci R (2000) Intramolecular dynamics and conformational transitions in proteins studed by biophysical labeling methods. Common and specific features of proteins from thermophylic microorganisms. Spectrochem Acta Part A Biomol Spectrosc 56:2011–2031
21. Likhtenshtein GI, Troshkina TV (1968) On the kinetics of the cooperative processes in proteins. Mol Biol (Moscow) 2:654–663
22. Likhtenshtein GI, Yamauchi J, Nakatsuji S, Smirnov A, Tamura R (2008) Nitroxides: application in chemistry, biomedicine, and materials science. Wiley-VCH, Weinhem
23. Likhtenshtein GI (1976) Spin labeling method in molecular biology. Wiley Interscience, New York
24. Likhtenshtein GI (2021) Biological water. Physicochemical aspects. Springer Nature, Cham
25. Constable FH (1925) The mechanism of catalytic decomposition. Proc R Soc Lond Ser A 1925(108):355–378
26. Dragan I, Read M, Crane-Robinson C (2017) Enthalpy–entropy compensation: the role of solvation. Eur Biophys J 46(4):301–308
27. Liu L, Yang C, Guo Q-X (2000) A study on the enthalpy–entropy compensation in protein unfolding. Biophys Chem 84:239–251
28. Lumry R (2003) Uses of enthalpy–entropy compensation in protein research. Biophys Chem 105(2–3):545–557
29. Starikov EB, Norden B (2007) Enthalpy–entropy compensation: a phantom or something useful? J Phys Chem B 111(51):14431–14435

30. Ulf R (2014) A fundamental view of enthalpy–entropy compensation. MedChemComm 5(9):1324–1336
31. Grunwald E, Steel C (1995) Solvent reorganization and thermodynamic enthalpy–entropy compensation. J Am Chem Soc 117(21):5687–5692
32. Khrapunov S (2018) The enthalpy–entropy compensation phenomenon. Limitations for the use of some basic thermodynamic equations. Curr Protein Peptide Sci 19(11):1088–1091
33. Starikov E (2020) Entropy–enthalpy compensation. Finding a methodological common denominator through probability, statistics, and physics. Jenny Stanford Publishing, New Delhi
34. Pan A, Biswas T, Rakshit AK, Moulik SP (2015) Enthalpy–entropy compensation (EEC) effect: a revisit. J Phys Chem B 119(52):15876–15884
35. Cornish-Bowden A (2017) Enthalpy–entropy compensation and the isokinetic temperature in enzyme catalysis. J Biosci 42:665–670. https://doi.org/10.1007/s12038-017-9719-0
36. Barrie P (2012) The mathematical origins of the kinetic compensation effect: 2. The effect of systematic errors. J Phys Chem Chem Phys 14(1):327–336. https://doi.org/10.1039/c1cp22667c
37. Starikov EB, Norden B (2012) Entropy–enthalpy compensation as a fundamental concept and analysis tool for systematical experimental data. Chem Phys Lett 2012(538):118–120
38. Freed KF (2011) Entropy–enthalpy compensation in chemical reactions and adsorption: an exactly solvable model. J Phys Chem B 115:1689–1692
39. Perez-Benito JF, Mulero-Raichs M (2016) Enthalpy–entropy compensation effect in chemical kinetics and experimental errors: a numerical simulation approach. J Phys Chem A 120:7598–7609
40. Krug RR, Grieger WG, Hunte RA (1976) Enthalpy–entropy compensation. 2 Separation of the chemical from the statistical effect. J Phys Chem 80:2341–2351
41. Ouvrard C, Berthelot M, Lamer T, Exner O (2001) A program for linear regression with a common point of intersection: the isokinetic relationship. J Chem Inf Comput Sci 41:1141–1144
42. Sharp K (2001) Entropy–enthalpy compensation: fact or artifact? Protein Sci 10:661–667
43. Piguet C (2011) Enthalpy–entropy correlations as chemical guides to unravel self-assembly processes. Dalton Trans 40:8059–8071
44. Griessen R, Dam B (2021) Simple accurate verification of enthalpy–entropy compensation and isoequilibrium relationship. Chem Phys Chem 22(17):1774–178. https://doi.org/10.1002/cphc.202100431
45. Olsson TSG, Ladbury JE, Pitt WR, Williams MA (2011) Extent of enthalpy–entropy compensation in protein–ligand interactions. Protein Sci 20(9):1607–1618
46. Khakhel OA, Romashko TP (2016) Extrathermodynamics: varieties of compensation effect. J Phys Chem A 120(12):2035–2040
47. Vlcek L, Chialvo AA (2016) Single-ion hydration thermodynamics from clusters to bulk solution. Fluid Phase Equilibria 407:58–75
48. Law K-Y (2014) Definitions for hydrophilicity, hydrophobicity, and superhydrophobicity: getting the basics right. J Phys Chem Lett 5(4):686–688. https://doi.org/10.1021/jz402762h
49. Schauperl M, Podewitz M, Waldner BJ, Liedl KR (2016) Enthalpic and entropic contributions to hydrophobicity. J Chem Theory Comput 2016(12):4600–4610
50. Gavrilov Y, Leuchter JD, Levy Y (2017) On the coupling between the dynamics of protein and water. Phys Chem Chem Phys 19:8243
51. Conti Nibali V, D'Angelo G, Paciaroni A, Tobias DJ, Tarek M (2014) On the coupling between the collective dynamics of proteins and their hydration water. J Phys Chem Lett 5:1181–1186
52. Battistuzzi G, Borsari M, Di Rocco G, Ranieri A, Sola M (2004) Enthalpy/entropy compensation phenomena in the reduction thermodynamics of electron transport metalloproteins. J Biol Inorg Chem 9(1):23–26
53. Fox JM, Zhao M, Fink MJ, Kang K, George M, Whitesides GM (2018) The molecular origin of enthalpy/entropy compensation in biomolecular recognition. Annu Rev Biophys 47:223–250
54. Fox JM, Kang K, Sastry M, Sherman W, Sankaran B, Zwart PH, Whitesides GM (2017) Water-restructuring mutations can reverse the thermodynamic signature of ligand binding to human carbonic anhydrase. Angew Chem Int Ed 56(14):3833–3837

55. Mazen A, Helms V, Lengauer T, Kalinina OV (2015) Enthalpy–entropy compensation upon molecular conformational changes. J Chem Theory Comput 11(4):1410–1418
56. Ryde U (2014) A fundamental view of enthalpy–entropy compensation. MedChemComm 5(9):1324–1336
57. Snyder PW, Mecinovic J, Moustakas DT et al (2011) Mechanism of the hydrophobic effect in the biomolecular recognition of arylsulfonamides by carbonic anhydrase. PNAS 108(44):17889–17894
58. Leuchter JD, Levy Y (2017) On the coupling between the dynamics of protein and water. Phys Chem Chem Phys 19:8243–8257
59. Singh RK, Suzuki T, Mandal T, Balsubramanian N, Haldar M, Mueller DJ, Strode JA, Cook G, Mallik S (2014) Thermodynamics of binding of structurally similar ligands to histone deacetylase 8 sheds light on challenges in the rational design of potent and isozyme-selective inhibitors of the enzyme. Biochemistry 53:7445–7458
60. Breiten B, Lockett MR, Sherman W et al (2013) Water networks contribute to enthalpy/entropy compensation in protein-ligand binding. J Am Chem Soc 135(41):15579–15584
61. King NM, Prabu MM, Bandaranayake R et al (2012) Extreme entropy–enthalpy compensation in a drug-resistant variant of HIV-1 protease. ACS Chem Biol 7(9):1536–1546
62. Krimmer SG, Betz M, Heine A, Klebe G (2014) Methyl, ethyl, propyl, butyl: futile but not for water, as the correlation of structure and thermodynamic signature shows in a congeneric series of thermolysin inhibitors. Chem Med Chem 9(4):833–846
63. Klebe G (2015) Applying thermodynamic profiling in lead finding and optimization. Nat Rev Drug Discov 14(2):95–110
64. Peccati F, Jiménez-Osés G (2021) Enthalpy–entropy compensation in biomolecular recognition: a computational perspective. ACS Omega 6(17):11122–1113
65. Reynolds CH, Holloway MK (2011) Thermodynamics of ligand binding and efficiency. ACS Med Chem Lett 2(6):433–433
66. Blasie CA, Berg JM (2004) Entropy–enthalpy compensation in ionic interactions probed in a zinc finger peptide. Biochemistry 43(32):10600–10604

Chapter 9
Artificial and Nanoenzymes

Abstract Artificial enzymes are a class of catalysts being potentially viable alternatives to natural enzymes. Enzyme mimics can have the important advantages including tunable structures and catalytic efficiencies, stability in experimental conditions, lower cost, and simple synthetic routes to their preparation. Cyclodextrins, metal complexes, porphyrins, polymers, dendrimers, membranes, nanoparticles, organic and metalorganic compounds and biomolecules can serve as the basis for an artificial catalyst. On the path of mimicry, seven stages have been formulated. Typical examples of enzyme mimetic catalysts, performance, the reactions they catalyze, and the main source of their catalytic efficiency were considered. Catalytically active nanomaterials (nanozymes) also show several advantages over natural enzymes. A wide range of materials are used as nanoenzymes, such as nanoparticles of Fe_3O_4, Prussian blue, Mn_3O_4, CeO_2, WO_2, $NiCo_2O_4@MnO_2$, graphene oxide, graphene-hemin nanocomposites, carbon nanotubes, carbon nanodots, mesoporous silica-encapsulated gold nanoparticles, gold nanoclusters, Pep–Au-NP = peptide–gold nanoparticle conjugate gold nanoparticles, ferrous ferrocyanide, nonheme iron, vanadium, and copper complexes bearing hemicryptophane. Metal atom (M = K, Ti, Fe, Co, Ni, Cu, Rh) doping on a Mo6S8 cluster and mononuclear rhodium species anchored on a zeolite or titanium dioxide illustrate single-atom catalysis. Nanozymes can exhibit activity of oxidase, catalase, superoxide dismutase, DNAse, alkaline phosphatase phospholipase, topoisomerase and can display antioxidant and biofilm activity. Molecular imprinting is a technique to create template (analogs of substrates, transition states, or products) shaped cavities in polymer matrices with predetermined selectivity and high affinity in a specific function. Schemes of the principle of molecular imprinting recognition, a molecularly imprinted catalyst scheme for selective catalysis, and, as an illustration, imprinted catalytic system based on methacrylic layer polymerization were presented. Despite the obvious advantages of artificial enzymes, it is still not possible to achieve the activity of natural catalysts.

G. I. Likhtenshtein, *Enzyme Catalysis Today and the Chemistry of the 21st Century*,
https://doi.org/10.1007/978-3-031-51552-1_9

9.1 Introduction

Natural enzymes, mediating every biological process in living organisms, are able to accelerate the rate of chemical reactions up to 10^{19} times for specific substrates and reactions. It is widely accepted that enzymes accelerate reactions through their steric and electronic complementarity to the reactants in the rate-determining transition states and conformational dynamics. The conformational diversities and complexities of natural enzymes exert in catalysis seriously restrict the detailed understanding of enzymatic mechanisms in molecular terms. The practical application of enzymes is often hampered by low operational stability, sensitivity of catalytic activity to environmental conditions, high costs in preparation and purification, etc.

Enzyme mimics or artificial enzymes are a class of catalysts that have heralded much interest as potentially viable alternatives to natural enzymes. Aside from having catalytic activities similar to their natural counterparts, enzyme mimics can have the desired advantages including tunable structures and catalytic efficiencies, stability in experimental conditions, lower cost, and simple synthetic routes to their preparation [1–27]. Enzyme mimics have exhibited high potential in the catalysis of a wide range of chemical and biological reactions and the development of chemical and biological sensing and anti-biofouling systems, the production of pharmaceuticals, clean fuels, and stable and low-cost alternatives to natural enzymes in a wide range of applications. A variety of materials including cyclodextrins, metal complexes, porphyrins, polymers, dendrimers, membranes, nanoparticles, and biomolecules have been designed and developed. Nevertheless, though synthetic catalysts offer an opportunity for more chemical versatility operating over a wider range of conditions currently, they do not reach the remarkable performance of natural enzymes.

The terms "mimicking enzymatic processes" or "chemical models of enzymes" or "artificial enzymes" have no monosemantic and exact definitions. This could mean the chemical model of the corresponding chemical reaction or the construction of chemical structures similar to an enzyme active site, or both. In the mimicking of an enzymatic process, there is no need to copy the structure of protein and coenzyme groups and all stages of this process. In the course of evolution, nature created enzymes in specific conditions, in certain media and utilized certain "building materials." Besides chemical functions, enzymes in a cell bear many other obligations, serving as units of complicated enzymatic and membrane ensembles. At present, on the bases of common features present in enzymes that derive from the study of natural enzymes, the following general steps of mimicking enzymatic processes may be formulated [6, 7]: (1) Previous detailed analysis of existing data on the structure and action mechanism of an enzyme on interest; (2) choose of basic (primary) catalytic groups directly involved in the catalytic process; (3) selection of secondary groups, which can regulate the reactivity of the attacking groups; (4) optimal disposition of primary and secondary catalytic groups on a matrix; (5) including in the catalytic system additional residues, which can form portions capable of bounding and precisely orienting the substrate molecule; (6) a matrix, carrying the model catalysis active site, should be dynamics, providing condition for dynamic adaptation (Chap. 4); (7) a pretransition state complex that activates cleavage or formation of chemical bonds should have to include optimal number nucleous moving to the transition state (the "principle of optimum motion" (Chap. 4).

9.2 Enzyme Mimetic Catalysts

9.2.1 Modeling Enzymes Active Centers Structure and Activity

Artificial catalysts can be based on natural molecules, such as cyclodextrins, catalytic antibodies, synthetic complex systems, functional macrocycles, imprinted polymers, dendrimers, and complex nanosystems, such as micellar and vesicular aggregates or nanoparticles [8, 9]. To give an idea of their typical performance and the source of catalysis as enzyme mimics, selected examples and kinetic results obtained with these catalysts are demonstrated in Figs. 9.1 and 9.2 and Table 9.1. Each system reported in Table 9.1 is based on a specific feature pertaining to enzyme catalysis.

Synthetic catalysts (Fig. 9.2) offer an opportunity for more chemical versatility operating over a wider range of conditions but currently do not reach the remarkable performance of natural enzymes [8]. With a focus on the enzyme as a whole catalytic construct, the translational impact of natural enzyme principles to synthetic enzymes, supramolecular capsules, and electrocatalytic surfaces was put forward.

Examples of chemical modeling of enzymes of particular interest such as nitrogenase, nitric oxide synthase, cytochrome P459, methane monooxydase, hydrolases, photosynthetic reaction centers, water oxidation complex, etc. are given in the following chapters.

9.2.2 Nanoenzymes

Nanozymes are catalytic nanomaterials with enzyme-like characteristics [14–47]. Recently, the merging of nanotechnology with biology has ignited extensive research efforts for designing functional nanomaterials that exhibit various properties intrinsic to enzymes. As a promising candidate for artificial enzymes, catalytically active nanomaterials (nanozymes) show several advantages over natural enzymes, such as controlled synthesis in low cost, tunability in catalytic activities, as well as high stability against stringent conditions. Nevertheless, nanozymes as artificial catalysts, having many advantages, still face many challenges (Fig. 9.3).

A comprehensive review [38] summarized and analyzed the progress, especially the achievements from more than 1100 research papers, published by 2019. The review covered all principle aspects in the area: development of nanotechnology,

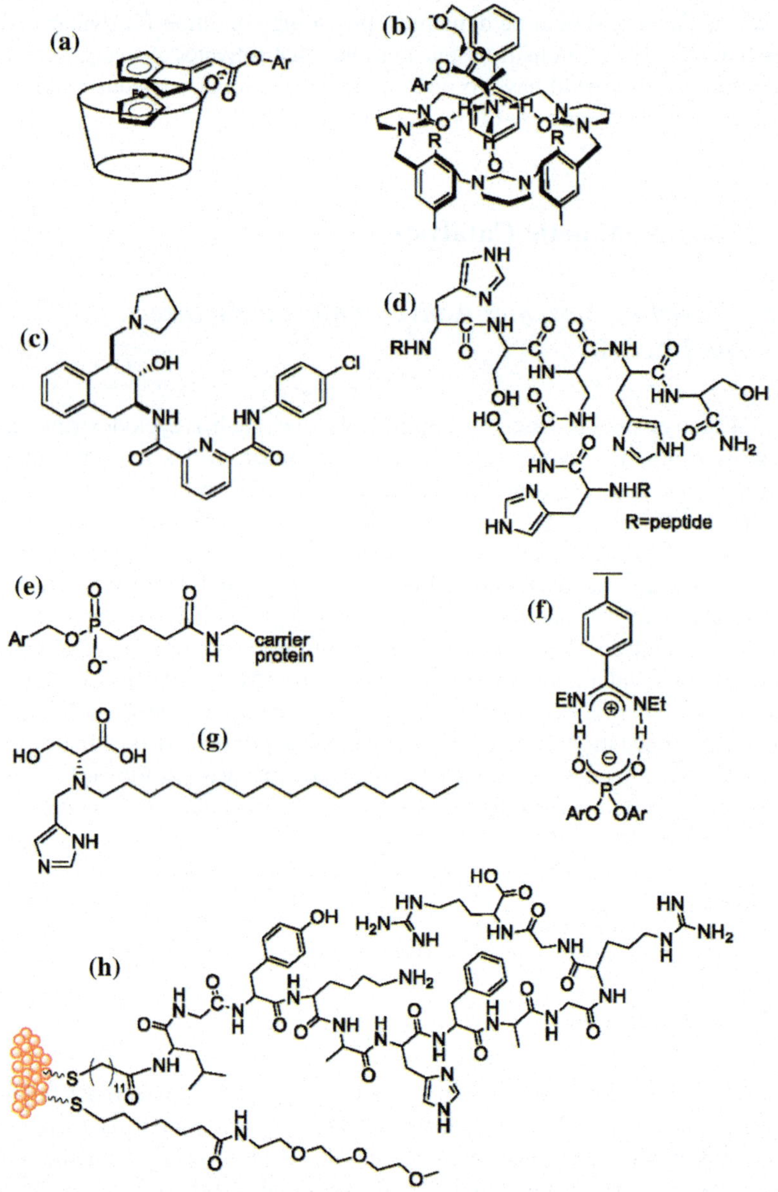

Fig. 9.1 Structure of enzyme mimetic catalysts: **a** β-cyclodextrin with ferrocenyl derivative; **b** macrocyclic host with a complexed protonated alanine *p*-nitrophenyl ester substrate; **c** rigid receptor for methanolysis; **d** dendrimers; **e** hapten used for the development of catalytic antibodies for ester hydrolysis; **f** catalytic site of an imprinted polymer with transition state analog for carbamate hydrolysis; **g** functional peptide for ester hydrolysis; **h** peptide-functionalized gold nanoparticle for ester hydrolysis [9]. Published under CC BY license from American Chemical Society

a

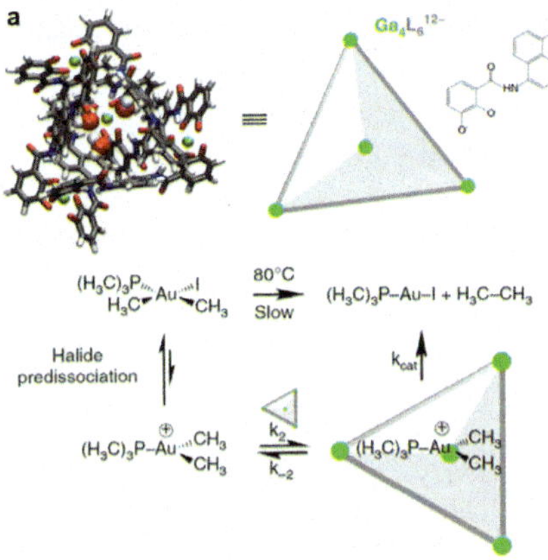

b

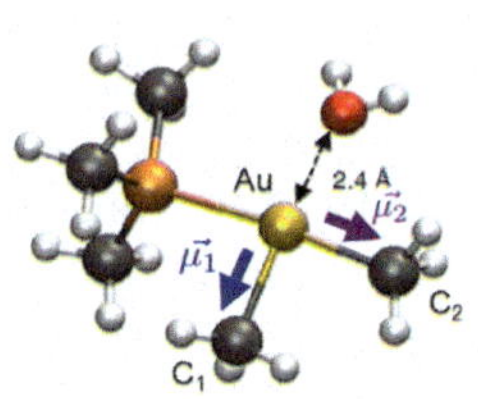

c

Activation free energy stabilization (kcal mol^{-1}):

$$\Delta G^{\ddagger}_{elec} = -\sum_i \left(\mu^i_{TS} \cdot E^i_{TS} - \mu^i_{RS} \cdot E^i_{RS}\right)$$

	Bulk water	Complexed water	Cage
Uncatalyzed	10.58	−11.29	N/A
Catalyzed	9.02	−13.62	−1.04

Fig. 9.2 $Ga_4L_6^{12-}$ supramolecular catalyst for reductive elimination from gold complexes. **a** Structure of $Ga_4L_6^{12-}$ and the proposed reaction mechanism. **b** The gold complex, the reactive Au–C_1 and Au–C_2 bonds, and the complexed water molecule position in the transition state; the nanocage and greater water environment are not shown for clarity. **c** The activation free energy stabilization by region for the complexed water molecule, the nanocage, and the remaining water solvent obtained by ensemble-averaged MD calculations [8]. Published under a license CC-BY from American Chemical Society

biotechnology, catalysis mechanism, computational design, imitating new enzymatic activities with high-performance nanomaterials, regulating the nanozyme activities, elucidating the catalytic mechanisms, and broadening potential applications (Fig. 9.4). Separate sections of the review are devoted to the following important subjects: (1) Iron, vanadium, noble metal, metal–organic framework (MOF), and other nanomaterial based on peroxidase mimics; (2) gold, cupper, molybdenum, and platinum compounds-based oxidase mimics; (3) catalase mimics and superoxide dismutase mimics; and (4) hydrolase mimics. Other relevant topics covered in the review are: multi-enzyme-mimicking nanozymes, multi-functional nanozymes, and engineering nanozyme activity.

In review [14], with the expanded types of nanozymes and engineered high performance, applications have been accomplished and discussed as follows: in vitro sensing H_2O_2 detection based on peroxidase mimics, detection of glucose and other oxidase substrates, and cancer markers on cell surface detection. For this purpose, magnetic resonance, computed tomography, and optical techniques, benefiting from the intrinsic properties (e.g., magnetism of Fe, X-ray absorption ability of Ir, and optics of Au) were utilized. Hundreds of nanomaterials have been investigated to mimic the catalytic activity of peroxidase, oxidase, catalase, haloperoxidase, glutathione peroxidase, uricase, methane monooxygenase, hydrolase, and superoxide dismutase [17]. For example, nanoparticles Fe_3O_4 NPs showed pH-dependent peroxidase-like and catalase-like activities. Prussian blue nanoparticles (NPs)

Table 9.1 Examples of enzyme mimetic catalysts, their performance, the reactions they catalyze, and the main source of their catalytic efficiency [9]

Entry	Catalyst/Substrate (panel in Fig. 9.1)[a]	Catalyzed reaction	Main source of catalysis[b]	$K_{rel} = k_{cat}/k_{uncat}$[c]	k_{cat}/K_M[d] or k_2[e] (s^{-1} M^{-1})	$K_{tr} = (k_{cat}/K_M)/k_{uncat}$[f]($M^{-1}$)
1	Cyclodextrin/ Carboxylate ester (A)	Transacylation	iii	5.9×10^6	16	1.0×10^9
2	Macrocycle/ Carboxylate ester (B)	Transacylation	ii, iii	1.7×10^3	2×10^9	2×10^{12}
3	Rigid molecular receptor/ Carboxylate ester (C)	Transacylation	iii	Very high[g]	3.2×10^{-4}[h]	nd
4	Dendrimer/ Carboxylate ester (D)	Hydrolysis	iii, iv	3.9×10^4	3.2×10^2	1.3×10^9
5	Catalytic antibody/ Carboxylate ester (E)	Hydrolysis	i	1.3×10^5	214	4.6×10^8
6	Imprinted polymer/ Diphenylcarbamate (F)	Hydrolysis	i	3.8×10^3	0.24	1.2×10^6
7	Micelles/ Carboxylate ester (G)	Hydrolysis	ii–iv	1.6×10^4	1.3×10^3	1.1×10^7
8	Gold nanoparticles/ Carboxylate ester (H)	Hydrolysis	ii–iv	1.2×10^5	6.2×10^5	4.8×10^9

a–h: see details in [9]

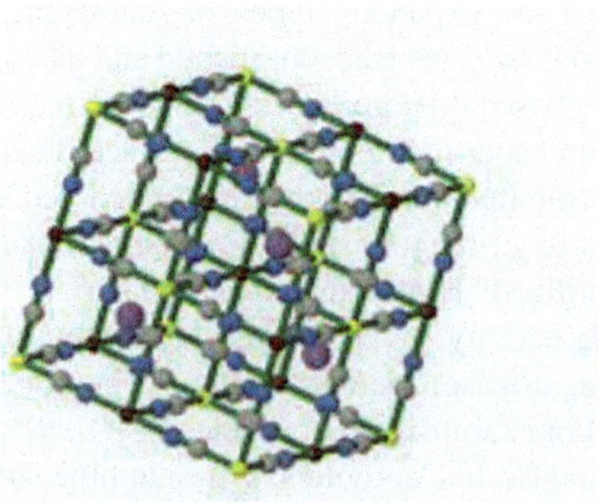

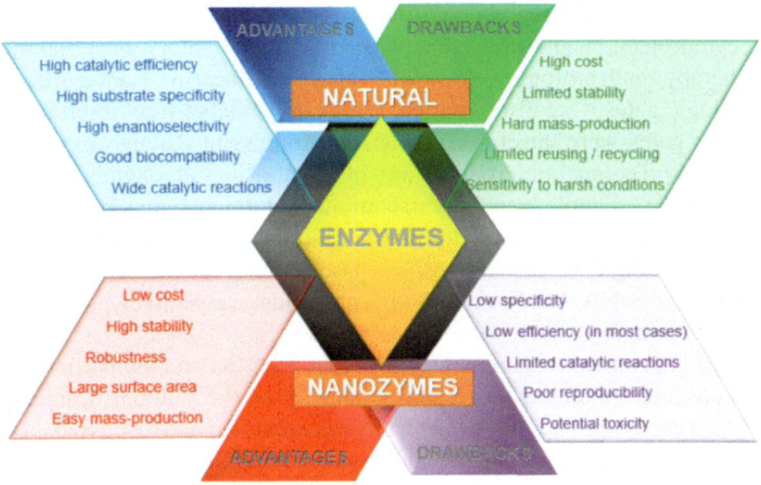

Fig. 9.3 Comparison of some characteristics between nanozymes and natural enzymes [37]. Published by MDPI

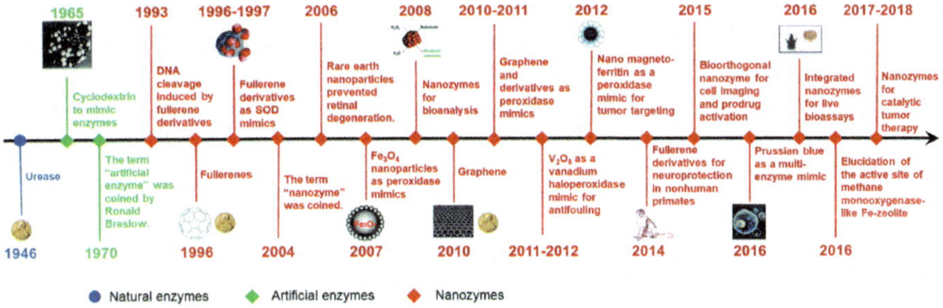

Fig. 9.4 Brief timeline for the development of nanozymes (natural enzymes and artificial enzymes are listed for comparison) [19]. With permission from RCS

simultaneously possess peroxidase-, catalase-, and superoxide dismutase-like activity. Mn_3O_4 NPs can also mimic all three cellular antioxidant enzymes including superoxide dismutase, catalase, and glutathione peroxidase. Targeting antibody-conjugated ferromagnetic nanozymes simultaneously provides three functions: target capture, magnetic separation, and nanozyme color development for target detection. Account [15] focused on progress in exploring and constructing artificial enzymes, including graphene oxide, graphene-hemin nanocomposites, carbon nanotubes, carbon nanodots, mesoporous silica-encapsulated gold nanoparticles, gold nanoclusters, and nanoceria. The biomimetic catalysts can serve as colorimetric probes for the detection of cancer cells, nucleic acids, proteins, metal ions, and other small molecules. For example, graphene oxide could be a modulator to greatly improve the catalytic activity of lysozyme-stabilized gold nanoclusters at neutral pH.

Nano enzymes have been widely used in therapeutic fields including neuroprotection, cytoprotection, environmental protection, inflammatory, bacterial infection, cancer therapy, and biofouling [39] and references therein. Nanozymes used to construct logic gates were also reported. For example, label-free, resettable, colorimetric logic gates were developed. Research progress in reactive oxygen species ROS-Mediated Therapeutic Applications of Composite Nanozymes was summarized in [42]. For example, nanomaterials mimic natural enzymes in action, display pro- or antioxidative catalytic activity, and also mimick hydrolase-like activity. Nanomaterials exhibiting DNAse and phospholipase activity, caused hydrolytic cleavage of extracellular DNA (eDNA) and phospholipids, respectively, and displayed antibiofilm activity.

A review [12] aimed to summarize the recent progress on nanoparticles in catalysis and provided potential theoretical guidance for the rational design of high-performance catalysts. The following results were reported and discussed: (1) A brief summary of the synthetic strategies and characterization methods of nanoclusters NCs; (2) discussion of the model catalyst role of NCs in catalysis as illustrated from electrocatalysis, photocatalysis, photoelectric conversion, and catalysis of organic reactions; and (3) examination of the main challenges and opportunities for a deep understanding of the key catalytic steps. In addition, some new design strategies based on the contributions of nonlocal electric fields and thermodynamic fluctuations to improve the catalytic step and turnover for rate acceleration in arbitrary synthetic catalysts were put forward [8].

A variety of nanomaterials has been studied as artificial enzyme mimics, including CeO_2 NPs, Fe_3O_4 NPs, Pt nanomaterials, Au NPs, bimetal and trimetal NPs, graphene oxide, carbon nanotubes, fullerene derivates, quantum dots, metal organic frameworks (MOFs) as some common examples (Fig. 9.5) [37].

Inspired by natural peroxidases [40], ferromagnetic nanoparticles (**Fe_3O_4**) were modified with histidine to form a similar microenvironment to natural hydrolases. It was found that such modification improved the affinity toward H_2O_2 by at least tenfold, as a hydrogen bond would form between the side-chain imidazole group of histidine and H_2O_2. A peroxidase mimic composed of iridium (Ir) nanoparticles with an average size of only 1.1 nm that are coupled to a tungsten suboxide (WO_2) nanorod as support was prepared [48]. Such a peroxidase mimic exhibited an area-specific catalytic efficiency as high as $1.5 \times 10^4\ s^{-1}\ nm^{-2}$. The peroxidase mimic was applied to colorimetric immunoassays of carcinoembryonic antigen (a cancer biomarker) and aflatoxin B1

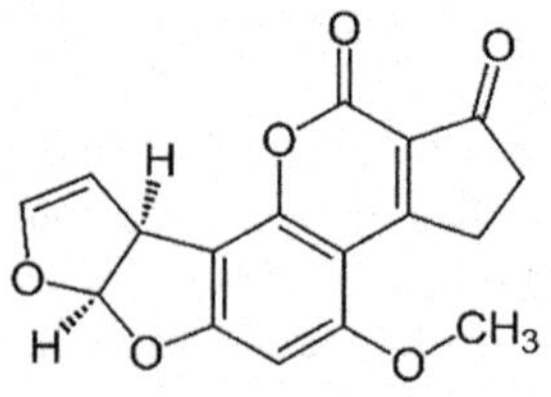

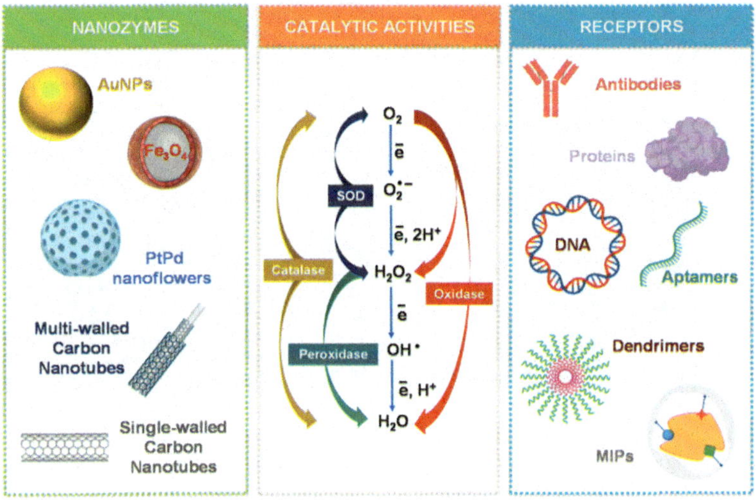

Fig. 9.5 Schematic illustration of some nanomaterials mimicking enzymes, the most studied enzyme-like activities, and common ligands to tune the catalytic efficiency and selectivity of nanozymes [37]. Published by MDPI

(a potent carcinogen), which provided substantially enhanced detection sensitivities compared to conventional enzyme-based assays.

Heme-based catalase enzymes and structurally related catalase mimics destroy H_2O_2 by a mechanism involving oxoferryl intermediates [47]. A large number of nanomaterials, including CeO_2, Fe_3O_4, Fe-doped carbon and carbon nitride, gold nanoparticles, and MOFs demonstrated their ability to catalyze H_2O_2 disproportionation. Composite nanozymes designed recently showed both intrinsic peroxidase- and oxidase-like activity to generate ROS $\cdot OH$, $\cdot O_2^-$, and H_2O_2, which can vividly damage more lipids, amino acids, and polysaccharides in the bacterial membrane and also kill cancerous cells due to their oxidative and electrophilic nature [46]. The scavenging reactions of reactive oxygen species (ROS) like H_2O_2 and O_2 bound by {111} planes on two facets of the nanooctahedra (Fig. 9.6) were examined [49]. Specifically, the reaction energy (Er) of the rate-determining step was calculated and more negative Er evidenced higher activity. As shown in Fig. 9.6, the scavenging abilities of H_2O_2 and O_2 on the {111} facet (Er equaled 2.81 and 0.60 eV, respectively) were stronger than those on the {100} facet (Er equaled 2.64 and 0.13 eV, respectively).

$NiCo_2O_4@MnO_2$ microcubes were synthesized to form numerous nanoscale p–n junctions between p-type $NiCo_2O_4$ and n-type MnO_2 [50]. The resulting nanozyme exhibits both oxidase- and peroxidase-like activities:

$$AH + O_2 \xrightarrow{\text{oxidase}} A + H_2O$$

$$AH + H_2O_2 \xrightarrow{\text{peroxidase}} A + H_2O$$

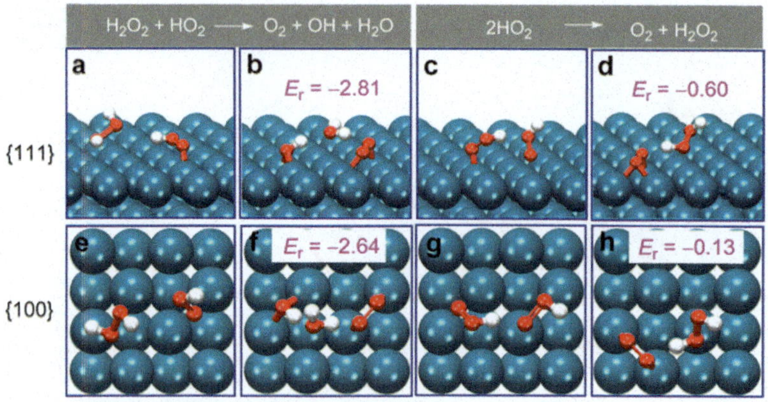

Fig. 9.6 Lowest-energy adsorption structures and reaction energies (in eV) for the reactions on structures having either Pd{111} or {100} facets [49]. With permission from American Chemical Society

where AH denotes substrates. The nanoenzyme efficiency can be enhanced under illumination (Fig. 9.7). Chromogenic substrate, 3,3′,5,5 tetramethylbenzidine dihydrochloride (TMB) was utilized in the investigation of the photoenhanced oxidase–peroxidase-like activity of $NiCo_2O_4$@ MnO_2 microcubes: The calculated K_m, V_{max}, and K_{cat} of $NiCo_2O_4$@MnO_2 were found to be 2.31×10^{-3} M^{-1}, 1.27×10^{-6} M s^{-1}, and 5.40×10^8 s^{-1}. When $NiCo_2O_4$@MnO_2 receives light irradiation, the photoenhanced oxidase–peroxidase-like activity occurs. The band gap of $NiCo_2O_4$ and MnO_2 are 1.77 and 2.2 eV, respectively. In addition, the p–n junction interface transports hot electrons from $NiCo_2O_4$ to MnO_2 and restrains the recombination of photogenerated electrons and holes.

Inspired by enzymatic cases, it was suggested that thermal motion can play a crucial role in influencing the reaction mechanism and determining the energetic stability of CO binding in heterogeneous catalysis of the electrochemical CO_2 reduction reaction (CO_2RR) [8]. In this case, a DFT functional gave a look at the CO binding motifs on the weak binding metal surfaces, which was found to be a mixture of chemisorbed and physisorbed species (Fig. 9.8). The importance of thermal fluctuations in a number of catalytic systems was stressed.

Sensitive and stable fluorometric sensors for the detection of hydrogen peroxide based on the polyoxotungstate intrinsic peroxidase ($Na_{10}[\alpha\text{-}SiW_9O_{34}]$) which induced fluorescent enhancement of benzoic acid (BA), thiamine (TH)

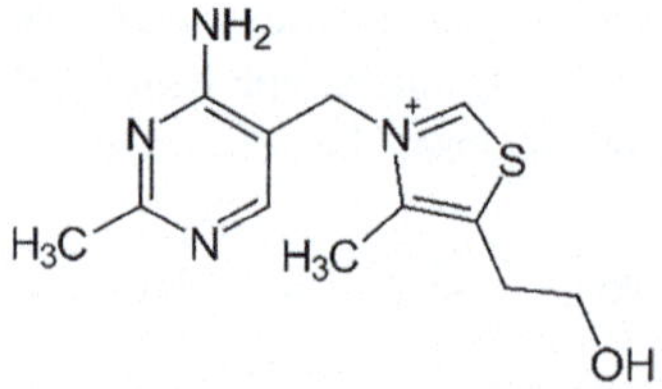

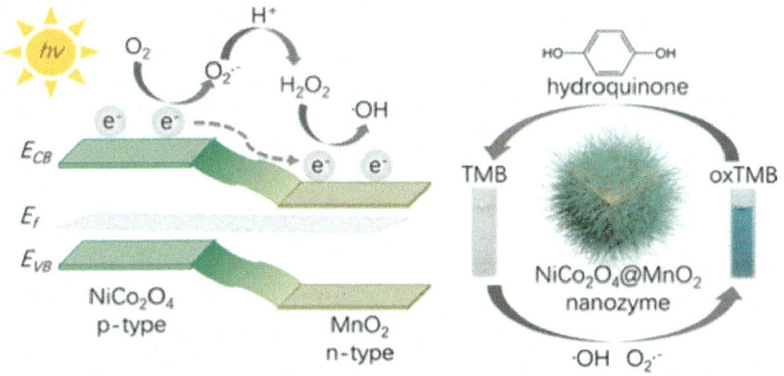

Fig. 9.7 Mechanism of the photoenhanced oxidase–peroxidase-like activity of $NiCo_2O_4@MnO_2$ microcubes. The p–n junctions in $NiCo_2O_4@MnO_2$ microcubes transfer the photogenerated hot electrons from $NiCo_2O_4$ to MnO_2. These electrons react with devolved O_2 or H_2O_2 to form $O_2^{\cdot -}$ and ·OH, which oxidize the colorless TMB to blue oxTMB [50]. With permission from American Chemical Society

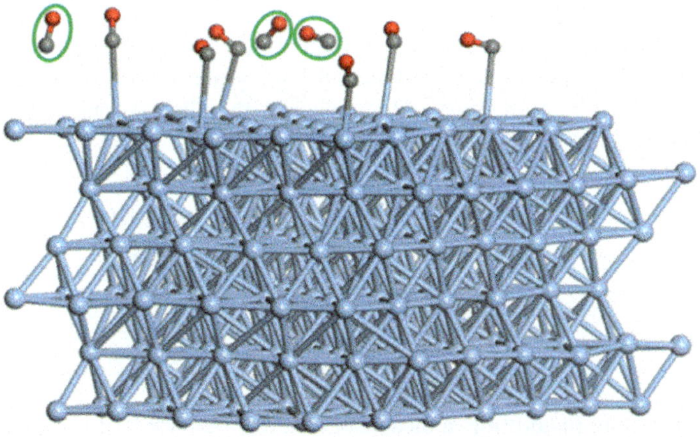

Fig. 9.8 Representative configuration showing the mixture of chemisorbed and physisorbed CO molecules (circled in green) at the last 2.0 ps time point of the AIMD simulation for CO binding to Ag(111) for B97M-rV. The statistical data are collected after 500 fs's pre-equilibration for the 2 ps trajectories [8]. Published under a license CC-BY from American Chemical Society

and 3-(4-hydroxyphenyl)propionic acid (HPPA)

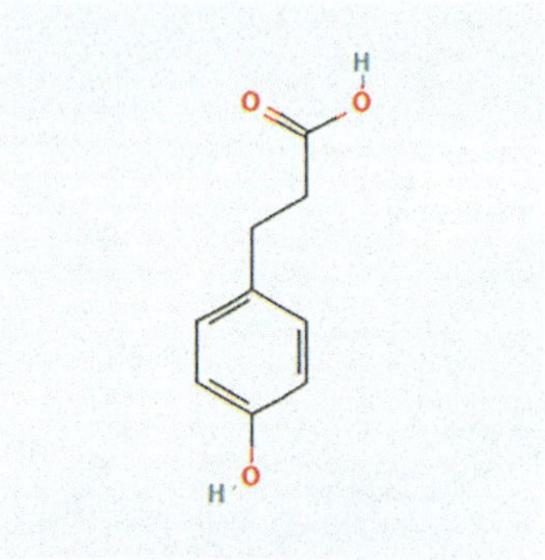

were designed [51]. The three nonfluorescent substrates BA, TH, and HPPA were oxidized with the OH radicals decomposed from $\mathbf{H_2O_2}$ under the catalysis of $Na_{10}[\alpha\text{-}SiW_9O_{34}]$. Detection limits for BA, TH, and HPPA were found to be 6.7×10^{-9}, 2.2×10^{-7} and 9.6×10^{-6} M, respectively.

Work [52] focused on a combination of metal–organic framework (MOF)-based nanozymes and a G-quadruplex (G4)-based catalytic system (G4-DNAzyme). On this pathway, MIL-53(Fe) MOF was linked covalently to a G4-forming sequence (F3TC), itself covalently linked to its cofactor hemin. These properties were exploited to devise a new biosensing system based on a cascade of reactions catalyzed by the nanozyme (ABTS oxidation) and an enzyme, the alkaline phosphatase (ALP), catalyzed ascorbic acid 2-phosphate

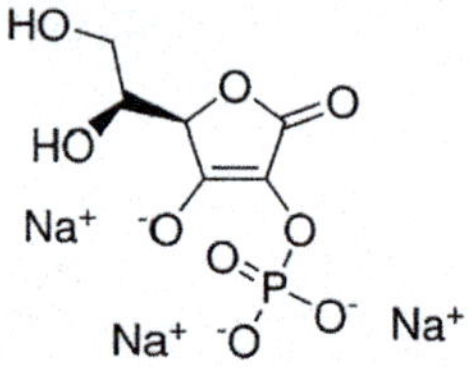

dephosphorylation. In addition, a biosensor for ALP (a marker of bone diseases and cancers), with a very low limit of detection (LOD, 0.02 U L^{-1}), which is operative in human plasma samples, was designed.

Prussian blue (PB) ferrous ferrocyanide nanoparticles were prepared by using mild synthetic conditions in a cellulose-based hydrogel network in the presence of glucose oxidase (GOx). This protocol resulted in the simultaneous immobilization of PB NPs and active GOx in the hydrogel [53]. This system displayed peroxidase-like activity relative to the PB NPs and revealed an enhanced cascade catalytic performance for the colorimetric detection of glucose in human serum samples. Compared to the analog mixture with GOx in solution, the composite hydrogel enhanced glucose detection and improved stability. Prussian Blue chitosan enzyme mimetic tubular micromotors were utilized for on-the-fly (bio)-sensing assays [54]. PB micromotors display enzyme mimetic capabilities with key functions: (1) The autonomous oxygen bubble propulsion with PB acting as a catalase mimic for hydrogen peroxide decomposition; (2) 3,3′,5,5′-tetramethylbenzidine (TMB)

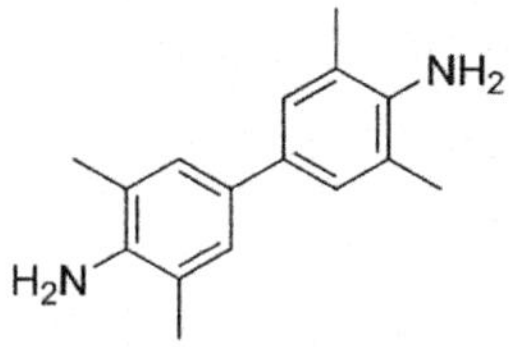

oxidation with PB acting as a peroxidase mimic for analyte detection; and (3) as a magnetic material to simplify the (bio)-sensing steps. After functionalization with acetylthiocholinesterase enzyme (ATChE), the micromotor was explored in fast inhibition assays for the colorimetric determination of the nerve agent neostigmine, with analytical performance in terms of quantification limit (0.30 μM) and concentration linear range (up to 500 μM).

It was reported that cysteine-derived chiral carbon dots (CDs)

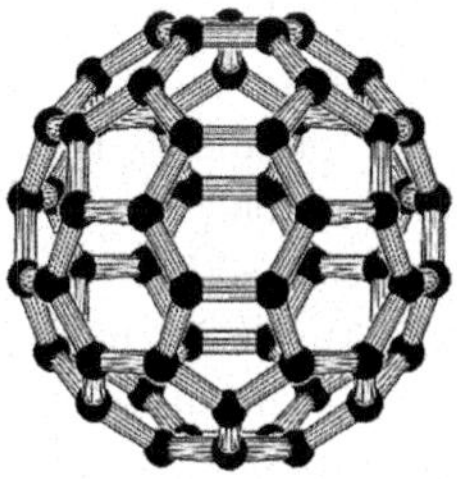

can mimic topoisomerase to mediate topological rearrangement of supercoiled DNA enantioselectively [55]. Experiments suggested the underlying mechanism, namely: (1) D-CDs intercalatively bind with DNA double helix more strongly than L-CDs; (2) the intercalative CDs can catalyze the production of hydroxyl radicals to cleave phosphate backbone in one strand of the double helix; and (3) this leads to topological rearrangement of supercoiled DNA. In addition, molecular dynamics simulation indicated that the affinity for hydrogen-bond formation and hydrophobic interaction between D-cysteine and DNA is stronger than that of L-cysteine.

Pengo et al. pioneered the field of catalytically active Pep–Au-NPs by invention a simple dipeptidic model system synthesized by a ligand-exchange protocol [56]. For example, spherical precursor Au-NPs capped by a stabilizing monolayer of a water-soluble *N*-(3,6,9-trioxadecyl)-8-sulfanyloctanamide (HS-C8-TEG)

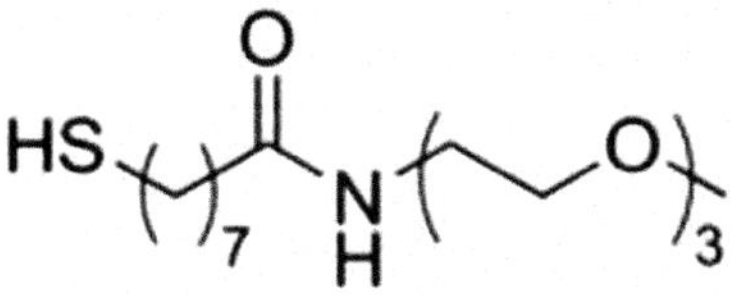

were conjugated to a His-Phe dipeptide **1** with an N-terminal undecanethiol and an unmodified C-terminus. Degradation of organophosphorus nerve agents is of primary importance for the severe toxicity of these compounds. Gold nanoparticles passivated with thiolated molecules, bearing 1,3,7-triazacyclononane and 1,3,7,10-tetraazacyclododecane

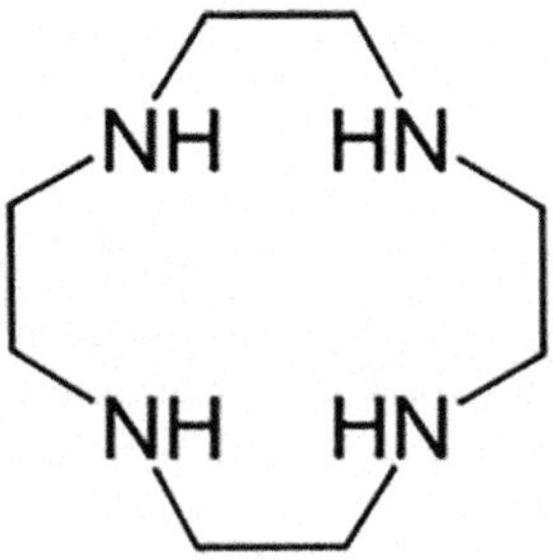

ligands and hydrolyzing nerve agents simulants *p*-nitrophenyl diphenyl phosphate and methylparaoxon, were prepared and examined [57]. The analysis was performed at 25 °C and pH 8 with half-lives of the order of a few minutes. The catalytic site appeared to involve a single-metal ion and its recognition of the substrates can be driven mostly by hydrophobic interactions.

Review [58] summarized progress made over the past 15 years in the field of catalytically active Au-NPs Pep–Au-NP = peptide–gold nanoparticle conjugate (Figs. 9.9 and 9.10). The choice of natural or unnatural amino acids allowed for the fine-tuning of reactivity and specificity of the ligand shell. The review covered the following topics: (1) The origin of distinct properties evoked by the conjugation of peptides to the Au-NPs and their similarities to natural enzymes; (2) the confinement of tailored peptides onto the surface of Au-NPs created dense peptide monolayers, in which novel hydrogen-bond networks and charge-relay interactions between individual functional groups appear; (3) carboxylate-imidazole-mediated ester hydrolysis showed Au@E1H8 functions as an efficient esterase mimic, based on a centrally located His functioning as the catalytic moiety; (4) a dipeptide can be replaced with rationally designed peptide consisting of 12 amino acids (Fig. 9.9); after immobilization onto the gold surface, His is buried inside the resulting monolayer analogous to the active sites of natural enzymes; (6) Lys and Arg were incorporated to stabilize the negatively charged transition state of ester hydrolysis, to enable (together with His) nucleophilic and/or general acid/base catalysis; and (7) the individual Pep–Au-NPs were stabilized due to electrostatic repulsion.

Thus, Pep–Au-NPs are able to generate improved substrate and transition state interactions, and the associated rate enhancements. Main advantages of catalytically active Pep–Au-NPs were illustrated in Fig. 9.10 [58]. Nevertheless, these systems still do not enable the high levels of catalytic efficiency achieved by natural enzymes.

Gold nanoparticles (AuNPs) functionalized with Zn(II) complexes, synthetic nanonucleases (Fig. 9.11) catalyzed metal (Zn)-aided hydrolytic reactions of nucleic acids and phosphate esters. The nanoparticles mimic metallonuclease enzymes which

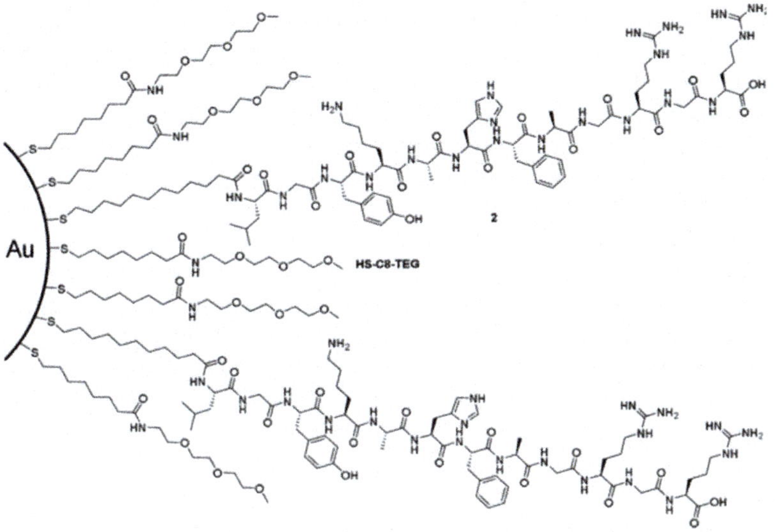

Fig. 9.9 Schematic prepresentation of Au@2 composed of mixed monolayer of HS-C8-TE G and peptide 2 synthesized by Pengo et al. [56]. From permission from Wiley

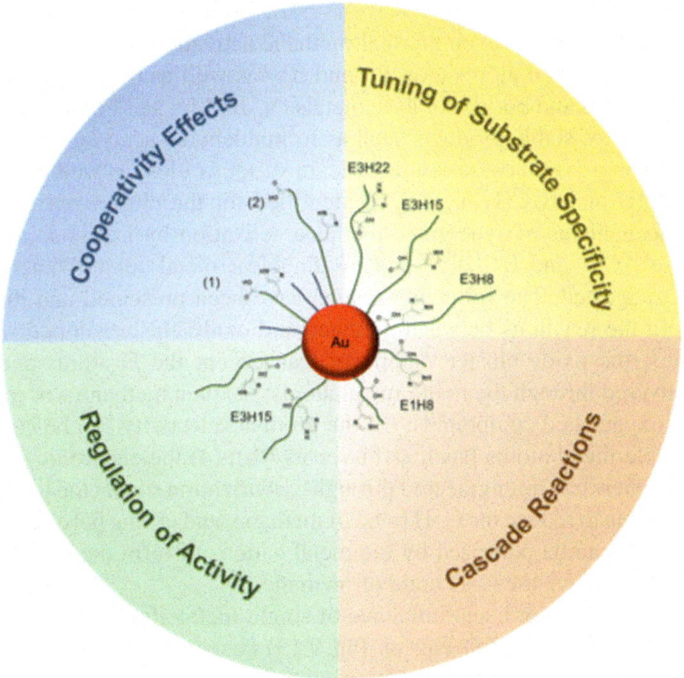

Fig. 9.10 Advantages of catalytically active Pep–Au-NPs [58]. Published under the Creative Commons CC BY license from Wiley

hydrolyzed of the 2-hydroxypropyl-4-nitrophenylphosphate (HPNP), a typical RNA model substrate [5]. The structure and dynamics of the outer coating monolayer of these nanoparticles, either alone or in complex with HPNP, were examined using microsecond-long atomistic simulations and NMR experiments.

Establishing whether a reaction is catalyzed by a single-metal catalytic center or cooperatively by a fleeting complex encompassing two such centers was a subject of the work [59]. Possibility of single-copper versus dicopper mechanisms of the "click" reaction was discussed [59]. "Click" chemistry is a class of biocompatible small molecule reactions commonly used in bioconjugation, allowing the joining of substrates of choice with specific biomolecules. The suggested method relied on geometrical considerations that can distinguish between cooperative and noncooperative mechanisms. The latter can be realized in a nanoparticle-based platform. In view of the potential practical significance of the oxidation of methane by atmospheric oxygen under relatively mild conditions, special attention was drawn to the corresponding catalysts' [60–68]. A strategy for the electrocatalysis of selective methane oxidation by immobilizing tricopper catalysts on the cathodic surface was developed (Fig. 9.12) [60]. In the presence of dioxygen and methane, above a threshold cathodic potential, oxygen atom transfers to methane. Turnover frequencies $> 40\ \text{min}^{-1}$ and $>$ 30,000 in 12 h were achieved for this chemical transformation in water under ambient conditions. This technology is suitable for on-site direct conversion of methane into methanol.

In account [61], progress on thermal methane activation by metal oxide clusters doped with noble-metal atoms (Au, Pt, and Rh) as well as by oxygen-free species including **carbides** and borides of base metals (V, Ta, Mo, and Fe) was summarized. The generations of stable products such as formaldehyde, acetylene, and syngas (a mixture of hydrogen and carbon monoxide, in various ratios) as well as closed-shell species $AuCH_3$ and B_3CH_3 have been identified for the cluster reaction systems. The new mechanisms of synergistic methane activation by Lewis acid–base pairs (such as $Au^{\delta+}$-$O^{\delta-}$ and $B^{\delta+}$-$B^{\delta-}$) and by dinuclear metal centers (such as Ta-Ta) have been suggested. The following results have been presented and discussed in detail: (1) In the reactions between methane and oxide clusters doped with noble-metal atoms, the oxide cluster "supports" can accept the H atoms and the CH_x species delivered through the noble-metal atoms; (2) then methane was transformed into stable oxygenated compounds; (3) the product selectivity can be controlled by different noble-metal atoms (such as Pt versus Rh); (4) the electronic structures of base metal centers can be engineered through carburization so that the low-spin states can be accessible to reduce the C–H bond of methane; and (5) the boron clusters (such as B_3 in VB_3^+) can be polarized by the metal cations to form the Lewis acid–base pair $B^{\delta+}$-$B^{\delta-}$ to cleave the C–H bond of methane.

In work [62], the significant influence of single-metal-atom (M = K, Ti, Fe, Co, Ni, Cu, Rh) doping on a Mo6S8 cluster (Fig. 9.13) was revealed for the direct methane to methanol conversion in water stream using density functional theory calculations. The direct $CH_4 \rightarrow CH_3OH$ conversion via the steam reforming of methane (SRM) and over the single-metal-atom (M = K, Ti, Fe, Co, Ni, Cu, Rh) on Mo6S8 clusters was explored using DFT. The following findings have been reported and discussed:

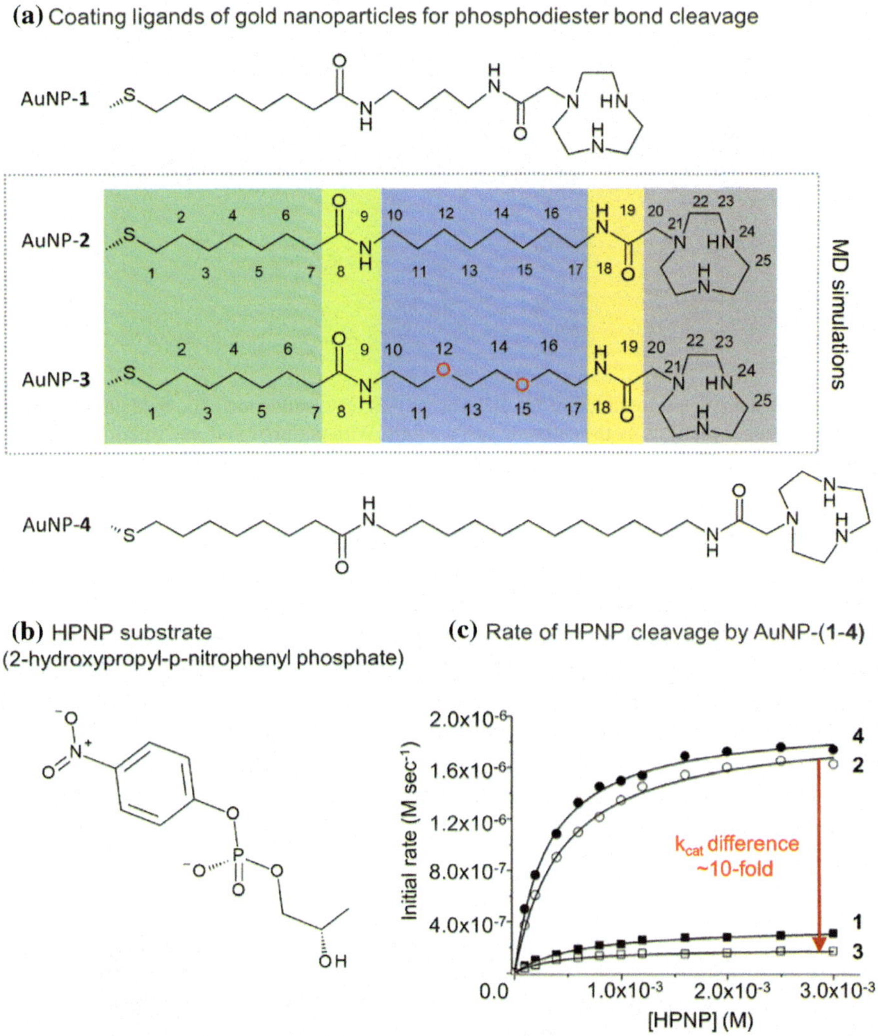

Fig. 9.11 **a** Coating ligands of four phosphodiester bond-cleaving nanozymes from ref (10a). The individual components of the coating ligands of AuNP-2 and AuNP-3, two nanoparticles (similar in structure and different in activity) selected for this study, are divided into five main segments: (i) the region most proximal to the gold core, which is a hydrophobic C7 alkyl chain (in green); (ii) an inner amide group (in yellow); (iii) a linker chain that differs in the two nanoparticles, being a hydrophobic alkyl chain in AuNP-**2** and a hydrophilic polyethylenglycol (PEG) in AuNP-**3** (in blue); (iv) an outer amide group (in orange); and (v) the terminal TACN crown which chelates Zn ions (in gray). **b** Chemical structure of the HPNP substrate. **c** Data from ref (10a), on the dependence of the initial rate of HPNP at fixed AuNP-**2**(**3**)/Zn concentration as a function of the substrate HPNP concentration, with an activity drop ($\sim$ tenfold, highlighted in red) between AuNP-**2** and AuNP-**3**. Specifically, $k_{cat} = 0.21\ s^{-1}$ for AuNP-**2** and $k_{cat} = 0.02\ s^{-1}$ for AuNP-**3** for the hydrolysis of HPNP, while K_M is 0.40 nM for AuNP-**2** and 0.38 nM for AuNP-**3** [5]. Published under the Creative Commons CC BY license from American Chemical Society

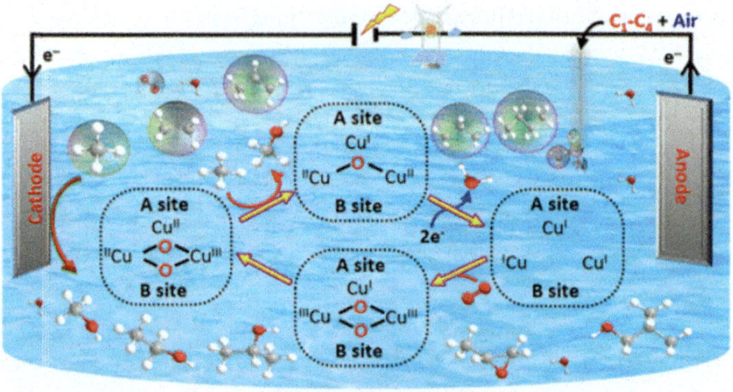

Fig. 9.12 Schematic presentation of the electrocatalysis of selective methane oxidation by immobilizing tricopper catalysts on the cathodic surface [60]. With permission from American Chemical Society

(1) In Mo6S8 cluster, each S atom tightly combines with three Mo atoms and a binding energy of – 5.90 eV; (2) for Mo6S8, the energy of adsorption of CH_4 E_{ads} = – 0.30 eV, and decreases in the order: Ni–Mo6S8 > Co–Mo6S8 > Ti–Mo6S8 > Fe–, Cu–Mo6S8 > Mo6S8 > K–Mo6S8 > Rh–Mo6S8; (3) the first C–H bond cleavage ($*CH_4 + * \rightarrow * CH_3 + *H$) corresponds to an energy barrier (E_a) of 1.45 eV; (4) the barrier for the second bond cleavage Ea = 1.07 eV, $*CH_3 + * \rightarrow *CH_2 + *H$; (5) the barrier for the third bond cleavage Ea = 1.17 eV, $*CH_2 + * \rightarrow *CH + *$; (6) the formation of $*C$ ($*CH + * \rightarrow *C + *H$) with the barrier as high as 2.00 eV is slower step; and (7) all doped single-atoms M show the enhancement in CH_3OH synthesis during the SRM on Mo6S8, where the promoting effect of Fe is the most significant.

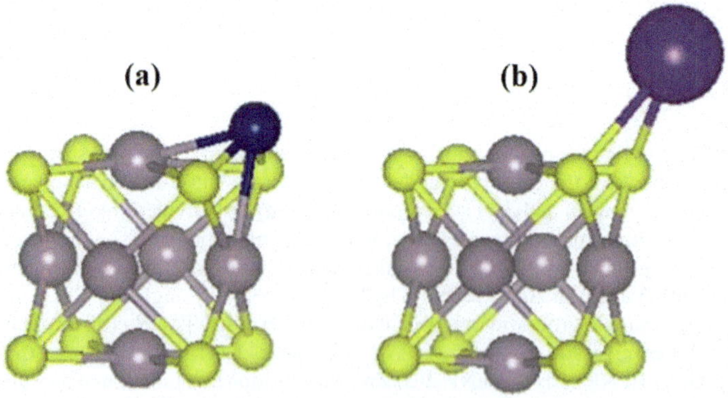

Fig. 9.13 Optimized M-Mo6S8 structure. **a** M atom anchored at S-Mo-Mo-S fourfold site; **b** M atom anchored at S-S twofold site (Mo: small purple, S: yellow, M: big purple or dark blue [62]). With permission from American Institute of Physics

High CH_4 oxidation activity of ZnO nanoparticles loaded with Ag nanoparticles under simulated sunlight was investigated using surface plasmon resonance, SPR [63]. SPR is a phenomenon where the electrons in the metal surface layer are excited by photons of incident light with a certain angle of incidence and then propagate parallel to the metal surface. The deposited Ag nanoparticles showed SPR, reduced recombination rate of the photogenerated charge carriers with faster surface reaction, and increased visible light absorption. The experiment on atmospheric methane oxidation indicated the high quantum yield of 8% at wavelengths < 400 nm and over 0.1% at wavelengths ~ 470 nm was achieved on the silver decorated zinc oxide nanostructures. To account for the methane photooxidation, a two-step photocatalytic reaction process was suggested. In the first step, CH_4 reacted with oxygen to produce H_2O. In the second step, a HCHO ($\mathbf{CH_4} + O_2 \rightarrow HCHO + H_2O$), as the intermediate, reacted with O_2 to produce CO_2 and H_2O ($HCHO + O_2 \rightarrow CO_2 + H_2O$) [63]. By means of DFT calculations, the ability of bidimensional (2D) and three-dimensional (3D) Cu_5 and Cu_7 clusters to oxidize partially methane to methanol was explored [64]. The calculation indicated that a highly selective Eley–Rideal pathway describes the reaction between an adsorbed molecule, AB^*, and the gaseous molecule, and involves homolytic C–H bond dissociation. The defining characteristic of an Eley–Rideal reaction is that one of the reactants is not chemisorbed locally, and hence, not in equilibrium with the surface temperature. Appearance of a nonadsorbed radical-like methyl intermediate was found to be favored when bicoordinated oxygen atoms, stabilizing at the edges of 2D clusters, are available. In addition, it was established that the reactivity of oxygen atoms adsorbed on copper clusters is closely linked to cluster size and morphology.

Density functional theory calculations were also performed to investigate the catalytic cycle of methane conversion to methanol in $Cu_2(O_2)^{2+}$ active sites of the Cu-ZSM-5 catalyst [65]. The calculation indicated that the oxidation has a higher energy barrier in the methane activation step, a very low energy barrier in the methanol formation step (36.3 kJ mol^{-1}) and a lower methanol desorption energy (52.5 kJ mol^{-1}). The methane activation step was found to follow the homolytic route and the heterolytic route for the $[Cu_2(O_2)]^{2+}$ and $[Cu_2(\mu\text{-}O)]^{2+}$ active sites, respectively, whereas the methanol formation step follows the direct radical rebound mechanism and the indirect rebound mechanism, respectively. The calculations further indicated that: (1) The electronic properties of the reactive O atoms in the active site influence their reactivity toward methane oxidation were evaluated; (2) the higher the spin density, the more negative the charge of the reactive O atom at the active site is; (3) the lower the energy barrier for methane activation, the more effective the catalyst can be; (4) the higher energy barrier of the methanol formation step can be in the triplet state; (5) a larger cluster model predicts the mechanism of the methane activation step and the effect of atomic charge of the O atom at the $[Cu_2(\mu\text{-}O)]^{2+}$; (6) the larger cluster model are consistent with those drawn using the smaller double-5T-ring cluster model; and (7) the partial oxidation of methane at $[Cu_2(O_2)]^{2+}$ can occur via a single step by direct insertion of one of the O atoms at the active site into the C–H bond of methane.

The review [67] summarized the main achievements obtained with photocatalyst alternatives to TiO_2 in the three main niches for this technology: water splitting for hydrogen production, decontamination and disinfection processes, and organic synthesis. For example, semiconductors based on cations with d0 configuration such Ta^{5+} or Nb^{5+}, and oxides or nitrides of d10 elements such as Bi^{3+}, In^{3+} or Ga^{3+} were found to be among the most successful novel catalysts [67]. Non-semiconductor solids like cation-interchanged zeolites and some semiconductors like ZnO or CdS also can produce promising results. The catalyst, having the coarse structure of the ZnO islands supported on a Cu_2O/Cu(111) substrate and providing a collection of catalytic multiple centers, was designed [68]. The catalyst was capable of methane activation at room temperature and transforms mixtures of methane and oxygen to methanol at 450 K with a selectivity of ~ 30%. ZnO–Cu_2O acted as active centers for methanol synthesis when exposed to CH_4 and O_2 due to an effective O–O bond dissociation, enabling a methane-to-methanol conversion. Upon addition of water, the defected O-rich ZnO sites, introduced by Zn vacancies, showed enhance the overall methanol selectivity to over 80%. Methane over-oxidation by dicopper $[Cu_2\ O]^{2+}$ and $[Cu_2\ O_2]^{2+}$ in zeolite mordenite (MOR) was examined sites using DFT [66]. The calculation showed that $[Cu_2\ O_2]^{2+}$ has a very large methane C–H activation barrier and stabilization of methanol and methyl at unreacted dicopper sites leads to over-oxidation via sequential hydrogen atom abstraction steps. For methanol, near 200 °C the process was initiated by abstraction of the CH_3 group followed by OH. The following reaction mechanism was suggested: (1) In the first step, methane is activated in the presence of O_2 on isolated Rh^+ cations under mild conditions to produce Rh–CH_3; (2) the latter can be functionalized via oxygen insertion to produce methanol, or CO insertion to produce acetic acid; (3) after the hydrolysis step, isolated Rh^{+1} species are available for the next catalytic cycle; and (4) the zeolite supported or the titanium-dioxide-supported catalyst during three-hour-long batch-reactor tests conducted at 150 °C, showed yield around 22,000 μmol of acetic acid per gram of catalyst, or around 230 μmol of methanol per gram of catalyst, respectively, with selectivities of 60–100%.

ZnO loaded with appropriate cocatalysts (Pt, Pd, Au, or Ag), enabling direct oxidation of methane to methanol and formaldehyde in water using only molecular oxygen as the oxidant under mild light irradiation at room temperature, was prepared [69]. It was found that: (1) Up to 250 μmol of liquid oxygenates with ~ 95% selectivity is achieved for 2 h over 10 mg of ZnO loaded with 0.1 wt% of Au; (2) molecular $^{18}O_2$, rather than $^{18}H_2O$, is the source of oxygen for direct CH_4 oxidation; and (3) ZnO and cocatalyst could concertedly activate CH_4 and O_2 into methyl radical and hydroperoxyl radical in water. The authors of work [70] underlined two equally significant aspects for efficient C–H bond activation of CH_4 over hematite and controllable activation of O_2. Hematite, iron oxide compound (Fe_2O_3) was found to be highly effective and thermally stable to catalyze methane combustion at temperatures (< 500 °C) with a low light-off temperature of 230 °C, a low apparent activation energy of 17.60 kcal mol^{-1} and 100% selectivity to a final product CO_2 [70]. Isotopic oxygen tracer experiments, diffuse reflectance infrared spectroscopy, Fourier transform spectroscopy (DRIFTS), and DFT calculations were performed.

The research indicated that the performance stems from a tetra-iron center with an antiferromagnetically coupled iron dimer on the hematite (110) surface. The results supported a Mars van Krevelen redox mechanism. According to this mechanism, CH_4 is activated by reaction with a hematite surface oxygen first, followed by a catalytic cycle through a molecular-dioxygen-assisted pathway. The evolution of reaction intermediates from a methoxy CH_3–O–Fe, to a bridging bidentate formate b-HCOO–Fe, to a monodentate formate m-HCOO–Fe was also revealed.

A review [71] discussed the advancements in the electrocatalytic and photocatalytic oxidation of methane at low temperatures with products containing oxygen atom(s). The reactions that occurred at the electrodes are given below

$$\text{Cathode: } O^2 + 4H^+ + 4e^- \rightarrow 2H_2O \text{ (1)}$$
$$\text{Anode: } CH_4 + H_2O \rightarrow CH_3OH + 2H^+ + 2e^-$$

CH_4 oxidation is thermodynamically favorable at modest potentials (e.g., CH_4 (g) + H_2O (l) $\rightarrow$ CH_3OH (aq) + $2H^+$ + $2e^-$ reaction has $E_0 = 0.586$ V versus normal hydrogen electrode (NHE) at pH = 0 and 298 K. It was found that a photocatalyst must have a conduction band minimum more negative than the potential of the O_2/O_2^- redox couple (− 0.16 V vs. normal hydrogen electrode, NHE) and a valence band maximum more positive than the potential of the ·OH/OH^- redox couple (+ 2.59 V vs. NHE).

At high overpotentials (at ≥ 0.8 V vs. reversible hydrogen electrode (RHE)), water oxidation (oxygen evolution) occurred at high rate [72]. Electrocatalytic CH_4 oxidation over multiwalled carbon nanotube Nafion/nickel hydroxide-modified nickel electrodes in 1.0 m NaOH solution was examined. A linear relationship between the oxidation peak current and the concentration of CH_4 was observed [72]. Thus, findings indicated that CH_4 was effectively oxidized to methanol by NiOOH electrocatalyst. A gas diffusion electrode (GDE) for the selective methanol electrosynthesis from CH_4 on a V_2O_5 incorporated TiO_2/RuO_2/polytetrafluoroethylene (PTFE) electrode was utilized [73]. Formaldehyde and formic acid were synthesized at the similar rate. In such oxidation reactions, electrochemically activated oxygen species (O*) act as oxidizing agent involving in several steps: (1) Cathodes can reductively activate O_2 molecules to generate O* ($O_2 + 2H^+ + 2e^- + Mn^+ \rightarrow$ O*–Mn^+ + H_2O), where Mn^+ is the active site on the cathode, such as Sm^{3+}, Fe^{3+}, and Cu^{2+}; (2) the O* can selectively oxidize light hydrocarbons to the corresponding oxygenates (O* + R–H $\rightarrow$ R–OH); (3) O* reacts with methane to produce methanol (O* + $CH_4 \rightarrow$ CH_3OH); and (4) different reactive oxygen species $O^{\cdot -}$ ($H_2O \rightarrow O^{\cdot -} + 2H^+ + e^-$) and $O_2^{\cdot -}$ ($2H_2O \rightarrow O_2^{\cdot -} + 4H^+ + 3e^-$) were formed electrochemically over $V4^+$ sites.

Mononuclear rhodium species, anchored on a zeolite or titanium dioxide support suspended in aqueous solution, can catalyze the direct conversion of methane to methanol and acetic acid, using oxygen and carbon monoxide under mild conditions [74]. Supported rhodium catalysts were prepared using a heat treatment protocol on the zeolite (ZSM-5) supports and by anchoring rhodium precursor species on

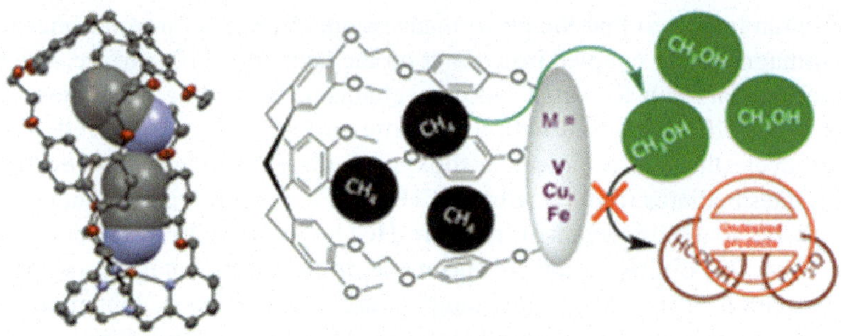

Fig. 9.14 Schematic presentation of nonheme iron, vanadium, and copper complexes bearing hemicryptophane cavities [75]. With permission from American Chemical Society

reduced titanium dioxide (titania) assisted by ultraviolet irradiation. The temperature at which the catalytic reaction was initiated was 110 °C in the case of Rh-ZSM-5. Results suggested that the conversion of methane to oxygenates may occur via a reaction pathway, in which methane is activated by formation M–CH_3, namely: (1) In the first step, methane is activated in the presence of O_2 on isolated Rh+ cations under mild conditions to produce Rh–CH_3; (2) the formed Rh–CH_3 can then be functionalized via two independent reaction pathways, namely oxygen insertion to produce methanol, or CO insertion to produce acetic acid; and (3) after the hydrolysis step, isolated Rh^{+1} species are available for the next catalytic cycle. The zeolites supported or the titanium-dioxide-supported catalyst, showing yield around 22,000 μmol of acetic acid per gram of catalyst, or around 230 μmol of methanol per gram of catalyst, respectively, with selectivities of 60–100%, were examined in three-hour-long batch-reactor tests150 °C. A batch reactor is a chemical reactor in which a noncontinuous reaction is conducted.

Nonheme iron, vanadium, and copper complexes bearing hemicryptophane cavities were evaluated in the oxidation of methane in water by hydrogen peroxide (Fig. 9.14) [75]. Cryptophanes

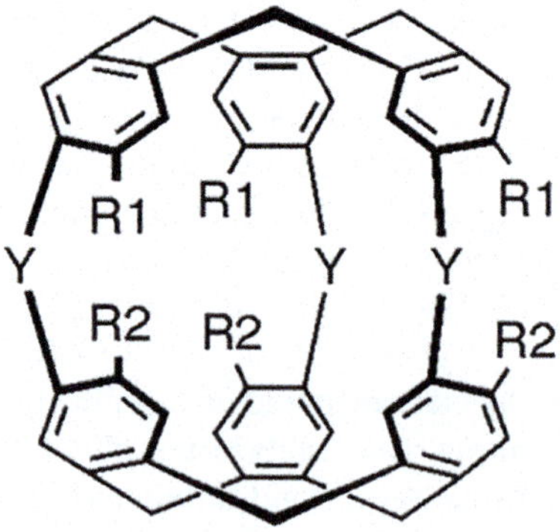

are a class of organic supramolecular compounds. To evaluate the importance of the supramolecular structure to catalytic efficiency, six supramolecular complexes, including the two novel tris(2-pyridylmethyl)amine-

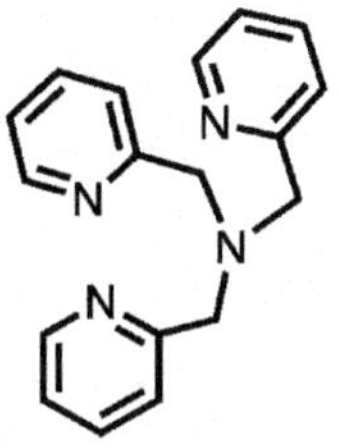

based hemicryptophane complexes CuII(Hm-TPA) and more effective FeII(Hm-TPA), displaying various first and second coordination spheres, were immobilized onto silica and examined. Several metal complexes bearing a hemicryptophane cavity and analogous complexes devoid of the cage were also prepared. The results obtained with the iron-(II)-based catalysts indicated the following: (1) The involvement of a high-valence metal-based oxidant occurs in a manner similar to that involved for enzymatic systems (MMO); (2) the catalytic performances of "encaged" and "naked" catalysts were evaluated by the direct comparison of the selectivity for primary oxidation products (CH_3OH and CH_3OOH) and the turnover number; and (3) copper complexes showed low catalytic efficiency; (4) vanadium and iron complexes exhibited higher turnover numbers, ≤ 13.2 and ≤ 9.2, respectively, providing target primary oxidation products (CH_3OH and CH_3OOH) as well as over-oxidation products (HCHO and HCOOH) (Fig. 9.14). Thus, this study highlighted the relevance of the hemicryptophane nanoreactors which can combine the efficiency of heterogeneous catalysis with the convenient character of homogeneous host–guest systems.

The **review** [76] accounted for recent progress in the field of single-atom catalysts (SACs) for the selective oxidation of CH_4 to C_{1-2} oxygenates. A single-atom catalyst (SAC) is the catalyst with a single atom on its surface. SACs with coordinated metal centers on active support have displayed high reactivity and selectivity for methane oxidation (Fig. 9.15). Specifically, SACs can significantly reduce the activation energy due to induced electrostatic polarization of the C–H bond. Suggestions that the distinct metal–support interaction can stabilize the intermediate, the chemical nature of catalytic sites, effects of metal–support interaction, and stabilization of intermediate species on catalysts to minimize over-oxidation were discussed in details.

In account [61], progress on thermal methane activation by metal oxide clusters doped with noble-metal atoms (Au, Pt, and Rh) as well as by oxygen-free species including carbides and borides of base metals (V, Ta, Mo, and Fe) was summarized. The generations of stable products such as formaldehyde, acetylene, and syngas as well as closed-shell species $AuCH_3$ and B_3CH_3 have been identified. The new mechanisms of synergistic methane activation by Lewis acid–base pairs (such as $Au^{\delta+}$-$O^{\delta-}$ and $B^{\delta+}$-$B^{\delta-}$) and by dinuclear metal centers (such as Ta-Ta) have been revealed. The new mechanisms suggested that: (1) In the reactions between methane

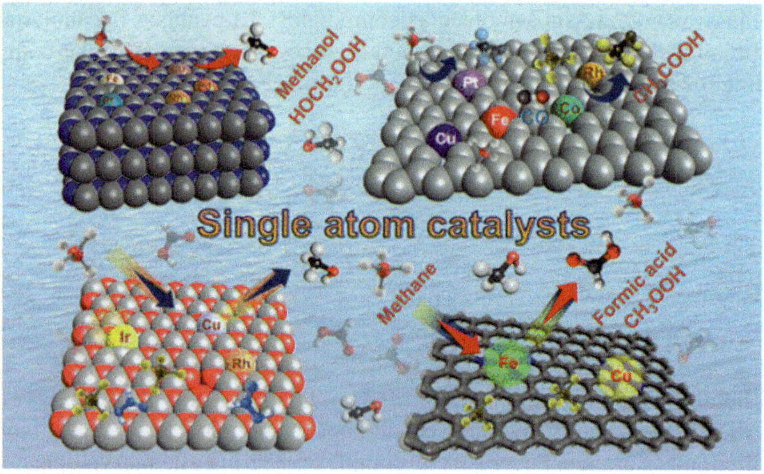

Fig. 9.15 Schematic presentation of the single-atom catalysts (SACs) for the selective oxidation of CH_4 to C_{1-2} oxygenates [76]. With permission from American Chemical Society

and oxide clusters doped with noble-metal atoms, the oxide cluster "supports" can accept the H atoms; (2) the CH species delivered through the noble-metal atoms; (3) then its transform methane into stable oxygenated compounds; (4) the product selectivity can be controlled by different noble-metal atoms (such as Pt vs. Rh); (5) the electronic structures of base metal centers can be engineered through carburization; and (6) the boron clusters (such as B_3 in VB_3^+) can be polarized by the metal cations to form the Lewis acid–base pair $B^{\delta+}$-$B^{\delta-}$ to cleave the C–H bond of methane.

The reconstitution of purified photosystem II (PSII) from *Thermosynechococcus elongatus BP*-1 into the membrane fraction containing particulate p-monooxygenase (pMMO) from *Methylosinus trichosporium OB3b* was performed [77]. This result indicated that the sequential redox chain from PSII through the quinone pool methane mono oxygenase (pMMO) can be constructed and that water can serve as the electron donor for methane hydroxylation under irradiation with light. When an electron donor was added for methane hydroxylation, pMMO in the membrane fraction produced hydrogen peroxide as a byproduct.

9.3 Molecular Imprinting

Molecular imprinting (MIP) is a technique to create template-shaped cavities in polymer matrices with predetermined selectivity and high affinity in a specific function [78–92]. Analogs of substrates, transition states, or products as template can be used in preparation of molecularly imprinted systems. Molecular imprinting on nanozymes provides a solution to substrate specificity problem, and the catalytic activity can be enhanced. The synthesis of molecular imprinting (MIP) materials can include the following steps (Figs. 9.16 and 9.17) [78]: (1) The template or print molecule is mixed with functional monomers that get positioned spatially around the template and guide the assembly; (2) the polymerization reaction creates a cross-linked matrix around the template; (3) the template is removed from the polymer network leaving a complementary cavity. A molecularly imprinted catalyst general scheme for selective catalysis is displayed in Figs. 9.16 and 9.17.

The process adsorption of substrates, reaction, and product release in printed nanozymes was studied using reaction kinetics measurement, dynamic light scattering, UV–vis spectrometry, and isothermal titration calorimetry [90]. Diffusion of the substrate across the imprinted gel layer was studied by a pre-incubation experiment. Difference between imprinted and nonimprinted gel layers was highlighted. Enrichment of local substrate concentration by imprinting and consequent increase activity was around eightfold and a substrate-imprinted sample had the lowest activation energy of 13.8 kJ mol^{-1}. Artificial enzyme with magnetic properties of F_3O_4 core, silica methacrylic layers, and peroxidase activity on indoleamine metabolite tumor marker was designed and characterized (Fig. 9.18) [85]. To protect the iron nucleus from oxidation and to provide anchoring for hydroxyl surface groups, the

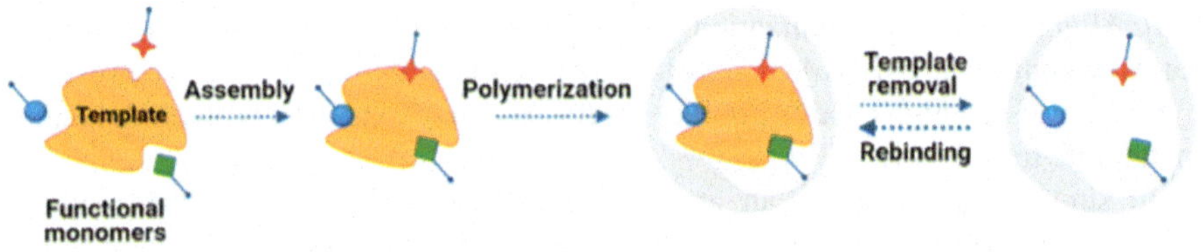

Fig. 9.16 Scheme of the principle of molecular imprinting recognition [78]. Published by MDPI

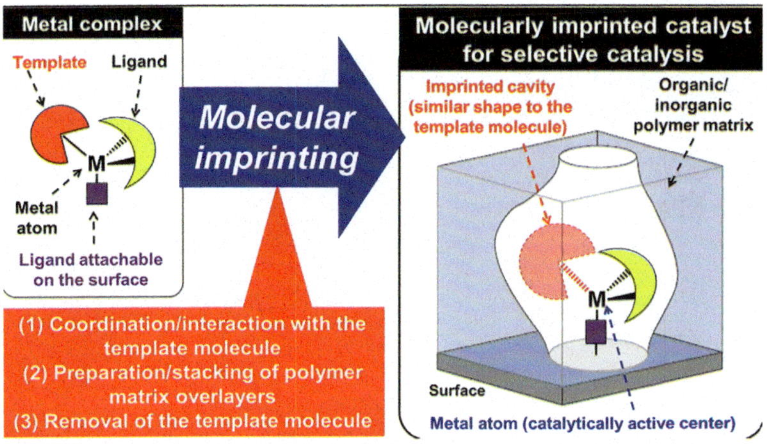

Fig. 9.17 Molecularly imprinted catalyst scheme for selective catalysis [3]. With permission from Elsevier

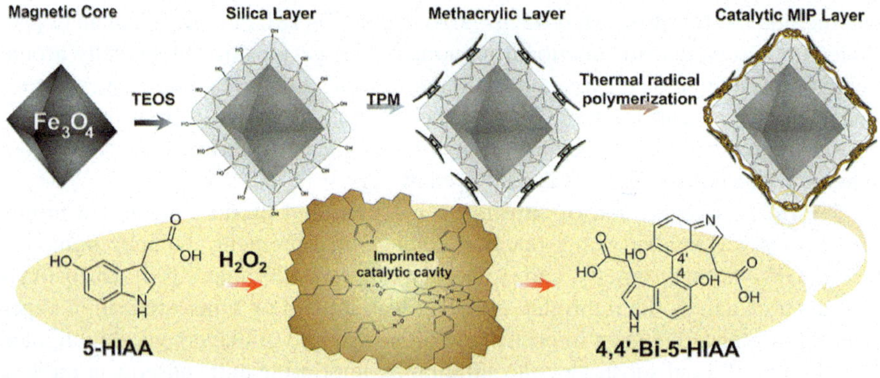

Fig. 9.18 Schematic presentation of imprinted catalytic system based on methacrylic layer polymerization [85]. With permission from Elsevier

magnetic molecularly imprinted catalytic stable methacrylic polymers that exhibit peroxidase-like activity toward 5-hydroxyindole-3-acetic acid (5-HIAA) oxidation were obtained from highly crystalline magnetite nuclei coated with a silica layer [85]. After acrylic functionalization, a molecularly imprinted polymer with hemin as catalytic center and 5-HIAA as template was attached to the structure. The resulting hybrid possessed high catalytic ability for the selective oxidation of the indoleamine metabolite tumor marker.

In study [91], to create an artificial hydrolase with specific selectivity, the molecular imprinting strategy was combined with the self-assembly of Fmoc-FFH

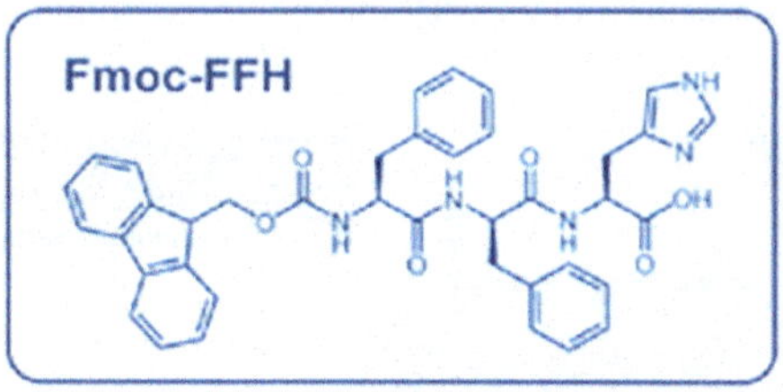

by using *p*-NPA

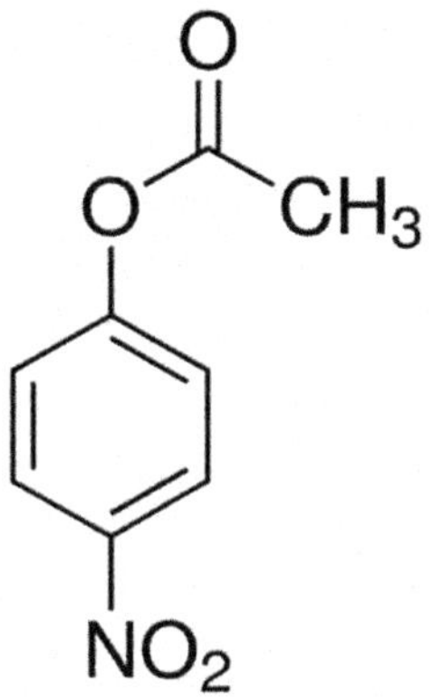

as the imprinting template. It was found that the imprinted polymer coating on the surface of the catalytic peptide nanofibers provided a specific binding site to *p*-NPA, enhancing the catalytic efficiency of the artificial hydrolase. The substrate selectivity of this artificial enzyme can be improved by changing the imprinting template.

Peroxidase (POD)-mimicking catalysts based on the strategy of peptide assembly Fmoc-FFH and molecular imprinting were constructed [86]. Specifically, CA-H/**Hemin** was fabricated and showed POD-like catalytic activity with 21-fold rate acceleration in the oxidation of 2,2′-azinobis-(3-ethylbenzthiazoline). Upon combining CA-H/Hemin with the 2,2′-azinobis-(3-ethylbenzthiazoline-6-sulphonate),

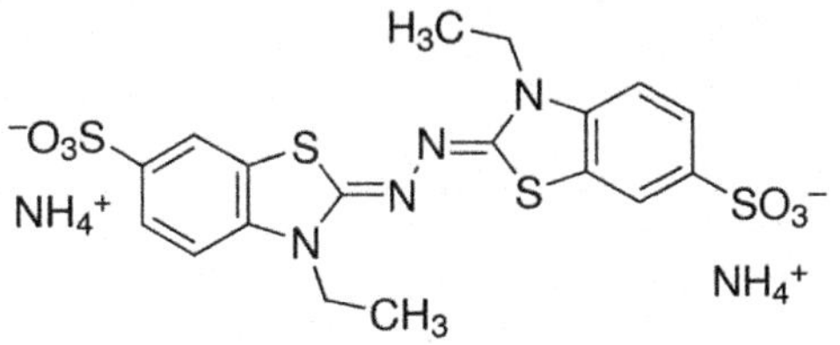

the obtained imprinted catalyst (MIP-H/Hemin) indicated 52-fold acceleration. The substrate specificity of MIP-H/Hemin and MIP$^+$-H/Hemin toward ABTS was also highlighted.

Molecular imprinting technology based on the possibility of imprinting within the nanospace of doubly cross-linked micelles was developed [87]. The obtained water-soluble MIP NPs have a hydrophobic/hydrophilic core–shell morphology, making this micellar imprinting useful for a variety of template molecules using molecular imprinting; a peroxidase-mimicking DNAzyme has achieved substrate specificity with enhanced activity in a nanoscale gel for three different substrates [88]. DNAzymes are DNA oligonucleotides were fund to be capable of performing a specific chemical reaction, similar to the action of ribozymes (enzymes composed of RNA). Specifically, a molecularly imprinted DNAzyme nanogel of high peroxidase-mimicking was prepared using Amplex Red

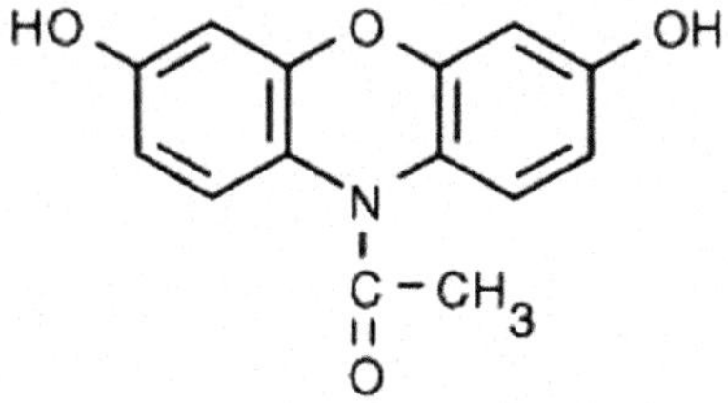

as the template [89]. The nanogel selectively oxidized Amplex Red in the presence of H_2O_2 to form a fluorescent product, while the oxidation of other substrates was inhibited. The activity of this imprinted DNAzyme was even higher than that of the free DNAzyme.

References

1. Reymond J-L, Darbre T (2012) Peptide and glycopeptide dendrimer apple trees as enzyme models and for biomedical applications. Org Biomol Chem 10:1483–1492
2. Zhang G, Mujahid Ali M, Feng X (2021) Mesoporous molecularly imprinted materials: from preparation to biorecognition and analysis. rAC Trends Anal Chem 144:116426
3. Muratsugu S, Shirai S, Tada M (2020) Recent progress in molecularly imprinted approach for catalysis. Tetrahedron Lett 61:151603
4. Geronimo I, Vidossich P, De Vivo M (2021) Local structural dynamics at the metal-centered catalytic site of polymerases is critical for fidelity. ACS Catal 11(22):14110–14121. https://doi.org/10.1021/acscatal.1c03840
5. Pecina A, Rosa-Gastaldo D, Riccardi L et al (2021) On the metal-aided catalytic mechanism for phosphodiester bond cleavage performed by nanozymes. ACS Catal 11(14):8736–8748
6. Likhtenshtein GI (2003) New trends in enzyme catalysis and mimicking chemical reactions. Kluwer Academic/Plenum Publishers, New York
7. Denisov ET, Sarkisov OM, Likhtenshtein GI (2003) Chemical kinetics. Fundamentals and recent developments. Elsevier Science
8. Li W-L, Head-Gordon T (2021) Catalytic principles from natural enzymes and translational design strategies for synthetic catalysts. ACS Cent Sci 7:72−80
9. Lyu Y, Scrimin P (2021) Mimicking enzymes: the quest for powerful catalysts from simple molecules to nanozymes cite. ACS Catal 11:11501–11509
10. Kirby AJ, Hollfelder F (2009) From enzyme models to model enzymes. RSC Publishing
11. Czescik J, Zamolo S, Darbre T, Mancin F, Scrimin P (2021) Factors influencing the activity of nanozymes in the cleavage of an RNA model substrate. Molecules 24:2814
12. Du Y, Sheng H, Astruc D, Zhu M (2020) Atomically precise noble metal nanoclusters as efficient catalysts: a bridge between structure and properties. Chem Rev 120:526–622
13. La Sorella G, Strukul G, Scarso A (2015) Recent advances in catalysis in micellar media. Green Chem 17:644–683
14. Wei H, Wang E (2013) Nanomaterials with enzyme-like characteristics (nanozymes): next-generation artificial enzymes. Chem Soc Rev 42(14):6060–6093
15. Lin Y, Ren J, Qu X (2014) Catalytically active nanomaterials: a promising candidate for artificial enzymes. Acc Chem Res 47(4):1097–1105
16. Wei H, Wang E (2008) Fe_3O_4 magnetic nanoparticles as peroxidase mimetics and their applications in H_2O_2 and glucose detection. Anal Chem 80(6):2250–2254. https://doi.org/10.1021/ac702203f. PMID 18290671

17. Liang M, Yan X (2019) Nanozymes: from new concepts, mechanisms, and standards to applications. Acc Chem Res 52:2190–2200
18. Scott S, Zhao H, Dey A, Gunnoe TB (2020) Nano-apples and orange-zymes. ACS Catal 10:14315−14317
19. Wu J, Wang X, Wang Q (2019) Nanomaterials with enzyme-like characteristics (nanozymes): next-generation artificial enzymes (II). Chem Soc Rev 48:1004–1076
20. Robert A, Meunier B (2022) How to define a nanozyme. ACS Nano 16(5):6956–6959. https://doi.org/10.1021/acsnano.2c02966
21. Ma Y, Zhu M, He Q, Zhao M, Cui H (2022) Photoenhanced oxidase–peroxidase-like $NiCo_2O_4@MnO_2$ nanozymes for colorimetric detection of hydroquinone. ACS Sustain Chem Eng 10(17):5651–5658. https://doi.org/10.1021/acssuschemeng.2c00621
22. Wu T, Huang S, Yang H, Ye N et al (2022) Bimetal biomimetic engineering utilizing metal–organic frameworks for superoxide dismutase mimic. ACS Mater Lett 4(4):751–757. https://doi.org/10.1021/acsmaterialslett.2c00075
23. Luo Q, Li J, Wang W et al (2022) Transition metal engineering of molybdenum disulfide nanozyme for biomimicking anti-biofouling in seawater. ACS Appl Mater Interfaces 14(12):14218–14225. https://doi.org/10.1021/acsami.2c00172
24. Lu Z, Lu N, Xiao Y et al (2022) Metal-nanoparticle-supported nanozyme-based colorimetric sensor array for precise identification of proteins and oral bacteria. ACS Appl Mater Interfaces 14(9):11156–11166. https://doi.org/10.1021/acsami.1c25036
25. Li S, Zhang Y, Wang Q, Lin A, Wei H (2022) Nanozyme-enabled analytical chemistry. Anal Chem 94(1):312–323. https://doi.org/10.1021/acs.analchem.1c04492
26. Wu J, Wang X, Wang Q et al (2019) Nanomaterials with enzyme-like characteristics (nanozymes): next-generation artificial enzymes (II). Chem Soc Rev 48:1004–1076
27. Liang M, Yan X (2019) Nanozymes: from new concepts, mechanisms, and standards to applications. Acc Chem Res 52(8):2190–2200
28. Liu M, Zhu Y, Jin D, Li L, Cheng J, Liu Y (2021) Hemin-caged ferritin acting as a peroxidase-like nanozyme for the selective detection of tumor cells. Inorg Chem 60(19):14515–14519. https://doi.org/10.1021/acs.inorgchem.1c01863
29. Zhao R, Liu H, Li Y, Guo M, Zhang X-D (2021) Catalytic nanozyme for radiation protection. Bioconjug Chem 32(3):411–429. https://doi.org/10.1021/acs.bioconjchem.0c00648
30. Ma M, Liu Z, Gao N et al (2020) Self-protecting biomimetic nanozyme for selective and synergistic clearance of peripheral amyloid-β in an alzheimer's disease model. J Am Chem Soc 142(52):21702–21711. https://doi.org/10.1021/jacs.0c08395
31. Han J, Yoon J (2020) Supramolecular nanozyme-based cancer catalytic therapy. ACS Appl Bio Mater 3(11):7344–7351. https://doi.org/10.1021/acsabm.0c01127
32. Gu C, Kong X, Yan S, Gai P, Li F (2020) Glucose dehydrogenase-like nanozyme based on black phosphorus nanosheets for high-performance biofuel cells. ACS Sustain Chem Eng 8(44):16549–16554. https://doi.org/10.1021/acssuschemeng.0c05743
33. Huang Y, Ren J, Qu X (2019) Nanozymes: classification, catalytic mechanisms, activity regulation, and applications. Chem Rev 119:4357–4412
34. Heckmann CM, Paradisi F (2020) Looking back: a short history of the discovery of enzymes and how they became powerful chemical tools. ChemCatChem 12:6082–6102
35. Jiang B, Duan D, Gao L et al (2018) Standardized assays for determining the catalytic activity and kinetics of peroxidase-like nanozymes. Nat Protoc 13:1506–1520
36. Gabrielli L, Prins LJ, Rastrelli F, Mancin F, Scrimin P (2020) Hydrolytic nanozymes. Eur J Org Chem 5044–5055
37. Cardoso AR, Frasco MF, Serrano V, Fortunato E, Goreti V, Sales F (2021) Molecular imprinting on nanozymes for sensing applications. Biosensors 11(5):152
38. Wu J, Wang X, Wang Q et al (2019) Nanomaterials with enzyme-like characteristics (nanozymes): next-generation artificial enzymes (II). Chem Soc Rev 48:1004
39. Lin YH, Xu C, Ren JS, Qu XG (2012) Using thermally regenerable cerium oxide nanoparticles in biocomputing to perform label-free, resettable, and colorimetric logic operations. Angew Chem Int Ed 51:12579–12583

40. Gao J, Zhuang L, Nie JB (2007) Intrinsic peroxidase-like activity of ferromagnetic nanoparticles. Nat Nanotechnol 2:577–583
41. Fan KL, Wang H, Xi JQ, Liu Q, Meng XQ, Duan DM et al (2017) Optimization of Fe_3O_4 nanozyme activity via single amino acid modification mimicking an enzyme active site. Chem Commun 53:424–427
42. Maddheshiya S, Nara S (2022) Recent trends in composite nanozymes and their pro-oxidative role in therapeutics. Front Bioeng Biotechnol Sect Nanobiotechnol. https://doi.org/10.3389/fbioe.2022.880214
43. Jiang D, Ni D, Rosenkrans ZT, Huang P, Yan X, Cai W (2019) Nanozyme: new horizons for responsive biomedical applications. Chem Soc Rev 48:3683–3704. https://doi.org/10.1039/C8CS00718G
44. Chen Z, Ji H, Liu C, Bing W, Wang Z, Qu X (2016) A multinuclear metal complex based DNase-mimetic artificial enzyme: matrix cleavage for combating bacterial biofilms. Angew Chem 128:10890–10894. https://doi.org/10.1002/ange.201605296
45. Khulbe K, Karmakar K, Ghosh S et al (2020) Nanoceria-based phospholipase-mimetic cell membrane disruptive antibiofilm agents. ACS Appl Bio Mater 3:4316–4328. https://doi.org/10.1021/acsabm.0c00363
46. Li J, Lei Y, Qing Z et al (2020) Al centre-powered graphitic nanozyme with high catalytic efficiency for pH-independent chemodynamic therapy of cancer. Chem Commun 56:6285–6628. https://doi.org/10.1039/D0CC01331E
47. Alfonso-Prieto M, Biarnes X, Vidossich P, Rovira C (2009) The molecular mechanism of the catalase reaction. J Am Chem Soc 131:11751−11761
48. Xi Z, Gao W, Biby A, Floyd A, Xia X (2022) Ultrasmall iridium nanoparticles as efficient peroxidase mimics for colorimetric bioassays. ACS Appl Nano Mater 5(5):6089–6093
49. Ge C, Fang G, Shen XM et al (2016) Facet energy *versus* enzyme-like activities: the unexpected protection of palladium nanocrystals against oxidative damage. ACS Nano 10:10436–10445
50. Ma Y, Zhu M, He Q, Zhao M, Cui H (2022) Photoenhanced oxidase−peroxidase-like $NiCo_2O_4@MnO_2$ nanozymes for colorimetric detection of hydroquinone. ACS Sustain Chem Eng 10:5651−5658
51. Tian R, Zhang B, Zhao M, Zou H, Zhang C, Qi Y, Ma Q (2019) Fluorometric enhancement of the detection of H_2O_2 using different organic substrates and a peroxidase-mimicking polyoxometalate. RSC Adv 9(22):12209–12217
52. Mao X, He F, Qiu D et al (2020) Efficient biocatalytic system for biosensing by combining metal-organic framework (MOF)-based nanozymes and G-quadruplex (G4)-DNAzymes. Anal Chem 94(20):7295–7302
53. Baretta R, Gabrielli V, Frasconi M (2022) Nanozyme–cellulose hydrogel composites enabling cascade catalysis for the colorimetric detection of glucose. ACS Appl Nano Mater, Article ASAP
54. María-Hormigos R, Molinero-Fernández A, Miguel Ángel López M, Jurado-Sánchez B, Escarpa A (2022) Prussian blue/chitosan micromotors with intrinsic enzyme-like activity for (bio)-sensing assays. Anal Chem 94(14):5575–5582
55. Li F, Li, S, Guo X et al (2020) Chiral carbon dots mimicking topoisomerase I to mediate the topological rearrangement of supercoiled DNA enantioselectively. Angew Chem Int Ed 59:11087–11092
56. Pengo P, Polizzi S, Pasquato L, Scrimin P (2005) Carboxylate-imidazole cooperativity in dipeptide-functionalized gold nanoparticles with esterase-like activity. J Am Chem Soc 127:1616–1617
57. Lyu Y, Morillas-Becerril L, Mancin F, Scrimin P (2021) Hydrolytic cleavage of nerve agent simulants by gold nanozymes. J Hazard Mater 415:125644
58. Mikolajczak DJ, Berger AA, Koksch B (2020) Catalytically active peptide–gold nanoparticle conjugates: prospecting for artificial enzymes. Angew Chem Int Ed 59:8776–8785
59. Quintana C, Ahumada JC, Ahumada G (2022) Proving cooperativity of a catalytic reaction by means of nanoscale geometry: the case of click reaction. J Am Chem Soc 144(25):11238–11245. https://doi.org/10.1021/jacs.2c02556

60. Tsai YF, Natarajan T, Lin ZH, Tsai IK, Janmanchi D, Chan SI, Yu SS (2022) Voltage-gated electrocatalysis of efficient and selective methane oxidation by tricopper clusters under ambient conditions. J Am Chem Soc 144(22):9695–9706
61. Zhao YX, Li ZY, Yang Y, He SG (2018) Methane activation by gas phase atomic clusters. Acc Chem Res 51(11):2603–2610
62. Zhang HT, Liu C, Liu P, Hu YH (2019) Mo_6S_8-based single-metal-atom catalysts for direct methane to methanol conversion. Chem Phys 151(2):024304
63. Chen X, Li Y, Pan X, Cortie D, Huang X, Yi Z (2016) Photocatalytic oxidation of methane over silver decorated zinc oxide nanocatalysts. Nat Commun 7:1
64. Gallego M, Corma A, Boronat M (2022) Sub-nanometer copper clusters as alternative catalysts for the selective oxidation of methane to methanol with molecular O_2. J Phys Chem A 126(30):4941–4951
65. Yu X, Zhong L, Li S (2021) Catalytic cycle of the partial oxidation of methane to methanol over Cu-ZSM-5 revealed using DFT calculations. Phys Chem Chem Phys 23(8):4963–4974
66. Panthi D, Adeyiga O, Odoh SO (2021) DFT analysis of methane C-H activation and overoxidation by $[Cu_2O]^{2+}$ and $[Cu_2O_2]^{2+}$ sites in zeolite mordenite: intra-versus inter-site overoxidation. ChemPhysChem 22(24):2517–2525
67. Wadak H, Yamada H, Watanabe Y, Mitsudo TA (1998) Development of alternative photocatalysts to TiO_2: challenges and opportunities. J Chem Soc Faraday Trans 94:1771
68. Huang E, Orozco I, Ramírez PJ (2021) Selective methane oxidation to methanol on $ZnO/Cu_2O/Cu(111)$ catalysts: multiple site-dependent behaviors. Am Chem Soc 143(45):19018–19032
69. Song H, Meng X, Wang S et al (2019) Direct and selective photocatalytic oxidation of CH_4 to oxygenates with O_2 on cocatalysts/ZnO at room temperature in water. J Am Chem Soc 141(51):20507–20515
70. He Y, Guo F, Yang KR (2020) Identification of reaction intermediates and mechanistic understandings of methane oxidation over hematite: a combined experimental and theoretical study. J Am Chem Soc 142(40):17119–17130
71. Sher Shah MSA, Oh C, Park H, Hwang YJ, Ma M, Park JH (2020) Catalytic oxidation of methane to oxygenated products: recent advancements and prospects for electrocatalytic and photocatalytic conversion at low temperatures. Adv Sci (Weinh) 7(23):2001946
72. Qiao J, Tang S, Tian Y et al (2009) A nano-nickel electrochemical sensor for sensitive determination of chemical oxygen demand. Sens Actuators B-Chem 138:402
73. Rocha RS, Camargo LM, Lanza RV, Bertazzoli R (2010) A feasibility study of the electrorecycling of greenhouse gases: design and characterization of a (TiO_2/RuO_2)/PTFE gas diffusion electrode for the electrosynthesis of methanol from methane. Electrocatalysis 1:224
74. Shan J, Li M, Allard LF, Lee S, Flytzani-Stephanopoulos M (2017) Mild oxidation of methane to methanol or acetic acid on supported isolated rhodium catalysts. Nature 551(7682):605–608
75. Ikbal SA, Colomban C, Zhang D, Delecluse M, Brotin T, Dufaud V, Dutasta JP, Sorokin AB, Martinez A (2019) Bioinspired oxidation of methane in the confined spaces of molecular cages. Inorg Chem 58(11):7220–7228
76. Kumar P, Al-Attas TA, Hu J, Kibria MG (2022) Single atom catalysts for selective methane oxidation to oxygenates. ACS Nano 16(6):8557–8618
77. Kondo R, Yoshimori K, Kamachi T (2018) Methane hydroxylation with water as an electron donor under light irradiation in the presence of reconstituted membranes containing both photosystem ii and a methane monooxygenase. ChemBioChem 19(20):2152–2155
78. Cardoso AR, Frasco MF, Serrano V et al (2021) Molecular imprinting on nanozymes for sensing applications. Biosensors 11(5):152
79. Resmini M (2012) Molecularly imprinted polymers as biomimetic catalysts. Anal Bioanal Chem 402:3021–3026
80. Wulff G (2002) Enzyme-like catalysis by molecularly imprinted polymers. Chem Rev 102:1–27
81. Mathew D, Thomas B, Devaky KS (2018) Transition state analogue imprinted polymers as artificial amidases for amino acid p-nitroanilides: morphological effects of polymer network on catalytic efficiency. Artif Cells Nanomed Biotechnol 46:1830–1837

82. Pan J, Liu S, Jia H (2019) Rapid hydrolysis of nerve agent simulants by molecularly imprinted porous crosslinked polymer incorporating mononuclear zinc (II)—picolinamine-amidoxime module. J Catal 380:83–90
83. Stepanova M, Solomakha O, Ten D, Tennikova T, Korzhikova-Vlakh E (2020) Flow-through macroporous polymer monoliths containing artificial catalytic centers mimicking chymotrypsin active site. Catalysts 10:1395
84. Zhang Z, Zhang X, Liu B, Liu JJ (2017) Molecular imprinting on inorganic nanozymes for hundred-fold enzyme specificity. Am Chem Soc 139(15):5412–5419
85. Antuña-Jiménez D, Blanco-López MC, Miranda-Ordieres AJ, Lobo-Castañón MJ (2014) Artificial enzyme with magnetic properties and peroxidase activity on indoleamine metabolite tumor marker. Polymer 55:1113–1119
86. Li J, Zhu M, Wang M et al (2020) Molecularly imprinted peptide-based enzyme mimics with enhanced activity and specificity. Soft Matter 16:7033–7039
87. Hu L, Zhao Y (2017) Cross-linked micelles with enzyme-like active sites for biomimetic hydrolysis of activated esters. Helv Chim Acta 100:1–10
88. Zhang Z, Liu B, Liu J (2017) Molecular imprinting for substrate selectivity and enhanced activity of enzyme mimics. Small 13:1–7. https://doi.org/10.1002/smll.201602730. Epub 2016
89. Zhang Z, Liu J (2018) Intracellular delivery of a molecularly imprinted peroxidase mimicking DNAzyme for selective oxidation. Mater Horiz 5:738–744
90. Zhang Z, Li Y, Zhang X, Liu J (2019) Molecularly imprinted nanozymes with faster catalytic activity and better specificity. Nanoscale 11(11):4854–4863
91. Zhu M, Wang M, Qi W, Su R, He Z (2019) Constructing peptide-based artificial hydrolases with customized selectivity. J Mater Chem B 7:3804–3810
92. Pengo P, Pasquato L, Scrimin P (2022) Substrate modulation of the activity of an artificial nanoesterase made of peptide-functionalized gold nanoparticles. Angevandte Chimi 119(3):404–408

Chapter 10
Artificial Photosynthesis

Abstract The studies of artificial photosynthesis use biomimetic techniques to replicate the process of natural photosynthesis, which employs abundant resources of sunlight, water, and carbon dioxide to produce oxygen and energy-rich carbohydrates. In this chapter, requirements for successive mimicking of these processes and means for fulfillment, these requirements were formulated. The compounds such as porphyrins or other cyclic tetrapyrroles, coratenoides, ruthenium and zinc complexes, fullerenes, borondipyrromethene, melanin, polyphenylene, go forth, can be light harvesters. Examples of charge separation in artificial donor–acceptor and pairs and photophysical and photochemical processes in dual flourophore-nitroxide molecules were presented and discussed in separate sections. The evolution dioxygen from water in a cluster of transition metals in the biological systems at the absorption of four light quanta of low energy occurs by a sequence of elementary steps: four one-electron steps of oxidation of the manganese complex and, most probably by one four-electron elementary step of O_2 evolution (Sect. 3.5.2). In approaching mimicking problem, a number of artificial manganese clusters and other transition metal clusters were synthesized and investigated. The reduction, or fixation, of carbon dioxide is a vital process associated with natural and artificial photosynthesis in which can generate hydrocarbon fuels such as formic acid, methanol, carbon monoxide, and methane, with the utilization of hydrogen. Employing of transition metal complexes catalyzed carboxylation reactions with CO_2 and carboxylation of preactivated substrates and examples of successive carboxylation were presented The chapter focused on advances in the field of direct carboxylation reactions of C(sp^3)–H and C(sp^2)–H bonds using CO_2 encompassing both transition metal catalysis and base-mediated approach and on applications of electrochemical and photochemical methods.

G. I. Likhtenshtein, *Enzyme Catalysis Today and the Chemistry of the 21st Century*,
https://doi.org/10.1007/978-3-031-51552-1_10

10.1 Introduction

Artificial photosynthesis is mimicking the process of natural photosynthesis, uses abundant resources of sunlight, water, and carbon dioxide to produce oxygen and energy-rich carbohydrates [1–49]. Fuels that are able to be produced by artificial photosynthesis include hydrocarbons such as formic acid (HCOOH), methanol (CH_3OH), carbon monoxide (CO), and methane (CH_4), or pure hydrogen fuel. The chemical processes involved in artificial photosynthesis include the splitting of water into oxygen and hydrogen, and the reduction of carbon dioxide into various hydrocarbons. In this way, energy can be being stored densely and cheaply in the form of a fuel (Fig. 10.1) [1–12].

There are three principal steps in artificial photosynthesis, which are similar to natural photosynthesis: the absorption of light to reach an excited state, charge generation and separation, and chemical conversion for the production of fuel. There is also the possible extension of carbon dioxide fixation for the further production of other fuels besides pure hydrogen.

Photoinduced charge separation is the process of an electron in an atom or molecule, being excited to a higher energy level by the absorption of a photon and

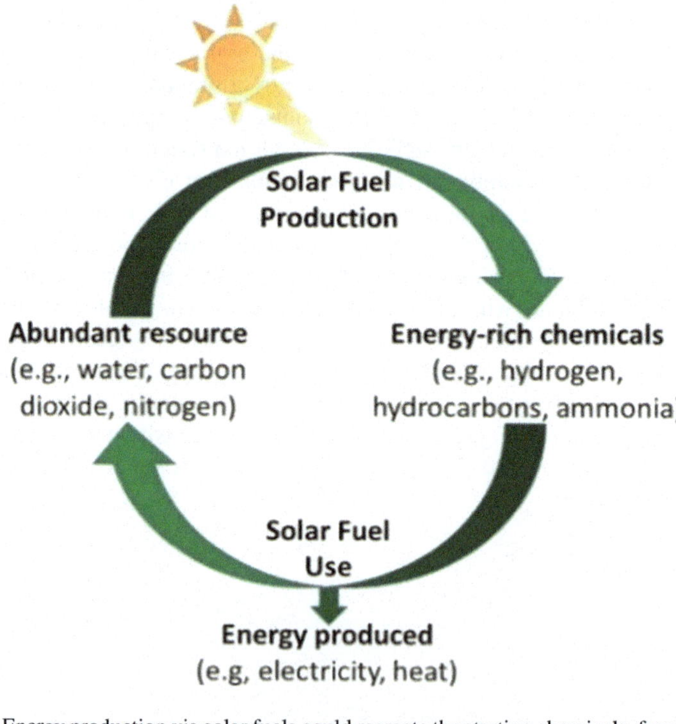

Fig. 10.1 Energy production via solar fuels could recreate the starting chemicals, forming a closed cycle that minimizes unwanted by-products. https://www.energy.gov/science/doe-explainssolar-fuels

then electron transfer to a nearby electron acceptor. Charge separation processes play key role in photovoltaics, natural and artificial photosynthesis, and photoluminescence sensors and switches [2–49]. There are two main objections to the mimicking of the photosynthetic process of light energy conversion in reaction centers via the mechanism of charge photoseparation on chemically active cation and anion radicals: (1) establishing factors affecting fast stages of primary electron transfer in donor–acceptor pairs using femto-, pico-, and nanosecond time domain techniques and (2) building donor–acceptor structures capable of retaining the photoseparated structures long enough for secondary chemical reactions of the charges to occur.

For successive mimicking the following requirements, sometimes contradictory ones, should be fulfilled: (1) Creation of light harvesting system with broad absorption spectrum covering sun light radiation spectra; (2) Construction of systems which have long charge separation state lifetimes; (3) Minimization of a loss of light energy during the multistep electron transfer processes; (4) Maintaining high reactivity of cation and anion segments after charge separation sufficient for consequent chemical and photochemical reactions; and (5) Prevention of side reactions including the donor exited state quenching.

The following means for fulfillment of these requirements can be formulated: (1) Chromophores in excited singlet state of high energy as electron donor should be preferably used; (2) The thermodynamic gap between ET intermediates should be optimum; (3) To avoid a competition with the spontaneous quenching of the primary donor exited state, the first steps of the charge separation (CS) by electron transfer should be fast; (4) The best way to prevent the fast charges recombination (CR) is a construction of a cascade system of intermediate acceptors separated by nonconductive bridge of 6–8 Å composed with nonsaturated chemical bonds [18–21]; (5) A promising way to slow down charges recombination (CR) is an organization of the spin catalysis which provide singlet-triplet conversion on the CS pair; and (6) A redox system can be protected from side reactions by using for example membranes or polymers, which separate initial and final reagents, or by incorporation of donor and acceptor groups in a protein or other matrix, as it is realized in natural photosynthesis.

10.2 Light Harvesting and Charge Separation in Artificial Photosynthetic Systems

10.2.1 Artificial Light Harvesting Antenna

The requirement of an antenna coupled to a photochemical reaction center, as occurs in natural photosynthetic constructs, is a necessary feature for efficient photocatalytic operation in artificial solar energy conversion systems. Another principal requirement is high quantum yield of the photo separation, which can be provided by optimum combination of fast electron transfer and slow recombination in photochemical cascade systems keeping high chemical redox reactivity of the photoseparated pair. The specific classes of compounds such as porphyrins or other cyclic tetrapyrroles, coratenoides, ruthenium and zinc complexes, fullerenes, borondipyrromethene, melanin, go forth, have been attracted special interest. A promising means to increase variety of effective and durable photochemical light conversion systems is the use of templates, for examples zeolites and mesoporous silica, TiO_2, nanofibers, polymers, liposomes, metal-protein complexes, fullerenes, go forth. It is necessary to stress an outstanding role of fullerenes as antennas, electron acceptors, and molecular wires.

A review [22] discussed self-assembly strategies for integrating light harvesting and charge separation in artificial photosynthetic systems. Self-assembly strategies involving π-stacking in covalent building blocks based on chemically robust arylene imide and diimide dyes, biomimetic porphyrins, and chlorophylls were developed [21]. The constructed blocks were used to integrate light harvesting with charge separation and transport. To elucidate the solution phase structures, monodisperse noncovalent aggregates systems small- and wide-angle X-ray scattering (SAXS/WAXS) from a synchrotron source were exploited. The author stressed that the central scientific challenge is to develop small, functional building blocks with a minimum number of covalent linkages. In the search for design principles of efficient artificial light harvesters, an approach based on self-assembly of pyropheophorbide-peptide conjugates was proposed and the solvent-driven and controlled aggregation of the peptide moieties was prepared [23]. The formation of a dense network of interacting pigments gave rise to an excitonic network. The obtained system was characterized by intense and spectrally wide absorption bands. The ultrafast dynamics of the nanosystems was studied using two-dimensional electronic spectroscopy. The experiment revealed that the excitation energy is funneled in an ultrafast time range (hundreds of femtoseconds).

Results from transient absorption spectroscopy on a series of artificial light harvesting dyads made up of a zinc phthalocyanine (**Pc**) covalently linked to carotenoids with 9, 10, or 11 conjugated carbon–carbon double bonds (dyads 1, 2, and 3), were presented [17]. The results indicated that: (1) The energy transfer and excited-state deactivation pathways followed excitation of the carotenoid S_2 state as a function of the conjugation length; (2) The S_2 state rapidly relaxes to the S^* and S_1 states; (3) In all systems a new pathway of energy deactivation within the carotenoid manifold in which the S^* state acts as an intermediate state was detected; (4) The $S_2 \rightarrow S_1$ internal conversion pathway on a sub-picosecond time scale was revealed; (5) In dyad **3**, a novel type of collective carotenoid–Pc electronic state that may correspond to a carotenoid excited state(s)–Pc Q exciplex was observed; (6) In dyad **1**, two carotenoid singlet excited states, S_2 and S_1, contribute to singlet–singlet energy transfer to Pc; (7) The process is very efficient ($> 90\%$) while for dyads **2** and **3** the S_1 energy transfer channel is precluded and only S_2 is capable of transferring energy to Pc; and (8) In the latter two systems, the lifetime of the first singlet excited state of Pc is shortened compared to the 9 double-bond dyad and model Pc, indicating that the carotenoid acts as a strong quencher of the phthalocyanine excited-state energy.

A biomimetic light harvesting antenna based on negatively charged poly(phenylene ethynylene) conjugated polyelectrolytes assembled within a positively charged lipid membrane scaffold, formed by the lipid 1,2-dioleoyl-3-trimethylammonium-propane (DOTAP), was constructed [22]. For a lipophilic energy acceptor (the cyanine dye DiI) effective molar absorption was enhanced by up to 18-fold due to the antenna effect. Absorption amplification of DiI was found to be due primarily to direct energy transfer from polymers. The membrane embedding fullerene derivative phenyl-C_{61}-butryic acid methyl ester (PCBM)

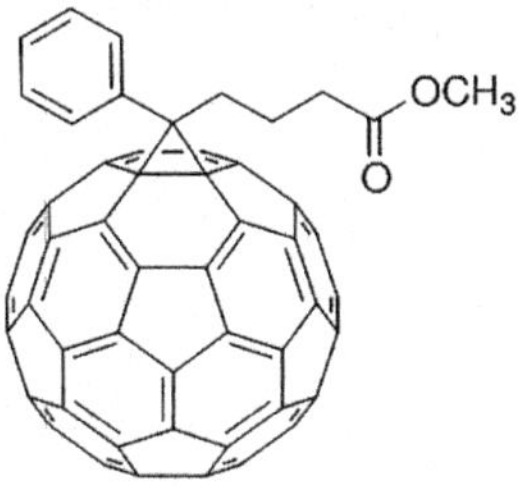

acts as an electron acceptor. It was suggested that the light harvested by the polymers was transferred via through-space mechanisms.

10.2.2 Charge Separation in Artificial Donor–Acceptor Pairs

In works [10, 22, 24–55] a congruent and systematic set of well-designed models has been synthesized and the photochemical and photophysical properties of these models were characterized. Below are a few typical examples. In the covalent donor–acceptor–acceptor molecule (D–A–A) in toluene at 295 K (Fig. 10.2), selective photoexcitation of with a 416 nm, 120 fs laser pulse initiates a series of photophysical and photo chemical processes [28]. The rate of intramolecular charge transfer from biphenyl to naphthalene was determined for the radical anions and radical cations of molecules with the general structure: (2-naphthyl)-(steroid spacer)-(4-biphenylyl). Degrees of unsaturation (one double bond, NSenB; two double bonds, $NSen_2B$; and the b-ring completely aromatized, NSarB) incorporated into the steroid spacer were varied. The several main results were reported: (1) The charge-transfer rate, the decay of the biphenyl radical ion absorption increased in all cases relative to the completely saturated 3-(2-naphthyl)-16-(4-biphenylyl)-5α-androstane (NSB) reference molecule; (2) For the anion charge transfer, the decay rates increased by factors of 1.4, 4.2, and 5.1, respectively; and (3) For the cation, the decay rates increased by factors of 5, 276, and 470. To explain these results, the charge-transfer process was viewed as a combination of two independent mechanisms: a single-step, superexchange mechanism, and a two-step, sequential charge. The critical variable for this analysis was the free energy of formation ($\Delta G_I°$) of the intermediate state: (2-naphthyl)-[spacer]$^1\pm$-(4-biphenylyl). It was concluded that superexchange

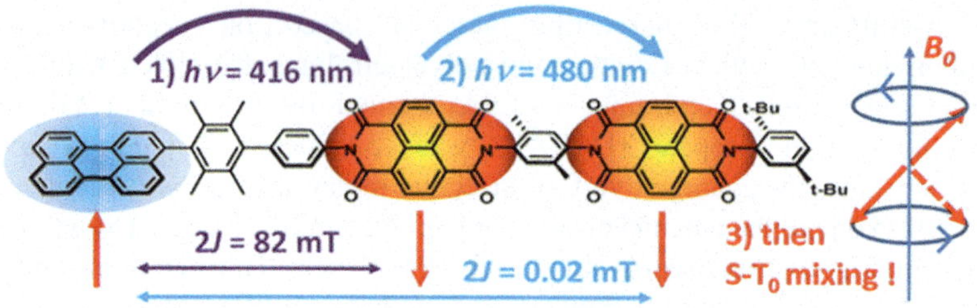

Fig. 10.2 Schematic illustration of photophysical and photo chemical processes in a triad [28]. With permission from American Chemical Society

is the dominant mechanism when $\Delta G_{\mathrm{I}}°$ is large, but at small $\Delta G_{\mathrm{I}}°$, the sequential mechanism will dominate.

Photoinduced intramolecular electron transfer (ET) and energy transfer (EnT) processes in two rotaxanes, one containing both zinc porphyrin and C60 fullerene moieties incorporated around the Cu(I) bisphenanthroline core [**(ZnP)$_2$**-Cu(I)(phen)$_2$-C60] and a second complex lacking the fullerene [(ZnP)$_2$-Cu(I)(phen)$_2$] [16] (Fig. 10.3) were studied by time-resolved electron paramagnetic resonance (TREPR) spectroscopy at 9.5 GHz (X-band) combined with a selective photoexcitation of the rotaxane moieties. The experiments were carried out in isotropic toluene and ethanol and in anisotropic nematic liquid-crystal (E-7) media over a wide range of temperatures. The results indicated that the efficiencies and pathways of the light-driven ET and EnT processes in both rotaxanes depend on the properties of their microenvironment, resulting in the formation of distinct charge-separated states under different experimental conditions.

A series of donor−bridge−acceptor (D–B–A) molecules was examined [29]. In the supermolecule, the photoexcited charge-transfer state of 3,5-dimethyl-4-(9-anthracenyl)julolidine (DMJ−An) was the donor, and naphthalene-1,8:4,5-bis(dicarboximide) (NI) was the acceptor, along with 1,1-diphenylethene, *trans*-stilbene, diphenylmethane, and xanthone bridges [29]. Time-resolved spectroscopy indicated that photoinitiated charge separation (CS) through the cross-conjugated 1,1-diphenylethene bridge and through the diphenylmethane bridge is about 30 times slower than through its linearly conjugated *trans*-stilbene counterpart. It was suggested that cross-conjugation decreased the π orbital contribution to the donor−acceptor electronic coupling electron transfer that most likely uses the bridge σ system as its primary CS pathway. The effect of relative orientation on the charge-transfer CT that excited-state energies of porphyrin-fullerene dyads (base- and Zn-tetraphenyl porphyrin coupled to C_{60} as the acceptor molecule in an end-on configuration) was examined [30]. The energetics of a few low-lying CT states for the end-on geometry to calculated CT energetics of a co-facial orientation were compared. The calculated CT excitation energies were obtained to be larger for the end-on orientation in comparison with the co-facial structure by about 0.7 eV. Furthermore, changes in relative donor–acceptor orientation have a larger impact on the CT energies than changes in donor–acceptor distance.

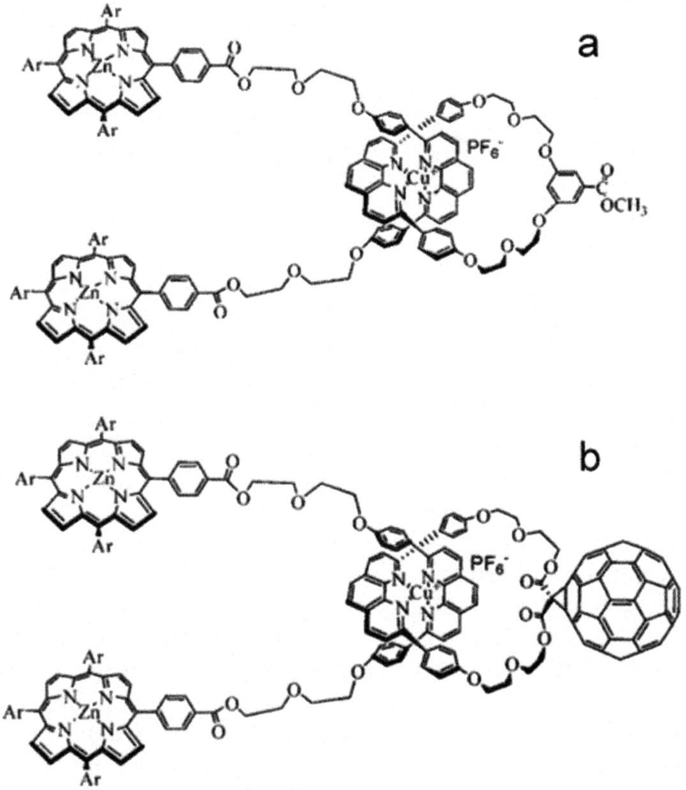

Fig. 10.3 Structure of a complex containing both zinc porphyrin and C60 fullerene moieties incorporated around the Cu(I) bisphenanthroline core [$(ZnP)_2$-Cu(I)$(phen)_2$-C60] and a second complex lacking the fullerene [$(ZnP)_2$-Cu(I)$(phen)_2$] [16]. With permission from American Chemical Society

Double excitation by nanosecond pump–pump–probe experiments was used to interrogate the photoinduced charge transfer and charge accumulation on a molecular dyad composed of a porphyrin chromophore and a ruthenium-based catalyst in the presence of a reversible electron acceptor [31]. The experiment indicated rapid reverse electron transfer to the photosensitizer upon the second excitation and the low driving force of the forward photodriven electron transfer reaction. Such a method allowed the understanding of the relaxation mechanism after two sequential photon absorptions, deciphering the undesired electron transfer reactions that limit the charge accumulation efficiency.

The effect of positioning an electron donor, ferrocene (Fc), and a charge-transfer complex, Fc–tetracyanobutadine (TCBD), at different locations of the BF_2-chelated dipyrromethene (BODIPY) ring

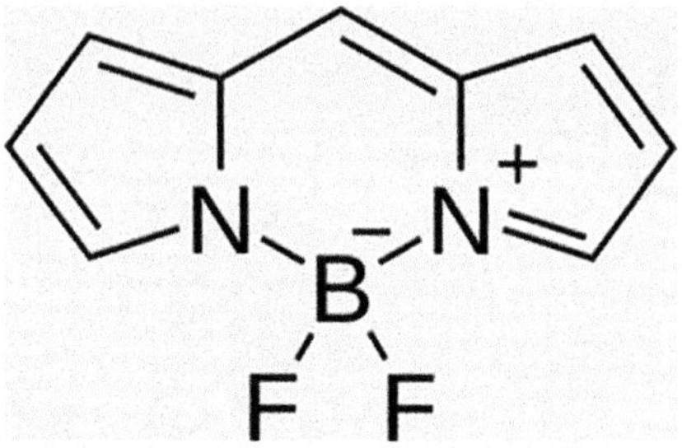

on governing excited-state charge separation was examined [32]. For this series, the Stokes shift followed the order β > meso > α-substitution for a given series of BODIPY derivatives, while $E_{0,0}$ followed the order meso > α- > β-substitution. BODIPY has been modified to contain an electron donor moiety, 8-(2,4,5-trimethoxyphenyl)-4,4-difluoro-1,3,5,7-tetramethyl-4-bora-3a,4a-diaza-*s*-indacene (MEOPHBDP) [33]. The photovoltaic measurements combined with femtosecond laser flash photolysis were performed using a standard two-employing electrode system consisting of a working electrode and a Pt sputtered electrode in methoxyacetonitrile containing 0.5 M iodide and 0.05 M I_2. The lifetime of the charge-separated state of MEOPHBDP was found to be 59 ps at 298 K. The dye-sensitized solar cells (DSSC) were prepared using MEOPHBDP with carboxylic acid (MEOPHBDP−COOH). The photovoltaic measurements were performed utilizing a standard two-electrode system consisting of a working electrode and a Pt sputtered electrode in methoxyacetonitrile containing 0.5 M iodide and 0.05 M I(2).

The photophysical properties of two pyrene–bodipy molecular dyads, composed of a phenyl-pyrene (Py-Ph) linked to the *meso* position of a bodipy (BD) molecule with either H-atoms (BD1) or ethyl groups (BD2) at the 2,6 positions, were investigated by stationary, nanosecond, and femtosecond spectroscopy [34]. Experiments revealed a weak coupling among the subunits in both dyads (Py-Ph-BD1 and Py-Ph-BD2) and showed that electron transfer occurs within 0.3–0.5 ps which is independent of the solvent and the type of the BD subunit. Excited state of Py-Ph-BD1 in ACN was quenched having a lifetime of 1 ns. Target analysis of the TA spectra has shown that the charge separation (CS) follows inverted kinetics and is slower than the recombination of the charge-separated state. Three bodipy-based (BDP = 4,4-difluoro-4-bora-3a,4a-diaza-*s*-indacene) donor–acceptor dyads were synthesized, and their ground-state and photophysical properties were characterized [35]. The electronic coupling between the BDP chromophore and an electron-donating carbazole occurs. By controlling the rate of photoinduced internal charge transfer (ICT), effect of variation of the linkage between the donor and acceptor moieties on the photophysical behavior of excited-state deactivation was examined. Density functional theory (DFT) and time-dependent DFT (TDDFT) showed that, when attachment occurred through the *beta* position of BDP, molecular orbital density spread across the HOMO. A Cu(I)-promoted oxidative dimerization of **BODIPY dyes** was developed to give a series of α,α-ethylene-bridged BODIPY dimers and trimers [48]. This methodology relies on the singlet electron transfer process between alkylated BODIPYs and Cu(I) salt to generate BODIPY-based radical species, which undergo a selective radical homocoupling reaction.

A series of 3,5-*bis*(hetero)arylethenyl-substituted BODIPY derivatives have been prepared by Knoevenagel-type condensation of alkyl-substituted BODIPY with the corresponding aldehydes [37]. Alkylation at the *N*-pyrrolic position of the ethenyl substituent allowed for the installation of the hydrophilic PEG group and afforded amphiphilic BODIPY derivatives. 2-pyrrolylethenyl-substituted derivatives featured near-IR emission (λ_{em} > 700 nm) with a high fluorescence quantum yield. Both the emission maxima and fluorescence quantum yields were insensitive to solvent polarity, contrary to the corresponding near-IR-emitting 4-(*N*,*N*-dimethylaminophenyl)ethenyl derivatives.

The photoinduced charge-transfer process in a hydrogen-bonded pair of melamine (donor)

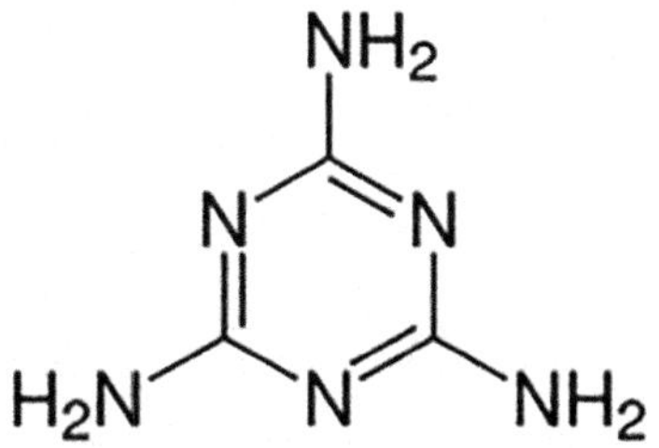

and isocyanuric acid (acceptor)

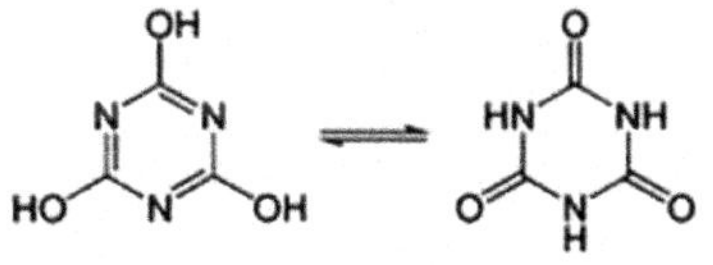

1 2

(Fig. 10.4) was investigated by using real-time quantum–classical simulations of the relaxation from the Franck–Condon point to the region of strong nonadiabatic coupling, where electron transfer occurs [38]. The simulation, suggesting the of strong nonadiabatic coupling, revealed how the charge transfer is coupled to specific vibrational modes and how it is affected by isotope substitution. The importance of resonance in nuclear and electron dynamics and the role of dynamic symmetry breaking was emphasized and exchange of angular momentum between nuclear and electronic subsystems in an electron–nuclear resonant process was inferred. Specifically, photoinduced electron transfer and vibronic coupling in a pair of melamine

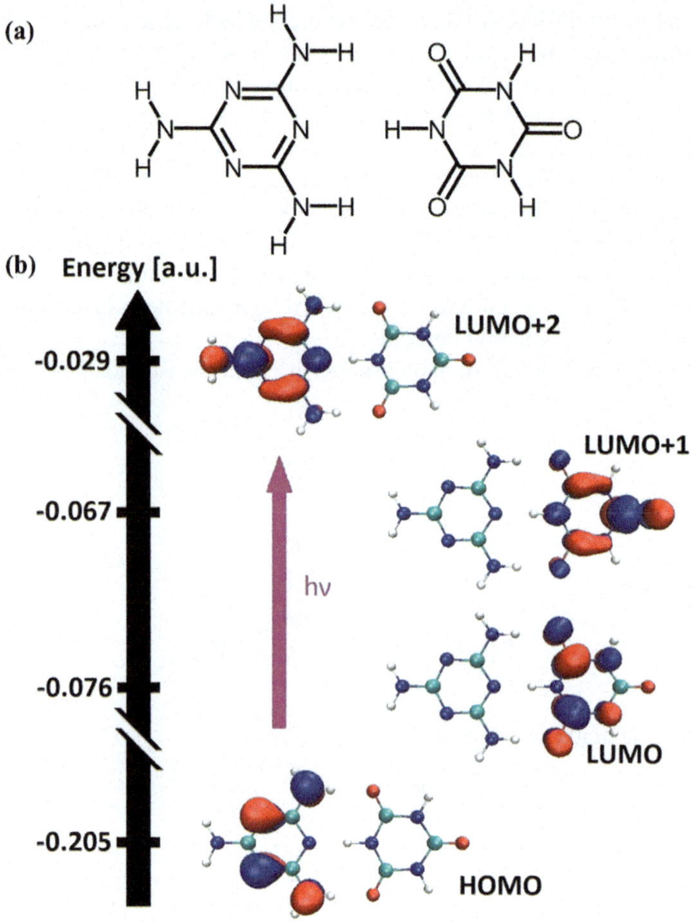

Fig. 10.4 **a** Chemical structure of the pseudo base pair melamine (left, donor) and isocyanuric acid (right, acceptor) and **b** Kohn–Sham orbitals relevant for the photoinduced charge transfer [38].

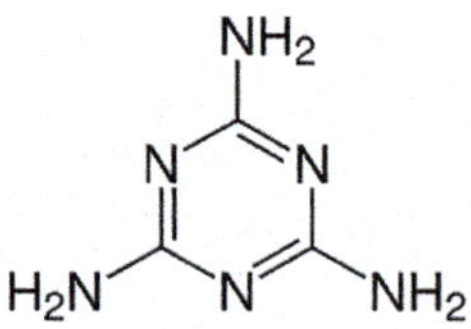

(donor) and isocyanuric acid

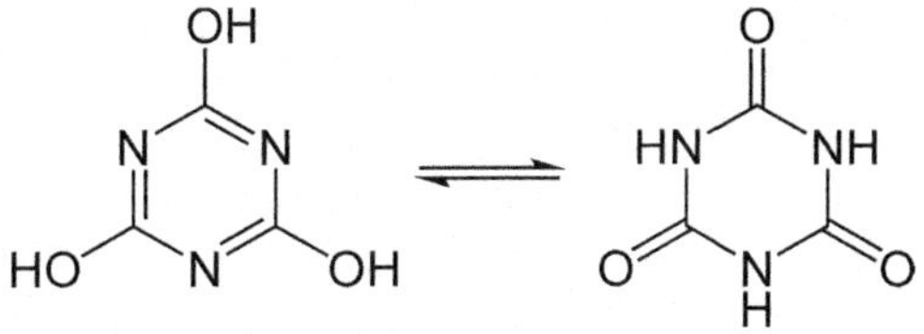

1 2

(acceptor) were examined. Figure 10.14 illustrates the relevant molecular orbitals involved in the photoinduced ET process obtained for an optimized geometry at the DFT/BLYP level. See details in [38]. According to suggested scheme: (1) The melamine is excited from its HOMO (HOMO of the total system) to its LUMO (the LUMO+2 of the total system), with the LUMO and LUMO+1 localized on the acceptor; (2) Upon excitation, an electron is transferred from the melamine highest occupied molecular orbital (HOMO) to its lowest unoccupied molecular orbital (LUMO); and (3) This excitation localized on the donor is higher in energy than the charge-transfer state from HOMO to LUMO and from HOMO to LUMO+1. Time-dependent density functional theory (TDDFT) (both with BLYP and CAM-B3LYP) calculations and Ehrenfest dynamics simulations with BLYP revealed some important details of the process including the follows: (1) The most relevant excitonic excitation has the highest oscillator strength in the energy range explored; (2) The charge-transfer state has a lower energy resulting in an energy gradient upon excitation from the excitonic to the charge-transfer state; (3) The β-HOMO occupation was changed to 0, while the β-LUMO+2 occupation was adapted to 1, representing a local excitation on the melamine; and (4) Upon instantaneous excitation in the Franck–Condon region, the system starts to relax toward a more preferable geometry. In the Ehrenfest dynamics, classical particles evolve on a mean potential energy surface which is averaged over all quantum states and weighted by the corresponding populations.

A schematic potential energy plot along a generic nuclear relaxation coordinate shown in Fig. 10.5 [38] indicated that: (1) In the deuterated case, D case, additional nuclear modes couple to the electronic motion in contrast to the ^{1}H case, where the N–H frequencies are energetically isolated from all other modes; (2) Upon excitation, the system relaxes from the Franck–Condon point (FC); (3) In the ^{1}H system, the first available high-energy vibrations are due to the N–H stretching and are well-separated from the other modes; (4) The isotope substitution moves the resonance condition to lower frequencies; (5) N–H modes are the first nuclear vibrations to match the energy difference between the electronic states during the relaxation process; and (6) In the D case, the N–D stretching has a vibrational energy similar to several other modes, coupling to the electronic motion.

Electron transfer quenching dynamics in nonpolar solvents was investigated using ultrafast spectroscopy and vibrational absorption spectroscopy with a series of six fluorophore/quencher pairs, covering a driving force range of more than 1.3 eV [39]. The intrinsic ET rate constants, k_0, deduced from the quenching dynamics in the

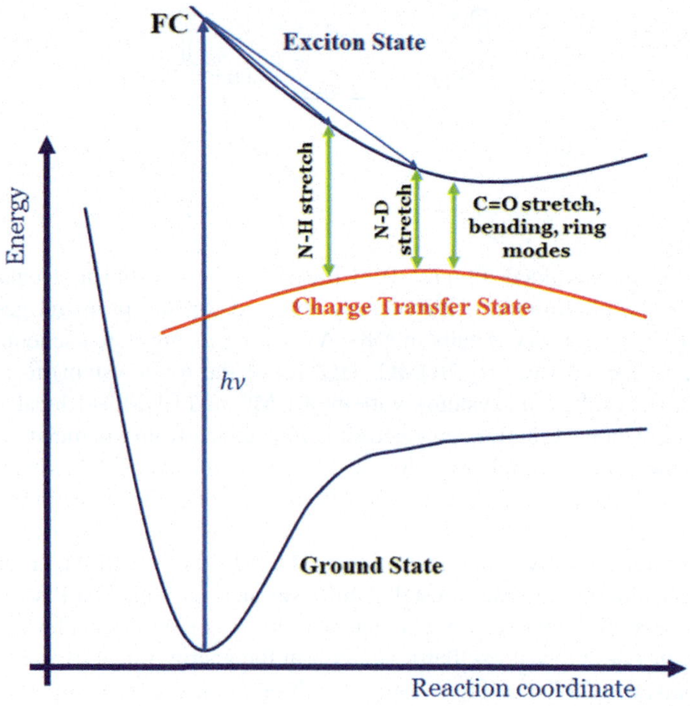

Fig. 10.5 Schematic representation of the coherent process in the investigated system. See details in the text [38]. Published under a Creative Commons public use license CC-BY from American Chemical Society

static regime were in nonpolar solvents beyond the diffusion limit. The measurements revealed that the primary product of static quenching is a strongly coupled exciplex that decays within a few picoseconds. Exciplexes with a longer lifetime were generated subsequently, during the dynamic, diffusion-controlled stage of the quenching. It was suggested that static ET quenching in nonpolar solvent scan can be an internal conversion from a locally excited state to a charge-transfer state of a supermolecule.

To study the relationship between the mutual orientation of the electron donor/acceptor in the dyads and the spin-orbit charge-transfer intersystem crossing (SOCT-ISC) efficiency, perylenemonoimide (PMI)–carbazole (Cz) compact electron donor/acceptor dyads were prepared and the photophysical properties of the dyads were examined with steady state and time-resolved femtosecond optical spectroscopies [40]. The fluorescence of the PMI unit in the dyads, in which the PMI and the Cz units were connected via either a C–C or C–N bond, or with an intervening phenyl moiety, was quenched due to photoinduced electron transfer. It was found that the charge separation takes 0.28 ps and the charge recombination takes 1.21 ns. It was also found that reversible photoreduction of the PMI–Cz dyads and generation of the near-IR-absorbing PMI radical anion ($PMI^{-\cdot}$) occurs in the presence of a sacrificial electron donor (triethylamine).

Photoselection and charge separation in donor–bridge–acceptor (DBA) (butadiyne-bridged donor–bridge–acceptor supermolecular) strongly depending on the structural and electronic properties of the bridge were examined using transient mid-IR absorption spectra of the C≡C bridge stretching modes, transient spectra in the visible range, and time depending TD-DFT calculations [42]. Constructed alkyne bridges found to be compact possess linear structure permitting access to high symmetry DBA molecules and allow torsional motion of donor (D) and acceptor (A), especially for longer bridges. Conformation-dependent electron transfer dynamics in a set of novel donor–bridge–acceptor DBA compounds featuring butadiene (C4) bridge, *N*-isopropyl-1,8-napthalimide (NAP) acceptors, and donors that span a range of reduction potentials (trimethyl silane (Si-C4-NAP), phenyl (Ph-C4-NAP), and dimethyl aniline (D-C4-NAP)) were investigated. The research revealed that: (1) Excitation of D-C4-NAP at 400 nm predominantly prepares an S_1 excited state in the planar conformations ($\theta \sim 0$) but selects an S_2 state with $\theta \sim 90°$; (2) The symmetry of the frontier orbitals of such DBA compounds defines the photo-selection outcome and determines the rate of the $S_2 \rightarrow S_1$ charge separation reaction; (3) Variation of the S_2–S_1 electronic coupling with θ by over four orders of magnitude resulted in slow ET at θ ca. 0° and 90° but extremely fast ET at θ of 20–60°; and (4) The electronic excited-state energies and, the transition dipoles ($S_0 \rightarrow S_n$) depend strongly on the dihedral angle (θ) between D and A and the frontier orbital symmetry.

A two-step charge separation process in a newly synthesized diketopyrrolopyrrole-pyrrolopyrrole (DPP-PP) dyad (AD) was examined employing transient absorption, broad-band fluorescence upconversion, and transient impulsive stimulated Raman spectroscopy [43]. The process starts from the initial photoexcited bright exciton and goes through a partial charge-transfer (CT) state before finally reaching the charge-separated (CS) state. Further results of this work included the following: (1) AD in THF evolves to the diabatic potential energy surface of the CS state, whose absorption spectrum converges to that of a DPP anion for which fluorescence is completely quenched; (2) The trend of shifting for certain vibrational frequencies supports the proposed CT dynamics and mechanism takes place; (3) This finding gives quantitative insight into the CT characters of the bright state (0.1 e) and intermediate partial CT state (0.5 e); (4) The linear relationship exists between the vibrational frequency of the marker modes and the CT character; (5) The structure of the bridge between donor and acceptor enables an intermediate level of electronic communication; and (6) The charge separation can rapidly occur through a distinct partial charge-transfer state. Finally, it was suggested that the strong electronic communication at positions 2 and 5 of the pyrrolo[3,2-b]pyrrole core is a crucial element for this charge-transfer mechanism, which could be applied to organic photovoltaics or light-emitting diodes requiring efficient charge separation.

Six new "axial-bonding" type "phosphorus(V) porphyrin–naphthalene" conjugates with deferent distance between the porphyrin and naphthalene (NP) varied using polyether bridges, consisting of octaethylporphyrinatophosphorus(V) ($POEP^+$)/ tetraphenylporphyrinatophosphorus(V) ($PTPP^+$) and (NP), have been prepared [44].

The steady-state and transient spectral properties were investigated as a function of redox properties, distance, and molecular topology. The main results of the research are: (1) Strong electronic interactions between the phosphorus(V) porphyrin and NP in directly bound conjugates take place; (2) Suggested energy diagrams predicted reductive electron transfer involving singlet excited phosphorus(V) porphyrin and NP to generate high-energy (~ 1.83–2.11 eV) charge-separated states $(POEP/PTPP)^{\cdot -}$–$(NP)^{\cdot +}$; (3) Femtosecond transient absorption spectral studies revealed rapid deactivation of singlet excited phosphorus(V) porphyrin due to charge separation wherein the estimated forward rate constants were in the range of 10^9–10^{10} s^{-1}; (4) The process dependents on the distance between the NP and porphyrins units, as well as the redox potentials of the type of the phosphorus(V) porphyrin; and (5) Due to high exothermicity and low-lying triplet states, the charge recombination process was found to be rapid, leading to populating the triplet states of phosphorus(V) porphyrins.

Two electronically diverse carboxy-terminated zinc porphyrins, BFBP(Zn)-COOH and TFP(Zn)-COOH, incorporating into the hexagonal pores of NU-1000 via solvent-assisted ligand incorporation (SALI), and resulting in BFBP(Zn)@NU-1000 and TFP(Zn)@NU-1000 compositions were prepared [45]. Experiments with the varied number of trifluoromethyl groups at the porphyrin e showed that: (1) The ground-state redox potentials of the porphyrins within ca. 0.1 V relative to that of NU-1000 were tuned; (2) For BFBP(Zn)@NU-1000, the relative ground- and excited-state redox potentials of the components facilitate an energy transfer (EnT) from NU-1000* to BFBP(Zn), forming $BFBP(Zn)_{S1}$; (3) A long-lived charge-separated complex was formed through an exciplex-like [$BFBP(Zn)_{S1}$*–TBAPy] intermediate; (4) Time-resolved spectroscopic data suggested that energy transfer (EnT) from NU-1000* occurs through a slow [NU-1000*–BFBP(Zn)] intermediate formation; and (5) TFP(Zn)@NU-1000 displays an efficient EnT from NU-1000* to exited [TFP(Zn)–TBAPy]*, formed at the ground state through electronic interaction.

The synthesis of a new fullerene–bis-Zn-porphyrin *e*-bisadduct by tether-directed functionalization of C_{60} via a multistep synthetic protocol was performed [46]. In the investigation, the transient pump–probe spectral technique covering the wide femtosecond-to-millisecond time scale, matrix-assisted laser desorption ionization-time of flight mass spectrometry (MALDI-TOF MS) MALDI-TOF spectrometry, and variable temperature NMR were utilized. Supramolecular oligomers were subsequently formed involving the two porphyrin-bearing arms embracing a fullerene cage of the vicinal molecule. The final supramolecular donuts revealed an inner cavity size estimated as 23 nm, close to that observed in photosynthetic antenna systems. The formation of long-lived charge-separated states was directly confirmed with the transient pump–probe method.

Quantum coherence deals with the idea that all objects have wave-like properties and **coherence** expresses the potential for two waves to interfere. Quantum coherence effects on charge transfer and spin dynamics in a system having two degenerate electron acceptors were studied employing femtosecond transient absorption spectroscopy and theoretical analysis [47]. A zinc 5,10,15-tri(*n*-pentyl)-20-phenylporphyrin (ZnP) electron donor covalently linked to either one or two naphthalene-1,8:4,5-bis(dicarboximide) (NDI) and electron acceptors with an anthracene (An) spacer, ZnP-An-NDI (1) and ZnP-An-NDI_2 (2), respectively were investigated. The research indicated that: (1) The electron transfer (ET) rate constant for 2 is about three times larger than that of 1; (2) The rate constant for charge recombination (CR) of 1 is about 25% faster than that of 2; (3) The electron on $NDI_2^{\cdot -}$ in 2 localizes onto one of the two NDIs prior to CR, thus precluding electronically coherent CR from NDI_2; (4) CR in both 1 and 2 is spin coherent; and (5) Observation of a resonance in the ^{3*}ZnP yield following CR as a function of applied magnetic field, gave spin–spin exchange interaction energies of $2J = 210$ and 236 mT, respectively.*

In work [49] it was demonstrated that donor–acceptor composition in complexes of electrolytic polythiophene (PTAK)

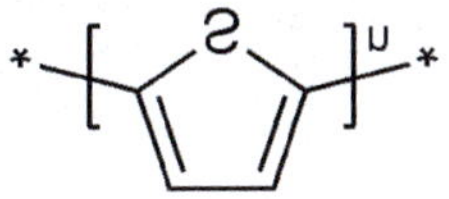

and naphthalene diimide (ENDI)

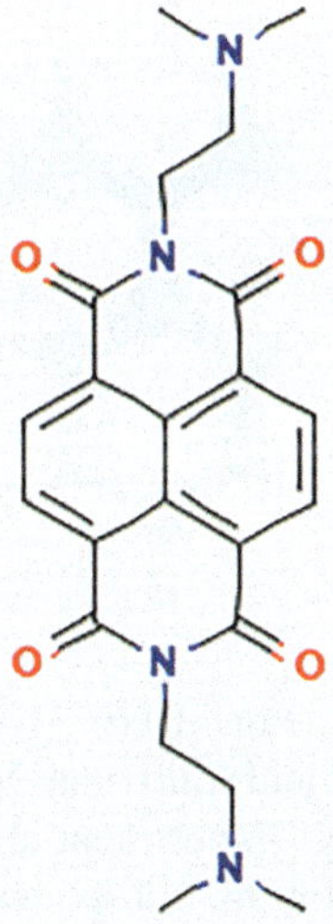

impacted the microstructure of the polymer and donor–acceptor interaction, with consequences for photoinduced charge-pair formation and recombination. Among the results of the study are the following: (1) PTAK evolved from microstructures showed increase of H-like to random-coil J-like excitonic coupling character with increasing ENDI/PTAK charge ratio; (2) ENDI exhibits weak J-like coupling at high acceptor densities; (3) Sub-100-fs charge separation between PTAK and ENDI implies close donor–acceptor orbital proximity and hot-exciton relaxation via charge-transfer exciton states; (4) Multiphasic recombination observed reflecting a distribution of charge-pair separation distances that is correlated with degree of exciton delocalization; and (5) Recombination timescales and the fraction of long-lived charge pairs increased with higher excitation energies, consistent with energy-dependent coupling between hot polymer and delocalized charge-transfer excitons.

A host–guest supramolecular charge-transfer (CT) assembly utilizing photoactive anthraquinone-

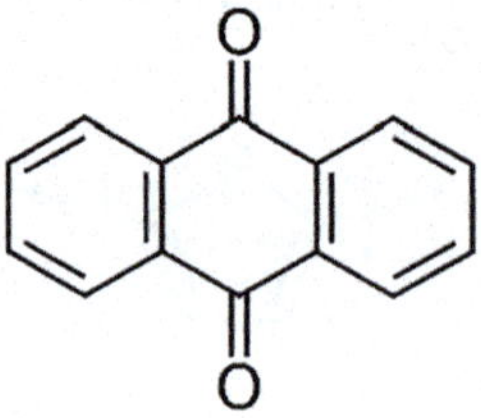

based porous 2D covalent organic frameworks (COF) as an accept and *N*,*N*-dimethylaniline (DMA)

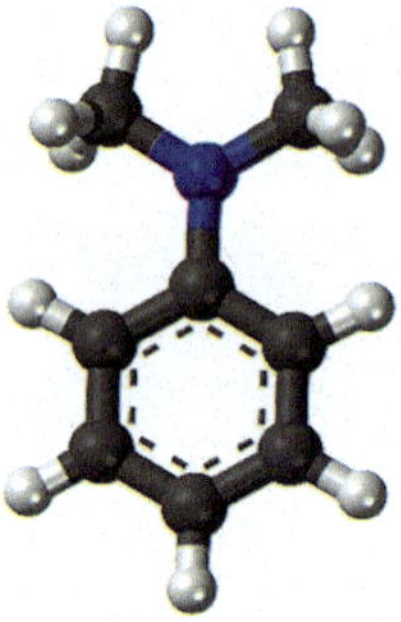

inside as donor, was designed (Fig. 10.6) [10]. The assembly was examined employing the photoinduced EPR and solid-state NMR and femtosecond broad-band transient absorption spectroscopy. Experiment showed that the CT occurs rapidly within < 110 fs after photoexcitation, subsequently leading to long-lived charge separation with 13% quantum efficiency at room temperature. ESR and solid-state NMR, using ^{1}H–^{13}C correlation experiments, confirmed appearing the long-lived confined DMA cation radical and the packing of the amines inside the host–guest COF assembly.

The intramolecular charge delocalization/localization character of a newly synthesized acceptor–donor–acceptor molecule (ADA) containing symmetric quadrupolar molecules was investigated, employing femtosecond stimulated Raman scattering (FSRS), Raman spectroscopy and femtosecond transient absorption (fs-TA) spectroscopy [50]. Properties of these systems can be explained by the excited-state evolution from initial charge delocalization over all molecules to localization on

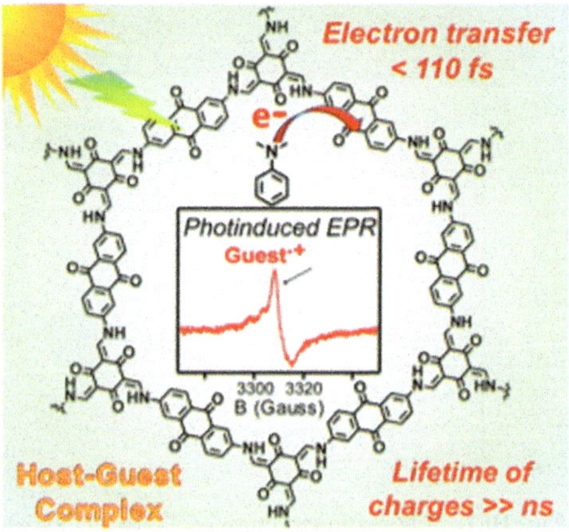

Fig. 10.6 Host–guest supramolecular charge-transfer assemble [10]. With permission from American Chemical Society

one branch of the molecules after a femtosecond pulse excitation. The excited-state Raman spectra of the specific alkynyl (−C≡C−) bonds at each branch of ADA indicated that the nature of the relaxed S_1 state is strongly governed by solvent polarity. Three types of charge separation modes based on their driving forces, namely: asymmetric energetics, asymmetric kinetics, and their hybrid were identified [51]. Physical parameters of the investigated system and the respective performance limitation relevant to all types of photocatalysts have been summarized. The charge separation strategies employed in the reported photocatalyst systems were elucidated, analyzed, and compared.

Due to their functional diversity, 1,2,3-triazoles (Fig. 10.7) have proofed to be versatile building blocks for a wide range of applications, as lgand in metal complexes, H-bond donor/acceptor, anion receptor, anchoring group on metal surfaces, or linker promoting preventing intramolecular charge transfer, in particular [52]. A synthetic strategy for the coupling of D- and A-units was the copper-catalyzed alkyne-azide cycloaddition (CuAAC), forming a 1,4-disubstituted 1,2,3-triazole ring.

The following criteria were derived for the design of triazole-bridged D-A dyads with emphasis on electron transfer properties: (1) The charge separation rate (k_{CS}) and the charge-recombination rate (k_{CR}) was higher for dyads with flexible linkers (through-space ET); (2) The charge-recombination rate (k_{CR}) is lower for rigid dyads (3) The energy barrier ($\Delta G^{\ddagger}$) for oxidative reaction is higher than for reductive photoinduced ET; (4) The charge separation rate (k_{CS}) for reductive photoinduced ET (acceptor excitation) is higher in 1-compared to 4-connected dyads; and (5) The electron transfer from the photoexcited D to the A unit is mediated by the bridge-centered unoccupied molecular orbitals. Structural and electronic properties of the bridge influence charge separation and recombination rates in triazole-bridged donor–acceptor dyads, the role of the triazole bridge energetics in thermal as well as oxidative and reductive photoinduced ET criteria for the design of dyads with different ET properties were discussed in details [52].

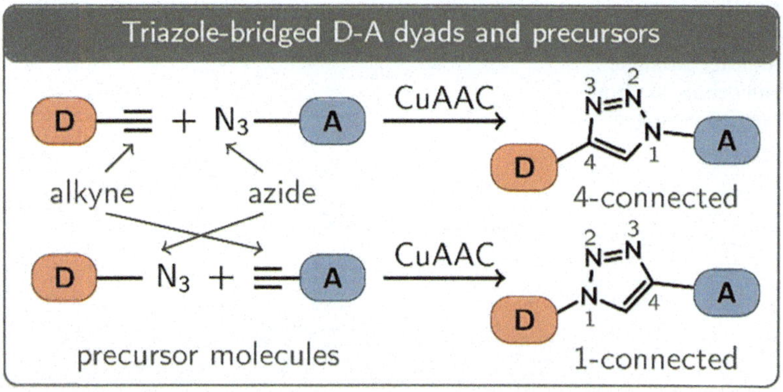

Fig. 10.7 Schematic structure of 1- and 4-connected triazole-bridged D–A dyads and the corresponding alkyne- and azide-substituted precursors. The building blocks are *clicked* together via a copper-catalyzed alkyne-azide cycloaddition (CuAAC) reaction [52]. With permission from Elsevier

10.2.3 Photophysical and Photochemical Processes in Dual Flourophore-Nitroxide Molecules

In pioneering work [53] it was demonstrated that irradiation of the chromophore segment of the dual compound, dansyl-TEMPO (FNO)

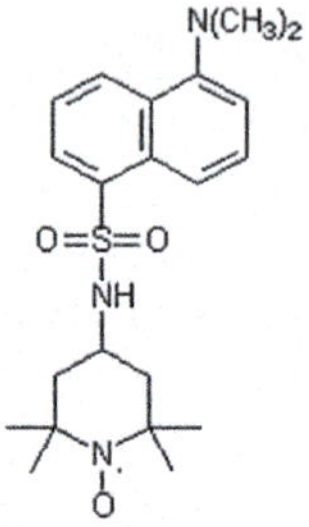

(FN1)

in a glassy liquid (glycerol, 75%, water 20%, ethanol 5%) invoked the production of the hydroxylamine derivative accompanying the decay of the electron spin resonance (ESR) signal from nitroxide and a parallel eightfold increase in fluorescence. Both processes occur with the same rate constant k_{red} under identical conditions. Subsequently, fluorophore nitroxide compounds found wide application as double fluorescent nitroxide labels, redox probes of high selectivity and sensitivity, and models for studying the mechanism of the conversion of light energy into chemical energy [54–64].

Dual FNO˙ as model systems for conversion of light energy to chemical energy were first proposed and developed [53]. [For example, in the dual compound (fluorophore)–spacer–nitroxide] the fluorophore (F) in an excited state is an electron donor and nitroxide fragment (NO˙) serves as an electron acceptor. In this system, photochemical and photophysical processes follow a scheme:

$$\mathbf{FNO^{\cdot}} + h\upsilon \rightarrow \mathbf{F^{*}NO^{\cdot}}$$
$$\mathbf{F^{*}NO^{\cdot}} \rightarrow \mathbf{FNO^{\cdot}} \quad (\mathrm{Q[I]})$$
$$\mathbf{F^{*}NO^{\cdot}} \rightarrow \mathbf{F^{+}NO^{-}} \quad (\mathrm{CS[I]})$$
$$\mathbf{F^{+}NO^{-}} \rightarrow \mathbf{FNO^{\cdot}} \quad (\mathrm{CR[I]})$$
$$\mathbf{F^{+}NO^{-}} + \mathrm{RH} \rightarrow \mathrm{D\text{-}}\mathbf{FNOH} + \mathrm{products} \quad (\mathrm{ChR})$$

where RH is a solvent (glycerol, ethanol, ethylene glycol, go forth) or an ingredient; Q[I], CS[I], CR[I], and ChR designate fluorescence quenching, charge separation, charge recombination, and chemical reaction, correspondingly and FNOH is hydroxylamine derivatives. Importantly, nitroxide photoreduction in the dual probes occurs without a violation of the fluorophore structure. Therefore, the photochemical reactions of the dual molecules in the system can be considered as processes of transfer of light energy to chemical one.

Fluorophore–nitroxide compounds of a system designed and investigated in [64] consist of triad tethered with nitroxide [64]: (fluorophore)–bridge–(acceptor)*n*–[nitroxide] (Fig. 10.8). The stable free radical 2,2,6,6-tetramethylpiperidinoxyl (TEMPO, T*) was covalently attached to the electron acceptor in a donor–chromophore–acceptor (D–C–A) system, MeOAn-6ANI-Phn-A-T*, having well-defined distances between each component, where MeOAn = p-methoxyaniline, 6ANI = 4-(N-piperidinyl)naphthalene-l,8-dicarboximide, Ph = 2,5-dimethylphenyl (n = 0,1), and A = naphthalene-1,8:4,5-bis(dicarboximide) (NI) or pyromellitimide (PI piperidinyl)naphthalene-l,8-dicarboximide, Ph = 2,5-dimethylphenyl ($n = 0{,}1$), and A = naphthalene-1,8:4,5-bis(dicarboximide) (NI) or pyromellitimide (PI). Using both time-resolved optical and EPR spectroscopy, it was showed that T* influences the spin dynamics of the photogenerated triradical states 2,4(MeOAn+*-6ANI-Phn-A-*-T*), resulting in modulation of the charge recombination rate within the triradical. In this system, the nitroxide segment effects on the separated charge recombination by mechanism of spin catalysis:

$$(\mathrm{D{-}A_1{-}A_2{-}}\mathbf{FNO^{\cdot}}) + h\upsilon \rightarrow (\mathrm{D^{*}{-}A_1{-}A_2{-}}\mathbf{FNO^{\cdot}})$$
$$(\mathrm{D^{*}{-}A_1{-}A_2{-}}\mathbf{FNO^{\cdot}}) \rightarrow (\mathrm{D{-}A_1{-}A_2{-}}\mathbf{FNO^{\cdot}}) \quad (\mathrm{Q[II]})$$
$$(\mathrm{D^{*}{-}A_1{-}A_2{-}}\mathbf{FNO^{\cdot}}) \rightarrow (\mathrm{D^{+}{-}A_1^{-}{-}A_2{-}}\mathbf{FNO^{\cdot}}) \quad (\mathrm{CS_1[II]})$$
$$(\mathrm{D^{+}{-}A_1^{-}{-}A_2{-}}\mathbf{FNO^{\cdot}}) \rightarrow (\mathrm{D^{+}{-}A_1{-}A_2^{-}{-}}\mathbf{FNO^{\cdot}}) \quad (\mathrm{CS_2[II]})$$
$$(\mathrm{D^{+}{-}A_1^{-}{-}A_2{-}}\mathbf{FNO^{\cdot}}) \rightarrow (\mathrm{D{-}A_1{-}A_2{-}}\mathbf{FNO^{\cdot}}) \quad (\mathrm{CR[II]a})$$
$$(\mathrm{D^{+}{-}A_1{-}A_2^{-}{-}}\mathbf{FNO^{\cdot}}) \rightarrow (\mathrm{D{-}A_1{-}A_2{-}}\mathbf{FNO^{\cdot}}) \quad (\mathrm{CR[II]b})$$

The presence of nitroxide allowed to control the lifetimes of photoinduced radical ion pairs. Owing to the spin catalysis effect, the charge recombination can be locked up and the ability of the system to retain the photoseparated state is long enough for secondary chemical reactions of these charges. Energy level diagram showing the spin manifolds triad after charge separation and charge recombination is displayed in Fig. 10.9. The authors concluded that the nitroxide tethering does not affect markedly the charge separation and the resonance integral but decreases the charge recombination time by factor 2.5.

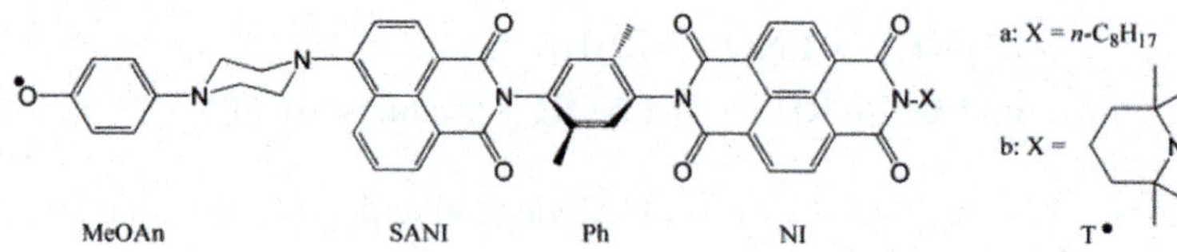

Fig. 10.8 Donor–chromophore–acceptor (D*–C–A) system, MeOAn–6ANI–Phn–A–T˙: used in [64]. With permission from American Chemical Society

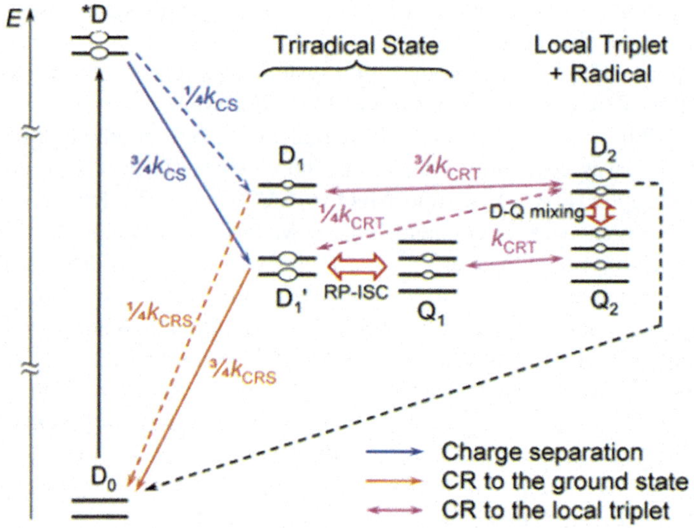

Fig. 10.9 Energy level diagram showing the spin manifolds of 1b4b after charge separation and charge recombination, in a magnetic field of ~ 0.35 T. Blue arrows denote charge separation forming the triradical state; orange arrows, charge recombination to the ground doublet state; and magenta arrows, reversible D-D and Q-Q charge recombination steps leading to the local triplet 3*NI, which is nearly isoenergetic with the RP state in toluene (see text). D1, D1′, and Q1 form a complete spin basis set of the triradical state. Dashed lines mean less probable transitions. Red double arrows stand for processes that mix certain Zeeman levels of the relevant D and Q states. The size of the ellipse on each spin level represents its population qualitatively [64]. With permission from American Chemical Society

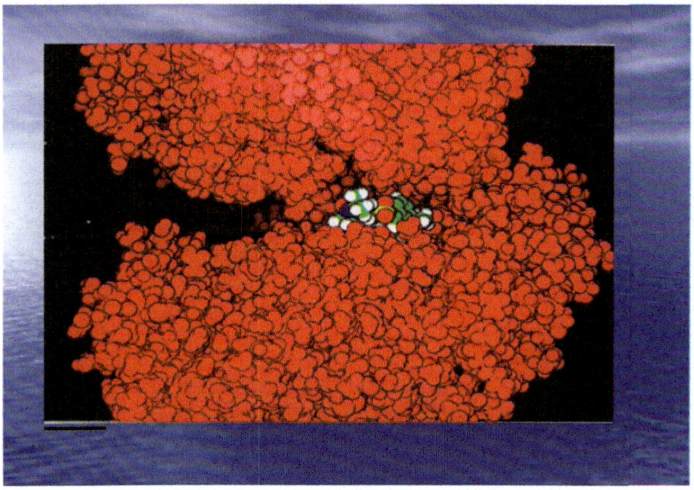

Fig. 10.10 BSA modified with the dual fluorophore nitroxide probe *x*. (Liktenshtein G. I., privet communication)

In order to model effects of protein on the ET in fluorophor-nitroxides, the dual probes FN1 was incorporated in bovine serum albumin (BSA) (Fig. 10.10) [57–61]. The photoreduction of the nitroxide fragment was monitored by ESR, and fluorescence quenching was measured by steady-state and picosecond time-resolved techniques. It is important that both fluorescent and nitroxide fragments allowed to investigate the factors affecting the ET to be measured, namely, the molecular dynamics and micropolarity of the medium in the vicinity of the donor (by fluorescence technique) and the same properties in the vicinity of nitroxide acceptor (by ESR).

The measurements of the dynamic shift of time-resolved spectra of the dual probe fluorescent fragment using picosecond fluorescence time-resolved technique (Fig. 10.11) allowed to obtain time of dynamic reorganization (relaxation) of the BAS binding which shows protein dynamics of the nanosecond scale. This conclusion was independently confirmed by analyzing ESR spectra of the nitroxide fragment. Diagram of the Gibbs energy levels for ET in the dual molecule FN1 in BSA is presented in Fig. 10.11 [57].

Based on experimental data on the local apparent dielectric constant ε_0 in the vicinity of the donor dansyl groups (fluorescence technique) and around the acceptor nitroxide segments (ESR), the following parameters of the Marcus–Levich theory (Chap. 4) were estimated for the ET at $T = 300$ K: Gibbs energy $\Delta G_0 = -1.7$ eV, reorganization energy $\lambda = 0.9$ eV, and activation energy $\Delta G^{\#} \approx 0.25$ eV. It should be emphasized that the process of thermodynamic charge photoseparation is thermodynamically allowed only in the case of stabilization of the charge separation pair due to the protein medium (Fig. 10.12). The latter, in turn, is possible only as

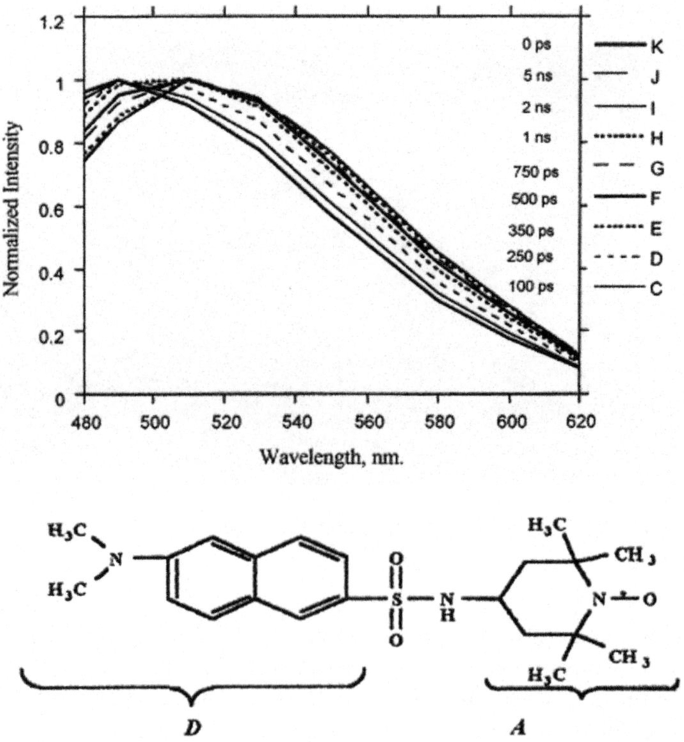

Fig. 10.11 Time-resolved fluorescence spectra of dansyl-TEMPO dual probe in human serum albumin [57]. With permission from Elsevier

a result of the nanosecond intramolecular dynamics of the protein recorded by ESR and fluorescence techniques.

10.2.4 Artificial Photoenzymes on the Base of Model Vesicles

The photosynthetic (thylakoid) membranes can contain pigments that participate in light absorption and charge separation and can act as a barrier to inhibit the undesirable recombination and side reactions of photogenerated oxidizing and reducing species [65–68]. Figure 10.13 illustrates a scheme of a liposome formed by phospholipids in an aqueous solution which is widely used in design and investigation of artificial photoenzymes. Works from several laboratories were devoted the problems of charge separation through the lipid bilayer [70–75]. In early work [65], electron transfer across membranes using vitamin K_1

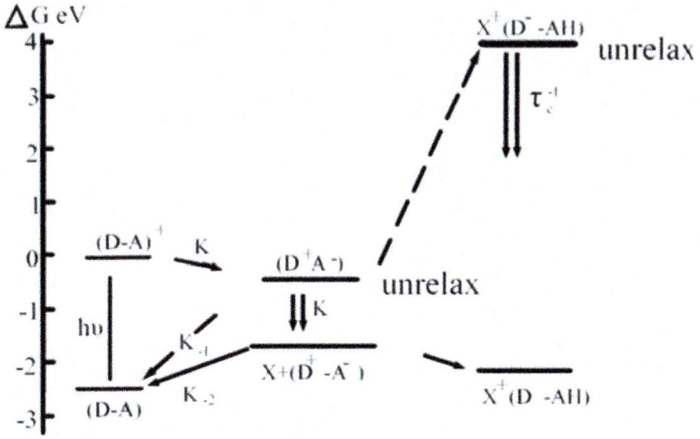

Fig. 10.12 Diagram of the Gibbs energy levels for ET in the dual molecule FN1 in BSA [57]. With permission from Elsevier

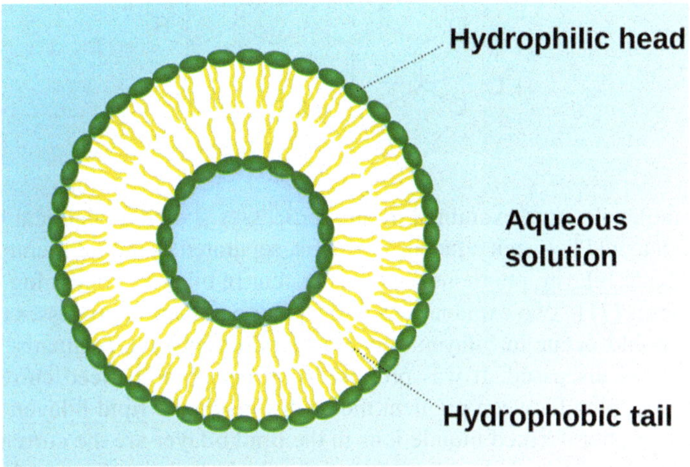

Fig. 10.13 Scheme of a liposome formed by phospholipids in an aqueous solution https://en.wikipedia.org/wiki/Vesicle_(biology_and_chemistry)

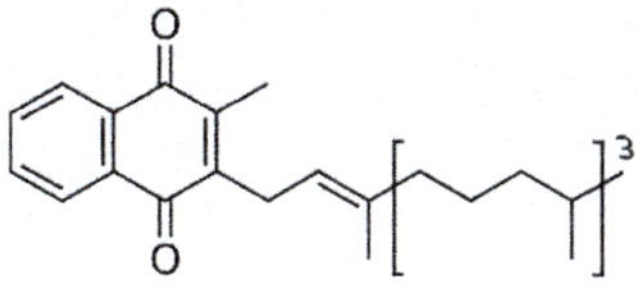

and coenzyme Q_{10}

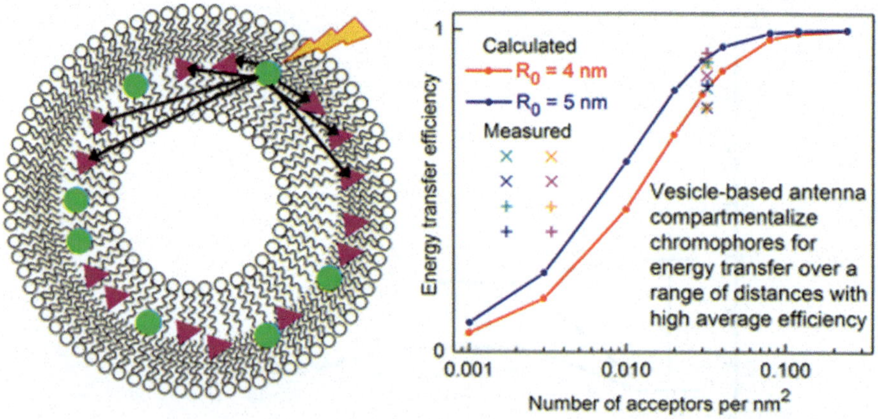

Fig. 10.14 Vesicle-based antennas used in [73]. With permission from American Chemical Society

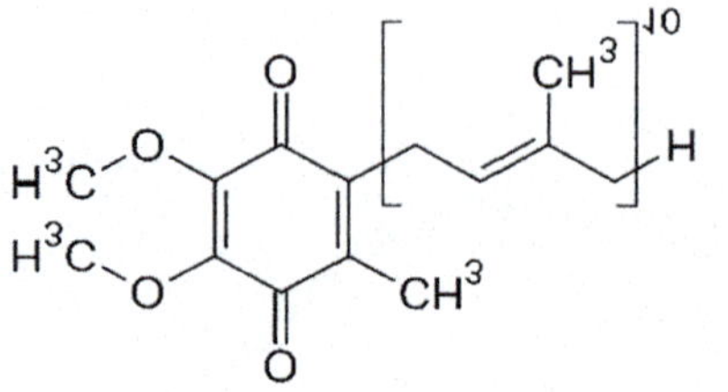

as carrier molecules was examined. Redox processes showing chemical specificity have been detected between aqueous reactants, separated by liquid membranes.

In the planar bilayer, photocurrents could occur in bilayers containing photosensitive pigments [71]. Two explanations of the photoconduction processes (photocurrents) that could occur in bilayers containing photosensitive pigments located in the bilayer was discussed. It was proposed that the photoinduced current occurs by electron hopping between pigment molecules inside the lipid bilayer. The other suggested that photoformed mobile ions in the lipid bilayer are the current carriers. Photoregulated transmembrane charge separation by linked piropyran-anthraquinone molecules was reported [67]. Experiment showed that amide-linked spiropyran-anthraquinone (SP-AQ) conjugates mediates $ZnTPPS_4^-$ photosensitized transmembrane reduction of occluded $Co(bpy)_3^{3+}$ within unilamellar phosphatidylcholine vesicles by external EDTA. The quantum yield decreased by 6-fold in the simple conjugate and 3-fold for an analogue containing a lipophilic 4-dodecylphenoxy substituent on the anthraquinone moiety. Rapid electroneutral charge transport across vesicle membranes using pyrylium cations

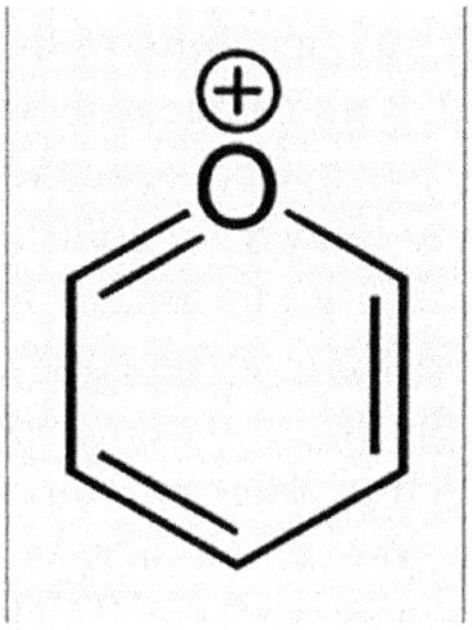

as the electron carrier, which undergo reversible ring-opening hydrolysis to form neutral diketones after deposition of the electron inside the vesicle, was investigated [70]. As the pyrylium cations are the primary acceptors for the photoproduced electrons, this approach simplifies the design of vesicle-based photocatalytic devices.

The photophysical properties of multichromophoric systems consisting of eight red or blue naphthalene diimide (NDIs)

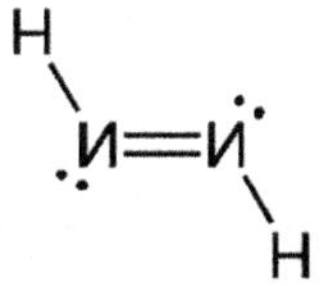

covalently attached to **a** *p*-octiphenyl scaffold, as well as a blue bichromophoric system with a biphenyl scaffold, have been investigated in detail using femtosecond time-resolved spectroscopy [72]. These systems formed self-assemble in lipid bilayer membranes and enable generation of a transmembrane proton gradient upon photoexcitation. The lifetime of the charge-separated state investigated using femtosecond time-resolved spectroscopy was found to increase from 22 to 45 ps by going from the bi- to the octachromophoric blue systems in methanol, while a 400 ps decay component was observed in the lipid membrane.

Endoergic electron transport across vesicle bilayers from ascorbate (Asc^-) in the inner waterpool to methylviologen (MV^{2+})

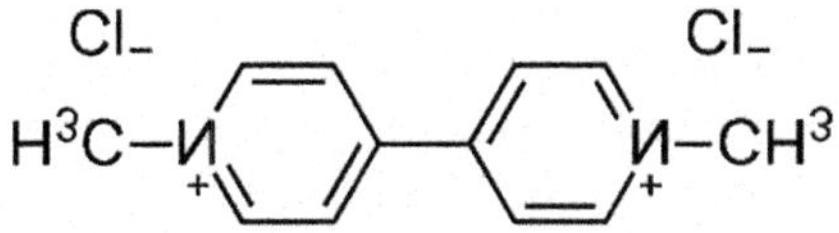

in the outer aqueous solution driven by the irradiation of pyrene derivatives embedded in the vesicle bilayers was examined [75]. The main results of the work are: (1) The initial rate of MV^{2+} reduction dependents on the substituent group of the pyrenyl ring; (2) A hydrophilic functional group linked with the pyrenyl ring by a short methylene chain acts as a sensitizer for the electron transport; and (3) The electron transport initiated by the reductive quenching of the singlet excited state of the (1-pyrenyl) alkanoic acids occurred. It was suggested that the latter process proceeds by a mechanism involving electron exchange between the pyrenes located at the inner and outer interface across the vesicle bilayer. In addition, unsymmetrically substituted pyrenes having both a hydrophilic group linked by a short methylene chain and a hydrophobic long alkyl group were designed and investigated. Thus, described system acted as sensitizers for the electron transport across vesicle bilayers.

Voltammograms for the electron transport through a bilayer lipid membrane, BLM, from one aqueous phase, W1, to another, W2, were recorded with the BLM containing 7,7,8,8-tetracyanoquinodimethane, TCNQ

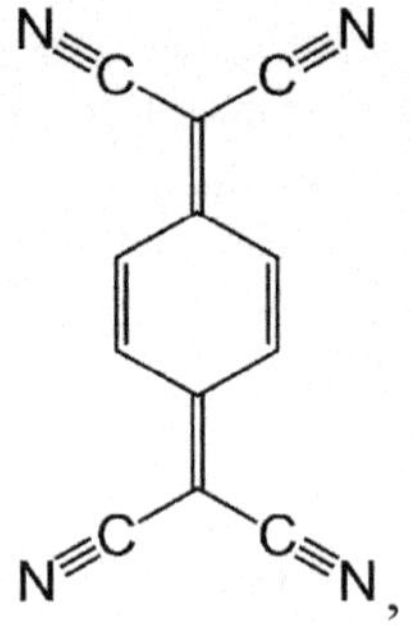

or decamethyl ferrocene, DMFc

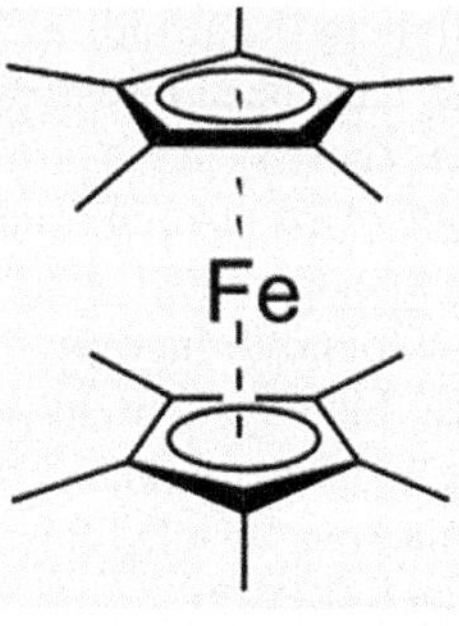

exposed to W1 containing $[Fe(CN)_6]^{4-}$ and W2 containing $[Fe(CN)_6]^{3-}$ [74]. The voltammogram experiments indicated that: (1) The formation of an anion radical of TCNQ or a cation of DMFc in the BLM followed by the interfacial redox reactions at both W1|BLM and BLM|W2 interfaces; (2) The BLM containing TCNQ gave a large current for the electron transport controlled by the electron exchange in the BLM; (3) The electron transport through the BLM requires the proximity between the redox potential of redox species in the BLM; and (4) In aqueous phases both the oxidized and reduced form in the BLM exist.

† Lipid vesicles were used as the organizational structure of self-assembled light harvesting systems (Fig. 10.14) [73]. Six chromophores were selected for inclusion in

vesicle-based antennas. Energy transfer was characterized in 13 representative binary mixtures using fluorescence excitation versus absorptance spectra, quenching of donor fluorescence, modeling emission spectra of a mixture versus components and time-resolved spectroscopy (fluorescence, ultrafast absorption). The main reported results of the experiments are as follows: (1) Binary donor–acceptor systems that employ a boron-dipyrrin donor

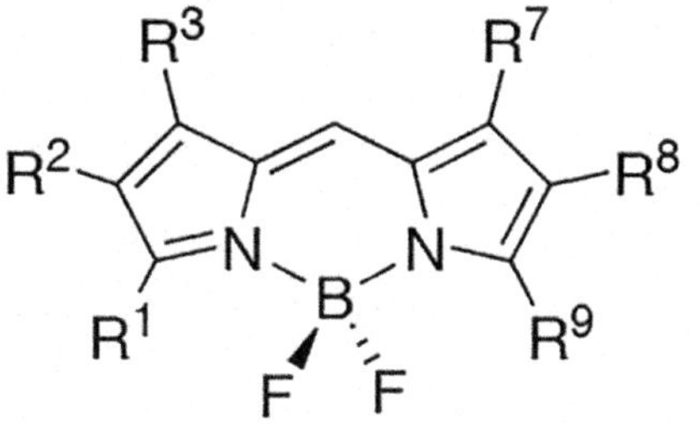

($S_0 \leftrightarrow S_1$ absorption/emission in the blue-green) and a chlorin or bacteriochlorin acceptor ($S_0 \leftrightarrow S_1$ absorption/emission in the red or near-infrared) have an average excitation-energy-transfer efficiency (Φ_{EET}) of $\sim$ 50%; (2) Binary systems with a chlorin donor and a chlorin or bacteriochlorin acceptor have Φ_{EET} $\sim$ 85%; (3) The differences in Φ_{EET} track the donor-fluorescence/acceptor-absorption spectral overlap within a dipole–dipole coupling (Förster) mechanism; and (4) The time profiles and resulting Φ_{EET} were modeled on the basis of Förster energy transfer between chromophores relatively densely packed in a two-dimensional compartment. The chemical structure of chlorine is:

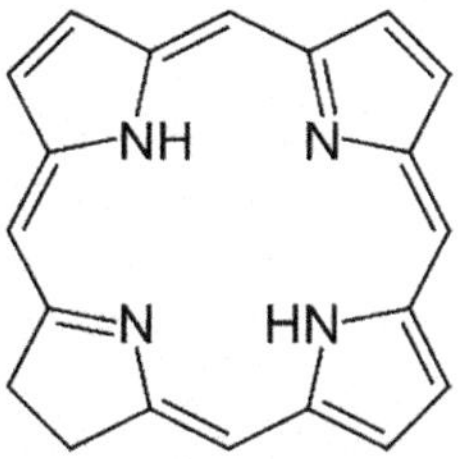

10.3 Oxygen Evolution Complex (OEC) Mimics

The lifegiving process of water decomposition in containing manganese photosynthetic systems and its chemical models still remains one of the most challenging mysterious problems to biochemists and chemists. The evolution dioxygen from water in a cluster of transition metals in the biological systems at the absorption of light quanta of low energy occurs by a sequence of elementary steps: four one-electron steps of oxidation of the manganese complex and, most probably by one four-electron elementary step of O_2 evolution (Sect. 3.5.2) [76]. In approaching this problem, a number of artificial manganese clusters and other transition metal clusters were synthesized and investigated [24, 77–97]. The synthesis of Mn cluster-based catalysts to mimic the structural and catalytic properties of the oxygen-evolving complex in photosystem II is a longstanding goal for researchers. Review [92] described a progress for three strategies which have been followed to prepare functional synthetic analogues of the oxygen evolution complex (OEC): (1) The synthesis of dinuclear manganese complexes designed to act as water oxidation catalysts in homogeneous solution; (2) Preparation of heterogeneous catalysts in the form of clay hybrids of such Mn_2-complexes; and (3) The preparation of manganese oxide particles of different compositions and morphologies. The prospect of designing efficient, robust and affordable catalysts for light-driven water oxidation photosynthesis was discussed.

Mn(IV) bound to lipid vesicles is an active catalyst for O_2 evolution in the presence of strong one-electron oxidants, such as $[Ru(bpy)_3]^{3+}$ and $[Fe(bpy)_3]^{3+}$ where bpy is 2,2′-bipyridyl [88]. The mechanism of evaluation of the O_2 forming center was discussed. It was speculated that if a lipid membrane is formed in the presence of Mn^{2+}, the Mn^{2+} may be incorporated into the membrane forming the catalyst for O_2 evolution from H_2O.

A model water oxidation complex $[H_2O(terpy)Mn(O)_2Mn(terpy)OH_2](NO_3)_3$ (terpy is 2,2′:6′,2″-terpyridine), containing a di-μ-oxo manganese dimer, was synthesized and structurally characterized [89]. This complex catalyzed O_2 evolution using strong oxidant either KHSO (potassium oxone) or NaOCl via an intermediate complex $[(terpy)(SO_4)MnIV(O)_2MnIV(SO_4)(terpy)]$. Isotope-labeling studies using $H_2{}^{18}O$ and $KHS^{16}O_5$ showed that O_2 evolution proceeds via an intermediate that can exchange with water. The rate-limiting step of O_2 evolution was proposed to be the formation of a formally MnV:O moiety which could then competitively react with either oxone or water/hydroxide to produce O_2. Dioxygen evolution in systems containing cubane-type tetramers, $[Ru_4(CO)_{12}(\mu 3\text{-}Se)_4]$ and $[(dpp)_6Mn_4O_4]$ (dpp- = diphenyl phosphinate anion) have been examined [79]. The system released O_2 under UV-light absorption in the gas phase. The structure of oxocubane core $Mn_4O_4{}^{6+}$ complex which mimics of the photosynthetic water oxidation system was revealed. A series of dimanganese complexes, $[Mn_2III,IV(\mu\text{-}O)_2(terpy)_2(H_2O)_2]^{3+}$ (1), $[Mn_2III,IV(\mu\text{-}O)_2(terpy)_2(CF_3CO_2)_2]^+$ (2), and $[Mn_2III,III(\mu\text{-}O)(terpy)_2(CF_3CO_2)_4]$ (terpy = 2,2′:6,2″-terpyridine) have been crystallographically characterized [87]. Two new terpyridine dimanganese complexes: a manganese(III, III) complex with a single unsupported oxo bridge and a manganese(III, IV) complex with a dioxo bridge were synthesized and its electrochemical behavior was examined.

The chemical principles that may govern the photosystem II water-oxidizing complex (PSII-WOC) by comparing the water oxidation capabilities of structurally related synthetic manganese-oxo complexes, particularly those with a cubical Mn_4O_4 core were formulated [82]. The following important aspects of the field were in the focus of the authors' attention: (1) The $[Mn_4O_4]^{6+}$ cubane core assembles spontaneously in solution from monomeric precursors or from $[Mn_2O_2]^{3+}$ core complexes in the presence of bidentate chelates, for example, diarylphosphinates which bridge pairs of Mn atoms on each cube face ($Mn_4O_4L_6$); (2) The $[Mn_4O_4]^{6+}$ core is enlarged relative to the $[Mn_2O_2]^{3+}$ core, resulting in considerably weaker Mn–O bonds; (3) Cubanes are ferocious oxidizing agents, stronger than analogous complexes with the $[Mn_2O_2]^{3+}$ core; (4) The cubane core topology is structurally suited to releasing O_2, with high yield; (5) The photoproduct, $[Mn_4O_2L_5]^+$, an intact nonplanar butterfly core complex, is poised for oxidative regeneration of the cubane core upon binding of two water molecules and coupling to an anode; (6) Catalytic evolution of O_2 and protons from water exceeding 1000 turnovers was achieved by suspending the oxidized cubane, $[Mn_4O_4L_6]^+$, into a proton-conducting membrane (Nafion); (7) Catalytic water oxidation can be also achieved using sunlight as the only source of energy by replacing the external electrical bias with redox coupling to a photoanode incorporating a Ru(bipyridyl) dye.

Two new polynuclear heterometallic cluster complexes with [MnIII3MIINa] (M = Mn, Ca) core were synthesized using two in situ formed Schiff bases and structurally characterized by single crystal X-ray analysis [90]. Schiff base is a compound with the general structure $R^1R^2C{=}NR^3$ (R^3 = alkyl or aryl). Utilizing Clark electrode and online mass spectrometry, it was shown that the compound with [MnIIICaIINa] appeared to catalyze water oxidation. As determined by electrochemistry and gas chromatography–mass spectrometry (GC-MS) the tetranuclear manganese complex $[Mn(IV)_4O_5(terpy)_4H_2O\ (2)](ClO_4))_6$ (1; terpy = 2,2′:6′,2″-terpyridine) gave catalytic water oxidation in aqueous solution [91]. Complex 1 exhibited catalytic water oxidation when adsorbed on kaolin clay, with Ce(IV) as the primary strong oxidant. The latter was characterized by diffuse reflectance UV/visible and EPR spectroscopy. DFT calculations showed that the oxygen in the mono-μ-oxo bridge, rather than Mn(IV), is oxidized after an electron removed from the Mn(IV,IV,IV,IV) tetramer. In addition, on the basis of the calculation, the formation of O_2 is proposed to occur by reaction of water with an electrophilic manganese-bound oxyl radical species, $(\cdot)O\text{-}Mn_2$ (IV/IV), produced during the oxidation of the tetramer.

The rates of water oxidation/O_2 evolution catalyzed by six cobalt–oxo clusters including the Co_4O_4 cubanes, $Co_4O_4(OAc)_4(py)_4$ and $[Co_4O_4(OAc)_2(bpy)_4]^{2+}$, using the common $Ru(bpy)_3{}^{2+}/S_2O_8{}^{2-}$ photo-oxidant assay, were compared (Fig. 10.15) [93]. The main experimental results of the study are as follows: (1) At pH 8, the first-order rate constants for these cubanes differ by 2-fold, 0.030 and 0.015 s^{-1}, respectively; (2) A deprotonation step and two electrons remove before O_2 evolution occur; (3) Incomplete cubane trimers were found not capable of evolving O_2 (Fig. 10.15); (3) Only the Co_4O_4 cubane core was shown to be the smallest catalytic unit for the intramolecular water oxidation pathway; (4) Oxidation of both cubanes to formally Co_4(3III,IV) (0.7 V vs. Ag/AgCl) occurs readily and neither dimers nor trimers are oxidized below 1.5 V pointing to appreciably greater charge delocalization in the $[Co_4O_4]^{5+}$ core. The origin of catalytic activity by Co_4O_4 cubanes illustrates three key features for water oxidation: (1) four one-electron redox steps, (2) efficient charge delocalization of the first oxidation step across the Co_4O_4 cluster, allowing for stabilization of higher oxidizing equivalents, and (3) terminal coordination site for substrate aquo/oxo formation. These conclusions are fully consistent with the concept of a four-electron mechanism for the oxidation of two water molecules in the complex of the Photosystem II predicted in [76].

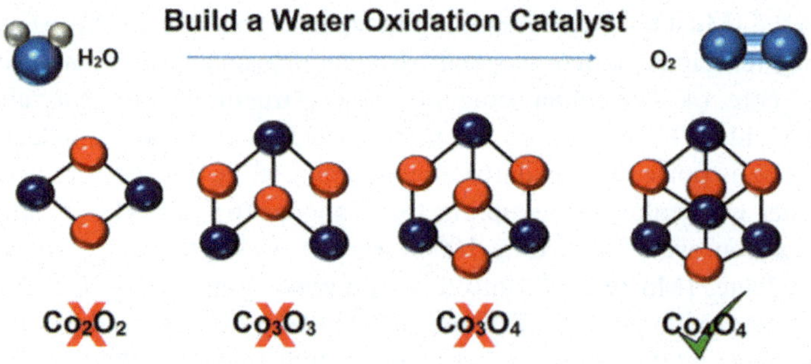

Fig. 10.15 Schematic illustration that only the Co_4O_4 cubane core is the smallest catalytic unit for the intramolecular water oxidation pathway [93]. With permission from American Chemical Society

A novel family of heteronuclear MnIVCa–oxido complexes containing MnIVCa–oxido cuboidal moieties and reactive water molecules on Ca^{2+} have been synthesized and characterized to mimic the oxygen-evolving complex (OEC) of photosystem II (PSII) in nature [92]. Complex A

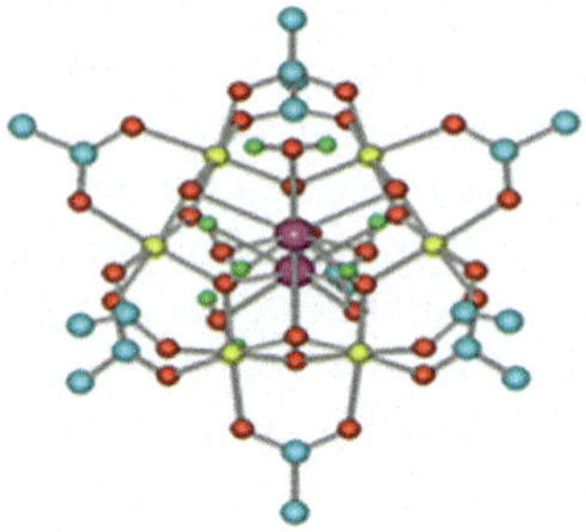

was synthesized in a reaction of Bun $4NMnO_4$, $Mn(ClO_4)_2 \cdot (H_2O)_6$ and $Ca(CH_3CO_2)_2$ with a molar ratio of 4:1:1 in boiling acetonitrile in the presence of an excess of pivalic acid. The bond lengths of Mn–oxido (1.80–1.86 Å), Ca^{2+}–oxido (2.42–2.64 Å), and Ca^{2+}–O water (2.38–2.44 Å) were found to be similar to those observed in reported multinuclear Mn oxido complexes and the OEC in PSII [92].

A novel family of heteronuclear Mn(IV)Ca-oxido complexes containing **Mn(IV)Ca**-oxido cuboidal moieties and reactive water molecules on Ca^{2+} have been synthesized and characterized to mimic the oxygen-evolving complex (OEC) of photosystem II (PSII) [94]. On the basis of hybrid density functional **c**alculations, a water oxidation mechanism for a bioinspired Mn catalyst, driven by the mild single-electron oxidant $[Ru(bpy)_3]^{3+}$ was proposed [95]. The calculation suggested that in the catalyst, the crucial O–O bond formation proceeds from the formal Mn_4(IV,IV,IVV) state by direct coupling of a Mn(IV)-bound terminal oxyl radical and a di-Mn bridging oxo group. The proposed mechanism exhibits some features similar to the presently leading suggestion for the natural system.

In work [96] the first asymmetric Mn_4Ca-cluster

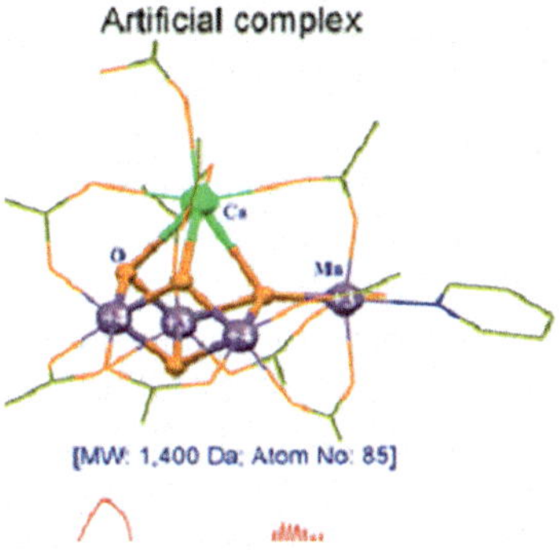

has been synthesized through a two-step procedure with a high yield (~ 50%) by using commercial chemicals Bun 4$NMnO_4$, $Mn(CH_3CO_2)_2 \cdot (H_2O)_4$, $Ca(CH_3CO_2)_2 \cdot H_2O$, and pivalic acid.

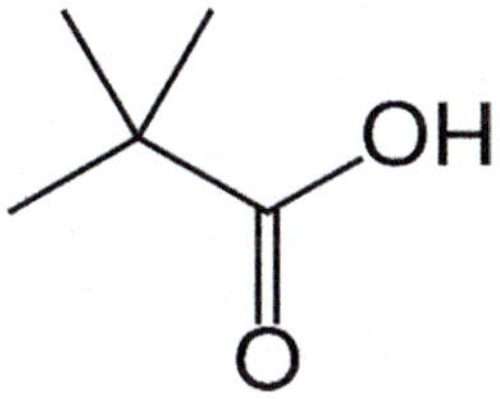

The new artificial Mn_4Ca-cluster displays structural similarity to that of the OEC in nature in respects of the asymmetric Mn_4Ca core structure and peripheral ligands [98]. Like the biological OEC, the artificial Mn_4Ca-cluster can undergo four redox transitions and display two low-temperature electron paramagnetic resonance (EPR) signals ($g = 2.0$ and $g = 4.9$) arising from two distinct ground states. A novel family of heteronuclear Mn(IV)Ca-oxido complexes containing Mn(IV)Ca-oxido cuboidal moieties and reactive water molecules on Ca^{2+} have been synthesized and characterized to mimic the oxygen-evolving complex (OEC) of photosystem II (PSII) in nature [98].

An account [99] described the results of research on manganese model chemistry relevant to PSII. The functional model $[Mn^{III/IV}{}_2(terpy)_2(\mu\text{-}O)_2(OH_2)_2](NO_3)_3$ complex (terpy = 2,2′;6′,2″-terpyridine), a mixed-valent di-μ-oxo Mn dimer with two terminal aqua ligands was designed. In the presence of strong oxo-donor oxidants such as HSO_5, this complex evolves O_2. In addition, after binding to the surface of TiO_2 via a chromophoric linker, the catalyst continued its activity and can be oxidized by a photoinduced electron-transfer mechanism, mimicking the natural PSII process. Highly active MnO_x catalysts was formed in situ under photocatalytic conditions [100]. The most active $^4MnO_x$ catalyst (~ 0.84 mmol O_2 mol Mn^{-1} s^{-1})

was prepared from a Mn_4O_4. $^4MnO_x$ was characterized by pair distribution function analysis (PDF), Raman spectroscopy, and high-resolution transmission electron microscopy (HR-TEM) as a disordered, layered Mn-oxide with high surface area ($216\ m^2\ g^{-1}$). HR-TEM is an imaging mode of specialized transmission electron microscopes that allows for direct imaging of the atomic structure of samples. In contrast, the $^SMnO_x$ formed from Mn^{2+} salt gave an amorphous species of lower surface area ($80\ m^2\ g^{-1}$) and lower activity (~ 0.15 mmol O_2 mol $Mn^{-1}\ s^{-1}$).

Crystalline oxides consisting of Mn^{3+} (manganite, γ-MnOOH; bixbyite, Mn_2O_3), Mn^{4+} (pyrolusite, β-MnO_2), and multiple monophasic fixed-valence manganese oxides, exhibiting catalytic activities for water oxidation, were investigated employing electrochemical measurements and ligand field theory [101]. The experiments and simulation showed that: (1) Pure β-MnO_2 has no detectable catalytic activity; (2) γ-MnOOH (tetragonally distorted $Mn^{3+}O_6$, D_{4h} symmetry) is significantly more active; (3) Mn_2O_3 (trigonal antiprismatic $Mn^{3+}O_6$, D_{3d} symmetry) is the most active; (4). In a comparison of 2D-layered crystalline birnessites (δ-MnO_2), the monovalent Mn^{4+} form was catalytically inert; (5) The hexagonal polymorph, containing few out-of-layer corner-sharing $Mn^{3+}O_6$, has $\sim$ 10-fold higher catalytic activity than the triclinic polymorph, containing in-plane edge-sharing $Mn^{3+}O_6$; (6) Electrochemical measurements showed that $Mn^{3+}O_6$ sites possess D_{3d} ligand field symmetry; and (7) In D_{4h} $Mn^{3+}O_6$ sites have weaker Mn–O bonding to the two axial oxygen ligands, and were catalytically less efficient.

- Efforts have been made toward understanding O–O bond activation pathways of Mn^{III}-peroxo adducts [102]. With this purpose, hydrogen-atom transfer reactivity of Mn^{IV}-oxo and Mn^{III}-hydroxo complexes was investigated. A variety of Mn^{III}-peroxo adducts with varied ligand environments was generated. A series of Mn^{IV}-oxo complexes supported by neutral, pentadentate ligands with perturbed equatorial donation were synthesized and examined using parallel-mode EPR and Mn K-edge X-ray absorption techniques. Monitored decay of one Mn^{III}-peroxo complex bearing a bulky macrocylic ligand yielded mononuclear Mn^{III}-hydroxo and Mn^{IV}-oxo products, resulting from O–O bond activation of the Mn^{III}-peroxo unit. Kinetic investigations of these complexes revealed a correlation between equatorial ligand field strength and hydrogen-atom and oxygen-atom transfer reactivity. The reactivity trend also correlated with variations in $Mn^{III/IV}$ reduction potential caused by changes in the ligand field.

A series of artificial Mn_4XO_4-clusters (X = Ca/Y/Gd) have been synthesized and mimic both the geometric structure and the electronic structure, as well as the redox property of the OEC [103]. As an example, critical comparison of the OEC and the artificial Mn_4CaO_4-complex is presented in Fig. 10.16.

Figure 10.17 illustrates proposed synthesizing mechanism of the artificial Mn_4CaO_4-cluster [103].

Based on the characterization of artificial Mn_4XO_4-clusters, it was found that all μ3- and μ4-oxide bridges are tightly bound to the cluster, supporting that they may play roles in maintaining the cluster's stability and integrity [103]. The oxidation states of the four Mn ions in all these Mn_4XO_4-clusters are (III, III, IV, IV), which

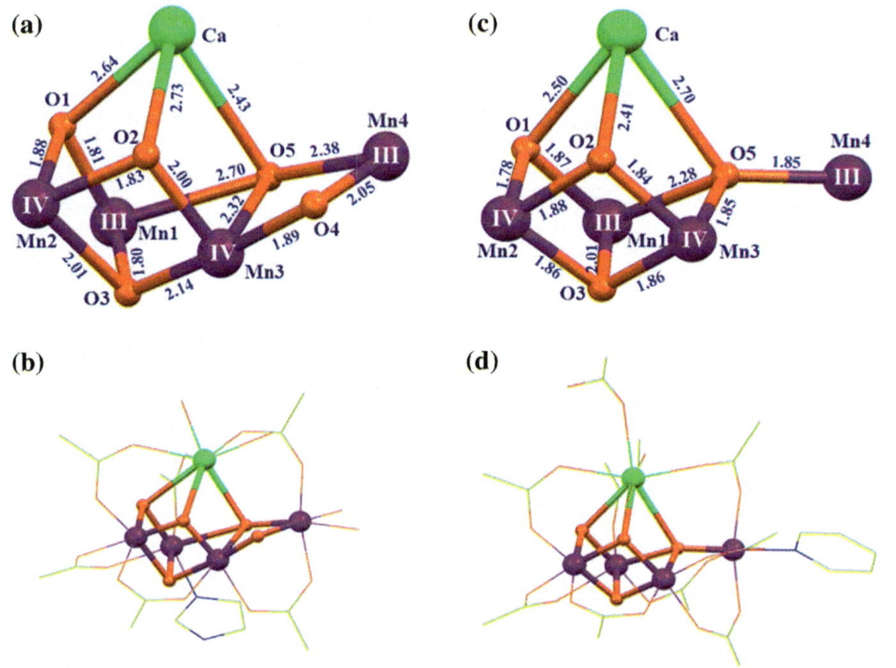

Fig. 10.16 Structural comparison of the OEC and the artificial Mn4CaO4-complex. **a** Core of the OEC; **b** structure of the OEC; **c** core of 4; **d** structure of 4. The data for the OEC is taken from the first monomer in the crystal structure data of PSII with the Protein Data Bank code 4UB6. For clarity, the methyl groups and the hydrogen atoms are not shown [103]. Published under the Creative Commons Attribution license CC-BY from Frontiers

provides the chemical evidence for the assignment of the four Mn ions in the S1 state of the OEC predicted in the frame four-electron mechanism of the water oxidation in four nuclear clusters [76].

In work [104] a hypothesis that Ca^{2+} fine-tunes the redox potential of synthetic oligonuclear transition metal complexes was put forward. By computing the relative redox potentials of substituted derivatives of the oxygen-evolving complex with the cations Sr^{2+}, Gd^{3+}, Cd^{2+}, Zn^{2+}, Mg^{2+}, Sc^{3+}, Na^{+}, and Y^{3+} for two sequential transitions of its catalytic cycle it was evaluated that the redox potential correlate with the Lewis acidity of a redox-inactive cation connected to the redox-active transition metals of the cluster via oxo or hydroxo bridges (Fig. 10.18) [104]. The primary role for Ca^{2+} in biological water oxidation was stressed.

The water oxidation mechanism of the highly active water oxidation catalyst $[Mn_4V_4O_{17}(OAc)_3]^{3-}$, a polyoxometalate catalyst with a $[Mn_4O_4]^{6+}$ cubane core modeling the natural oxygen-evolving complex was unraveled [105]. A proposed water oxidation mechanism for the bioinspired water oxidation catalyst $[Mn_4V_4O_{17}(OAc)_3]^{3-}$ includes the following steps: (1) The process starts from the activated species $[Mn_4{}^{4+}V_4O_{17}(OAc)_2(H_2O)(OH)]^{1-}$ (**1**); (2) After activation, the

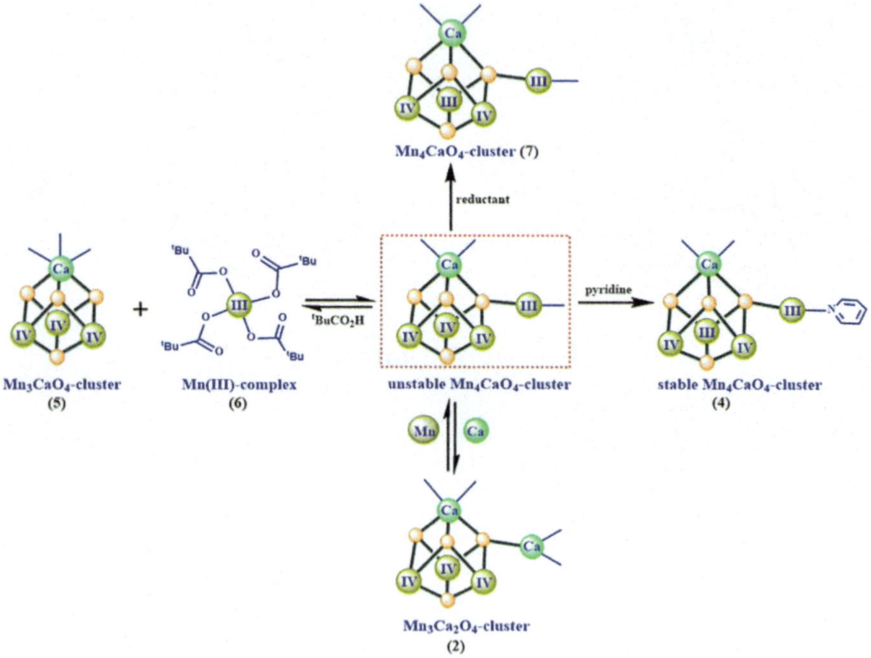

Fig. 10.17 Possible synthesizing mechanism of the artificial Mn_4CaO_4-cluster [103]. Published under the Creative Commons Attribution license CC-BY from Frontiers

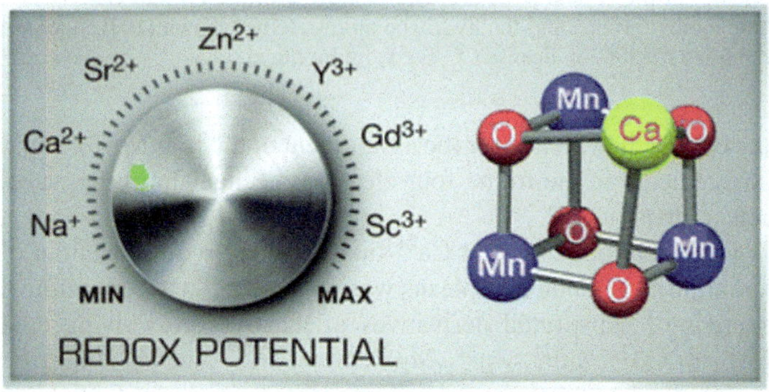

Fig. 10.18 Schematic illustration of the relative redox potentials of substituted derivatives of the oxygen-evolving complex with the cations Sr^{2+}, Gd^{3+}, Cd^{2+}, Zn^{2+}, Mg^{2+}, Sc^{3+}, Na^{+}, and Y^{3+} [104]. With permission from Royal Chemical Society

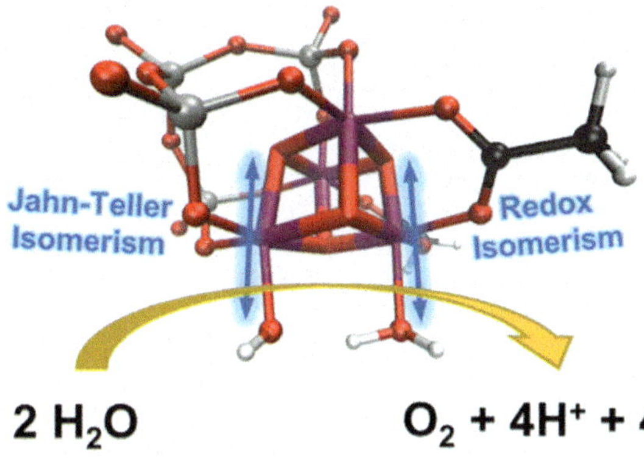

Fig. 10.19 Illustration of the role played by redox isomerism and JT effects in the natural OEC [105]. Published under a license CC-BY from American Chemical Society

catalyst holds four redox equivalents in the form of four Mn^{4+} centers; (3) The catalyst binds co-facial OH and H_2O ligands that are positioned in close proximity, allowing them both to participate in water oxidation; (4) Water oxidation proceeds by an intramolecular ET from the OH ligand of Mn_B to that metal center, coupled with deprotonation of the neighboring H_2O ligand; (5) This is followed by another proton-coupled electron transfer (PCET) and O–O bond formation, with a peroxo species formed on Mn_B; and (6) An intramolecular ET and final proton-coupled electron transfer (PCET) lead to the evolution and subsequent dissociation of O_2, resulting in the deactivated catalytic species $[Mn_4{}^{3+}V_4O_{17}(OAc)_2]^{4-}$ with one open coordination site on Mn_A and Mn_B each. It was also found that redox isomerism and Jahn–Teller (JT) effects play a prominent role in the water oxidation cycle of $[Mn_4V_4O_{17}(OAc)_3]^3$, namely, in the O_2 release step, JT distortions occurs (Fig. 10.19). The Jahn–Teller effect is a mechanism of spontaneous symmetry breaking in molecular and solid-state system.

The recent progress in understanding a half-reaction of electricity-driven water splitting, the oxygen evolution reaction (OER) mechanisms were summarized [81]. The suggested mechanism includes the conventional adsorbate evolution mechanism (AEM) (Fig. 10.20) and lattice-oxygen-mediated mechanism (LOM) (Fig. 10.21) considered from both theoretical and experimental aspects. According to AEM, the oxygen-containing adsorbates undergo catalytic reactions on the surface-active TM cations of the catalyst. The strategies to reduce overpotential based on the AEM were introduced. LOM involves the participation of lattice oxygen for oxygen generation, and is able to rationalize the high intrinsic activity and surface reconstruction issues in some catalysts. Strategies of stabilization of adsorbed HOO*, proton acceptor functionality, and switching the OER pathway to LOM were discussed.

In concluding the section, it should be noted that the release of oxygen under the action of relatively weak oxidizing agents, as in the case of OEC, occurs only in the case of catalysts, four nuclear clusters of manganese, most likely according to the four-electron mechanism predicted in [76]. While the catalysis under the action of mono and two nuclear complexes requires significantly strong oxidizing agents.

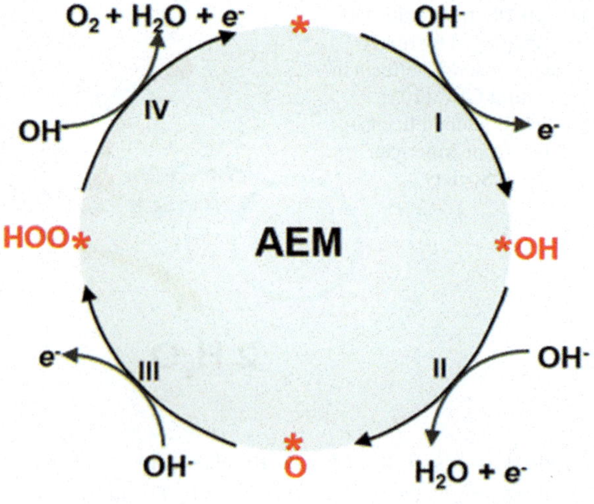

Fig. 10.20 OER catalytic cycle based on AEM. Here, * represents the surface-active TM cation. The four elementary steps I, II, III, IV are indicated [81]. With permission from Springer Nature

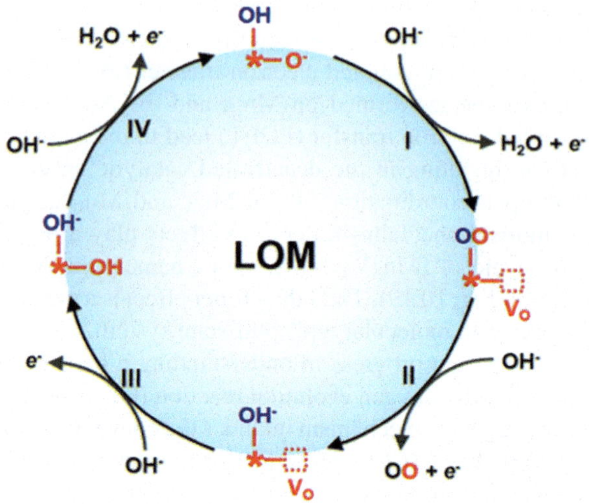

Fig. 10.21 OER catalytic cycle based on LOM. Here, * represents the surface-active TM cation, and the red dotted square represents the oxygen vacancy VO. The four elementary steps I, II, III, IV [81]. With permission from Springer Nature

10.4 Carbon Dioxide Utilization

The reduction, or fixation, of carbon dioxide is a vital process associated with artificial photosynthesis. In addition to the production of oxygen and hydrogen derived from water, there is a great interest in generating other hydrocarbon fuels by chemically reducing CO_2 with the utilization of hydrogen. Examples of fuel compounds that can be produced are formic acid (HCOOH), methanol (CH_3OH), carbon monoxide (CO), and methane (CH_4), for which the reduction reactions were presented in [106–125]. According to the Latimer-Frost diagram for the multi-multi-proton reduction of electron, CO_2 in one-electron reduction of CO_2 to $CO_2^{\cdot -}$ can be affected by only an extremely negative $E^{\circ\prime}$ redox potential of − 1.90 V versus NHE [106]. Latimer diagrams are the way to represent electrochemical equilibria for substances that have multiple oxidation states, while in a Frost diagram, $\Delta G^{\circ}/F$ (= nE°) versus oxidation number is plotted. Therefore, the one-electron process is thermodynamically forbidden in ambient condition. Nevertheless, two-electron and two-proton reduction of CO_2:

$$CO_2 + 2H^+ + 2e^- \rightarrow CO + H_2O \quad E^{\circ\prime} = -0.53\ V$$

$$CO_2 + 2H^+ + 2e^- \rightarrow HCOOH \quad E^{\circ\prime} = -0.61\ V$$

are affected by a lower thermodynamic barrier. In particular, this means that the hydride can participate in reactions as a hydrogenating agent. Candidates for corresponding catalysts are rhenium-based complexes, as well as more abundant cobalt-based, nickel-based, iron-based, and zinc-based complexes.

A review [117] focused on advances in the field of direct carboxylation reactions of C(sp^3)–H and C(sp^2)–H bonds using CO_2 encompassing both transition metal catalysis and base-mediated approach. The review described a number carboxylation reactions, namely: (1) Transition metal-catalyzed carboxylation of preactivated substrates including allylstannanes, organoboronic esters, organozinc reagents and aryl halides; (2) Transition metal-catalyzed hydrocarboxylation of unsaturated compounds (olefins, allenes, alkynes) requiring $AlEt_3$, $ZnEt_2$ coreactants to generate a metal-hydride species undergoing insertion of the unsaturated substrate and subsequent carboxylation of the organometallic intermediate; (3) The use of CO_2 in conjunction with alkenes, alkynes, dienes, allenes, diynes, in oxidative cycloaddition reaction of low-valent metal complexes to form five-membered metallacycles; and (4) Direct carboxylation of C–H bonds with CO_2 avoiding C–H prefunctionalization, hydrocarboxylation or metallacycle formation. In addition, several examples of base-promoted reactions integrating base recycling or base regeneration (through electrosynthesis)" steps have been reported. A representative example of synthetically efficient, base-promoted processes is presented in Fig. 10.22.

Substantial progress reviewed in [111] has recently been provided the general understanding of the interfacial CO_2 reduction reaction (CO_2RR) in electro and photocatalysis. A classification scheme to group different aspects influencing the interfacial CO_2RR thermodynamics and kinetics was introduced. A survey of the literature was focused on the local chemical environment encompassing surface effects (adsorbates, support), solution interactions (electrolyte constituents) and three-dimensional chemical surroundings (polymers, metal organic frameworks (MOFs), covalent organic frameworks (COFs)).

The use of carbon dioxide as a building block in organic synthesis from an industrial perspective was demonstrated [115].

The utilization of bases such as KOtBu, Cs_2CO_3, aluminum-based Lewis acids, and transition metals, as a building block in an organic synthesis from an industrial perspective to promote C–H carboxylation was reported [115]. According to recently significant results in the process improvements with respect to synthetic scope,

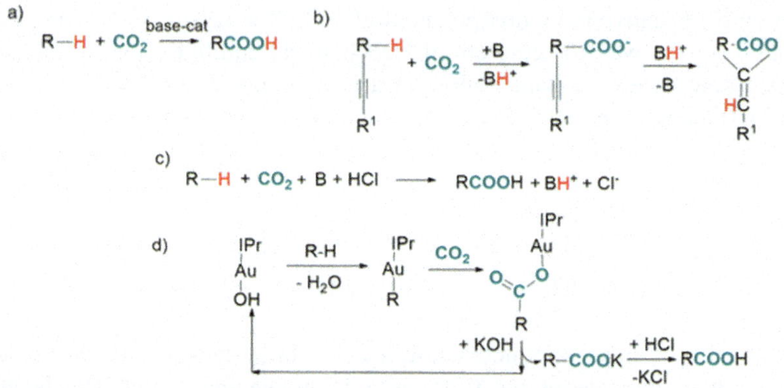

Fig. 10.22 General stoichiometry of base-catalyzed reactions (**a**, **b**); **c** general stoichiometry of base-promoted reactions (under the assumption that $pK_{BH^+} > pK_{C\text{–}H}$); **d** carboxylation of heteroarenes catalyzed by IPrAu(OH) catalyst [118]. With permission from American Chemical Society

yields, and green credentials were achieved. At temperature 200–350 °C, molten salts containing caesium or potassium cations enable carbonate ions to deprotonate very weakly acidic C–H bonds (pKa > 40), generating carbon-centered nucleophiles. The latter reacts with CO_2 to form carboxylates [116]. In particular, C–H carboxylation is followed by protonation to convert 2-furoic acid into furan-2,5-dicarboxylic acid (FDCA)

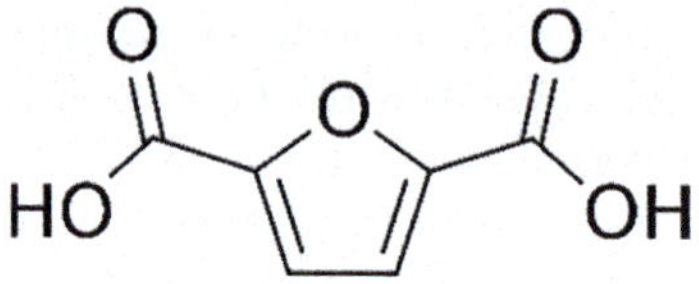

which was used for the synthesis of polyethylene furandicarboxylate.

It was shown that intermediate-temperature (200–350 °C) molten salts containing caesium or potassium cations enable carbonate ions (CO_3^{2-}) that can deprotonate very weakly acidic C–H bonds ($pK_a > 40$), generating carbon-centered nucleophiles that react with CO_2 to form carboxylates [116]. As an example, C–H carboxylation is followed by protonation to convert 2-furoic acid

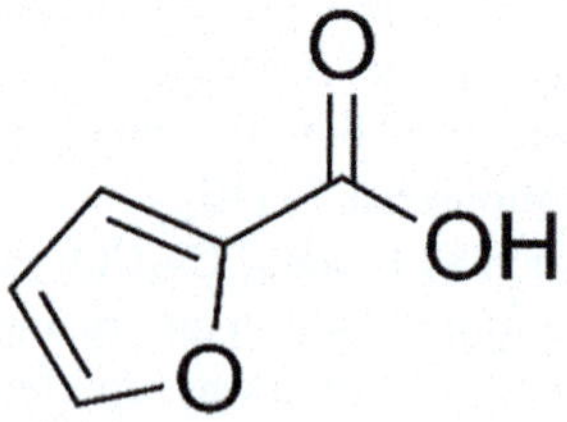

into furan-2,5-dicarboxylic acid. Since 2-furoic acid can be made from lignocellulose, CO_3^{2-}-promoted C–H carboxylation thus reveals a way to transform inedible biomass and CO_2 into a valuable feedstock chemical.

A strategy based on a monolithic nanoporous silver-based membrane electrocatalyst (npm-Ag) to increase the density of active sites was developed [120]. The effective thickness of the catalyst was improved from tens of nanometers to several micrometers. The npm-Ag showed selective reduction of CO_2 to CO at low overpotentials (ca. 80%@40 mV, ca. 100%@90–290 mV). Combined with a NiFe-based O_2-evolving anode, a stable CO_2-reducing photosynthetic system with an η_{STF} of ca. 20.1% was active for 28 h in the absence of a DC/DC converter A DC-to-DC converter is an electronic circuit or electromechanical device that converts a source of direct current (DC) from one voltage level to another.

A 3D mass transport model utilizing the accurate electrode geometry to calculate local concentration distributions of $CO_2(aq)$, OH^-, HCO_3^-, and CO_3^{2-} was designed [121]. The model based on Butler–Volmer correlations which described how the electrical current through an electrode depends on the voltage difference between the electrode and the bulk electrolyte for an unimolecular redox reaction, considered that both a cathodic and an anodic reaction occur on the same electrode. The CO evolution reaction with a proportional dependency on the carbonate concentration was also observed. The developed design showed high electrode roughness per film thickness, which translated to smaller pore size and allowed to increase the CO selectivity.

An example of a nonaqueous flow cell electrolyzer with $[Ni(cyclam)]^{2+}$ as the homogeneous electrocatalyst for CO_2 reduction was presented [122]. Using ferrocene as a sacrificial electron donor and ammonium salts as both electrolyte and proton donor, efficient catalytic CO_2 reduction was achieved. The design showed high selectivity for the reduction of CO_2 to CO (> 80%) and reached high current densities with a graphite felt working electrode (up to 50 mA cm^{-2} with 0.5 M proton donor in MeCN solution), producing > 40 mL h^{-1} of CO, continuous process. By taking advantage of the high solubility of CO_2 and CH_4 at low temperature, the electrochemical conversion of these gases in aqueous media down to – 40 °C was performed [123]. Specifically, the use of brines (a high-concentration solution of a salt in water) electrolytes was shown to be an efficient route to perform electrochemistry experiments at low temperatures. Immobilized CO_2 reduction enzymes—such as formate dehydrogenase—can operate with high turnover and selectivity at a minimal overpotential and are therefore effective model catalysts [124]. Through the coimmobilization of carbonic anhydrase on thin indium oxide, the effect of CO_2 hydration on the local environment and performance of a range of disparate CO_2 reduction systems from the enzymatic to heterogeneous systems was achieved. The experiments indicated that the coimmobilization of carbonic anhydrase increases the kinetics of CO_2 hydration at the electrode.

References

1. Wakerley D, Lamaison S, Christine A (2019) Caputo C: highlights from faraday discussion: artificial photosynthesis. Chem Commun 55:8351–8358
2. Anthony F. Collings AF, Christa Critchley C (2007) Artificial photosynthesis: from basic biology to industrial application. Wiley, Hoboken
3. Bruno R (2016) Artificial photosynthesis, 1st edn. Advances in botanical research, vol 79. ISBN: 9780128032893
4. Ye M, Yuan Y, Lin Z (2020) Artificial photosynthesis: from materials to devices. NOVA
5. Fukuzumi S (2023) Principles and applications of artificial photosynthesis. Royal Society of Chemistry
6. Razeghifard R (2013) Natural and artificial photosynthesis: solar power as an energy source. Wiley, Hoboken
7. Rozhkova EA, Ariga K (2015) From molecules to materials pathways to artificial photosynthesis. Springer, Berlin
8. Bachmeier ASJL (2017) Metalloenzymes as inspirational electrocatalysts for artificial photosynthesis from mechanism to model devices. Springer, Berlin
9. Bozal-Ginesta C, Durrant JR (2019) Artificial photosynthesis—concluding remarks. Faraday Discuss 215:439–451
10. Jha A, Mote KR, Chandra S, Madhu PK, Dasgupta J (2021) Photoactive Anthraquinone-Based Host–Guest Assembly for Long-Lived Charge Separation. J Phys Chem C 125(20):10891–10900
11. Likhtenshtein GI, Syrtsova LA, Samuilov VD et al (1975) XII International botanic congress: theses of reports. Nauka, Leningrad, p 429
12. Likhtenshtein GI, Kotel'nikov AI, Kulikov AV, Syrtsova LA et al (1979) Some peculiarities of the electronic transfer in redox enzymes. Int J Quantum Chem 16(3):419–435
13. Likhtenshtein GI, Kulikov AV, Kotelnikov AI, Bogatyrenko VR (1982) Structure and action mechanism of reaction centers of photosynthetic bacteria. Photobiochem Photobiol 3:178–182
14. Likhtenshtein GI (2003) New trends in enzyme catalysis and mimicking chemical reactions. Kluwer Academic/Plenum Publishers, Dordrecht
15. Gratzel M (2005) Solar energy conversion by dye-sensitized photovoltaic cells. Inorg Chem 44(20):6841–6851
16. Jakob M, Berg A, Rubin R, Levanon H, Li K, Schuster DI (2009) Photoinduced electron transfer in porphyrin- and fullerene/porphyrin-based rotaxanes as studied by time-resolved EPR spectroscopy. J Phys Chem A 113(20):5846–5854
17. Gust D, Moore TA, Moore AL (2009) Solar fuels via artificial photosynthesis. Acc Chem Res 42(12):1890–1898
18. Berera R, Herrero C, van Stokkum IHM, Vengris M et al (2007) Energy transfer, excited-state deactivation, and exciplex formation in artificial caroteno-phthalocyanine light-harvesting antennas. J Phys Chem B 111(24):6868–6877
19. Mozer AJ, Griffith M, Tsekouras G (2009) Zn−Zn porphyrin dimer-sensitized solar cells: toward 3-D light harvesting. J Am Chem Soc 131(43):15621–15623
20. Park M, Im D, Rhee YH, Joo T (2014) Coherent and homogeneous intramolecular charge-transfer dynamics of 1-tert-butyl-6-cyano-1,2,3,4-tetrahydroquinoline (NTC6), a rigid analogue of DMABN. J Phys Chem A 118:5125–5134. https://doi.org/10.1021/jp411227r
21. Wasielewski MR (2009) Self-assembly strategies for integrating light harvesting and charge separation in artificial photosynthetic systems. Acc Chem Res 42(12):1910–1921
22. Calver CF, Schanze KS, Cosa G (2016) Biomimetic light-harvesting antenna based on the self-assembly of conjugated polyelectrolytes embedded within lipid membranes. ACS Nano 10(11):10598–10605. https://doi.org/10.1021/acsnano.6b07111
23. Meneghin E, Biscaglia F, Volpato A et al (2020) Biomimetic nanoarchitectures for light harvesting: self-assembly of pyropheophorbide-peptide conjugates. J Phys Chem Lett 11(19):7972–7980. https://doi.org/10.1021/acs.jpclett.0c02138

24. Likhtenshtein GI (2012) Solar energy conversion. Chemical aspects. Wiley-VCH, Weinheim
25. Sessler JL (1992) Photosynthetic model systems. Israel J Chem 32:449–468; McLendon G, Hake R (1992) Interprotein electron transfer. Chem Rev 92:481–490
26. Gust D, Moore TA, Moore AL (1999) An artificial photosynthetic membrane. Z Phys Chem 213:149–155
27. Wasielewski MR (2002) High time resolution Q-band EPR study of sequential electron transfer in a triad oriented in a liquid crystal. J Phys Chem A 106:1933–1937
28. Paulson BP, Miller JR, Gan W-X, Closs G (2005) Superexchange and sequential mechanisms in charge transfer with a mediating X state between the donor and acceptor. J Am Chem Soc 127:4860–486837
29. Ricks AB, Solomon GC, Colvin MT et al (2010) Controlling electron transfer in donor−bridge−acceptor molecules using cross-conjugated bridges. J Am Chem Soc 132(43):15427–15434
30. Olguin M, Zope RR, Baruah T (2013) Effect of geometrical orientation on the charge-transfer energetics of supramolecular (tetraphenyl)-porphyrin/C_{60} dyads. J Chem Phys 138:074306
31. Ha-Thi MH, Pham VT, Pino T (2018) Photoinduced electron transfer in a molecular dyad by nanosecond pump—pump—probe spectroscopy. Photochem Photobiol Sci 17:903–909
32. Shinde J, Thomas MB, Madhurima Poddar M et al (2021) Does location of BF2-chelated dipyrromethene (BODIPY) ring functionalization affect spectral and electron transfer properties? Studies on α-, β-, and meso-functionalized BODIPY-derived donor-acceptor dyads and triads. J Phys Chem C 43:23911–23921
33. Hattori S, Ohkubo K, Urano Y, Sunahara H, Nagano T, Wada Y, Tkachenko NV, Lemmetyinen H, Fukuzumi S (2005) Charge separation in a nonfluorescent donor−acceptor dyad derived from boron dipyrromethene dye, leading to photocurrent generation. J Phys Chem B 109(32):15368–15375
34. Fakis M, Beckwith JS, Seintis K et al (2018) Energy transfer and charge separation dynamics in photoexcited pyrene–bodipy molecular dyads. Phys Chem Phys 20:837–849
35. Popere BC, Khomein P, Pointer CA et al (2020) Modulating absorption and charge transfer in bodipy-carbazole donor–acceptor dyads through molecular design. Dalton Trans 48:8488–8501
36. Wei Z, Philip AM, Jager WF, Grozema FC (2022) Fast charge separation in distant donor-acceptor dyads driven by relaxation of a hot excited state. J Phys Chem C Nanomater Interfaces 126(45):19250–19261
37. Ansteatt S, Meares A, Ptaszek M (2021) Amphiphilic near-IR-emitting 3,5-bis(2-pyrrolylethenyl)BODIPY derivatives: synthesis, characterization, and comparison with other (hetero)arylethenyl-substituted BODIPYs. J Org Chem 86(13):8755–8765. https://doi.org/10.1021/acs.joc.1c00586
38. Menzel JP, de Groot HJM, Francesco Buda F (2019) Photoinduced electron transfer in donor–acceptor complexes: isotope effect and dynamic symmetry breaking. J Phys Chem Lett 10(21):6504–6511
39. Nançoz C, Rumble C, Rosspeintner A, Vauthey E (2020) Bimolecular photoinduced electron transfer in non-polar solvents beyond the diffusion limit. J Chem Phys 152:244501
40. Zhang X, Elmali A, Duan R, Liu Q, Ji W, Zhao J, Li C, Karatay A (2020) Charge separation, recombination and intersystem crossing of directly connected perylenemonoimide–carbazole electron donor/acceptor dyads. Phys Chem Chem Phys 22:6376–6390
41. Yang X, Keane T, Delor M et al (2017) Identifying electron transfer coordinates in donor-bridge-acceptor systems using mode projection analysis. Nat Commun 8:14554. https://doi.org/10.1038/ncomms1455
42. Li X, Valdiviezo J, Banziger SD et al (2020) Symmetry controlled photo-selection and charge separation in butadiyne-bridged donor–bridge–acceptor compounds. Phys Chem Chem Phys 22:9664–9676
43. Kim W, Vakuliuk O, Gryko DT et al (2020) Two-step charge separation passing through the partial charge-transfer state in a molecular dyad. J Am Chem Soc 142(3):1564–1570

44. Poddutoori PK, Bayard BJ (2021) (V) Porphyrin-naphthalene conjugates: triplet formation via rapid charge recombination. Inorg Chem 60(23):17952–17965
45. Li X, Yu J, Lu Z (2021) Photoinduced charge transfer with a small driving force facilitated by exciplex-like complex formation in metal–organic frameworks. J Am Chem Soc 143(37):15286–15297
46. Caballero R, Barrejón M, Cerdá J et al (2021) Self-assembly-directed organization of a fullerene–bisporphyrin into supramolecular giant donut structures for excited-state charge stabilization. J Am Chem Soc 143(29):11199–11208
47. Bancroft L, Zhang J, Harvey SM et al (2021) Charge transfer and spin dynamics in a zinc porphyrin donor covalently linked to one or two naphthalenediimide acceptors. J Phys Chem A 125(3):825–834
48. Gong Q, Cheng K, Wu Q et al (2020) One-pot access to ethylene-bridged BODIPY dimers and trimers through single-electron transfer chemistry. J Org Chem 86(21):15761–15767
49. Clark-Winters TL, Bragg AE (2021) Control of photoinduced charge separation in conjugated polyelectrolyte complexes through microstructure-dependent exciton delocalization. J Phys Chem C 125(42):22982–22997
50. Zhang W, Xu W, Zhang G, Kong J, Niu X, Chan JMW, Liu W, Xia A (2021) Direct tracking excited-state intramolecular charge redistribution of acceptor–donor–acceptor molecule by means of femtosecond stimulated Raman spectroscopy. J Phys Chem B 125(17):4456–4464
51. Yanagi R, Zhao T, Solanki D, Pan Z, Hu S (2022) Charge separation in photocatalysts: mechanisms, physical parameters, and design principles. ACS Energy Lett 7(1):432–452
52. Müller C, Bold S, Chavarot-Kerlidou M, Dietzek-Ivanšić B (2022) Photoinduced electron transfer in triazole-bridged donor-acceptor dyads—a critical perspective. Coord Chem Rev 472(1):214764
53. Bystryak IM, Likhtenshtein GI, Kotel'nikov AI, Hankovsky O, Hideg K (1986) The influence of the molecular dynamics of the solvent on the photoreduction of nitroxyl radicals. Russ J Phys Chem 60:1679–1683
54. Vogel R, Rubtsova ET, Kotel'nikov AI, Likhtenshtein GI (1986) Study of protein collisions in solutions by triplet labels. Biofizika 31:152–154
55. Vogel R, Rubtsova ET, Likhtenshtein GI, Hideg K (1994) Factors affecting photoinduced electron transfer in a donor-acceptor pair (D-A) incorporated into bovine serum albumin. J Photochem Photobiol A Chem 83:229–236
56. Lozinsky E, Martin VV, Berezina TA, Shames, Weis AL, Likhtenshtein GI (1999) Dual fluorophore-nitroxide probes for analysis of vitamin C in biological liquids. J Biochem Biophys Meth 38:29–42
57. Likhtenshtein GI, Febbrario F, Nucci R (2000) Intramolecular dynamics and conformational transitions in proteins studied by biophysical labeling methods. Common and specific features of proteins from thermophylic microorganisms. Spectrochem Acta Part A Biomol Spectrosc 56:2011–2031
58. Likhtenshtein GI, Pines D, Pines E, Khutorsky V (2009) Effect of a protein on the light energy conversion in dual fluorophore-nitroxide probes studied by ESR and fluorescence spectroscopy. Appl Magn Reson 35(3):459–472
59. Likhtenshtein GI (2009) Appl Biochem Biotechnol 152(1):13587
60. Likhtenshtein GI (2020) Nitroxides: brief history and recent developments. Springer Nature, Berlin
61. Likhtenshtein GI, Yamauchi J, Nakatsuji S, Smirnov A, Tamura R (2008) Nitroxides: application in chemistry, biomedicine, and materials science. Wiley-VCH, Weinheim
62. Blinco JP, Hodgson J, Morrow BJ et al (2008) Experimental and theoretical studies of the redox potentials of cyclic nitroxides. J Org Chem 73:6763–6771
63. Gerlock JL, Zacmanidis PL, Bauer DR et al (1990) Fluorescence detection of free radicals by nitroxide scavenging. Free Radical Res Commun 10(1–2):119–121
64. Qixi M, Chernick ET, McCamant DW, Weiss EA, Ratner MA, Wasielewski MR (2006) Spin dynamics of photogenerated triradicals in fixed distance electron donor-chromophore-acceptor-TEMPO molecules. J Phys Chem A 110:7323–7333

65. Anderson SS, Lyle IG, Paterson R (1976) Electron transfer across membranes using vitamin K_1 and coenzyme Q_{10} as carrier molecules. Nature 259:147–148. https://doi.org/10.1038/259147a0
66. Gust D, Moore TA, Moore AL (2005) In: Collings AF, Critchley C (eds) Artificial photosynthesis. Wiley, Weinheim, p 187
67. Zhu L, Khairutdinov RF, Cape JL, Hurst JK (2006) Nanoporous block copolymer micelle/micelle multilayer films with dual optical properties. J Am Chem Soc 108:825
68. Balzani V (2001) Electron transfer in chemistry. Wiley-VCH Verlag GmbH
69. Akbarzadeh A, Rezaei-Sadabady R, Davaran S et al (2013) Liposome: classification, preparation, and applications. Nanoscale Res Lett 8(1):102
70. Khairutdinov RF, Hurst, JK (1999) Cyclic transmembrane charge transport by pyrylium ions in a vesicle-based photocatalytic system. Nature 402(6761):509–511
71. Mauzerall D, Sun K (2003) Photoinduced charge separation across membranes and interfaces is the basis of photosynthesis and of artificial solar energy conversion. Photoinduced charge separation in lipid bilayers, chap 34. Membrane science and technology, vol 7, pp 963–979
72. Banerji N, Fürstenberg A, Bhosale S, Sisson AL, Sakai N, Matile S, Vauthey E (2008) J Phys Chem B 112(30):8912–8922
73. Sahin T, Harris MA, Vairaprakash P et al (2015) Self-assembled light-harvesting system from chromophores in lipid vesicles. J Phys Chem B 119(32):10231–10243
74. Hiroyuki S, Kohji M, Nobuyuki I (2003) Voltammetric study on the electron transport through a bilayer lipid membrane containing neutral or ionic redox molecules. J Electroanal Chem 556(1):1–11
75. Tadashi M, Yoshida A, Harada A et al (2006) Pyrene-sensitized electron transport across vesicle bilayers: dependence of transport efficiency on pyrene substituents. Org Biomol Chem 4(23):4336–4344
76. Semenov NN, Shilov AE, Likhtenshtein GI (1975) Doklady Akademii Nauk SSSR 221(6):1374–1377
77. Shafirovich VY (1995) Chemical models of reaction centers for photoinduced charge separation and photosynthetic enzyme systems. Ross Khim Zh 39:80–88
78. Shilov AE (1997) Metal complexes in biomimetical reactions. N_2 fixation, activation and oxidation of alkanes, chemical models of photosyntheses. CRC Press, Boca Raton, p 302
79. Rüttinger W, Yagi M, Wolf K, Bernasek S, Dismukes GC (2000) O_2 evolution from the manganese-oxo cubane core $Mn_4O_4^{6+}$: a molecular mimic of the photosynthetic water oxidation enzyme? J Am Chem Soc 122:10353–10357
80. Mayfield JR, Grotemeyer EN, Jackson TA (2020) Concerted proton-electron transfer reactions of manganese-hydroxo and manganese-oxo complexes. Chem Commun (Camb) 56(65):9238–9255
81. Li J (2022) Oxygen evolution reaction in energy conversion and storage: design strategies under and beyond the energy scaling relationship. Nanomicro Lett 14(1):112
82. Dismukes GC, Brimblecombe R, Felton GA et al (2009) Development of bioinspired Mn_4O_4-cubane water oxidation catalysts: lessons from photosynthesis. Acc Chem Res 42(12):1935–1943
83. Brieskorn V, González L (2021) Flexibility enhances reactivity: redox isomerism and Jahn-Teller effects in a bioinspired Mn_4O_4 cubane water oxidation catalyst. ACS Catal 11(21):13320–13329
84. Song J, Wei C, Huang ZF, Liu C, Zeng L, Wang X, Xu Z (2020) A review on fundamentals for designing oxygen evolution electrocatalysts. J Chem Soc Rev 49(7):2196–2214
85. Likhtenshtein GI (2016) Electron spin in chemistry and biology: fundamentals, methods, reactions mechanisms, magnetic phenomena, structure investigation. Springer Nature, Berlin
86. Likhtenshtein GI (1988) Chemical physics of redox metalloenzymes. Springer, Heidelberg
87. Baffert C, Collomb MN, Deronzier A et al (2002) Two new terpyridine dimanganese complexes: a manganese(III,III) complex with a single unsupported oxo bridge and a manganese(III,IV) complex with a dioxo bridge. Synthesis, structure, and redox properties. Inorg Chem 41:1404

88. Luneva NP, Knerel'man EI, Shafirovich VYa, Shilov AE (1987) A manganese cluster bound to a bilayer membrane: a chemical model for the oxygen-forming center of photosynthesis. J Chem Soc Chem Commun (19):1504–1505
89. Limburg J, Vrettos JS, Liable-Sands LM et al (1999) A functional model for O-O bond formation by the O_2-evolving complex in photosystem II. Science 283:1524–1527
90. Nayak S, Nayek HP, Dehnen S, Powell AK, Reedijk J (2011) Trigonal propeller-shaped [Mn(III)$_3$M(II)Na] complexes (M = Mn, Ca): structural and functional models for the dioxygen evolving centre of PSII. Dalton Trans 40(12):2699–2702
91. Gao Y, Crabtree RH, Brudvig GW (2012) Water oxidation catalyzed by the tetranuclear Mn complex [Mn(IV)4O$_5$(terpy)$_4$(H$_2$O)$_2$](ClO$_4$)$_6$. Inorg Chem 51(7):4043–4050
92. Chen C, Zhang C, Dong H, Zhao J (2015) Artificial synthetic MnIVCa–oxido complexes mimic the oxygen-evolving complex in photosystem II. Dalton Trans 44:4431. https://doi.org/10.1039/c4dt03459g
93. Smith PF, Kaplan C, Sheats JE, Robinson DM, McCool NS, Mezle N, Dismukes CG (2014) What determines catalyst functionality in molecular water oxidation? Dependence on ligand and metal nuclearity in cobalt clusters. Inorg Chem 53:2113–2121
94. Marang C, Dong H, Zhao J (2015) Artificial synthetic Mn(IV)Ca-oxido complexes mimic the oxygen-evolving complex in photosystem II. Dalton Trans 44(10):4431–4435
95. Liao RZ, Kärkäs MD, Lee BL, Åkermark B, Siegbahn PE (2015) Photosystem II like water oxidation mechanism in a bioinspired tetranuclear manganese complex. Inorg Chem 54(1):342–351
96. Zhang C, Chen C, Dong H, Shen JR, Dau H, Zhao JA (2015) Synthetic Mn$_4$Ca-cluster mimicking the oxygen-evolving center of photosynthesis. Science 348:690–693
97. Laine TM, Kärkäs MD, Liao RZ et al (2015) A dinuclear ruthenium-based water oxidation catalyst: use of non-innocent ligand frameworks for promoting multi-electron reactions. Chemistry 21(28):10039–10048
98. Zhang C, Dong H, Zhao J (2015) Artificial synthetic Mn(IV)Ca-oxido complexes mimic the oxygen-evolving complex in photosystem II. Dalton Trans 44(10):4431–4435
99. Young KJ, Brennan BJ, Tagore R, Brudvig GW (2015) Photosynthetic water oxidation: insights from manganese model chemistry. Acc Chem Res (3):567–574
100. Deibert B, Zhang J, Smith PF et al (2015) Surface and structural investigation of a MnOx birnessite-type water oxidation catalyst formed under photocatalytic conditions. Chem Eur J 21(40):14218–14228
101. Smith PF, Deibert B, Kaushik S et al (2016) Coordination geometry and oxidation state requirements of corner sharing MnO_6 octahedra for water oxidation catalysis: an investigation of manganite (γ-MnOOH). ACS Catal 6:2089–2099
102. Rice DB, Massie AA, Jackson TA (2017) Manganese-oxygen intermediates in O-O bond activation and hydrogen-atom transfer reactions. Acc Chem Res 50(11):2706–2717
103. Chen Y, Xu B, Yao R, Chen C, Zhang C (2022) Mimicking the oxygen-evolving center in photosynthesis. Front Plant Sci 13:929532
104. Krewald NF, Pantazis DA (2016) Redox potential tuning by redox-inactive cations in nature's water oxidizing catalyst and synthetic analogues. Phys Chem Chem Phys 18:10739–10750. https://doi.org/10.1039/c5cp07213a
105. Schwiedrzik L, Brieskorn V, González L (2021) Flexibility enhances reactivity: redox isomerism and Jahn-Teller effects in a bioinspired Mn_4O_4 cubane water oxidation catalysis. ACS Catal 11(21):13320–13329
106. Schneider J, Hongfei J, Muckerman JT, Fujita E (2012) Thermodynamics and kinetics of CO_2, CO, and H^+ binding to the metal centre of CO_2 reduction catalysts. Chem Soc Rev 41:2036–2051
107. New nature-inspired concepts for efficiently turning CO_2 into clean, sustainable fuels. University of Cambridge, SciTechDaily (2022). https://www.cam.ac.uk/research/news/new-nature-inspired-concepts-for-turning-co2-into-clean-fuels
108. Oneill M (2022) Artificial photosynthesis: new device advances commercial viability of solar fuels. SciTechDaily

109. Floating "artificial leaves" generate clean fuels from sunlight and water. SciTech-Daily (2022). https://scitechdaily.com/floating-artificial-leaves-generate-clean-fuels-from-sunlight-and-water/
110. Artz J, Müller TE, Thener T et al (2018) Sustainable conversion of carbon dioxide: an integrated review of catalysis and life cycle assessment. Chem Rev 118:434–504
111. Wagner A, Sahm CD, Reisner E (2022) Towards molecular understanding of local chemical environment effects in electro- and photocatalytic CO_2 reduction. Nat Catal 3:775–786
112. Zheng Y, Yao H, Zhang Z (2022) Water coordinated on Cu(I)-based catalysts is the oxygen source in CO_2 reduction to CO. Nat Commun 13(1):25772022
113. Ma Y, Yi X, Wang J (2022) Selective photocatalytic CO_2 reduction in aerobic environment by microporous Pd-porphyrin-based polymers coated hollow TiO_2. Nat Commun 13:1400
114. Wagner A, Sahm CD, Reisner E (2020) Towards molecular understanding of local chemical environment effects in electro- and photocatalytic CO_2 reduction. Nat Catal 3:775–786
115. Dabral S, Schaub T (2019) The use of carbon dioxide (CO_2) as a building block in organic synthesis from an industrial perspective. Adv Synth Catal 361:223–246
116. Banerjee A, Dick GR, Yoshino T, Kanan MW (2016) Carbon dioxide utilization via carbonate-promoted C-H carboxylation. Nature 531:215–219
117. Tommasi I (2017) Direct carboxylation of C(sp^3)-H and C(sp^2)-H bonds with CO_2 by transition-metal-catalyzed and base-mediated reactions. Catalysts 7:380
118. Boogaerts IIF, Nolan SP (2010) Carboxylation of C-H bonds using N-heterocyclic carbene Gold(I) complexes. J Am Chem Soc 132:8858–8859
119. Zhang Y, Zhang T, Das S (2020) Catalytic transformation of CO_2 into C1 chemicals using hydrosilanes as a reducing agent. Green Chem 22:1800–1820
120. Xiao Y, Qian Y, Chen A et al (2020) An artificial photosynthetic system with CO_2-reducing solar-to-fuel efficiency exceeding 20%. J Mater Chem A 8:18310–18317
121. Suter S, Haussener S (2019) Optimizing mesostructured silver catalysts for selective carbon dioxide conversion into fuels. Energy Environ Sci 12:1668–1678
122. Jiang C, Nichols AW, Walzer JF, Machan CW (2020) Electrochemical CO_2 reduction in a continuous non-aqueous flow cell with $[Ni(cyclam)]^{2+}$. Inorg Chem 59(3):1883–1892
123. Sargeant E, Kolodziej A, Le Duff CS, Rodriguez P (2020) Electrochemical conversion of CO_2 and CH_4 at subzero temperatures. ACS Catal 10:7464–7474
124. Cobb SJ, Vivek M, Badiani VM, Dharani AM et al (2022) Fast CO_2 hydration kinetics impair heterogeneous but improve enzymatic CO_2 reduction catalysis. Nat Chem 14:417–424
125. Yu D, Teong SP, Zhang Y (2015) Transition metal complex catalyzed carboxylation reactions with CO_2. Coord Chem Rev 93(294):279–291

Chapter 11
Modeling Nonhemin Iron Proteins and Enzymes

Abstract Iron–sulfur clusters occur in many biological systems, often as components of electron transfer proteins and as the H-cluster in hydrogenase and the P-cluster and FeMo-cofactor in nitrogenase. This chapter describes significant progress made in synthetical, compositional and functional aspects of the nitrogenase-like clusters through the recent decades. New synthesized clusters were found to react with dihydrogen and proton mimicking the hydrogenase reaction S ($H_2 + A_{ox} \rightarrow 2H^+ + A_{red}$ and $2H^+ + D_{red} \rightarrow H_2 + D_{ox}$). Number of synthetic rhombic and cuboidal clusters could convert N_2 to hydrazine and ammonium modeling the nitrogenase reaction. As an example of four electron mechanism of dinitrogen reduction, four nuclear complex V(II)-pyrocatechol ($V^{(II)}{}_4$) was utilized in a homogeneous reaction $V^{(II)}{}_4N_2 + V^{(II)}{}_4 + 8H^+ \rightarrow 8V^{(III)} + 2NH_3 + H_2O$. The Chatt Cycle, a hypothetical model for the mechanism of nitrogenase action at the atomic level based upon reactions of molybdenum compounds and scheme of the multi-stage nitrogenase process were considered.

11.1 Iron–Sulfur Clusters

Iron–sulfur clusters occur in many biological systems, often as components of electron transfer proteins and are proved to be the catalysts for the generation of biologically relevant molecules such as ammonia and hydrogen [1–20]. Versatile structural and chemical characteristics biological Fe/S clusters have allowed their widespread use in virtually all living organisms. In nature, Fe–S clusters are structured in three main shapes: rhombic [Fe_2S_2] clusters, cuboidal [Fe_3S_4] clusters, and cuboidal [Fe_4S_4] clusters (Fig. 11.1) [8, 9]. Other more complex clusters exist, such as the H-cluster in hydrogenase and the P-cluster and FeMo-cofactor in nitrogenase. Fe–S clusters have been widely investigated specifically for electron transfer, catalysis, sulfur donation, and sensing (Fig. 11.2).

The paper [9] provided an overview of Fe–S clusters to highlight the growing *versatile* uses of these compounds. The discussion was divided into sections that describe the structure–function relationship of Fe–S clusters in five research areas:

G. I. Likhtenshtein, *Enzyme Catalysis Today and the Chemistry of the 21st Century*,
https://doi.org/10.1007/978-3-031-51552-1_11

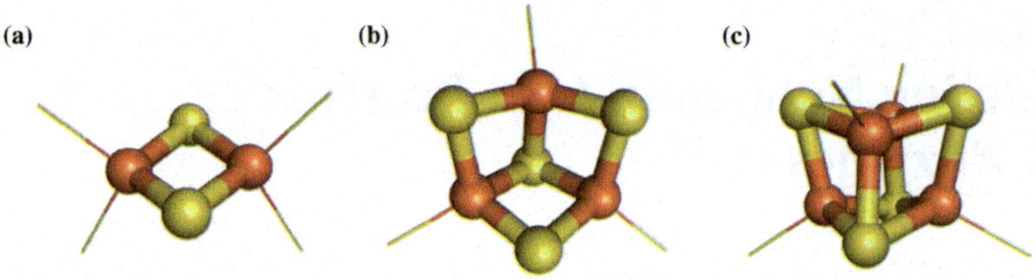

Fig. 11.1 Most common Fe–S cluster types found in nature are the **a** $[Fe_2S_2]$, **b** $[Fe_3S_4]$, and **c** $[Fe_4S_4]$ clusters. In these figures, Fe and S atoms are displayed as orange and yellow spheres, respectively. Bonds within the cluster are shown as sticks, and bonds from Fe atoms to their coordinating ligands are shown as lines [9]. With permission from Elsevier

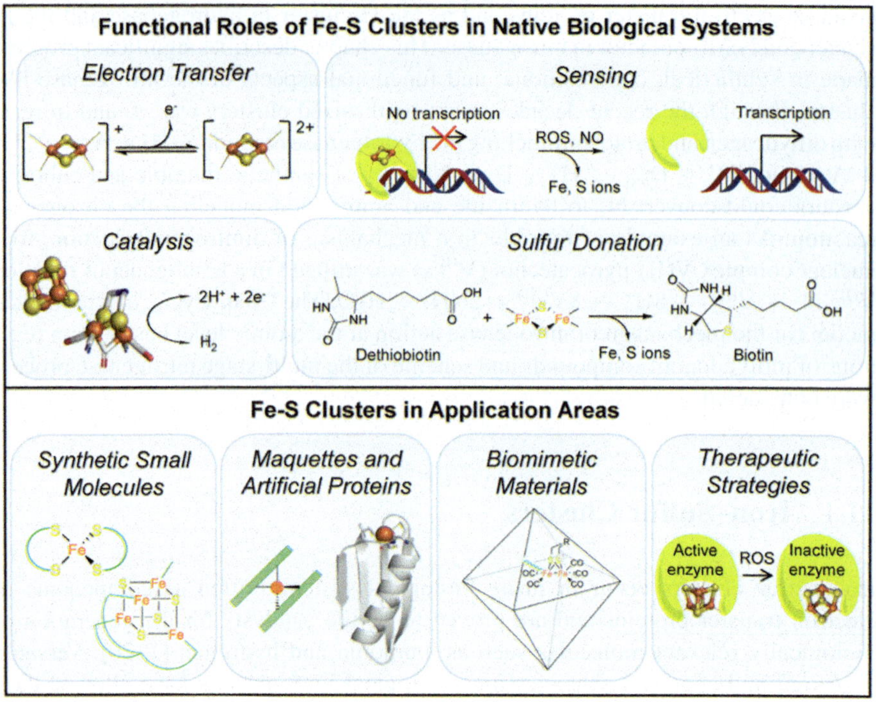

Fig. 11.2 Top: In biology, Fe–S clusters participate in four main functional roles: electron transfer, catalysis, sulfur donation, and sensing. Bottom: Depiction of Fe–S clusters in four application areas discussed in this review: synthetic small molecules, maquettes and artificial proteins, biomimetic materials, and therapeutic strategies [8, 9]. With permission from Elsevier

biology, synthetic small molecules, maquettes and artificial proteins, biomimetic materials, and therapeutic development (Fig. 11.2, Table 11.1). The clusters redox-active and catalytic properties, structural diversity and modular nature, abundance of Fe and S, and stability under mild conditions were described in details. The hydroxylase-like photocatalytic systems and electrocatalytic systems were discussed separately. Of particular interest are mimic clusters with the corresponding nitrogenase and hydrogenase catalytic activity. Synthetic catalysts are compared by their turnover frequencies (TOF) and turnover numbers (TON) are presented in Fig. 11.2.

11.2 Nitrogenase Chemical Models

Of particular interest are mimic clusters with the corresponding nitrogenase and hydrogenase catalytic activity. The process of assimilation of atmospheric nitrogen by microorganisms was known since 1838. Involvement of these molecules in a chemical reaction under mild conditions seemed to be disagreed with traditional theoretical considerations and experimental evidence of the extraordinary chemical inertness of dinitrogen. Since the discovery of discrete iron–sulfur metalloclusters within the nitrogenase MoFe protein, synthetic inorganic chemists have sought to reproduce the structural features of these clusters in order to understand how they facilitate the binding, activation and hydrogenation of N_2. Through the recent decades, significant progress has been made to synthetically replicate certain compositional and functional aspects of the nitrogenase-like clusters [1–50]. Beyond approximating the architectures of nitrogenase clusters with synthetic models, research efforts have also focused on reproducing nitrogenase functionality, i.e., substrate binding, activation, and hydrogenation. Although much work remains to generate synthetic iron–sulfur clusters that can reduce N_2 to NH_3 to reach the level of biological fixation of atmospheric nitrogen.

The mechanism of reducing N_2 to NH_3 by nitrogenase has been be modeled in work [51]. The Fe-protein has a single [4Fe4S] cluster that bridges its two subunits, one of which has one binding site that is the only known reductant of the MoFe-protein [51]. In this paper, the literature on alternative, available, and sustainable NH_3 production processes in terms of the scientific aspects of the spatial structures of nitrogenase metalloclusters, the mechanism of reducing N_2 to NH_3 catalyzed by nitrogenase, the synthetic analogues of nitrogenase metalloclusters, and the opportunities for continued research were reviewed.

Transition metal dinitrogen chemistry began in 1965 with the discovery of a complex $[Ru(NH_3)_5(N_2)]^{2+}$, which was prepared by treating Ru salts with hydrazine [31]. An unstable binuclear titanium complex with molecular nitrogen has been isolated at low temperatures and its composition has been determined [32]. The first break through problem of extraordinary stability of N_2 was made in the pioneering work of Volpin and Shur [33, 34]. These authors showed the first reactions of dinitrogen reduction by strong reducing agents such as $Al–AlBr_3$, $LiAlH_4–AlBr_3$ in aprotic media in the presence of transition metals [33, 34]. In the 1970s, on the basis of the concept of the elementary multielectron mechanism of dinitrogen reduction in polynuclear transition metal complexes, reduction of dinitrogen to hydrazine and ammonia in protic media (methanol, water) involving relative weak reducing agents was demonstrated [25, 38–40]. The first systems discovered were heterogeneous and included metal hydroxides $Mo(OH)_3–Ti(OH)_3$ or $V(OH)_2–Mg(OH)_2$, which can be

Table. 11.1 Characteristics of Fe–S clusters in proteins and enzymes [8, 9]

Fe centers or Fe–S cluster	Protein	Ligands	Function
$[Fe(S\text{-}Cys)_4]$	Rubredoxin	$(Cys)_4$	Electron transport
$[Fe_2S_2]$	Ferredoxin	$(Cys)_4$	Electron transport
	Rieske centers	$(cys)_2(His)_2$	Electron transport
	Biotin synthase	$(Cys)_3(Arg)_1$	Sulfur donation
	SoxR	$(Cys)_4$	Sensing
	IscR	$(Cys)_3(His)_1$	Sensing
	RsrR	$(Cys)_2(His)_1(Glu)_1$	Sensing
	MitoNEET	$(Cys)_3(His)_1$	Sensing
$[Fe_3S_4]$	Ferredoxin	$(Cys)_3$	Electron transport
	RirA	$(Cys)_3/(Cys)_3(Unknown)_1$	Sensing
$[Fe_4S_4]$	Ferredoxin	$(Cys)_4$	Electron transport
	High potential iron proteins (HiPIPs)	$(Cys)_4$	Electron transport
	Radical SAM enzymes	$(Cys)_3(AdoMet)_1$	Catalysis
	Lipoyl synthase	$(Cys)_3(Ser)_1$	Sulfur donation
	NsrR	$(Cys)_3(Asp)_1$	Sensing
	Endolll, MutY	$(Cys)_4$	Sensing/ Structural
$[Fe_4S_4]$, $[Fe_2S_2]$	FNR	$(Cys)_4$	Sensing
$[Fe_4S_4]$ component of H-cluster	[FeFe] hydrogenase	$(Cys)_4$	Electron transport
$[Fe_2S_2]$ component of H-cluster	[FeFe] hydrogenase	$(Cys)_1(CO)_3(CN)_2(\text{-}SCH_2NHCH_2S\text{-})$	Catalysis
$[Fe_4S_4]$	Nitrogenase Fe protein	$(Cys)_4$	Electron transport
$[Fe_8S_7]$ P-cluster	Nitrogenase MoFe protein	Reduced: $(Cys)_6$ Oxidized: $(Cys)_6X_1$ (X = Ser or Tyr)	Electron transport
FeMo-cofactor	Nitrogenase MoFe protein	$(Cys)_1(His)_1(Homocitrate)_1$	Catalysis

With permission from Elsevier

considered as giant clusters of transition metals. In this system, N_2 was reduced to hydrazine by Ti^{3+}, Cr^{2+}, or V^{2+} in the presence of Mo compounds in aqua and alcohol solutions, while N_2 was reduced to NH_3 at higher temperatures.

One of homogenic system, complexes of V(II) and catechol

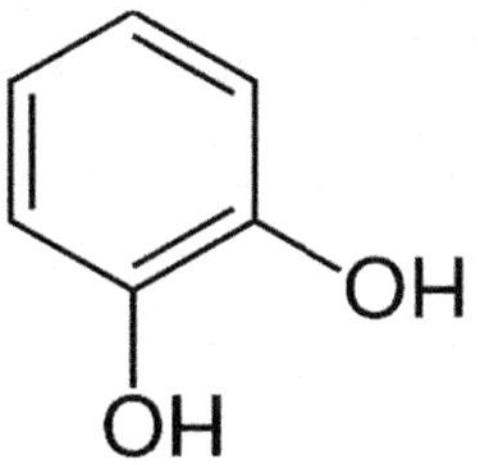

in the protic media, including water, turned out to be homogeneous [40].

In the presence of this complex the following reactions occur:

$$V^{(II)}{}_4 + N_2 = V^{(II)}{}_4N_2$$

$$V^{(II)}{}_4N_2 + V^{(II)}{}_4 + 8H^+ \rightarrow 8V\left({}^{III}\right) + 2NH_3 + H_2O$$

The authors suggested that the vanadium four-nuclear complex reduces dinitrogen by the four-electron mechanism. Later, detailed analysis of the reduction of dinitrogen in solution catalyzed by the V(II)-pyrocatechol was performed [41]. The hyperfine structure of the EPR spectrum of the complex was explained as a consequence of strong exchange interactions. Conclusions were reached on the tetranuclear character of the active complex of the centrosymmetric structure with the romboid disposition of the vanadium atoms. A hydrazine derivative and hydrogen were shown to be formed in the coordination sphere of tetramers, while ammonia is formed in the coordination sphere of an octamer.

The Chatt Cycle is a hypothetical model for the mechanism of nitrogenase action at the atomic level and is based upon reactions of certain well-defined molybdenum compounds (Fig. 11.3) [20, 26, 43]. Chatt's team established N_2 binding and activation at Mo (and W) and showed that the coordinated N_2 ligand could be protonated to release NH_3 (and hydrazine) in variable yield (as high as 90% for $W(PMe_2Ph)_4(N_2)_2$ with H_2SO_4 [43]. Protonation of an $M(NHNH_2)$ or $M(NH_2NH_2)$ intermediate lead to release hydrazine or ammonia. On the basis of the classic Chatt-type bis(dinitrogen) Mo/W complexes containing diphosphine coligands (either of two or more different ligands in a compound) the first mechanism for the transition metal-mediated conversion of N_2 to NH_3 was formulated.

Complex $[Mo(N_2)(P_2{}^{Me}PP_2{}^{Ph})]$

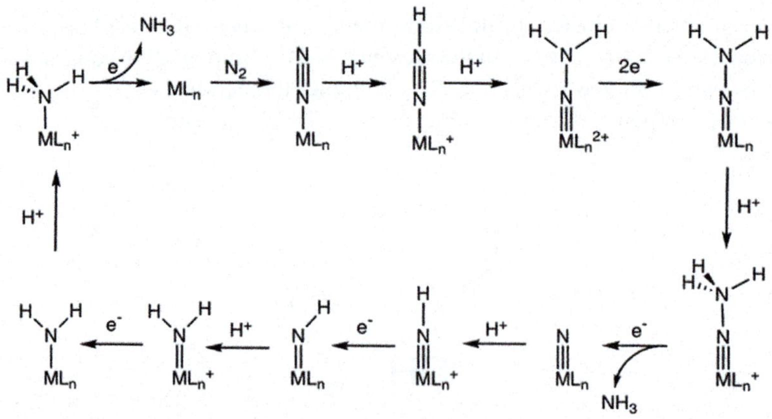

Fig. 11.3 Hypothesized cycle for M-catalyzed nitrogen fixation according to Chatt et al. https://en.wikipedia.org/wiki/Abiological_nitrogen_fixation_using_homogeneous_catalysts#/media/File:ChattCycle.svg

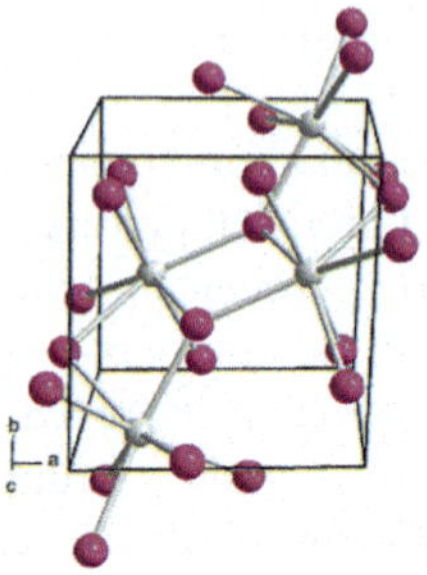

was found to be the first Chatt-type complex with one coordination site that catalytically converts N_2 to ammonia (26 equivalents) and the corresponding catalytic cycle follows a mononuclear pathway [25, 26]. Proton and electron transfer to N_2 are mediated by the proton-coupled electron transfer (PCET) reagent SmI_2/H_2O as protonating agent and strong reductant. Three structurally related Mo-N_2 complexes with pentaphosphine

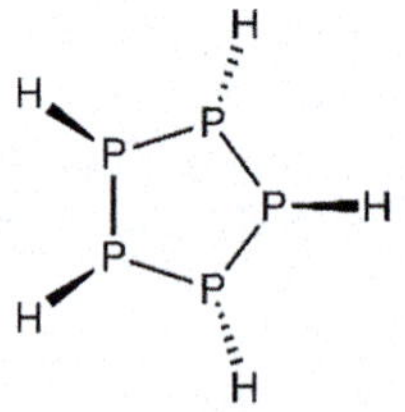

environment have been investigated as catalysts for the conversion of N_2 to NH_3, using SmI_2 as a strong reductant [25, 26]. Only the complex $[Mo(N_2)(P_2^{Me}PP_2^{Ph})]$

was found to be catalytically active, exhibiting the highest activation of N_2 and the pentadentate tetrapodal (**pentapod**) coordination. Protonation of the dinitrogen **complex** 3 leads to the hydrazido(2−) complex **4** which was isolated and spectroscopically characterized. In the Chatt scheme, binuclear complex of molybdenum

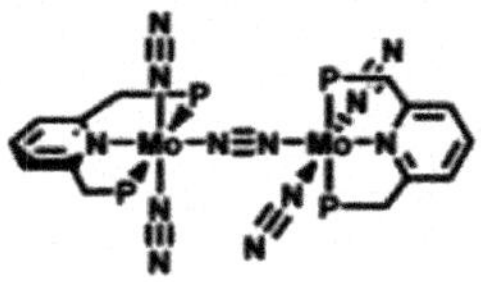

can serve as a catalysis, as well as mononuclear complexes with one N_2

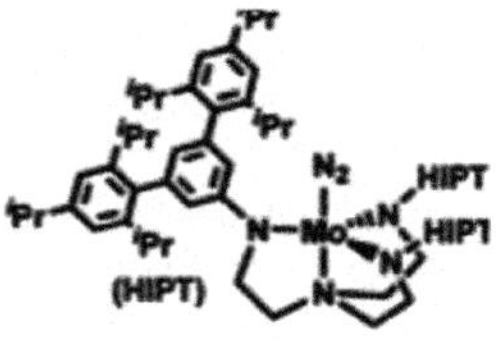

or complex with two N_2.

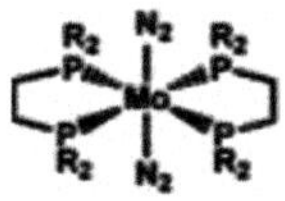

Several Fe catalysts for N_2-to-NH_3 conversion that operate at low temperature (−78 °C) and atmospheric pressure while relying on a very strong reductant (KC_8) and acid ($[H(OEt_2)_2][BAr^F{}_4]$) were designed [36]. It was shown that original catalyst system, $P_3{}^BFe$, when employing a combination of reductant ($Cp^*{}_2Co$) and acid ($[Ph_2NH_2][OTf]$ or $[PhNH_3][OTf]$), achieved efficiency for NH_3 formation (up to 72% for e^- delivery) and a high turnover number for a synthetic molecular Fe catalyst (84 equivalents of NH_3 per Fe site) [36]. The results of the study are presented in Figs. 11.4 and 11.5.

Electrocatalytic N_2 reduction in the presence of metal–sulfur clusters has been investigated [42]. The reduced species were electrochemically generated by controlled potential electrolysis of homo and heterometallic cubane clusters $[Fe_4S_4(SPh)_4]^{2-}$ and $[\{MoFe_3S_4(SPh)_3\}_2(\mu SPh)_3]^{3-}$, and then tested the N_2-fixation reactions. A report [44] suggested that solutions containing clusters such as $[Fe_4S_4(SPh)_4]_2$ and $[Mo_2Fe_6S_8(SPh)_9]_3$ could convert N_2 to NH_3 with very low rates and yields of these reactions. Reactivity studies of clusters containing the MFe_3S_4 cores (M = Mo, V) with catecholate, multicarboxylate, **or** dimethylformamide ligands coordinated to the Mo (or V) atoms, and Cl ligands coordinated to the Fe atoms have been carried out [10]. The studies showed that the M/Fe/S single cubane clusters are effective catalysts in the reduction of nitrogenase substrates such

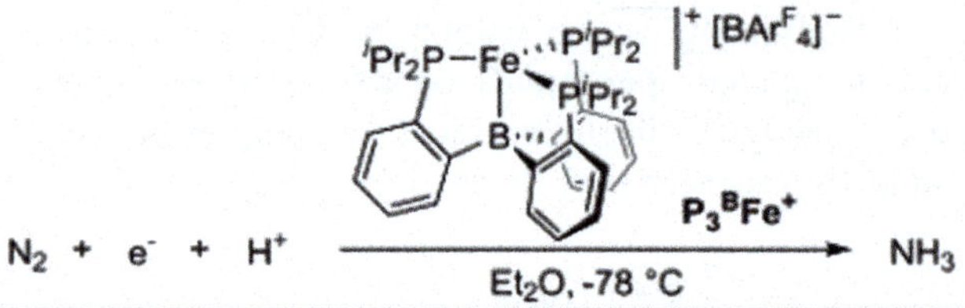

e^- (E^0 vs $Fc^{+/0}$)	H^+ (pK_a in THF)	$\Delta\Delta H_f$ (kcal/mol)	% yield NH_3
Cp^*_2Co (-1.96 V)	$[PhNH_3][OTf]$ (7.6)	59	38
Cp^*_2Co (-1.96 V)	$[Ph_2NH_2][OTf]$ (3.2)	77	72
KC_8 ($\leq$ -3.0 V)	$HBAr^F_4$ ($\leq$ 0)	$\geq$ 156	45

Fig. 11.4 Summary of conditions used for catalytic N_2-to-NH_3 conversion by $P_3{}^BFe^+$ highlighting the estimated enthalpic driving force ($\Delta\Delta H_f$) [36]. Published under the Creative Commons public use license CC-BYF from American Chemical Society

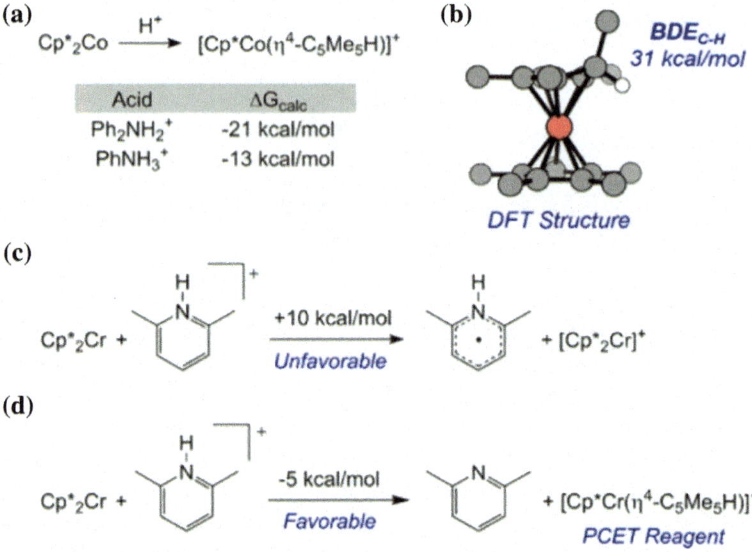

Fig. 11.5 **a** Calculated free-energy changes for the protonation of Cp^*_2Co. **b** DFT-optimized structure of *endo*-$Cp^*Co(\eta^4-C_5Me_5H)^+$ (methyl protons omitted for clarity). **c** Unfavorable reduction of 2,6-lutidinium by Cp^*_2Cr with the calculated free energy change. **d** Favorable protonation of Cp^*_2Cr by lutidinium with the calculated free energy change [36]. Published under the Creative Commons public use license CC-BYF from American Chemical Society

as hydrazine, acetylene giving ammonia, ethylene and dihydrogen, respectively. The same molecules did not activate or catalyze the reduction of dinitrogen. A study [45] was conducted to elucidate the mechanism of FeMo-cofactor catalytic activity in non-enzymic conditions and to compare its catalytic behavior with that of the nitrogenase. The kinetics were investigated of C_2H_2 reduction by multielectron donors, Zn and Eu amalgams, catalyzed by isolated FeMo-cofactor. Inhibition of this process by CO was also examined. Results indicated that the FeMo-cofactor can realize of some of its functions without the protein.

Complex [IPTN$_3$N]Mo(N_2) (**Mo**N_2) ([HIPTN$_3$N]$^{3-}$ = [(HIPTNCH$_2$CH$_2$)$_3$N]$^{3-}$ where HIPT = 3,5-(2,4,6-*i*-Pr$_3$C$_6$H$_2$)$_2$C$_6$H$_3$) was found to react with dihydrogen slowly (days) at 22 °C to yield [HIPTN$_3$N]MoH$_2$ (**Mo**H_2) [11]. The following results were reported: (1) **Mo**N_2, **Mo**NH_3, and **Mo**H_2 are interconverted rapidly in the presence of H_2, N_2, and NH_3 (Fig. 11.6); (2) The process runs with **Mo**H_2 as a catalyst and is competent for reduction of N_2 with protons and electrons under standard conditions: and (3) Reduction of dinitrogen to ammonia in the presence of 32 equivalents of added hydrogen (vs. Mo) is not catalytic, consistent with dihydrogen being an inhibitor of dinitrogen reduction. Suggested catalytic N_2RR cycle depicting (HIPTN3N)Mo intermediates and their efficacy as (pre)catalysts for N2RR using [LutH]BArF 4 (48 equivalents) and Cp*2Cr (36 equiv.); [Mo] = (HIPTN3N)Mo is presented in the figure.

A series of nitrogenase-mimicking complexes have been synthesized and studied by Peters' group [46]. It was reported that a tris-(isopropylphosphino(o-phenylene)borane iron complex, (P3 B) Fe, was capable to catalyze nitrogen fixation. The process included the following steps: (1) The P3 B ligand reacts with $FeBr_2$ in the presence of Fe powder to form (P3 B)Fe(Br), a valuable synthon (a hypothetical

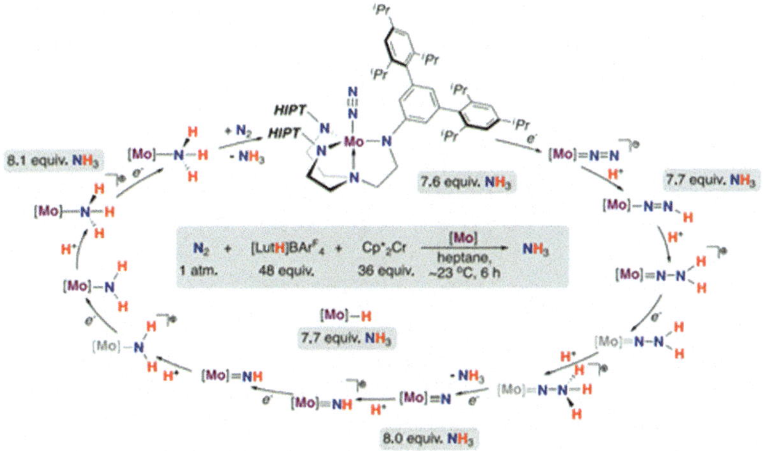

Fig. 11.6 Catalytic N_2RR cycle depicting (HIPTN$_3$N)Mo intermediates and their efficacy as (pre)catalysts for N2RR using [LutH]BAr$^F{}_4$ (48 equiv.) and Cp*2Cr (36 equiv.); [Mo] = (HIPTN$_3$N)Mo [10, 11]. With permission from American Chemical Society

unit within a target molecule that represents a potential starting reagent in the retroactive synthesis of that target molecule); (2) The complex is reduced by stoichiometric sodium naphthalenide

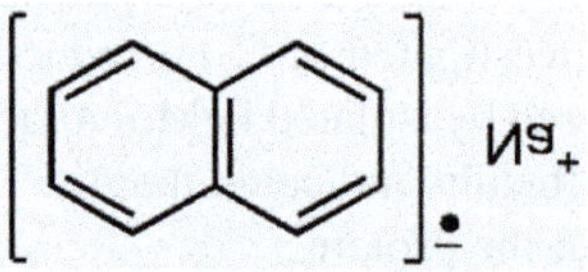

to give the $Fe(N_2)$ complex ($v_{NN} = 2011\ cm^{-1}$); (3) Reaction with excess sodium naphthalenide led to formation of the anionic $Fe(N_2)$ complex (vNN $= 1877\ cm^{-1}$); (4) Encapsulation of the sodium with 12-crown-4 afforded a separated ion pair (vNN $= 1918\ cm^{-1}$); and (5) Activating π-accepting ligands, (P3 B)Fe which is capable of stabilizing π-donating imido ligands occurs.

In 2011, Nishibayashi and co-workers introduced a dinuclear molybdenum system as a N_2 fixation (pre)catalyst [47]. Specifically, a PNP-based (PNP = 2,6-bis(di-tert-butylphosphinomethyl)pyridine) complex $(PNP)MoCl_3$

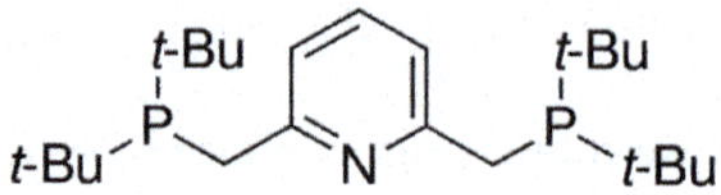

was prepared and reduced to the corresponding dinitrogen Mo(0) complex, $[(PNP)Mo(N_2)_2]2(\mu\text{-}N_2)$, through reaction with a strong reductant (NaHg amalgam). In 2017, the same group proposed that a reactive terminal (PNP)Mo(N) could be formed via cleavage of a bridging dinitrogen unit, thus circumventing the formation of early protonation intermediates such as diazenido, hydrazido, and hydrazidium [7, 8]. Use of $(PNP)Mo(I)_3$ as a precursor that afforded access to (PNP)Mo(N)(I) following reduction by Cp*2Co. PCP-type pincer ligands,

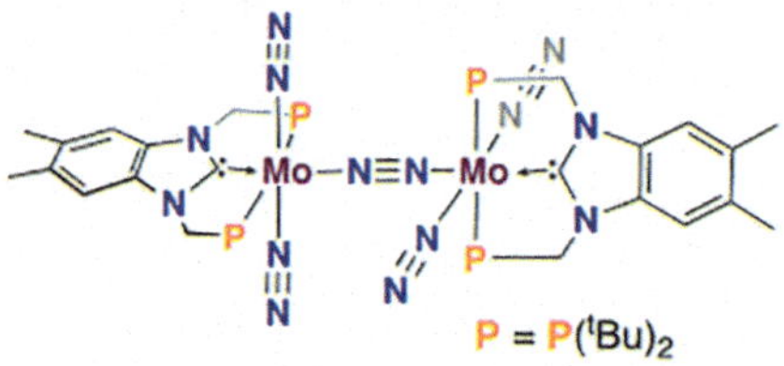

comprising of both N-heterocyclic carbene and diphosphine donors, have also been explored in an effort to evaluate N2RR catalysis performance [7, 8].

The first example of the Fe-catalyzed transformation of N_2 into $N(SiMe_3)_3$ I under ambient conditions was reported [48]. The Fe-catalyzed reduction of N_2 into **1** under ambient conditions, where up to 34 equivalents of **1** occurred. In this reaction system, Fe complexes such as $[Fe(CO)_5]$ and ferrocenes have been found to work as effective catalysts. With the aid of density-functional-theory calculations, a plausible reaction pathway was proposed, where Fe(II)-N_2 complex bearing two Me(3)Si groups as ancillary ligands has an important role as a key reactive intermediate. A tris(phosphine)borane-supported iron complex that catalyzes the reduction of N_2 to NH_3 under mild conditions was synthesized [49]. In the complex, more than 40% of the proton and reducing equivalents are delivered to N_2. The results indicated that a single iron site may be capable of stabilizing the various N_xH_y intermediates generated during catalytic NH_3 formation and geometric tunability at iron. For example, to explore the potential of the iron catalyst 2

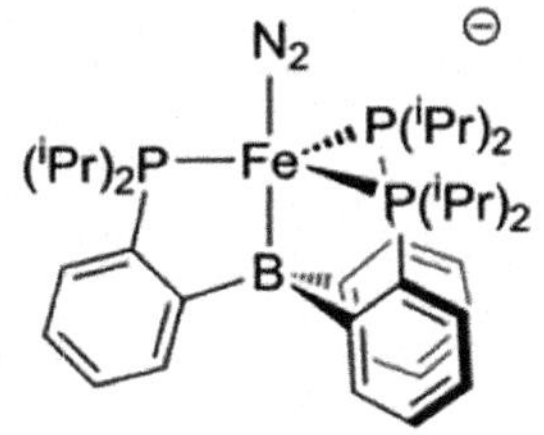

2

a range of different proton sources (HCl, HOTf, $HBAr^F{}_4 \cdot 2Et_2O$, [LutH]$BAr^F{}_4$) and electron sources from strong reductant (Na/Hg, KC8, CoCp*2, CrCp*2, potassium) were used as indicated in (Figs. 11.7 and 11.8). The reaction can run in tetrahydrofuran, dimethoxyethane, Et_2O and toluene. Catalytic conversion of dinitrogen into ammonia yielding up to 8.5 equivalents NH_3 per Fe was achieved with use of 58 equiv. of KC8 and 48 equiv. of $HBAr^F{}_4 \cdot 2Et_2O$ in diethyl ether at −78 °C [49].

The positively charged catalytic component of nitrogenase ($NifDK^{Fe}$) can work together with the reductase component of nitrogenase to reduce C_2H_2 in an ATP-dependent reaction [15, 50]. It also acted as an enzyme on its own in the presence of Eu^{II}DTPA (DTPA is diethylenetriaminepentaacetic acid).

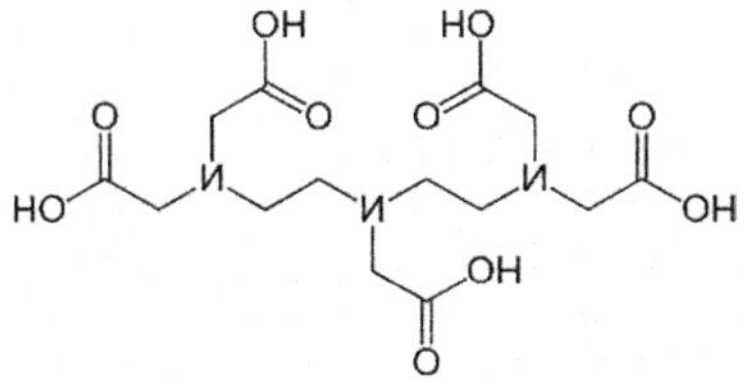

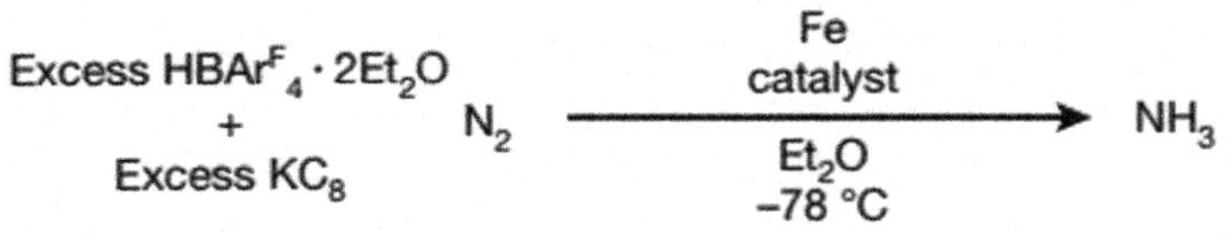

Fig. 11.7 Schematic depiction of postulated N_2 reduction at an Fe site investigated in [49]. With permission from Springer Nature

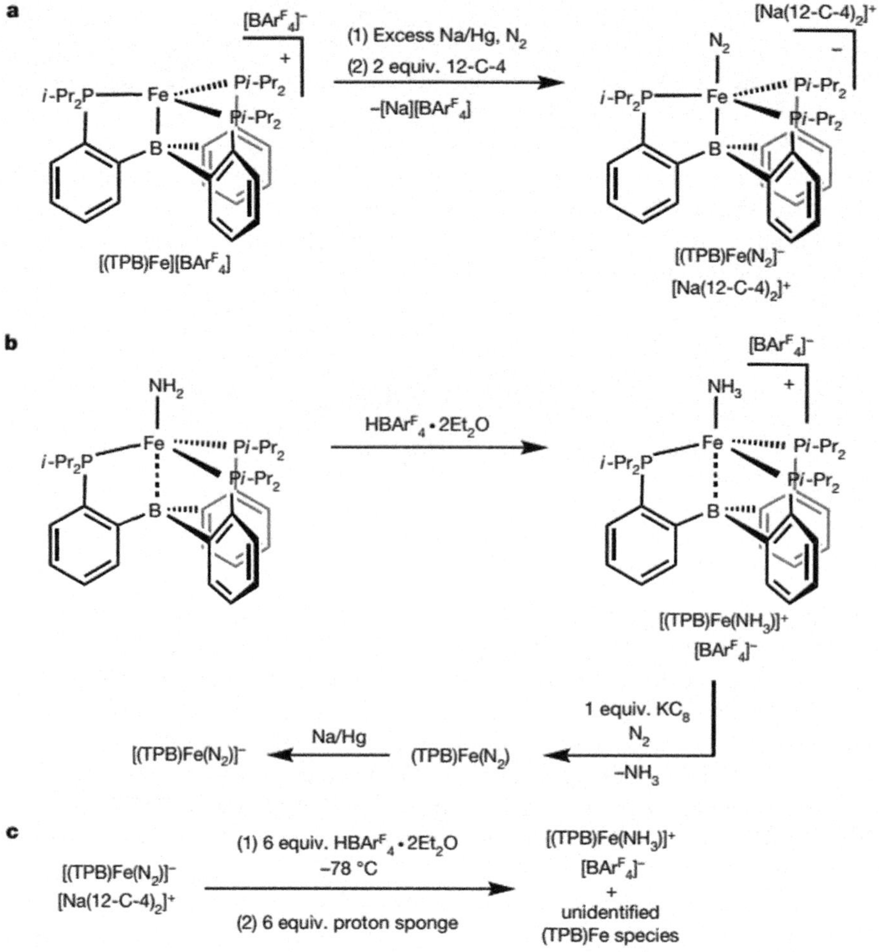

Fig. 11.8 Stoichiometric (TPB)Fe–N_2 model reactions. **a** N_2 binding to [(TPB)Fe][$BAr^F{}_4$] under electron loading to generate [(TPB)Fe(N_2)][Na(12-crown-4)$_2$]. **b** Reductive protonolysis of (TPB)Fe(NH_2) to release NH_3 with concomitant N_2 uptake. **c** Generation of [(TPB)Fe(NH_3)][$BAr^F{}_4$] and other (TPB)Fe species on addition of acid to [(TPB)Fe(N_2)][Na(12-crown-4)$_2$] at low temperature, followed by warming and then addition of base. 12-C-4, 12-crown-4 [49]. With permission from Springer Nature

The above system exhibited an ability to reduce CN^- to C_1–C_3 hydrocarbons in an ATP-independent manner. As an example, a complex bis(thiolato)sulfido-bridged double cubane [Mo_2Fe6S9(SEt)$_8$] of two [$MoFe_3S_4$] units bridged via the Mo centers by inorganic sulfide and two thiolate ligands was synthesized [15]. In this work three main biosynthetic platforms facilitate nitrogenase M-cluster biosynthesis were designed, namely: (1) Assembly of mobilized Fe and S atoms on NifU to generate Fe_2S_2 and Fe_4S_4 building blocks; (2) Coupling of two Fe_4S_4-like clusters (K-cluster)

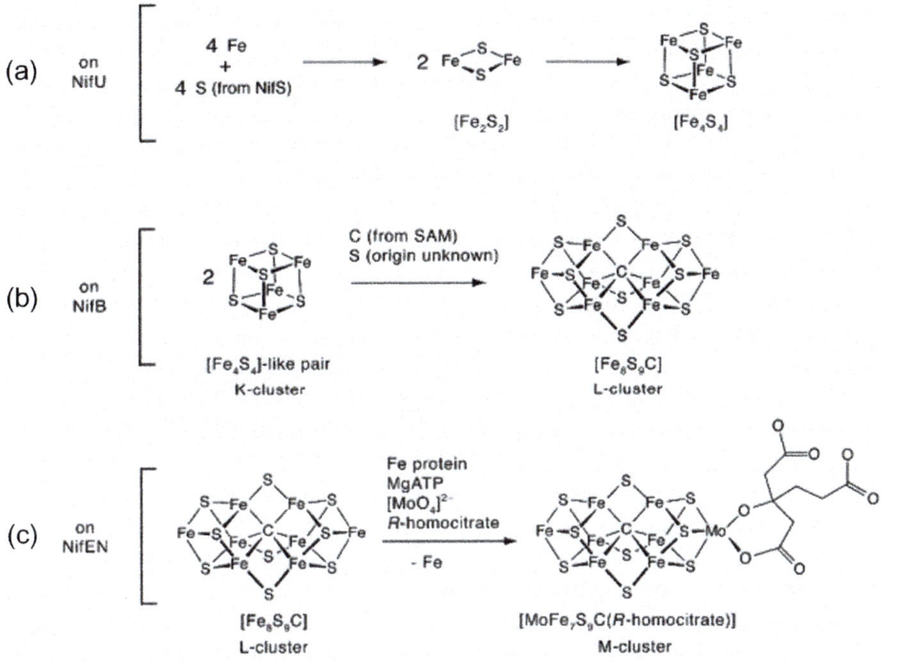

Fig. 11.9 Synthetic Fe clusters relevant to nitrogenase metallocofactors [15]. With permission from Wiley

on NifB to generate the precursor cluster through the addition of C and S atoms; and (3) Maturation of L-cluster to M-cluster on the scaffold NifEN, with Fe protein providing Mo and R-homocite [15]. Figure 11.9 displays elected synthetic Fe clusters relevant to nitrogenase metallocofactors.

The first example of N_2 binding at a mononuclear iron site supported by only S and C donors was described [18]. The mechanism of N_2 binding to the iron(II) tris(thiolate)

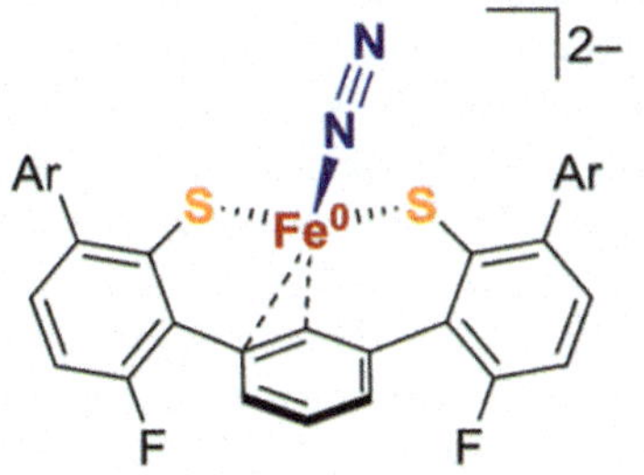

(complex **2**-K) was examined. The following main results of the research were reported: (1) The reduction of an iron(II) tris(thiolate) complex with 1 equivalent of strong reductant KC_8 leads to a thermally unstable intermediate; (2) A combination of Mössbauer, EPR, and X-ray absorption spectroscopies identified the complex as a high-spin ($S = 3/2$) iron(I) species that maintains coordination of all three sulfur atoms; (3) DFT calculations suggested that this iron(I) intermediate has a pseudotetrahedral geometry that resembles the S_3C iron coordination environment of the belt iron sites in the resting state of the FeMoco; (4) Further reduction to the iron(0) oxidation level under argon causes the dissociation of one of the thiolate donors and gives an η^6-arene species which reacts with N_2; (5) Additional reduction of the iron(0)–N_2 complex gives a reactive, formally iron(−1) species; (6) Treatment of the putative iron(−I) complex with weak acids gives low yields of ammonia and hydrazine, demonstrating that these nitrogenase products can be generated from N_2 at a synthetic Fe–S–C site; and (7) Catalytic N_2 reduction was not observed.

The formation of a nitrogenase-mimicking enzyme upon insertion of a negatively charged synthetic model complex ($[Fe_6S_9(SEt)_2]^{4-}$, designated Fe_6^{RHH}, into the positively charged catalytic component of nitrogenase ($NifDK^{apo}$) was observed [50]. The article provided a comprehensive **review** of the chemistry of M–S compounds related to the conversion of N_2 and the structures/functions of the nitrogenase cofactors. About 150 clusters have been considered. Three classes of M–S compounds described are: (1) Multinuclear M–S clusters structurally or functionally relevant to the nitrogenase cofactors; (2) Mono and dinuclear transition metal complexes supported by sulfur-containing ligands; and (3) Metal sulfide-based solid materials employed in the reduction of N_2.

In review [21], comprehensive insights into synthetic nitrogen fixation assisted by molecular catalysts under ambient or mild condition was described. Recent advances in the development of molecular catalysts for synthetic N_2 fixation under ambient or mild conditions were summarized, and potential directions for future research were discussed. The review focused on the advances on the catalytic conversion of N_2 into NH_3, N_2H_4, and silylamines mediated by molecular catalysts under ambient or mild conditions. Among them are catalytic conversion of N_2 into NH_3 or N_2H_4 mediated by Mo catalysts under ambient conditions without using H_2O or with using H_2O. Other catalytic systems (Fe, Ru, Os, Co, Ti, V, Re); and the catalytic conversion of N_2 into silylamines were also considered.

Recently it was shown that the Fe atoms in synthetic $[Mo_3S_4Fe]$ cubes can capture a N_2 molecule and catalyze N_2 silylation to form $N(SiMe_3)_3$ under treatment with excess sodium and trimethylsilyl chloride (Figs. 11.10 and 11.11) [22]. The treatment of the clusters with strong reductants (Potassium graphite, KC8 or sodium naphthalenide $Na(C_{10}H_8)$ under N_2 in tetrahydrofuran led to the formation of the corresponding N_2 clusters. It was found that $[Mo_3S_4Fe]$ dimer bridged by N_2, $[\{Cp^*3Mo_3S_4Fe\}_2(\mu\text{-}N_2)]^{2-}$ and monomers each bearing a terminal N_2 ligand. Catalytic reduction of N_2 into $N(SiMe_3)_3$ promoted by $[Mo_3S_4Fe]$ clusters occurred with high yield. The results demonstrated the N_2-reduction capability of Fe atoms in a sulfur-rich environment, which is reminiscent of the ability of FeMoco to bind and activate N_2.

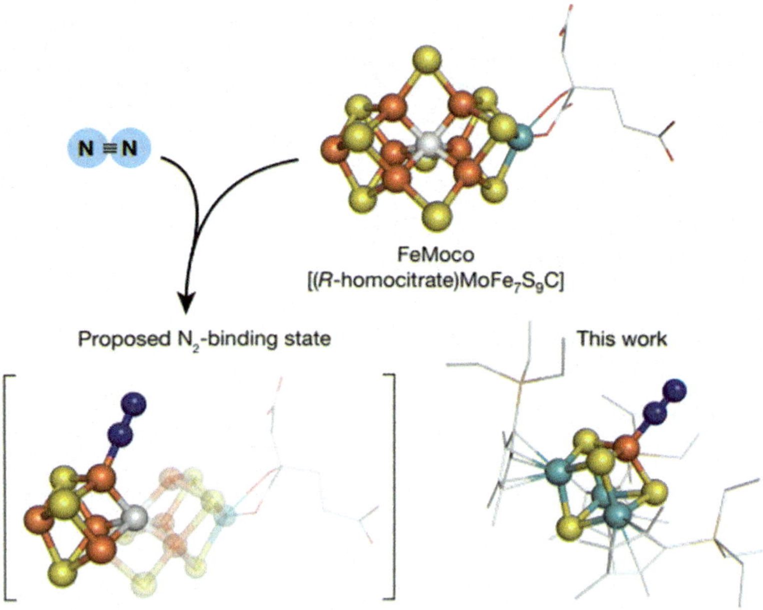

Fig. 11.10 Proposed N_2-binding state for FeMoco and a N_2-bound Mo–Fe–S cluster in this work. Orange, iron; teal, molybdenum; yellow, sulfur; grey, carbon; blue, nitrogen [21, 22]. With permission from Springer Nature

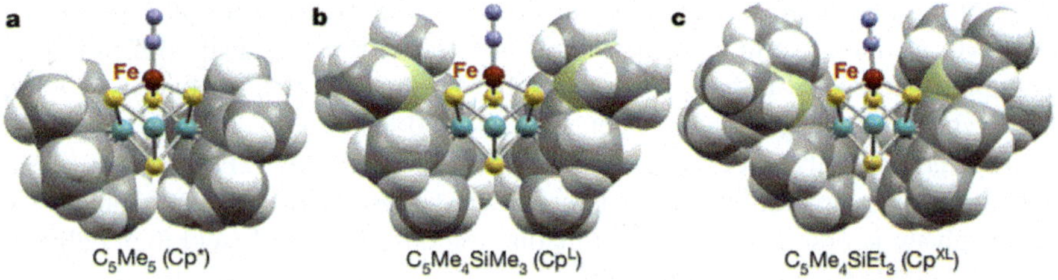

Fig. 11.11 Structural models of N_2-bound [Mo_3S_4Fe] clusters. A space-filling model is applied to two CpR ligands, whereas the other CpR is omitted to show the [Mo_3S_4Fe]–N_2 moiety. **a** C_5Me_5 (Cp*), **b** $C_5Me_4SiMe_3$ (CpL), **c** $C_5Me_4SiEt_3$ (CpXL). Panels **a** and **c** were prepared from the crystal structures of 2a and 2c; panel **b** was prepared via software using the Cl-bound cluster 1b [21, 22]. With permission from Springer Nature

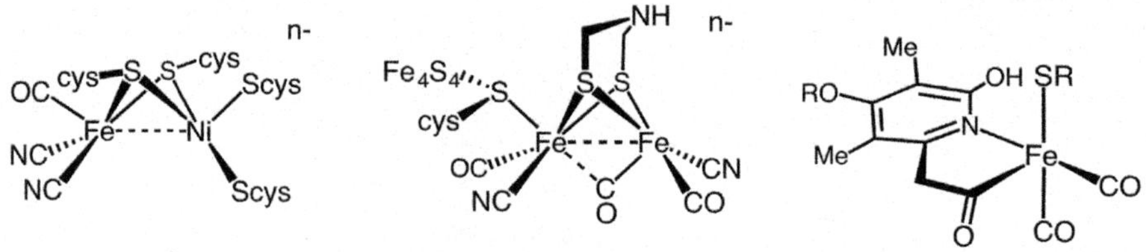

Fig. 11.12 Structures of the active sites of the three types of hydrogenase enzymes [53]. With permission from Springer Nature

11.3 Hydrogenases Chemical Models

11.3.1 General

Hydrogenases which catalysis the reversible oxidation of molecular hydrogen (H_2)

$$H_2 + A_{ox} \rightarrow 2H^+ + A_{red}$$
$$2H^+ + D_{red} \rightarrow H_2 + D_{ox}$$

were discovered in the 1930s by Marjory Stephenson [51]. Since then synthetic chemistry has been used to prepare a number of similar mimics of hydrogenase subsites, but it has commonly failed to reproduce the natural enzymatic activities [51–79]. Synthesis of, free-standing analogue of the hydrogenase active cite may enable to develop useful electrocatalytic materials for a practical application, for example, the reversible hydrogen fuel cells. Structures of the active sites of the three types of hydrogenase enzymes are indicated in (Fig. 11.12) [53]. In [FeFe]-hydrogenases, catalysis occurs at a unique diiron center (the [2Fe] subsite), which contains a bridging dithiolate ligand, three CO ligands and two CN(−) ligands. This [2Fe] subsite is first assembled on a maturation enzyme, HydF, and then delivered to the apo-hydrogenase for activation.

11.3.2 [Fe–Fe] Hydrogenase Models

The enzyme [FeFe]-hydrogenase (HydA1) contains a unique 6-iron cofactor, the H-cluster, that has unusual ligands to an Fe–Fe binuclear subcluster: CN^-, CO, and an azadithiolate (adt) ligand that provides 2 S bridges between the 2 Fe atoms. HydF is a scaffold protein in [FeFe] hydrogenase H-cluster biosynthesis. A semisynthetic form of HydF was generated by loading it with a synthetic low-valent dinuclear Fe complex [61]. Employing the semisynthetic form of HydF and FTIR spectroscopy, it was verified that the pre-catalyst is truly a four CO species, and that transfer of the pre-catalyst from HydF to the hydrogenase enzyme resulted in the release of CO.

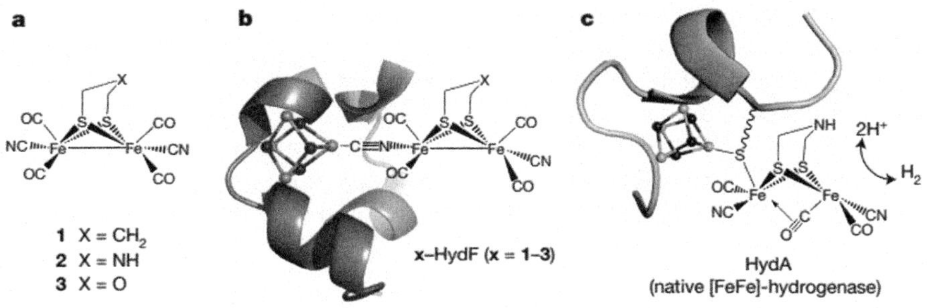

Fig. 11.13 Structures of the diiron clusters discussed in the study. **a** Synthetic mimics 111–13, 214 and 315. **b** Proposed structure for the x–HydF (x 5 1–3) hybrid proteins. **c** H-cluster (active site) of [FeFe]-hydrogenase. The protein ribbon and the [4Fe-4S] clusters (shown as balls and sticks with Fe shown as white spheres) are shown only schematically [53]. With permission from Springer Nature

Combined FTIR and EPR study of the H-cluster assembly reaction revealed that the [4Fe4S] cluster presented in HydF is responsible for the binding of the pre-catalyst to HydF and directly involved in H-cluster assembly by acting as an electron acceptor in the reaction. According to a proposed mechanism of the reaction an electron transfer occurs during H-cluster assembly, resulting in an oxidation of the [2Fe] subsite with concomitant reduction of the [4Fe4S] cluster present on the HydF protein.

Three synthetic mimics, containing different bridging dithiolate ligands, were loaded onto bacterial *Thermotoga maritima* hydrogenase HydF and then transferred to apo-HydA1, one of the hydrogenases of *Chlamydomonas reinhardtii* algae (Fig. 11.3) [53]. Full activation of HydA1 was achieved using the HydF hybrid protein containing the mimic with an azadithiolate ($NH(CH_2S)_2^{2-}$) bridge. As a model of the active site of [FeFe]-hydrogenase, complexes $(Et_4N)_2[Fe_2(adt)(CO)_4(CN)_2]$ (2, adt2– = 2-azapropanedithiolate), $(Et_4N)_2[Fe_2(pdt)(CO)_4(CN)_2]$ (1, pdt2– = propanedithiolate), $(Et_4N)_2[Fe_2(pdt)(CO)_4(^{13}CN)_2]$, and $(E_{t4}N)_2[Fe_2(odt)(CO)_4(CN)^2]$ (3, odt2– = 2-oxopropanedithiolate) were prepared [53]. A maturation enzyme TmHydF (HydF) was overexpressed, purified and its [4Fe–4S] cluster was reconstituted. Apo-CrHydA1 (apo-HydA1) was also overexpressed in *Escherichia coli* BL21 DE3 ΔiscR using growth conditions. The following spectra were recorded: Fourier-transform infrared spectroscopy (FTIR) spectra of protein samples, X-band EPR spectra, hyperfine sublevel correlation (HYSCORE) spectra and their combination frequency (CF)–nuclear frequency (NF) variants. Specific hydrogenase activity of reconstituted HydA1 (μmol H_2 min^{-1} mg $HydA1^{-1}$ was measured in the presence of methyl viologen (10 mM) and sodium dithionite (100 mM)

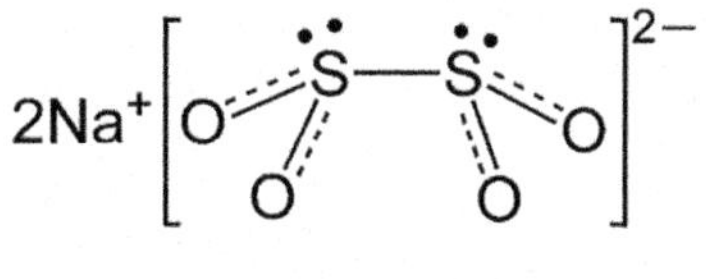

after in vitro maturation of apo-HydA1 for 30 min at 37 °C with 10 equivalents of x-HydF (x = 1–3), HydF, or CaHydF. The obtained three synthetic mimics with different bridging dithiolate ligands) was loaded onto bacterial *T. maritima* HydF and then transferred to apo-HydA1. Full activation of HydA1 was achieved when using the HydF hybrid protein containing the mimic with an azadithiolate bridge.

According to work [52], the [FeFe]-hydrogenase reaction runs in several steps [52]: (1) The dimeric Mj [Fe] hydrogenase binds of the second substrate methenyl-tetrahydromethanopterin (methenyl-H4MPT+)

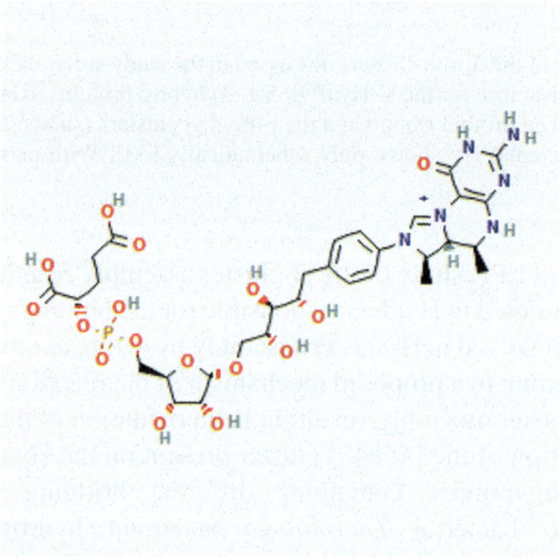

to the open form of the enzyme; (2) The Fe-GP cofactor is activated and a hydrogen molecule binds to the open coordination site of the iron; (3) Methenyl-H4MPT+ binding triggers the breaking of the Fe–Hδ· · · Hδ+ –O (pyridinol oxygen) dihydrogen bond; (4) The hydrogen molecule is heterolytically cleaved; (5) The proton is transferred to the thiol of Cys176; and (6) The adjacent C14a carbocation of the methenylH4MPT+ accepts the hydride.

It was reported that chitosan improves the efficiency and durability of a typical mimic of the diiron subsite of [FeFe]-hydrogenase for photocatalytic hydrogen evolution (Fig. 11.14). The turnover number of the self-assembling system increases ~ 4000-fold compared with the same system in the absence of chitosan. The diiron subsite of [FeFe]-H_2ases, [$Fe_2(CO)_6$(μ-adt)$CH_2C_6H_5$] [μ-adt = $N(CH_2S)_2$] [27, 36], was selected as a catalyst. The present artificial system using chitosan-confined environment is reminiscent of the [Fe_2S_2] subcluster of natural [FeFe]-H_2a. The assembly of the iron–sulfur framework of the active site of iron-only hydrogenase (the H-cluster) was synthesized and its functions as an electrocatalyst for proton reduction were examined [69]. Through linking of a diiron subsite to a {4Fe4S} cluster, the first synthesis of a metallosulfur cluster core involved in small-molecule catalysis was achieved.

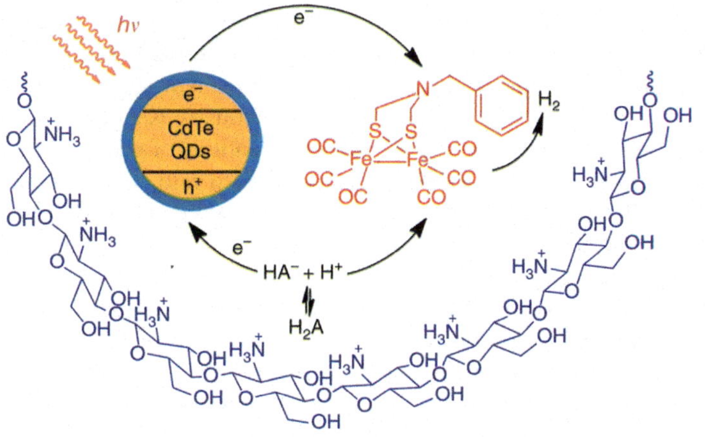

Fig. 11.14 Schematic describing the H_2 photogeneration of a chitosan-confined mimic of the diiron subsite of [FeFe]-H_2ase in the presence of CdTe quantum dots and H_2A [72]. With permission from Springer Nature

Photocatalytic hydrogen evolution from rhenium(I) complexes to [FeFe] hydrogenase mimics in aqueous sodium dodecylsulfate (SDS) micellar systems was investigated [71]. Electrochemical measurements were conducted in standard three-electrode systems with a working electrode, counter electrode, and reference electrode. The working electrodes directly incorporate Fe–S functionalized materials. A common working electrode setup consists of a glassy carbon base material with a spin-coated polymeric film. To offer an access to photocatalytic H_2, generation in an aqueous solution, the hydrophobic photosensitizers, Re(I)(4,4′-dimethylbpy)$(CO)_3$Br (**1**) or Re(I)(1,10-phenanthroline)$(CO)_3$Br (**2**), and [FeFe] H_2ases mimics, $[Fe_2(CO)_6(\mu\text{-adt})CH_2C_6H_5]$ (**3**) or $[Fe_2(CO)_6(\mu\text{-adt})C_6H_5]$ (**4**) [μ-adt $= N(CH_2S)_2$], have been incorporated into sodium dodecyl sulfate (SDS)

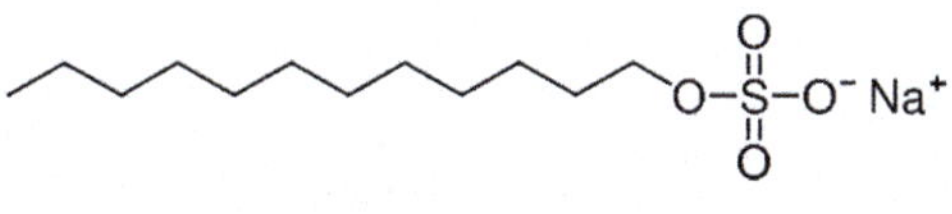

micelle solution. Ascorbic acid was used as a sacrificial electron donor and proton source.

To give a complete model for the [FeFe]-hydrogenase, the function of the ferredoxins with the redox-active ligand Cp*Fe(C(5)Me(4)CH(2)PEt(2)) (FcP*) was examined combining acid–base, redox, and substrate binding functionalities [73]. FcP* was oxidized at mild potentials of a mimic of ferredoxin cofactors. The specific model, Fe(2) [(SCH(2))(2)NBn](CO)(3)(FcP*)(dppv) (1), contains the three functional components of the active site: a reactive diiron cite, an amine proton relay and a one-electron redox module was designed. A covalently linked photosensitizer-catalytic center dyad Ps-Hy, consisting of two bis(2-phenylpyridine)(2,2′-bipyridine)iridium(iii) chromophores (Ps) and a diiron hydrogenase mimic (Hy) was constructed by using click reaction [65]. Click chemistry is a class of simple reactions commonly used for joining two molecular entities of choice for example by joining small modular units. Ps-Hy was incorporated into K(+)-exchanged mesoporous molecular sieve MCM-41 to form a composite (Ps-Hy@MCM-41). MCM-41 is a synthetic mesoporous material with cylindrical pores suitable for catalytic cracking, separations, dye removal, and as an adsorbent. The photo catalytic activity of Ps-Hy@MCM-41 was found to be ~ threefold higher when compared with that of Ps-Hy in the absence of MCM-41. The incorporation of Ps-Hy into MCM-41 stabilizes the catalyst, and consequently, advances the photocatalysis.

Terminal hydride complexes, as the diiron carbonyl models, containing azadithiolate, and the dithiolate cofactor $[HN(CH_2S)_2]^{2-}$, (Fig. 11.15) were designed and examined [58]. The complex was prepared by alkylation of $Fe_2(SH)_2(CO)_6$ with formaldehyde in the presence of ammonia (to give adt^H derivatives) or amines (to give adt^R derivatives). The crystal structure of the doubly protonated model $[(term\text{-}H)Fe_2(Hadt^H)(CO)_2(dppv)_2]^{2+}$ confirmed the presence of both ammonium and terminal hydrido centers, which interact through a dihydrogen bond (dppv = cis-$C_2H_2(PPh_2)_2$). The amine in the azadithiolate transfers protons to and from the iron. Both $[(term\text{-}H)Fe_2(Hadt^H)(CO)_2(dppv)_2]^{2+}$ and $[(term\text{-}H)Fe_2(adt^H)(CO)_2(dppv)_2]^+$ were highly active electrocatalysts for hydrogen evolution reaction **(HER)** which was initiated by reduction of the diferrous center. This finding indicated inducing coupling of the protic ammonium center and the hydride ligand. Complex $[(term\text{-}H)Fe_2(Hadt^H)(CO)_2(dppv)_2]^{2+}$ in CF_3CO_2H showed $k = 58{,}000\ s^{-1}$ and overpotential of 0.51 V. Figure 11.15 illustrates similarity of active site of the [FeFe]-Hydrogenase and given synthetic model.

It was reported that the biomimetic $[Fe_2(adt)(CO)_4(CN)_2](2-)$ (adt(2−) = azadithiolate) complex resembling the diiron sub-cluster was inserted in vitro into the apo-protein of [**FeFe**] hydrogenase from *C. reinhardtii*, which contains only

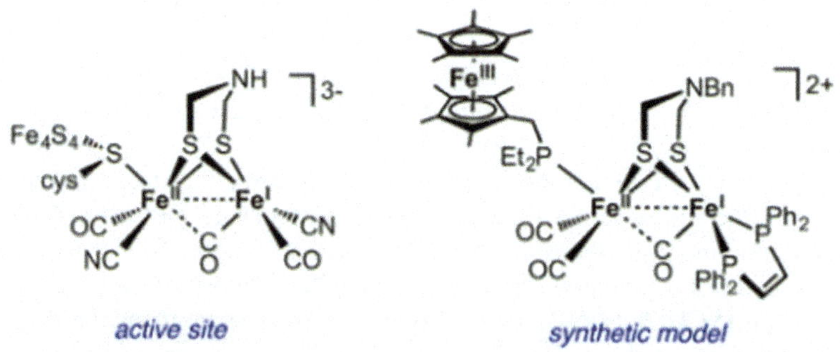

Fig. 11.15 Active site of the [FeFe]-hydrogenase and its synthetic model [58]. With permission from American Chemical Society

the [4Fe–4S] part of the H-cluster [59]. This process resulted in a fully active enzyme. Studies on the mixed valence Fe(I)Fe(II) state of the diiron sub-cluster of this hybrid protein were performed employing ^{15}N and ^{13}C isotope labeled CN^- ligands, hyperfine sublevel correlation (HYSCORE), and electron-nuclear double resonance (ENDOR). Detailed selective ENDOR and HYSCORE experiments at multiple field positions enabled to know the relative orientations of the nitrogen and carbon hyperfine tensors. These finding were consistent with the crystal structure assuming a g-tensor orientation. To probe the influence of phosphine ligand substitution on the [2Fe2S] model, two new [FeFe]-hydrogenase model complexes with the phosphine ligands, PMe_3 or $P(CH_3O)_3$, such as μ-($SCH(CH_2CH_3)CH_2S$)-$Fe_2(CO)_5PMe_3$ 1, and μ-($SCH(CH_2CH_3)CH_2S$)-$Fe_2(CO)_5P(CH_3O)_3$, were synthesized [60]. Confirmation of structures was provided by FTIR, ^1H NMR, ^{13}C NMR, ^{31}P NMR, elemental analyses and single-crystal X-ray analysis. Experiments indicated that in the presence of acetic acid in CH_3CN solution, the hydrogen evolution overpotentials of complexes 1 and 2 were 0.91 V and 0.81 V, respectively. In addition, the electrocatalytic study showed the maximum charges for complexes 1 and 2 were 31.3 mC and 56.3 mC at − 2.30 V for 10 min, respectively, while complex 2 exhibited better electrocatalytic ability than complex 1.

An iron-based catalyst covalently attached to silicon (Si) was constructed for the investigation of activity and electrochemical decomposition pathways of diiron hydrogenase enzyme mimics [64]. The employed catalyst, $Fe_2(CO)_6(\mu$-S-C_6H_4-p-$OH)_2$ ([FeFe]), mimics the active sites of this enzyme. [FeFe] modification improves the overpotential required to produce 10 mA cm^{-2} by 100 mV, with a hydrogen evolution rate of 2.31×10^{-5} mol h^{-1} cm^{-2} (− 0.78 V vs. RHE). Structural rearrangement transpires within 1 h of electrolysis, where Fe–S bond dissociates at the catalytic center, resulted in an aromatic linkage modified Si interface. The [FeFe] hydrogenase from *C. reinhardtii* (*Cr*HydA1) containing a propane-1,3-dithiolate (PDT) ligand instead of the native 2-azapropane-1,3-dithiolate (ADT) ligand was investigated using time-resolved infrared (TRIR) spectroscopy [62]. In this experiment, the natural redox-active dithiolene ligand ADT, which contains a nitrogen as the bridgehead atom, was exchanged for a propane-1,3-dithiolate (PDT) ligand. The following experimental results were reported: (1) The PDT substitution presumably prevented protonation of the bridging ligand in $[2Fe]_H$; (2) Electron transfer from $[4Fe–4S]_H$ to $[2Fe]_H$, trapping the H-cluster in a one-electron reduced state occurred; (3) This effect abolished the catalysis and allowed to monitor the first reduction of $[4Fe–4S]_H$ going from the oxidized (H_{ox}) state to the one-electron reduced (H_{red}) state; (3) The PDT modification does not affect the electron transfer step to $[4Fe–4S]_H$ but prevents the enzyme from proceeding further through the catalytic cycle; (4) The rate of the first electron transfer step is independent of the pH, supporting a simple electron transfer rather than a proton-coupled event; (5) A proton-coupled electronic rearrangement (PCER) took place, basing on the fact that the H_{red} and $H_{red}H^+$ states, can vary their populations with pH; (6) H_{red} can be prevalent at high pH; (7) $H_{red}H^+$ can be prevalent at low pH; (8) Catalytically important protonation occurs at the nitrogen of the ADT bridge, leading to PCER of the H-cluster to give a protonated, one-electron reduced state ($H_{red}H^+$); and (9) This state can accept an additional electron on [4Fe–4S].

Two series of silicon-supported [(μ-SCH$_2$)$_2$N^{R}Fe$_2$(CO)$_6$] (R = *p*-OHC$_6$H$_4$, **3**, and (CH$_2$)$_2$OH, **4a**, and (CH$_2$)$_5$OH, **4b**)] hydrogenase mimics, as well as octamethylsilsesquioxane (POSS)-based nanostructures containing eight methylsilsesquioxane (POSS)-based and nanostructures containing eight [(μ-SCH$_2$)$_2$N^R**Fe$_2$**(CO)$_6$] moieties, have been prepared and characterized [63]. The following chemical processes were observed: (1) Reaction of the corresponding silyl and silyloxy chlorides and compounds **3** and **4** P; (2) NNCuAAC reaction between propargyl derivatives of silyl and silyloxy chlorides and azide; (3) Efficient cycloaddition between POSS-derived azide; and (4) Formation [(μ-SCH$_2$)$_2$N^{R}Fe$_2$(CO)$_6$] (R = alkyne) complexes, leading to POSS-based nanostructures containing eight [(μ-SCH$_2$)$_2$N^{R}Fe$_2$(CO)$_6$] units. All of the complexes prepared through this work were characterized by spectroscopic and electrochemistry methods, that is: (1) The electrochemistry of the linear siloxanes proceeded with a reduction wave around − 1.7 V, which is characteristic of these [(μ-SCH$_2$)$_2$N^{R}Fe$_2$(CO)$_6$] entities; (2) Complexes lacking 1,2,3-triazole moieties were electrochemically stable in the presence of acetic acid (AcOH), showing a strong electrocatalytic wave at − 2.2 V; and (3) Complexes having the 1,2,3-triazole ring were electrocatalytically active in the wave at − 2.2 V. In addition, POSS-[(μ-SCH$_2$)$_2$N^{R}Fe$_2$(CO)$_6$] derivatives indicated an irreversible reduction event and were deposited in the electrode. Experiments employing Si NMR and using H addition indicated that the integrity of the silicon cage is not affected by time or acetic acid experiments. Suggested synthetic protocols and electrochemical studies were applied in the design of silicon-supported [(μ-SCH$_2$)$_2$N^{R}Fe$_2$(CO)$_6$] mimics.

The synthesis of hydrogenase mimics has been based on the preparation of complexes having [(μ-SCH$_2$)$_2$N^{R}Fe$_2$(CO)$_6$] [63]:

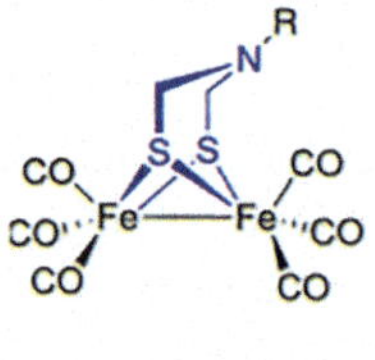

Fe$_2$(μ-adt)(CO)$_6$

or [Fe$_2$(μ-bdt)^R(CO)$_6$] (bdt = benzene-1,2-dithiolate) cores.

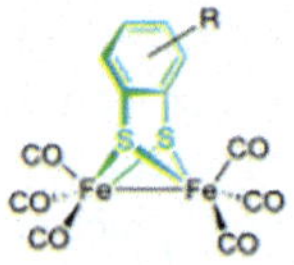

Fe$_2$(μ-bdt)(CO)$_6$

Inclusion of mimics of [FeFe]-hydrogenases into TiO_2 surfaces, nanotubes nanocrystals, metaorganic frameworks, dendrimers, polymers, sodium dodecyl sulfate micellar systems, nanocrystals, and other supramolecular assemblies have been carried out [63]. And references therein of experiments showed that the silicon photoelectrode interface having covalently bonded [FeFe]-hydrogenase mimics photoelectrochemically generate hydrogen. Several other silica-based heterogeneous systems can be used to immobilize [2Fe2S], photosensitizer-[2Fe2S], and other diiron complexes. In all cases, an improvement in the stability of the complexes and in their catalytic activity was achieved.

Novel biomimetic model of [FeFe]-hydrogenase prepared using the reaction of ferrocenyl α-thienyl thioketone with $Fe_3(CO)_{12}$ was introduced [68]. Heating of $Fe_3(CO)_{12}$ with three mol-equivalents of ferrocenyl α-thienyl thioketone in boiling THF for 4 h led to a mixture of amounts of the *ortho*-metalated complex **2** and also formed complex **3** (Fig. 11.16).

The influence of the substitution pattern in ferrocenyl α-thienyl thioketone used as a proligand in complexation reactions with $Fe_3(CO)_{12}$ was examined. Two new sulfur–iron complexes, considered [FeFe]-hydrogenase mimics, were characterized by spectroscopic techniques (^1H, $^{13}C\{^1H\}$ NMR, IR, MS), cyclic voltammetry elemental analysis and X-ray single crystal diffraction methods. The analysis of cyclic voltammograms of both complexes indicated that **2** reacts via two successive one-electron processes (EE mechanism) and in the case of **3**, transfer of two electrons in a single step via the ECE mechanism. ECE is an electrolysis reaction with any electrode steps being labeled E and any chemical steps labeled C:

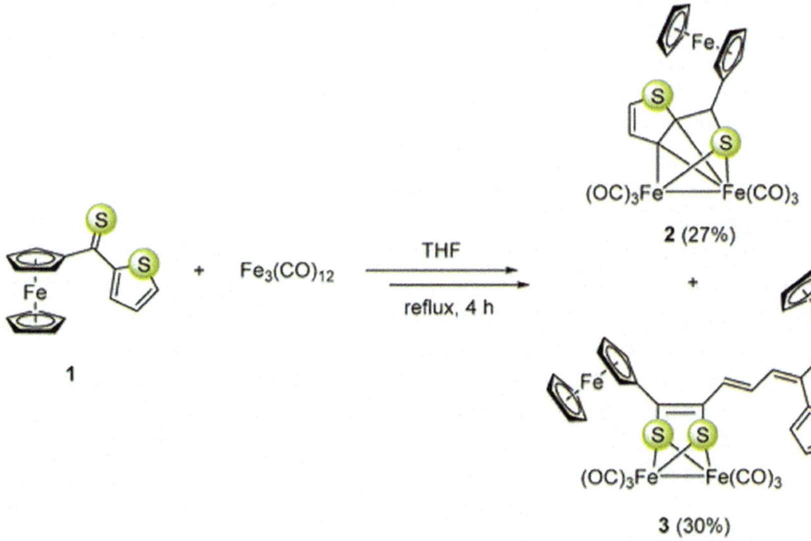

Fig. 11.16 Formation of a mixture of complexes **2** and **3** starting with ferrocenyl α-thienyl thioketone (**1**) and $Fe_3(CO)_{12}$ [68]. Published by MDPI

$$\text{E} \quad \text{O} + \text{e} \rightleftharpoons \text{R}$$

$$\text{C} \quad \text{R} \xrightarrow{k_{ECE}} \text{S}$$

$$\text{E} \quad \text{S} + \text{e} \rightleftharpoons \text{T}$$

The performed measurements demonstrated that both complexes can catalyze the reduction of protons to molecular hydrogen H_2.

11.3.3 Nickel–Iron and Iron Hydrogenases

[NiFe] hydrogenases catalyze the reversible oxidation of dihydrogen. The [NiFe] hydrogenase contains several metal centers, including the bimetallic Ni–Fe active site, iron–sulfur clusters and a Mg(2+) ion and the Ni–Fe active site is located in the inner part of the protein molecule (Fig. 11.12) [74, 75]. The crystal structure of the membrane-associated [NiFe] hydrogenase from *Allochromatium vinosum* has been determined to 2.1 Å resolution. The bimetallic [NiFe] active site is located in the large subunit of the heterodimer and possesses three diatomic non-protein ligands coordinated to the Fe (two CN(−), one CO). Ni is bound to the protein backbone via four cysteine thiolates; two of them also bridge the two metals. The additional metal site located in the large subunit appears to be an Mg^{2+} ion. The catalytic reaction route consists of an electron transfer pathway, a proton transfer pathway, and a gas-access channel [76]. The mechanism for H_2 cleavage in NiFe-hydrogenase has been reinvestigated with large models using hybrid DFT by itself, or in a QM/MM scheme following the ONIOM approach. In the suggested mechanism the key stages are: heterolytic H_2 cleavage occurs as an initial step to generate a complex with nickel in oxidation state Ni(I); and in the following cycles, H_2 is cleaved on nickel using an oxidative addition mechanism with a lower barrier.

Studies on the iron(II)–thiolate cyanocarbonyls, the mononuclear/dinuclear nickel(III)–thiolate complexes and the mononuclear nickel(II)–thiolate complexes with pendant thiol have led to the results related to the structure, reactivity and spectroscopic properties of the iron and nickel centers of the active site of [NiFe]/[Fe] hydrogenases [77]. Separate sections are devoted to a detailed description of the following topics related to the chemical compounds which model the structure and mechanism of [NiFe] hydrogenase structure and activity: (1) Iron–thiolate complexes with mixed CO/CN^- ligands mimic the iron part of [NiFe]/[Fe] hydrogenases; (2) Dinuclear iron(II)–cyanocarbonyl complexes with bridging ethylthiolates were prepared; and (3) Nickel–thiolate complexes (a) heterobimetallic nickel–iron complexes, (b) mononuclear nickel(II) carbonyl complexes, (c) mononuclear nickel(II)–thiolate complexes with pendant thiol, (d) mononuclear nickel(III)–thiolate complexes; and dinuclear nickel–thiolate complexes were synthetized.

The [**NiFe**]-hydrogenase model complex NiFe(pdt)(dppe)$(CO)_3$ (1) (pdt = 1,3-propanedithiolate) has been synthesized and characterized [76]. One CO ligand in [1H]BF_4 was substituted by organophosphorus ligands to afford the substituted derivatives [HNiFe(pdt)(dppe)(PR_3)$(CO)_2$]BF_4, where PR_3 = P$(OPh)_3$ ([2H]BF_4); PPh_3 ([3H]BF_4); PPh_2Py ([4H]BF_4, where Py = 2-pyridyl). In MeCN solution, complex 3, which was prepared by deprotonation of [3H]BF_4 with NaOMe, is about 10^4 stronger base than is 1. X-ray crystallographic analysis of [3H]BF_4 revealed a highly unsymmetrical bridging hydride for which the Fe–H bond being 0.40 Å shorter than the Ni–H distance. Experiments also indicated that: (1) Complexes [2H]BF_4, [3H]BF_4, and [4H]BF_4 underwent a one-electron reductions near − 1.46 V versus $Fc^{0/+}$; (2) The dependence of i_c/i_p on the concentration of the acid showed that H_2 evolution entails protonation of a reduced hydride; (3) For $[2H]^+$, $[3H]^+$, and $[4H]^+$, the acid-independent rate constants are 50–75 s^{-1}; (4) For $[2H]^+$ and $[3H]^+$, the over-potentials for H_2 evolution were estimated to be 430 mV; (5) The overpotential for the N-protonated pyridinium complex $[4H_2]^{2+}$ was found to be 260 mV; and (6) On the basis of their values for its pK_a and redox potentials, $\Delta G_{H^\cdot}$ and ΔG_{H^-} are 57 and 79 kcal/mol for [**1**H]$^+$ and [**1**]$^{2+}$, respectively. The mechanism of H_2 evolution was proposed to follow an ECEC sequence, where E and C correspond to one-electron reductions and protonation, respectively.

[Fe]-hydrogenase, which contains a unique iron cofactor (FeGP cofactor), catalyzes the reversible hydrogenation of methenyltetrahydrometanopterin (methenyl-H_4MPT^+) with H_2 to methylene-H_4MPT iron-chromophore [78]. Circular dichroism (CD) data indicated the binding of methenyl/methylene-H_4MPT induces a conformational change that closes the active site cleft of [Fe]-hydrogenase to form the intact active site. A crystal structure of [Fe]-hydrogenase at 1.75 Å resolution, showing a mononuclear iron coordinated by the sulfur of cysteine 176, two carbon monoxide molecules, and the sp^2-hybridized nitrogen of a 2-pyridinol compound with back-bonding properties was obtained in work [79]. The three-dimensional arrangement of the ligands was found to be similar to that of thiolate, CO, and cyanide ligated to the low-spin iron in binuclear [NiFe]- and [FeFe]-hydrogenases, although the CO and cyanide ligands are not found in any other metalloenzyme. In study [57], the electronic and molecular structure of the iron center by X-ray absorption spectroscopy (XAS) analysis of [Fe]-hydrogenase from *Methanocaldococcus jannashii* and five model complexes (Fe(ethanedithiolate)$(CO)_2(PMe_3)_2$, [K(18-crown-6)]$_2$[Fe$(CN)_2(CO)_3$], K[Fe(CN)$(CO)_4$], K_3[Fe(III)$(CN)_6$], K_4[Fe(II)$(CN)_6$]) were designed and examined. The experiment indicated that: (1) The different electron donors have a strong influence on the iron absorption K-edge energy position, which is used to determine the metal oxidation state; (2) The K-edges of Fe(II) complexes, achieved with low-spin ferrous thiolates, are consistent with a ferrous center in the [Fe]-hydrogenase; and (3) The metal geometry also strongly influences the X-ray absorption near edge structure (XANES). In addition, using in silico simulation, it was able to reproduce the main features of the XANES spectra and describe the effects of individual donor contributions on the spectra.

References

1. Lill R (2015) Issue of iron-sulfur protein. Biochem Biophys Acta 1853(6):1251–1252
2. Nicolet Y, Fontecilla-Camps JC (2017) Iron-sulfur clusters and molecular oxygen: function, adaptation, degradation, and repair. In: Characterization, properties and applications. Walter de Gruyter GmbH, pp 359–385. https://doi.org/10.1515/9783110480436-012
3. Boyd ES, Schut GJ, Shepard EM et al (2017) Origin and evolution of Fe-S proteins and enzymes. In: Biochemistry, biosynthesis and human diseases: biochemistry, biosynthesis, and human diseases. Walter de Gruyter GmbH, pp 445–461. https://doi.org/10.1515/9783110479850-017
4. Johnson DC, Dean DR, Smith AD, Johnson MK (2005) Structure, function, and formation of biological iron-sulfur clusters. Annu Rev Biochem 74:247–281. https://doi.org/10.1146/annurev.biochem.74.082803.133518
5. Meyer J (2008) Iron-sulfur protein folds, iron-sulfur chemistry, and evolution. J Biol Inorg Chem 13:157–170. https://doi.org/10.1007/s00775-007-0318-
6. Ichiye T (2017) Chemistry of iron-sulfur clusters. In: Characterization, properties and applications. Walter de Gruyter GmbH, pp 11–19. https://doi.org/10.1515/9783110480436-002
7. Eizawa A, Arashiba K, Tanaka H et al (2017) Remarkable catalytic activity of dinitrogen-bridged dimolybdenum complexes bearing NHC-based PCP-pincer ligands toward nitrogen fixation. Nat Commun 8:14874–14886
8. Boncella AE, Sabo ET, Santore RM et al (2020) The expanding utility of iron-sulfur clusters: their functional roles in biology, synthetic small molecules, maquettes and artificial proteins, biomimetic materials, and therapeutic strategies. Coord Chem Rev 453:214229
9. Coucouvanis D (1996) Functional analogs for the reduction of certain nitrogenase substrates. Are multiple sites within the Fe/Mo/S active center involved in the 6e$^-$ reduction of N_2? J Biol Inorg Chem 1:594–600
10. Hetterscheid DGH, Hanna BS, Schrock RR (2009) Molybdenum triamidoamine systems. Reactions involving dihydrogen relevant to catalytic reduction of dinitrogen. Inorg Chem 48:8569–8577
11. Lee SC, Holm RH (2004) The clusters of nitrogenase: synthetic methodology in the construction of weak-field clusters. Chem Rev 104:1135–1157
12. Rittle J, Peters C (2013) Synthetic analogs of nitrogenase metallocofactors: challenges and developments. Proc Natl Acad Sci USA 110:15898–15903. S15898/15891–S15898/15814
13. Coric BQ, Mercado Q, Bill E, Vinyard DJ, Holland PL (2015) Binding of dinitrogen to an iron–sulfur–carbon site. Nature 526:96–99
14. Burges BK, Lowe DJ (1996) Mechanism of molybdenum nitrogenase. Chem Rev 96:2983–3012
15. Sickerman NS, Tanifuji K, Hu Y, Ribbe MW (2017) Synthetic analogs of nitrogenase metallocofactors: challenges and developments. Chemistry 23(51):12425–12432
16. Jasniewski AJ, Sickerman NS, Hu Y, Ribbe MW (2018) The Fe protein: an unsung hero of nitrogenase. Inorganics 6:25
17. Speelman AL, Čorić I, Van Stappen C et al (2019) Nitrogenase-relevant reactivity of a synthetic iron–sulfur–carbon site. C J Am Chem Soc 141(33):13148–13157
18. Zhang X, Ward BB, Sigman DM (2020) Global nitrogen cycle: critical enzymes, organisms, and processes for nitrogen budgets and dynamics. Chem Rev 120(12):5308–5351
19. Chalkley MJ, Drover MW, Peters JC (2020) Catalytic N_2-to-NH_3 (or -N_2H_4) conversion by well-defined molecular coordination complexes. Chem Rev 120:5582–5636
20. Tanabe Y, Nishibayashi Y (2021) Comprehensive insights into synthetic nitrogen fixation assisted by molecular catalysts under ambient or mild con. Chem Soc Rev 50:5201–5242
21. Ohki Y, Munakata K, Matsuoka Y et al (2022) Nitrogen reduction by the Fe sites of synthetic [Mo_3S_4Fe] cubes. Nature 607:86–90. https://doi.org/10.1038/s41586-022-04848-1
22. Broere DLJ (2022) Synthetic molecular cluster hints at mechanism of nitrogen fixation. Nature 607:37–38

23. Hara R, Kachi M, Uchida K et al (2022) Nitrogen reduction by the Fe sites of synthetic [Mo_3S_4Fe] cubes. Nature 607:86–90
24. Shilov AE (1997) Metal complexes in biomimetical reactions. N_2 fixation, activation and oxidation of alkanes, chemical models of photosyntheses. CRC Press, Boca Raton, p 302
25. Engesser TA, Kindjajev A, Junge J et al (2020) Chatt-type catalyst with one coordination site for dinitrogen reduction to ammonia. Chemistry 26(65):14807–14812
26. Likhtenshtein GI (1988) Chemical physics of redox metalloenzymes. Springer, Heidelberg
27. Likhtenshtein GI (2003) New trends in enzyme catalysis and mimicking chemical reactions. Kluwer Academic/Plenum Publishers, New York
28. Denisov ET, Sarkisov OM, Likhtenshtein GI (2003) Chemical kinetics. Fundamentals and recent developments. Elsevier Science, Amsterdam
29. Likhtenshtein GI (2016) Electron spin in chemistry and biology: fundamentals, methods, reactions mechanisms, magnetic phenomena, structure investigation. Springer Nature, Berlin
30. Stripp ST, Duffus BR, Fourmond V et al (2022) Second and outer coordination sphere effects in nitrogenase, hydrogenase, formate dehydrogenase, and CO dehydrogenase. Chem Rev 122(14):11900–11973. https://doi.org/10.1021/acs.chemrev.1c00914
31. Allen AD, Senoff CV (1965) Nitrogenopentammineruthenium(II) complexes. Chem Commun (London) 621–622
32. Shilov AE, Shilova AK, Kvashina EF, Vorontsova TA (1971) Isolation of a binuclear complex intermediate in the reaction of molecular nitrogen with titanium compounds. J Chem Soc D 1590–1591
33. Volpin ME, Shur VB (1966) Nitrogen fixation by transition metal complexes. Nature 209:1236
34. Volpin ME, Shur VB (1970) Nitrogen fixation by transition—metal complexes in aprotic media. Organomet React 1:50–78
35. Likhtenshtein GI, Shilov AE (1970) On the thermodynamics of fixation of molecular nitrogen. Zhurnal Fiz Khem (Russ J Phys Chem) 44:849–856
36. Chalkley MJ, Drover MW, Peters JC (2017) Catalytic N_2-to-NH_3 (or -N_2H_4) conversion by well-defined molecular coordination complexes. ACS Cent Sci 3(3):217–223. https://doi.org/10.1021/acscentsci.7b00014
37. Semenov NN, Shilov AE, Likhtenshtein GI (1975) Dokl Akad Nauk SSSR 221(6):1374–1377
38. Denisov NT, Shuvalov VF, Shuvalova NI, Shilova AK, Shilov AE (1970) Catalytic reduction of molecular nitrogen in proton media. Kinet Katal 11:813–814
39. Shilov AE, Denisov NT, Efimov NO et al (1971) New nitrogenase model for reduction of molecular nitrogen in protonic media. Nature 231:460–461
40. Nikonova LA, Shilov AE (1977) Dinitrogen fixation in homogeneous protic media. In: Newton, Posgate JR, Rodrogues-Barrueco G (eds) Recent developments in nitrogen fixation. Academic Press, New York, pp 41–52
41. Shestakov AF, Shilov AE (2001) On the nuclearity of the vanadium(II)-pyrocatechol complex active in the reaction of molecular nitrogen reduction. Kinet Catal 42:653–656
42. Tanaka K, Hozumi, Y Tanaka T (1982) Dinitrogen fixation catalyzed by the reduced species of $[Fe_4S_4(SPh)_4]^{2-}$ and $[Mo_2Fe6S_8(SPh)]^{3-}$. Chem Lett 11:1203−1206
43. Buscagan TM, Ree DC (2019) Rethinking the nitrogenase mechanism: activating the active site. Joule 3:2662–2678
44. Hozumi Y, Imasaka Y, Tanaka K, Tanaka T (1983) Reduction of isocyanomethane and acetonitrile by the reduced species of $[Fe_4S_4(SPh)_4]_2$- and $[Mo_2Fe_6S_8(SPh)_9]_3$-: model reactions to nitrogenase. Chem Lett 12:897–900
45. Bazhenova TA, Bazhenova MA, Petrova GN, Shilov AE (2000) Catalytic behavior of isolated FeMo-cofactor of nitrogenase in non-protein surroundings. In: Current plant science and biotechnology in agriculture, vol 38. Nitrogen fixation: from molecules to crop productivity, pp 49–50
46. Moret M-E, Peters JC (2011) Terminal iron dinitrogen and iron imide complexes supported by a tris(phosphino)borane ligand. Angew Chem Int Ed 50:2063−2067
47. Arashiba K, Miyake Y, Nishibayashi Y (2011) A molybdenum complex bearing PNP-type pincer ligands leads to the catalytic reduction of dinitrogen into ammonia. Nat Chem 3:120–125

48. Yuki M, Tanaka H, Sasaki K et al (2012) Iron-catalysed transformation of molecular dinitrogen into silylamine under ambient conditions. Nat Commun 3:1254
49. Anderson JS, Rittle J, Peters JC (2013) Catalytic conversion of nitrogen to ammonia by an iron model complex. Nature 501:84–87
50. Tanifuji K, Ohki Y (2020) Metal–sulfur compounds in N_2 reduction and nitrogenase-related chemistry. Chem Rev 120:5194–5251
51. Yang J (2019) Progress in synthesizing analogues of nitrogenase metalloclusters for catalytic reduction of nitrogen to ammonia. Catalysts 9(11):939. https://doi.org/10.3390/catal9110939
52. Lubitz W, Ogata H, Rüdigger O, Reijerse E (2014) Hydrogenases Rev 114:4081−4148
53. Berggren G, Adamska A, Lambertz C et al (2013) Biomimetic assembly and activation of [FeFe]-hydrogenases. Nature 499(7456):66–69. https://doi.org/10.1038/nature12239G
54. Cao Z, Hall MB (2001) Modeling the active sites in metalloenzymes. 3. Density functional calculations on models for [Fe]-hydrogenase: structures and vibrational frequencies of the observed redox forms and the reaction mechanism at the diiron active center. J Am Chem Soc 123(16):3734–3742
55. Hinnemann B, Moses PG, Bonde J et al (2005) Biomimetic hydrogen evolution: MoS_2 nanoparticles as catalyst for hydrogen evolution. J Am Chem Soc 127(15):5308–5309
56. Liu JJ, Long MN (2005) Recent advances on the structure and catalytic mechanism of hydrogenase. Chin J Biotechnol 21(3):348–353
57. Salomone-Stagnia M, Stellatob F, Whaleyc CM et al (2010) Model systems: an X-ray absorption near edge spectroscopy study. The iron-site structure of [Fe]-hydrogenase. Dalton Trans 39(12):3057–3064
58. Rauchfuss TB (2015) Diiron azadithiolates as models for the [FeFe]-hydrogenase active site and paradigm for the role of the second coordination sphere. Acc Chem Res 48(7):2107–2116
59. Adamska-Venkatesh A, Simmons TR, Siebel JF et al (2015) Artificially maturated [FeFe] hydrogenase from *Chlamydomonas reinhardtii*: a HYSCORE and ENDOR study of a non-natural H-cluster. Phys Chem Chem Phys 17(7):5421
60. Zhang X, Ma X, Zhang T (2018) The influence of phosphine ligand substituted [$2Fe_2S$] model complexes as electro-catalyst on proton reduction. RSC Adv 8(74):42262–42268
61. Németh B, Esmieu CM, Redman HJ, Berggren G (2019) Monitoring H-cluster assembly using a semi-synthetic HydF protein. Dalton Trans 48(18):5978–5986
62. Sanchez MLK, Wiley S, Reijerse E, Lubitz W (2022) Time-resolved infrared spectroscopy reveals the pH-independence of the first electron transfer step in the [FeFe] hydrogenase catalytic cycle. J Phys Chem Lett 13(25):5986–5990
63. Aguado S, Diego J, Vicent DJ et al (2022) Two complementary approaches to silicon-supported soluble [FeFe]-hydrogenase mimics. Organometallics 41(22):3281–3329
64. Williams NB, Nash A, Yamamoto N et al (2021) Unraveling activity and decomposition pathways of [FeFe] hydrogenase mimics covalently bonded to silicon photoelectrodes. Adv Mater Interfaces 8:2001961. https://doi.org/10.1002/admi.202001961
65. Wang W, Yu T, Zeng Y et al (2014) An [Fe-Fe]-hydrogenase mimic immobilized on MCM-41 for the photochemical production of hydrogen in pure water. Chin J Chem 32(6):479–484. https://doi.org/10.1002/cjoc.201400083
66. Wang W, Yu T, Zeng Y et al (2014) Enhanced photocatalytic hydrogen production from an MCM-41-immobilized photosensitizer-[Fe-Fe] hydrogenase mimic dyad. Photochem Photobiol Sci 13(11):1590–1597. https://doi.org/10.1039/C3PP50446H
67. Kluwer M, Kapre R, Hartl F et al (2009) Self-assembled biomimetic [$2Fe_2S$]-hydrogenase-based photocatalyst for molecular hydrogen evolution. Proc Natl Acad Sci USA 106:10460–10465. https://doi.org/10.1073/pnas.0809666106
68. Daraosheh AQ, Abul-Futouh H, Murakami N et al (2022) Novel [FeFe]-hydrogenase mimics: unexpected course of the reaction of ferrocenyl α-thienyl thioketone with $Fe_3(CO)_{12}$. Materials (Basel) 15(8):2867
69. Tard C, Liu X, Ibrahim SK et al (2005) Synthesis of the H-cluster framework of iron-only hydrogenase. Nature 433(7026):610–661

70. Barton BE, Rauchfuss TB (2010) Hydride-containing models for the active site of the nickel−iron hydrogenases. J Am Chem Soc 132:14877
71. Wang H-Y, Wang WG, Si G et al (2010) Photocatalytic hydrogen evolution from rhenium(I) complexes to [FeFe] hydrogenase mimics in aqueous SDS micellar systems: a biomimetic pathway. Langmuir 26:9766–9771
72. Jian JX, Liu Q, Li ZJ et al (2013) Chitosan confinement enhances hydrogen photogeneration from a mimic of the diiron subsite of [FeFe]-hydrogenase. Nat Commun 4:2695
73. Camara JM, Rauchfuss TB (2012) Combining acid-base, redox and substrate binding functionalities to give a complete model for the [FeFe]-hydrogenase. Nat Chem 4:26
74. Ogata H, Kellers P, Lubitz W (2010) The crystal structure of the [NiFe] hydrogenase from the photosynthetic bacterium *Allochromatium vinosum*: characterization of the oxidized enzyme (Ni-A state). J Mol Biol 402(2):428–444
75. Ogata H, Lubitz W, Higuchi Y (2009) [NiFe] hydrogenases: structural and spectroscopic studies of the reaction mechanism. Dalton Trans 37(37):7577–7587. https://doi.org/10.1039/b903840j. PMID19759926
76. Lill SON, Siegbahn PEM (2009) An autocatalytic mechanism for NiFe-hydrogenase: reduction to Ni(I) followed by oxidative addition. Biochemistry 48(5):1056–1066. https://doi.org/10.1021/bi801218n.PMID19138102
77. Chiou T-W, Liaw W-F (2008) Nickel–thiolate and iron–thiolate cyanocarbonyl complexes: modeling the nickel and iron sites of [NiFe] hydrogenase. C R Chim 11(8):818–833. https://doi.org/10.1016/j.crci.2008.04.003
78. Hiromoto T, Warkentin E, Moll J et al (2009) Iron-chromophore circular dichroism of [Fe]-hydrogenase: the conformational change required for H^2 activation. Angew Chem Int Ed 49(51):9917–9921
79. Shima S, Pilak O, Vogt S et al (2008) The crystal structure of [Fe]-hydrogenase reveals the geometry of the active site. Science 321(5888):572–575

Chapter 12
Artificial Hydroxylases and Hydrolases

Abstract In hydroxylation reactions: $2R_3C\text{–}H + O_2 \rightarrow 2R_3C\text{–}OH$ and $R_3C\text{–}H + O_2 + 2e^- + 2H^+ \rightarrow R_3C\text{–}OH + H_2O$ a substrate C–H bond is converted to an alcohol group. A mechanism of abstracting a hydrogen atom from the substrate by a high active intermediate, an iron(IV)-oxo porphyrin radical intermediate (Compound I) (rebound mechanism) was proposed for cytochrome P450. A similar rebound mechanism was suggested for the methane monooxygenase (MMO). The large number of complexes mimicking cytochromes P450 including iron protoporphyrin IX were synthesized and their hydroxylation activity was tested. To avoid problems associated with indirect pathway for conversion methane to methanol, which widely used in chemical industry using 1000 °C and elevated pressure, numerous complexes mimicking the structure and enzymatic activity of MMO were prepared. Among them are the following compounds: (μ-oxo)(μ-hydroxo)diiron (III) core $[Fe_2(O)(OH)(6TLA)_2(ClO_4)_3]$, $[Fe_2(O)_2(6TLA)_2(ClO_4)_2]$ (II), a (-1,2-peroxo)bis(-carboxylato)diiron(III). According to the Shilov Cycle, CH_4 oxidation in mild condition is catalyzed by $PtCl_2$ in an aqueous solution, where $PtCl_6{}^{2-}$ acts as the ultimate oxidizing agent. Artificial hydrolases was utilized to catalyze the cleavage a number of substrates such as p-nitrophenyl acetate, glycerophospholipids, the phosphodiester bond of DNA, phosphate diester model of RNA, RNA model substrate 2-hydroxypropyl-4-nitrophenyl phosphate (HPNP). As a successful modeling nature's catalytic triad, consisting of a hydroxyl group of a serine residue, an imidazole group of a histidine residue, and a carboxyl group of an aspartic acid, the miniature organic model of chymotrypsin, was designed on the basis of cyclodextrin. A water-soluble fullerene functionalized with carboxylic acid moieties, C60-1, mono- and binuclear metalloenzymes, carbon-based nanozymes, peptide-based compounds are examples of successful mimicry of hydrolases.

G. I. Likhtenshtein, *Enzyme Catalysis Today and the Chemistry of the 21st Century*,
https://doi.org/10.1007/978-3-031-51552-1_12

12.1 Introduction

In biochemistry, hydroxylation reactions

$$2R_3C - H + O_2 \rightarrow 2R_3C - OH \quad (12.1)$$

$$R_3C - H + O_2 + 2e^- + \ 2H^+ \rightarrow R_3C - OH + H_2O \quad (12.2)$$

are facilitated by enzymes hydroxylases [1–6]. A substrate C–H bond is converted to an alcohol by insertion of an oxygen atom into a C–H bond. The driving force for the four-electron reduction of oxygen to water is 80 kcal mol^{-1}, thus, showing large favorable enthalpic change. Nevertheless, the one-electron reduction of $^3O_2$ to the superoxide ion is endergonic by 7.8 kcal mol^{-1} and therefore thermodynamically forbidden in ambient biological conditions (see also Eq. 12.5). In contrast, as follows from Eq. (12.4), the two-electron reduction of the oxygen molecule is energetically favorable.

$$O_2(g) + 4H^+(aq) + 4e^- \rightarrow 2H_20 \quad E^0 = +1.229 \quad (12.3)$$

$$O_2(g) + 2H^+(aq) + 2e^- \rightarrow H_2O_2 \quad E^0 = +0.670 \quad (12.4)$$

$$O_2(g) + H^+(aq) + e^- \rightarrow HO_2^{\bullet} \quad E^0 = -0.34V^- \quad (12.5)$$

Consequently, two-electron reduction of complexed O_2 can lead to the formation of a peroxide intermediate, which in turn generates a highly active an iron(IV)-oxo porphyrin radical intermediate (Compound I) capable of abstracting a hydrogen atom from the substrate (rebound mechanism Figs. 12.1 and 12.2). A similar mechanism with the formation of a high active intermediate with $[Fe^{IV}{}_2(\mu\text{-}O)_2]$ diamond core was proposed for the methane monooxygenase (MMO)

A survey on the range of O_2 activation processes mediated by heme proteins and model compounds with a focus on recent progress in the characterization and reactivity of important iron–oxygen intermediates was reported [20]. Along this way, a number of aspects in this area have been presented and analyzed in detail: (1) Representative reactions initiated by these reactive intermediates; (2) The fundamental mechanistic features of these transformations; and (3) The recent developments in biocatalysis for nonnatural chemistries and the advantages of radical chemistry in organic synthesis. It was stressed that new enzymatic and synthetic transformations will emerge based on the radical processes mediated by metalloproteins and their synthetic analogs. Examples of typical reactions illustrating the diverse and wide possibilities of cytochrome models were described.

It is obvious that the creation of effective models of hydroxylases should take into account the principal stages of enzymatic processes outlined above, including the two-electron reduction of O_2 and the rebound mechanism. This section briefly

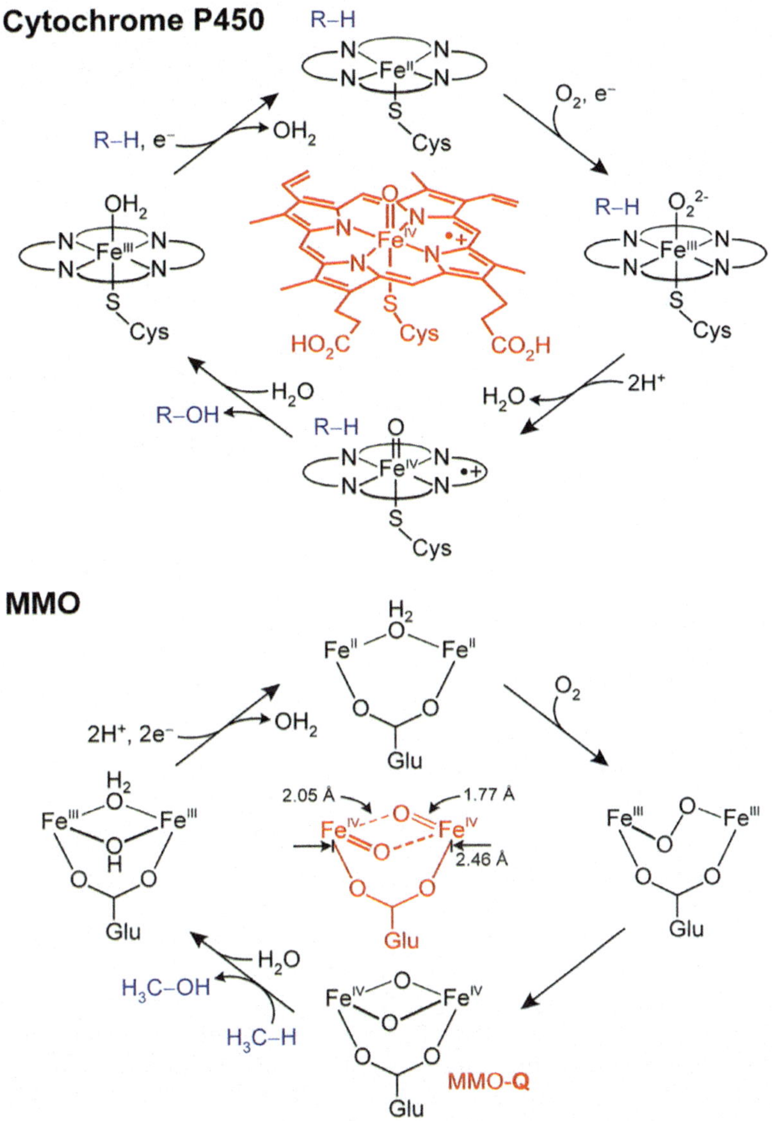

Fig. 12.1 Principal steps in the O_2 activation and substrate (ROH) oxidation mechanisms used by cytochrome P450 (upper) and MMO (lower), along with schematic representations of the corresponding key oxidizing species (shown in red). In the cytochrome P450 catalytic cycle, binding of substrate to the resting enzyme triggers the reduction of the FeIII center of the heme cofactor to the reactive FeII state, which binds and activates O_2 to generate a high-valent iron(IV)-oxo porphyrin radical intermediate that is capable of oxidizing a wide range of organic substrates (a notable exception being methane, the native substrate of MMO). In the MMO catalytic cycle, the diiron(III) cluster of the resting enzyme is reduced by two electrons to yield the reactive diiron(II) species, which binds O_2 to generate a peroxodiiron(III) intermediate. Further reduction of the O_2 moiety in this species yields the oxodiiron(IV) key oxidizing intermediate (MMO-Q), which converts the substrate methane to methanol [2]. With permission from PNAS

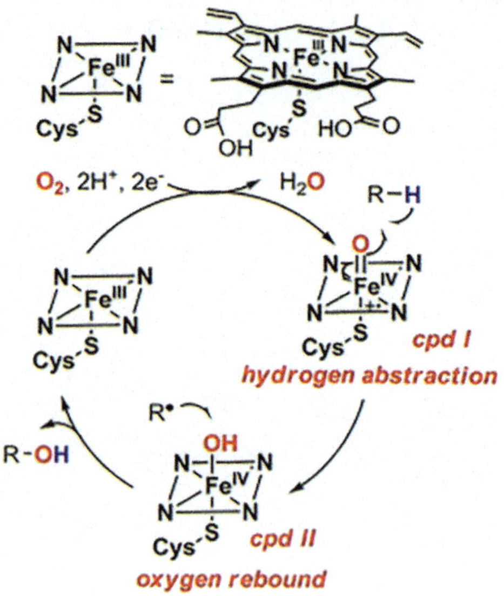

Fig. 12.2 Mechanism of aliphatic C–H hydroxylation catalyzed by cytochrome P450 (the oxygen rebound mechanism) [6]. With permission from Spinger Nature

reviews advances in chemical modeling of important reactions catalyzed by the enzymes cytochrome P450 and methane monooxygenase.

12.2 Cytochrome P450

The large number of studies have been performed during the last three decades both on cytochromes P450 themselves and on iron porphyrin models [2–28]. The synthesis of three successive generations of porphyrin ligands eventually led to metalloporphyrin catalysts more efficient for oxygenation of substrate. Some examples of porphyrin metal porphyrin complexes of are presented in references [3, 4]. Iron porphyrins derived from the heme: iron protoporphyrin IX, Fe(PPIX), $R_1 = CH\,CH_2$, $R_2 = CH_3$, $R_3 = CH_2CH_2COOH$; iron protoporphyrin IX dimethylester, Fe(PPIXDME), $R_3 = CH_2CH_2COOCH_3$; iron deuteroporphyrin dimethylester, Fe(DPDME), $R_1 = H$, $R_2 = CH_3$, $R_3 = CH_2CH_2COOCH_3$; iron beta-octaethylporphyrin, Fe(OEP), $R_1 = R_2 = R_3 =$ Et; on the right, synthetic iron *meso*-tetraarylporphyrins. Fe(TPP), Ar $= C_6H_5$; Fe(TMP), Ar = mesityl(2,4,6-trimethylphenyl); Fe(TDCPP), Ar = 2,6-dichlorophenyl; Fe(TF$_5$PP), Ar $= C_6H_5$[3]. The high-valent iron (IV)-oxo intermediate of the P450 catalytic cycle was formed by the reaction of iron(III) porphyrins with different monooxygen donors and used for the hydroxylation of hydrocarbons, epoxidation of olefins, oxidation of heteroatoms, and the cleavage of C–C bonds in organic substrates [4].

According to the oxygen rebound mechanism (Figs. 12.1 and 12.2) suggested and established in a pioneering work of Groves and Van der Puy [5]. Stereospecific aliphatic hydroxylation by iron-hydrogen peroxide occurs in wide range of oxygen- and heteroatom-transfer reactions. In this process, a high-valent metal-oxo species (Compound I) abstracts a hydrogen atom from the substrate to generate a carbon-centered radical, which recombines with the hydroxometal intermediate with rate constants that can be in the nanosecond to picosecond regime. In addition to catalyzing C–O bond formation, it was found that high-valent metal-oxo porphyrins can directly catalyze C–H halogenations and pseudohalogenations, including chlorination, bromination, and fluorination. Review [6] recount the 40 years history of radical rebound mechanisms. The following aspects of this process were in the focus of the authors attention: (1) The mechanisms general features; (2) Key intermediates involved in the process; (3) The factors that affect the behavior of the initial caged radical pair; and (4) The lifetimes of the incipient substrate radicals. To illustrate how the behaviors of the radical pair $[Fe^{n-1} - OH \cdot R]$ determine the eventual reaction outcome several representative examples of enzymatic C–H transformations were selected.

In 1978, Groves et al. reported the first catalytic alkane hydroxylation and alkene epoxidation catalyzed by a synthetic iron(III) porphyrin complex, Fe(TPP)Cl using an iodosylbenzene (PhIO) oxidant [7] and references within. In the reactions, alkenes and alkanes were oxidized to the corresponding epoxides and alcohols, respectively. The first use of a chiral iron porphyrin to carry out asymmetric epoxidation was reported in 1983 by Groves and Myers. In 1980, Groves group ([7] and **references within**) reported the **first manganese** porphyrin, (Mn(TPP)Cl), catalyzed hydrocarbon oxidation reaction. In the study, oxidation of cyclohexane with iodosylbenzene catalyzed by Mn(TPP)Cl gave mixture of cyclohexanol and cyclohexyl chloride in a total 70% yield. Oxidation of a radical clock substrate, norcarane,

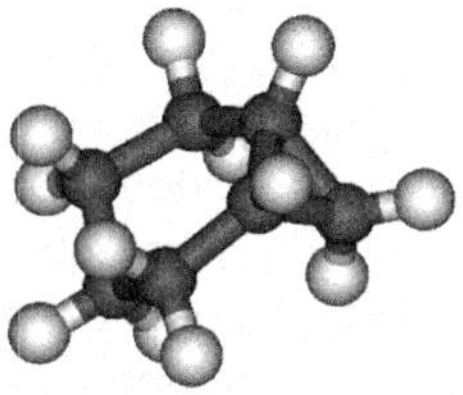

produced significant amounts of radical rearranged products. The radical-clock compound itself reacts at a known rate, which provides a calibration for determining the rate of another reaction.

The detection and kinetic characterization of a cytochrome P450 model $OFe^{IV}-4\text{-}TMPyP]^{+}$ were carried out [8]. This compound aqueous solution showed reaction rates for C−H hydroxylations (Fig. 12.6) [8]. Stopped-flow spectrophotometric monitoring of the oxidation of $Fe^{III}-4\text{-}TMPyP$ with meta-chloroperoxybenzoic acid (*m*CPBA)

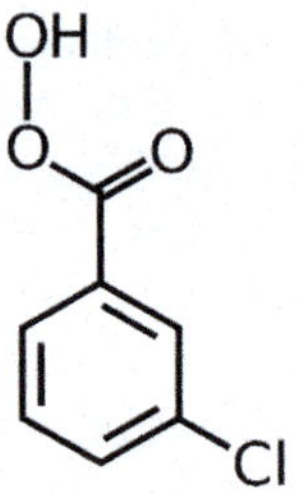

revealed the intermediate, typical of a porphyrin π-radical cation which was subsequently transformed into the OFe^{IV}–4-TMPyP. Global analysis afforded a second-order rate constant $k_1 = (1.59 \pm 0.06) \times 10^7\ M^{-1}\ s^{-1}$ for the formation of **1** followed by a first-order decay with $k_2 = 8.8 \pm 0.1\ s^{-1}$. ^1H and ^{13}C NMR determined 9-xanthydrol to be the major product (~90% yield) of xanthene

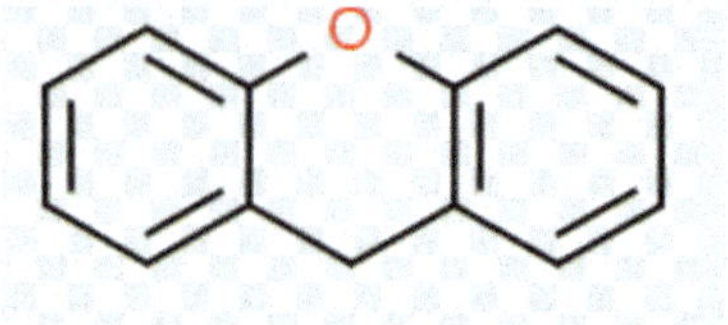

oxidation by **1**. The following results were reported: (1) Electrospray ionization mass spectrometry carried out in 47.5% $^{18}OH_2$ indicated 21% ^{18}O incorporation; (2) Xanthene/xanthene-d_2 showed a modest kinetic isotope effect, $k_H/k_D = 2.1$; (3) Xanthene hydroxylation by **1** occurred with a second-order rate constant $k_3 = (3.6 \pm 0.3) \times 10^6\ M^{-1}\ s^{-1}$ (Fig. 12.3); (4) Similar reactions of fluorene-4-carboxylic acid and 4-isopropyl- and 4-ethylbenzoic acid gave high rates for C–H hydroxylation that correlated well with the scissile C–H bond energy, indicating a homolytic hydrogen abstraction transition; (5) The observed rate constants for C–H bond cleavage onto the Brønsted–Evans–Polanyi relationship for similar substrates were mapped; and 5) The H–OFe^{IV}–4-TMPyP bond dissociation energy was determined as ~100 kcal mol^{-1}. It was suggested that the high kinetic reactivity observed for **1** is result from a high porphyrin redox potential and spin-state-crossing phenomena. Spin crossover was referred to as spin transition or spin equilibrium behavior when the change in spin state usually involves interchange of low-spin (LS) and high-spin (HS) configuration [2].

The first data on iron porphyrins to mimic the $Fe(III)OO^-$, Fe(III)OOH and iron–oxo intermediates of the P450 were published by Groves et al. [9]. The experiments showed the formation of a species, with proposed an oxo–ferryl porphyrin ***Π***-cation radical structure, formed by the reaction of Fe(TMP)Cl with *meta*-chloroperbenzoic acid at – 78 °C for process:

$$Fe^{III}(TMP)Cl + ArCO_3H \xrightarrow[-78^\circ C]{CH_2Cl_2} (TMP^{+\cdot})Fe^{IV} = O$$

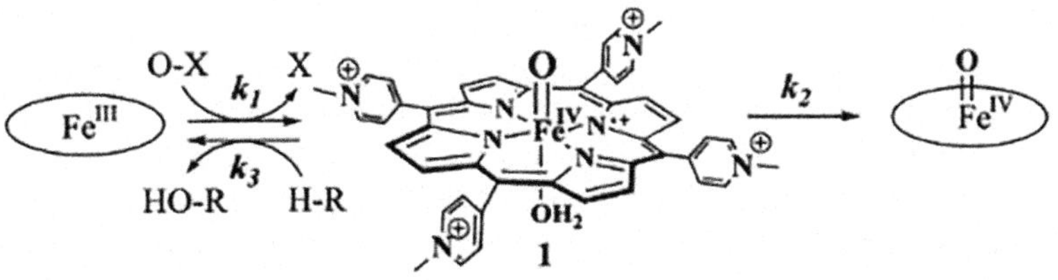

Fig. 12.3 Kinetic partitioning of [OFe^{IV}-4-TMPyP$^+$] (**1**) between spontaneous decay to OFe^{IV}-4-TMPyP (k_2) and substrate hydroxylation (k_3) resulting in the formation of Fe^{III}-4-TMPyP [8]. With permission from American Chemical Society

The product exhibited spectral characteristics analogous to those of peroxidase compound ***I***. Further studies on the reactions of iron(III)porphyrins with various oxygen atom donors such as peracids, iodosylarenes or O_3, at low temperature, indicated the formation of porphyrin cation radical Fe(IV)O and porphyrin Fe(IV)O complexes. Electronic function of the porphyrin and axial ligands of iron were established on the basis of the detailed UV–vis, ^{1}H NMR, EPR, Mössbauer and resonance Raman characteristics.

It was shown [10] that in case **of** Tetraphenylporphyrin (TPP) FeCl/PhIO system, only one product of oxidation of nicotine to cotinine

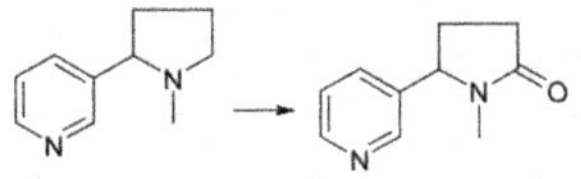

was obtained. When using TPPMnCl as catalyst with PhIO, the yield of cotinine was substantially increased and cotinine can be also oxidized to 3-OH cotinine. Mass spectrometry (**MS**) and NMR data indicated that it is isomeric with the keto-amide, which has been isolated as a urinary metabolite from Rhesus monkey.

The improving efficiency of the third generation metalloporphyrins can be illustrated on the example of the hydroxylation of heptane, by PhIO [11]. The yield of oxygenation of heptane to heptanols and heptanones heptane + PhIO → heptanols + heptanones + PhIwas found to increase from 38% when utilizing Fe(TDCPP)Cl as catalyst and to to 78% with the third-generation catalyst Fe(TDCPCl$_8$P)Cl

The aim of work [12] was to evaluate whether **metalloporphyrin models** could mimic the action of cytochrome P-450 in the oxidation of atrazine, a herbicide

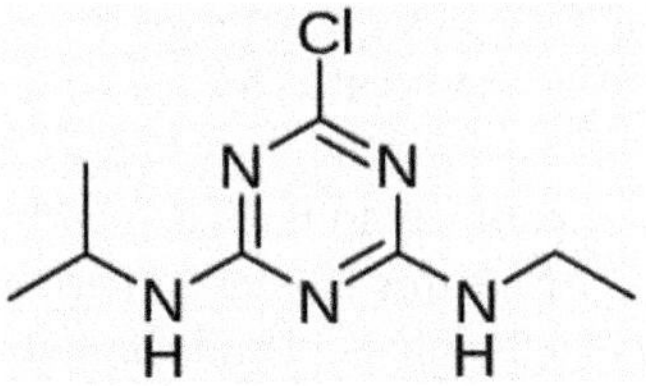

employed second-generation metalloporphyrins 5,10,15,20-tetrakis(2,6-dichlorophenyl)porphyrin metal(III) chloride [M(TDCPP)Cl] and 5,10,15,20-tetrakis(pentafluorophenyl)porphyrin metal(III) chloride [M(TFPP)Cl] (metal = Fe or Mn) and the oxidants iodosylbenzene and metachloroperbenzoic acid, it was found that the metalloporphyrins can oxidize atrazine. Yields as high as 32% were obtained for the Mn(TFPP)Cl/PhIO system, indicating that these catalysts can mimic action of cytochrome P-450.

The biomimetic oxidation of metribuzin

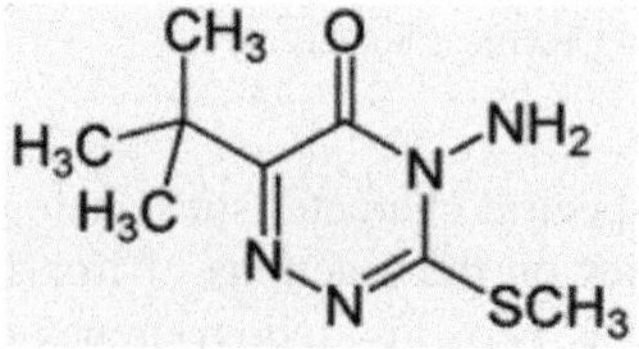

a pre- and post-emergence herbicide, with hydrogen peroxide catalyzed by 5,10,15,20-tetraarylporphyrinatoiron(III) chlorides [TAPFe(III)Cl], has been studied [13]. The reaction yielded 6-*t*-butyl-3-methylthio-1,2,4-triazine-5(4*H*)-one, 4-amino-6-*t*-butyl-3,5(2*H*,4*H*)-dione and 6-*t*-butyl-1,2,4-triazin-3,5(2*H*,4*H*)-dione. Oxidative degradation of the herbicides atrazine (**1**), **atraton** (**2**), ametryn (**3**), and mecoprop (**4**) was carried out with hydrogen peroxide and metalloporphyrins as catalysts [14]. Two different reaction conditions were studied, the first involving Mn(TDCPP)Cl in an aprotic solvent with buffer (**S-I**), and the second using Fe(TPFPP)Cl in a protic solvent (**S-II**). The reaction was studied in two conditions, the first involving Mn(TDCPP)Cl in an aprotic solvent with buffer (**S-I**), and the second using Fe(TPFPP)Cl in a protic solvent (**S-II**). The experiments showed the following results: (1) In the case of the **S-I** conditions, the *s*-triazines were oxidized through hydroxylation of the alkyl side chains followed by dealkylation, while **S-II** was ineffective for these reactions; (2) Mecoprop was oxidized with high efficiency by **S-II**, leading to decarboxylation and further oxidation, while in the presence of **S-I**, low substrate conversion was observed, and reaction resulted mainly from oxidation at the benzyl position; (3) Sulfoxidation of ametryn was observed with both systems; (4) Involvement of different reactive species assigned to the oxo and hydroperoxy complexes occurred; (5) These routes show similarities with metabolic pathways, with the reactivity pattern of **S-I** analogous; and (6) **S-II** catalyzes mecoprop

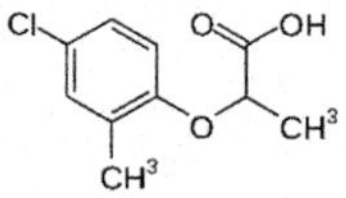

decarboxylation *via* a similar pathway to that seen with peroxidase catalyzed reactions. Biomimetic oxidation of a *trans* isomer of permethrin, fenvalerate, together with the corresponding chrysanthemic acid and alcohol moieties in the presence of hydrogen peroxide or *meta*-chloroperbenzoic acid as an oxidant was conducted using catalytic amounts of six iron-porphyrins [15]. The weak electron donor as the axial ligand of iron was found to increase the oxidative reactivity of porphyrin. The introduction of electron-withdrawing substituents at β- and *meso*-positions of a porphyrin ring resulted in higher conversion.

Using 2,4,6-tri-tert-butylphenol

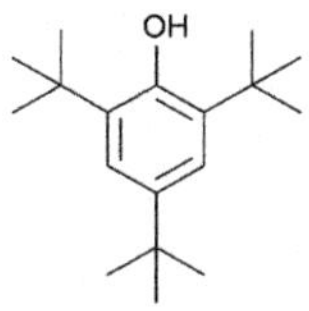

as the reductant, the direct ultraviolet-visible spectroscopic observation of the soluble iodosylarene-manganese porphyrin adduct following catalytic oxidation was reported [16]. Manganese(V)-oxo porphyrin species

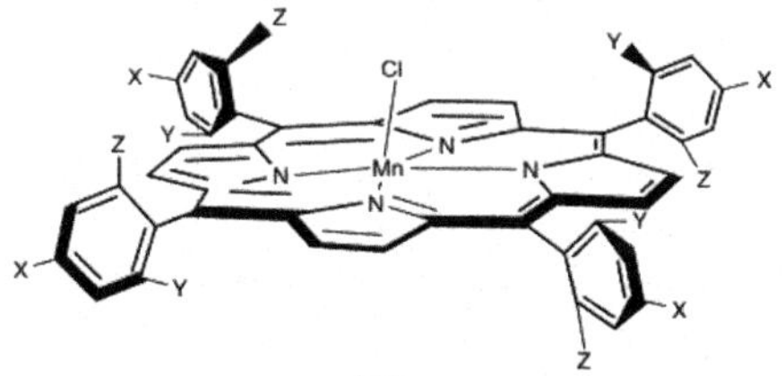

2a-e

2a X=Y=Z=H, [Mn(TPP)Cl]
2b X=OMe, Y=Z=H, [Mn(T(*p*-OMe)PP)Cl]
2c X=Cl, Y=Z=H, [Mn(T(*p*-Cl)PP)Cl]
2d X=Y=H, Z=Cl, [Mn(T(*o*-Cl)PP)Cl]
2e X=H, Y=Z=Cl, [Mn(TDCPP)Cl]

were utilized *using 2,4,6-tri-tert-butylphenol as the reductant* [16]. Both the iodosylarene-manganese porphyrin adduct and [(porphyrin)Mn(V) O] species were *simultaneously* observed. In the absence of reductant, the adduct of iodosylarene with sterically hindered [Mn(meso-tetrakis(2,6-dichlorophenyl)porphinato)Cl] was formed, and converted into a high-valent [(porpyrinato)MnO]. On the basis of kinetic data, obtained by ultraviolet–vis spectroscopy and electrospray ionization mass spectrometry (ESI-MS) analysis, a detailed mechanism of the reaction was proposed.

A reactive hydroxoferric porphyrazine complex, $[(PyPz)Fe^{III}(OH)\ (OH_2)]^{4+}$ (**1**, PyPz = tetramethyl-2,3-pyridino porphyrazine) has been prepared via one-electron oxidation of the corresponding ferrous species $[(PyPz)Fe^{II}(OH_2)_2]^{4+}$ (**2**) (Fig. 12.7) [25]. Electrochemical analysis revealed high Fe^{III}–OH/Fe^{II}–OH_2 reduction potential of 680 mV versus Ag/AgCl at pH 5.2 and indicated a one-proton, one-electron interconversion from pH 2 to pH 8. The O–H bond dissociation energy of the Fe^{II}–OH_2 complex was estimated as 84 kcal mol^{-1}. Experiments showed that compound 1 reacts rapidly with substrates via C-H hydrogen atom transfer (HAT), reducing 1 to $[(PyPz)Fe^{II}(OH_2)_2]^{4+}$ (2) (Fig. 12.4). The second-order rate constant for the reaction of $[(PyPz)Fe^{III}(OH)\ (OH_2)]^{4+}$ with xanthene was found to be $2.22 \times 10^3\ M^{-1}\ s^{-1}$ that 5–6 orders of magnitude faster than other reported Fe^{III}-OH complexes.

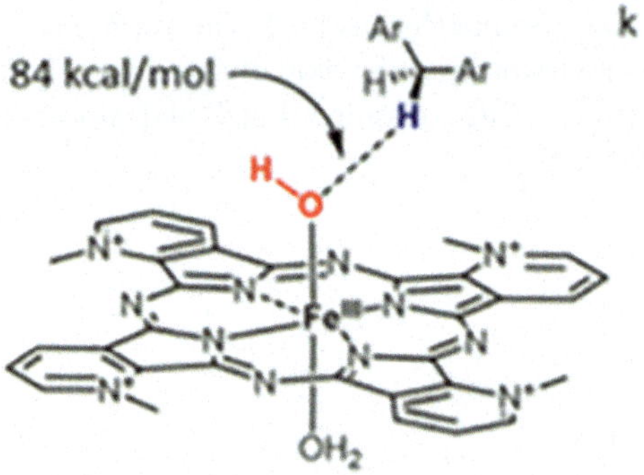

Fig. 12.4 Structure of hydroxoferric porphyrazine complex $[(PyPz)Fe^{III}(OH)(OH_2)]^{4+}$ [25]. With permission from American Chemical Society

Specifically, compound **1** reacts rapidly with a panel of substrates via C–H hydrogen atom transfer (HAT), reducing **1** to $[(PyPz)Fe^{II}(OH_2)_2]^{4+}$ (**2**) (Fig. 12.5).

Main experimental results of study of the C–O bond-formation step in metalloporphyrin-catalyzed C–H oxygenation reactions were formulated as follows [17]: (1) Fluoride, hydroxide, and oxo ligands all slow down the oxygen rebound rate by factors of 10–40-fold; (2) The oxidation of norcarane

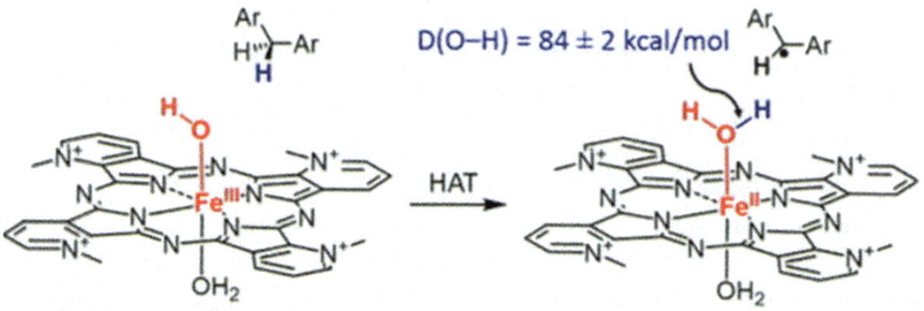

1 2

Fig. 12.5 Conversion of $[(PyPz)Fe^{III}(OH)(OH_2)]^{4+}$ (1) to $[(PyPz)Fe^{II}(OH_2)_2]^{4+}$ (2) via HAT [25]. With permission from American Chemical Society

Fig. 12.6 Block flow diagram of conversion methane to methanol [27]. Published under Creative Commons Attribution 3.0 license with permission from IOSCIENCE

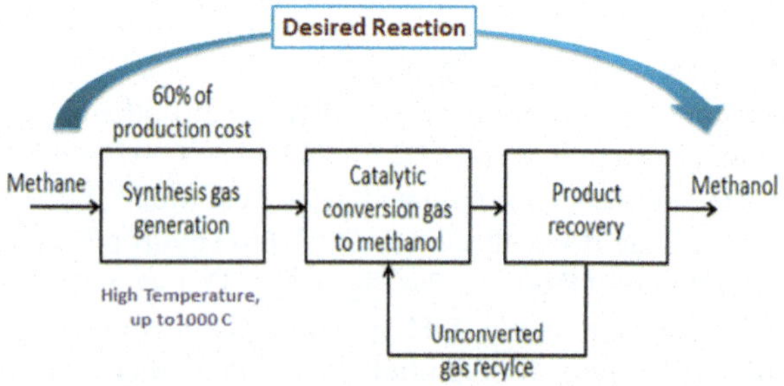

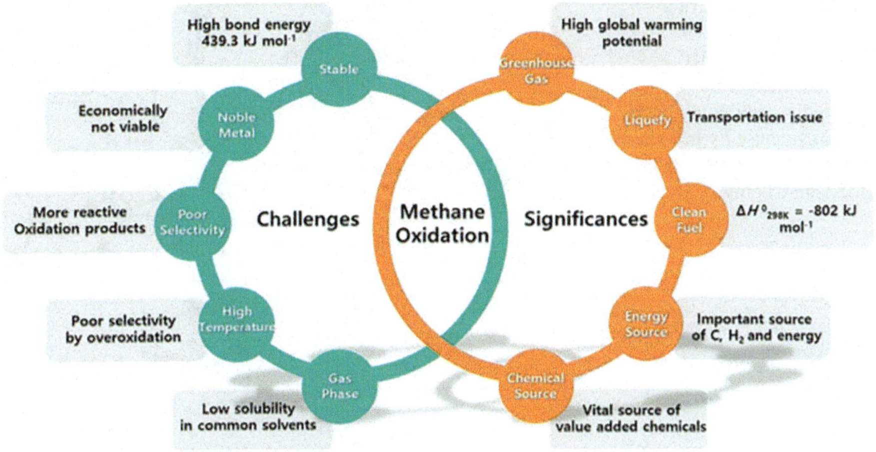

Fig. 12.7 Challenges and significances of CH_4 oxidation [28]. Published under the terms of the http://creativecommons.org/licenses/by/4.0/ License' from John Wiley and Son

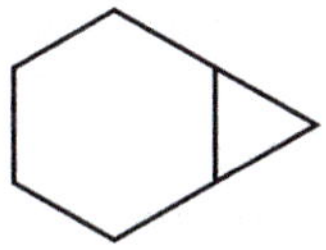

by a manganese porphyrin coordinated with fluoride or hydroxide leads to the formation of significant amounts of radical rearranged products; (3) *cis*-Decalin

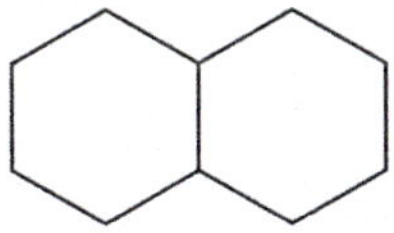

oxidation afforded both *cis*- and *trans*-decalol; (4) Xanthene afforded dioxygen trapped products and the radical dimer product bixanthene, under aerobic and anaerobic conditions, respectively; and (5) Axial ligands of metalloporphyrin-catalyzed C–H oxygenation reactions significantly affect the oxygen rebound rate. In addition, DFT calculations probing the rebound step indicated that the rebound barrier increases significantly (by 3.3, 5.4, and 6.0 kcal mol^{-1}, respectively) with fluoride, hydroxide, and oxo as axial ligands.

A series of biomimetic catalysts consisting of iron(III)/porphyrin complexes and biocompatible polyamidoamine dendrimers incorporated on mesoporous silica MCM-41 has been synthesized [18]. The structural and surface characteristics of catalysts were characterized by transmission electron microscopy (TEM), field-emission scanning electron microscopy, N_2 adsorption–desorptionisotherm (BET specific surface area), Fourier transform infrared spectroscopy, X-ray diffraction, diffuse reflectance DR UV–vis spectroscopy, thermal gravimetric and elemental

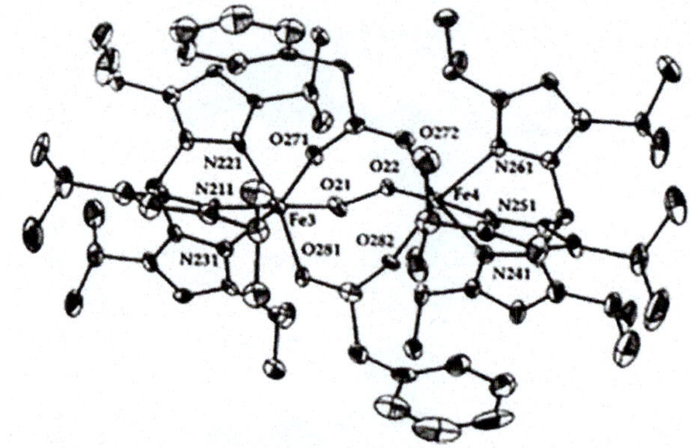

Fig. 12.8 Structure of a (-1,2-peroxo)bis(-carboxylato)diiron(III) model for the peroxo intermediate in the methane monooxygenase hydroxylase [29]. With permission from American Chemical Society.

analyses, gas chromatography and GC-mass spectrometry. The catalytic properties of the prepared catalysts were evaluated. The oxidation of 2,3,6-trimethylphenol to 2,3,5-trimethyl-*p*-benzoquinone with hydrogen peroxide was studied by varying the structure of catalysts, generation numbers of dendrimers, temperature, solvent, and reaction time. Role of dendritic functionality at the peripheral positions of a metalloporphyrin in regioselective catalytic performance of the hybrid catalysts was discussed.

The first direct C–H ^{18}F fluorination reaction of unactivated aliphatic sites using no-carrier-added [^{18}F]fluoride was realized [19]. In suggested protocol (Fig. 12.8), ^{18}F fluoride captured on an anion exchange cartridge was eluted using an acetone/acetonitrile solution of Mn^{III}(TPFPP)Ots

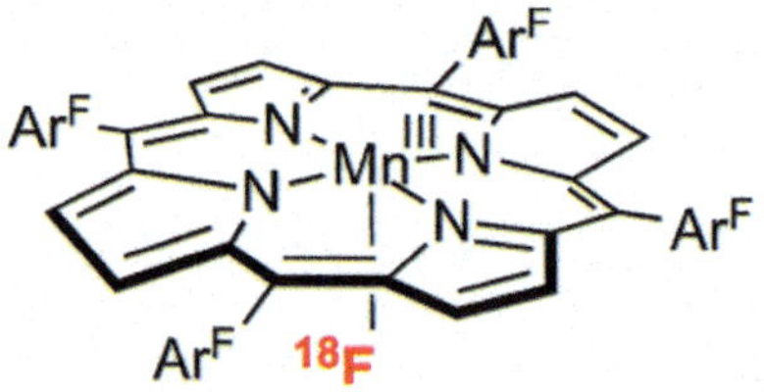

with over 90% elution efficiency of the ^{18}F fluoride. Radiolabeling of protected FACPC to afford 3-[^{18}F]-FACPC

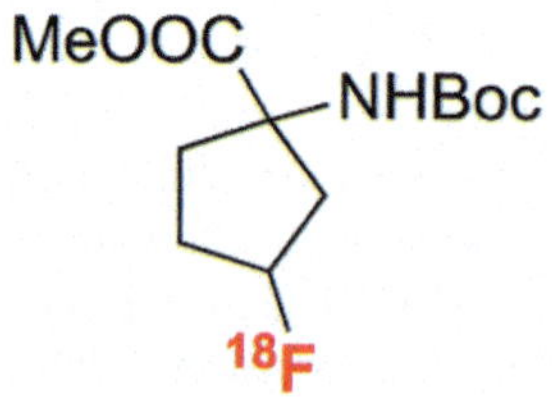

was accomplished within 10 min using the eluted Mn(TPFPP)^{18}F solution in conjunction with the terminal oxidant, PhIO, with microwave irradiation.

In order to evaluate whether metalloporphyrins were good biomimetic models for the oxidation of herbicides [21], the oxidation metalloporphyrin-mediated oxidation of a herbicide of the triazine class, simazine,

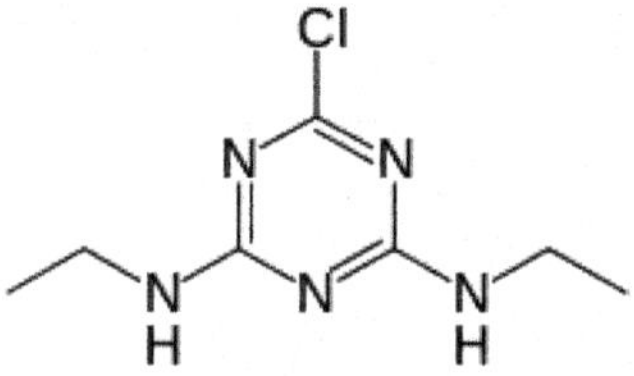

was examined. *In order to evaluate whether metalloporphyrins were good biomimetic models for the oxidation of this herbicide* [21], the following second-generation metalloporphyrins were subjects of the research: 5,10,15,20-tetrakis(2,6-dichlorophenyl)porphyrin metal(III) chloride, [M(TDCPP)]Cl, 5,10,15,20-tetrakis(pentafluorophenyl)porphyrin metal(III) chloride, [M(TFPP)]Cl; and 5,10,15,20-tetrakis(N-methyl-4-pyridyl)porphyrin metal(III) chloride, [M(TMPyP)]Cl_5 (metal = Fe or Mn). The iodosylbenzene (PhIO), hydrogen peroxide, and 3-chloroperoxy benzoic acid (*m*-CPBA) were utilized as oxidants. Results indicated that these metalloporphyrins can mimic both the *in vivo* and *in vitro* action of cytochrome P450 for simazine oxidation. Among the systems, [Fe(TDCPP)]Cl/*m*-CPBA was the most efficient for simazine oxidation, giving up to 49% total conversion. Degradation of atrazine (ATZ), a herbicide, catalyzed by manganese porphyrins as biomimetic cytochrome P450 models, using PhIO, $PhI(OAc)_2$, H_2O_2, t-BuOOH, m-CPBA, or Oxone® as oxidant was evaluated [22]. Oxidant, iodosylbenzene provided the best result of ATZ degradation which was as high as 47%. The same catalyst/oxidant systems in natural water from a Brazilian river as solvent were studied and indicated yield up to 100% ATZ degradation when iodobenzene diacetate was the oxidant.

12.3 Methane Monooxygenase Mimicking

Indirect pathway for conversion methane to methanol, which was widely used in chemical industry, run at 1000 °C and elevated pressure divided to three steps (Fig. 12.6) [27]: (1) Synthesis gas by steam reforming of methane; (2) High pressure catalytic conversion syngas to methanol using Cu–Zn–O-based catalyst such as Cu/ZnO/Al_2O_3; and (3) Separation of crude methanol from the unconverted synthesis gas and purified, while the unconverted syngas was recycled. Direct routes involving the partial oxidation methane to methanol, or one-step catalytic oxidation, can be chosen as an alternative process [26–40].

Review [28] discussed the advancements in the electrocatalytic and photocatalytic oxidation of methane at low temperatures with products containing oxygen atom(s). The main challenges and significance of CH_4 oxidation has been summarized in Fig. 12.7.

First two complexes with a (μ-oxo)(μ-hydroxo)diiron (III) core[$Fe_2(O)(OH)(6TLA)_2(ClO_4)_3$] (I) and [$Fe_2(O)_2(6TLA)_2(ClO_4)_2$] (II) were isolated and characterized in work [29]. Structure of a (-1,2-peroxo)bis(-carboxylato)diiron(III) model for the peroxo intermediate in the methane monooxygenase hydroxylase reaction cycle is presented in Fig. 12.8. Kim and Lippard [26] have synthesized and characterized a model for the peroxo intermediate in the methane monooxygenase hydroxylase reaction cycle, the complex $[Fe_2(\mu - 1,2 - O_2)\mu - O_2CCH2Ph)_2 - \{HB(pz')3\}2$ where pz′ = 3.5-bis(isopropyl)-pyrazol. In the complex, the two iron atoms are linked by two bridging phenyl acetate ligands and a peroxoligand is coordinated in a cis-$\mu - \eta 1 : \eta 1$ fashion [26].

The Shilov Cycle is a way to realize CH_4 oxidation in mild condition [30]. The Shilov system is catalyzed by $PtCl_2$ in an aqueous solution, where $PtCl_6{}^{2-}$ acts as the ultimate oxidizing agent. The cycle includes the following steps: (1) Electrophilic addition of CH_4 (or an alkane) to the Pt(II) center of chloroplatinate occurs in an aqueous medium; (2) Simultaneous deprotonation which leads to the formation of a Pt(II)–CH_3 complex; (3) Oxidation of the Pt(II)–CH_3 complex forms the Pt(IV)–CH_3 complex; and (4) Subsequent nucleophilic attack by H_2O molecules or Cl^- ions at the methyl group of the Pt(IV)–CH_3 complex which leads to the formation of methanol or methyl chloride (which is hydrolyzable to methanol), with simultaneous regeneration of the Pt(II).

To mimic dinuclear active sites of some nonheme diiron proteins, ten new polydentate and potentially dinucleating ligands containing a carboxylate moiety designed to bridge two metal atoms were synthesized [31]. The reaction of these ligands with $Fe(ClO_4)_3 \cdot 9H_2O$ leads to ferric μ-oxo-μ-carboxylato iron complexes [$Fe_2O(L)_2$ $(H_2O_2)_2](ClO_4)_2$ and [$Fe_2O(L)(BzO)](ClO_4)_2$ (L = ligand), containing one or two immobilized bridging carboxylates, respectively. Later it was shown [32] that iron complexes formed during the thermal treatment of FeZSM-5 zeolite perform single-turnover cycles of methane oxidation to methanol at ambient conditions when nitrous oxide can be used as a source of oxygen. The long-living active intermediate is capable of transferring an accepted O atom into a C-H bond of methane to produce methanol at 100% selectivity. The synthesis and dioxygen reactivity **of** diiron(II) tetracarboxylate complexes [Fe_2(mu-O_2CAr(Tol))(2)(O_2CAr(Tol))(2)(N,N-Me_2en)(2)] (2) and [Fe_2(mu-O_2CAr(Tol))(2)(O_2CAr(Tol))(2)(N,N-Bn(2)en)(2)] (6), where Ar(Tol)CO_2(-) = 2,6-di(p-tolyl)benzoate was described [33]. The experiment indicated that compound 6 reacted with dioxygen to afford PhCHO in 60(5)% yield, following oxidative N-dealkylation of the pendant benzyl group on the diamine ligand. The 4.2 K Mössbauer spectrum of the diiron(III) complex [Fe_2 (mu-OH)$_2$(mu-O_2CAr(Tol))(O_2CAr(Tol))(3)(N-Bnen)(N,N-Bn(2)en)] (8) displayed a single quadrupole doublet with parameters delta = 0.48(2) mm s^{-1} and $\Delta E(Q)$ = 0.61(2) mm s^{-1}. It is important that the $[Fe_2(mu\text{-}OH)(2)(mu\text{-}O_2CR)]^{3+}$ core structure in the diiron(III) complex $[Fe_2$ $(mu\text{-}OH)_2(mu\text{-}O_2CAr(Tol))(O_2CAr(Tol))(3)(N\text{-}Bnen)(N,N\text{-}Bn(2)en)]$ matches to the fully oxidized form of methane mono oxidase fragment MMOH.

The reaction between an Fe(III) complex and O_2 to afford a stable catalytically active diiron(IV)-mu-oxo compound was designed [35]. Phosphonium salts of orange five-coordinated Fe^{III}–TAML complexes with an axial aqua ligand ([PPh_4]**1**–H_2O, tetraamidato macrocyclic Fe^{III} species derived from 3,3,6,6,9,9-hexamethyl-3,4,8,9-tetrahydro-1*H*-1,4,8,11-benzotetraazacyclotridecine-2,5,7,10(6*H*,11*H*)-tetraone) reacted rapidly with O_2 in CH_2Cl_2 or other weakly coordinating solvents to produce black μ-oxo-bridged diiron(IV) complexes, **2**, in high yields. Main results of this work included the following: (1) Phosphonium salts of five-coordinated Fe(III)-TAML complexes with an axial aqua ligand ([PPh_4)]1-H_2O, tetraamidato macrocyclic Fe(III) species derived from 3,3,6,6,9,9-hexamethyl-3,4,8,9-tetrahydro-1H-1,4,8,11-benzotetraazacyclotridecine-2,5,7,10(6H,11H)-tetraone) reacted rapidly with O_2 in CH_2Cl_2; (2) Weakly coordinating solvents black mu-oxo-bridged diiron(IV) complexes, 2, were produced in high yields; (3) The complexes were characterized by X-ray crystallography microanalytical data, mass spectrometry, UV/Vis, Mossbauer, and (1)H NMR spectroscopies, showing the diamagnetic Fe-O-Fe unit contains antiferromagnetically coupled $S = 1$ Fe(IV) sites; (4) The oxidation of PPh_3 to $OPPh_3$ by 2 was confirmed by UV/Vis and gas chromatography–mass spectrometry (GC-MS); (5) The bridging oxygen atom of 2 that derives from O_2 was established in the labeling experiments with $^{18}O_2$ and $H_2{}^{18}O$; and (6) Complexes 2 catalyzed the selective oxidation of benzylic alcohols into the corresponding aldehydes and bleach rapidly organic dyes, such as Orange II in MeCN-H_2O mixtures.

Using electrochemical methods, Xue et al. have succeeded in preparing the first example of a synthetic complex possessing an $[FeIV_2(\text{-}O)_2]$ diamond core [36]. Specifically, a synthesis of an $[Fe(IV)_2(mu\text{-}O_2)]])$ has been performed in MeCN at – 40 °C $[Fe(III)Fe(IV)(mu\text{-}O)_2(L(b))_2]^{3+}$ complex exhibited potential $E_{(1/2)} =$ 760 mV versus Fc(+/0). Information about the structure and physical properties of this complex was obtained using a number of physical methods, namely: (1) A product of oxidation of 1b exhibited a visible spectrum with lambda(max) at 485 nm (9800 M^{-1} cm^{-1}) and 875 nm (2200 $M^{-1}cm^{-1}$); (2) The 4.2 K Mössbauer spectrum of 2b showed parameters typical of an iron(IV) center; (3) The Mössbauer patterns observed in strong applied fields indicated that 2b is an antiferromagnetically coupled diiron(IV) center; (4) Resonance Raman studies revealed the diagnostic vibration mode of the $[Fe_2(mu\text{-}O)_2]$ core; (5) Extended x-ray absorption fine structure (EXAFS) analysis confirmed structure of the complex $[Fe(IV)_2(mu\text{-}O)_2(L(b))_2{}^{4+}]$; and (6) A valence-delocalized $[Fe(III)Fe(IV)(mu\text{-}O)_2(L(b))_2]^{3+}$ complex exhibited potential $E_{(1/2)} = 760$ mV versus Fc(+/0).

The design and synthesis of metal organic framework (MOF) catalysts inspired by pMMO for selective methane oxidation to methanol was reported [37]. The structure of the catalysts active sites was confirmed using solid-state UV–vis, resonance Raman, and X-ray absorption spectroscopies complemented by density functional theory (DFT) calculations. Compound MOF-808 (Fig. 12.9) was utilized as a scaffold to host and stabilize highly active copper oxygen complexes, bis(μoxo) **dicopper** species. The complex was ligated to imidazole moieties including L-histidine (His), 4-imidazoleacrylic acid (Iza), and 5-benzimidazolecarboxylic acid (Bzz). Methane oxidation was conducted with an isothermal series of treatments at 150 °C., the average methanol productivity corresponded to 31.7 ± 13.0, 61.8 ± 17.5, and 71.8 ± 23.4 μmol gMOF-808-L-Cu^{-1} for MOF-808-His-Cu, MOF-808-Iza-Cu, and MOF-808-Bzz-Cu, respectively. This finding indicated that the MOF808-Bzz-Cu has the high methanol productivity. Combined spectroscopies and density functional theory calculations suggested that bis(μ-oxo) dicopper species as probable active site of the catalysts. The authors concluded that the resulting catalysts are capable of highly selective oxidation of methane to methanol under isothermal conditions. A series of iron(IV) oxo complexes, which differ in the donor groups (CH_2py or CH_2COO^- have been prepared and investigated in water at pH 2 and 7 [38]). Experiments revealed the following properties of the system under study: (1) The $\nu_{Fe=O}$ values of 832 ± 2 cm^{-1} indicate similar Fe^{IV}=O bond strengths; (2) Different reactivities toward C–H substrates in water; (3) Hydrogen atom transfer (HAT) occurs at rates that differ by one order of magnitude with KIEs (k_H/k_D = 30–66) consistent with hydrogen atom tunneling; (4) Higher KIEs correlation with faster reaction rates as well as a greater thermodynamic stability of the iron(III) resting states took place; (5) A doubling in rate from pH 7 to pH 2 for substrate C–H oxidation by the most potent complex, that with a *cis*-carboxylate donor, $[Fe^{IV}O(Htpena)]^{2+}$ was observed; (6) Supramolecular assistance by the first and second coordination spheres in activating the substrate was proposed; (7) The lifetime of this complex in the absence of a C–H substrate is the shortest; and (8) The iron(IV)=O complex bearing an alcohol moiety in the second coordination sphere *which* displayed significantly shorter lifetimes.

A computational design screen of over 2500 mid-row 3d transition-metal complexes with four metals in numerous spin and oxidation states **to** design strategies for spin-allowed methane-to-methanol catalysis was carried out [39]. The reaction energies for the radical rebound mechanism of methane-to-methanol conversion were calculated for mononuclear transition metal complexes with four mid-row 3d metals (M = Cr, Mn, Fe, Co) in two oxidation states (M(II)/M(III)). The computation demonstrated the importance of spin/oxidation state. Obtained weak coupling of energetics and design principles between reactions steps (e.g., oxo formation vs. methanol release) indicated that that linear free energy relationships LFERs are not fulfilled across the catalyst set. Suggested catalytic cycle for radical rebound oxidation of methane to methanol is displayed in Fig. 12.10.

In a recent review [40] simultaneously optimization methane activation and methanol release (Fig. 12.11, 1–4), calculating with machine learning-accelerated density functional theory in a space of 16 M candidate catalysts including novel macrocycles was performed. The large-scale search showed a compromise *that low-spin Fe(II) compounds paired with strong-field* between hydrogen atom transfer (HAT) and methanol release. By decoupling equatorial and axial ligand effects, it was determined that negatively charged axial ligands are critical for more rapid release of methanol. For example, M(III) versus M(II) are likely to be rate-limited

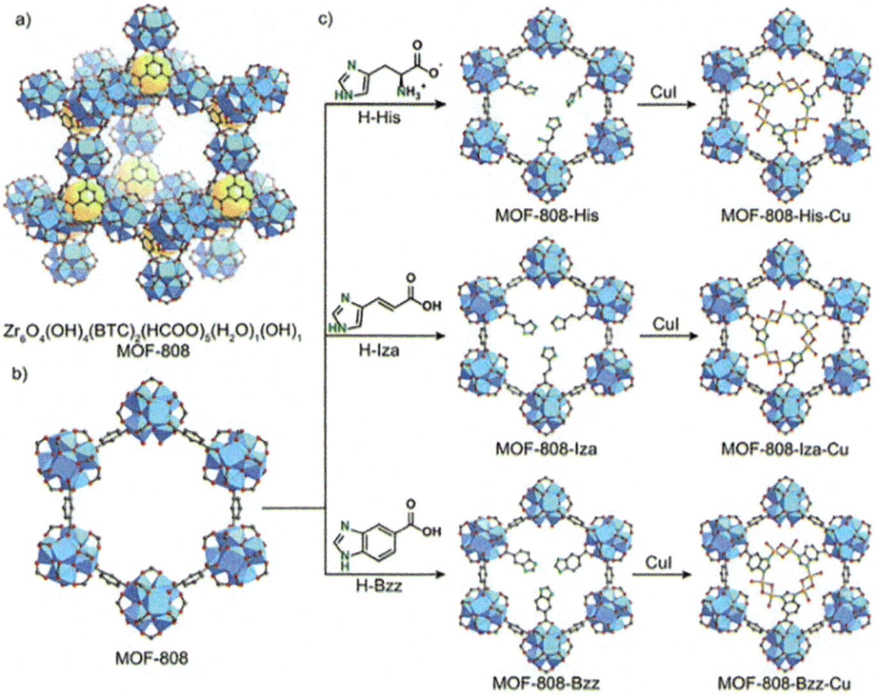

Fig. 12.9 Design and synthesis of the catalysts bearing copper–oxygen complexes in MOF-808 for methane oxidation to methanol. **a** Structure of MOF-808. **b** Pseudohexagonal pore opening of MOF-808. **c** Synthesis of the catalysts comprising the replacement of formate with imidazole-containing ligands and metalation with Cu(I). Atom labeling scheme: C, black; O, red; N, green; Cu, orange; Zr, blue polyhedra. H atoms are omitted for clarity. Orange spheres represent the space in the tetrahedral cages [37]. With permission from American Chemical Society

by slow methanol release. Energetic span analysis compared for an intermediate-spin Mn(II) catalyst and a low-spin Fe(II) catalyst.

A cyclophane

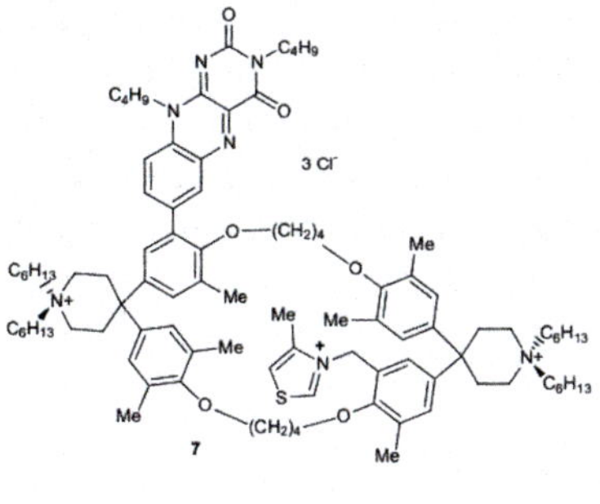

7

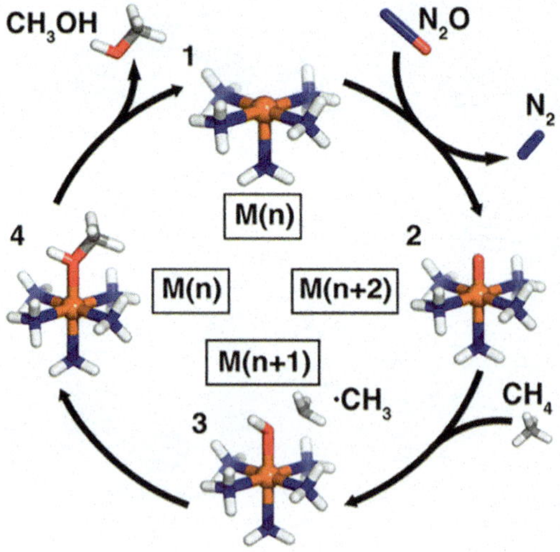

Fig. 12.10 Catalytic cycle for radical rebound oxidation of methane to methanol. From the resting state (**1**, top) in oxidation state $n = \text{II/III}$, the cycle proceeds clockwise (intermediate formal oxidation state indicated in adjacent boxes): metal–oxo (**2**, right) formation with N_2O oxidant, hydrogen atom transfer to a metal–hydroxo (**3**, bottom), and rebound to form methanol (**4**, left). A representative $Fe(II)(NH_3)_5$ catalyst is shown in ball and stick colored as follows: Fe in brown, N in blue, O in red, C in gray, and H in white [39]. Published under Creative Commons public use license CC-BY from American Chemical Society

is a hydrocarbon consisting of an aromatic unit (typically a benzene ring) and a chain that forms a bridge between two nonadjacent positions of the aromatic ring. Cyclophanes have been used as enzyme mimics; *because their geometric and electronic properties can be easily controlled.* Diederich and coworkers have designed a cyclophane-based pyruvate oxidase model [42]. Flavo-thiazolio-cyclophane was prepared by an 18-step synthesis. This pathway involved the following steps: (1) Preparation of bromo-cyclophane (37% yield over 13-steps); (2) The bromide was subsequently converted into the corresponding boronic acid and connected to the 7-bromoflavin via Suzuki coupling to give flavo-cyclophane; and (3) The thiazolium unit was introduced after quaternization of the tertiary amino groups; forming flavo-thiazolio-cyclophane , with both prosthetic groups attached in proximity to the cyclophane binding site; and (4) In basic methanolic solution, Cyclophane 7 catalyzed the oxidation of aromatic aldehydes to their corresponding methyl esters, indicating the turnover number calculated for the oxidation of naphthalene-2-carbaldehyde to methyl naphthalene-2-carboxylate ($k_{cat} = 0.22\ s^{-1}$).

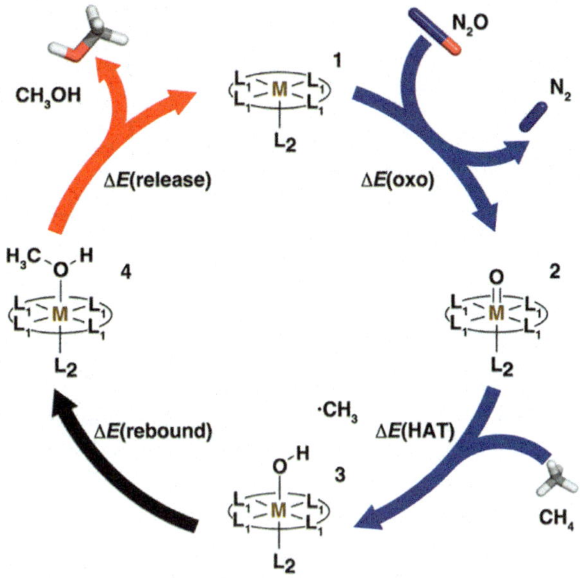

Fig. 12.11 Radical rebound mechanism for the partial oxidation of methane to methanol. The cycle proceeds clockwise from the resting state (**1**) in oxidation state n = II/III to the metal-oxo intermediate (**2**) formed by two-electron oxidation with N_2O, followed by HAT to form a metal-hydroxo intermediate (**3**) and rebound to form a metal-bound methanol intermediate (**4**). A representative catalyst is shown, with the metal (M) shown in brown corresponding to Mn or Fe in this work. All catalysts have a tetradentate equatorial ligand (L_1) and a monodentate axial ligand (L_2). We color the arrows of steps that have been observed to be turnover-determining [40].

12.4 Artificial Hydrolases

The most promising direction for enzyme modeling is synthetically mimetic nature of the binding site and the active site in terms of the close similarity of catalytic groups, stereochemistry, interatomic distances and the mechanism of the action of the enzyme [41–61].

Review [49] explored research in the development of synthetic catalysts based on the chemistry of hydrolytic enzymes published by 2019. Active site chemistry, tuning transition state interactions, and establishing reactive nanoenvironments conducive to attracting, binding, and releasing target molecules were in the review focus. In a typical early example, mimicking of the "proton-transfer relay", proposed for the mechanism of the action of chymotrypsin, was performed. The miniature organic model of chymotrypsin was designed on the basis of cyclodextrin (CD) which is cyclic oligosaccharide consisting of α-1,4-linked (Fig. 12.12) [41]. The catalytic activity of "artificial chymotrypsin" (Fig. 10.12) in the hydrolysis of m-tert-butylphenyl acetate ($k_{cat} = 2.8 \times 10^2\ s^{-1}$, $K_M = 13 \times 10^5$ M) was found to be close to the activity of chymotrypsin in the hydrolysis of p-nitrophenyl acetate (k_{cat}

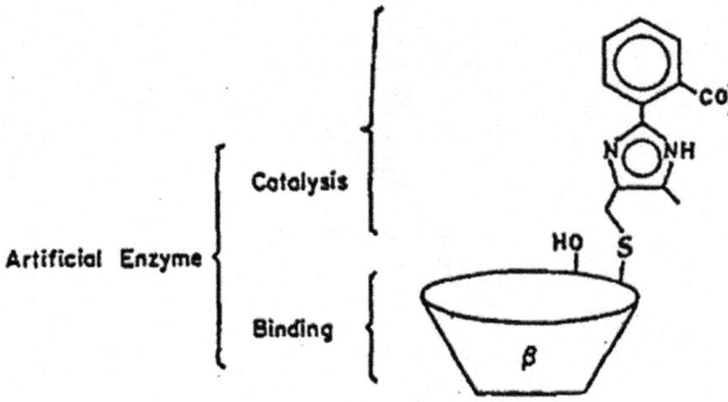

Fig. 12.12 Mechanism of action of "artificial chymotrypsin" (**a**); complete model of acylchymotripsin (**b**) and miniature organic model of chymotrypsin [41]. With permission from American Chemical Society

$= 1.1 \times 10^2$ s^{-1}, $K_M = 4 \times 10^5$ M). Another example of mimicking enzyme catalysis by β-cyclodextrin is general acid-base-catalyzed hydrolysis and nitrosation of amines by alkyl nitrites [60]. Addition of β-CD strongly inhibits yв the acid hydrolysis of alkyl nitrite. The retardation of the reaction was attributed to a separation of the reagents: β-CD and alkyl nitrites form host–guest 1:1 inclusion complexes. The strong catalysis observed with 1-phenyl-1-propyl nitrite upon the addition of DTABr was indicative of an example of allosteric activation (Fig. 12.12).

A nucleosidase hydrolyzes the bonds of nucleotides T. Carbon-based nanozymes were firstly discovered to mimic natural nucleases in the 1990s [61]. Water-soluble fullerene

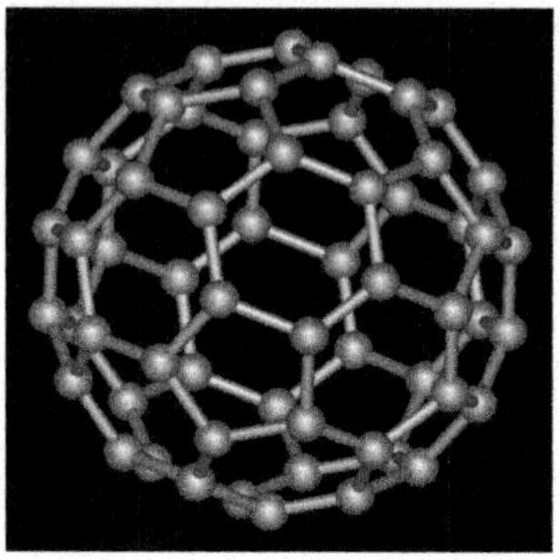

functionalized with carboxylic acid moieties, C60-1, was utilized to catalyze the cleavage of the phosphodiester bond of DNA when irradiated by light.

A *Pyrococcus furiosus* Argonaute (*Pf*Ago)-based platform for generating artificial restriction enzymes (AREs) capable of recognizing and cleaving DNA sequences was developed [48]. The Argonaute protein family plays a central role in RNA silencing processes as essential components of the RNA-induced silencing complex. Short DNA guides were used to direct *Pf*Ago to target sites for cleavage at high temperatures (> 87 °C) followed by reannealing of the cleaved single stranded DNAs. Peptide dendrimers with one to three amino acids between branches were prepared with up to 38 amino acids (MW ~ 5000 Da) [47]. Structural studies of peptide dendrimers by circular dichroism (CD), Fourier transform infrared (FT-IR), NMR and molecular dynamics revealed molten globule states containing up to 50% of α-helix. The Fourier transform (FT) is a transform that converts a function into a form that describes the frequencies present in the original function. Mimicking esterase and aldolase peptide dendrimers displayed enzyme kinetics ($k_{cat}/k_{uncat} \sim 10^5$) were screening from large combinatorial libraries. Combinatorial chemistry comprises chemical synthetic methods in which compound libraries can be made as mixtures, sets of a large number individual compounds or chemical structures generated by computer software. Efficient ligands for cobalamin, cytotoxic colchicine conjugates and antimicrobial peptide dendrimers were developed. In study [50], a peptide-based artificial hydrolase with photoswitchable activity was fabricated by the introduction of a catalytic histidine residue and a photoresponsive azobenzene group into the peptide chain. The activity of the hydrolase mimic was controlled using UV and visible light. An enhanced catalytic activity on *p*-nitrophenyl acetate was observed due to the proximity effect of the histidine residues and the hydrophobic microenvironment in the supramolecular assemblies.

The functioning of two chemically promiscuous mono- and binuclear metalloenzymes (insulin degrading enzyme and glycerophosphodiesterase) and two synthetic analogues was examined [52]. Several studies in which a plethora of computational techniques have been successfully employed were carried out. To model mechanism of the activity of hydrolytic enzymes, the conformation of an unstructured bimetallic Zn(II) complex was controlled by its inclusion in the cavity of a γ-cyclodextrin [54]. For this purpose, a catalytic bimetallic site for the hydrolytic cleavage of the RNA model substrate 2-hydroxypropyl-4-nitrophenyl phosphate (HPNP), whose reactivity is 30-fold larger with respect to the unstructured complex, was constructed (Fig. 12.13). Competitive inhibition with 1-adamantanecarboxylate

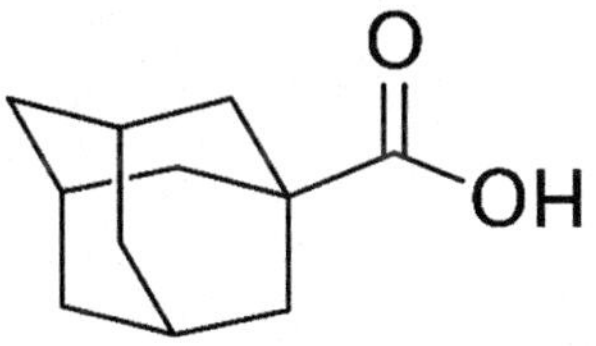

displaces the metal complex from the cyclodextrin decreasing the reactivity.

A three-dimensional geometry analysis of the key catalytic residues in natural hydrolases has been exploited to design and synthesize small-molecule artificial enzymes which mimic the active centers of chymotrypsin and N-terminal hydrolases (Fig. 12.14) [55]. DFT studies and X-ray diffraction analysis of the catalyst bound to a transition state analogue proved the similarity with the geometry of natural hydrolases. The optimized prototype catalyzed the methanolysis of the acyl enzyme mimic with a half-life of 3.7 min at 20 °C and also perform the transesterification of vinyl acetate at room temperature.

A few examples in which the performance of a metallocatalyst for the cleavage of a phosphate diester model of RNA (2-hydroxypropyl nitrophenyl phosphate, HPNP) [56]:

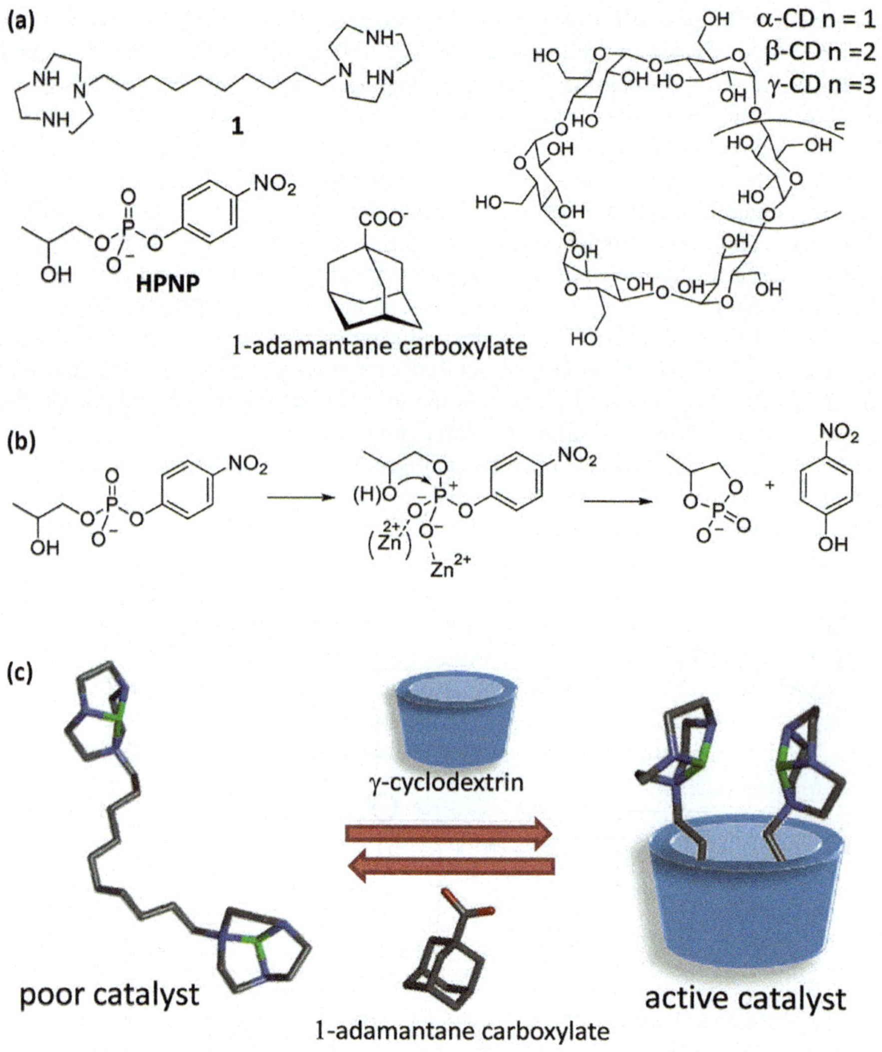

Fig. 12.13 **a** Ligand **1** and other compounds used in this work; **b** general mechanism for the HPNP transesterification catalyzed by mono- and dimetallic Zn(II) complexes; **c** proposed mechanism for the allosteric modulation of the activity of catalyst studied (the structure drawn are molecular models with hydrogens omitted) [54]. Published under a Creative Commons public use license CC-BY from American Chemical Society

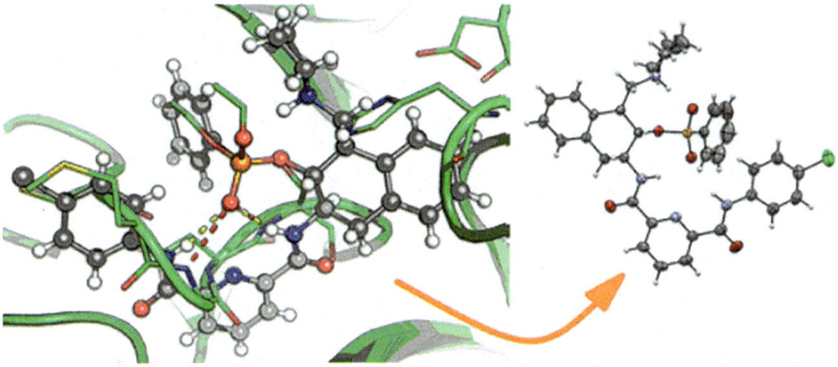

Fig. 12.14 Enzyme model which mimics chymotrypsin and N-terminal hydrolases [55]. Published under a Creative Commons public use license CC-BY from American Chemical Society

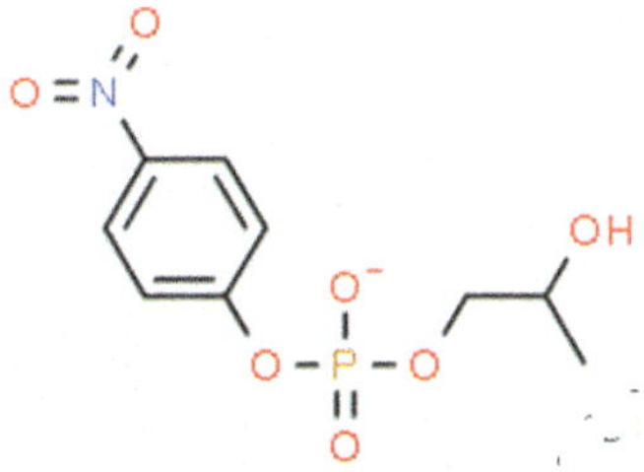

were reported. The number of metal ions present in the catalyst, their precise orientation to elicit cooperativity, the introduction of hydrogen bond donors, and the control of the solvating medium properties were taking in consideration. An important role of these metal ions in decreasing the pKa of the nucleophilic species was stressed. A comparison of catalysts 1 and 2 (Fig. 12.15), red arrow shows that the dinuclear catalyst is 34- or 200-fold better than the mononuclear one when k_{cat} or k_2 (k_{cat}/K_M) is considered, respectively.

Selected examples of structure of enzyme mimetic catalysts are indicated in Fig. 12.16 [56].

A molecular device that can selectively hydrolyze glycerophospholipids (GPLs) under physiologically relevant conditions was engineered [57]. A chemically modified α-cyclodextrin bearing amine functional groups was shown to hydrolyze lyso-GPLs, generating free fatty acids (Fig. 12.17). Lyso-GPLs are preferentially hydrolyzed when part of a mixture of GPL lipid species, and reaction efficiency was dependent on lyso-GPL chemical structure.

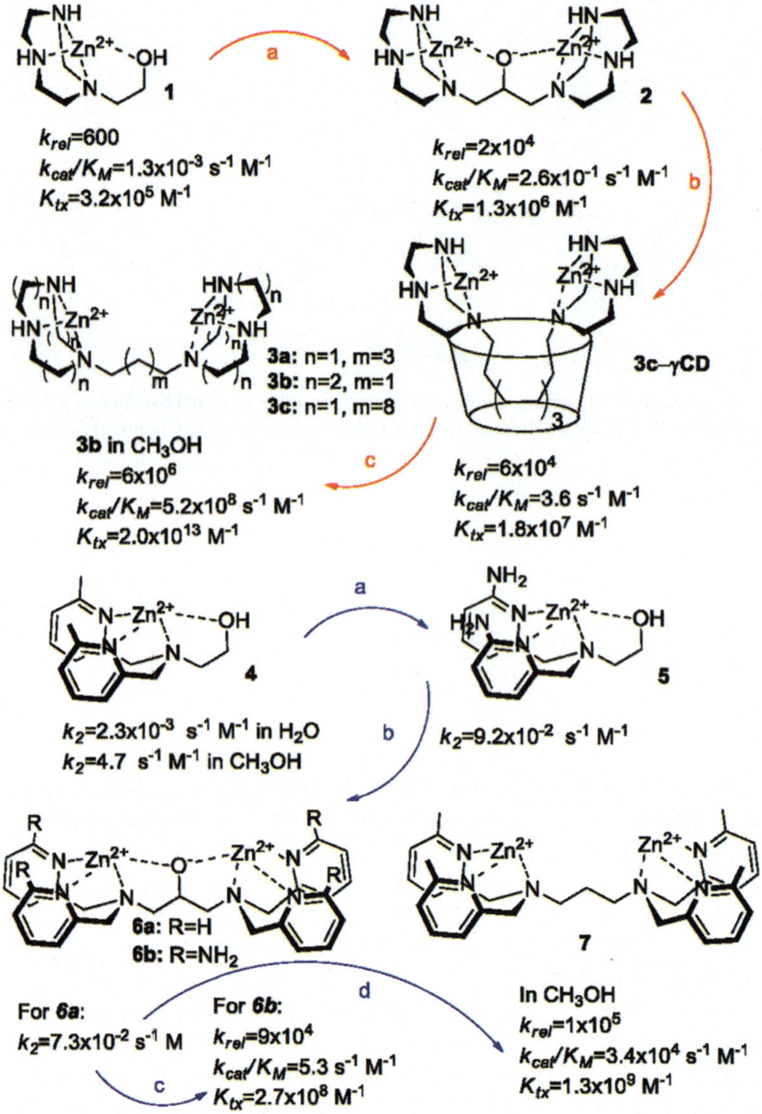

Fig. 12.15 Evolution of the efficiency of simple synthetic metallocatalysts for the cleavage of the RNA model substrate HPNP. The red and blue arrows are meant to guide the eye in following the progressive improvement of the catalysts. In red: **a** from one to two metals; **b** the need for structure organization; **c** solvent effect. In blue: **a** the role of proton donors; **b** from one to two metal ions; **c** merging (**a**) and (**b**); **d** solvent effect [56]. Published under a Creative Commons public use license CC-BY from American Chemical Society

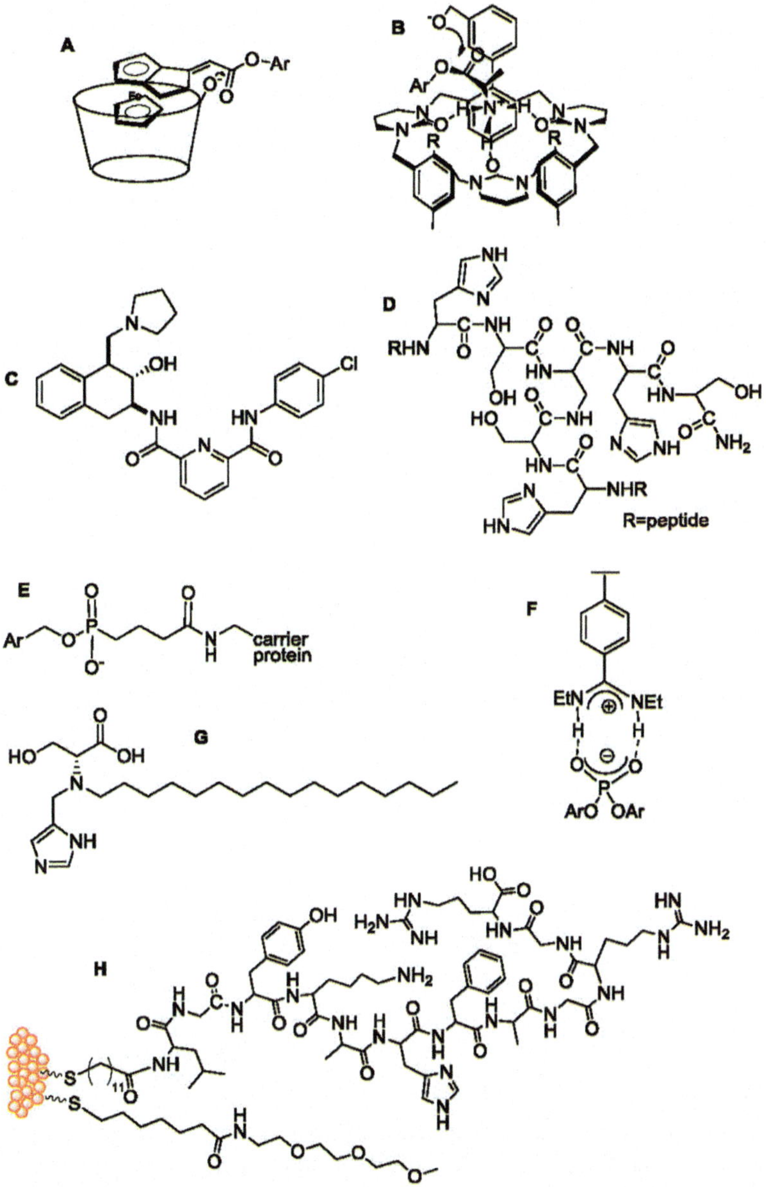

Fig. 12.16 Structure of enzyme mimetic catalysts: **a** β-cyclodextrin with ferrocenyl derivative; **b** macrocyclic host with a complexed protonated alanine *p*-nitrophenyl ester substrate; **c** rigid receptor for methanolysis; **d** dendrimers; **e** hapten used for the development of catalytic antibodies for ester hydrolysis; **f** catalytic site of an imprinted polymer with transition state analogue for carbamate hydrolysis; **g** functional peptide for ester hydrolysis; **h** peptide-functionalized gold nanoparticle for ester hydrolysis [56]. Published under a Creative Commons public use license CC-BY from American Chemical Society

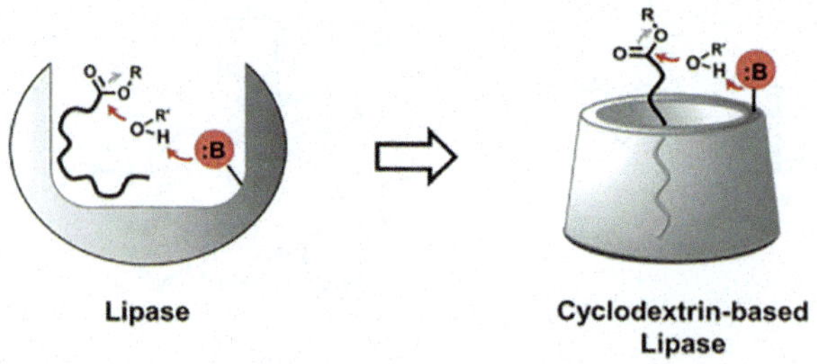

Fig. 12.17 Schematic representation showing the development of lipase mimetic cyclodextrins (CDs) inspired by the mechanism of base promoted hydrolysis in the active site of natural lipases [B: base, R: alkyl group, R′: alkyl group or proton] [57]. Article is licensed under a Creative Commons Attribution 3.0 Unported Licence

Fig. 12.18 Natural and artificial catalytic triads [58]. With permission from Elsevier

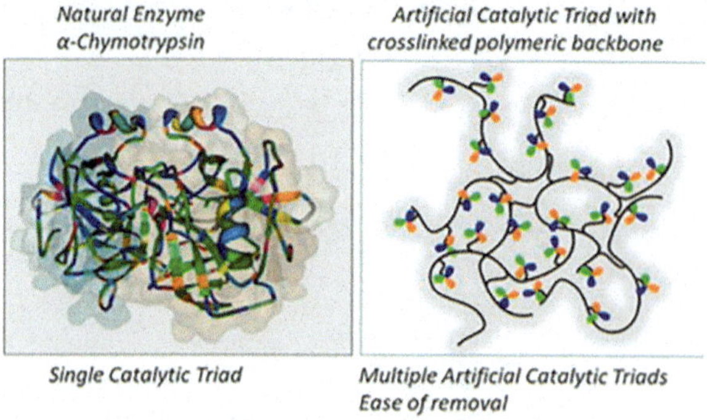

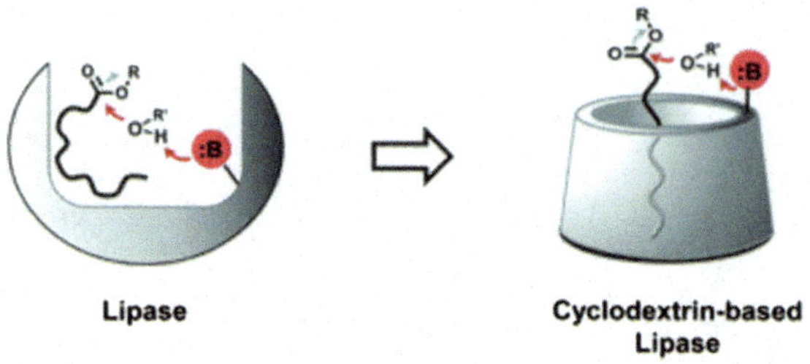

To modeling nature's catalytic triad, consisting of a hydroxyl group of a serine residue, an imidazole group of a histidine residue, and a carboxyl group of an aspartic acid, a polymer scaffolded enzyme mimic was incorporated with artificial catalytic triad units (Fig. 12.18) [58]. The experiments were supported by molecular dynamics simulations. Compared to a water soluble artificial catalytic triad, enhanced esterolytic activity of p-nitrophenyl benzoate was observed for the polymeric catalyst. It was suggested that this effect appears due to the hydrophobic microenvironment formed by the tertiary structure of the polymer backbone.

References

1. Esteves F, Rueff J, Kranendonk V (2021) The central role of cytochrome P450 in xenobiotic metabolism—a brief review on a fascinating enzyme family. J Xenobiot 11(3):94–114. https://doi.org/10.3390/jox11030007
2. Brunold TS (2007) Synthetic iron-oxo "diamond core" mimics structure of key intermediate in methane monooxygenase catalytic cycle. PNAS 104(52): 20641–20642
3. Mansuy D (2007) A brief history of the contribution of metalloporphyrin models to cytochrome P450 chemistry and oxidation catalysis. C R Chim 10(4–5):392–413. https://doi.org/10.1016/j.crci.2006.11.001
4. Balogh GT, Keserű GM (2004) Metalloporphyrin mediated biomimetic oxidations. A useful tool for the investigation of cytochrome P450 catalyzed oxidative metabolism. ARKIVOC 2004, pp 124–139
5. Groves JT, Van der Puy M (1976) Stereospecific aliphatic hydroxylation by iron-hydrogen peroxide. Evidence for a stepwise process. J Am Chem Soc. 98:5290–5297
6. Huang X, Groves JT (2016) Beyond ferryl-mediated hydroxylation: 40 years of the rebound mechanism and C-H activation. J Biol Inorgan Chem
7. Groves JT, Ortiz de Montellano PR (ed) (2005) Cytochrome P450: structure, mechanism and biochemistry (third ed.). Kluwer Academic/Plenum Publisher, New York, p 1
8. Seth R, Bell SR, Groves JT (2009) A highly reactive P450 model compound I. J Am Chem Soc 131(28):9640–9641
9. Groves JT, Haushalter RC, Nakamura M et al (1981) The related to peroxidase and cytochrome P-450. J Am Chem Soc 102:2884
10. Nwosu CG, Crooks PA (1988) Metalloporphyrin mediated biomimetic oxidations. A useful tool for the investigation of cytochrome P450 catalyzed oxidative metabolism. Xenobiotica 31:637
11. Bartoli JF, Brigaud O, Battioni P, Mansuy D (1991) Hydroxylation of linear alkanes catalysed by iron porphyrins: particular efficacy and regioselectivity of perhalogenated porphyrins. J Chem Soc Chem Commun 440–442
12. Gotardo MCAF, de Moraes LAB, Assis MD (2006) Metalloporphyrins as biomimetic models for cytochrome P-450 in the oxidation of atrazine. J Agric Food Chem 54(26):10011–10018
13. Shive MS (2007) Chauhan, Pratibha Kumari: Biomimetic oxidation of metribuzin with hydrogen peroxide catalyzed by 5,10,15,20-tetraarylporphyrinatoiron(III) chlorides. Tetrahedron Lett 48(29):5035–5038
14. Rebelo SLH, Pereira MM, Monsanto PV, Burrows HD (2009) Catalytic oxidative degradation of s-triazine and phenoxyalkanoic acid based herbicides with metalloporphyrins and hydrogen peroxide: Identification of two distinct reaction schemes. J Mol Catal A 297(1):35–43
15. Fukushima M, Katagi T (2009) Iron-Porphyrin catalyzed oxidation of permethrin and fenvalerate. J Pestic Sci 34(4):241–252
16. Guo M, Dong H, Li J, Cheng B et al (2012) Spectroscopic observation of iodosylarene metalloporphyrin adducts and manganese(V)-oxo porphyrin species in a cytochrome P450 analogue. Nat Commun 3:1190
17. Liu W, Cheng M, Nielsen RJ, Goddard WA, Groves JT (2017) Probing the C–O bond-formation step in metalloporphyrin-catalyzed C–H oxygenation reactions. ACS Catal 7:4182–4188
18. Fallah AH, Endud S, Alizadeh A, Chien LS (2018) Metalloporphyrin/dendrimer-decorated MCM-41 biomimetic hybrid catalysts: high stability combined with facile catalyst recyclability. J Porous Mater 25(6):1813–1823
19. Liu W, Huang X, Placzek MS et al (2018) Site-selective F fluorination of unactivated C–H bonds mediated by a manganese porphyrin. Chem Sci 9:1168–1172
20. Huang X, Groves JT (2017) Oxygen activation and radical transformations in heme proteins and metalloporphyrins. Chem Rev. https://doi.org/10.1021/Acs.Chemrev.7B00373
21. Dos Santos JS, Palaretti V, De Faria AL, Crevelin EJ, De Moraes LA, das Dores Assis M, (2021) Biomimetic simazine oxidation catalyzed by metalloporphyrins. Appl Catal A General 408(1–2):163–170. https://doi.org/10.1016/j.apcata.2011.09.023

22. Lage AL, Ribeiro JM, de Souza-Fagundes EM et al (2019) Efficient atrazine degradation catalyzed by manganese porphyrins: determination of atrazine degradation products and their toxicity evaluation by human blood cells test models. J Haz Mater 378:120748
23. Li Z, Jiang Y, Peter Guengerich XF et al (2020) Engineering cytochrome P450 enzyme systems for biomedical and biotechnological applications. J Biol Chem 295(3):833–849
24. Porter TR, Mayer JM (2014) Radical reactivity of the fetetramesitylporphyrin couple: hydrogen atom transfer, oxyl radical dissociation, and catalytic disproportionation of a hydroxylamine. Chem Sci 5(1):372–380. https://doi.org/10.1039/C3SC52055B
25. Gao H, Groves JT (2017) Fast hydrogen atom abstraction by a hydroxo iron(iii) porphyrazine. J Am Chem Soc. https://doi.org/10.1021/Jacs.6B13091. Jalalvand AR, Gholivand M-B, Goicoechea HC, Thomas Skov T, Kamran Mansouri K (2015) Mimicking enzymatic effects of cytochrome P450 by an efficient biosensor for in vitro detection of DNA damage Int J Biol Macromol 79:1004–1010
26. Huang X, Groves JT (2018) Oxygen activation and radical transformations in heme proteins and metalloporphyrins. Chem Rev. https://doi.org/10.1021/Acs.Chemrev.7B00373.. Kim K, Lippard SJ (1996) Structure and Mössbauer spectrum of a (μ-1,2-Peroxo)bis(μ-model for the peroxointermediate in the methan monooxygenase hydroxylase reacton circle. J Am Chem Soc 118:4914-4915
27. Anggoro DD, Chamdani ET, Buchori L (2021) One step catalytic oxidation process of methane to methanol at low reaction temperature: a brief review. IOP Conf Ser Mater Sci Eng 1053:012056
28. Sher Shah MS, Oh C, Park H, Hwang YJ, Ma M, Park JH (2020) Catalytic oxidation of methane to oxygenated products: recent advancements and prospects for electrocatalytic and photocatalytic conversion at low temperatures. Adv Sci 7(23):2001946
29. Zang Y, Dong Y, Que L Jr., Kauffmann K, Münck E (1995) The first Bis(m-oxo)diirin complex. Structure and magnetic properties of $[Fe_2(\mu\text{-}O)_2(TLA)_2(ClO_4)_2$. J Am Chem Soc 117:1169–1170
30. Shilov AE, Shul'pin GB (1997) Activation of C–H bonds by metal complexes. Chem Rev 97:2879
31. Trukhan VM, Polukhov VV, Sulimenkov IV et al (1998) First structural-functional model of methane monooxygenase. Kinet Catal 39:788–791
32. Ovanesyan NS, Dubkov KA, Pyalling AA, Shteinman AA (2000) The Fe active sites in FeZSM-5 catalyst for selective oxidation of CH_4 to CH_3OH at room temperature. J Radioanal Nucl Chem 246:149–152
33. Lee D, Lippard SJ (2002) Synthetic analogue of the $[Fe_2(\mu\text{-}OH)_2(\mu\text{-}O_2)CR_{3+})$ core of soluble methane monooxygenase hydroxylase via synthesis and dioxygen reactivity of carboxylate-bridged diiron(II) complexes. Inorg Chem 41(4):827–837
34. Collins TJ (2002) TAML oxidant activators: a new approach to the activation of hydrogen peroxide for environmentally significant problems. Acc Chem Res 35:782–790
35. Ghosh A, de Oliveira FT, Yano T et al (2005) Catalytically active mu-oxodiiron(IV) oxidants from iron(III) and dioxygen. J Am Chem Soc 127:2505–2513
36. Xue G, Wang D, De Hont R, Fiedler AT, Shan X, Münck E, Que L (2007) A synthetic precedent for the $[FeIV_2(\mu\text{-}O)_2]$ diamond core proposed for methane monooxygenase intermediate Q. Proc Natl Acad Sci USA 104:20713–20718
37. Baek J, Rungtaweevoranit B, Pei X (2018) Bioinspired metal-organic framework catalysts for selective methane oxidation to methanol. J Am Chem Soc 140(51):18208–18216
38. Wegeberg C, Skavenborg ML et al (2021) Engineering the oxidative potency of non-heme iron(IV) oxo complexes in water for C–H oxidation by a cis donor and variation of the second coordination sphere. Inorgan Chem 60(3):1975–1984
39. Nandy A, Kulik HJ (2020) Why conventional design rules for C–H activation fail for open-shell transition-metal catalysts. ACS Catal 10(24):15033–15047
40. Nandy A, Duan C, Goffinet C, Kulik HJ (2022) New strategies for direct methane-to-methanol conversion from active learning exploration of 16 million catalysts. JACS 5:1200–1213

41. D'Souza VT, Bender ML (1987) Miniature organic models of enzymes. Acc Chem Res 20:146–152
42. Mattei P, Diederich F (1997) Catalytic Cyclophanes. Part XI. A flavo-thiazolio-cyclophane as a biomimetic catalyst for the preparative-scale electro-oxidation of aromatic aldehydes to methyl esters. Helv Chim Acta 80:1555–1588
43. Breslow R, Dong SD (1998) Biomimetic reactions catalyzed by cyclodextrins and their derivatives. Chem Rev 98(5):1997–2012
44. Bruice TC, Benkovic S (2000) Chemical basis for enzyme catalysis. Biochemistry 39:6267–6274
45. Breslow R (ed) (2005) Artif enzymes. Wiley-VCH Verlag GmbH & Co. KGaA, Weinheimm. Print ISBN: 3-527-31165-1
46. Chin J, Kim H (2005) Artificial hydrolytic metalloenzymes in artificial enzymes. Edited by Ronald Breslow. Wiley-VCH Verlag GmbH & Co, KGaA, Weinheim. Print ISBN: 3527-31165-1A
47. Reymond J-LL, Darbre T (2012) Peptide and glycopeptide dendrimer apple trees as enzyme models and for biomedical applications. Org Biomol Chem 10:1483–1492
48. Enghiad B, Zhao H (2017) Programmable DNA-guided artificial restriction enzymes. ACS Synth Biol 6(50):752–757
49. Nothling MD, Xia Z, Bhaskaran A et al (2019) Synthetic catalysts inspired by hydrolytic enzymes. ACS Catal 9:168–187. https://doi.org/10.1021/acscatal.8b03326REVIEW
50. Zhao Y, Lei B, Wang M, Wu S, Qi W, Su R, He Z (2018) A supramolecular approach to construct a hydrolase mimic with photo-switchable catalytic activity. J Mater Chem B 6(16):2444–2449
51. Mathew D, Thomas B, Devaky KS (2019) Design, synthesis and characterization of enzyme-analogue-built polymer catalysts as artificial hydrolases. Artif Cells Nanomed Biotechnol 47(1):1149–1172
52. Hu Q, Jayasinghe-Arachchige VM, Sharma G et al (2020) Mechanisms of peptide and phosphoester hydrolysis catalyzed by two promiscuous metalloenzymes (insulin degrading enzyme and glycerophosphodiesterase) and their synthetic analogues. Wiley Interdiscip Rev Comput Mol Sci 10:e1466
53. Astruc D, Boisselier E, Ornelas C (2020) Dendrimers designed for functions: from physical, photophysical, and supramolecular properties to applications in sensing, catalysis, molecular electronics, photonics, and nanomedicine. Chem Rev 110:1857–1959
54. Czescik J, Lyu Y, Neuberg S, Scrimin P, Mancin F (2020) Host–guest allosteric control of an artificial phosphatase. J Am Chem Soc 142(15):6837–6841
55. Garrido-Gonzalez JJ, Iglesias Aparicio MM, García MM, et al (2020) An enzyme model which mimics chymotrypsin and n-terminal hydrolases. ACS Catal 10(19):11162–11170
56. Lyu Y, Scrimin P (2021) Mimicking enzymes: the quest for powerful catalysts from simple molecules to nanozymes. ACS Catal 11(18):11501–11509
57. Lee Y, Devaraj NK (2021) Lipase mimetic cyclodextrins. Chem Sci 12(3):1090–1094
58. Bhaskaran A, Aitken HM, Xiao Z et al (2021) Enzyme inspired polymer functionalized with an artificial catalytic triad. Polymer 225:123735
59. Lee Y , Devaraj NK (2021) Lipase mimetic cyclodextrins. Chem Sci
60. Iglesias E (1998) Cyclodextrins as enzyme models in nitrosation and in acid-base-catalyzed reactions of alkyl nitrites. J Am Chem Soc 120:13057–13069
61. Tokuyama H, Yamago S, Nakamura E, Shiraki T, Sugiura Y (1993) Photoinduced biochemical activity of fullerene carboxylic acid. J Am Chem Soc 115:7918–7919. https://doi.org/10.1002/3527606645.ch5

Chapter 13
Miscellaneous Biomimetic Processes

Abstract The key role of nitric oxide (NO) in numerous vital biological processes, including the human body in normal and pathological conditions, poses an urgent problem of dosed and regulated flow of objects of interest. *S*-Nitrosothiols have received much attention in the chemistry of NO-based signaling in living systems, because they serve as donors of both the nitrosonium ion NO+ and of NO. S-Nitrosoglutathione (GSNO) mimics the beneficial activity of endothelial Nitric Oxide Synthase-derived nitric oxide in organisms. For example, administration of exogenous of GSNO or enhancing the level of endogenous GSNO protects against neurovascular injury (endothelial dysfunction) of mouse models, reduces brain infarctions and edema, improves neurobehavioral functions, attenuates cerebral ischemia-reperfusion (IR) injury in a mice, exhibits anti-ischemic and anti-inflammatory functions, and stimulates neurorepair and functional recovery following experimental stroke in rats. Due to the possibility of synthesizing almost unlimited number of the iron–sulfur–nitrosyl complexes (ISNC) of various structures, stability, and bioavailability, they can be used for predictable regulation of synthesis of nitric oxide in biological systems. The complexes showed encouraging results and their therapeutic potential are subject of discussion. Principles and recent developments of catalysis by abzyme, a monoclonal antibody with catalytic activity of a corresponding enzyme and their potential applications of catalytic antibodies in reactive immunization, therapy, biochemical analysis, and biotechnology were described. Data on artificial enzymes of common interest not included in previous chapter are discussed in a separate section.

13.1 Nitric Oxide Producing Activity

Nitric oxide (NO) is a vertebrate biological messenger, playing a key role in a variety of biological processes being a bioproduct in almost all types of organisms, including bacteria, plants, fungi, and animal cells [1–3]. Nitric oxide synthase (NOS) is an enzyme catalyzing the formation of L-citrulline and nitric oxide from L-arginine (Section X). Small molecular NO donors usually show lack stability and have a short

G. I. Likhtenshtein, *Enzyme Catalysis Today and the Chemistry of the 21st Century*,
https://doi.org/10.1007/978-3-031-51552-1_13

half-life in human tissues, limiting the therapeutic application. The dosed production of NO and its effect on physiological processes is an important factor and an actual challenge in modern biomedicine (Section X). The past decade has seen the advent of a new paradigm in controlled delivery of NO, namely localized bioconversion of the endogenous prodrugs of NO, specifically by enzyme mimics.

13.1.1 S-Nitrosoglutathione as NO Donor

S-Nitrosoglutathione (GSNO)

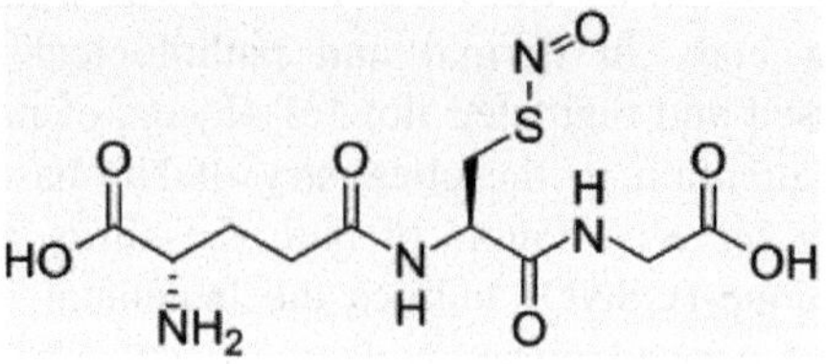

is an endogenous *S*-nitrosothiol (SNO) that plays a critical role in nitric oxide signaling and is a source of bioavailable NO. *S*-Nitrosothiols serve as a good source of nitric oxide due to the ease of cleavage of the S-N bond which consequently produces NO [4–15].

The reductive decomposition of GSNO by L-ascorbic acid yields NO which was monitored both electrochemically and spectrophotometrically [4]. The kinetics experiments showed several important results, namely: (1) The decomposition rate increases with pH demonstrating that the L-ascorbic acid dianion (A(2-)) is the most reactive species of L-ascorbic acid (H(2)A); (2) the derived rate expression, measuring the disappearance of the absorption at ca. 336 nm due to GSNO), was established as rate = $-d[GSNO](t)/dt$ = $((k(a)[H^+]^2 + k(b)[H^+]K(1) + k(c)K(1)K(2))/([H^+]^2 + K(1)[H^+] + K(1)K(2)))[GSNO](t)[H(2)A](t)$ where k(a), k(b), and k(c) are second-order rate constants via the H(2)A, HA(−), and A(2−) pathways, respectively. And K(1) and K(2) represent the first and second equilibrium dissociation constants of L-ascorbic acid; (3) at 25 °C, k(b) = $5.23 \pm 1.47 \times 10^{-3}$ $dm^3\ mol^{-1}\ s^{-1}$ and k(c) = $1.22 \pm 0.04 \times 10^3$) $dm^3\ mol^{-1}\ s^{(-1)}$, activation parameters ΔH (b) = 54.4 ± 4.3 kJ mol^{-1}, ΔS (b) = -106 ± 16 J $K^{-1}\ mol^{-1}$, ΔH (c) = 80.5 ± 7.5 kJ mol^{-1}, ΔS (c) = 84 ± 7 kJ mol^{-1}; (4) induced catalytic decomposition of GSNO becomes significant above ca. 10 μM Cu(2+); and (5) to nullify any catalysis by Cu^{2+}, EDTA was added to all experimental reactions except those where catalysis by Cu^{2+} was studied.

Studies on brain and spinal cord (a long, thin, tubular structure made up of nervous tissue) trauma using a rat model of cerebral ischemia and reperfusion (IR) showed that exogenous treatment with the S-nitrosylating agent S-nitrosoglutathione (GSNO)—a nitric oxide and glutathione metabolite of the human body—stimulates neurorepair and aids functional recovery [9]. The testing of a hypothesis that GSNO invokes the neurorepair process and improves neurobehavioral functions through the angiogenic HIF-1α/VEGF pathway indicated that GSNO stimulates the process of neurorepair, promotes angiogenesis, and aids functional recovery through the HIF-1α-dependent pathway, showing therapeutic and translational promise for stroke. In work [13], experiments were performed to determine whether the nitric oxide donor, S-nitrosoglutathione, may exert different effects on diabetic and nondiabetic myocardial ischemia/reperfusion (MI/R) injury. Diabetic state was induced in mice by multiple intraperitoneal injections of low-dose streptozotocin (STZ).

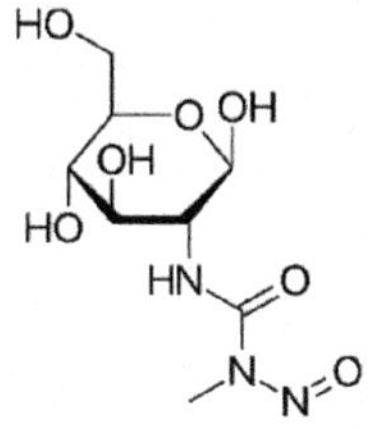

Diabetic mices were subjected to 30 min ischemia for 3 or 24 h eperfusion. The experiments indicated that GSNO attenuated MI/R injury in nondiabetic mice, as measured by improved cardiac function, reduced infarct size, and decreased cardiomyocyte apoptosis. GSNO aggravated the MI/R injury in diabetic mice. Thus, diabetes may cause superoxide overproduction, increase NO inactivation and peroxynitrite formation, and convert GSNO from a cardioprotective molecule to a cardiotoxic molecule. An administration of exogenous of GSNO or enhancing the level of endogenous GSNO protected against neurovascular injury in wild-type (WT) and eNOS-(endothelial dysfunction) mouse models of cerebral ischemia-reperfusion (IR) [14] Reduced brain infarctions and edema, and improved neurobehavioral functions by pre- or post-injury GSNO treatment of eNOS indicated that GSNO can attenuate cerebral ischemia-reperfusion (IR) injury. It was suggested that this effect is caused by mimicking the eNOS-derived NO-dependent anti-ischemic and anti-inflammatory functions. The hypothesis that treatment with S-nitrosoglutathione is sufficient to revert insulin resistance induced by a high-sucrose diet was tested utilizing Male Wistar rats [15]. The animals were divided in a control or high-sucrose group, and insulin sensitivity was determined: (i) in the fast state; (ii) after a standardized test meal; (iii) after GSH + NO; and after (iv) S-nitrosoglutathione (administration). It was found that GSNO was able to revert insulin resistance induced by sucrose feeding that they have an insulin-sensitizing effect *in vivo* and that GSNO promotes insulin sensitivity in a sucrose-induced insulin-resistant animal model.

13.1.2 NO Complexes Donors

The iron–sulfur–nitrosyl complexes (ISNC) are actively studied compounds due to their applications as nitric oxide (NO) carriers and donors [16–29]. The purpose of review [22] was to summarize the recent advances that contributed in unveiling the intricate relationships between the structural modifications on iron complexes and their reactivity, cellular trafficking, and global mechanisms of action. These advantages allowed the iron complexes to be used use as anticancer drugs in clinics, due to influencing on iron homeostasis, the antiproliferative activity, and elevating safety index. The finding showed encouraging results as per their therapeutic potential.

Paper [19] described the comparative study of the decomposition of two iron nitrosyl complexes (NICs) with a cysteamine thiolate ligand $\{Fe_2[S(CH_2)_2NH_3]_2(NO)_4\}SO_4{\cdot}2.5H_2O$ (**I**) and a glutathione (GSH)-ligand, $[Fe_2(SC_{10}H_{17}N_3O_6)_2(NO)_4]SO_4{\cdot}2H_2O$ (**II**), which spontaneously evolve NO in aqueous medium. NO formation was measured by using a spectrophotometric method by the formation of a hemoglobin (Hb)–NO complex. In work [20], the natural utilization of the $[Fe(NO)_2]$ motif for translocation of endogenous NO and the translational development of synthetic dinitrosyl iron complexes (DNICs) for biomedical applications were performed. It was demonstrated that mechanistic study of NO-release and NO-transfer reactivity of structure-characterized DNICs promoted the discovery of cell-penetrating and *in vivo* NO-delivery reactivity. These compounds were used for treatment of cancer and wound healing in diabetes. Activation of soluble guanylate cyclase sGC and vasodilation (the widening of blood vessels), phase I/II clinical trials of glutathione-bound DNICs against hypertension were observed. On this base, bioinorganic engineering of DNICs into scaffolds for tissue regeneration and repair relying on antibacterial, anti-inflammation, cytoprotective, and proliferative effects of NO was achieved. The antibacterial activity of [1Fe-2S]- and [2Fe-2S]-nitrosyl complexes as nitric oxide donors on the ability of *M. luteus* to form biofilms of *Micrococcus luteus* bacteria was examined [27]. The parameters of NO liberation of the complexes were found to be higher than 20 nmol L^{-1} during the first 500 s for the binuclear complexes and 17–30 nmol L^{-1} for the mononuclear complexes.

Two {FeNO}complexes $[Fe(NO)(^{TMS}PS2)(^{TMS}PS2H)]$ (1, $^{TMS}PS2H_2$ = 2,2′-dimercapto-3,3′-bis(trimethylsilyl)diphenyl)phenylphosphine; H is a dissociable proton) with a pendant thiol and $[Fe(NO)(^{TMS}PS2)(^{TMS}PS2CH_3)]$ (2) bearing a pendant thioether, were designed and spectroscopically and structurally characterized [23]. Both complexes were found to be highly sensitive to visible light (Fig. 13.1). The experiment revealed the following sequence of the process steps: (1) Upon photolysis, complex 2 undergoes NO dissociation to yield a mononuclear Fe(III) complex, $[Fe(^{TMS}PS2)(^{TMS}PS2CH_3)]$; (2) the pendant SH of 1 can act as a trap for the departing NO radical upon irradiation; (3) subsequent photolysis of the intermediate A resulted in HNO production; (4) the photogenerated HNO can further react with $[Mn^{III}(^{TMS}PS3)(DABCO)]$ ($^{TMS}PS3H_3$ = (2,2′2″-trimercapto-3,3′,3″-tris(trimethylsilyl)triphenylphosphine; DABCO = 1,4-diazabicyclo[2.2.2]octane) in organic media to yield anionic $[Mn(NO)(^{TMS}PS3)]^-$ (5^-) with a {MnNO} electronic configuration; and 5) effective differentiation of the formation of HNO from complex 1 with the pendant SH versus NO from 2 with the pendant SMe was achieved by the employment of $[Mn^{III}(^{TMS}PS3)(DABCO)]$.

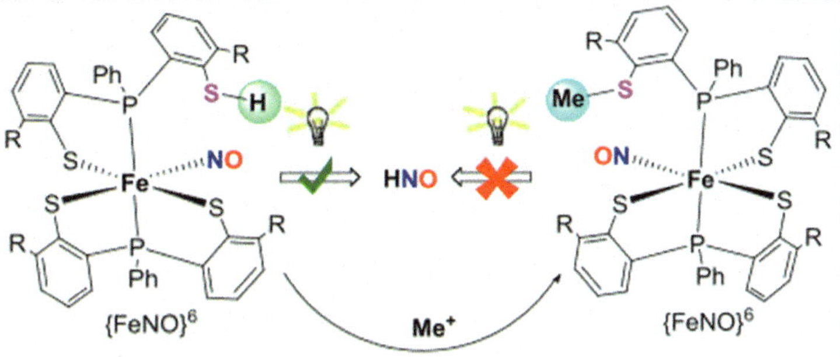

Fig. 13.1 Schematic illustration photochemical release HNO from $[Fe(NO)(^{TMS}PS2)(^{TMS}PS2CH_3)]$ [23]. With permission from American Chemical Society

Synthesis of a series of nontoxic nanoparticles based on chitosanhyaluronic acid (HA)

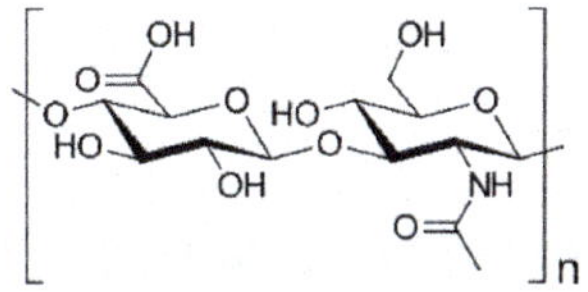

encapsulating dinitrosyl iron complexes (DNICs, donors NO) into the nanoparticles to increase the stability and effectiveness of their action was performed [26]. The effect of the nanoparticle-DNIC complex on the cell viability of cardiomyocytes was assessed. Specifically, nanoparticles were obtained from chitosan-HA using the ionotropic gelation technology, and its morphology size and stability were characterized by the electrochemistry and electron microscopy. It was emphasized that the incorporation of the DNICs into the composition of the nanoparticles prolonged the generation of NO, enhanced the yield of nitrogen monoxide, and also significantly increased the stability of the DNICs. The anionic tetranitrosyl iron complex with thiosulfate ligands (TNIC) was characterized as one of the most promising NO donors [24].

It was reported that bovine serum albumin (BSA) can effectively stabilize the anionic tetranitrosyl iron complex with thiosulfate ligands $(Na_2[Fe_2(S_2O_3)_{2(}NO)_4]\cdot 4H_2O$ TNIC (Fig. 13.2), under physiological conditions [24]. This event contributed to its prolonged action as NO donor. The experimental part of the work included the following sections: synthesis of TNIC, preparation of BSA-bounded TNIC, employing UV–Vis absorption and ESR spectroscopies. The experiment indicated that TNIC, being in a protein-bound form, has an inhibitory effect on the formation of thiobarbituric acid reactive substance (TBARS) in a mouse brain homogenate and influence on the catalytic activity of mitochondrial membrane-bound enzymes such as cytochrome *c* oxidase and monoamine oxidase A. Thus,

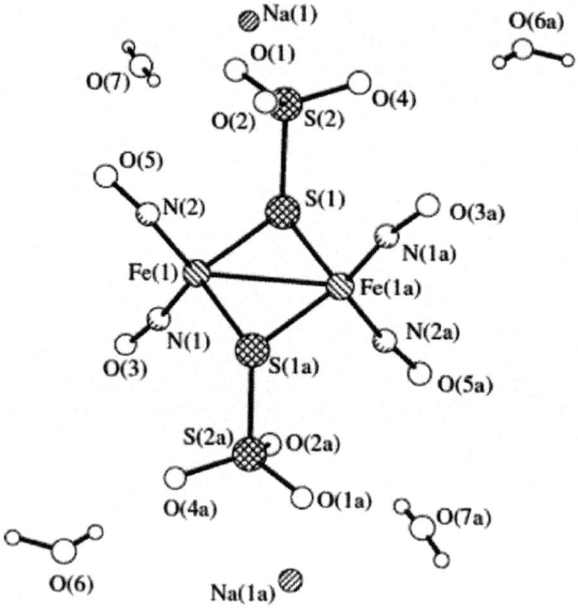

Fig. 13.2 Molecular structure of the anionic binuclear complex $Na_2[Fe_2(S_2O_3)2(NO)_4]$ [24]. With permission from Elsevier

stabilization of TNIC with BSA opens up the possibility of its practical application in chemotherapy.

The synthesis, structure, and properties of a new series of DNIC dimers (viz dimers of dinitrosyl iron complexes (DNICs) of $[Fe_2(SR)_2(NO)_4]$ composition, where R is 5-(2-nitrophenyl)-4-*H*-1,2,4-triazole-3-thiolyl (**I**), 5-(3-nitrophenyl)-4-*H*-1,2,4-triazole-3-thiolyl (**II**), 5-(4-nitrophenyl)-4-*H*-1,2,4-triazole-3-thiolyl (**III**), were examined employing NMR ^{1}H, IR, Mossbauer spectroscopies, and X-ray structural analysis as well as quantum-chemical calculations in the frame of DFT (NBO and QTAIM) methods and monitoring antibacterial activity [25]. DNIC dimers belong to μ–N–C–S structural type that are functionalized with 5-(nitrophenyl)-4-H-1,2,4-triazole-3-thiolyls. Several new compounds were prepared including dimers of dinitrosyl iron complexes (DNICs) of $[Fe_2(SR)_2(NO)_4]$ composition, where R is 5-(2-nitrophenyl)-4-*H*-1,2,4-triazole-3-thiolyl (**I**), 5-(3-nitrophenyl)-4-*H*-1,2,4-triazole-3-thiolyl (**II**), *5-(4-nitrophenyl)-4-*H*-1,2,4-triazole-3-thiolyl (III)*. The complex-(4-nitrophenyl)-4-*H*-1,2,4-triazole-3-thiolyl (III) was obtained by replacement of thiosulfate ligands in $[Fe(S_2O_3)_2(NO)_2]^{2-}$ anion. NO generation was determined using amperometric method. Data on molecular interactions stabilizing DNIC dimer of μ–N–C–S structural type functionalized by 5-(2-nitrophenyl)-4-*H*-1,2,4-triazole-3-thiolyl were revealed based on investigation of the structures of I (Fig. 13.3). Assessment of antibacterial activity of **I–III** with respect to gram-negative (*Escherichia coli*) and gram-positive (*Micrococcus luteus*) bacteria was also performed and the ability of the compounds to inhibit growth of bacteria was determined by the serial dilution method. A serial dilution is the stepwise dilution of a substance in solution.

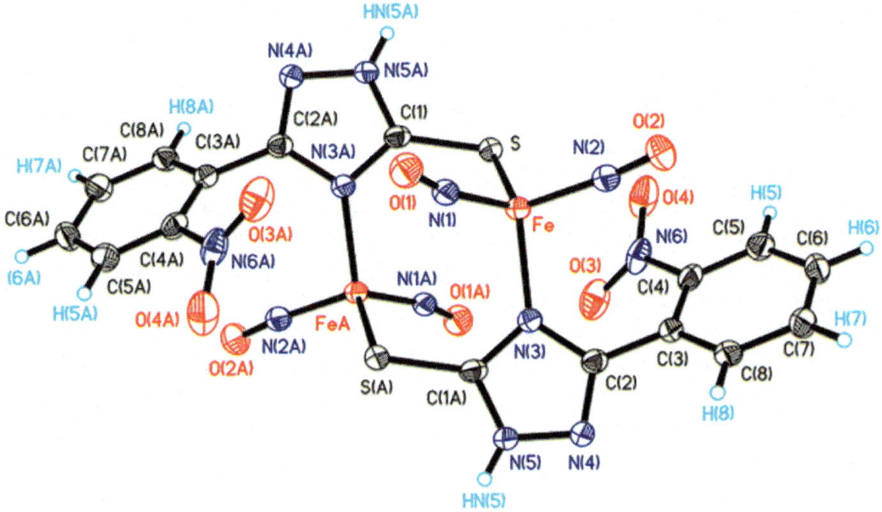

Fig. 13.3 Molecular structure of complex I [25]. With permission from Elsevier

The mechanisms of hydrolysis of a model cationic dinitrosyl iron complex with a prototypic thioformaldehyde ligand have been examined using the density functional theory and polarizable continuum water model [18]. The free energy calculations with the implicit aqueous medium demonstrated exothermic character of hydrolysis (about $-$ 70 kJ mol^{-1}) and preference for the associative mechanism of this reaction. Other results of this work included the following: (1) The dissociative mechanism involves ~ 34 kJ mol^{-1} higher activation barrier than the associative one; (2) the reaction intermediates with unfilled vacant position after removal of SCH_2 at the central iron are less stable than the intermediates where the positions of the SCH2 ligand are filled with water molecules; and (3) the exchange of NO· for H_2O is unfavorable by several tens kJ mol^{-1}. In work [28], transformation mechanism of complex DNIC $[Fe(SC(NH_2)(NHC_2H_5))_2(NO)_2]^+Cl^-\cdot[Fe(SC(NH_2)(NHC_2H_5))Cl(NO)_2]^0$ (complex 1) as cardioprotective and cytostatic agent has been studied in anaerobic aqueous solution (pH = 7.0), DMSO, and ethanol. According to EPR-spectroscopy and UV-spectra, in anaerobic aqueous solution (pH = 7.0), DMSO, and ethanol, the solvents have a significant effect on the decomposition of complex. From the experimental and theoretical data obtained, it followed that complex 1 is the most stable in DMSO. The complex can be served as a promising prodrug for the treatment of diseases. Solutions of complex in a DMSO-water mixture can be used to improve its bioavailability in further *in vitro* and *in vivo* studies.

13.2 Catalytic Antibodies

13.2.1 Fundamentals

An abzyme is a monoclonal antibody with catalytic activity [31–55]. A monoclonal antibody is an antibody produced from a cell lineage made by cloning a unique white blood cell. Recent developments of catalysis by antibodies are based on fundamental ideas advanced many years ago. Pauling [30] proposed that the lowering of the activation energy in enzyme catalysis results from the enzymes affinity for the transition state (TS) exceeding its affinity for the substrate. Developing this idea, Jencks [31] suggested that stable molecules, analogs of the transition state of a reaction could be used as haptens for the production of antibodies, which catalyze these reactions. Transition state analogs have been used for generating catalytic antibodies (Abs) abzymes that catalyze chemical reactions [32–40]. If an antibody can bind the transition state tightly like an enzyme, it can also catalyze the reaction just like an enzyme. The TS stable mimics, the transition state analogs, which have long half-lives and could be synthesized chemically, may function as an antigen to generate the antibody. Thus, antibodies and enzymes use the similar set of chemical interactions. The discovery of catalytic Abs has opened the door to a new form of targeted-enzyme therapy referred to as antibody-directed abzyme prodrug therapy (ADAPT).

At present, a significant number of reactions have been successfully performed with the use of catalytic antibodies [32–70]. Among them are reactions such as sigmatropic rearrangements, triterpen

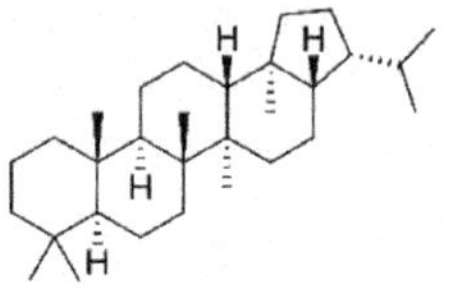

synthesis, hydrolysis, deprotonation, complexation of bivalent metals by protoporphyrin IX, acyl transfer and retroaldol reactions, the regio- and stereochemical of reactions, cleavage of acetals and glycosides, phosphate ester cleavage, amides and organophosphorus compounds hydrolysis, removal of the p-nitrobenzyl ester protecting group, the derivatization of primary amines with naphthalene-2,3-dicarboxaldehyde, hydrolysis of organophosphorus, hydrolyzing nerve agents such as Sarin and Soman *in vivo*, oxidosqualene-cyclase action in steroid ring poisons, formation of steroid ring A of the lanosterol nucleus, cationic cyclization of an oxidosqualene derivative., immunization with an 4-aza-steroid aminoxide. Success in the synthesis of new catalytic antibodies (CAs) depends on the efficiency of each of the following steps: (1) hapten design, (2) immunogen synthesis, (3) preparation of the enzymatic tracer; (4) generation and purification of antibodies; and (5) kinetic assays. Though the values of the Michaels complex ($K_m \approx 10^{-4}$–10^{-5} M) for reactions catalyzed by natural enzymes and catalytic antibodies were found to be of the same order of magnitude, the catalytic constants (k_{cat}) for CAs commonly 10^4 fold lower than that for correspondent enzymes. Experimental ratio values (k_{cat}/k_{uncat}) for abzymes range within 10^2–10^6, while these values for similar enzymatic reactions can reach 10^{17}.

The purpose of review [69] was to summarize the recent advances that contributed in unveiling the intricate relationships between the structural modifications on iron complexes and their reactivity, cellular trafficking, and global mechanisms of action. These advantages allowed the iron complexes to be used use as anticancer drugs in clinics, due to influencing on iron homeostasis, the antiproliferative activity, and elevating safety index. The finding showed encouraging results as per their therapeutic potential. Interaction network of articles within the research topic "Autoantibodies" was summarized [69]. Keywords of all 88 articles were downloaded from the website of the research topic "Autoantibodies." These articles were selected from the theme of this topic, clustered around "Autoantibodies" and "Autoimmunity." Clustering was observed for specific diseases, namely, pemphigus and pemphigoid, lupus, arthritis, and neuroimmunology. Cytokines, B cells, cell signaling. The complement cascade was also the focus of many manuscripts. Differences and similarities between enzymes and antibodies including different mechanisms of enzyme inhibition by antibodies were discussed in detail using data on typical examples [62]. Recent review [65] summarized the up-to-date application of monoclonal antibodies mAb-based approaches in assisting.to invasive aspergillosis (IA) diagnosis and therapy. It was demonstrated that monoclonal antibodies (mAbs) with high specificity in targeting the corresponding antigen(s) may have the potential to improve diagnostic tests. This work forms the basis for novel treatments of invasive aspergillosis which is a life-threatening fungal disease that causes high morbidity and mortality in immunosuppressed patients.

13.2.2 Typical Examples

Ferrochelatase catalyzes the eighth and terminal steps in the biosynthesis of heme, converting protoporphyrin IX

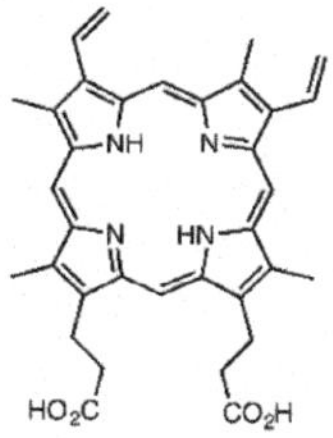

into heme B: protoporphyrin IX + $Fe^{+2} \rightleftharpoons$ heme B + 2 H^+, Ferrochelatase antibody 7G12 was generated against *N*-methylmesoporphyrin (NMP) [53]. The crystal structures of the Michaelis complex between the Fab fragment of **7G12** and its substrate mesoporphyrin (MP) as well as the Fab fragment of the germ-line precursor antibody to 7G12 and its complex with the hapten *N*-methylmesoporphyrin have been solved to 2.6-Å resolution (Fig. 13.4) [53]. The structures showed that the antibody bound mesoporphyrin *clearly* adopts a nonplanar conformation and revealed that the antibody catalyzes the porphyrin metalation reaction by distorting the bound substrate toward the transition state configuration. When *N*-methylmesoporphyrin was injected to animals as the antigen, monoclonal tightly binding antibodies were produced. Using a similar technique, antibodies that catalyze ester and amide hydrolysis, transesterification, and photoinduced cleavage, among other reactions, have been prepared.

Murine catalytic antibody 7A1 was generated against a phosphonate mono-ester transition state analog (TSA) of cocaine 5 (GNL)

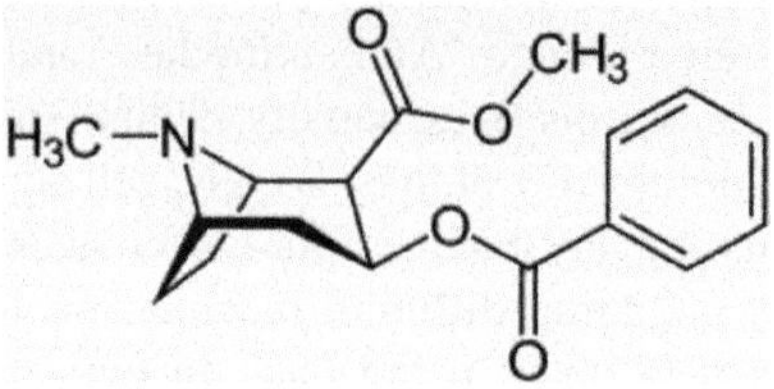

and hydrolyzes GNL to produce nonpsychoactive metabolites ecgonine methyl ester

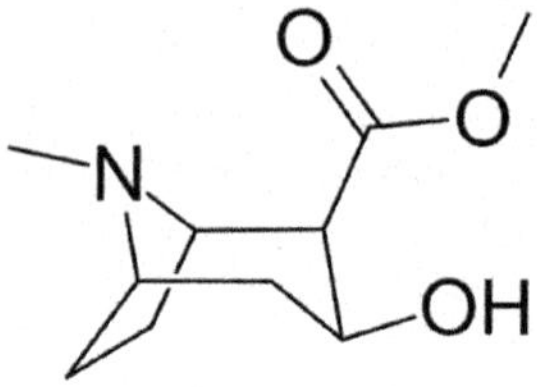

and benzoic acid [68]. Seven structures of Cocaine Catalytic Antibody 7A1 Fab0 (Fig. 13.5) were determined to high resolution in the apo form (1.5 Å) and as complexes with six ligands: (1) substrate cocaine (1.5 Å); (2) the transition state analog (1.85 Å); (3) two products, ecgonine methyl ester and benzoic acid (2.1 Å); (4) product ecgonine methyl ester (2.3 Å); (5) product benzoic acid (1.85 Å); and heptaethylene glycol (1.70 Å). 7A1 Fab0 has an immunoglobulin (Ig) g1 heavy chain (H) and a light chain (L).

The results of the activity assay of antibody 7A1 are shown in Table 13.1. As can be seen from the Table, the use of a catalytic antibody increases the reaction rate by 4.6×10^3 times. It was concluded that the potential therapeutic application for cocaine catalytic antibodies would be enhanced by improved kinetic parameters, manifested either by a lower K_M and/or by a greater k_{cat} (Table 13.2).

Examining noncatalytic and catalytic vaccines for cocaine treatment was reported [69]. A cocaine-like hapten GNE

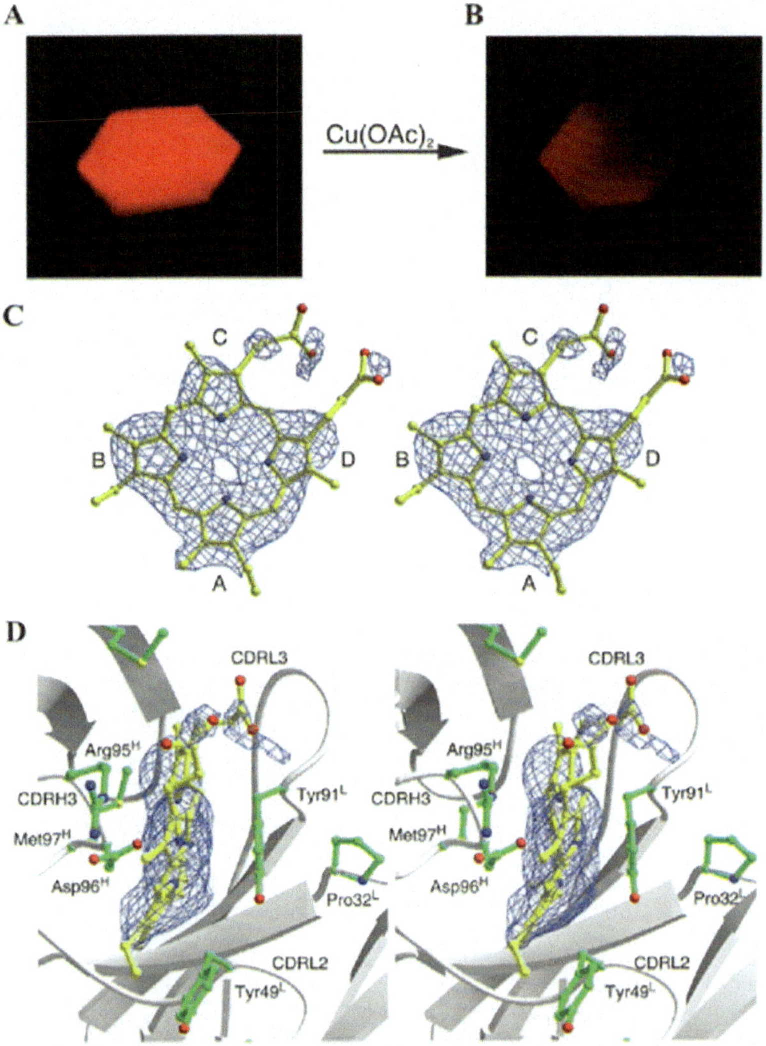

Fig. 13.4 Fluorescent micrographs of a single crystal of 7G12 Fab–MP complex before (*A*) and after (*B*) the addition of copper acetate to a final concentration of 1 mM. Photographs were taken on a Zeiss Axioskop fluorescence microscope equipped with a 546-nm exciter filter. The exposure time was 0.25 s for both pictures (Magnifications, _40). (*C* and *D*) *F*o_*F*c electron density maps contoured at 2.0 for the substrate MP molecule bound with 7G12 Fab. (*D*) Residues making important packing interactions with MP in the antibody-combining site. *C* and *D* were drawn with BOBSCRIPT and RASTER3D [53]. With permission from PNAS

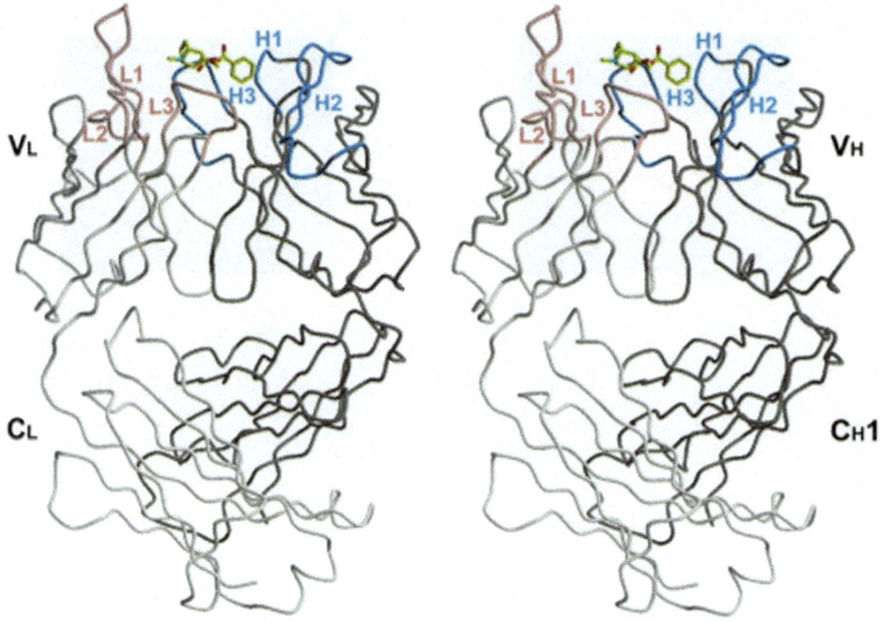

Fig. 13.5 Crystal structure of the 7A1 Fab0 cocaine complex with the Ca Trace of the Antibody Light and Heavy Chains Colored in Light and Dark Gray, Respectively CDRs L1, L2, and L3 are colored brown, while CDRs H1, H2, and H3 are colored blue. Substrate cocaine is also shown with yellow carbons in the active site [68]. With permission from Elsevier

Table 13.1 Kinetic parameters for antibody 7A1-catalyzed cocaine hydrolysis [68]. With permission from Elsevier

Buffer	100 mM PB (pH7.4)	100 mM Bicine (pH 8.0)	100 mM MOPS (pH 7.4)
K_M (mM)	0.75	3.47	2.77
k_{cat} (min^{-1})	0.025	0.087	0.035
k_{uncat} (min^{-1})	9.37×10^{-6}	2.93×10^{-5}	7.65×10^{-6}
k_{cat}/k_{uncat}	2.63×10^3	2.98×10^3	4.63×10^3

With permission from Elsevier

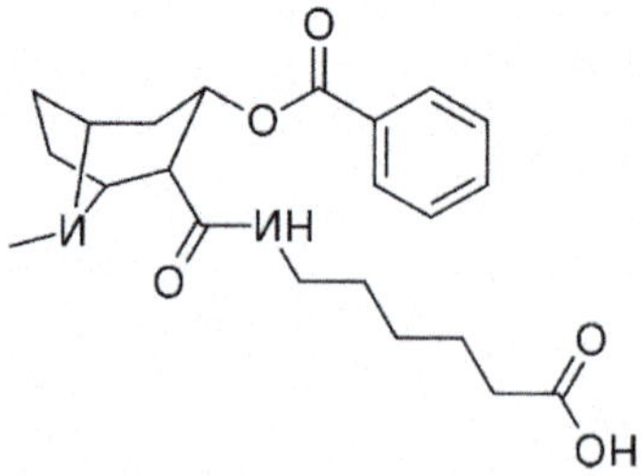

Table 13.2 Kinetic efficiency of disease catalytic antibodies [50]

Entry	Catalyst/substrate (panel in Fig. 13.1)[a]	Catalyzed reaction	Main source of catalysis[b]	$k_{rel} = k_{cat}/k_{uncat}$	k_{cat}/K_M[d] or k_2[e] (s^{-1} M^{-1})	$K_{tr} = (k_{cat}/K_M)/k_{uncat}$[f] ($M^{-1}$)	Refs.
1	Cyclodextrin/ carboxylate ester (A)	Transacylation	iii	5.9×10^6	16	1.0×10^9	[34]
2	Macrocycle/ carboxylate ester (B)	Transacylation	ii, iii	1.7×10^3	2×10^9	2×10^{12}	[30]
3	Rigid molecular receptor/ carboxylate ester (C)	Transacylation	iii	Very high[g]	3.2×10^{-4}[h]	nd	[35]
4	Dendrimer/ carboxylate ester (D)	Hydrolysis	iii, iv	3.9×10^4	3.2×10^2	1.3×10^9	[36]
5	Catalytic antibody/ carboxylate ester (E)	Hydrolysis	i	1.3×10^5	214	4.6×10^8	[37]
6	Imprinted polymer/ diphenylcarbamate (F)	Hydrolysis	i	3.8×10^3	0.24	1.2×10^6	[38]
7	Micelles/ carboxylate ester (G)	Hydrolysis	ii–iv	1.6×10^4	1.3×10^3	1.1×10^7	[39]
8	Gold nanoparticles/ carboxylate ester (H)	Hydrolysis	ii–iv	1.2×10^5	6.2×10^5	4.8×10^9	[32]

and a cocaine transition state analog GNT were used to generate the active vaccines and GNE–KLH, keyhole limpet hemocyanin (Fig. 13.6). KLH is a large, multisubunit, oxygen-carrying, metalloprotein. The latter it was found to elicit persistent high-titer, cocaine-specific antibodies, and blunt cocaine-induced locomotor behaviors. Catalytic antibodies induced by GNT–KLH were shown to produce potent titers and suppress locomotor response in a mice.

Catalytic antibodies with a metalloporphyrin cofactor, hemoabzymes, were used as models for hemoproteins like peroxidases and cytochrome P450s, representing a promising route to catalysts tailored for selective oxidation reactions [39]. Research in this direction has led to the following results: (1) Monoclonal antiporphyrin antibodies, raised against various N-substituted- and meso-carboxyaryl-porphyrins showed, in the presence of the corresponding iron-porphyrin cofactor,

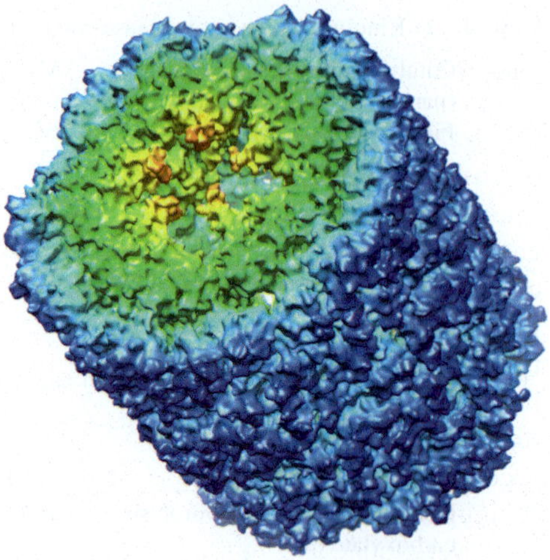

Fig. 13.6 Keyhole limpet hemocyanin. https://en.wikipedia.org/wiki/Keyhole_limpet_hemocyanin

a significant peroxidase activity; (2) an artificial hemoprotein obtained by associating a monoclonal antibody, 13G10, and its iron(III)-α,α,α,β-meso-tetrakis(ortho-carboxyphenyl)porphyrin (Fe(ToCPP)) hapten, also exhibited this activity; (3) in this antibody, a carboxylic acid side chain of the protein participated in the catalysis; and (4) the 3A3–MP8 complex can catalyze the selective oxidation of substrates, the regioselective nitration of aromatics by NO_2^-/H_2O_2 and the stereoselective oxidation of sulfides like thioanisole by H_2O_2. Data on kinetic efficiency of disease-related catalytic antibodies were summarized in a comprehensive review [50]. Kinetic studies collected in a table indicated that natural catalytic antibodies have measures of K_m, k_{cat}, and catalytic efficiency is found to be similar to those shown by their synthetic counterparts. The metal-dependence of catalytic antibodies (catabody) construct 2E6, an engineered heterodimer of immunoglobulin light chain variable domains that specifically hydrolyzes amyloid β peptides (Aβ) were designed [69]. The studies revealed the following features of the process under study: (1) The electrophilic phosphonate inhibitor of serine proteases, the metal chelators ethylenediaminetetraacetic acid (EDTA) and 1,10-phenanthroline

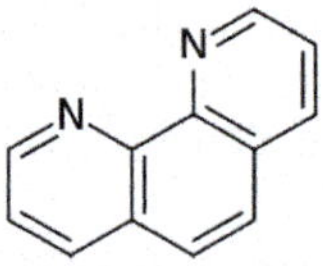

completely inhibited the hydrolysis of Aβ by catabody 2E6; (2) formation of catabody-electrophilic phosphonate inhibitor adducts was unaffected by EDTA; (3) the EDTA-inactivated catabody failed to disaggregate fibrillar Aβ, indicating the functional importance of the Aβ hydrolytic activity; (4) treating the EDTA-inactivated catabody with Zn^{2+} or Co^{2+} restored the Aβ hydrolytic activity; and

(5) Zn^{2+}-induced catabody conformational transitions were evident by fluorescence emission spectroscopy.

Problems of the design of artificial hemoproteins that could catalyze selective oxidations using oxidants O_2 or H_2O_2 under ecocompatible conditions were discussed [61]. Several new artificial hemoproteins that mimic enzymes as heme-thiolate proteins, cytochromes P450 were introduced. Artificial hemoproteins catalyzed the oxidation of substrates by dioxygen in the presence of electrons delivered from NADPH by cytochrome P450 reductase. The first strategy of the design involved the noncovalent association of synthetic hemes with monoclonal antibodies raised against these cofactors. Obtained the first generation of artificial hemoproteins, hemoabzymes, displayed a peroxidase activity, and catalyzed the regioselective nitration of phenols by H_2O_2/NO_2 and the stereoselective oxidation of sulfides by H_2O_2. A second generation of hemozymes involved the noncovalent association of affordable proteins with metalloporphyrin derivatives. In addition, several artificial hemoproteins were found able to catalyze the stereoselective oxidation of organic compounds such as sulfides and alkenes by H_2O_2.

The 18B7 monoclonal antibody against glucuronoxylomannan (GXM), the major component of the *Cryptococcus neoformans* polysaccharide capsule, hydrolyzed a peptide antigen mimetic [66]. Using mass spectrometry and Forster resonance energy transfer techniques, the hydrolytic activity of 18B7 against peptide mimetics was characterized. Specifically, it was found that: (1) 18B7 is able to hydrolyze an oligosaccharide substrate, providing the first example of a naturally occurring catalytic antibody for polysaccharides; (2) the catalytic 18B7 antibody increased release of capsular polysaccharide from fungal cells; (3) a serine protease inhibitor blocked peptide and oligosaccharide hydrolysis by 18B7; and (4) a putative serine protease-like active site was identified in the light chain variable region of the antibody. In addition, an algorithm was developed to detect similar sites present in unique antibody structures in the Protein Data Bank. Programmed cell death 1 (PD-1) is an immune checkpoint molecule regulating T-cell function. Preventing cell PD-1 binding to its ligand PD-L1 has emerged as an important tool in immunotherapy and molecule regulating T-cell function [64]. An unique human catalytic antibody light chain, H34, which mediates enzymatic degradation of human PD-1 peptides and recombinant human PD-1 protein, was examined. This abenzyme prevented the binding of PD-1 with PD-L1. Investigating the acquisition of the catalytic function by H34 revealed the importance of this sequence. EDTA inhibited the catalytic activity of H34, which works as a metalloprotease, while Zn^{2+} or Co^{2+} ions may work as a cofactor.

Paper [70] presented a comprehensive overview of bioanalytical tools based on quinone methide (QM)

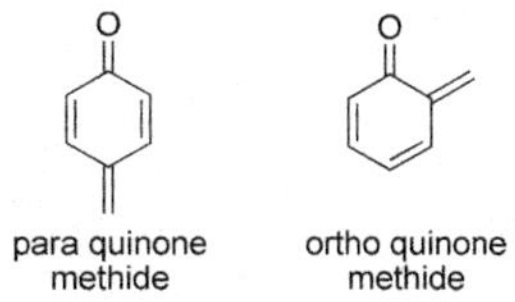

chemistry, focusing on self-immolative linkers and the highly electrophilic property of QMs enabling them to react rapidly with neighboring nucleophiles. The authors established a chemical selection method to select catalytic antibodies with superior turnover. The hydrolysis of galactopyranoside was used as a model reaction to select catalytic antibodies that catalyze this reaction (Fig. 13.7). The selection process included the following stages: (1) The hapten, which mimics the transition state of glycosidic bond cleavage, was synthesized; (2) about 100 clones with the ability to bind to this hapten using conventional hybridoma methodology (a method for producing large numbers of identical antibodies) were selected; (3) after several steps, the heavy chain and light chain sequences derived from these clones were combinatorially subcloned into the pComb3H vector to form a Fab-phage library with approximately 9000 clones in total; (4) the cycle of selection and amplification of phage in *E. coli* was repeated four times; (5) the selected vectors were transformed into β-galactosidase-deficient XL1-Blue cells to express the selected Fab; 6) these transformed cells were cultured on LB agar plates containing X-gal, which reacts with β-galactosidase to form blue precipitates; (7) further sorting was performed based on the formation of blue precipitates; (8) one of the blue-colored clones named Fab 1B was selected and its kinetics was analyzed; and (9) the kinetics of Fab 1B and the monoclonal antibody (1F4) selected by the hybridoma methodology, targeting the originally used hapten against *p*-nitrophenyl-β-galactopyranoside, was measured, and experiments showed $k_{cat} = 0.007\ \text{min}^{-1}$, $K_M = 530\ \mu\text{M}$ (rate enhancement $= 7 \times 10^4$), whereas 1F4 showed $k_{cat} = 10^{-5}\ \text{min}^{-1}$, $K_M = 330\ \mu\text{M}$.

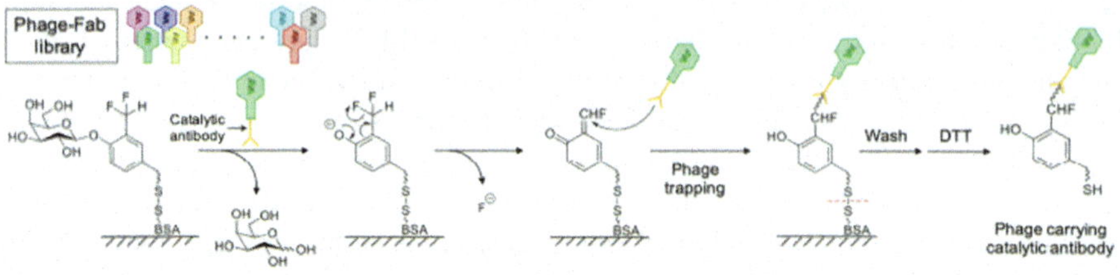

Fig. 13.7 Mechanism-based panning for the selection of antibodies that can catalyze the hydrolysis of a galactopyranoside substrate. Each phage-Fab is illustrated in a variety of colors. The trapping device was first coupled to bovine serum albumin (BSA) through a disulfide exchange reaction. The phage-Fab library was then applied to the plate coated with the trapping device. Only clones that catalyzed hydrolysis of the glycosidic bond could cause quinone methide formation and be trapped. Nonbinders and phage that did not bind covalently were removed by washing with buffer and acid, respectively. After that, trapped phage-Fab was eluted with dithiothreitol (DTT) and amplified in *E. coli* [70]. With permission from Elsevier

13.3 Miscellaneous Artificial Enzymes [70–88]

Review [76] concerned the development of various types of enzyme mimics, namely polymeric and dendrimeric, supramolecular, nanoparticulate, and proteinic enzyme mimics, with an emphasis on their synthesis, catalytic properties, and technical applications. The following requirements to artificial catalysts and biomacromolecule hybrid catalysts as models toward the development of competitive artificial enzymes should be [87]: (1) The confinement of the substrates and the active site within a catalytic pocket; (2) the creation of a hydrophobic pocket in water; (3) self-replication and allosteric properties; (4) stabilization of the transition state; and (5) favorable conformations of the active site and the product(s). The development contributions from 2005 to 2014 of catalytic systems in water in the presence of micelles obtained by addition of surfactants for the detection of hydrogen peroxide were summarized [82]. The effects of these systems on simple, economic, and green reaction media aspects like on recyclability, activity, product, and substrate selectivity were in focus of the investigation. Surfactants enable formations of nanosized apolar aggregates in bulk water, where the catalysts and reagents can be dissolved, which were observed. Specific interactions between substrates, catalysts, and micelles were examined by means of NMR and other techniques. Below are some typical examples from this area.

An artificial enzyme composed of hydrophobic thiazolium and imidazolium coenzyme in the presence of modified-polyethylenimine enzyme mimics catalyzes the benzoin

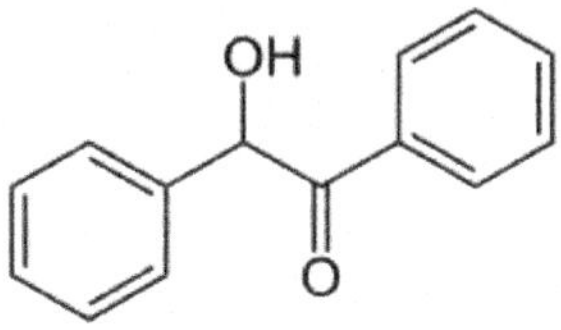

condensation 2300–3300 times faster than the coenzyme mimics alone [71]. Polycationic enzyme mimics provided a hydrophobic binding domain for coenzyme and substrate and electrostatic stabilization of anionic species. Transamination to form phenylalanine and alanine from their related keto acids produced enantioselectivities was examined, utilizing Poly(amidoamine), PAMAM [72]. The latter is a class of dendrimer made of repetitively branched subunits of amide and amine functionality. Series of PAMAM dendrimers have been synthesized from two opposite sites on a pyridoxamine core and capped with chirally terminal amino groups. The observed catalysis and chiral induction demonstrated that a dendrimer with catalytic groups remoting from the pyridoxamine unit is a good mimic of transaminase enzymes.

To investigate transamination by polymeric enzyme mimics, pyridoxamine

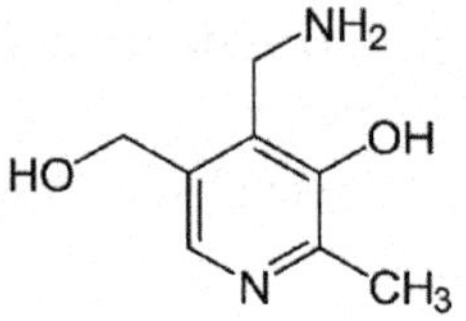

linked to a series of permethylated polyethylenimines ($M_n = 600$, 1800, 10,000, and 60,000)

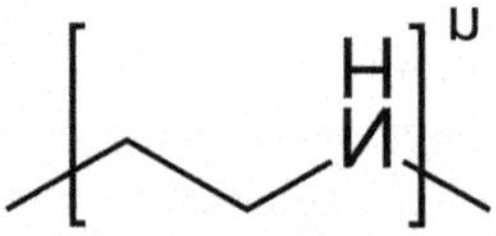

with additional attached dodecyl chains was examined [73]. The transamination kinetics of pyruvic acid and of phenylpyruvic acid showed Michaelis–Menten behavior and only small variations of the values of k_2 and of K_M with polymer size. A modified polyethylenimine can be utilized also for decarboxylation, oxidation-reduction, or hydrolysis reactions.

A restriction endonuclease is a protein produced by bacteria that cleaves DNA at specific sites along the molecule and can be essential tool for recombinant DNA technology. A *Pyrococcus furiosus* Argonaute (*Pf*Ago), a protein playing a central role in RNA silencing processes basing platform for generating artificial restriction enzymes (AREs) capable of recognizing and cleaving DNA sequences was designed [75]. Short DNA guides were used to direct *Pf*Ago to target sites for cleavage at high temperatures (> 87 °C). Reannealing of the cleaved single-stranded DNAs was found to be next step of the process. This finding was used to generate over 18 AREs for DNA fingerprinting and molecular cloning of PCR-amplified or genomic DNAs. Simple, sensitive, and stable fluorometric sensors for the detection of hydrogen peroxide based on the polyoxotungstate intrinsic peroxidase model ($Na_{10}[\alpha\text{-}SiW_9O_{34}]$), inducing fluorescent enhancement of benzoic acid (BA), thiamine (TH) and 3-(4-hydroxyphenyl)propionic acid (HPPA) were fabricated [80]. The three nonfluorescent substrates BA, TH, and HPPA were oxidized with the ·OH radicals decomposed from H_2O_2 under the catalysis of cluster $Na_{10}[\alpha\text{-}SiW_9O_{34}]$. Laccases are multicopper proteins for dioxygen-involved oxidation of a broad spectrum of organic compounds. Recently, novel amyloid-like phenylalanine-Cu (F-Cu(II)) fibrils as a chemical model of laccases were developed [85]. The model was obtained via supramolecular self-assembly of Cu^{2+} and phenylalanine (F)

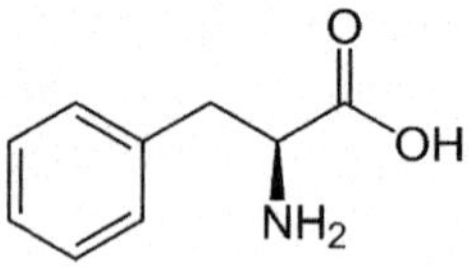

under basic conditions via alternating hydrophobic (aromatic environment) and hydrophilic (both hydrogen bonding and Cu(II) coordination) interactions [85]. The amyloid-like F-Cu(II) architecture exhibited laccase-like affinity (Michaelis constant, $K_m = 0.75$ mM) and higher catalytic efficiency ($k_{cat}/K_m = 773.33 \times 10^{-3}\ g^{-1}\ min^{-1}\ L$), *t* in pH (4 ~ 10), temperature (room temperature ~ 200 °C), organic solvent, and long-term storage (> 15 days). In addition, F-Cu(II) fibrils can catalyze the oxidation of dopamine (DA)

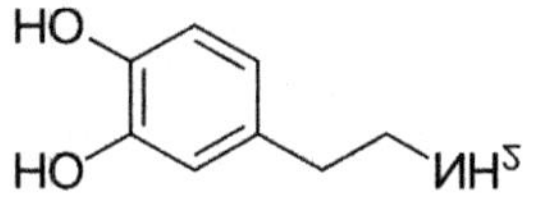

to a product with absorption band at 470 nm.

To improve the catalytic step and turnover for rate acceleration in arbitrary synthetic catalysts, through bioinspired studies of natural enzymes new design strategies based on the contributions of nonlocal electric fields and thermodynamic fluctuations were suggested [86]. The translational impact of natural enzyme principles to synthetic enzymes, supramolecular capsules, and electrocatalytic surfaces was in focus of consideration. In study [83], peptide fibril-based artificial hydrolase with photoswitchable activity was fabricated. For this purpose, catalytic histidine residue and a photoresponsive azobenzene group into the peptide chain were introduced in the peptide fibril. An enhanced catalytic activity on *p*-nitrophenyl acetate and the positive proximity effect of the histidine residues and the hydrophobic microenvironment in the supramolecular assemblies were observed. The activity of the hydrolase mimic can be reversibly controlled using UV and visible light. Opportunities to employ peptide-based catalysis to solve stereochemical problems in asymmetric synthesis are illustrated in Fig. 13.8 [84]. It was found that peptides provide enantiotopic group differentiation in catalytic desymmetrization reactions, that is, remote desymmetrizations of diarylmethane. In the obtained structure, the enantiotopic groups are separated from the prochiral center by ~6 Å and from one another by nearly 1 nm. The peptide also catalyzed brominations of pharmaceutically relevant biaryl, nonbiaryl, and hetero-biaryl atropisomers, hydroxyl group functionalizations, and electrophilic brominations, as well as C–C, C–O, and C–N cross-coupling reactions using peptidic ligands on copper(I).

Key features regulating supramolecular order and disorder in peptide-based supramolecular systems (Fig. 13.9) were formulated [88]. On the basis highlight examples, data on the development of active and adaptive systems by considering kinetic and thermodynamic design aspects and combinatorial dynamic covalent and noncovalent interactions in the systems peptide-based supramolecular systems were presented and discussed in detail. Preorganization and induced fit, which are critical to developing materials with adaptive and tunable photonic, electronic, and catalytic properties were considered. Open Access license

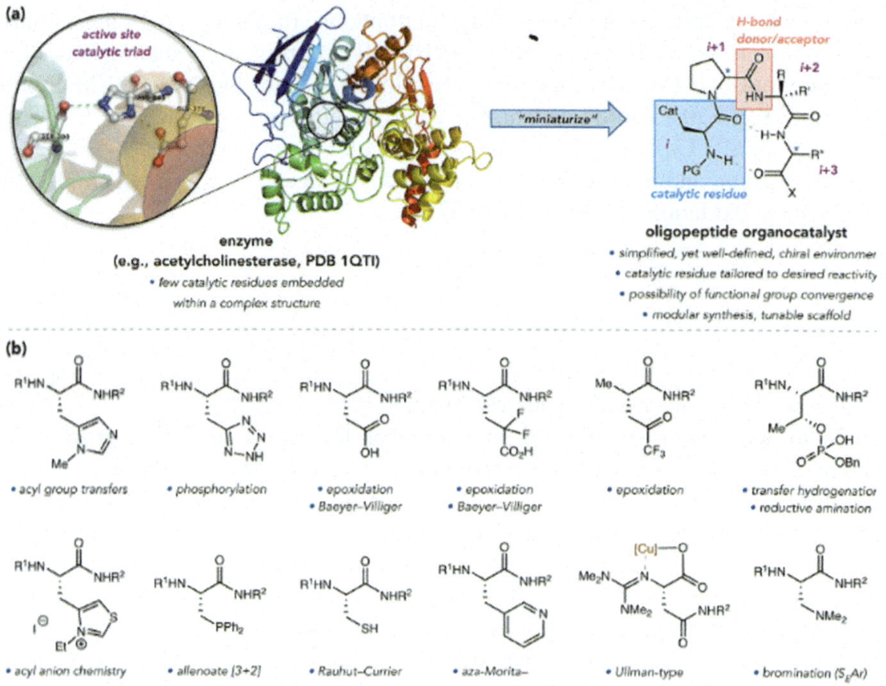

Fig. 13.8 (**a**) Peptide catalysis summarized. (**b**) Catalytic residues targeting specific transformations [84]. Published under Creative Commons public use license CC-BY from American Chemical Society

- The heterochiral tripeptide that forms a robust heterogeneous short peptide catalyst similar to the "histidine brace" active site of lytic polysaccharide monooxygenases was designed [79]. The "histidine brace" is a result of the spatial interaction between two histidine residues at the 36th and 135th positions in a typical lytic polysaccharide monoxoygenases (LPMO). To create the T-shaped 3N geometry at the reaction center, the "histidine brace" formed a conserved divalent copper ion-binding motif that comprises two histidine side chains and an amino group was constructed (Fig. 13.10). Data on UV–Vis, vis-circular dichroism, Raman, solution- and solid-state NMR, and electron paramagnetic resonance spectroscopic analyses suggested the following properties of the synthesized complex: (1) Distorted square-pyramidal geometry with a 3N coordination at pH 7; (2) the 3N coordination in the copper center of the complex; (3) the complex is pH-dependent and could catalyze the oxidation of benzyl alcohol in water to benzaldehyde with yields up to 82% in 3 h at pH 7 and above at 40 °C; (4) the catalyst achieved 100% selectivity for benzaldehyde compared to conventional copper catalysis; and (4) the geometry parameters, including a large twist angle (73°) between the two imidazole rings of the model complex, were found to be identical to those of native lytic polysaccharide monooxygenases (72.61°)

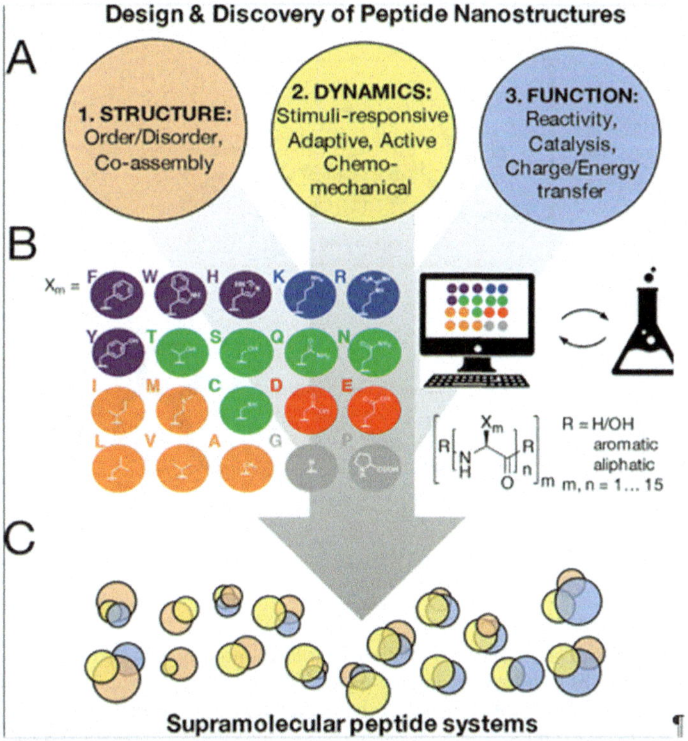

Fig. 13.9 Peptide-based supramolecular systems chemistry. (**A**) Outline of the topics covered in this review. (**B**) Color-coded amino acid building blocks which build up to functional peptides, discovered through computation and experimentation. (**C**) Schematic representation of peptide systems that combine structure, dynamics, and function [88]. Published under Creative Commons public use license CC-BY from American Chemical Society

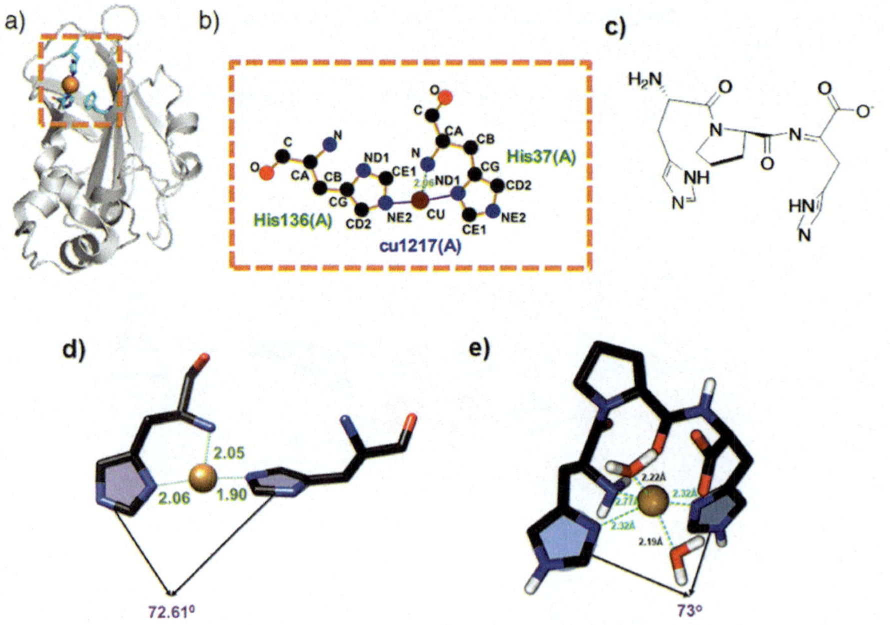

Fig. 13.10 (**a**) 3D structure of the lytic polysaccharide monooxygenase (PDB: 5FJQ) and active site residues (cyan). (**b**) LigPlot showing the ligand interactions with the histidine residues in the LPMO active site. (**c**) Chemical structure of the designed peptide mimicking the histidine brace. (**d**, **e**) Comparison between the structural parameters of the copper-chelated modeled peptide and enzyme active site, respectively [79]. Published under Creative Commons public use license CC-BY from American Chemical Society

References

1. Ignaro LJ, Freemen BA (2017) Nitric oxide: biology and pathobiology, 3rd edn. Elsevier, Amsterdam
2. Bryan NZ, Bill Gottlieb B (2010) The nitric oxide (NO) solution. NEOGENIS Labs
3. Likhtenshtein GI (2018) Nitric oxide in chemistry and biology—physicochemical aspect: fundamental and recent developments. NOVA Science Publishers
4. Smith JN, Dasgupta TP (2000) Kinetics and mechanism of the decomposition of S-nitrosoglutathione by L-ascorbic acid and copper ions in aqueous solution to produce nitric oxide. Nitric Oxide 4(1):57–66
5. Xu A, Vita JA, Keaney JF Jr (2000) Ascorbic acid and glutathione modulate the biological activity of S-nitrosoglutathione. Hypertension 36(2):291–295
6. Chiueh CC, Rauhala P (1999) The redox pathway of S-nitrosoglutathione, glutathione and nitric oxide in cell to neuron communications. Free Radic Res 31(6):641–650
7. Do KQ, Benz B, Grima G, Gutteck-Amsler U, Kluge I, Salt TE (1999) Nitric oxide precursor arginine and S-nitrosoglutathione in synaptic and glial function. Neurochem Int 29(3):213–224
8. Khan M, Qiao F, Islam SMT, et al (2021) GSNOR and ALDH2 alleviate traumatic spinal cord injury. Brain Res 1758:147335. Sousa-Lima I, Fernandes AB, Patarrão RS, Kim YB, Macedo MP (2020) S-nitrosoglutathione reverts dietary sucrose-induced insulin resistance. Antioxidants (Basel) 9(9):870

9. Khan M, Dhammu TS, Qiao F, Kumar P, Singh AK, Singh IJ (2019) S-nitrosoglutathione mimics the beneficial activity of endothelial nitric oxide synthase-derived nitric oxide in a mouse model of stroke. Stroke Cerebrovasc Dis 28(12):104470
10. Yin XH, Yan JZ, Hou XY, Wu SL, Zhang GY (2013) Neuroprotection of S-nitrosoglutathione against ischemic injury by down-regulating Fas S-nitrosylation and downstream signaling. Neuroscience 248:290–298. https://doi.org/10.1016/j.neuroscience.2013.06.012. Epub 2013 Jun 20
11. Khan M, Kumar P, Qiao F, Islam SMT, Singh AK, Won JS, Feng W, Singh I (2020) Targeting GSNOR for functional recovery in a middle-aged mouse model of stroke. Brain Res 1741:146879
12. Sakakima H, Khan M, Dhammu TS, Shunmugavel A, Yoshida Y, Singh I, Singh AK (2012) Stimulation of functional recovery via the mechanisms of neurorepair by S-nitrosoglutathione and motor exercise in a rat model of transient cerebral ischemia and reperfusion. Restor Neurol Neurosci 30(5):383–396
13. Liu Y, Xia C, Wang R et al (2017) The opposite effects of nitric oxide donor, S-nitrosoglutathione, on myocardial ischaemia/reperfusion injury in diabetic and non-diabetic mice. Clin Exp Pharmacol Physiol 44(8):854–861
14. Khan M, Dhammu TS, Qiao F et al (2019) S-nitrosoglutathione mimics the beneficial activity of endothelial nitric oxide synthase-derived nitric oxide in a mouse model of stroke. J Stroke Cerebrovasc Dis 28(12):104470
15. Sousa-Lima I, Fernandes AB, Patarrão RS, Kim YB, Macedo MP (2020) S-nitrosoglutathione reverts dietary sucrose-induced insulin resistance. Antioxidants (Basel) 9(9):870
16. Vanin AF (2021) Physico-chemistry of dinitrosyl iron complexes as a determinant of their biological activity. Int J Mol Sci 22:10356
17. Lu TT, Wang YM, Hung CH et al (2018) Bioinorganic chemistry of the natural [$Fe(NO)_2$] motif: evolution of a functional model for NO-related biomedical application and revolutionary development of a translational mode. Inorg Chem 57:12425
18. Krapivin VB, Luzhkov VB, Saninaa NA, Aldoshina SM (2020) Decomposition of dinitrosyl iron complex with thioformaldehyde ligands in water: reaction mechanisms and the role of chemical hardness of ligands. Mendeleev Commun 32:457–459
19. Syrtsova LA, Sanina NA, Kabachkov EN, Shkondina NI, Kotelnikov AI, Aldoshin SM (2014) Exchange of cysteamine, thiol ligand in binuclear cationic tetranitrosyl iron complex, for glutathione. RSC Adv 4:24560
20. Hsiao HY, Chung CW, Santos JH et al (2019) Fe in biosynthesis, translocation, and signal transduction of NO: toward bioinorganic engineering of dinitrosyl iron complexes into NO-delivery scaffolds for tissue engineering. Dalton Trans 48:9431
21. Lu T-T, Lai S-H, Li Y-W et al (2011) Discrimination of mononuclear and dinuclear dinitrosyl iron complexes (DNICs) by S K-edge X-ray absorption spectroscopy: insight into the electronic structure and reactivity of DNICs. Inorg Chem 50(12):5396–5406
22. Bouché M, Hognon C, Grandemange S, Monari GP (2020) Recent advances in iron-complexes as drug candidates for cancer therapy: reactivity, mechanism of action and metabolites. Dalton Trans R Soc Chem 49(33):11451–11466
23. Chiang C-K, Chu K-T, Lin C-C et al (2020) Photoinduced NO and HNO production from mononuclear $\{FeNO\}^6$ complex bearing a pendant. Thiol J Am Chem Soc 142(19):8649–8661
24. Faingold II, Poletaeva DA, Soldatova YV et al (2021) Effects of albumin-bound nitrosyl iron complex with thiosulfate ligands on lipid peroxidation and activities of mitochondrial enzymes in vitro. Nitric Oxide 117(2021):46–52
25. Sanina NA, Yakuschenko IK, Svyatoslav Y, Gadomskii SY (2022) Synthesis, structure and antibacterial activity of dinitrosyl iron complexes (DNICs) dimers functionalized with 5-(nitrophenyl)-4-H-1,2,4-triazole-3-thiolyls. Polyhedron 220:115822
26. Akentieva NP, Gizatullin AR, Sanina NA et al (2020) Fabrication of chitosan-hyaluronic acid nanoparticles and encapsulation into nanoparticles of dinitrosyl iron complexes as potential cardiological drugs. Nanomed J 7(3):199–210

27. Mumyatova GI, Kozub A, Kondratéva AA, Terentév NA, Sanina NA (2019) Antibacterial activity of [1Fe-2S]- and [2Fe-2S]-nitrosyl complexes as nitric oxide donors. Russ Chem Bull Int Ed 68(5):1025–1030
28. Pokidova OV, Emel'yanova NS, Psikha BL et al (2022) Effect of solvents and glutathione on the decomposition of the nitrosyl iron complex with N-ethylthiourea ligands: an experimental and theoretical study. J Inorg Biochem 235:111926
29. Yang T, Zelikin AN, Chandrawati R (2020) Enzyme mimics for the catalytic generation of nitric oxide from endogenous prodrugs. Small 16:1907635. https://doi.org/10.1002/smll.201907635(2020)
30. Pauling L (1948) Nature of forces between large molecules of biological interest. Nature 161:707–709
31. Jenks WP (1969) Catalysis in chemistry and enzymology. McGraw-Hill, New York
32. Tramontano A, Janda KD, Lerner RA (1986) Antibodies. Science 234:1566–1570
33. Keinan E (ed) (2005) Catalytic antibodies. Wiley, New York
34. Abe A, Kamiya M (2021) A versatile toolbox for investigating biological processes based on quinone methide chemistry: from self-immolative linkers to self-immobilizing agents. Bioorg Med Chem 44:116281. https://doi.org/10.1016/j.bmc.2021.116281
35. Hilvert D (2000) Critical analysis of antibody catalysis. Ann Rev Biochem 69:751–793
36. Blackburn GM, Garcon A (2000) Catalytic antibodies. In: Kelly DR (ed) Biotechnology (2nd Edition). Wiley-VCH Verlag GmbH, Weinheim, pp 403–490
37. Tanaka F (2002) Catalytic antibodies as designer proteases and esterases. Chem Rev 2002(102):4885–4906
38. Carl Veith Hanson V, Sudhi P (2005) Catalytic antibodies and their applications. Curr Opin Biotechnol 16(6):631–636
39. Ricoux R, Raffy Q, Mahy J-P (2007) New biocatalysts mimicking oxidative hemoproteins. Hemoabzym Chimi 10(8):684–702
40. Li J-W, Xia J, Sub Y et al (2012) Molecular imprint of enzyme active site by camel nanobodies. Rapid and efficient approach to produce abzymes with alliinase activity. J Biol Chem 287:13713–13721
41. Carter PJ, Lazar GA (2018) Next generation antibody drugs: pursuit of the 'high-hanging fruit.' Nat Rev Drug Discov 17(3):197–223. https://doi.org/10.1038/nrd.2017.227x
42. Elgundi Z, Reslan M, Cruz E, Sifniotis V, Kayser V (2017) The state-of-play and future of antibody therapeutics. Adv Drug Deliv Rev 122:2–19. https://doi.org/10.1016/j.addr.2016.11.004
43. Tiller KE, Tessier PM (2015) Advances in antibody design. Annu Rev Biomed Eng 17(1):191–216
44. Pérez de la Lastra JM, Baca-González V, González-Acosta S et al (2021) Antibodies targeting enzyme inhibition as potential tools for research and drug development. Biomol Concepts 12(1):215–232
45. Mahy JP, Maréchal J-D (2015) Ricoux R" From "hemoabzymes" to "hemozymes": towards new biocatalysts for selective oxidations. Chem Commun 51:2476–2494
46. Lian X, Scott-Thomas A, Lewis JG, Bhatia M, MacPherson SA, Zeng Y, Chambers ST (2022) Monoclonal antibodies and invasive aspergillosis: diagnostic and therapeutic perspectives. Int J Mol Sci 23(10):5563. https://doi.org/10.3390/ijms23105563
47. Bolon DN, Voigt CA, Mayo SL (2002) De novo design of biocatalysts. Curr Opin Chem Biol 6(2):125–129. https://doi.org/10.1016/s1367-5931(02)00303-4
48. Barrera GJ, Portillo R, Mijares A et al (2009) Immunoglobulin A with protease activity secreted in human milk activates PAR-2 receptors, of intestinal epithelial cells HT-29, and promotes beta-defensin-2 expression. Immunol Lett 123(1):52–59
49. Xu Y, Janda KD (2004) Catalytic antibodies: hapten design strategies and screening methods. Bioorg Med Chem 12(20):5247–5268
50. Bar-Even A, Noor E, Savir Y, Liebermeister W, Davidi D, Tawfik DS, Milo R (2011) The moderately efficient enzyme: evolutionary and physicochemical trends shaping enzyme parameters. Biochemistry 50:4402–4410. https://doi.org/10.1021/bi2002289

51. Hanson CV, Nishiyama Y, Sudhir P (2005) Catalytic antibodies and their applications. Curr Opin Biotechnol 16(6):631–636
52. Padiolleau-Lefèvre S, Naya RB, Shahsavarian MA et al (2014) Catalytic antibodies and their applications in biotechnology: state of the art. Biotechnol Lett 36:1369–1379. https://doi.org/10.1007/s10529-014-1503-8
53. Yin J, Andryski SE (2003) Structural evidence for substrate strain in antibody catalysis. PNAS 100:856–861
54. Schultz P, Lerner R (2002) Antibody catalysis: completing the circle. Nature 418:485. https://doi.org/10.1038/418485a
55. Motherwell WB, Bingham M, Six Y (2001) Recent progress in the design and synthesis of artificial enzymes. Tetrahedron 57(22):4663–4686. https://doi.org/10.1016/S0040-4020(01)00288-5
56. Witte M, Zillikens D, Schmidt E (2018) Diagnosis of autoimmune blistering diseases. Front Med 5:296. https://doi.org/10.3389/fmed.2018.00296
57. Zong S, Hoffmann C, Mané-Damas M, Molenaar P, Losen M, Martinez-Martinez P (2017) Neuronal surface autoantibodies in neuropsychiatric disorders: are there implications for depression. Front Immunol 8:752. https://doi.org/10.3389/fimmu.2017.00752
58. Tipton CM, Fucile CF, Darce J, Chida A, Ichikawa T, Gregoretti I et al (2015) Diversity, cellular origin and autoreactivity of antibody-secreting cell population expansions in acute systemic lupus erythematosus. Nat Immunol 16:755–765. https://doi.org/10.1038/ni.3175
59. Vayron P, Renard P-Y, Taran F et al (2001) Toward antibody-catalyzed hydrolysis of organophosphorus poisons. Proc Natl Acad Sci USA 97:7058–7063
60. Hasserodt J, Janda KD, Lerner RA (2000) Anibodies mimic natural oxidosqualene-cyclase action in steroid ring A formation. J Am Chem Soc 122:40–45
61. Mahy J, Maréchal J-D, Ricoux R (2015) From "hemoabzymes" to "hemozymes": towards new biocatalysts for selective oxidations. Chem Commun 51:2476–2494
62. Pérez de la Lastra JM, Baca-González V, González-Acosta S, Asensio-Calavia P, Otazo-Pérez A, Morales-delaNuez A (2021) Antibodies targeting enzyme inhibition as potential tools for research and drug development. Biomol Concepts 12(1):215–232
63. Zhu X, Dickerson TJ, Rogers CJ et al (2006) Complete reaction cycle of a cocaine catalytic antibody at atomic resolution. Structure 14(2):205–216
64. Hifumi E, Taguchi H, Nonaka T, Harada T, Uda T (2021) Finding and characterizing a catalytic antibody light chain, H34, capable of degrading the PD-1 molecule. RSC Chem Biol 2:220–229
65. Lian X, Scott-Thomas A, Lewis JG, Bhatia M, MacPherson SA, Zeng Y, Chambers ST (2020) Int J Mol Sci 23(10):5563. https://doi.org/10.3390/ijms23105563
66. Bowen A, Wear M, Casadevall A (2017) Antibody-mediated catalysis in infection and immunity. Infect Immun 85(9):e00202-17
67. Matsushita M, Hoffman TZ, Ashley JA et al (2001) Cocaine catalytic antibodies: the primary importance of linker effects. Bioorg Med Chem Lett 11:87–90
68. Zhu X, Tobin J et al (2005) Complete reaction cycle of a cocaine catalytic antibody at atomic resolution. Structure 14:205–216
69. Holmdahl R, Nimmerjahn F, Ludwig RJ (2019) Editorial: Autoantibodies. Front Immunol. Immunological Tolerance and Regulation. https://doi.org/10.3389/fimmu.2019.0048
70. Abe Atsuki, Kamiya Mako (2021) A versatile toolbox for investigating biological processes based on quinone methide chemistry: from self-immolative linkers to self-immobilizing agents. Bioorg Med Chem 44:116281. https://doi.org/10.1016/j.bmc.2021.116281
71. Zhao H, Foss FW, Breslow R Jr (2008) Artificial enzymes with thiazolium and imidazolium coenzyme mimics. J Am Chem Soc 130(38):12590–12591
72. Wei S (2011) Synthetic pyridoxamine models as transaminase enzyme mimic. Proquest, Umi Dissertation Publishing, Charleston, SC
73. Zhou W, Liu L, Breslow R (2003) Transmission by polymeric enzyme mimics. Helvetica 86:3560–3567
74. Klotz IM (2006) Synthetic polymers with enzyme-like activities. Ann NY Acad Sci 434(1):302–320

75. Enghiad B, Zhao H (2017) Programmable DNA-guided artificial restriction enzymes. ACS Synth Biol 6(5):752–757
76. Kuah E, Toh S, Yee J, Ma Q, Gao Z (2016) Enzyme mimics: advances and applications. Chemistry 22(25):8404–8430
77. Dong Z, Luo Q, Liu J (2012) Artificial enzymes based on supramolecular scaffolds. J Chem Soc Rev 41(23):7890–7908
78. Torres Castillo NE, Melchor-Martínez EM, Ochoa Sierra JS et al (2021) Enzyme mimics in-focus: redefining the catalytic attributes of artificial enzymes for renewable energy production. R Int J Biol Macromol 179:80–89
79. Saikia J, Bhat VT, Potnuru LR et al (2022) Minimalist de novo design of an artificial enzyme. ACS Omega 7(23):19131–19140
80. Tian R, Zhang B, Zhao M, Zou H, Zhang C, Qi Y, Ma Q (2019) Fluorometric enhancement of the detection of H_2O_2 using different organic substrates and a peroxidase-mimicking polyoxometalate. RSC Adv. 9(22):12209–12217
81. Song Z, Zhou H, Tian H et al (2020) Unraveling the energetic significance of chemical events in enzyme catalysis via machine-learning based regression approach. Commun Chem 3:134
82. La Sorella G, Strukul G, Scarso A (2015) Recent advances in catalysis in micellar media. Green Chem 17:644–683
83. Zhao Y, Lei B, Wang M et al (2018) A supramolecular approach to construct a hydrolase mimic with photo-switchable catalytic activity. J Mater Chem B 6:2444–2449
84. Metrano AJ, Miller SJ (2019) Peptide-based catalysts reach the outer sphere through remote desymmetrization and atroposelectivity. Acc Chem Res 52:199–215
85. Liu M, Yin JH, Lan C, Meng L, Xu N (2022) Supramolecule self-assembly synthesis of amyloid phenylalanine-Cu fibrils with laccase-like activity and their application for dopamine determination. Mikrochim Acta 189(3):98
86. Li W-L, Head-Gordon T (2021) Catalytic principles from natural enzymes and translational design strategies for synthetic catalysts. ACS Cent Sci 7:72–80. https://doi.org/10.1021/acscentsci.0c01556
87. Raynal M, Ballester P, Vidal-Ferran A, van Leeuwen PWNM (2014) Supramolecular catalysis. Part 2: Artificial enzyme mimics. Chem Soc Rev 43:1734–1787
88. Sheehan F, Sement D, Jain A et al (2021) Peptide-based supramolecular systems. Chem Rev 121(22):13869–13914

The manufacturer's authorised representative in the EU is Springer Nature Customer Service Centre GmbH, Europaplatz 3, 69115 Heidelberg, Germany. If you have any concerns regarding our products, please contact ProductSafety@springernature.com

Printed and bound by CPI Group (UK) Ltd, Croydon, CR0 4YY

01/10/2026

02225870-0017